ELECTRIC RELAYS

PRINCIPLES AND APPLICATIONS

ELECTRICAL AND COMPUTER ENGINEERING

A Series of Reference Books and Textbooks

FOUNDING EDITOR

Marlin O. Thurston
Department of Electrical Engineering
The Ohio State University
Columbus, Ohio

1. Rational Fault Analysis, *edited by Richard Saeks and S. R. Liberty*
2. Nonparametric Methods in Communications, *edited by P. Papantoni-Kazakos and Dimitri Kazakos*
3. Interactive Pattern Recognition, *Yi-tzuu Chien*
4. Solid-State Electronics, *Lawrence E. Murr*
5. Electronic, Magnetic, and Thermal Properties of Solid Materials, *Klaus Schröder*
6. Magnetic-Bubble Memory Technology, *Hsu Chang*
7. Transformer and Inductor Design Handbook, *Colonel Wm. T. McLyman*
8. Electromagnetics: Classical and Modern Theory and Applications, *Samuel Seely and Alexander D. Poularikas*
9. One-Dimensional Digital Signal Processing, *Chi-Tsong Chen*
10. Interconnected Dynamical Systems, *Raymond A. DeCarlo and Richard Saeks*
11. Modern Digital Control Systems, *Raymond G. Jacquot*
12. Hybrid Circuit Design and Manufacture, *Roydn D. Jones*
13. Magnetic Core Selection for Transformers and Inductors: A User's Guide to Practice and Specification, *Colonel Wm. T. McLyman*
14. Static and Rotating Electromagnetic Devices, *Richard H. Engelmann*
15. Energy-Efficient Electric Motors: Selection and Application, *John C. Andreas*
16. Electromagnetic Compossibility, *Heinz M. Schlicke*
17. Electronics: Models, Analysis, and Systems, *James G. Gottling*
18. Digital Filter Design Handbook, *Fred J. Taylor*
19. Multivariable Control: An Introduction, *P. K. Sinha*
20. Flexible Circuits: Design and Applications, *Steve Gurley, with contributions by Carl A. Edstrom, Jr., Ray D. Greenway, and William P. Kelly*

ELECTRIC RELAYS

PRINCIPLES AND APPLICATIONS

VLADIMIR GUREVICH

Israel Electric Corporation
Haifa, Israel

Taylor & Francis
Taylor & Francis Group
Boca Raton London New York

A CRC title, part of the Taylor & Francis imprint, a member of the
Taylor & Francis Group, the academic division of T&F Informa plc.

Published in 2006 by
CRC Press
Taylor & Francis Group
6000 Broken Sound Parkway NW, Suite 300
Boca Raton, FL 33487-2742

International Standard Book Number-10: 0-8493-4188-4 (Hardcover)
International Standard Book Number-13: 978-0-8493-4188-5 (Hardcover)
Library of Congress Card Number 2005049392

Library of Congress Cataloging-in-Publication Data

Gurevich, Vladimir, 1956-
   Electric relays : principles and applications / Vladimir Gurevich.
      p. cm.
   Includes bibliographical references and index.
   ISBN 0-8493-4188-4 (alk. paper)
   1. Electric relays. I. Title.

TK7872.R38.G87 2005
621.31'7--dc22   2005049392

---

informa
Taylor & Francis Group
is the Academic Division of Informa plc.

Visit the Taylor & Francis Web site at
http://www.taylorandfrancis.com

and the CRC Press Web site at
http://www.crcpress.com

In memory of my beloved father Igor Gurevich who tragically perished in January 2003

If we do not find anything very pleasant,
at least we shall find something new
Voltaire

Introduction

This book contains a description of electrical relays, their principles of operation, and applications for all basic types, for as widespread as knowledge of the subject is, it is still not abundant.

The scope of this book is very broad and unique in the sense that this book represents the first illustrated encyclopedia of electrical relays.

The historical background of the design of many different types of relays, not always known even to specialists, has been included and given much attention not only because it is interesting, but more importantly because of the frequent need for a display of expertise on the subject, enhancing the perception of competency of the specialist.

In describing some of the complicated types of relays (for example, electronic relays), the related issues of design and principles of operation of the relay components are discussed too (in our case vacuum, gas discharge, and semiconductor devices), which allows the reader to better understand the principles of operation of the described relays, without having to refer to additional information sources.

The book is written in a clear and easy-to-understand language, without mathematical treatment, and includes numerous illustrations, making it attractive not only for specialists in relays, but also for a wide range of engineers, technicians, and students interested in extending their knowledge in electric relays. Lecturers and university teachers will also find a lot of valuable material for their lectures in the book.

Acknowledgments

I wish to thank Mary Malinkovich for her enormous assistance in translating this manuscript to English and Haim Ron for reading and editing the final manuscript.

Acknowledgment is also made to the following firms and organizations for their kind permission to allow me to use various information and illustrations:

ABB Oy
ABB Switzerland Ltd — Semiconductors
Airpax Corp.
AutomationDirect
Behlke Electronic GmbH
CIGRE Central Office
Comus Group Companies
Dowetron GmbH
Eaton Corp.
EHG Elektroholding GmbH (AEG)
General Electric Co.
Gigavac
Gruner AG
Hermetic Switch, Inc.
Instrumentation & Control System, Inc.
Jennigs Technology
Leach International
LEM Components
Magnecraft & Struthers-Dunn
Maxim Integrated Products, Inc.
Moeller
NxtPhase Corp.
Olympic Controls Corp.
OPTEK Technology
Phoenix Contact GmbH & Co. KG
Powerex, Inc.
Precision Timer Co., Inc.
Rockwell Automation
Ross Engineering Corp.
Siemens AG
Smithsonian Institution Archives (Joseph Henry Papers)
Teledyne Technologies
Texas Instruments
Toshiba Corp.
Tyco Electronics
Yaskawa Electric America

Finally and most important, my utmost thanks and appreciation to my wife Tatiana, who has endured the writing of this (my third!) book, for her never-ending patience.

Preface

The electric relay is one of the most frequently used devices in modern technological systems. It can be found in cars, washing machines, microwave ovens, and medical equipment, as well as in tanks, aircraft, and ships. Practically no industry would function without relays. In some complex automatic control systems in industry, the number of relays is estimated in hundreds and even thousands. In the power-generation industry, no power device is allowed to operate without special protection relays. Certain electrical equipment, such as power transformers, may be protected by several different kinds of relays, each controlling different functions.

Because relays are so widely used and there are so many types, the broad population of engineers is unfamiliar with most of them. Generally speaking, engineers in a specific technical field are usually only familiar with relays that are applicable for specific devices. The same is true of specialists involved in the design and production of relays. Therefore, obtaining information on relays is a problem both for students whose future profession involves relay application, and for teachers in technical colleges or extension courses, who need up-to-date information about relays for their students.

Where can we find extensive publications that equally meet the needs of engineers, teachers, and students?

Various publications and books about relays currently on the market can be divided into two groups. One is generally called "Low Power Relays" or "Power Relays" (both terms mean the same, that is, a low-power electromagnetic relay with a switching current not exceeding 30 A). The second group is "Protective Relaying" (protective relays for protection of power networks), where the emphasis is placed not on a description of the principles and construction of relays, but on schematic principles of protection of electrical networks and calculation of their operating modes.

On the one hand, dividing the entire "world of electric relays" into two groups excludes some important relay implementations, for example relays with a switching current of hundreds of amperes, high-voltage relays, mercury relays, reed switch relays, solid-state relays, electric thermal relays, time-delay relays, safety relays, and many others. On the other hand, such an artificial division within the same field frequently results in separate treatment of common questions regarding relays which may be of different kinds, but are actually related and should be dealt with together. Experience accumulated for one type of relay is not always taken into account regarding other types of relays, even if the analogy is obvious. Moreover, modern protection relays usually contain electromagnetic, reed switch, or solid-state relays as output elements and experts in relay protection must be aware of their idiosyncrasies. In addition, in many particularly powerful and high-voltage modern electronic systems (power supplies, powerful lasers, radars, etc.) experts face challenges of providing protection against emergency states (overload, overcurrent, etc.), similar to challenges encountered by specialists in relay protection.

Another disadvantage of current publications is that they rarely meet the full range of engineering requirements. Some are intended mainly for experts and are abundant in equations and calculations for relays; others emphasize standards, methods of quality control, and other issues concerning production of relays; and still others are for engineers and technicians who are not experts in relays but only use relays in their equipment. Most of these publications provide the information in such a simplified and limited way that they are of little practical benefit, as they do not give simple and understandable answers to many questions concerning the implementation of relays, such as the following:

- Is it possible to switch on an electric light bulb having a nominal current of 0.3 A with the assistance of a reed switch relay with a nominal switching current of 1 A? (The correct answer is NO!)
- Why does a relay, which has worked well for a year, begin to drone and to malfunction? (The reason is that the relay has been incorrectly installed with respect to the vertical line.)
- Why does the ground fault relay ("residual current device") malfunction? Does it mean that the relay is out of order? (Not necessarily. Most often the reason is changes in insulation resistance of the equipment under exposure to moisture or high temperature.)

To answer these questions, it is essential to have a clear understanding of how relays function. That brings us to the question of what is necessary for effective study of the basic principles of relays of certain types. Is it enough just to analyze the specific construction of a certain relay? The author is convinced that it is not. The reason is that when a relay of a similar type but with a different construction is next encountered, the learning process must begin all over again.

For each type of relay, this book includes descriptions of several types of relay constructions, each functioning on a different basis. Moreover, readers will find full coverage here of the historical development of relay construction — from the earliest to modern times. The author is convinced that only such an approach can ensure understanding of principles applicable of all types of relays.

The author aimed to write a comprehensive book about relays without the disadvantages of other books and publications listed above. This book covers the diversity of the "world of electric relays" and reveals the dynamics of their development — from the earliest ideas to modern constructions and applications. In order to make the book understandable, not only for experts but also for laymen, the author utilizes the "picture-instead-of-formula" principle. Such an approach enables engineers, technicians, teachers, and students who are interested in relay construction to use the book as an encyclopedia of electric relays.

Furthermore, general readers who are interested in the history of engineering will discover many interesting historical facts about the invention of relays. Inquisitive readers will be able to enrich their knowledge in the field of electronics by reading the chapters devoted to electronic relays.

It is for the readers to decide whether the author has succeeded in attaining his objective.

This book consists of 16 chapters. The first four chapters cover the basic principles of relay construction and its major functional parts, such as contact systems, magnetic systems, etc. The following 12 chapters are devoted to various specific types of relays. Each of these chapters includes a description of the principles of relay functioning and construction as well as features of several different relays belonging to a certain type, but operating on different principles and developed at different times.

The information in the book is arranged such that the reader can work with any specific part without having to refer to another part of the book. It is also structured to function as an encyclopedia of relays by facilitating consultation when the need arises. It helps the reader find answers to particular questions, and avoids the pitfall of forcing

1

History

1.1 Relays and Horses

What is a relay?

There is probably no engineer or technician who would admit to his colleagues that he does not know what a relay is. It is an element so widely used in engineering that every engineer has had an opportunity to deal with it to some extent. Just for the sake of experiment try to define the notion "relay"... I don't think you will be able to do so at the first attempt, nor the second, and if you try to look up the word in a dictionary you will be puzzled even more.

See for yourself:

Relay

1. A fresh team of horses, posted at intervals along a route to relieve others
2. A series of persons relieving one another or taking turns; shift
3. To work in shifts
4. A sport race
5. A system of shifts at an enterprise
6. To provide with or replace by fresh relays
7. To retransmit
8. A switch.

Quite unexpected, so many different definitions of a word so widely used in the field of engineering. What's the explanation?

Let us start from the very beginning...

America's first "railway line" from Baltimore to Ellicott's factory was constructed in 1830. Its rail mileage was 13 miles. Trains consisted of several vans with wooden wheels pulled on wooden rails by a team of horses. Before long, such trains began to take people to far more distant towns. At the same time there had to be stopovers, for breaks to feed and rest horses. Such breaks considerably prolonged the journey until it occurred to someone that horses could be relieved at midpoint so that the journey could proceed with hardly any breaks at all. Each fresh team of horses was known as a "relay," from the French word *"relais,"* which means replacement. The same name was given to the small town located at the "relay" point where the horses were changed for the first time.

Although horsed vans may have little to do with modern trains, the date of that event, August 28, 1830, is considered to be the beginning of the era of railroads in the U.S.A. It

FIGURE 1.1
(a) Railroad station "Viaduct Hotel" and (b) the Town Hall in Relay.

was on that day that horsed vans began to circulate regularly via Relay station. In 1872, a special railroad station (a retransmitting station) was built in Relay (Figure 1.1). It had comfortable rest rooms for passengers along with a view of a viaduct, Viaduct Hotel, and the Town Hall.

At about the same time, some other remarkable events occurred, mostly in the same country but in quite a different field of human activity.

1.2   From Oersted to Henry

In 1820, the Danish physicist Hans Christian Oersted demonstrated for the first time that the interaction between a magnetic field and an electric current shows a slight impact of a single conductor on a compass needle. A few months later, during his experiments with a compass, the German scientist Schweigger, Professor of Chemistry at the University of Halle, noticed the fact that it is impossible to strengthen that influence by lengthening the conductor, because the compass will only interact with the nearest part of the wire. At that point it occurred to him to try to create a structure that would enable all the sections of the long wire to interact with the compass needle. He wound the long wire on a mandrel consisting of two studs, Aa and Cc, with the slots t and d in the form of several coils, attached outputs K and Z to a galvanic battery, and inserted the coils into the compass. He called this device a galvanic multiplier (Figure 1.2).

This was how the first prototype of an electromagnet was created, and if we put the compass in the B area as Professor Schweigger did, the multiplier becomes a galvano-

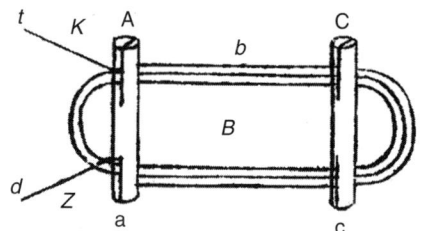

FIGURE 1.2
Schweigger's galvanic multiplier (From *Journal für Chemie und Physik* 31, Neue Reihe, Bd. I, 1821.)

FIGURE 1.3
William Sturgeon (1783–1850).

meter, enabling us to measure current strength and voltage. But this fact was unknown at that time, even to the inventor himself, the creator of this idea.

Unfortunately such a considerable decrease in the deflection of the needle was observed, even within 200 lb, that it was obvious that the scheme was inadequate.

At that point it seemed that Barlow's merciless verdict had put paid to the new telecommunication system suggested by Ampere.

The idea was appreciated to some extent by the outstanding French physicist Andre-Marie Ampere, who suggested applying Schweigger's multiplier to something similar to a telegraph or telephone system, where every letter and figure was transmitted through a separate circuit, with the compass needle becoming an indicator of current in circuits corresponding to the letters and figures.

Ampere declared that his experiments were successful, although he did not provide any commentary. Perhaps there was no need to comment. At any rate, in 1824 the English scientist Peter Barlow, commenting on Ampere's experiments, wrote that the components of the device were so obvious and the principle on which it was based so clear, that the only discovery needed was to find out whether electric current would be able to deflect the needle after passing through a long wire.

Fortunately the English scientist William Sturgeon (Figure 1.3), was unaware of Barlow's point of view. He did not give up his research on electromagnetism. On the contrary, he made every possible effort to try to find a way to increase the power of the electromagnet. Success was immediate. In 1824 he published the description of his new electromagnet, consisting of an iron core and a coil of bare copper wire. In order to enable winding a large number of turns, Sturgeon coated the surface of the horseshoe-shaped iron core with varnish, and then wrapped the spiral coil of bare copper wire, with the turns not touching each other (Figure 1.4).

At that point a new personage appeared in this tale: Joseph Henry, Professor of Mathematics and Natural Philosophy at the Albany Academy in New York (Figure 1.5).

Even from the point of view of modern science Henry's idea could be considered brilliant. He suggested insulating wire for an electromagnet by wrapping it with silk, and thus electrical wire was invented. From then on electromagnet coils were wound in hundreds of turns of insulated wire, and electromagnets were to become powerful devices widely used in many different experiments (Figure 1.6).

Soon after, Henry constructed the most powerful electromagnet in the world up to that time, which carried the weight of 750 lb. He sent descriptions of his experiments to Benjamin Silliman, Professor of Chemistry and Natural History at Yale College and editor of *The American Journal of Science*. Silliman appreciated Henry's work and in January 1831 he published in *The American Journal of Science* an article titled "Henry's Albany magnet

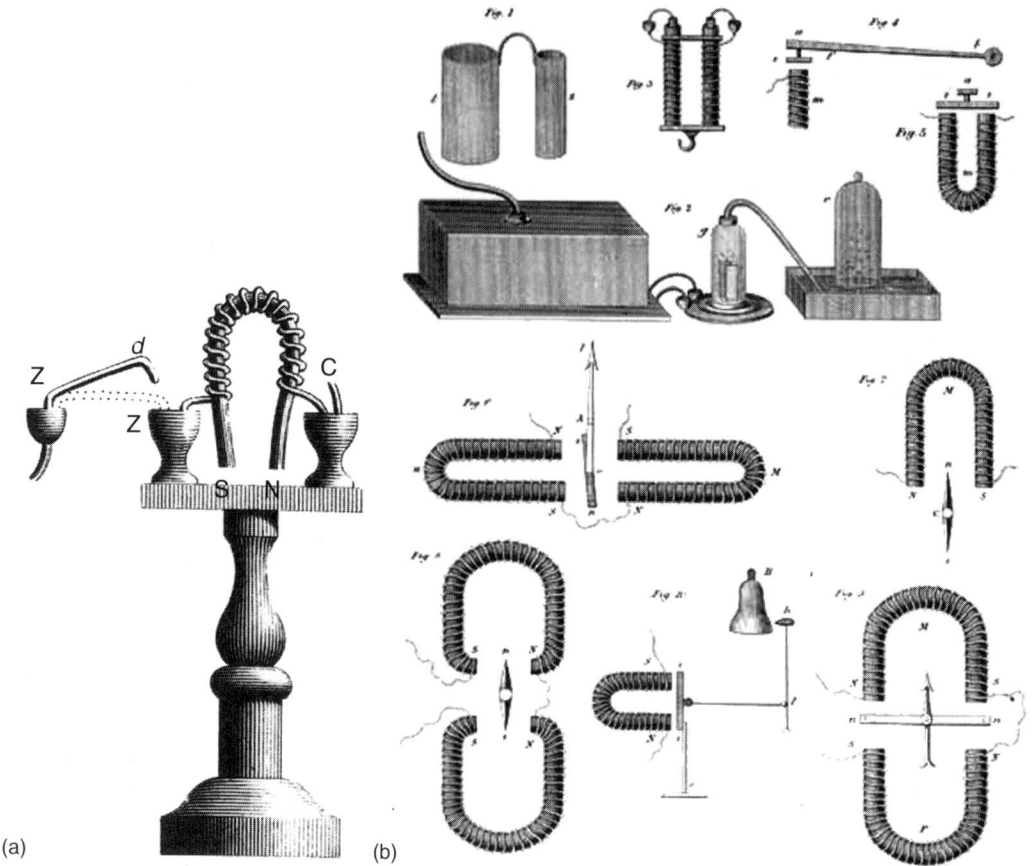

FIGURE 1.4

(a) Sturgeon's electromagnet. The horseshoe-shaped core with winding is at the top of the construction. (From *Transactions of the Society for the Encouragement of the Arts*, 43, 1824.) (b) Sturgeon's electromagnets. (From William Sturgeon, *Scientific Researches, Experimental and Theoretical, in Electricity, Magnetism, Galvanism, Electro-Magnetism and Electro-Chemistry*, Bury, T. Crompton, 1850.)

with its battery and apparatus for measuring its strength." In addition to his report, Henry proposed making a demonstrational electromagnet for his future experiments and lectures, which would lift 1000 or 1200 lb.

Benjamin Silliman accepted his proposal and in a few months Henry constructed his magnet, which surpassed even his own expectations.

This "Yale magnet" with an iron core weighing 59 lb could carry the unprecedented weight of 2063 lb. As a token of gratitude Benjamin Silliman published a detailed description of Henry's latest and most powerful magnet. In his editor's notes he mentioned that Henry had managed to create an electromagnet eight times more powerful than ever before.

Some time later, in another one of his articles, Henry put forward the idea of making a machine that could be moved by an electromagnet, an idea closely connected with the future idea of transmission of power at a distance with the help of an electromagnet.

In the summer of 1831, Henry described technical solutions for these problems in a short article titled, "On a reciprocating motion produced by magnetic attraction and repulsion." This was a simple device with a straight electromagnet rocking on a horizontal axis (Figure 1.7). Its motion automatically reversed its polarity as two pairs of wires

FIGURE 1.5
Professor Joseph Henry.

projecting from the ends made connections, alternately, with two electrochemical cells. Two vertical permanent magnets, C and D, alternately attracted and repelled the ends of the electromagnet, making it rock back and forth at 75 vibrations per minute. In fact, this device already comprised all of the basic components of an electric device known today as a polarized electromagnetic relay: a coil, a ferromagnetic core, a permanent magnet, and contacts for switching the electric circuit. Unfortunately, Henry failed to distinguish that he had a prototype for a modern relay in his device. He considered it merely a "philosophical toy," although useful for demonstrating the principles of electromagnetism to his students.

He continued improving it. In particular, instead of an iron core and two vertical magnets, his "motor" used one straight magnet with windings for the oscillating electromagnet. The description of this device was not published, but we have models based on these principles, which have lasted.

Before long Henry discovered that it was impossible to increase the power of his electromagnet due to the growth of the windings. He then tried dividing the windings into separate coils and studied the influence of parallel and series connections of coils on

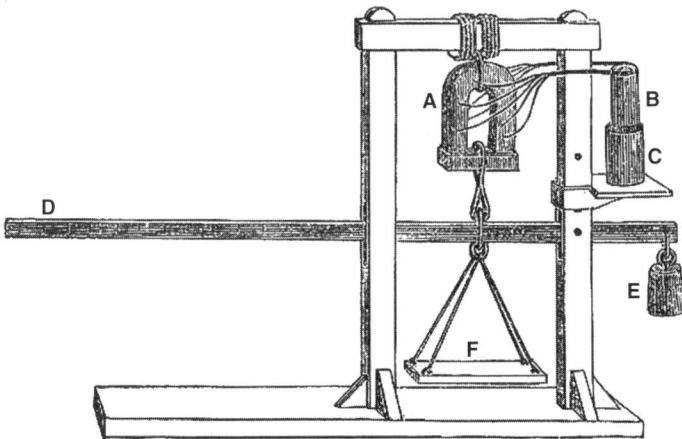

FIGURE 1.6
Henry's Albany magnet with battery and device for measuring its strength. (From J. Henry, *Silliman's American Journal of Science*, v. 19, 1831, 408.)

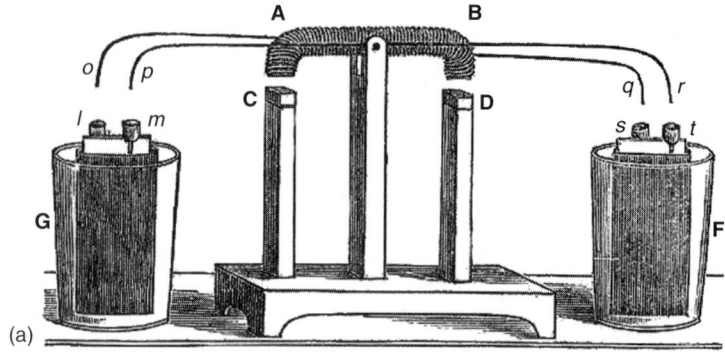

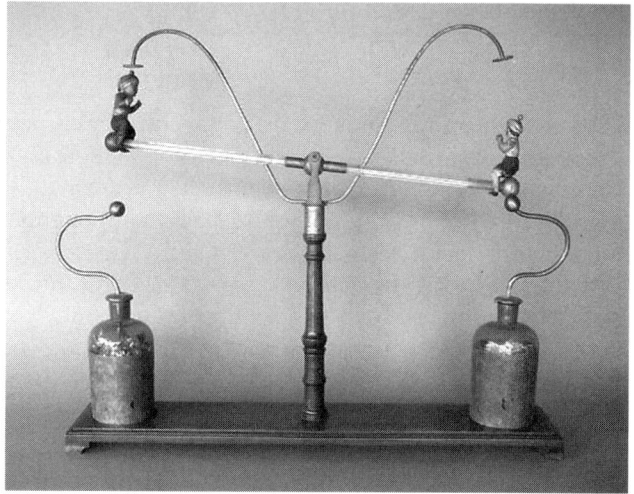

FIGURE 1.7
(a) Henry's oscillating electromagnet motor (From J. Henry, *Silliman's American Journal of Science*, v. 20, 1831, 342). (b) A toy of that time based on the principle of Henry's oscillating electromagnet motor.

the power of the electromagnet (Figure 1.8). He established an important interaction between the way the coils are connected and the number of galvanic cells placed in series, although some of the results he came up with in his experiments were quite astonishing and inexplicable. For example, the first experiments verified Barlow's conclusions that sensitivity of the compass needle decreases considerably when the wire connecting a galvanic battery with the electromagnet is lengthened, but further experiments showed an absolutely abnormal increase of sensitivity of the compass needle to the electromagnet attached not to one, but to a group of 25 galvanic cells connected in series. At the same time it was possible to transmit a distinguishable signal through wire that was hundreds of feet in length. Henry thought that the reason for this was that chemical qualities of galvanic cells change in such connections, but he did reach the proper conclusion that connections of galvanic batteries in series can compensate for lengthening of the wire connecting the electromagnet and the battery, thus creating a working telegraph.

Henry published the results of his research in the *American Journal of Science* in 1831 and made a model of a telegraph, which he demonstrated to students at his lectures until 1832. In his model Henry used an electromagnet with a horseshoe-shaped iron core, and a coil that perfectly matched the galvanic cell in the number of turns. Between the ends of the horseshoe he installed a permanent magnet on an axle swinging at the stimulation of the coil.

In fact he had produced that same Schweigger multiplier, but a much more powerful one. In addition, Henry placed a small bell near the turning magnet (Figure 1.9). The bell

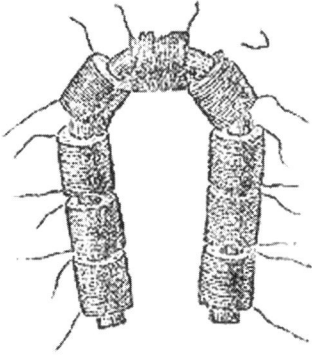

FIGURE 1.8
Henry's electromagnet with separate windings connected in series or parallel.

rang every time the magnet hit it. The electromagnet was attached to the battery with the help of copper wire about a mile in length, which was drawn through the lecture hall.

Joseph Henry became more and more popular in the American scientific community. In 1832 he was named Professor of Natural Philosophy at Princeton College. He restored his model of the telegraph, but this time the wire was drawn not in a lecture hall, but between campuses. Considering his teaching as foremost to conducting research, Henry created more and more new models for his lectures.

In 1835 he decided to combine his sensitive telegraph electromagnet. Working from a remote battery his superpower magnet lifted a record weight supplied by a powerful battery. In this new construction, instead of using a bell as in his telegraph, a permanent turning magnet closed contacts and connected the powerful electromagnet into the circuit.

As you may have guessed, this was *the very first relay in the world.* But neither Henry nor the others knew that it was a *relay.* Professor Henry enthusiastically demonstrated his new "toy" to his students: first he would turn on the whole system and fix a heavy weight on the electromagnet, then he turned it off from a long distance by means of the sensitive telegraph electromagnet. The turning electromagnet broke the supply circuit of the powerful electromagnet and the heavy weight came hurtling down followed by students' enthusiastic exclamations. Being far from practical, Professor Henry told students about prospects of his new device to be used for remote control of bells in churches, but it was not only his students who knew about his advances. Some famous and less famous

(a)

(b)

FIGURE 1.9
(a) A demonstrational model and (b) a later model of the receiving apparatus of Henry's telegraph, constructed in 1831.

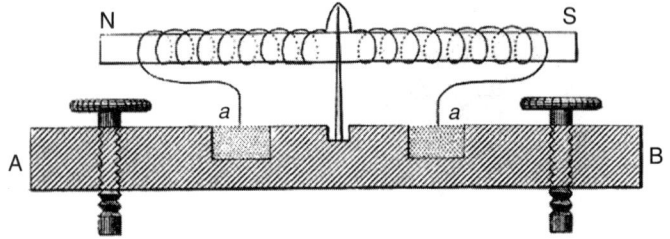

scientists and engineers followed his footsteps, applying his ideas to their own scientific and technical research and endeavors. In 1833, only 2 years after the publication of Henry's description of the oscillating electromagnet motor with a rocking electromagnet, an English clergyman unknown at that time, William Ritchie, published in "Philosophic Notes" an article titled, "Experimental researches in electro-magnetism and magneto-electricity" (*Philosophical Transactions*, 1833, v. 123) in which he describes a device with a continuously revolving electromagnet (Figure 1.10).

In this device the electromagnet moved not on a vertical axis as in Henry's model, but on a horizontal one. He caused its polarity to reverse twice in each revolution by an arrangement of wires grazing across two semicircular troughs filled with mercury. The mercury in both troughs was electrically connected to the poles of the galvanic cell with the help of additional wires.

It is still unknown whether Ritchie had heard of Henry's researches while inventing his oscillating electromagnet motor, since he made no reference to Henry's publications. Henry refused to accept Ritchie's ideas, as he considered himself to be the discoverer.

In the ensuing years Henry kept vigilant watch on his colleague's advances, and time and again engaged in controversy with them, even on pages of scientific journals contesting his pioneer status. The most notable instance was his longstanding dispute with Samuel Finley Breese Morse who used a rocking electromagnet and some other of Henry's ideas in his telegraph apparatus, without any references to Henry's publications.

Ritchie's device, like Henry's other devices, was in the first place just a didactic instrument and could not be used in practice.

1.3   Art Professor Samuel Morse

Meanwhile Henry continued his experiments on enhancement of electromagnets, in spite of his bitter dispute with his colleague Samuel Morse.

Professor Henry never "sank to the level of commercial profit" based on his inventions, and that is why he showed no interest to patenting his devices and apparatus. Samuel Morse (Figure 1.11), a professional portrayer, from 1832 Professor of Painting and Sculpture at New York University, with little formal grounding in technical sciences (on the contrary) never really gave his mind wholly to scientific research. He was a much more pragmatic man, with a remarkable capacity for work.

Morse just continued to construct and produce new and improved apparatus for his telegraph, while simultaneously patenting them. He consulted with famous scientists of that time, including Henry, but it was he who patented the devices. He believed that patents were to be granted not for fine theories but for constructing practical applications and it was he who applied and joined theory with practice. Many court hearings regarding

FIGURE 1.11
Samuel Finley Breese Morse.

Morse's numerous patents based on Henry's ideas and inventions made Henry's life a misery, and continued until his death. But Professor Henry remained firm, determined, and virile, continuing his research even under such pressure.

It was only from 1832 on that Samuel Morse really became interested in the electric telegraph, and by that time Joseph Henry had already created all of the necessary prerequisites for his successful construction of a telegraph. Morse's tenacious mind immediately grasped the enormous commercial potential hidden in the concept of instantaneous transmission of messages over large distances. He took off his coat and went to work. Not only did he have to be a scientist, but also an engineer, as he was forced to work even harder since he did not have enough money to purchase many of the necessary components.

He could not even afford insulated wire for coils. In spite of his high capacity for work it still took him 5 years to construct a first model of a telegraph that could be demonstrated to the public, and more important, to Congress, which could allocate money for further research. Rich men to whom he demonstrated his apparatus considered it as just an interesting toy, and were in no haste to invest in the project. It was only Alfred Vail, a student, who became interested in the invention. Vail's father and brother were proprietors of an iron and copper manufactory and were quite wealthy. Alfred promised to raise the necessary funds for an improved model of the apparatus, and Morse was forced to take him on as a partner. From then on Alfred Vail and William Baxter, another Morse assistant, took an active part in the construction of some Morse apparatus. Some sources even alleged that it was Morse's assistants who were the real inventors of many of Morse's devices, including his famous Morse Code alphabet. At any rate, by 1838 the apparatus had been demonstrated to Congress, which in fact took no interest in the invention. That failure did not put off Morse, who decided to prepare a new demonstration. Using pitch, tar, and rubber he made 2 miles of waterproof insulated wire. He planned to lay a submarine cable by means of which messages could be transmitted between two ships, but he failed again. During the demonstration one of the ships touched the cable and broke it.

At the same time Karl Gauss and Verner Weber in Germany, and Shilling von Kapstatt in Russia, were also attempting to design and construct a telegraph. That stimulated Morse's interest even more. Between 1839 and 1842, he frequently consulted Henry regarding technical problems, seeking his support. Henry readily helped Morse as he considered Morse's device to be just a practical application of his own research. Commercial profit did not attract him much; he just desired that his scientific achievements be implemented in practice and that is what he expected from Morse. In February 1842, making use of his reputation and influence, he appealed to Congress to assist Morse in funding his project.

Later on that same year (1842), after Henry's appeal, Morse succeeded in gaining the support of Congress and was able to obtain $30,000. In May 1844, he successfully demonstrated his telegraph to the public. "What hath God wrought?" was the first message officially transmitted by telegraph. The daughter of the Head of the Patent Authority had chosen that message.

Twelve years of hard work were crowned with triumph and international fame, and Morse was acknowledged as the inventor of this new means of communication. Unfortunately, he did not acknowledge Henry's contribution to the invention in his publications and patents, and this greatly soured the relations between them.

As a result, Morse and Henry became involved in a long series of litigations, and were to continue struggling for what each considered his rights, until their deaths.

Practically all components of Morse's apparatus were only slightly improved Henry's demonstrational models. For example the so-called sounder was a prototype of a loud-speaker (see Figure 1.12) used for sounding of Morse Code (dots and dashes) conveyed by the key. It helped in reception of vocal messages encoded by Morse Code.

Each key had a normally closed (NC) and a normally open (NO) contact (Figure 1.13). Every time the key at one end of the line (Station 1) was pressed, the tip of the turning beam of the sounder at the other end of the line (Station 2) was gravitated to the core of the vertically installed coil.

It simultaneously hit the metal element, producing a signal. Later, one more resonator was added to the sounder to intensify that signal (Figure 1.14).

As one can observe the sounder comprised all components of Henry's electromagnetic apparatuses: a coil, an iron core, a beam rocking on a vertical axis, and even a sound indicator, but Morse used two coils instead of one. It made the apparatus and the pole terminals of the beam more sensitive and considerably improved the entire construction. This scheme proved to be so effective that from then on it was used in all similar devices, produced by many different companies over the years (Figure 1.15).

Later Morse (and perhaps one of his numerous assistants) inserted a pencil in the sounder and attached to it a spring winding mechanism, which stretched paper tape under the pencil. As the telegraph became easier to use, its use quickly spread throughout the world. At first, telegraph lines in the U.S.A. were built only alongside railroad lines, because the land belonged to the railroad authorities. It was only natural that the first telegraph services were provided to railroads. By 1854, 20,000 miles of telegraph lines had been constructed in the U.S.A. The German engineer Siemens, who made his career and fortune in the construction of telegraph lines, was the founder of the company that later became a super concern of the same name.

FIGURE 1.12
Sketch of Morse sounder used in his telegraph apparatus. (From Prescott George B. *History, Theory, and Practice of the Electric Telegraph,* Boston, 1860.)

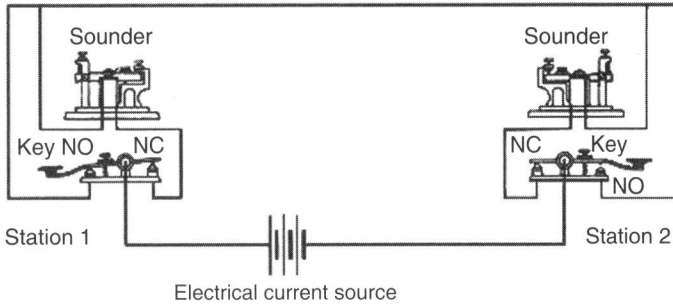

FIGURE 1.13
Scheme of Morse's telegraph.

As the length of telegraph lines constantly increased, the signal reaching the other end of the line became weaker, until someone from Morse's team recalled Henry's demonstrational experiments on remote control of a powerful electromagnet with the help of an interjacent sensitive electromagnet with contacts. Here was the solution! Technically the idea could be easily implemented: the well developed and reliable construction of the sounder included practically all the necessary components (see Figure 1.16) required to make an auxiliary element that would repeat signals from the transmission key and connect a subsidiary power supply (another galvanic battery) placed at the midpoint of

(a)

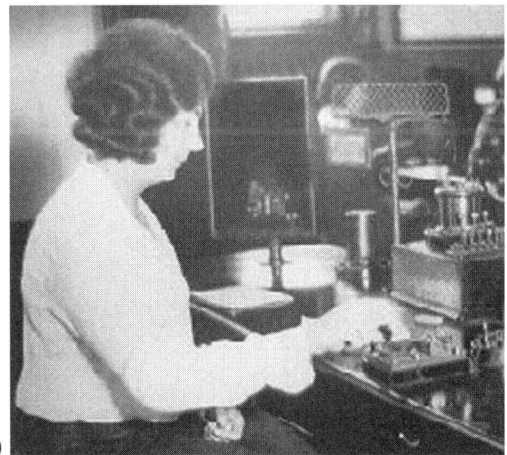

(b)

FIGURE 1.14
(a) Sounder supplied with a big wooden resonator. (b) Operator at work on telegraph key. On the rear plan a visible sounder with wooden resonator can be seen, standing on the table.

FIGURE 1.15
(a), (b), (c), and (d) Sounders produced by different companies in different years.

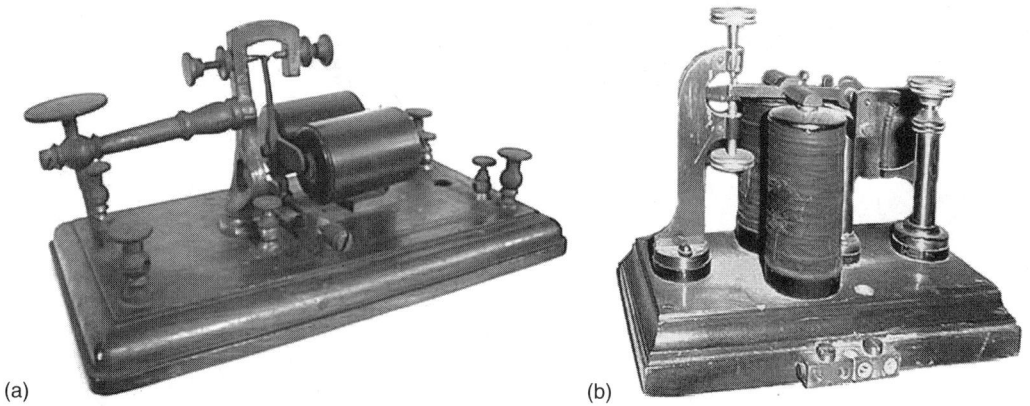

FIGURE 1.16
(a) and (b) Industrial models of sounders used as first electromagnetic relays.

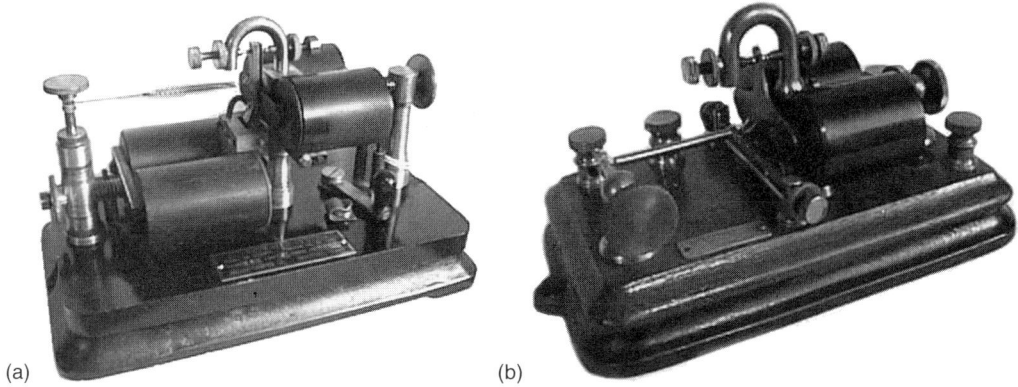

FIGURE 1.17
(a) and (b) Multi-wound and polarized (with an additional permanent magnet) relay produced in the 19th century.

the line, in time with the signals. Long distances between transmitting and receiving stations were no longer a problem as one or even several repeaters of the signal with "full" batteries could be installed on different telegraph stations.

At first such devices were called "repeaters" and "registers" but then someone noticed that the functions of such devices in the telegraph were similar to those of "relay stations" where horses were relieved. Such devices replaced a weak signal (run-down horses) with a more powerful one, connecting a "full" battery (another horse) at the midpoint of the line. As the term "relay" gradually caught on, it replaced all of the previously used terms.

Relays began to develop rapidly. New brands appeared, multiplying like clones. More and more companies began to specialize in the construction and production of relays (Figure 1.7) but for quite a long time relays remained just a part of the telegraphic system.

1.4 Edison's Relay

The outstanding inventor Thomas Alva Edison also could not help but contribute to this new field of science (Figure 1.18).

FIGURE 1.18
Thomas Alva Edison.

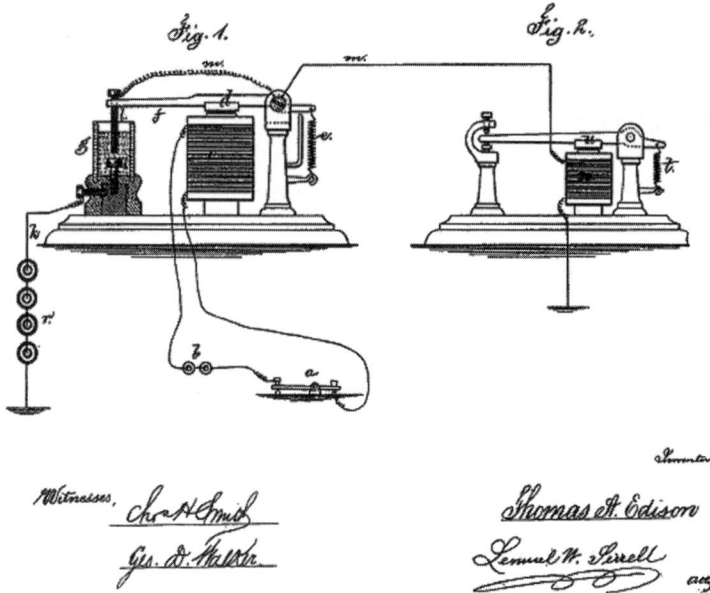

T. A. EDISON.
Relay Magnets.

Case 73.

No. 141,777.   Patented August 12, 1873.

FIGURE 1.19
Page copied from one of Edison's 200 patents relating to relays.

More than 200 of Edison's patents were devoted to relays and to other electromagnetic switching centers of telegraph apparatuses (Figure 1.19). By Edison's time the term "relay" had gained wide acceptance as the term most commonly used to denote this class of electric devices, and was the only term that Edison used. The relays designed and constructed by Edison gradually began to resemble the devices we use today in most industrial applications (Figure 1.20).

The invention of the telephone and further development of manually operated exchanges with hand-operated connectors; The American company Western-Electric used an electric relay in such switches for the first time in 1878.

1.5   The First Industrial Relays in Russia

At the end of the 19th century, the first manually operated telephone exchanges began to appear in the larger cities of Russia. As there were not yet domestically produced telephone systems, the construction and operation of such exchanges were carried out by foreign companies.

The Swede, Lars Eriksson, the proprietor of one of such companies, founded the first telephone factory in Russia in 1897, in Petersburg. With a staff of 200 people it produced

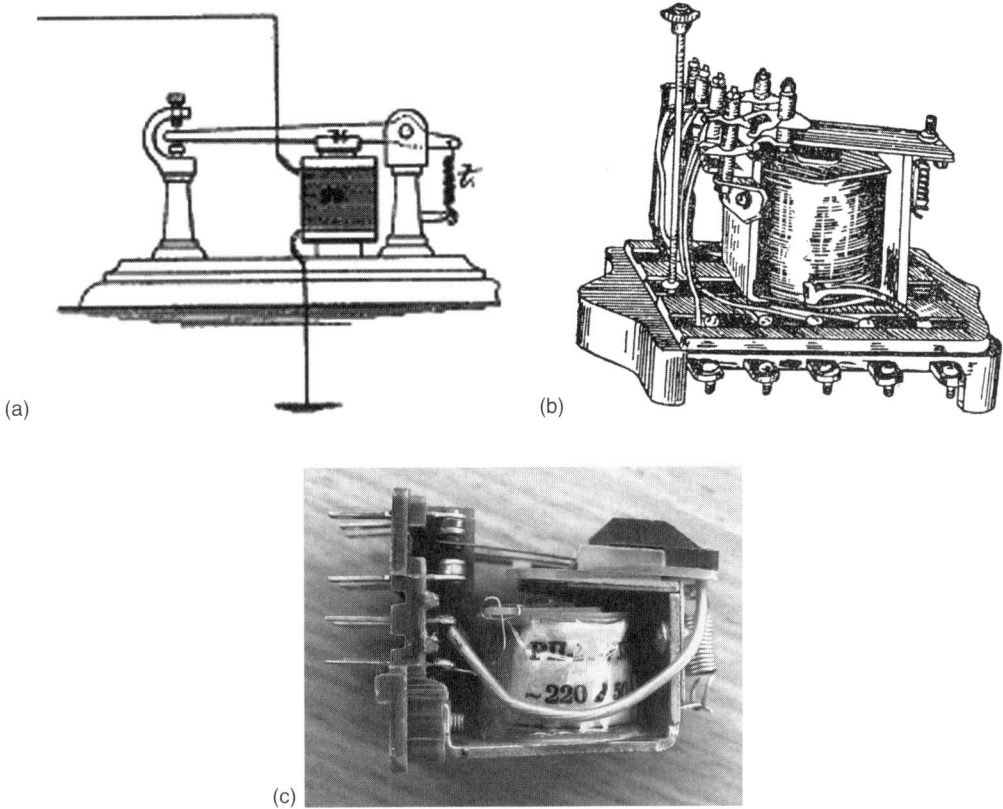

FIGURE 1.20

(a) Diagram of a relay from Edison's patent of 1873 and (b) and (c) its "descendants": Russian relays ER-100 (produced in 1940–1950s) and the modern one RP-21.

How little relays have actually changed in the course of more than a century of development. Nowadays we use the same principles and the same design schemes (RP-21 type).

12,000 telephone sets and about 100 switches over a period of 4 years. Later on it increased its capacity to produce over 60,000 telephone sets and a few hundreds of switches annually.

During the Russian–Japanese war the factory first began to produce military goods: field stations with phonic calls and outpost telephones. On January 1, 1905, it was renamed "The Russian Incorporated Company, L. Eriksson and Co."

When the First World War broke out, two more departments — naval and technical — were opened there for scientific and technical research. By 1915 the factory had evolved into a large plant, with a staff of more than 3000 people. The plant used mostly Swedish components for assembly of the various items it produced, among which were also relays.

In 1919, the plant was nationalized and transferred to the state-owned enterprises of weak-current electric industry. In 1922 the plant was renamed the Petrograd Telephone plant "Krasnaya Zarya" and joined the State Trust of weak-current electric plants (called the 9th Central Administrative Board of Ministry of Communication Industry in the closing stages of the Soviet Union), which also included the other 11 companies of the U.S.S.R. This marked the beginning of production of telephone relays in Russia.

The first models of relays, produced by "Krasnaya Zarya" until 1925, were of the clapper type, and almost exact reproductions of the "Eriksson" relays (Figure 1.21). Still,

FIGURE 1.21
Eriksson relay, produced by the "Krasnaya Zarya"
plant until 1925.

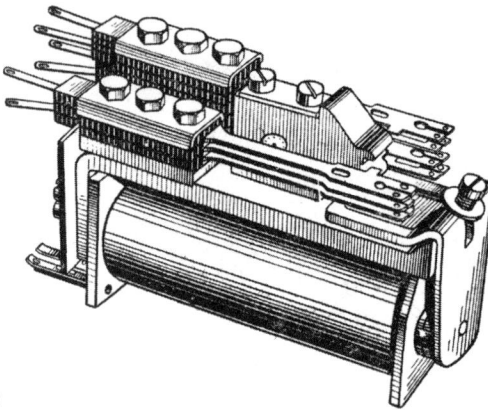

(a)

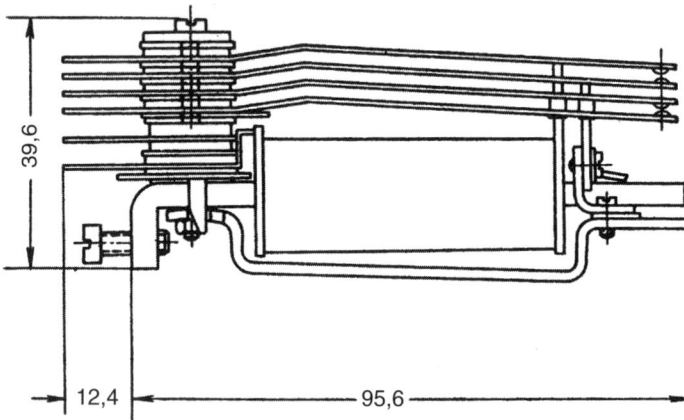

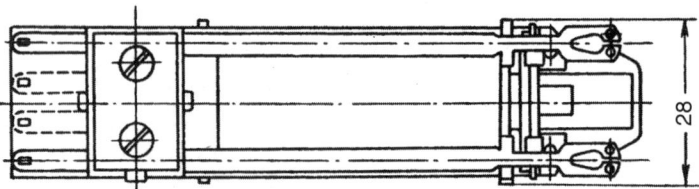

(b)

FIGURE 1.22
(a) PKH type relay (with cylindrical core). (b) PΠH relay (flat-core).

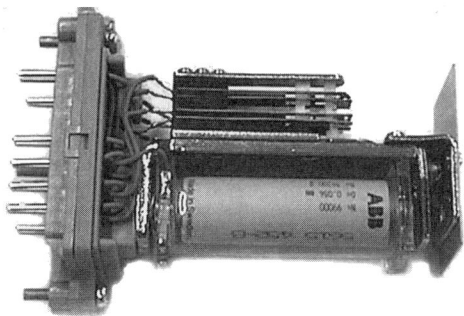

FIGURE 1.23

The modern relay RXMA-1 produced by the concern ABB, one of the world leaders in the relay field, is a practically identical copy of the PKH type relay and its "predecessors."

they were produced manually, in small series, and the materials for their production came from abroad.

Due to their lack of experience, Russian scientists at first attempted to obtain general knowledge and experience regarding the most common relay designs in use at that time around the world.

In 1934, Professor Matov of The Moscow Institute of Energy published the book *Telephone Relays, their Construction and Design*, which was the first Russian publication to present general knowledge and experience from around the world in the design of electromagnetic relays.

Being aware of Russia's technological gap, and realizing the importance of development of relay devices for modern communication and automation systems, the People's Commissariat of the Defense Industry created the Scientific-Research Institute of Electromechanics.

One of its main tasks was the invention and development of new types of relays for different fields of application, including defense technology. The Head of research was B. Sotskov. At the same time, in 1928, the Kharkov Electromechanical Plant began to produce relays for the power-generation industry and the electric drive industry. In a few years it was able to produce several different types of such relays, though still reproducing the best examples of foreign equipment. That policy lasted many long years. Copies of new German relays taken from captured German V-2 rockets were especially popular at that time. PKH and РПН type relays (Figure 1.22) produced in millions since 1946 were only copies of German and English relays.

It is interesting to note that although those relays were not used in telephone and telecommunication equipment, they "were revived" in systems of relay protection of electric networks. One of the biggest and most famous concerns producing all kinds of technical equipment for the power-generation industry still produces and utilizes analogs of RKH relays (Figure 1.23). Of course these relays are made today using new materials (they have a winding for 220 V of continuous current and contain 99,000 turns of 0.056 mm wire with a coil resistance of 39 kΩ). Also, today's model comes with a nice-looking transparent case, but that does not change the fact that it is still quite an ancient device.

The number of relay plants grew rapidly. Weak-current relays were then produced in Kharkov (by the production association "Radiorele"), S. Petersburg (by the research-and-production association "Severnaya zarya"), Irkutsk, Alatyr, Porhov, Krasnodon, Penza, and in many other cities.

About that same time there were many plants producing larger industrial relays (the electric equipment plant in Cheboksary, the production association of "Automation and Relays" in Kiev, the production association of electric equipment in Tiraspol, electric equipment plants in Moscow and Yerevan, etc.) and some plants that produced aircraft

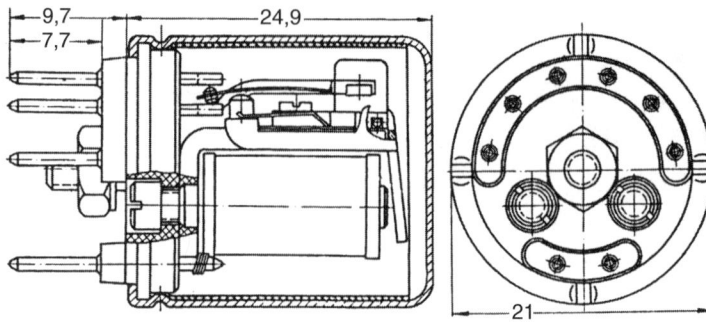

FIGURE 1.24
The design of a compact relay REC-9 produced by Kharkov Production Association "Radiorele" and some other plants in Russia is very much like that of an ancient sounder.

J. H. BUNNELL.
Telegraph-Sounder.

No. 159,894   Patented Feb. 16, 1875.

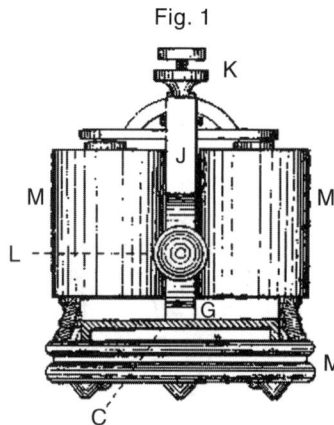

Fig. 1

FIGURE 1.25
Diagram of a sounder from a patent of J.H. Bunnell (1875), a famous inventor of electromagnetic devices at that time.

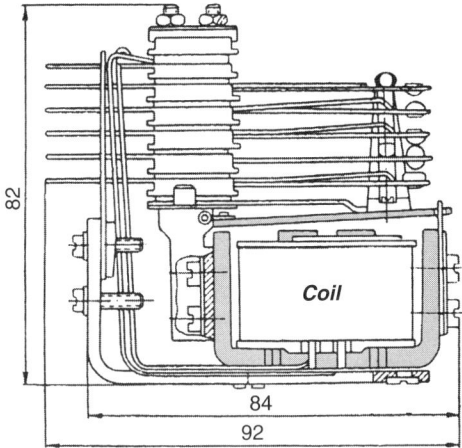

FIGURE 1.26
Out-of-date MKU-48 type relay for automation industry systems still produced by the Irkutsk Relay Plant.

switching equipment. In order to coordinate scientific activities in the field and to support the production of relays, the Scientific Research Institute of Switching Equipment, specializing in weak-current relays, was created in Leningrad, the All-Union Scientific Research Institute of the Relay Construction specializing in industrial automation relays and protective relays for the power-generation industry, in Cheboksary, and the Central Scientific Research Institute Number-22, coordinating researches in relay equipment for defense industry in Mytishi, situated near Moscow.

In the final stages of the U.S.S.R., relays were produced by dozens of big plants belonging to three Ministries: the Ministry of Communication Industry, the Ministry of Electrical Industry, and the Ministry of Aircraft Industry.

Today the range of types of electromagnetic relays produced by plants of the former Soviet Union is quite wide. It is possible, for instance, to come across REC-9 type relays, which used to be regularly produced by Kharkov Production Association "Radiorele" (Figure 1.24). The magnetic system of this two-coil relay is quite similar to many ancient sounders of the 19th century (Figure 1.25). Nevertheless, it was quite a reliable universal relay, applied in all fields of engineering. The author even happened to come across such relays in a reentry missile vehicle 9K21, fitting all types of charges.

The Irkutsk Relay Plant still continues producing such "veteran" relay devices, as the MKU-48 (Figure 1.26), which are in fact to be placed in a museum of technical equipment history.

Today Russian plants also produce many quite up-to-date relays, corresponding to their foreign analogs (Figure 1.27).

FIGURE 1.27
Modern Russian relays.

2

Magnetic Systems of Relays

2.1 Basic Components of an Electromagnetic Relay

An electromagnetic neutral relay is the simplest, most ancient, and widespread type of relay. What are its basic elements? As a rule, most people asked this question would probably name the following: a winding, a magnetic core, an armature, a spring, and contacts.

This all is true of course, but if you begin to analyze how a relay works, it might occur to you that something is missing. What is the purpose of a magnetic system? Apparently it is used to transform input electric current to the mechanical power needed for contact closure. And what does a contact system do? It transforms the imparted mechanical power back to an electric signal.

Don't you think that something is wrong here?

Everything will become more obvious if the list of basic components of a relay includes one more element, which is not so obvious from the point of view of the construction of a relay, for example, a coil or contacts. Very often, it is not just one element, but several small parts, that escape our attention. Such parts are often omitted on diagrams illustrating the principle of relay operation (Figure 2.1). I am referring to an insulation system providing galvanic isolation of the input circuit (winding) from output one (contacts). If we take such an insulation system into account, it becomes clear that an input signal at the relay input and the output signal at the relay output are not the same. They are two different signals that are completely insulated from each other electrically. Note that if you use Figure 2.1, which is often used to illustrate principles of relay operation, as the

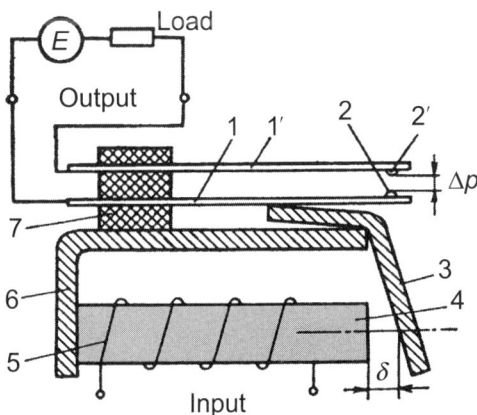

FIGURE 2.1
Construction of a simple electromagnetic relay.
1 — springs; 2 — contacts; 3 — armature; 4 — core;
5 — winding; 6 — magnetic core; 7 — insulator.

FIGURE 2.2
High-voltage relay from a Japanese patent, No. 62-32569. 3 — control winding; 4 — solenoid-type armature; 7 — insulation rod; 8, 9 — contacts; 5 — spring; 18 — outlets of input circuit; 19 — outlets of output circuit.

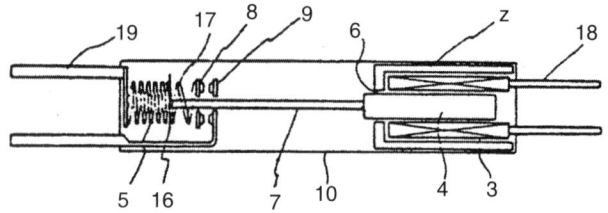

FIGURE 2.3
Construction of a relay with a noninsulated contact system. 1 — coil with an insulating bobbin; 2 — contacts; 3 — spring; 4 — armature; 5 — iron circuit.

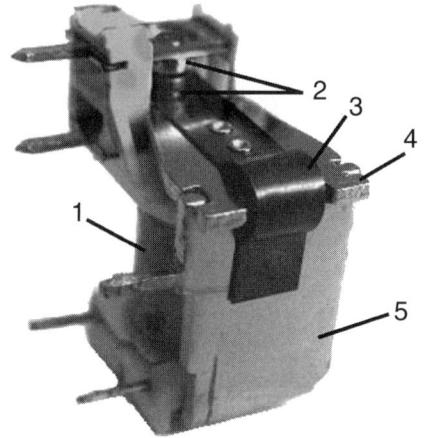

only guide while constructing a relay, the relay will not operate properly since its input circuit (the winding) is not electrically insulated from the output circuit (the contacts).

In simple constructions used for work at low voltage, insulating bobbins with winding (not shown in Figure 2.1) provide basic insulation (apart from an insulator). In a relay with a free bobbin coil, it is necessary to use a special insulating baffle pin between the armature (3) and the contacts (not shown in Figure 2.1, but it can be seen on the blueprint of the construction of an MKU-48-type relay (see Figure 1.26).

In more expensive constructions used for work at higher voltages (over 300 to 500 V) both elements are included. In high-voltage relays the insulating baffle pin usually comes in the form of a long rod (7) linking the armature with the contacts (Figure 2.2).

In relatively low-voltage relays (up to 220 V), in order to simplify the construction and to make it more compact, a coil from a magnetic core is insulated with the help of an insulating bobbin and connects a movable contact directly to the armature (Figure 2.3). Note that the contact spring and the restoring spring of the armature are the same component; a flat bent beryllium bronze plate. Since the magnetic core of the relay is not insulated from the contacts and is alive, the relay is put into a hermetic plastic case.

Let us consider the basic types of electromagnetic relays.

2.2   Hysteresis and Coercitive Force

The magnetic system of a typical low-voltage electromagnetic relay comprises first of all a control winding (1) made in the form of a coil with insulated wire, a magnetic core (2), and a movable armature (3) (see Figure 2.5). Elements of the magnetic circuit of the relay

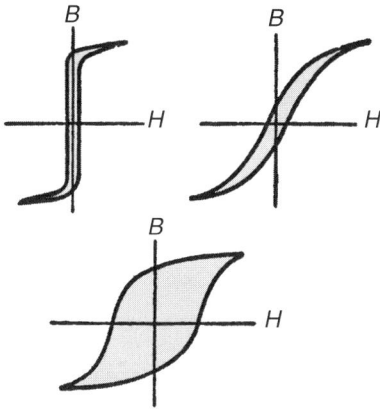

FIGURE 2.4
Hysteresis loops for soft and hard magnetic materials. B — flux density; H — magnetic field strength.

are usually made of soft magnetic steel, which is a type of steel that has a small hysteresis loop (Figure 2.4). "Hysteresis" can be translated as a "lag." A hysteresis loop is formed from a magnetization curve and a demagnetization curve. These curves are not superimposed on each other because the same magnetic field strength is not enough to demagnetize the magnetized material. An additional magnetic field is required for demagnetization of the material to its initial (nonmagnetized) state. This happens because the so-called residual induction still remains within the previously magnetized model, even after removal of the external magnetic field. Such phenomenon takes place because magnetic domains (crystal structures of ferromagnetic material) expanding along the external magnetic field lose their alignment when the external magnetic field is removed. The demagnetizing field necessary for a complete demagnetization of the previously magnetized model is called the *coercitive force*. The coercitive force of soft magnetic materials is small, which is why their hysteresis loop is also small. This means that when the external magnetic field is removed (deenergizing the winding of the relay), the magnetic core and armature do not remain magnetized, but return to their initial stage. Apparently this lack of retentivity is a very important requirement for materials used in magnetic circuits of typical neutral relays. Otherwise, relay characteristics would not be stable and the armature might seal.

Electric steel used for the production of electric motors, transformers, and relays is composed of this very soft magnetic material. This is steel with a lean temper (and other admixtures like sulfur, phosphor, oxygen, nitrogen) and rich silicon (0.5 to 5). Apart from enhancing the qualities of the steel, the silicon makes it more stable and increases electrical resistance, which considerably weakens eddy currents (see below). The hardness and the fragility of the steel are mostly dependent on the silicon content. With a content of 4 to 5% of silicon, steel usually can withstand no more than 1- to 2-fold of 90°. For a long time only the hot rolling method was used for the production of such steel, until in 1935 Goss discovered the superior magnetic properties of cold-rolled electric steel (but only along the direction of rolling). This gave such steel a magnetic texture and made it anisotropic. The utilization of anisotropic cold-rolled steel requires such construction of the magnetic core, to enable magnetic flux to pass only in the direction of rolling. Machining of parts of a magnetic core entails high internal stress, and consequently higher coercitive force. That is why after processes of forming, whetting, and milling, parts must be annealed at 800 to 900 °C, with further gradual reduction of temperature to 200 to 300 °C.

Sometimes Permalloy is used for magnetic cores of highly sensitive relays. It is a ferro-alloy with nickel (45 to 78%) alloyed with molybdenum, chrome, copper, and other elements. Permalloy has better magnetic properties in weak magnetic fields than electric steel; however, it cannot be used for work at great magnetic fluxes since its saturation induction is only half as much as that of electric steel. Magnetic cores for big AC relays are made of sheet electric steel, which is 0.35 to 0.5 mm thick.

2.3 Types of Magnetic Systems

Several types of the magnetic systems are used in modern constructions of relays (Figure 2.5).

2.3.1 The Clapper-Type (Attracted-Armature) Magnetic System

This is the most ancient type of magnetic systems. Its construction was already described in Edison's patents. It was used first in telephone relays, and later in industrial and compact covered relays. Today this type of magnetic system is widely used in constructions of middle and small-size relays, with a plastic rectangular cover that is often transparent. They are mostly designed for work in systems of industrial automation and the power-generation industry (Figure 2.6); and also in some relatively prominent electric open-type equipment (Figure 2.7). The disadvantage of relays with this type of magnetic system is sensitivity to external mechanical effects.

Considerable acceleration in certain directions can cause the armature of a relay to move spontaneously, thus provoking switching of external circuits by the contacts of the relay. When this happens, the magnetic system is called *unbalanced*. When this unpleasant characteristic of such types of relays was discovered scientists began to seek new

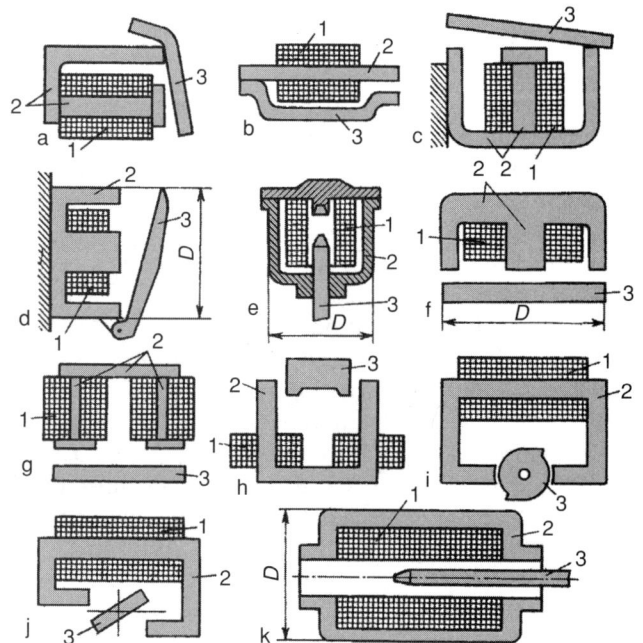

FIGURE 2.5
Types of the magnetic system of modern electromagnetic relays. 1 — control coil; 2 — magnetic core; 3 — armature; (a–d) — clapper-type (attracted-armature) systems; (f, g) — direct motion magnetic systems; (e, h, k) — systems with a retractable armature (solenoid type); (i, j) — systems with a balanced turning armature.

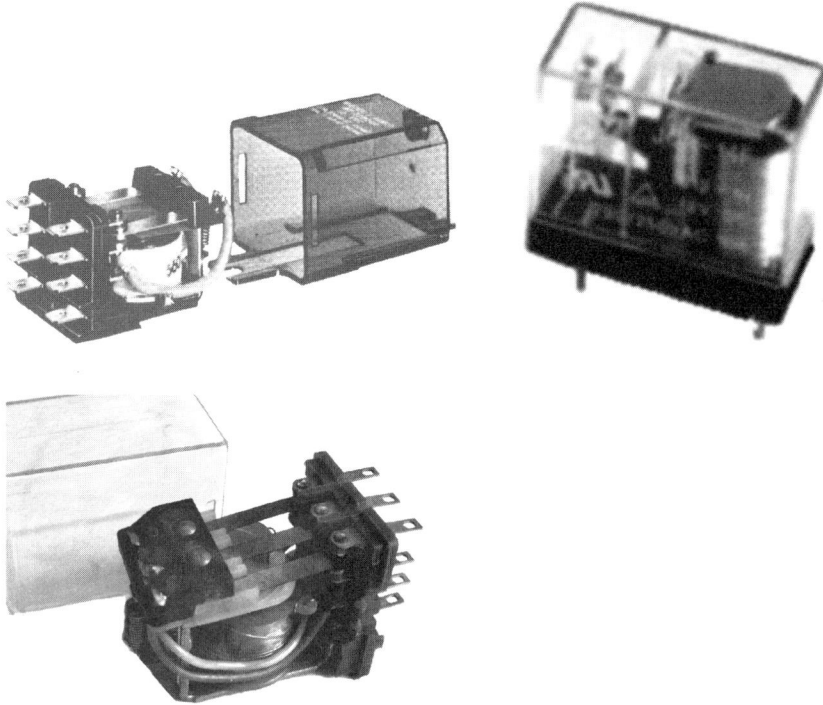

FIGURE 2.6
Modern attracted-armature relays with a transparent plastic rectangular cover for industrial usages.

technological solutions, which would enable the relay to work correctly within transportable equipment (mostly military equipment).

At first they tried the so-called "frontal solution": if there is some unbalanced weight it should be balanced, *but how?* The obvious solution: with the help of an additional weight — a counterweight placed in a certain position (Figure 2.8).

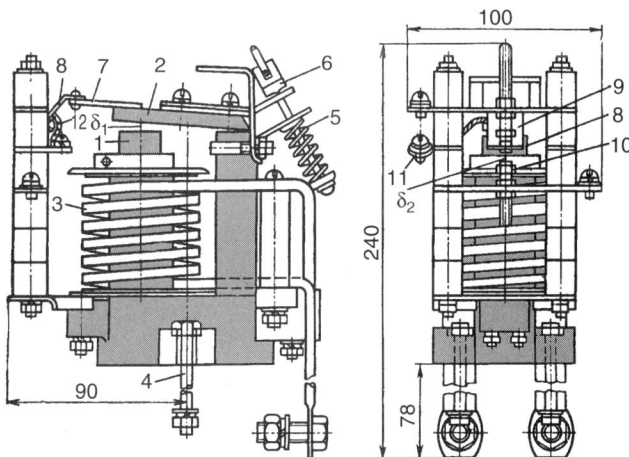

FIGURE 2.7
Large open relay (of REV-300 type) with clapper-type magnetic system. 1 — U-shaped magnetic core; 2 — flat armature rotating on the prism; 3 — winding (in this instance current winding wound by copper wire); 4 — fastening pins; 5 — spring; 6 — screw-nut regulating spring constrictor; 7 — insulating plate; 8 — movable contact; 9, 10 — stationary contacts; 11 — contact outlet; 12 — flexible copper conductor linking the contact with the outlet.

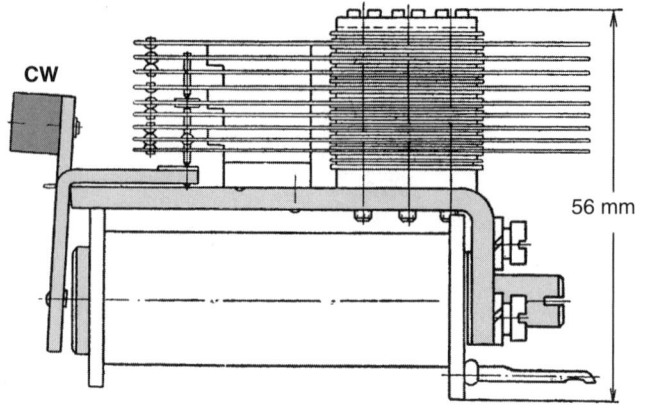

FIGURE 2.8
Multi-contact RKMP- type relay (Russia)
with an attracted-armature magnetic system balanced by a counterweight (CW)
fixed on the armature.

2.3.2   Systems with a Balanced Armature

Further research in this direction led to the invention of a new J-type magnetic system. In this system, a balanced turning armature was widely used in miniature hermetic relays for radio-electronic equipment, for work on movable units. In this magnetic system the rotation axis of the armature goes through its center of mass. As a result the relay becomes resistant to external mechanical impact and can sustain linear accelerations of 100 to 500 g, reiterated shocks at an acceleration of 75 to 150 g, and separate shocks at an acceleration rate of up to 1000 g. There are a lot of variants of this J-type magnetic system (Figure 2.9 and Figure 2.10).

Not only most modern miniature relays in plastic shells but also hermetic relays in metal shells (Figure 2.11), which have been produced for ages, have similar magnetic systems.

Such magnetic systems are typical of Russian relays RES47 (Figure 2.12), RES48, RES54, RES60, RES77, ... , RES80, etc.

Miniature and micro-miniature relays of this type are really very small in size: for example, the RES49-type relay is $10.45 \times 5.3 \times 23.2$ mm with a weight of 3.5 g, or the RES80-type relay is $10.4 \times 10.8 \times 5.3$ mm with a weight of 2 g.

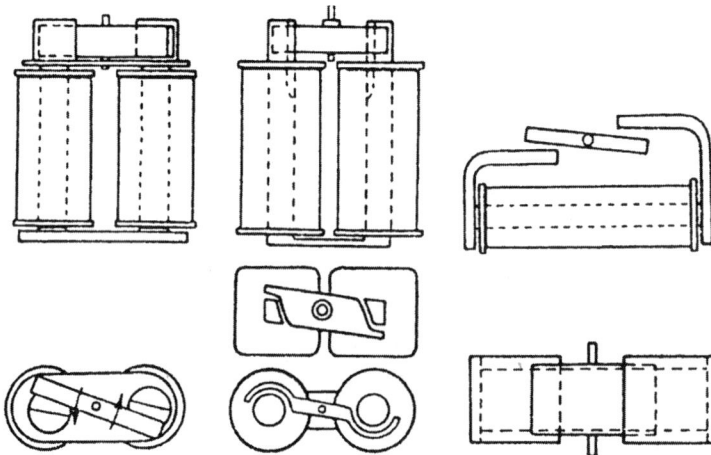

FIGURE 2.9
Some variants of construction of the J-type magnetic system.

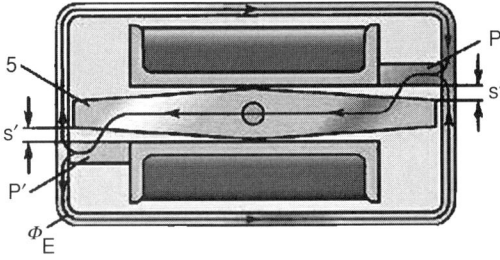

FIGURE 2.10
Another variant of construction of a magnetic system with a balanced turning armature placed inside the winding. P — stationary poles of the magnetic core; S — armature (magnetic pole) gape; F — magnetic flux in the magnetic core; 5 — turning armature.

Even power relays with switching currents of tens of amperes designed for use on movable units are often supplied with a magnetic system of this type. For example, the 8E-123M-type relay, Figure 2.13, designed by the All-Russian Scientific Research Institute of Electromechanics, and produced by one of the Moscow plants for many years for military applications only, has characteristic meeting the highest international standards.

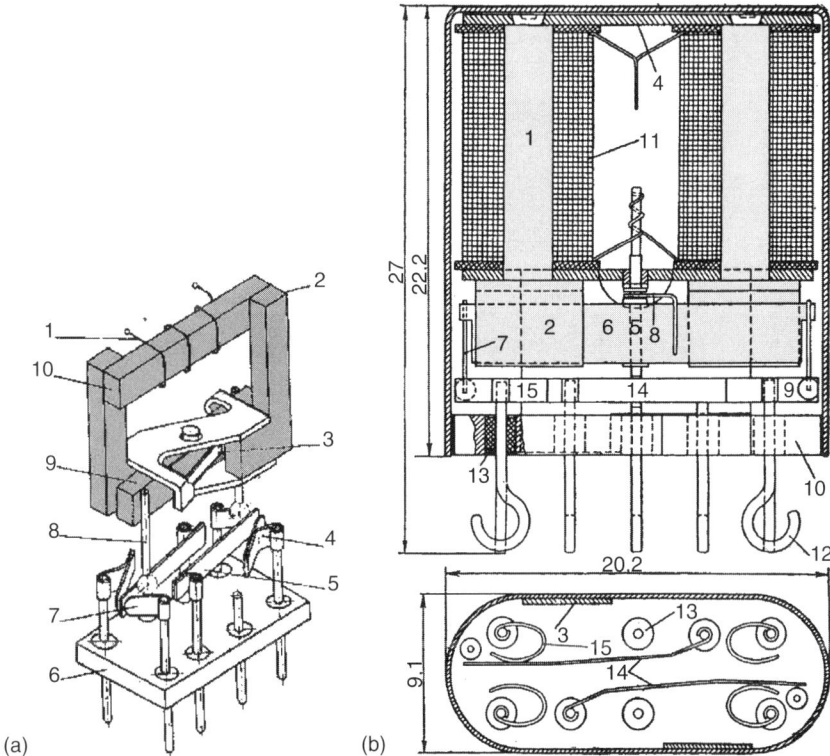

FIGURE 2.11
(a) One of the most popular variants of construction of the turning-type magnetic system, widely used in miniature relays. 1 — winding; 2 — pole of the magnetic core; 3 — restoring spring; 4 — fixed normally open contact; 5 — moving contact; 6 — heel piece of relay; 7 — fixed normally closed contact; 8 — pusher with a small insulating ball, weighted at the end; 9 — turning armature; 10 — core. (b) Miniature MV-type relay (weighing 14 g) produced by "Elgin National Watch Co" in the 1960s. 1 — core; 2 — pole; 3 — supporting bracket of the magnetic system; 4 — connecting plate of the magnetic core; 5 — rotary axis of the armature; 6 — armature; 7 — pusher of the armature with a small insulating ball, weighted at the end; 8 — restoring spring; 9 — restraining arm; 10 — heel piece of relay; 11 — winding; 12 — hook-shaped outlets of the relay (for point-to-point wiring by soldering); 13 — hermetic bushing glass insulators; 14 — movable contacts; 15 — fixed contacts.

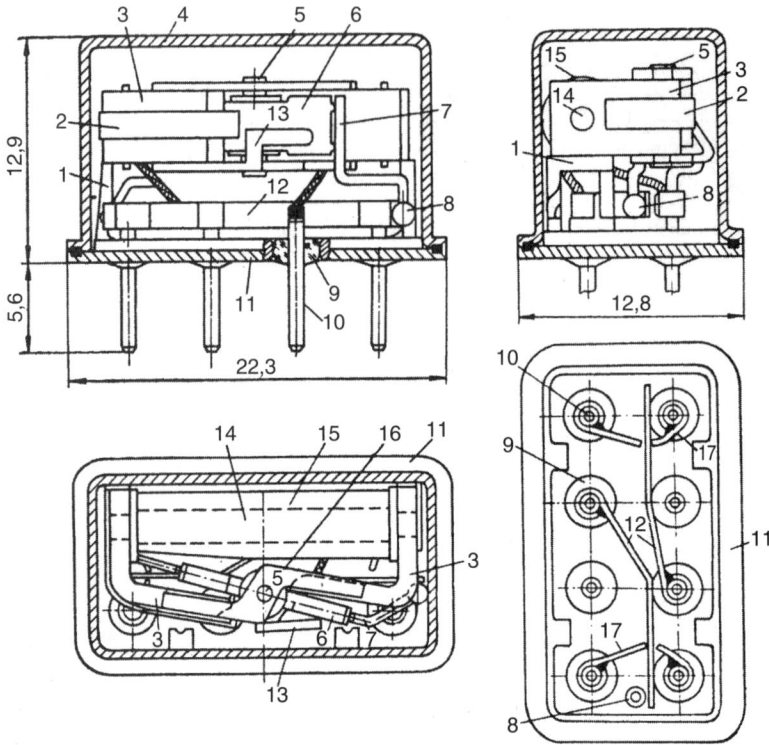

FIGURE 2.12
Miniature hermetic RES47-type relay (Russia) with two switching contacts at 1 A, 34 V, weight 9 g. 1 — post made from nickel silver; 2 — flat spring; 3 — Γ-shaped pole pieces; 4 — seamless copper case; 5 — axle journals (semiaxis); 6 — turning armature; 7 — pushers with small glass balls; 8 — weighted at the ends; 9 — glass insulators; 10 — outlet pins; 11 — heel piece of the relay; 12 — movable contact points; 13 — armature stop; 14 — iron core; 15 — free-bobbin coil; 16 — straps from nickel silver; 17 — stationary contact points (silver).

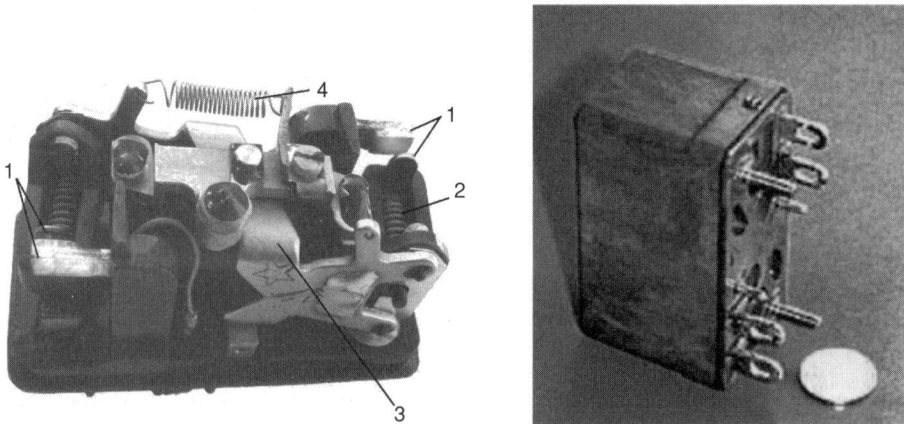

FIGURE 2.13
Relay 8E-123M type for military applications (Russia). 1 — closing contacts; 2 — spring for adjusting contacts; 3 — turning armature; 4 — restoring spring.

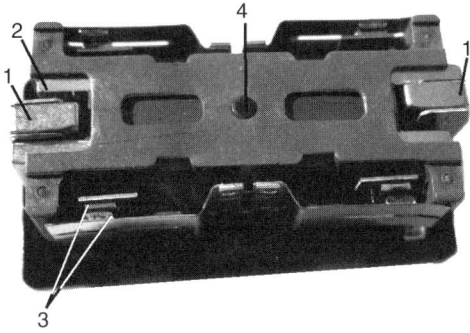

FIGURE 2.14

SP2-P-type relay produced by the SDS Relais company (16 A, 250 V). 1 — stationary core fixed coaxially inside the coil; 2 — movable poles of the armature; 3 — contacts; 4 — rotation axis if the armature.

This relay, with a switching current of 40 A (on each contact) can be used in temperatures ranging from –60 to +85 °C, on shock exposure of up to 100 g, and can also withstand exposure to dew, hoar-frost, salt, mist, mould, and other environmental factors, according to the Russian military standard B.20.39.404–81. At the same time the relay is of quite a small size (57 × 28 × 45 mm) and weight (80 g). Its service life is 20 years, with a storage life of 25 years. The core of such relay is fixed inside the coil with its ends going beyond the coil. A Π-shaped armature with movable contacts covers the coil. Such construction is quite typical of relays of this kind (Figure 2.14).

Magnetic systems with a turning armature are also widely used in protective (measuring) relays (current and voltage relays), which are often used in the electric power industry because one can accurately adjust the trip level by regulating volute spring force of the clockwork type (Figure 2.15). The Russian RES-8-type relay (Figure 2.16) also has a very interesting and original construction. Having one winding, the magnetic system of this relay has four poles (3), and an armature (2) in the shape of a cross, formed by four plates. The butt-ends of the plates are linked by means of a hollow muff with a molded axle. Another peculiarity of such relays is a protective case used as a magnetic core through which the magnetic flux is closed. This case is made of steel, making such relays quite heavy (110 g). One more interesting element is a magnetic armature damper consisting of two small stationary magnets (5) holding the armature in the initial position during vibrations that may affect the relay.

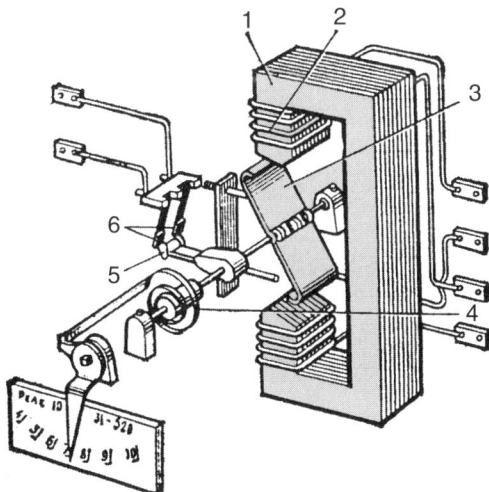

FIGURE 2.15

Magnetic system with a turning armature, used in protective relays. 1 — magnetic core; 2 — winding; 3 — turning armature; 4 — spring; 5,6 — elements of the contact system.

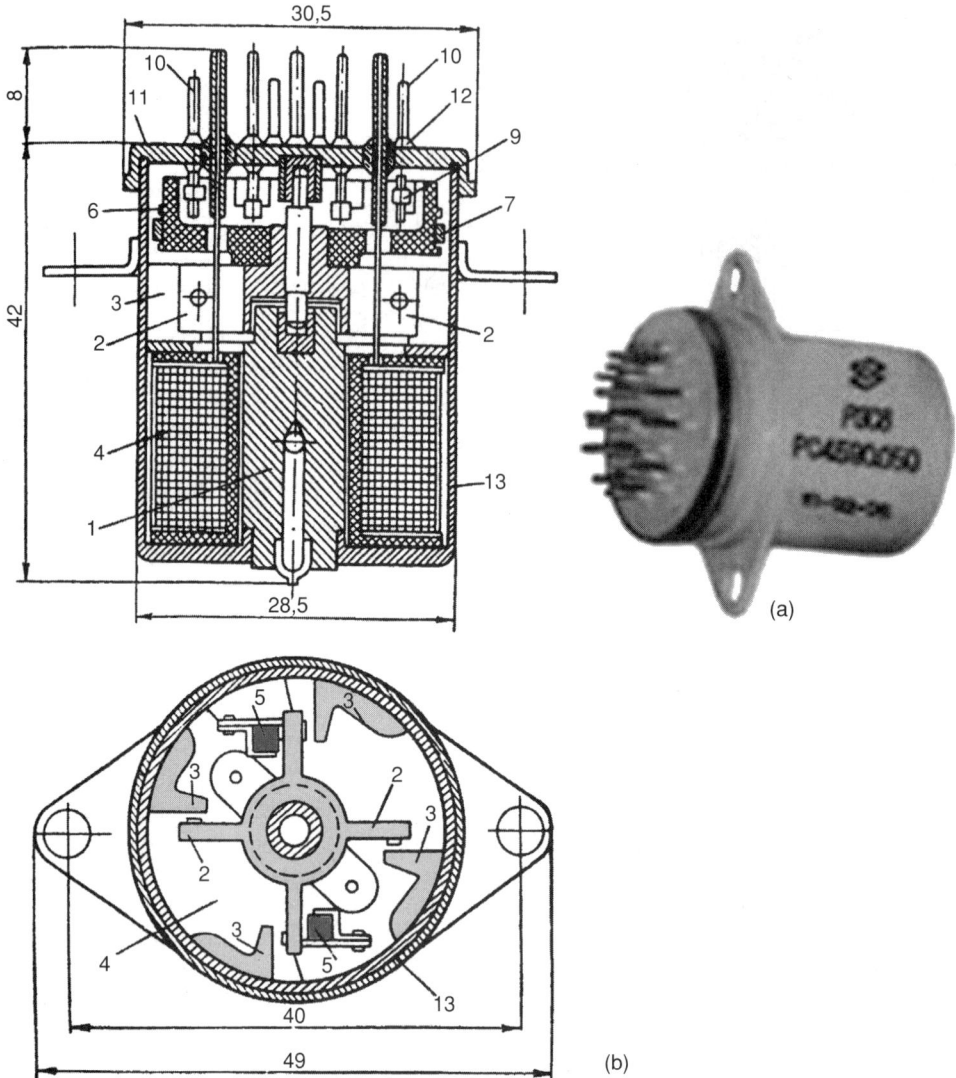

FIGURE 2.16
External view (a) and construction (b) of hermetic multi-contact RES-8 type relay. 1 — core; 2 — armature; 3 — poles; 4 — coil; 5 — stationary magnet; 6 — movable Teflon cup; 7 — damper circle; 8 — movable contact spring; 9 — stationary contact; 10 — outlet pin; 11 — heel piece from Kovar; 12 — glass insulators; 13 — steel case.

2.3.3 Direct Motion (Solenoid-Type) Systems and their Peculiarities

Direct motion systems of the "f" and "g" types are used in power relays, so-called "magnetic starters." They have been so named because originally such relays were designed for direct start up of asynchronous motors, and were also used in time delay relays with an airtime delay mechanism, described below.

Solenoid-type (e, h, k type) systems are used both in miniature relays (Figure 2.17), and in large devices with greater mechanical force.

In 1952, the Bell Company together with Western Electric Company designed an original Ш-shaped magnetic system (Figure 2.18), with a П-shaped armature (contacts were also located on both sides of the winding — Figure 2.19). The ends of this armature

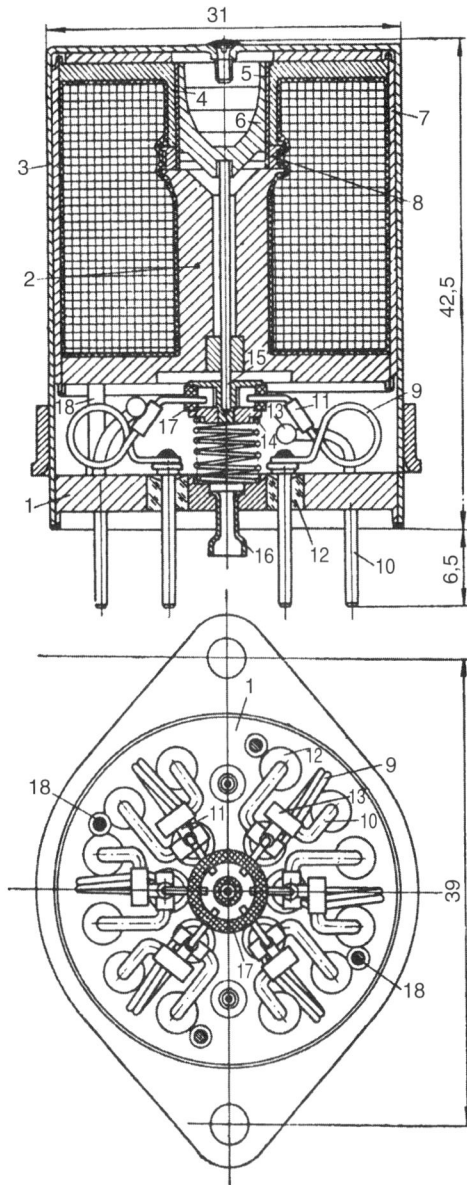

FIGURE 2.17

Miniature Mark-II-type relay produced by "Electro Tech Co." and an exact replica RES-39-type relay (Russia) with a retractable solenoid-type armature. 1 — heel piece; 2 — core; 3 — steel case; 4 — steel cup; 5 — guide brass tube; 6 — hollow retractable armature with a glass-shaped conical stopper; 7 — iron body; 8 — nonmagnetic bushing; 9 — movable contact springs; 10 — small pin; 12 — glass insulator; 13 — stationary contacts; 14 — cylindrical spiral spring; 15 — rod; 16 — union through which the relay is filled with dry air and sealed; 17 — pusher; 18 — steel posts with the magnetic system fixed.

were fixed on the free ends of a П-shaped flat spring. The core of the relay was Ш-shaped, its middle rod inside the coil and the outward ones outside of it. When the relay is energized, the П-shaped core lies level on the outward rods and the end of the internal rod sticks out of the winding, closing the magnetic circuit.

The advantages of this magnetic system are simplicity of production and a very slight magnetic resistance due to the absence of joints, the wide coupling surfaces of the armature and the core, which are the poles of this magnetic system. This relay was quite large, measuring 48.4 × 37.2 × 115 mm. (The year 2004 marks 52 years since it was invented.) It is all the more surprising to come across the descendant of this relay in a printed circuit card on a modern electric device. This is an A2440-type relay produced by

FIGURE 2.18
A magnetic Ш-shaped system for AF-type relay. 1 — winding; 2 — Ш-shaped core; 3 — П-shaped armature; 4 — flat П-shaped spring; 5 — pusher of c ontacts.

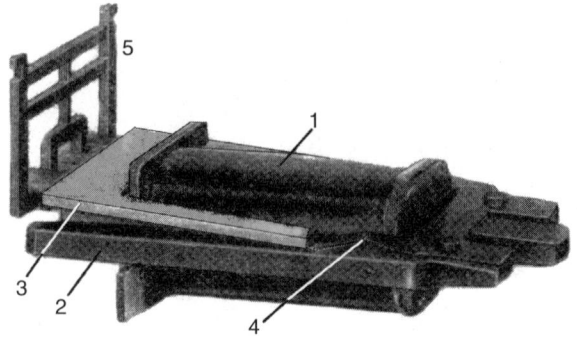

FIGURE 2.19
Relay produced by the "Western Electric Company" with a Ш-shaped magnetic system.

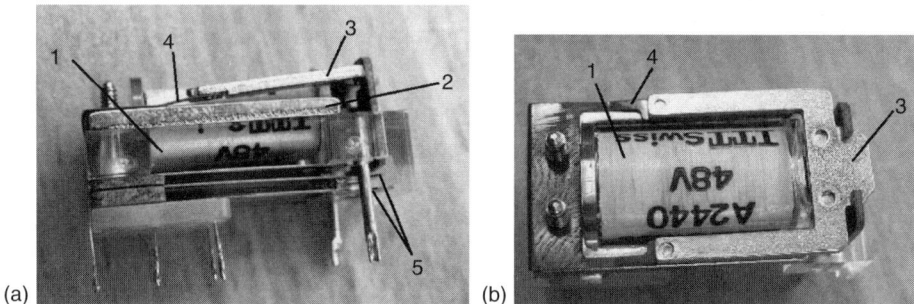

FIGURE 2.20
(a) Modern miniature A2440-type relay, (b) along with the original 52-year old design of the magnetic system. 1 — coil; 2 — Ш—shaped magnetic core; 3 — П-shaped armature; 4 — П-shaped spring linking the armature and the magnetic core; 5 — contact springs.

the firm ITT Swiss (Figure 2.20). At first sight it is difficult to recognize a construction with a 52-year history in this miniature device with a transparent plastic case whose size is $28 \times 15 \times 14$ mm, but that is really so.

Some types of magnetic systems are rather sensitive to their position. Relays with unbalanced and heavy armatures without a restoring spring may not work properly when incorrectly located in space. Among these are, for example, RPN-type relays,

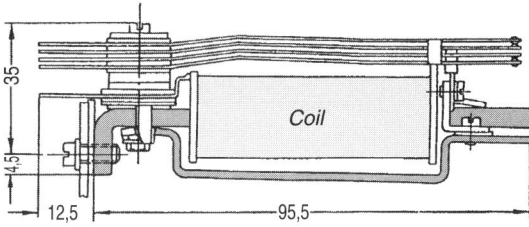

FIGURE 2.21
RPN-type relay.

which used to be quite popular with their analogs, produced for a long time in many different countries (Figure 2.21).

First, such relays can be used only in stationary equipment and second, they should be specifically located in space. For example, RPN-type relays should be located vertically. Some modern relays, especially relays with magnetic systems of alternating current, also have that same disadvantage, for example, popular relays of the DIL EM type, produced by the Klockner Moeller Company, have a forbidden position when the armature is located vertically at the bottom (Figure 2.22). This forbidden position is intentionally marked on the box (Figure 2.23). But then most people are reluctant to read instructions on boxes when dealing with such a "simple" and well-known device as a relay. If assembled incorrectly,

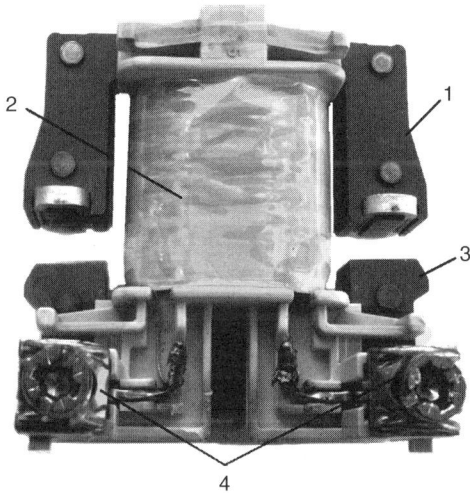

FIGURE 2.22
Ш-shaped magnetic system with a retractable armature of a DIL EM-type relay. 1 — stationary part of the magnetic core; 2 — coil; 3 — retractable armature; 4 — outlets of the coil (A1, A2).

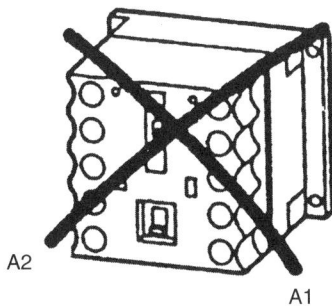

FIGURE 2.23
Notice on box warning of forbidden position when assembling a DIL EM-type relay.

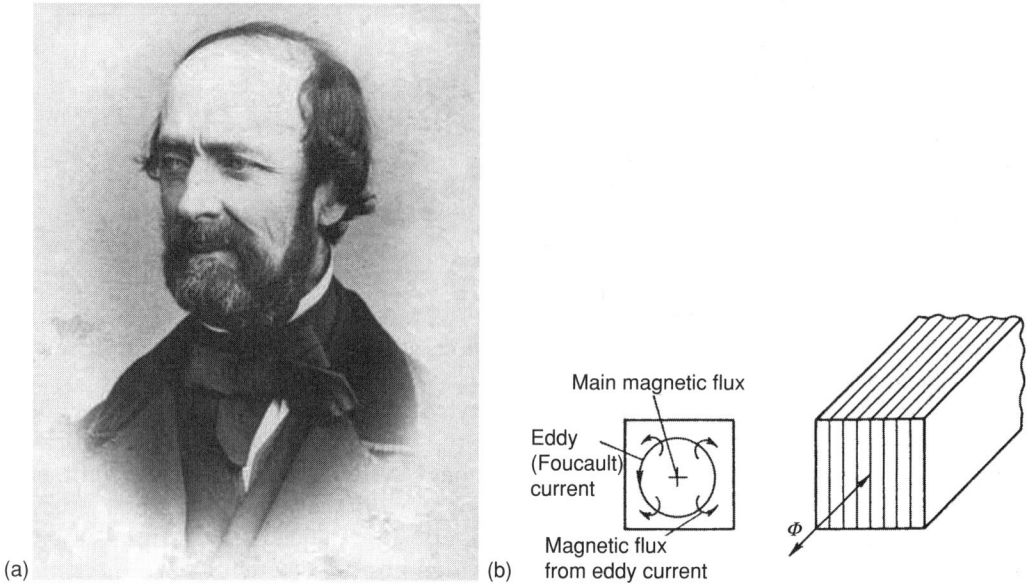

(a)   (b)

FIGURE 2.24
(a) Foucault Jean-Bernard-Léon; (b) eddy currents in an AC magnetic core.

the relay is not energized at all when switched on (the electromagnetic force created by the winding is not enough to lift the heavy armature located in the extreme bottom position of the magnetic system, with a big air-gap). It starts to chatter, or is not energized at all. If left switched on in that situation its winding will quickly burn down. The author himself experienced such an incident when such an automation system suddenly broke down, although it had worked properly for many years. Analysis of the reason for the malfunction showed that the relay had been improperly located. It was rather difficult to convince colleagues to believe it because the system had worked properly for a few years. This could be explained by the fact that at the initial stage of operation the relay was clean and friction between movable units weak. The relay became energized even in the incorrect position because its magnetic system was constructed with some reserve. When that reserve was overcome by growing friction, the relay malfunctioned. This was easily proven by the fact that when the malfunctioning relay was rotated 180°, the automation system started working properly again. Unfortunately we often hear people talking about bad relays, but not about bad engineers constructing automation systems without taking into account those peculiarities of the relay.

2.4   Differences between AC and DC Relays

It is well known that magnetic cores of any AC devices are laminated, that is, they are assembled from separate slim plates with a specific coating of high resistance. That is why *eddy currents* or *Foucault currents* (Foucault Jean-Bernard-Léon — French scientist, Figure 2.24),   producing additional heating of the magnetic core, have small value.

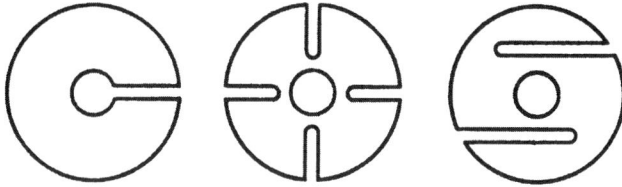

FIGURE 2.25
Core with slits.

In constructions of existing relays, one can observe quite a strange phenomenon: some AC relays have solid magnetic cores (for instance, MKU-48-type relay) and other DC relays, quite on the contrary, have laminated ones.

In the first case, in order to reduce the price of such relays additional heating from eddy currents is simply neglected. Such an approach is used in relatively small constructions with a small section of the magnetic core (for instance, in the MKU-48-type relay mentioned above the magnetic core is made of pressed iron about 4 mm thick. It is only natural that they failed to take on the additional cost of a laminated magnetic core).

In the second case, a laminated magnetic core is used in large constructions of DC relays in order to boost the speed of operation. During the transient process of switching large relays on and off, the magnetic flux in the magnetic core, changing rapidly as it does in AC relays, can cause eddy currents to appear. The point of the laminated core is to prevent that. Of course this process is time-bound so the eddy currents will not produce considerable heating. The problem is in another area here. Eddy currents weaken the main magnetic flux and enlarge the make delay and release time of the relay.

In DC relays with a massive round core, which is difficult to laminate, the core is sometimes simply cut up (Figure 2.25). We can almost always see a copper turn (ring) on the pole of the laminated magnetic core of AC relays, and two similar turns on boundary rods on Ш-shaped magnetic cores (Figure 2.26c).

Calculations show that area covered by the ring should constitute 0.7 to 0.85 of the total area of the pole. That proportion typifies correctly designed relays, but sometimes one can also come across simplified constructions with a ring covering the half of the pole. Such a ring, covering part of the area of the pole, indicates that such a relay is meant for alternating current. Sinusoidal alternating current in the winding of a relay, I_m $\sin(\omega t + t)$, produces a magnetic flux, $\Phi_m \sin \omega t$, lagging from the current by the angle φ_m affected by the eddy currents (see Figure 2.27).

Magnetic flux of alternating current produces an electromagnetic force F_m fluctuating at double frequency that can take on only positive values, from zero to its maximum, even at the negative sign of the magnetic flux. At the same time a permanent counterforce of the restoring spring F_S affects the armature of the relay. When the fluctuating electromagnetic force is weaker than the counterforce of the restoring spring, the armature of the relay loses contact with the pole of the magnetic core, and is pulled back when the level of the electromagnetic force is restored (a deenergized armature is in the interval A–A). Thus, vibration will appear in the armature of an AC relay.

In a magnetic system with a shading ring the main flux Φ_0 branches out to two fluxes Φ_1 and Φ_2. The flux Φ_2 passing through the ring produces current which in turn produces its own magnetic flux Φ_r lagging by one phase from the main flux. Thus the total magnetic flux $\Phi_\Sigma = \Phi_2 + \Phi_r$ will pass through that part of the pole covered by the ring. Throughout the covered part of the pole, and the uncovered one, these fluxes produce the electromagnetic forces F_1 and F_2 (see Figure 2.28), each joined to the other by a specific angle.

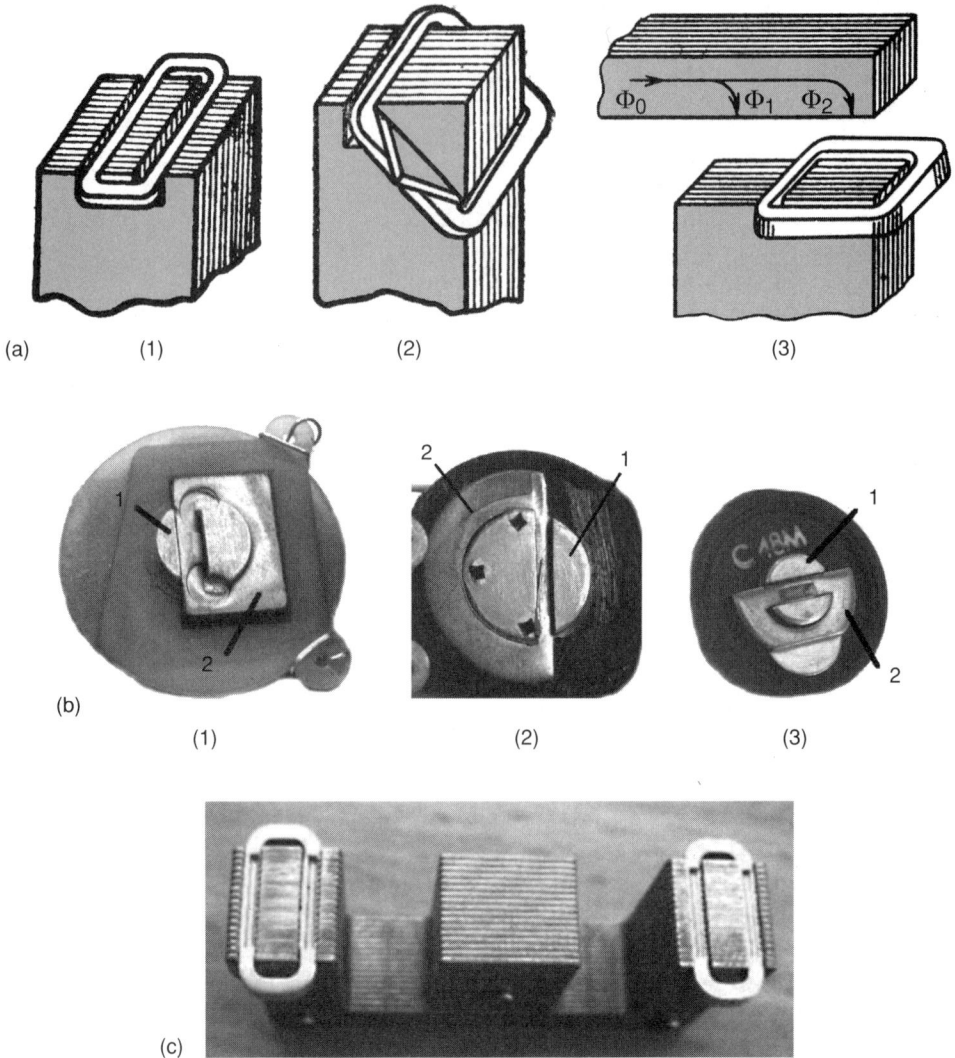

(a)   (1)   (2)   (3)

(b)

(1)   (2)   (3)

(c)

FIGURE 2.26
(a) Types of shading rings on the pole of the laminated magnetic core of an AC relay. (b) External design of shading rings of different form. 1 — pole of the core; 2 — shading ring; (c) Ш-shaped magnetic core with two shading rings.

Obviously, that angle can be no more than $90°$. The lower the pure resistance of the turn (it is usually pressed in the form of a ring of tough-pitch copper), and the closer the proportion between that part of the pole uncovered by the ring and the covered one, to the optimal value 0.7 to 0.85, the less the lagging from this angle will be. In these cases the total electromagnetic force F_Σ fluctuates not from zero, as in a magnetic core without a ring, but from a certain minimum value F_0 up to the maximum one F_m (Figure 2.28). If the ring is installed properly, F_0 will always be greater than the counterforce of the spring, and the armature of the relay will not fall away. However, one should bear in mind

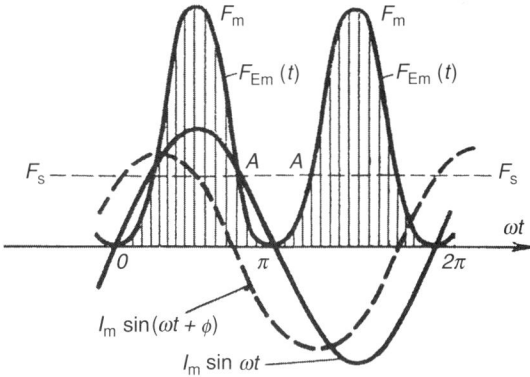

FIGURE 2.27
Fluctuations of magnetic flux and electromagnetic force in the magnetic system of an AC relay.

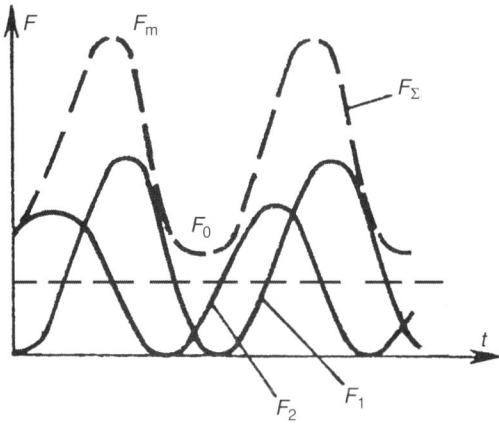

FIGURE 2.28
Electromagnetic forces in magnetic systems of relay with a shading ring.

that the use of a shading ring may lead to a certain attenuation of the average value of the electromagnetic force. This fact must be taken into consideration because due to the sinusoidal character of the changes in the current the average value of the electromagnetic force in an AC relay will be half as much as that in a DC relay, other conditions being equal.

2.5. Some Auxiliary Elements Improving the Relay Operation

In some *DC relays*, like the popular MKU-48-type relay, one may come across *a shading ring on the pole, made of electric steel*. Several questions emerge here. First, what is the purpose of such a ring in a DC relay? Second, since the resistance of steel is greater than that of copper, the effectiveness of the surge control of a magnetic flux with the assistance of such a ring will be less than if a copper ring were used, but what then is the effect of the use of such a ring?

You may have guessed that this ring has nothing to do with surge control of altering current, and if you guessed this, you are right! The point is that the core and the magnetic

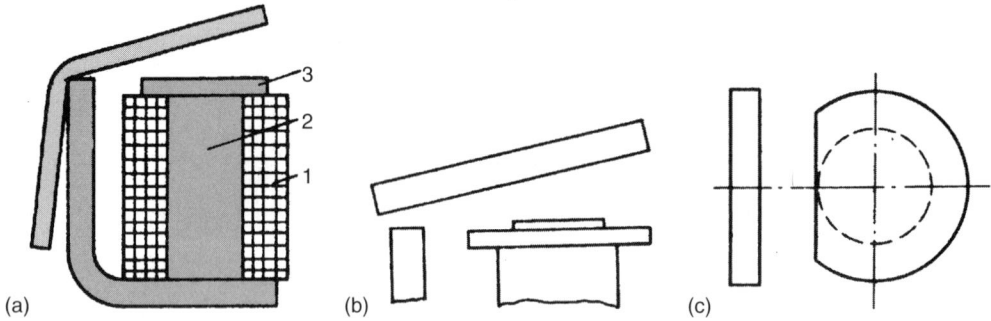

FIGURE 2.29
(a) Construction of the pole of a core. 1 — winding; 2 — core; 3 — pole packing; (b) lug on the pole packing; (c) notch on the pole packing.

core of the MKU-48-type relay are made of one piece of steel duly pressed and curved. Such a construction makes it difficult to create a pole lug, reinforcing the electromagnetic force of attraction of the armature. On the other hand there is already a notch for a copper shading ring at the end of the core of an AC relay, so that one needs a die for the production of copper rings and a press for molding these rings onto the pole of the core.

It occurred to some design engineers that one can already use existing equipment and technology to produce a pole lug. All that is needed to be done is to replace the copper with electric steel. This approach was applied in the production of the MKU-48-type relays.

Another component typifying a magnetic system whose purpose is not immediately obvious is a thin-walled copper tube placed directly on the core along the full length, between the core and the winding in some constructions of large relays. Such tubes are approximately 1.2 mm thick (sometimes these are separate shorted-circuit windings) enabling reduction of inductance 5 to 10 times more when a relay is supplied by alternating current of higher frequency (400 to 1000 Hz). This reduction of inductance increases relay performance and facilitates the switching conditions of the winding, along with other switching devices. The pole of the core is usually supplied by a lug (Figure 2.29), which increases magnetic flux in the gap, and increases pulling power as well.

Increased diameter of the core end (because of the pole lug) leads to an increase in tractive effort, and at the same time to contraction of the gap between the opposite polar elements of the magnetic system (that increase of leakage flux). That is why there is an optimal diameter of pole lug for each particular construction of relays (Figure 2.30). To weaken the negative effect caused by this increase of leakage flux, the edge of the pole lug is cut off.

In many relays, a nonmagnetic layer made of thin lamina is fixed to the end of the core touching the movable armature. As this layer is often transparent, sometimes it can only be found when the relay is taken to pieces. Sometimes instead of a layer a hole is drilled in the center of the armature, into which a thin copper pin, sticking out a bit from the armature, is molded. This too may be hard to notice as well.

The function of that small nonmagnetic component is the opposite of the function of the lug on the pole packing. It slightly weakens the main magnetic flux but also reduces the time for the drop out of the armature when the winding is deenergized. In small relays with a weak restoring spring, this will also prevent sticking of the relay due to residual magnetization of elements of the magnetic circuit.

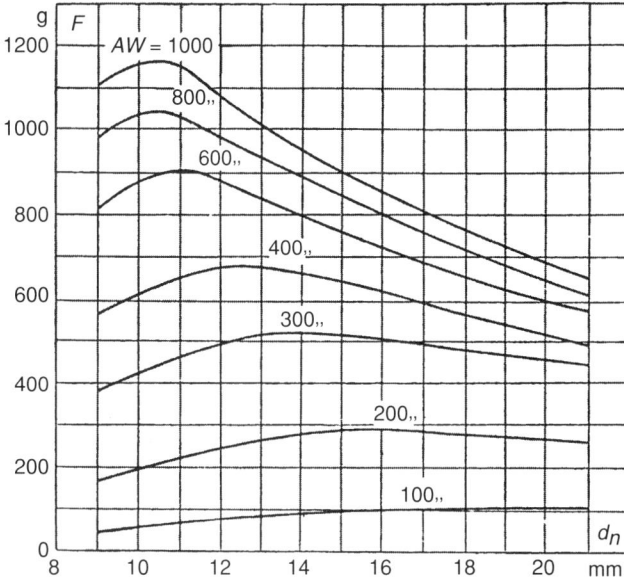

FIGURE 2.30
Curves illustrating the dependence of tractive effort of an RKN-type relay (in grams) from the diameter of the pole packing, with different ampere-turns of the winding (AW = 100–1000).

2.6. What Happens When a Relay is Energized

When a relay is energized one can observe some interesting variations of current in its winding (Figure 2.31). Just after the winding of a relay is connected to a power source the current increases according to the pattern of the 0–*a* curve, according to the exponential law as in a circuit with inductance. One may then observe a slight decrease of current on the interval *a–b* after which the current continues to increase until it reaches the steady-state value, but along another curve *b–c*. All of this becomes increasingly clear when one

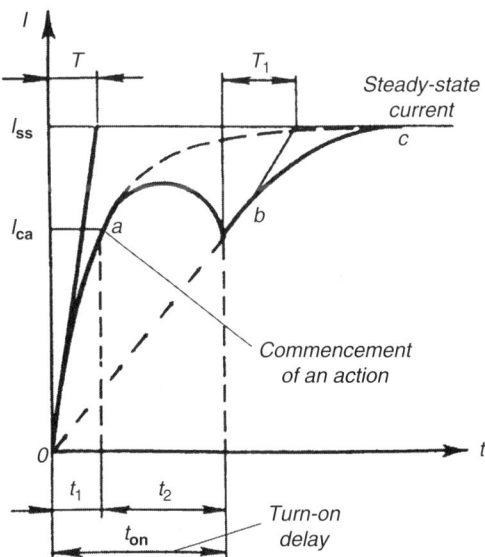

FIGURE 2.31
Current variation in the winding of a relay, when connected to a DC circuit.

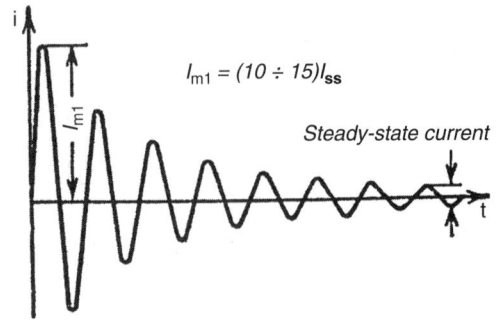

FIGURE 2.32
Transient process of switching on an AC relay.

takes into account the fact that during the energizing of a relay the inductance of the winding does not remain stable, but changes drastically as the armature is attracted to the core. If we place a nonmagnetic stopper between the armature and the core, preventing movements of the armature, and then switch the relay on, the current in the winding will vary according to the pattern of the 0–a–c curve which corresponds to the magnetic system with small inductivity and a rapid time constant T. If we press the armature against the core and then switch on the relay, the current will vary according to the pattern of the 0–b–c curve, corresponding to the magnetic system with greater inductance and a greater time constant T_1. When a relay is energized naturally the change from one curve pattern to the other is automatic which is why the current curve may look so strange. Figure 2.31 illustrates that the make delay of a relay t_{on} consists of two constituents t_1 and t_2. The first, t_1, is the current-rise time in the inductive circuit (winding), up to the value at which the electromagnetic force becomes greater than the counterforce of the spring, and the armature begins to move towards the core (I_{CA} is the pickup current) and the second, t_2, is the time of motion of the armature. This second constituent is also called the *travel time* of the armature. At point "b" the relay has already been energized and further current rise in the process of winding up to the steady-state value will not tell on the relay. Here a question may arise: If the relay has already been energized why does the current in the winding continue to rise, if it leads to further heating of the winding and increases electrical energy consumption, etc. Could it be that such relay was improperly designed and that point "c" should have been located on *the curve, much closer to the points* "a" *and* "b?"

In fact the nominal voltage (current) of a relay will always be greater than its pickup voltage (current) because of the following three factors:

First, we must take into account the technological diversity of the values existing in relay production. It is almost impossible to tell the exact operating value. For example, for the same construction of a relay with a nominal voltage of 12 V, actual pickup voltage value may vary in the range of 6 to 9 V. The relay will always work at the 12 V mentioned in the specification, and will save a lot of trouble for its manufacturer.

Second, differences in characteristics of relays may emerge in the course of their operation: adjustment changes and sometimes new gaps in the magnetic system of the relay may appear.

Third, the relative height of the nominal voltage, compared with the actual pickup voltage, enables enhancement of the reliability of the relay operation. Just imagine that you have a relay of an unknown type and you want it to be energized from a power source of 6 V. A simple experiment will show that the relay is energized at 5.8 V. Is not that great? Now try to imagine that for some reason (and there can be really dozens of reasons) the power source voltage goes down by just 0.3 V. This is enough to prevent the relay from being energized. That same situation will occur if the temperature goes up in the course of its usual operation. Then, resistance of the winding and a voltage of 6 V may not be enough for the relay to be energized. Imagine how resistance of the winding will change in real conditions of exploitation of a relay in the standard range of temperatures varying from −20 to +40 °C (not to mention the military standard, which is −60 to +85 °C).

Fourth, an increase of voltage imposed on a relay above and beyond its operational voltage may considerably reduce the actuation time of the relay (that is, it enhances the operating speed of the relay). All this must be taken into account when relays are designed. There is a special coefficient called the "safety factor for pick-up" which characterizes the ratio of nominal values of a relay (current, voltage) to its actual pickup values. Usually this factor varies in the range of 1.2 to 1.8.

When an AC relay is energized one can observe a damping aperiodic process (see Figure 2.32), characterized by a strong initial current (the armature is not attracted and impedance of the winding is low) exceeding the steady-state value by 10 to 15 times. Because of this peculiarity of an AC relay, the operational mode of such a relay is almost optimal: at large gaps, with the great magnetic resistance of the magnetic circuit (low winding inductance and impedance), the winding produces a great electromagnetic force when the relay is energized (because of the great initial current). After that (when gap is zero and magnetic circuit closed) there is no need for such a great current and electromagnetic force. The winding automatically reduces it because its inductance (impedance) is high in this position.

Such operational mode enhances the operating speed of the relay (the so-called *self-forcing*) and provides the possibility of extension of a specific gap, but frequent switching on may lead to overheating of the winding by strong starting currents, and if the armature sticks in the initial or intermediate position the winding may quickly burn down.

It is clear even from speculative reasoning that the voltage (current) at which the armature of a relay is attracted to the core will always be a bit stronger than voltage (current) needed for its retention, let alone the voltage at which the armature falls off from the core. The ratio of dropout value to operating value is called the *release factor of relay*. The value of this factor is always less than 1 and can fluctuate in the range of 0.1 to 0.99 for real relay constructions. The less this release factor is, the more stable and effective the function of the relay is; however, this cannot be applied to measuring relays. In such relays, which are energized at predetermined values of current and voltage, the release factor is frequently very high, depending on the design values of the relays and the

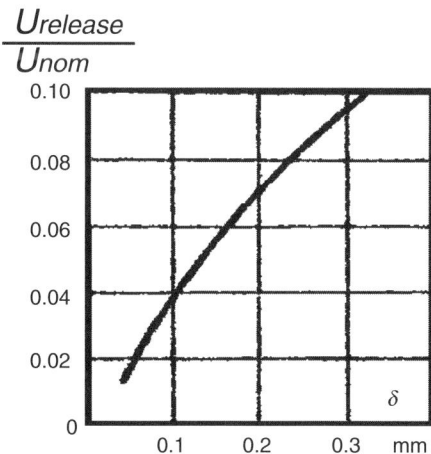

FIGURE 2.33
Typical dependence of the release factor from the value of the terminal nonmagnetic gap remaining in the magnetic system, after the relay has been energized.

character of the input value. The release factor of AC relays is usually higher than that of DC relays because two times a period ripple of magnetic flux creates conditions for easier separation of the armature from the core. Changes in the release factor in particular constructions of relays may be caused by modification of the terminal gap in the magnetic system (Figure 2.33 — thickness of nonmagnetic layer between the armature and the core, see above) or change of the restoring spring force.

There are also circuit means of increasing release factors of relays (Figure 2.34). After it has been energized one can connect in series, or parallel to the winding, an additional resistor or a semiconductor element with nonlinear voltage–current characteristics (a Zener diode, for instance) which rapidly changes its conductivity at the relay energizing point.

2.7   Windings of Relays: Types and Design Features

One of the most important elements of a magnetic system is winding. It is made in the form of a coil wound around with wire 0.02–2 mm in diameter (for power relays and contactors the diameter may be even larger). In the 1940–60's they mostly used wire with textile and silk insulation, in spite of their hygroscopic and a greater diameter. It was a forced measure because varnish of that time did not have enough mechanical strength especially on wire of greater diameter. At present, windings of relays are wound with wire with enamel or glass insulation. The latter has far better electric characteristics, not involving gas while being heated (which is essential for vacuum-processed relays), but it is more expensive.

In the course of operation of a relay, electromagnetic forces of interaction between turns and current cause mechanical stress in the coil, and heating and cooling of the coil causes thermal stress. Electric operating voltage is applied between turns of the coil, between the layers of winding, and also between the coil and the core. In transient processes of switching one must also deal with over voltage. Coils of relays can be affected by numerous negative environmental factors such as increased humidity of the atmosphere, salt spray, dew, mould, etc. The design of the coil should provide for safe performance of the relay under such circumstances.

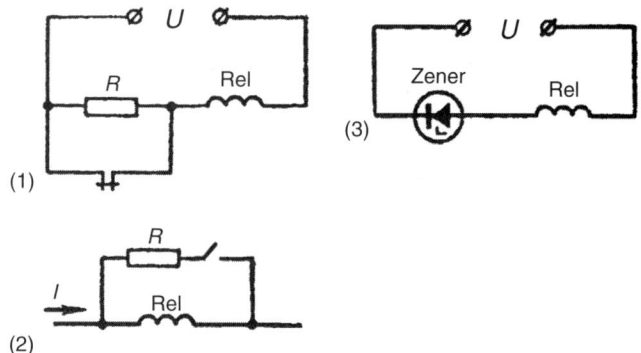

FIGURE 2.34
Circuit means of increasing release factor of relay. Variants 1 and 2 can be applied to AC and DC relays. Variant 3 is only for DC relay;

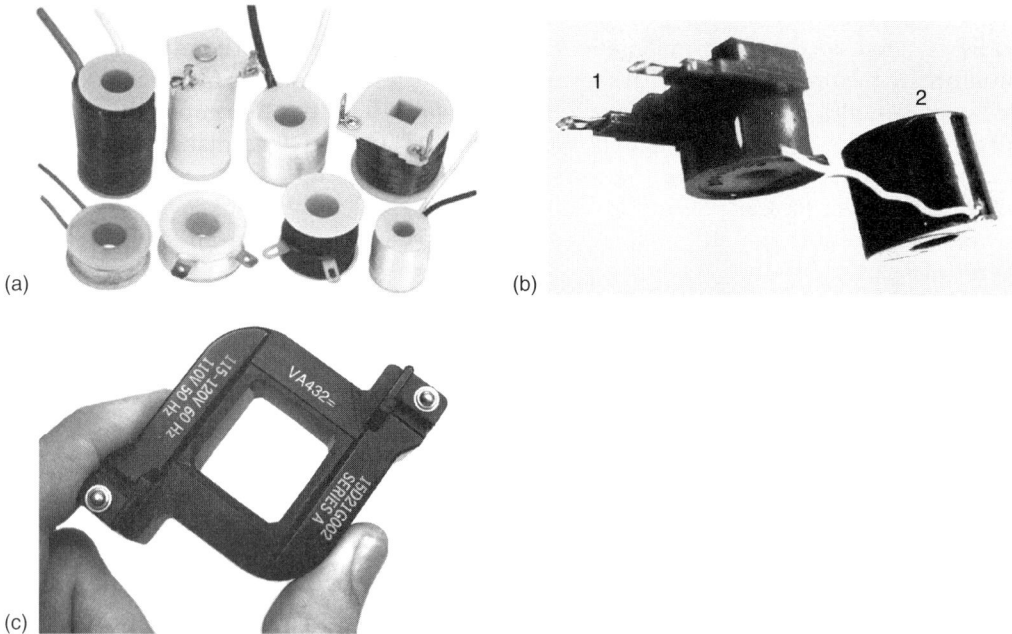

FIGURE 2.35
Constructions of coils of relays: (a) unimpregnated coils covered by protective film; (b) encapsulated coils (1 — coil molded with plastic; 2 — coil encapsulated by epoxide resin); (c) rectangular shape coil molded with plastic.

Coils can be bobbin and free-bobbin. In bobbin coils the winding is wound around a bobbin (made of insulating material or of metal with insulating lacquer). In free-bobbin coils, the wire is wound directly around the core of the magnetic system. In the past, in bobbin coils, paper impregnated with varnish and cardboard was widely used as a bobbin, but now it is usually replaced with plastic. Free-bobbin coils are rarely used in modern relays.

Depending on function and service conditions coils may also be covered by protective film and varnish, or impregnated with varnish or epoxy encapsulant (Figure 2.35). As the coils are encapsulated by epoxide resin, vacuum and increased pressure alternately affects them. This enables removal of unnecessary air bubbles from the winding and helps the encapsulant pervade between the turns of the coil. For impregnation epoxy encapsulant with low viscosity (for example, STYCAST 2651–40 type) is normally used. It is obvious that encapsulated coils provide better reliability of relays than unimpregnated ones, but they also increase the cost of the relay.

Coils may be square or round-shaped (in section) and very different in size (see Figure 2.35). There are special methods for optimal designing of relays, establishing

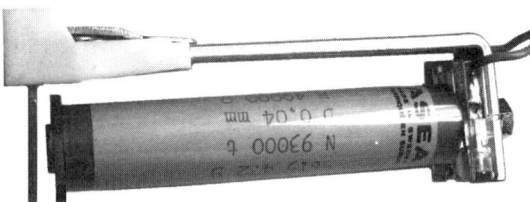

FIGURE 2.36
Elongated coil of a highly sensitive clapper armature relay with winding, containing 93,000 turns of wire with a diameter of 0.04 mm (resistance of 40 kΩ).

certain proportions between geometric sizes of coils. Obviously, it is impossible to endlessly increase the electromagnetic force of the coil with a core of small diameter by multiplying the number of turns of the coil (i.e., by increasing the external diameter of the coil or its length), because the core will quickly reach the saturation mode. In this case, further strengthening of the magnetic flux will not lead to strengthening of the electromagnetic effort of attraction of the armature to the end of this core; however, in some cases, when it is necessary to fulfill high requirements for relay sensitivity (i.e., in case of extremely small current in the coil) such a small working current must be compensated by considerable multiplication of turns.

Multiplication of layers of winding will not lead to considerable increase of sensitivity of a relay, because each following layer will be farther from the core than the previous one and its effect will be weaker than that of the previous layer. That is why in this case, the coils usually "grow" lengthwise and not in width (Figure 2.36).

The number of turns of a coil can be limited not only by the size of the bobbin, but also by a number of technical values characterized by *coil fill factor K1, coil correction factor K2, and winding fill factor K3.*

The coil fill factor is the ratio of copper section of the coil to the entire section of the coil. This factor depends on the type of winding, on the thickness and form of the bobbin of the coil, and on the thickness of the wire insulation (also on the form of wire section which is usually round for relays).

The coil correction factor is determined by the type of the winding of a coil with different winding densities. There can be the following types of winding:

- *Layer winding*: when turns of the same layer are placed densely to each other and turns of a higher layer are placed just above the turns of a lower layer (Figure 2.37)
- *Quincunx winding*: when turns of a higher layer are placed in gaps between turns of a lower layer
- *Pile winding*: when turns are placed without any specific order, in layers

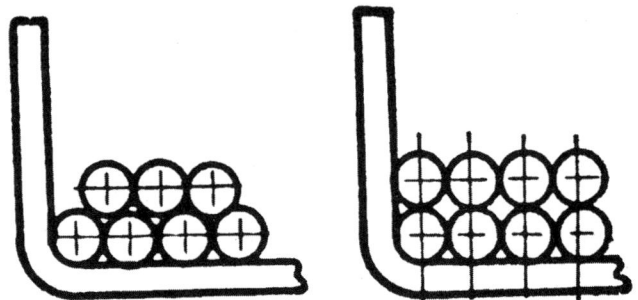

FIGURE 2.37
"Quincunx" and "layer" winding of coils.

FIGURE 2.38
Laying of winding wire without spacers and with a dielectric spacer of thickness δ between layers.

Quincunx winding enables us to obtain the maximum value of coil correction factor (16% more than layer winding); however, it is rarely used because of difficulties of automatic winding for thin wire.

The winding fill factor is taken into account because in some types of power relays (though very rarely) dielectric spacers are used between some layers of winding (Figure 2.38).

In the course of operation of a relay, its coil is heated by the Joule effect of current passing through copper wire, and due to other reasons mentioned below. Since the coil of a relay is a heterogeneous body consisting of materials with different thermal conduction: copper, varnish, plastic, and air layers, it is only natural that there will be different temperatures in different layers of the coil. The more monolithic the coil, the less the differential temperature will be between the outer surface and the internal layers (Figure 2.39).

Impregnation and compounding increase total thermal conductivity of the coil, thus increasing its heat emission by 5 to 10%.

Distribution patterns of temperature in coils of AC and DC relays are different. This is because the iron core in a DC relay where the coil is placed plays the part of a heat sink, bringing down the temperature of the adjoining layer of winding. In an AC relay, the core is a source of heat emerging due to the action of eddy currents (see above). That is why while constructing a DC relay an effort is usually made to try to reduce thermal resistance (or intensify thermal contact) between the winding and the core, and in fact on the contrary, to insulate the winding from the core for AC relays.

According to experimental data for mid-sized relays, an air gap of 0.25 mm on the side between the bobbin of the coil and the core reduces heat transfer of the coil by

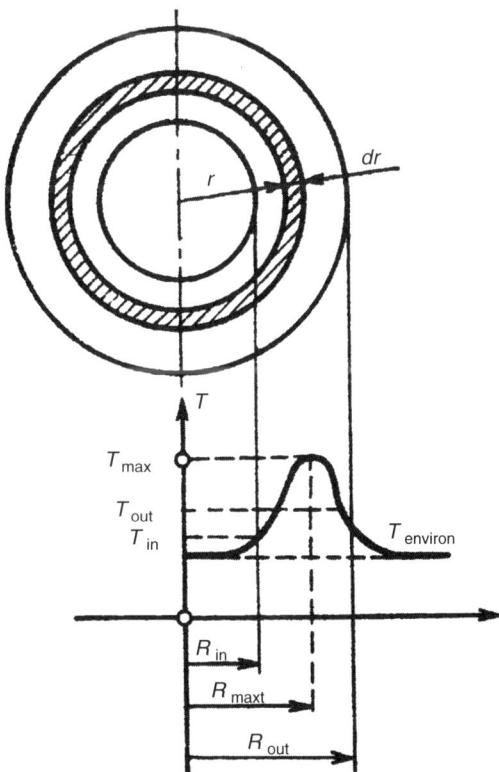

FIGURE 2.39
Radial distribution of temperature in winding of a DC relay. R_{in} — inner radius of the coil; R_{out} —, outer radius of the coil; R_{maxt} — zone of maximum heat; T_{max} — maximum heat temperature; T_{out} — temperature of the outer surface of the coil; T_{in} — temperature of the inner surface of the coil (adjoining to the core); $T_{environ}$ — ambient temperature.

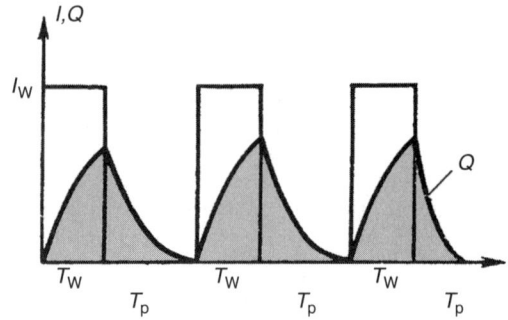

FIGURE 2.40
Diagram illustrating variations of temperature of the coil (Q) in the short-term mode when working current is passing through it. T_w — working time; T_P — pause time.

approximately 8%, and a gap of 0.5 mm — by 11%. A free-bobbin coil with direct winding on the core increases heat transfer by 10%. Maximum temperature of coil heating is limited mostly by the thermal endurance of the winding wire. Long-term effects of increased temperature may result in its rapid deterioration and decay. There are several classes of thermal-life characteristics of insulating materials (Table 2.1):

For copper winding wire insulating varnish and enamel corresponding to the classes A, B, and F is used. If high-temperature winding wire is employed, one also uses bobbins from heat-resistant plastic of the corresponding class, and the outer outlets of the coil are made of wire with Teflon insulation.

The heating of coils depends directly on the operational mode of the relay. One can distinguish the following modes of operation:

TABLE 2.1

Classes of Thermal-Life Characteristics of Insulating Materials

Classes of Thermal-Life Characteristics	Temperature Characterizing the Given Class, t (°C)	Materials Corresponding to the Given Class
Y	90	Unimpregnated fibrous materials made from cellulose and silk, and certain types of plastic: polyethylene, perspex (polyacrylate), polyurethane foam, and some others)
A	105	Impregnated fibrous materials from cellulose and silk or some types of plastic, varnish, enamel immersed into liquid dielectric material (transformer oil, for instance)
E	120	Some organic polyethylene films, plastic (polypropylene), and others
B	130	Materials based on mica, asbestos, glass fiber used with organic cohesive and impregnating encapsulate, varnish, enamel, plastics (glass fiber laminate, polystyrene, polycarbonate, etc.)
F	155	Materials based on mica, asbestos, glass fiber used with organic cohesive and impregnating encapsulate, varnish, enamel, and some brands of Teflon (polytetra-fluoroethylene)
H	180	Materials based on mica, asbestos, glass fiber used with organic-silicon cohesive and impregnating encapsulates, organic-silicon elastomer, poliarylate, polysulphone, etc.
C	>180	Mica, ceramic solids, glass, quartz, Teflon, polyamides, polyphenylene

- *Long-term mode:* when winding is under current for a long period of time
- *Short-term mode:* when winding is switched on for the period until its temperature reaches a steady-state value (this usually happens in a matter of seconds), and the pause between repeated switching is so long that temperature of the winding has time to come down to the ambient temperature (Figure 2.40)
- *Intermittent (cyclic) mode:* when performance periods of winding under current alternate repeatedly, with periods of silent state of the coil.

The latter mode is very popular and quite typical of relays used in technological automation and control systems.

For quantitative characteristics of such a mode we use the value called *"duty cycle."* It is a ratio of duration of the "on" state to duration of the whole cycle. Sometimes in technical literature it is given in terms of percentage.

On the face of it the first (long-term) operation mode of a relay may seem the most difficult and the two others easier, but in practice this is not always so. It depends on the value of the duty cycle and the ambient temperature. If the duty cycle is more than 50% this means that the coil is in the on state for a longer period of time than it in the off state.

Depending on thermal balance in a particular construction of a relay, such conditions can be created that the total temperature of the winding will increase with each following cycle, and eventually the temperature of the winding will reach a steady-state value (Figure 2.41).

If we take into account the fact that the transient process of switching on an AC relay will be accompanied by great starting currents (see above), frequent switching of the relay may make the short-term mode even more difficult than the long-term is.

The heating of insulating materials may accelerate their deterioration, causing mechanical decay and electric breakdown. For example, according to M. Vitenberg an RES-6-type relay already has frequent breakdowns after 350 h of operation at a temperature of +85 °C and within 700 h 50% of such relays go out of order. That is why industrial relays are designed in such a way that temperature of the winding does not exceed normally ambient temperatures by more than 15 to 20 °C. In some cases, however, while constructing miniature relays it may be necessary to provide the desired value of electromagnetic effort of the armature for small-sized coils and lightweight relays. In such cases the manufacturer has nothing else to do but increase current density in the winding wire of the coil, in other words to let the big current pass through a small coil wound around by thin wire. In modern miniature hermetically sealed relays for military applications winding temperature may reach as much as to 180 to 200 °C. Usually such relays are not designed for long-term

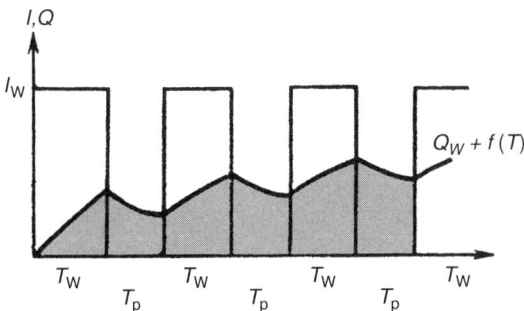

FIGURE 2.41
Diagram illustrating variations of temperature of the winding (Q) in the short-term mode when working current is passing through it. T_w — working time; T_p — pause time.

energization, and are used in on-board systems of single-mission weapons such as missiles, torpedoes, etc.

When browsing through specifications of such relays (only for relays produced in Russia) one may come across a value called "continuous (or *total*) time of a relay under working voltage (current) at working ambient temperature." For some types of light-weight miniature hermetic relays this time may only be some hundreds of hours. Unfortunately, most of the western relay manufacturers do not usually mark this parameter in their catalogs.

3

Contact Systems

3.1   Designs of Basic Types of Contacts

A contact system is the second essential component of any relay. Usually it consists of several elements (Figure 3.1). Current-conducting elements are made of elastic materials (as a rule, beryllium bronze or phosphor bronze) providing not only current supply to contacting surfaces but also the necessary contact pressure. Contact straps (which are actually contacts) are made of materials of high electric conductivity and resistance to electric erosion. They are riveted, soldered, or welded on by silver solder to current-conducting springs. Contact straps are usually made in the form of rivets or pins (Figure 3.2). Riveted attachment of contact straps is less reliable than welded attachment, because of the considerable increase in transient resistance of a rivet at the point of its joint to the contact spring, caused by heat cycling in course of its exploitation. Such straps, which are often bimetallic (two-layer), consist of a copper base and contact material, an alloy based on silver. Bimetallic contacts of a relay were invented long time ago.

Strange as it may seem, it was radio engineering that made the greatest contribution to the creation of powerful bimetallic contacts resistant to an arc. This field of technology is usually associated in our minds with miniature electronic elements or radio sets. It may seem that radio engineering as a wireless communication facility must have replaced a telegraph with its posts, wire, and sounders and in the long run put an end to wide use of relays in telecommunication systems, but the point is that early radio communication systems differed greatly from those we are familiar with today. Remember, the first radio set was a "storm indicator," that is, a device registering electric strikes of lightning, but in order to establish radio contact it is necessary to have not only a radio set, but also a radio-transmitter, and what can replace lightning? The answer is obvious: an electric spark, which may be regarded as miniature lightning.

Indeed, an electric spark turned out to be a source of electromagnetic radiation in a wide range of frequencies, including high frequencies, which are quickly spread over

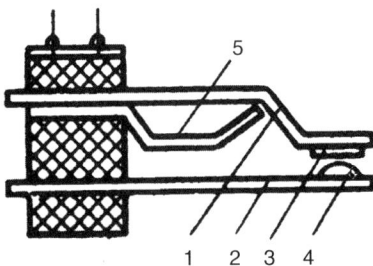

FIGURE 3.1
Popular types of relay contact systems. 1,2, — current-conducting elements; 3,4 — contact straps; 5 — stop.

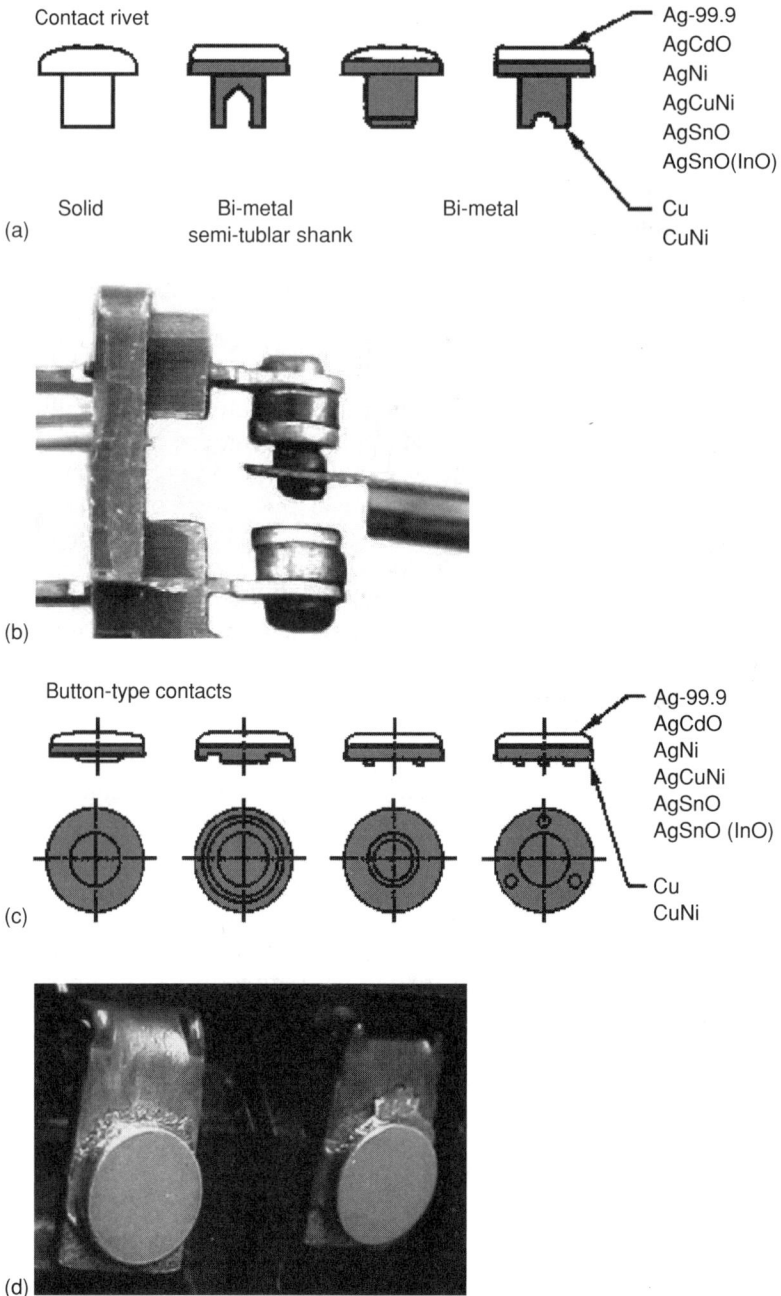

(a) Contact rivet

Solid

Bi-metal
semi-tublar shank

Bi-metal

Ag-99.9
AgCdO
AgNi
AgCuNi
AgSnO
AgSnO(InO)

Cu
CuNi

(b)

(c) Button-type contacts

Ag-99.9
AgCdO
AgNi
AgCuNi
AgSnO
AgSnO (InO)

Cu
CuNi

(d)

FIGURE 3.2
(a) Rivet-type contact straps; (b) rivet-type contact straps in an actual relay; (c) button-type contact straps;
(d) button-type welded contact straps in an actual power relay.

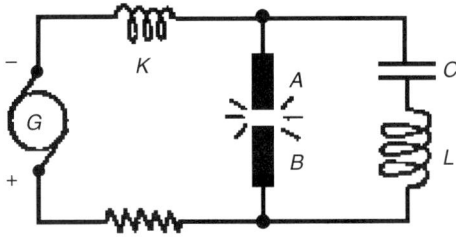

FIGURE 3.3
An oscillatory circuit with an arc is a power source of radio waves.

long distances. A continuous pulsating arc strikes an electric circuit consisting of a coil, a capacitor, active resistance, an DC power source, and two carbon electrodes (Figure 3.3), and is a power source of radio waves.

The first development of the oscillating arc was done by the Danish scientist, Valdemar Poulsen. Australian born and Stanford educated Cyril Elwell came across Poulsen's experiments and at once realized their commercial potential. In 1909, with the support of investors, he founded Federal Telegraph in San Francisco, in order to commercially exploit arc technology.

One technical drawback of that type of arc transmitter was that it was impossible to connect a Morse key directly to the circuit of the electric arc in order to receive a controlled (modulated) radio-signal, because the arc ran continuously and it was impossible to switch high powers (tens and hundreds of kilowatts) by means of a manual switch.

Here a relay came in useful. Being connected to a DC source, its winding was also in series with a Morse key. Referring to the schematic, note that the arc, denoted by "X," is switched by relay contacts, "K," between the antenna circuit and the dummy load tuned to the same frequency as the antenna circuit (Figure 3.4).

In 1913 Elwell developed and demonstrated this technology to the US Navy. By 1921, 80% of all commercial and military transmitters were of the arc variety. The capacity of such transmitters was continuously growing and eventually reached one million watt

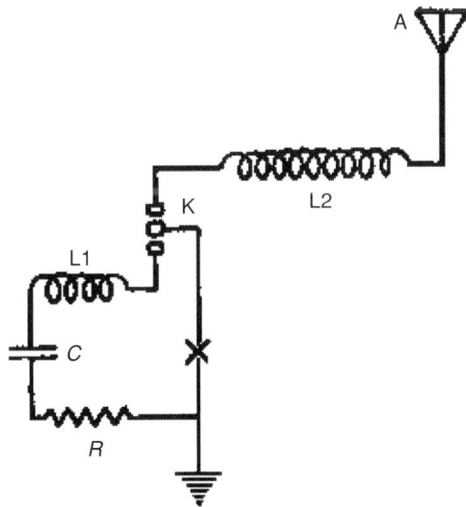

FIGURE 3.4
Schematic illustrating basic principles of operation of an arc transmitter of the early 19th century. X — arc; K — contacts of a modulator.

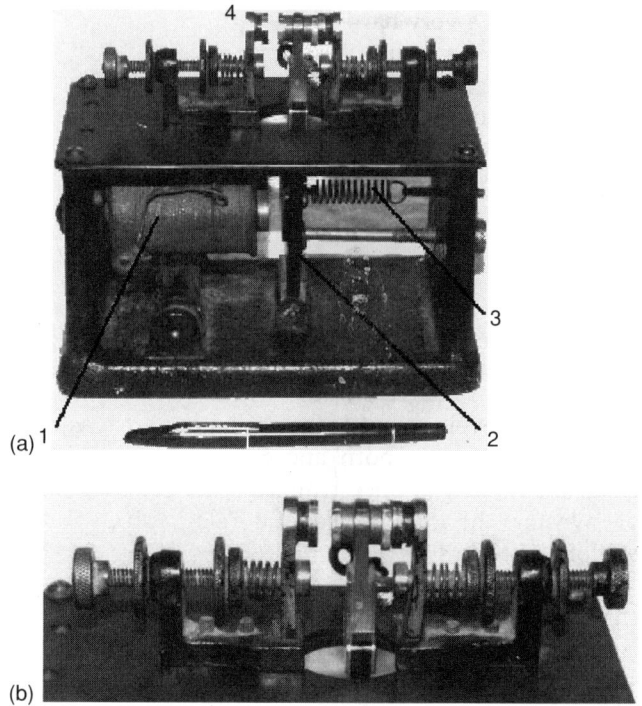

FIGURE 3.5
(a) A modulator relay for an arc transmitter control and its contact system. 1 — coil; 2 — armature; 3 — spring; 4 — contact system (increased magnified in (b)).

(a transmitter installed in Bordeaux, France). Obviously, to produce modulation arcs of such great power, relay transmitters must contain powerful wear-resistant contacts. As can be seen in photos and samples of relays of this type (not the most powerful ones) that have survived until today (Figure 3.5), such contacts consist of two layers of different metals and are quite wide in diameter (3/4 in.). Stationary contacts are supplied with spring-coupler dampers, extinguishing impact energy of a switching contact.

3.2   Silver, Gold, Platinum...

Silver (Ag) is the cheapest material used for this purpose. Due to its ductility, it is readily formed into various contact shapes. As a result of its high electrical and thermal conductivity, transient resistance of contacts is rather low. Silver's tendency to erode and be affected by arcing and its low hardness can be regarded as its main drawbacks. Sulfur-containing atmospheres will produce silver sulfide that increases contact resistance. That is why silver contacts are not recommended in constructions containing ebony, black rubber, or wire insulated with rubber, producing sulfuretted hydrogen (hydrogen sulfide) when heated. Silver contacts are used for switching of small and greater currents at low and medium voltages.

   Platinum (Pt) has a quite high corrosion resistance, being much less prone to erosion in comparison with silver. In pure form it is low in hardness and therefore generally alloyed with other materials. Platinum–iridium alloy (PtIr) combines hardness with excellent resistance in the formation of arcs, and provides excellent performance in corrosive environments. It is widely used for low-power or average-power contacts.

Tungsten is a very hard and refractory material. It is resistant to mechanical wear and welding, but as a result of its tendency to form thick oxide surface films it also requires high contact forces. Its high resistance limits its area of application. Tungsten is usually used as an auxiliary contact in contact systems containing both main and auxiliary (arcing) contacts.

Silver–tungsten alloy (AgW) is hard, with high resistance to contact welding and a high melting temperature. Its drawbacks are a tendency to oxidation and elevated resistance.

Silver–nickel alloy (AgNi) possesses good arc extinguishing properties and low oxidation, as well as high electrical conductivity similar to silver; however, it tends to form sulfide surface films. It is widely used for average-power contacts.

Silver–copper alloy ($AgCu_3$) possesses high resistance to mechanical wear and a weaker tendency to welding when compared to silver. Contact resistance is higher than that of silver.

Silver–palladium (AgPd) has quite high corrosion and sulfitation resistance, and is not prone to contact welding. Among its drawbacks is a tendency to absorb organic gases and to form a polymeric surface film. To prevent this it should be overlaid with gold, which is quite expensive.

Gold–silver (AuAg) possesses quite low and stable contact resistance, even with low currents and voltages and low resistance to welding; therefore, this alloy is used for contacts of measuring circuits with low currents and voltages.

Silver–cadmium oxide (AgCdO) is not an alloy, but rather a ceramic–metal composition. Contacts are pressed from powder, then heated to high temperatures for sintering of the components, calibrated by additional swaging in molds and then annealed for relief of work hardening. This material possesses high resistance to arcing and to mechanical wear and is prone to welding. It has stable properties but its resistance is higher than that of pure silver and it tends to form sulfide surface films. It is widely used for average and power contacts.

Silver–tin oxide (AgSnO) has come to be considered a good alternative to the AgCdO contacts described above. Over the past years the use of cadmium in contacts as well as in batteries has been restricted in many areas of the world; therefore, tin oxide contacts (10%) which are about 15% harder than AgCdO are a good alternative. Also, the above contacts are good for high inrush loads, like tungsten lamps, where the steady-state current is low.

To save space in micro-miniature relays with gaps of hundredths of a millimeter, contacts are formed by pressing-out protrusions on the ends of a flat spring. Sometimes slightly deflected ends of such springs also serve as contacts. Obviously in such cases contact springs should be made of specific materials. As a rule those materials are complex alloys based on silver with admixtures of magnesium (0.15 to 0.3%), nickel (0.1 to 0.25%), gold (1.5 to 2%), or zirconium (0.1 to 0.4%). Specific resistance of such alloys is half as much as that of typical contact springs with a beryllium bronze base, and that is why they can pass greater current. It is very essential when working with micro-miniature relays, to take into account the small size of their springs.

To enhance surface behavior of contacts, they are sometimes covered by gold or rhodium. Gold cover provides a clean surface, low resistance, and high stability of weak-current contacts. Rhodium cover is much heavier than gold and provides higher resistance to mechanical wear, but due to its tendency to absorb gases and form polymeric surface films it can only be used in hermetic relays.

Contacts with a thin cover are not reliable because the thinnest layer protecting a contact from oxidation can easily be damaged in the course of exploitation, when contacts are stripped. Stripping without cover may be expedient only in cases of considerable surface erosion. Stripping of unworn contacts with the help of a file or sandpaper only damages and contaminates the contact surface. Even washing the contact with ethyl alcohol or carbon tetrachloride leaves deposits after drying up. The contact surface can only be cleaned by a hardened iron polished plate, degreased in alcohol, and wiped clean with a dry and clean chamois.

3.3   Contacts with Two-Stage Commutation

As it can be seen from properties of the materials mentioned above, there are no perfect materials possessing low resistance (like silver), high ductility, and at the same time high resistance to arcing at switching (like tungsten). That is why designers invented two-stage contact systems, in which the switching is carried out by tungsten contacts, with the current flowing through silver contacts in a stationary mode. Contact systems of this type have been known for a long time. They are used in average-power relays switching currents of 20 to 50 A, and in power contactors switching currents of hundreds of Amperes (Figure 3.6).

Open RKS-3 type relays with a two-stage contact system shown in Figure 3.6a, have been produced in Russia by the Irkutsk Relay Plant for a long period of time. At present some western companies produce enhanced constructions of relays supplied with a contact system of this type.

When a RKS-type relay is energized, first tungsten contacts 1 close and then, after further spring flexure, silver contacts 2 close. Opening is carried out in a reverse order. A similar algorithm is applied to the contact system of a contactor.

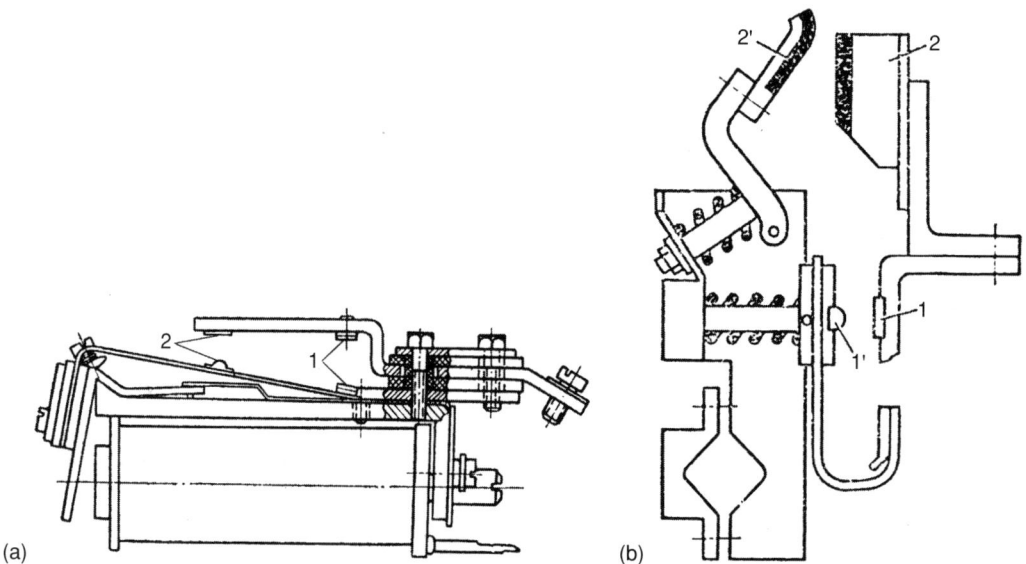

(a)   (b)

FIGURE 3.6
Two-stage contact systems used in (a) relays and (b) contactors. 1 — auxiliary tungsten contact producing switching; 2 — main silver contact.

3.4   What is the Purpose of "Contact Pressure?"

Relay contacts can have different forms. The most widespread contacts are of a flat, conical, and semicircular form (Figure 3.7) and can be applied in the same contacting pair in different combinations.

It may seem that the more contact area there is, the greater the current that can be passed through it, and that therefore a plane-to-plane contacting pair should be the most favorable contact, but in fact, things are much more complex.

As mentioned above, atmospheric oxygen, ozone, and sulfuretted hydrogen can provoke formation of oxide surface films of high resistance; however, it would be wrong to claim that such films cause only damage. Such films considerably constrain forces of intermolecular cohesion between surfaces from adjoining to each other under contact pressure, prevent interdiffusion of the contact materials, and assume the role of lubricant. Nevertheless, such films should be removed in order to provide more reliable contacts at relay energizing. Without a high-power arc on, contact film deteriorates under the heavy mechanical force caused by contact pressure.

In addition, the contact surface turned out to be very rough (Figure 3.8), and that is why electric contacts operate not along the entire osculant surface, but in only those points through which the current passes. Strengthening of contact pressure leads to an increase in the number of such points.

In spite of that, the bigger the contact area, the greater the effort that a magnetic system should develop in order to provide the required contact pressure, and vice versa. Pricking of this dense texture by a needle has been known to cause pressures equivalent to $1\,t/cm^2$ at that point. It is clear then that in small relays with weak effort developed by the magnetic system, one must reduce the contact area by changing forms of contacting surfaces.

With small switching currents and efforts not exceeding 25 to 40 g that are developed on contacts of a magnetic system that widely uses flat-pointed contacting pairs, increased specific pressure is provided. Such combination of contacts makes assembling of relays easier as there is no need for installation of contacting pairs along the common axis. However, when a point is used, electric field intensity between contacts rises sharply, which is why the contact gap should be increased. To switch voltages of hundreds of volts, greater distances are required. In earlier constructions of large relays this presented no difficulty. In modern small-sized and miniature relays with distances between contacts less than a millimeter, two hemispheres are used instead of flat-pointed contacts.

In systems with pressure of more than 50 g, hemisphere-plane and plane-to-plane contact layers are used. In older large relays, contact pressure was adjusted within some limits by proper tension and compression of the springs, and contact pressure was measured with the help of a special dynamometer (Figure 3.9). Such tools cannot

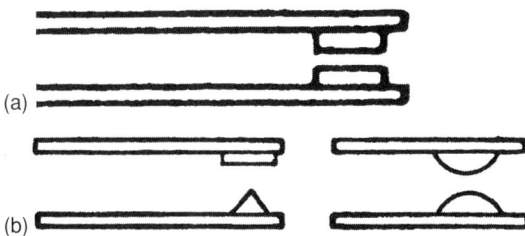

FIGURE 3.7
Forms of contacts of relays.

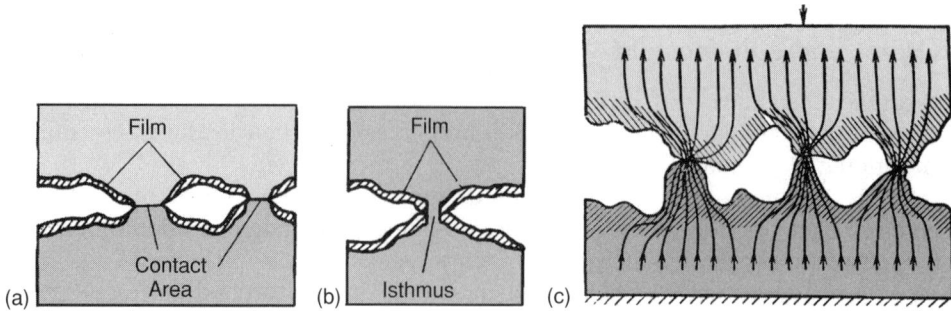

FIGURE 3.8
Surface structure of osculant contacts.

FIGURE 3.9
Dynamometer for measurement of contact pressure.

be applied to modern miniature relays, and therefore such relays cannot be adjusted in the course of operation.

3.5 Self-Cleaning Contacts

Another unconventional way to overcome resistance of films covering contacts is to wipe them when the contacts close. For that it is necessary that the contacts must move with respect to each other.

In some cases mutual micro-displacements of contacts are carried out automatically as they close, as for example in console contacts of relays with an attracted-armature magnetic system (Figure 3.10), where the contacts, which are already touching one another, continue moving. Due to the fact that the console springs of the closing contacts are placed at a certain distance from each other, they move along different radiuses when the contacts touch each other, leading to a certain amount of contact displacement. In other cases, specific, more complex constructions for wiping of oxide films are used for the same purpose.

In power equipment with large and heavy contacts, drive mechanisms enable a certain amount of rolling of the contacts with some mutual slip, causing them to self-clean (Figure 3.11). In such constructions the contacts fixed to a bridge are shifted beforehand with regard to the stationary contacts. At closure they take their place, wiping oxide films as they move as a result of slippage of the bridge along an inclined guide. In more powerful equipment the contacts are rolled during their closure, with a slight degree of slippage of the movable contact with regard to the stationary one (Figure 3.12).

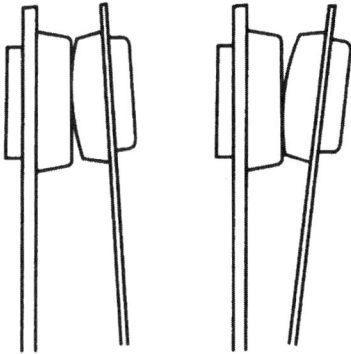

FIGURE 3.10
Mutual contact displacements during their closure, providing destructive effect to oxide films.

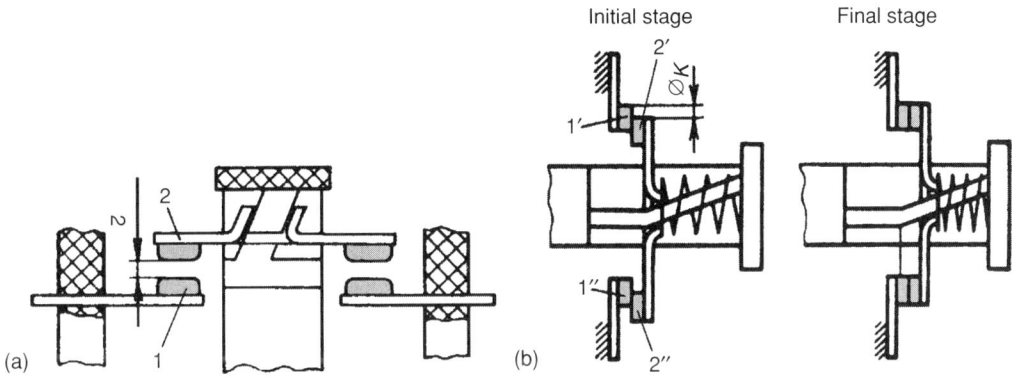

FIGURE 3.11
Construction of a bridge contact enabling the process of mutual contact displacement at closure. 1 — stationary contact; 2 — moving contact.

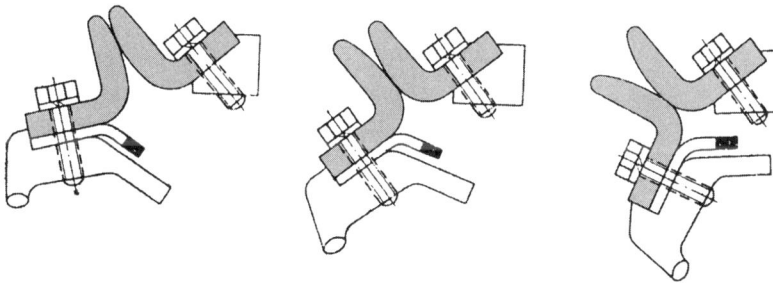

FIGURE 3.12
Contact positions of a more powerful contactor at closure.

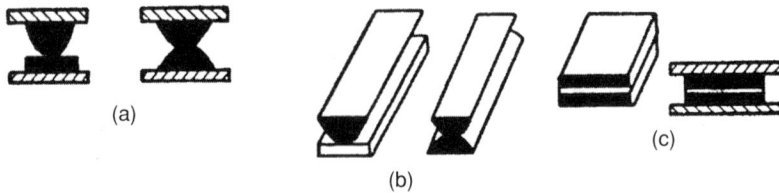

FIGURE 3.13

Various types of contacts used in relays. 1 — point contacts (for currents up to 20–40 A); 2 — line contacts; 3 — plane contacts.

Depending on the type of contacting surfaces, different kinds of contacts are used in different relays: point contacts, plane contacts, and line contacts (Figure 3.13). Each of these contact types is implemented in different construction diagrams of a contact system (Figure 3.14). In sensitive polarized relays (referred to below) a contact system usually looks as it appears in Figure 3.15.

3.6   Self-Adjusting Contacts

In order to enhance the switching capability of the relay, the contact system is made in the form of a bridge (Figure 3.16). Power, partitioned by the contact, is shared by the two serial contact gaps. In this case an open relay of quite a small size ($49 \times 36 \times 41$ mm) switches current up to 30 A, voltage up to 600 V, and power up to 1 kW.

Unlike contacts fixed on long and flexible console springs, a bridge contact is a hard component and requires additional elements to provide for reliable compression of the movable contact into the stationary one, for compensation of shock at closure, and also for automatic installation of the bridge in case there are technological variations of sizes, assembling inaccuracies and additional gaps emerging in the course of exploitation of the relay. All of these requirements are in fact fulfilled in the simplest constructions by means of a spring abutting from the central part of the bridge (Figure 3.17).

In some constructions of power relays designed in the 1950s and 1960s, self-installing bridge contacts were made as separate modules of a single-contact (Figure 3.17b) or multi-contact (Figure 3.17c) relay system. Despite the fact that the relays shown in Figure 3.17b and c look quite big and seemingly old-fashioned, General Electric Co., continues to produce them and include them in some types of power equipment it supplies today. The author came across relays of this type, for example, on a new gas-turbine power station still in the running-in process in 2004.

3.7   When Power does not Equal Multiplication by Current and Voltage

If on the miniature relay it is written: 250 V, 5 A it yet does not mean, that this relay is able to switch load 1250 W, that is, $250\,V \times 5\,A \neq 1250\,W$. Why? Because 250 V in our sample is the maximal switching voltage and 5 A is the maximal switching current. It has not been

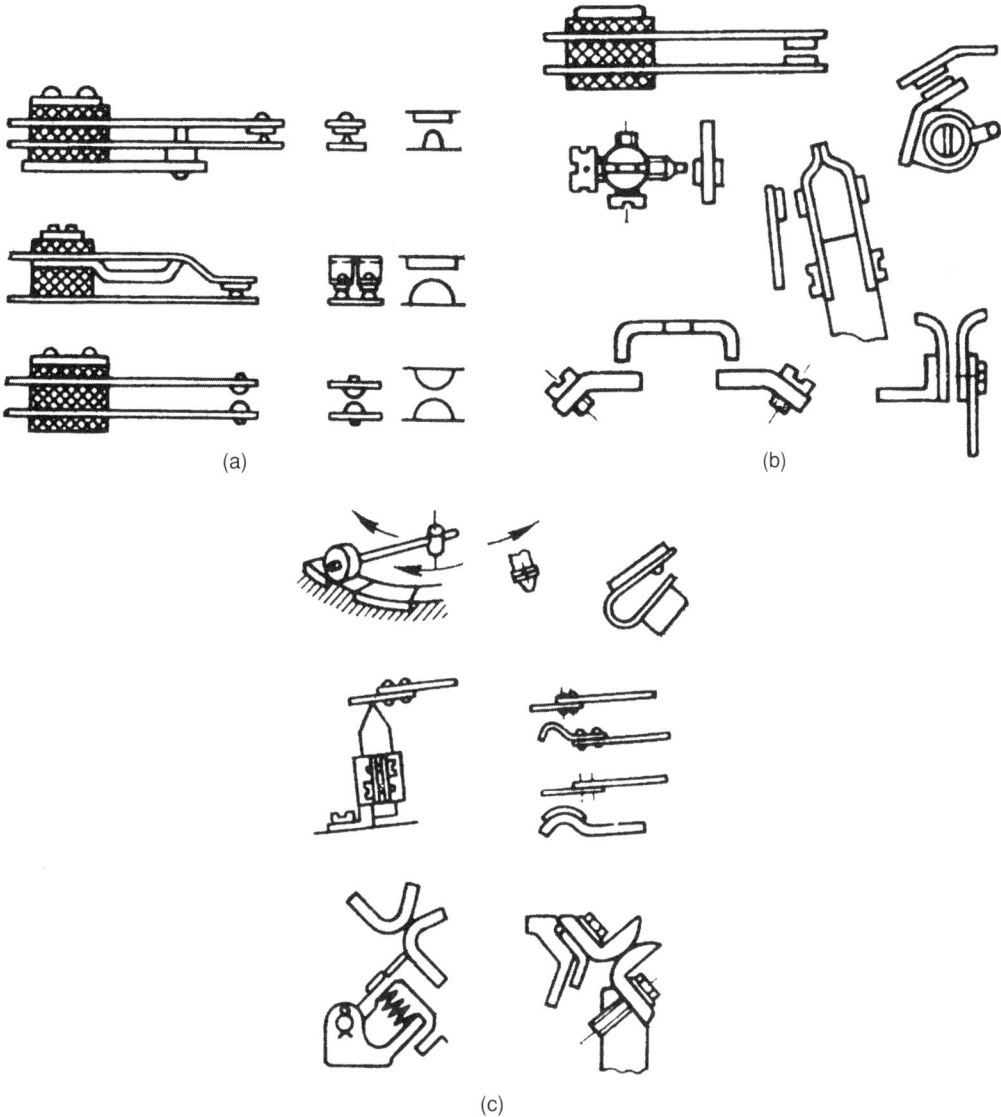

FIGURE 3.14
Different construction diagrams of a contact system with contacts of point (a), plane (b), and line (c) type.

said that both these maximal values can be applied simultaneously. In other words, our relay can switch a voltage of 250 V at current up to 1 or 5 A at voltage of not more than 50 V, that is, the relay has a maximal switching power of 250 W only.

For subminiature relays (which is frequently named as "power relays"), maximal switching power usually equal multiplication by switching voltage (250 V, for example) and switching current (8 A, for example) signed on relay, but also for these relays it is not so simple.

Power switching apparatus (contactors, starters) must have massive contacts, large in size and heavy in weight, supplied with additional powerful springs to provide the needed contact pressure, and with flexible band wireways through which the current is

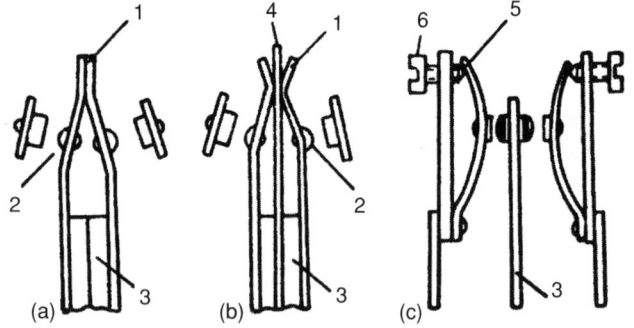

FIGURE 3.15
Contact systems of highly sensitive polarized relays. 1 — current-carrying springs; 2 — contacts; 3 — armature end; 4 — additional spring with silver cover; 5 — unfixed spring ends; 6 — adjusting screw.

fed to the contacts (Figure 3.18). As opposed to this, miniature relays contain very small contacts placed on flexible miniature contact springs (Figure 3.19).

As a rule, subminiature relays which are named power relays, provided with a disproportionately large single contact, can switch quite strong currents (Table 3.1). As has been mentioned above, increase in contact area leads to reduction of contact pressure needed for punching of oxide films. At low voltages on contacts and at low switching currents, this can cause an interruption of the switching process; however, if such relays are not used for switching low voltages and small currents, the oxide films will deteriorate because of thermal impact of arcing on contacts at the time of switching. In addition, massive contacts contribute to better cooling of the arc and to its quick starvation.

Sometimes one can find relays with relatively small dimensions that are really capable of switching high currents (20 to 30 A) at 250 V. One good example is the DI1U-112DMP-type relay, produced by Dongguan Sanyou Electrical Appliances Co. (Figure 3.20). This relay is made as a hybrid device, containing a magnetic system from the miniature relay and a single, normally open, large powerful contact usually used in relays of medium sizes.

When reducing the quantity of contacts to a minimum (that is, down to one) there is sufficient contact pressure even by means of a miniature magnetic system. Naturally, with such a small magnetic system, it is very difficult to provide the large gap necessary between the magnetic core and the armature; therefore, the contact gap in this relay is less (0.3 mm) than with larger relays designed for switching of the same currents. Nevertheless this relay does switch a 20-A current at a voltage of 250 V AC (for resistive load only). Of course, in the switching process a powerful spark between the contacts occurs and the contacts are strongly heated up; however, the author did not find any traces of

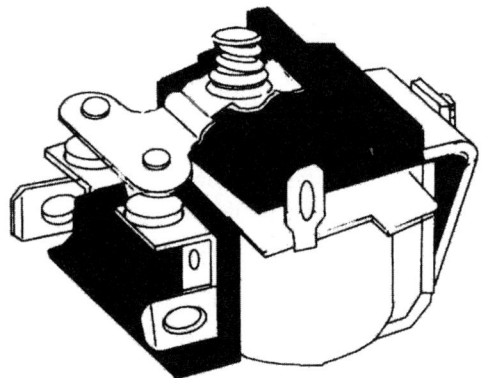

FIGURE 3.16
Open W88UK-type relay with a bridge-type contact, produced by Magnecraft.

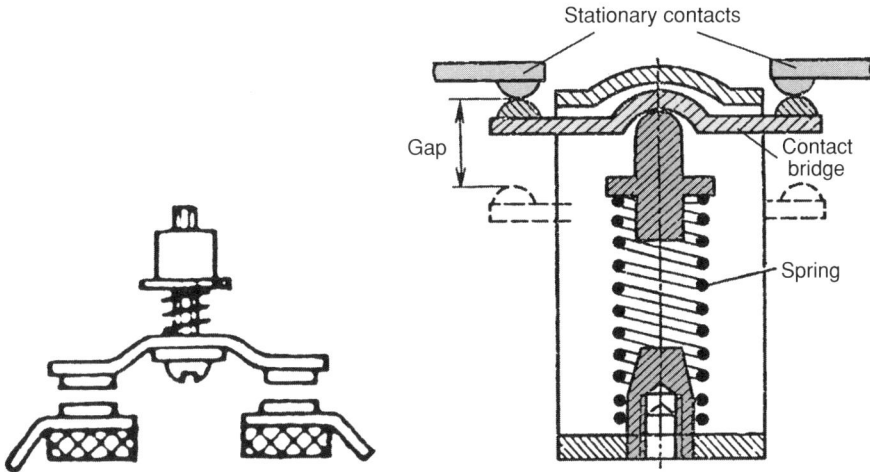

(a)

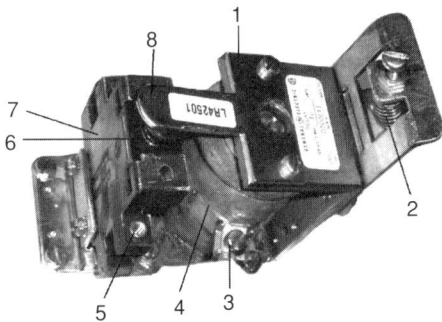

(b)

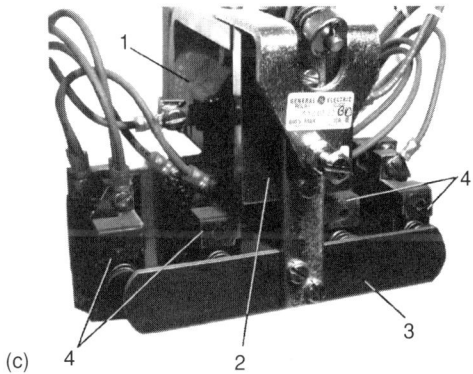

(c)

FIGURE 3.17

(a) Self-adjusting bridge contacts. (b) D300B11-type power relay (600 V, 10 A) with one switched bridge contact made as a separate module (General Electric Co.). 1 — armature; 2 and 6 — springs; 3 — coil outlet; 4 — coil; 5 — one of the outlets of the contact module; 7 — contact module with a switched bridge contact; 8 — contact pusher. (c) Contact system of A100BB3-type relay (600 V, 10 A) consisting of four modules with bridge contacts (General Electric Co.). 1 — coil; 2 — armature; 3 — plate transmitting the force from the armature to contact modules; 4 — contact modules.

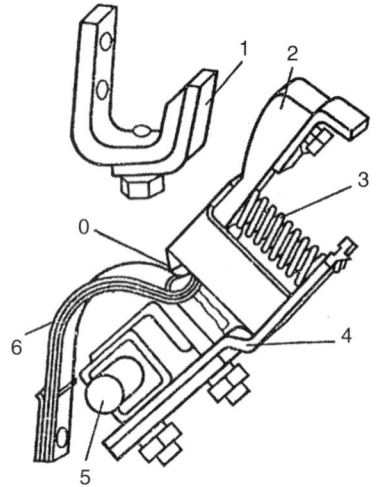

FIGURE 3.18
Construction of a contact system of greater power. 1 — stationary contact; 2 — movable contact; 3 — spring; 4 — contact lever; 5 — contact lever shaft; 6 — flexible band wireways; 0 — rotation axis of the movable contact.

electric erosion on the contact surface after several hundred switching cycles at a current of 22 A.

Switching capabilities of the relay for DC load are approximately ten times worse, than for AC load (Table 3.1). Values specified on the case of the relay usually refer only to AC. Further reduction in switching ability of the relay occurs with inductance in load, even in small ratios (Figure 3.21). Increased inductance of load decreases switching capability of the relay drastically, especially DC. Therefore, all data for switching capabilities of the relays are indicated, usually, only for resistive load.

Additional confusion in the choice of the relay type cause an inrush of current for many kinds of loads (electric motors, transformers, coils of other relays and solenoids, incandescent lamps). All this is inconvenient when it comes to choosing a relay for particular use, and frequently misleads the consumer.

To avoid misunderstanding and to bring different types of relays into sync, special standards (IEC 60947) have been introduced with all types of electric loads divided into so-called "utilization categories." In Table 3.2 there is some data for relay contacts switching auxiliary circuits. As it can be seen, in normal switching mode a relay of an AC-15 (AC-11) category (the most popular one for automation relays) can switch ten times more making current than its stated current and in the mode of infrequent switching 11 times more! For same loads on DC applications (DC-13) switching capacity is in ten times less. This implies that one industrial class relay with switching capacity 2000 VA (200 W)

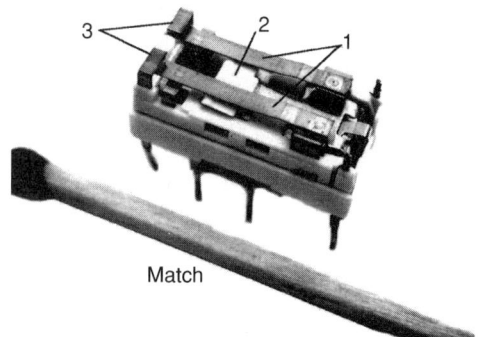

FIGURE 3.19
Miniature relay with weak-current contacts near a match.
1 — contact springs; 2 — pusher; 3 — contacts.

TABLE 3.1

Switching Parameters of Some Types of Subminiature Power Relays

Relay Type (Manufacturer)	Maximal Switching Power (for Resistive Load)		Rated Current and Voltage (for Resistive Load)		
	AC (VA)	DC (W)	AC	DC	For 250 V DC
ST series (Matsusita)	2000	150	8 A; 250 V	5 A; 30 V	0.40 A
JS series (Fujitsu)	2000	192	8 A; 250 V	8 A; 24 V	0.35 A
RT2 (Schrack)	2000	240	8 A; 250 V	8 A; 30 V	0.25 A
RYII (Schrack)	2000	224	8 A; 240 V	8 A; 28 V	0.28 A
G6RN (Omron)	2000	150	8 A; 250 V	5 A; 30 V	–
G2RL-1E (Omron)	3000	288	12 A; 250 V	12 A; 24 V	0.30 A

suitable to AC-11 can switch at once ten lamps on 200 W each on AC, and unable to switch even one such lamp on a DC, because inrush current for incandescent lamps is 6 to 11 times of that of the steady-state current. Unfortunately, miniature and subminiature relays are usually not intended for work under standard utilization categories, therefore, sometimes it is very difficult to choose such relays correctly.

Similar problems arise when one tries to determine areas of working voltages for relays to be used. Originally, miniature relays designed only for use within electronic equipment had less insulation reserve for switching loads than industrial relays. For example, if a voltage of 300 V is indicated as nominal on a miniature relay it is possible to use it also at voltages of 100 and 250 V in low voltage electronic equipment, but this does not mean that the same relay can be used in industrial automation systems with a nominal voltage of 220 V without checking its withstanding voltage. The level of this voltage is determined by the kind of relay: contactor, electromechanical elementary relay, measuring relay, etc. There are many standards for different kinds of relay:

IEEE C37.90-1989. Relays and relay systems associated with electric power apparatus

IEC 60664. Insulation coordination within low voltage systems, including clearances and creepage distance for equipment

IEC 60947-4-1. Low-voltage switchgear and controlgear — Part 4: Contactors and motor-starters — Section 1: Electromechanical contactors and motor-starters

FIGURE 3.20

Relay DI1U-112DMP type without cover (30.2 × 16.2 × 26 mm). 1 — coil; 2 — upper part of armature; 3 — stationary contact; 4 — heat sink; 5 — movable contact.

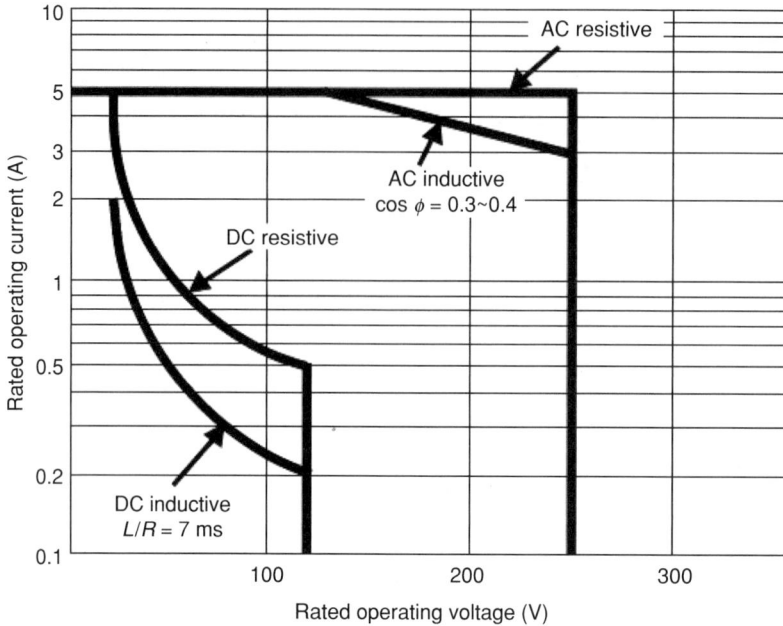

FIGURE 3.21
Typical relationship between switching current, voltage, load type, and electrical life (number of operation cycles) of a miniature relay.

IEC 60947-5-1. Low-voltage switchgear and controlgear. Part 5: Control circuit devices and switching elements. Section 1: Electromechanical control circuit devices

IEC 60947-6-2. Low-voltage switchgear and controlgear — Part 6: Control and protective switching devices

IEC 60255-5. Electrical relays — Part 5: Insulation coordination for measuring relays and protection equipment. Requirements and tests

IEC 61810-1. Electromechanical nonspecified time all or nothing relays — Part 1: General requirements

TABLE 3.2

Switching Capacity of Contacts Depending on the Type of Load for Control Electromagnets, Valves and Solenoid Actuators

Utilization Category IEC 60947–4	Type of Current	Make (Switching ON)			Break (Switching OFF)		
		Current	Voltage	cos φ	Current	Voltage	cos φ
Switching Capacity of Contacts in the Mode of Normal Switching							
AC-15	AC	10 I_N	U_N	0.3	10 I_N	U_N	0.3
DC-13	DC	I_N	U_N	–	I_N	U_N	–
Switching Capacity of Contacts in the Mode of Infrequent Switching							
AC-15	AC	10 I_N	1.1 U_N	0.3	10 I_N	1.1 U_N	0.3
DC-13	DC	1.1 I_N	1.1 U_N	–	1.1 I_N	1.1 U_N	–

I_N and U_N rated values of currents and voltages of electric loads switched by relay contacts.

It is sometimes very difficult to determine relay kind, but in most cases it is possible to use the simple equation for the minimal required value of dielectric withstanding voltage (DWV) on open contact of industrial relays: $U_{DWV} = 2U_N + 1000$. According to this equation for relays with nominal voltage 250 V, the minimal DWV value should not be less then 1500 V. Industrial relays have the DWV usually in the range of 1500 to 2500 V, and miniature relays not more than 1000 V.

3.8   Split, Make-Before-Break, High-Frequency Contacts

Let us continue our survey of contact constructions of relays. In many measuring relays (current and voltage relays) the armature turns at a greater angle. Its moment of rotation is very small, and in order to provide a reliable contacting line, contacts of specific construction are used (Figure 3.22 and Figure 3.23). The angle at which the movable contact touches the stationary one is 45 to 70°. When the surfaces of these contacts touch, the movable contact in the form of a pin (4) continues sliding on the stationary contact (1), drawing the contact spring (2) towards the backstop (3).

In point contact systems and plane contact systems of relays one may come across split contacts instead of the more common twin ones (Figure 3.24). Split (twin) contacts are more reliable than a single one because the switching is carried out simultaneously by both contacts. If there is a problem with switching, one contact will duplicate and back up the other. Another advantage of split contacts are their high resistance to vibration. This can be explained by the lower weight of each contact, fixed to a quasi-individual spring and by the high resonance frequency of this type of contact system. The vibration time of the contacts at closure is also reduced; however, small contacts do have lower thermal capacity and lower thermal resistance to arcing, so as a result of this the rated switching voltage of the relay may be reduced several times.

The construction and functions of the first three types of contacts (Figure 3.25) are quite obvious and require no additional comments.

As far as a "make-before-break" contact is concerned, it is a variant of a changeover contact in which the relay is energized first, the normally open contact (NO) closes, and only after that does the normally closed contact (NC) open (Figure 3.26). When the relay is energized the movable spring (2), compressed by the pusher, moves in the direction of the contact spring (1) (up) until closure. When contacts c and d close, the pusher continues to compress the contact spring (2), and the entire contacting mechanism ($c, d,$ and the contact spring (1)), moves up to break the a–b contact.

To switch high-frequency circuits (of hundreds of kHz) in radio equipment typical relays are of little use because of the high capacitance between contacts. Long current-

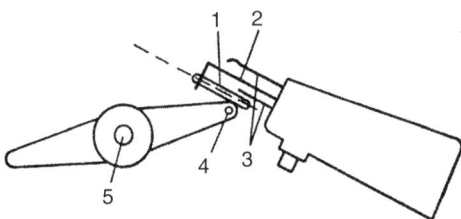

FIGURE 3.22
Diagram of a turntable contact system of a measuring relay with line contact; 1 — stationary contact in the form of a silver cylinder; 2 — contact spring; 3 — fixed stops; 4 — movable (turntable) contact in the form of a pin; 5 — rotating axis.

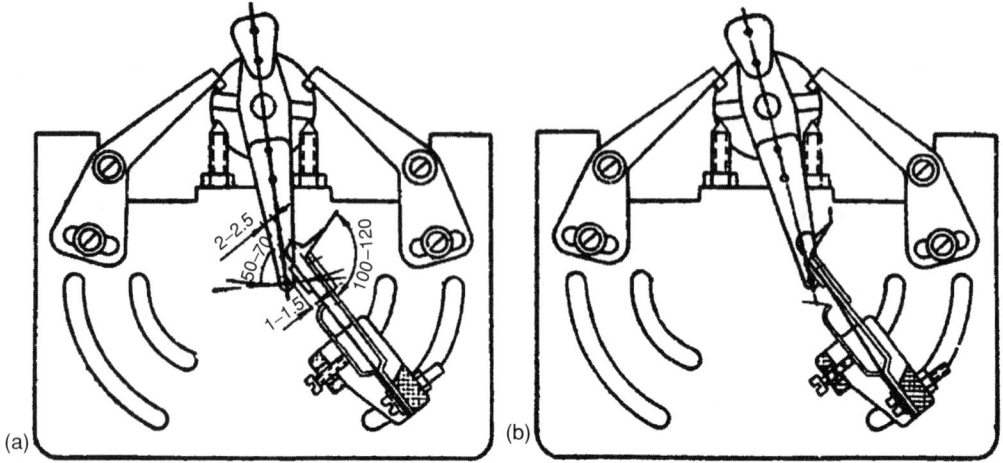

FIGURE 3.23
Construction of a turntable contact system of a measuring relay with a line contact. (a) before closure and (b) after closure.

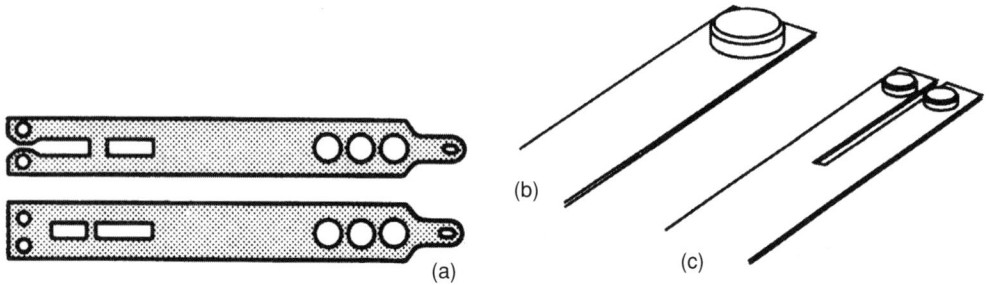

FIGURE 3.24
Single (a), twin (b), and split (c) contacts of relays.

FIGURE 3.25
Schematic of three types of contacts: type a — opening or normally closed contacts (NC), type b — closing or normally open contacts (NO), type c — changeover contact (CO); another type known as "make-before-break" contacts are also available. a, b and c — are standard types of contacts.

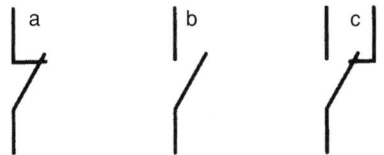

conducting springs with contacts form capacitance condensers, causing considerable leakage of a high-frequency signal (when contacts are open) and impact on switched circuit.

To reduce capacitance of a contact system, current-conducting springs of stationary contacts are made Γ-shaped and movable contacts are made flat in such a way that the intersection of their projections occurs only at the point where the contact is fixed (Figure 3.27). For switching of even higher frequencies (up to 1 GHz), special coaxial relays with outlets for a coaxial connector with a high-frequency cable are used (Figure 3.28). Stationary contacts (1 and 3) in such a relay are welded on to holders fixed to the inner ends of pins of the two coaxial connectors, insulated with glass.

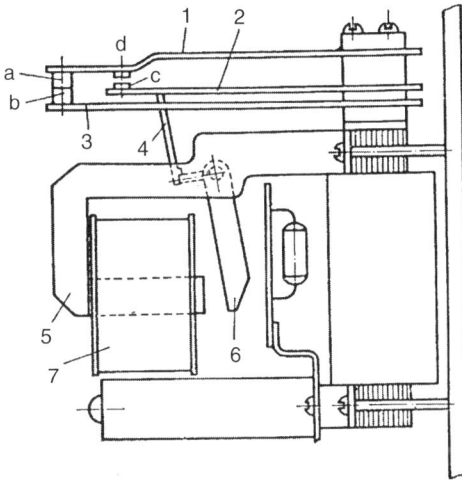

FIGURE 3.26
RP-341-type relay (Russia) with a changeover contact of the make-before-break type. a, b, c, d — contacts; 1, 2, 3 — contact springs, 4 — pusher, 5 — magnetic core; 6 — armature; 7 — coil.

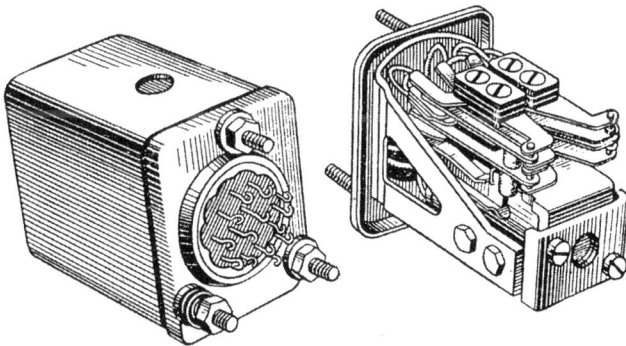

FIGURE 3.27
Hermetically sealed high-frequency relay type RMYG (Russia).

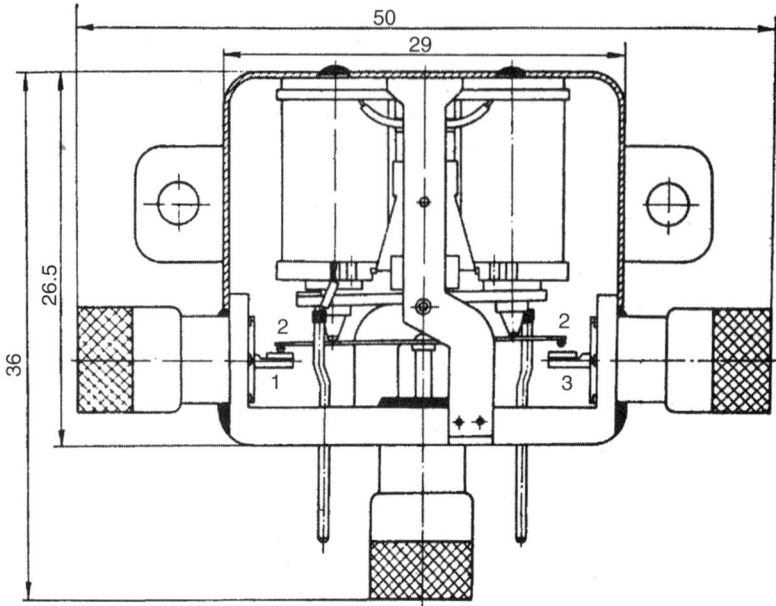

FIGURE 3.28
High-frequency coaxial RPV-5-type relay (Russia) with switching current of 8 A. 1, 3 — stationary contacts, linked with the inner part of the coaxial connector; 2 — movable contacts of the swinging bridge type forming a switching contact.

3.9   Compensation for Shock and Electrodynamic Forces in Contacts

In order to reduce erosion of contacts (failure caused by electric arc) and to enhance their switching capacity, the opening and closing velocity of the contacting pairs must be increased as much as possible, but high velocity may lead to collisions, rebounds, vibration causing new arcs, and other contact damage. In addition, in many cases, oscillations in the switched circuit resulting from contact vibration are inadmissible.

To prevent or at least to weaken such effects, one must choose the appropriate spring rate at the design stage, and produce the stationary current-conducting springs with built-in pretension. To prevent stiffness of the contact system of power relays and contactors, contacts are designed and constructed with a so-called armature overtravel or contact wear allowance (Figure 3.29). Values of the gap between the contacts (contact gap) and the amount of armature overtravel are assigned when the relay is designed. For power relays and contacts there is often a possibility to adjust the contacts during the process of exploitation. This is what many popular cheap relays often contain.

In complex measuring relays costing hundreds of dollars, additional complications of the contact system may not lead to a considerable rise in price, which is why special contact vibration suppressors are used. Contacts absorbing kinetic energy from shock have become especially popular (Figure 3.30). When a movable contact bumps into the element (7), all the energy of motion is transmitted through a flexible membrane to the ball (2), which rebounds, absorbing the energy of the shock and, due to a slight tilt of the tube, and returns to the initial position (as it is shown in the Figure). Adjustment of the initial gap between the movable and stationary contacts is carried out by moving tube

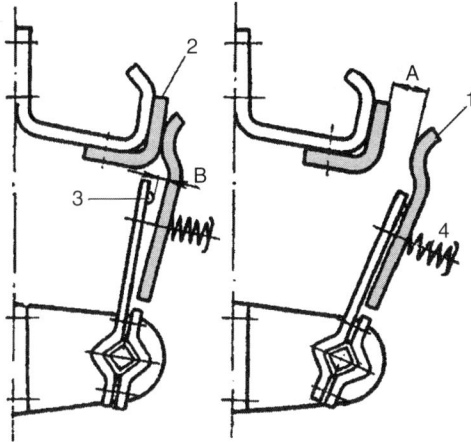

FIGURE 3.29
Contact gap (A) and armature overtravel (B) of power relays. 1 — movable contact; 2 — stationary contact; 3 — stop; 4 — loading spring.

1 in the holder, with the help of an external screw thread. Such contacts have been used for tens of years in different protective relays of the General Electric Co. (Figure 3.31).

At the same time, there still exist absolute hard-type contact systems, which resemble ancient electric arc transmitter relays (see above). They are even harder since they do not even contain damping springs (Figure 3.32). Such construction is possible only with very small gaps between contacts (in this particular relay the gap is 0.09 mm) and, short armature travel when kinetic energy from the shock of movable and stationary contact is so small that it does not lead to rebound of the contacts. On the other hand, it is possible to adjust and fix contacts with such small gaps only if the construction itself is hard enough.

In spite of such a small gap between contacts, it is able to withstand a test voltage of 500 V AC. The sensitivity of this relay is also quite remarkable: operating power, 13 mW; operating current, 2.5 mA. Apart from extremely small gaps, the contact system of such a relay is also remarkable because its NOs and NCs are made of different materials: the former are made of silver, and the latter of tungsten. During frequent switching of such contacts, the tungsten oxide film is impregnated with silver. This considerably lowers transient resistance and increases the reliability of the contact. Generally speaking, different (by size or by material) contacts in the same relay are not that rare.

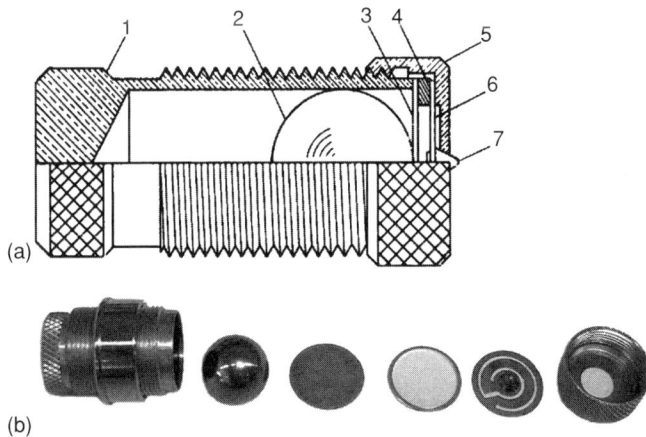

FIGURE 3.30
Stationary contact with compensation of kinetic energy as a result of shock. 1 — case (a hollow tube with external thread); 2 — metal ball; 3 — internal membrane; 4 — washer; 5 — case cover; 6 — second membrane; 7 — element receiving shock from the movable contact.

FIGURE 3.31
Protective relay with compensators of kinetic energy from shock of a movable contact, produced by the General Electric Co. 1 — stationary contacts with compensators with a movable contact between; 2 — stationary contact posts; 3 — screw setting position of the stationary contact.

As far as gaps between contacts are concerned, they may vary from 0.05 to 10 mm and more in more powerful and high-voltage equipment. The lesser the gap, the higher the operational speed of mechanical parts of a relay, but the voltage that the contact can switch is low, and its resistance to arcing is even lower. The electric strength of a gap between contacts depends on numerous conditions rather than just on the contact gap alone. In particular it depends on the shape of the contacts, the materials that they are made of, severity of erosion, the strength-to-weight ratio of the gas filling the gap between the contacts, frequency of applied voltage, etc., which is why it is possible to say that a certain gap corresponds to a certain voltage only after having taken into account all of the above conditions.

The closed contact of a relay warms up due to the current passing through it, because the resistance between the closed contacts does not equal zero. In old constructions of relays this level bound the amount of current passing through the contacts, and the temperature of the contacts could not exceed 50 to 70 °C (when the external temperature was +40 °C). In modern small-sized constructions working at their highest efficiency, the temperature of contact heating can reach 100 to 120 °C. In very powerful switching devices with currents up to 2000 A, the temperature of the contacts may reach 200 ° (for silver contacts). In such cases, sometimes, liquid cooling of contacts is used, thus one can make a simpler contact system, and the overall size of the device smaller. It is obvious that the less transient the resistance of the contacts, the greater the current that can pass through them without fear of exceeding maximum permissible temperatures. It is clear then why designers do their best to increase contact pressure between the contacts after they close. Theoretically, the greater the current passing through the contacts, the greater the pressure between them should be.

However, this is true only in theory. In practice, instead of snuggling up, they repel one another. This happens under electrodynamic forces tending to repel one contact from the

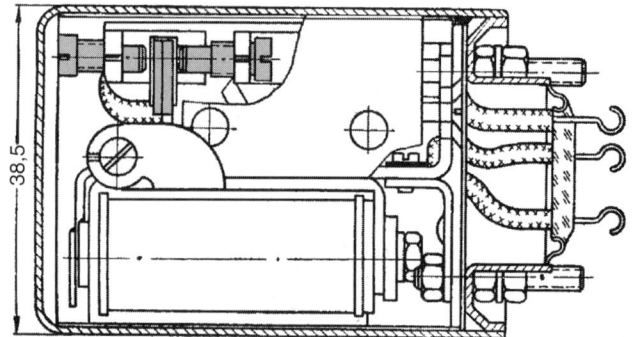

FIGURE 3.32
Construction of a Russian hermetic RDCHG-type relay of heightened sensitivity with an absolute hard-type contact system.

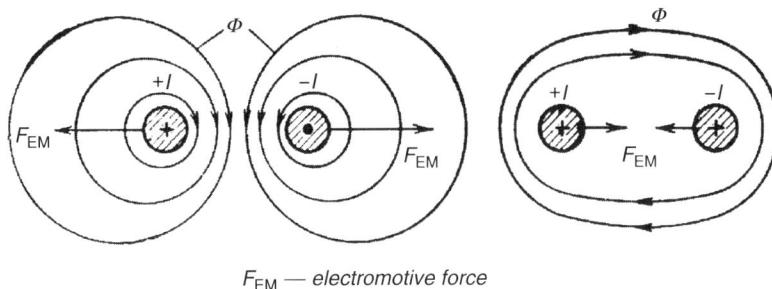

F_{EM} — electromotive force

FIGURE 3.33
Direction of electrodynamic forces between conductors. Left: Repulsion of conductors with opposite current direction. Right: Attraction of conductors with the same current direction.

other (Figure 3.33). At nominal working currents these forces are not appreciable but with short-circuit current passing through contacts, can increase so much that closed contacts may be thrown away from each other. What happens when a contact through which a short-circuit current passes and which is not designed for switching off short-circuit currents, is broken? In most cases it leads to welding of contacts or to significant damage of contacts.

To prevent spontaneous opening of closed contacts under electrodynamic forces of short-circuit currents in power switching devices, sometimes special compensators are used (Figure 3.34). In such devices short-circuit current I causes an attractive force F (1–3) between the elements 1 and 3 and a repulsive force F (2–3) between the elements 2 and 3. As a result, the element 3 where a contact is placed is affected by the sum of these two forces, causing additional compression of the contacts. Repulsive force F_2 between elements 1 and 2 creates efforts amplifying the pressing effort P. Joint coupling between elements 1 and 2 is usually shunted with flexible copper ties (Figure 3.35).

Knowledge of contact peculiarities allows us to design a contact system correctly without any specific constructions. For example, in the bridge contact shown in

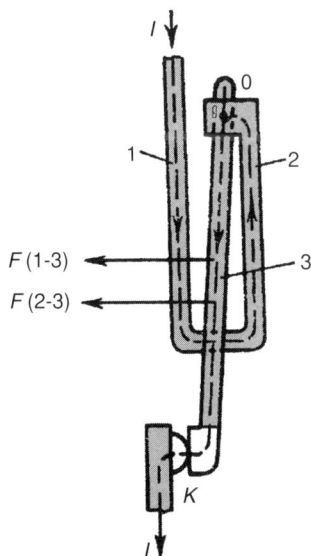

FIGURE 3.34
Compensators of electrodynamic forces.

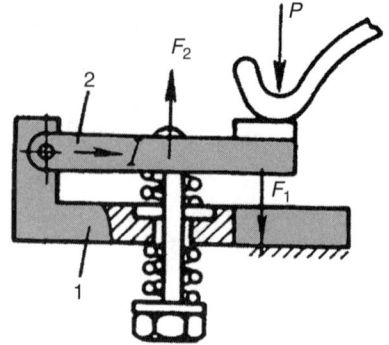

FIGURE 3.35
One of the practical constructions of compensators.

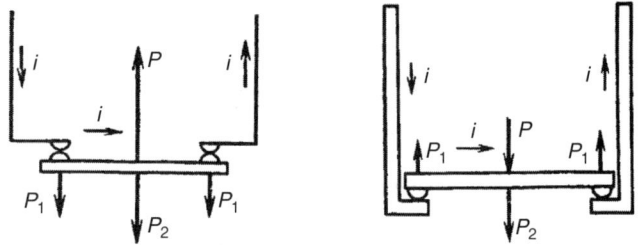

FIGURE 3.36
The effect of a bridge arrangement on
the distribution of forces, which are
operational on contact.

Figure 3.36 it is enough just to put the bridge form in a lower position relative to the top so that electrodynamic force P_2 can change the direction of its impact on the contacts. In the first case (Figure 3.36, left), this force is subtracted from force P of contact pressure provided by the springs, while in the second case (Figure 3.36, right) this force is added to the force P increased by the springs.

Moreover, due to the additional effort P_2, one can select a pressing spring with a smaller force leading to a diminution of the coil size and to facilitation of the magnetic system.

3.10   Sparking Contacts and their Control

As contacts to which working load voltage is applied close, they draw closer to each other. When the minimum distance (hundredth parts of a millimeter for low-voltage relays) is reached, electric breakdown of the gap between the contacts takes place. Discharge arising from that does not pass to an arc because a movable contact can pass distances of hundredth parts of a millimeter very quickly, and when the contacts open, the discharge disappears. However, the process of closing does not come to an end at this stage. Elastic collision of the contacts is accompanied by rebounding, with repeated closings, followed by additional rebounding and closing. Sparking contacts entail transfer of material from one contact to the other (the so-called electric erosion). As continuous current of certain polarity is switched, a bulge arises on one of the contacts and a crater on the other one (Figure 3.37). The direction of erosion depends on the type of discharge and current value.

However, due to the high speed of contact attraction the period of existence of this discharge is very short (up to the moment the contacts touch each other). When contacts open, contact pressure decreases, transient resistance proportionally rises and the temperature of the points of contact touch goes up considerably. At the moment of opening the contacts may be heated to melting point (if the amount of current in the circuit is great

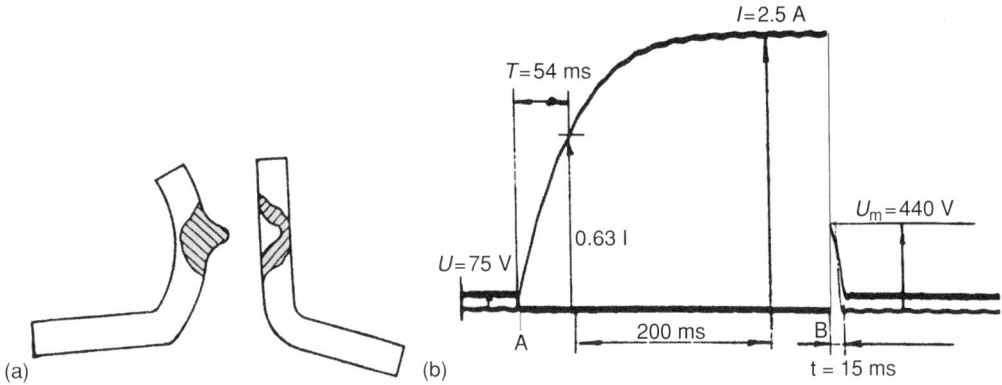

FIGURE 3.37
(a) Bulges and craters on contacts caused by electric spark. (b) Oscillogram of the process of switching ON and OFF of the inductive load with time constant $T = 54$ ms in DC current with nominal voltage $U = 75$ V; A — switch-on point; B — switch-off point.

enough) and when that happens a bridge of melded metal is formed between them. At further separation of contacts that bridge stretches and breaks turning into an arc discharge. Such a discharge will burn until the contacts separate at a distance where further burning of the arc is not possible.

On alternating current and at resistive load arc extinction occurs at the moment when the AC sinusoid passes through zero. If the contacts have separated at a distance where the electric strength of the gap between the contacts exceeds voltage restoring on contacts the switching process comes to the end (since repeated break-down of the gap is impossible). If not, another breakdown of the gap between the contacts will occur.

The basic requirement for normal circuit opening is excess of electric strength restoring during the switching process over restoring voltage. Restoring electric strength of a gap between contacts depends on the speed at which the contacts are separating, on the insulating environment of the gap between contacts (air, vacuum, sulfur-hexafluoride (SF_6), oil, etc.), and on the type of switching element (mechanical contact, semiconductor structure, etc.), with all of the above determined by the construction of switching equipment. Restoring voltage in circuit with resistive load equals the resistance of the power source.

In DC current with reactive load (containing considerable inductance or capacitance) restoring voltage depends mostly on load parameters and the rate of restoring electric strength rise.

At abrupt breakdown of current in a circuit with greater inductance energy accumulated in the form of a magnetic field is released on separating contacts with quite high intensity in the form of impulses of high-voltage exceeding voltage of the power source by 5 to 10 times (Figure 3.37b). In this particular case power source voltage of 75 V amplitude of over-voltage impulse reaches 440 V and may be subject to quite a great duration of this impulse (about 15 ms).

But restoring electric strength may be not enough to withstand such over-voltages and the following arcing process may break the contact gap jump-in changes in arc parameters during arcing may lead to self-oscillating processes in circuits with high inductance. Arcing continues on fully opened contacts until they completely burn out. The author witnessed such complete meltdown of a contact system of a hermetic relay with maximum switched current of 10 A and maximum working voltage of 350 V when coil of a power DC contactor (working as a contact load) with a nominal current in the circuit of this coil of 2 A and a nominal voltage of 100 V was disconnected.

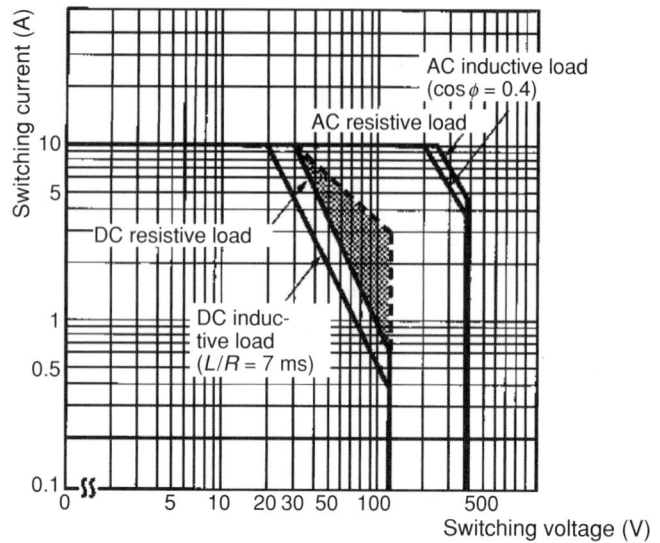

FIGURE 3.38
Typical relation between commutation parameters (voltage, current) and load character for relay contacts.

In AC circuits current passing through the contacts equals zero twice a period. It may seem that at that point the arcing processes on the contacts must cease. Indeed, switching of AC circuit is a simpler mode for contacts than DC switching (Figure 3.38). However, if there is high inductance in the switched circuit current is not in phase with the applied voltage and when the current equals zero, the voltage on the contacts can be quite high, thus maintaining an ionization state of the contact gap. Such conditions may lead to arcing even in AC circuits. Experimental investigations have shown that during switching of currents up to 6 A and voltages up to 380 V an arc goes out at first current passing through zero value in a wide range of contact separating rates (in the experiments rates from 8 up to 280 mm/s were studied). Problems in AC circuits usually arise when one has to deal with currents of a few tens of amperes. In this case specific technological solutions described below should be applied.

Repeated closing and opening leads to electric spark or arcing causing considerable wear of contacts due to fusion and dusting of contact material. Bulges and craters on contacts are dangerous because not only they destroy contacts but also they lead to frequent sealing of contacts caused by jamming of spurs in the craters, and eventually to malfunction of the relay. Contact rebounds can be controlled by special dampers and springs while electric spark must be controlled by circuit means using different protective circuits switched parallel to a contact or a load (Table 3.3 and Figure 3.39).

FIGURE 3.39
Protective RC type element produced by the firm RIFA consisting of a 0.25 μF, 630 V DC capacitor and a 100 Ohm resistor.

TABLE 3.3

Types and Features of Arc Protective Circuits

Protective Circuits	Commentary
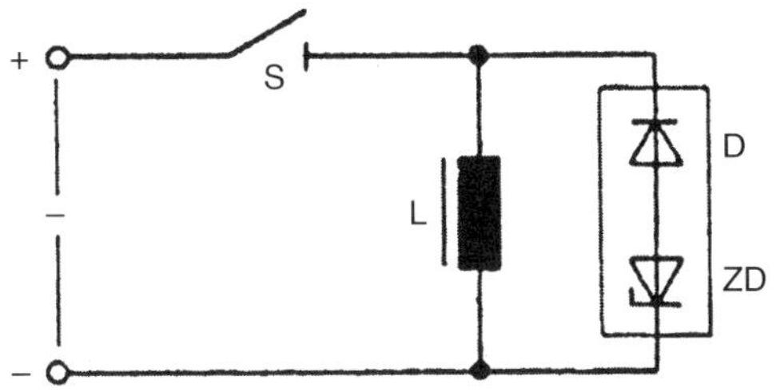	At circuit opening energy accumulated in inductance is discharged through resistance R Drawback of the circuit design: increase in current load of a contact
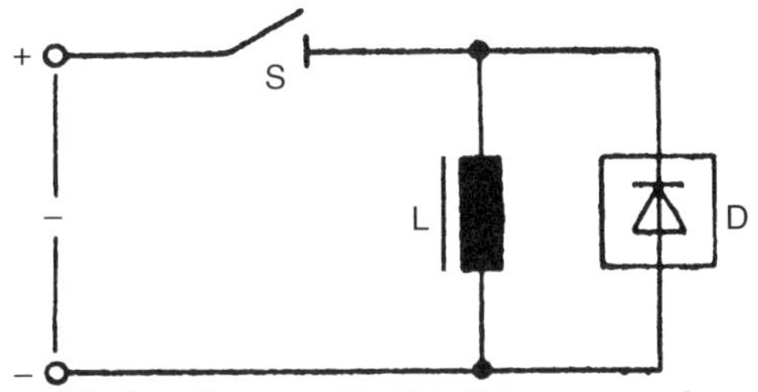 Diode voltage must be 3 to 5 times as much as circuit voltage, maximum pulse current must not be less than load current	For DC circuits only EMF of load self-induction arising at circuit opening has direction opposite to the source, which is why the diode is blocked in the normal mode and unblocked only at the moment of contact opening and shunt inductance Drawback: increase in time of current drop in inductance. When dealing with a relay winding or a contactor — increase in time of relay drop-out
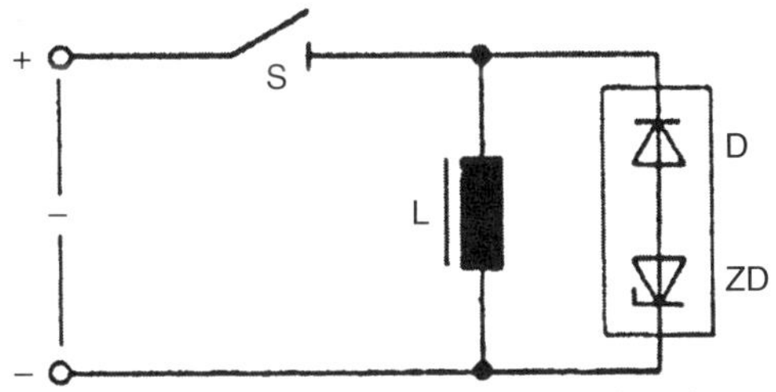 Voltage of Zener diode must be not less than voltage of power source, Zener current must not be less than 0.5 to 0.7 of load current	This circuit design, when compared to the previous one, does not have much effect on time of current drop in load because Zener diode is blocked, thus preventing shunting of the load by diode when voltage rating value in circuit decreases Drawback: high cost of a Zener diode for power loads
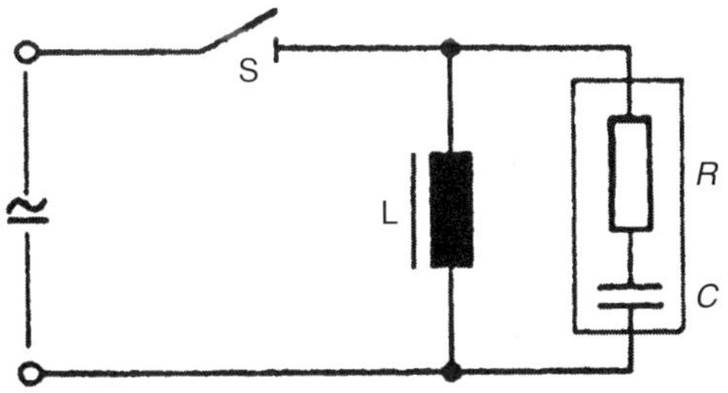	Popular type of protective circuits Does not have much influence on time of current drop in inductance. Energy of spark is used for condenser (C) charge. Resistance (R) restricts discharge of current of charged condenser at repeated contact closing Capacitance chosen is 0.5 to 1.0 µF for each ampere of switched current

(Continues)

TABLE 3.3 (*Continued*)

Spark Protective Circuits and its Characteristics

Protective Circuits	Commentary
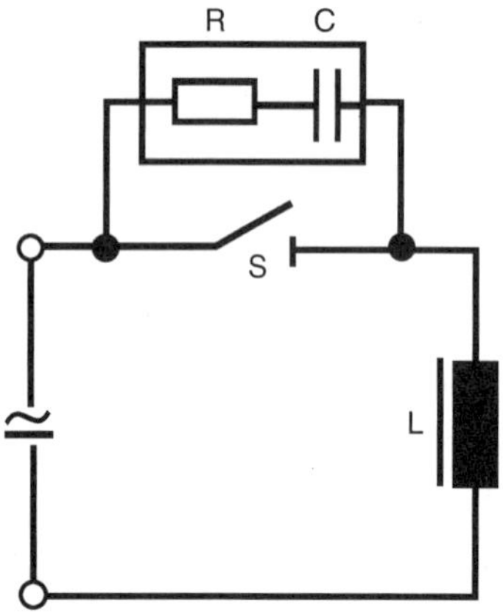	Resistance is 0.5 to 1 Ω for each Volt of working voltage The condenser should be designed for work in AC circuit with voltage not less than that exceeding rating voltage by 1.5 to 2 times In the schematic below on AC current there is a leakage of current through RC circuits that can have an influence on the load
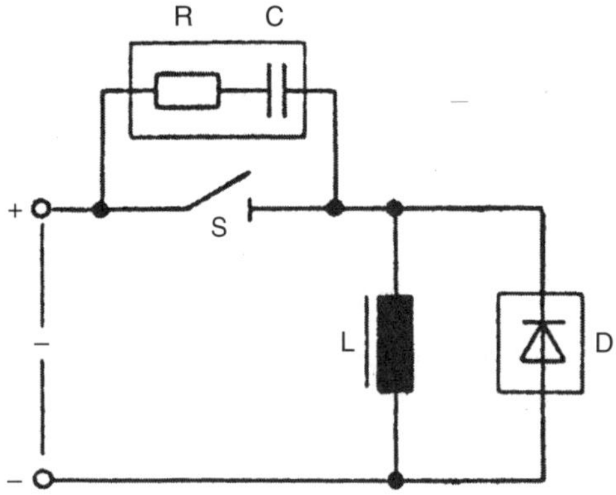	For DC circuits only Compound schematic combining both advantages and disadvantages of the variants mentioned above
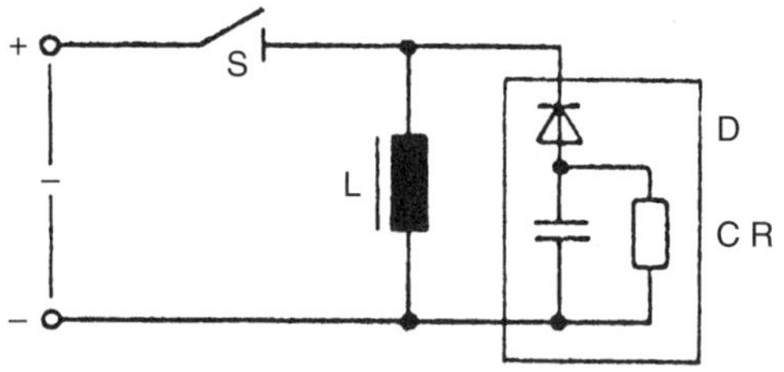	For DC circuits only Very effective circuit design, which practically does not affect current drop in a load. Resistance connected parallel to a condenser does not make it less effective for absorbing of spark energy, and discharges quickly after voltage surge
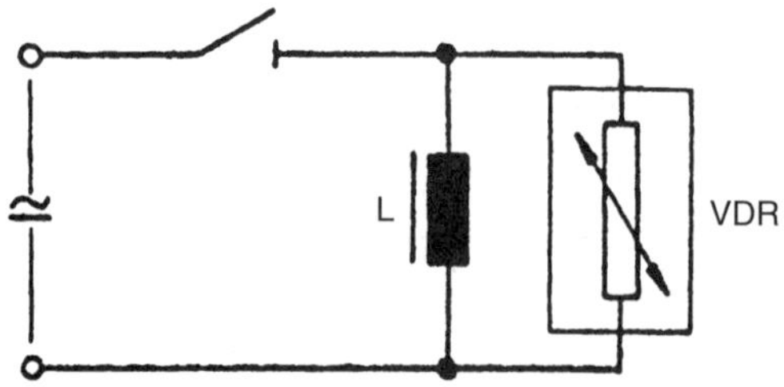 Classification voltage of a varistor must not be less than rayed voltage of circuit	Popular variant A varistor (VDR) is used. A varistor is a resistor with nonlinear resistance. Being affected by over-voltage its resistance considerably goes down Effectiveness depends on the proper choice of the varistor (voltage, dissipation energy) Has insignificant influence on time of current drop in the inductive load

(*Continues*)

TABLE 3.3 (*Continued*)

Spark Protective Circuits and its Characteristics

Protective Circuits	Commentary
	Combined schematic combining both advantages and disadvantages of the variants described above

Some companies produce protective circuits as separate articles. Companies, manufacturers of relays, produce protective circuits in special cases for easy mounting on their relays (Figure 3.40).

It must be mentioned that gas-discharge processes arising on opening contacts are quite complex and this book is not aimed at a detailed consideration of these processes, which are described in extensive monographs, and although the author does not want to simplify these problems it is still possible to draw a generalization digressing from the complex theory: when current switched by the contacts and the voltage applied to the contacts exceed certain threshold values necessary for maintaining arcing, the electric-spark discharge turns into an arc and the means described above are ineffective for extinguishing it. Conditions for arcing are ambiguous and may depend on many factors (Table 3.4).

Moreover, different researchers have provided considerably different data, which nevertheless because it gives the reader some idea regarding the conditions, we will cite here concerning critical current of arcing (that is current exceeding the arcing point, which

TABLE 3.4

Critical Currents of Arcing for Different Materials of Contacts and Different Voltages on the Contacts

Material of Contacts	Critical Current of Arcing A for Voltage on Contacts, V			
	25	50	110	220
Copper	—	1.3	0.9	0.5
Silver	1.7	1.0	0.6	0.25
Gold	1.7	1.5	0.5	0.5
Platinum	4.0	2.0	1.0	0.5
Nickel	—	1.2	1.0	0.7
Zinc	0.5	0.5	0.5	0.5
Iron	—	1.5	1.0	0.5
Tungsten	12.5	4.0	1.8	1.4
Molybdenum	18.0	3.0	2.0	1.0

(a)   LA4 DA 2U   (b)   LA4 DA 1U

(c)   LA4 DC 3U

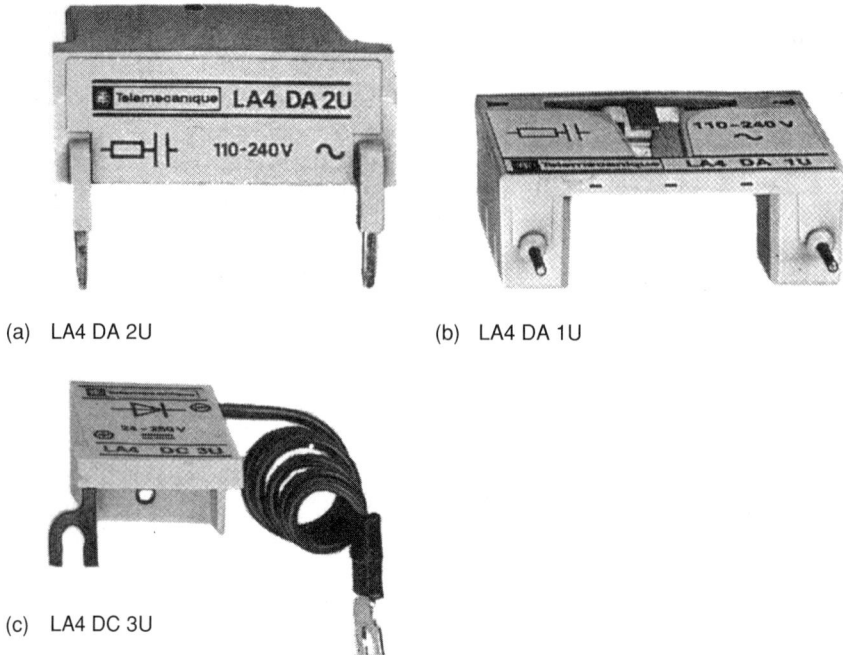

FIGURE 3.40
Spark proof elements produced by the firm Telemechanique. (a) LA4 DA 2U (b) LA4 DA 1U — RC-type element;
(c) LA4 DC 3U — diodes.

causes arcing) for different materials of contacts and different voltages on the contacts. The information is taken from research literature.

Basic means of arc affecting switching devices:

- Expansion of an arc channel by separating contacts
- Arc partition on small sections by metallic plates making the arc cool
- Displacement of arc in the zone between contacts by magnetic field
- Increase in pressure of gaseous atmosphere where there is arcing
- Arc extinction in vacuum
- Arc extinction in insulating liquid or gases.

With small cutoff currents and low voltages permanent magnets are often installed near the contacts in the relays (Figure 3.41). Electric arcs, which are at the same time flexible conductors of current, interact with the magnetic field of the permanent magnets and are forced out by this field from the area of the contacts.

3.11   High-Power Contact Systems

In power AC devices the electric arc is affected by a magnetic field created by working current passing through conducting bars and additional amplified ferromagnetic

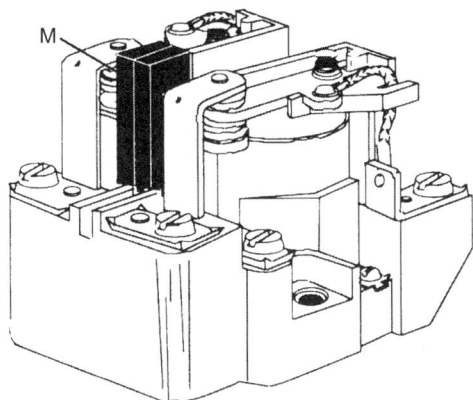

FIGURE 3.41
W199BX-14 type relay with an arc-suppressing magnet (M) designed for switching circuits of continuous current up to 30 A (produced by Magnecraft).

elements (2, Figure 3.42). As can be seen in Figure 3.42, contacts of this device are supplied not only with additional steel inserts but also with deflected metal plates. As we are now discussing an arc-suppressing chamber, it is obvious that these plates are essential for arc extinction. Any arc arising when the contact openings are supplied with such additional plates will move along these plates under electromagnetic forces of interaction between current fields in the plates, with the arc current rate reaching tens of meters per second.

If these plates are fixed at an angle forming horns (Figure 3.43), during arc movement, the arc will stretch. This will result in an increase in resistance and decrease in temperature, bringing about extinction of the arc.

One of the most popular types of arc-suppressing devices is a lattice placed near the contacts (Figure 3.44). This lattice can be made of metal (steel, copper) and of high-temperature insulating material. In the first case, the arc drawn by electromagnetic forces in such a lattice is divided into a number of short arcs with small potential difference (20 to 30 V) between the adjacent plates (anode and cathode). Thus conditions for extinction of these short arcs are created. In addition, in a metal lattice the arc cools down, which is also another significant factor in bringing about its extinction. When ferromagnetic (steel) plates are used, additional forces drawing an arc into a lattice are created. At great currents, due to the high resistance of steel, the plates of the lattice heat up to high temperatures during arcing.

In order to lower temperatures and to reduce electric erosion, the plates are covered with copper. In the second case arc extinction occurs due to its extension (Figure 3.45 (2)). This method of arc extinction by means of an arc-suppression lattice was invented more than a hundred years ago. In Russia it was implemented for the first time by M. Dolivo-Dobrovolsky. Later, more complex devices appeared; for example, in 1927 in Germany, an

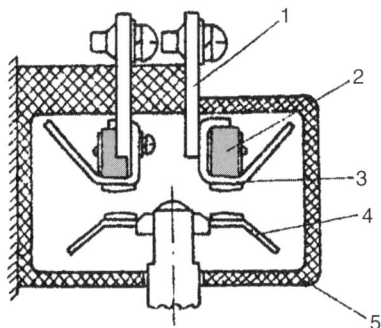

FIGURE 3.42
Arc-suppressing device of an AC contactor with rated currents of 50 to 150 A. 1 — current contact jaw; 2 — steel insert; 3 — stationary contact; 4 — movable bridge contact; 5 — vacuum chamber.

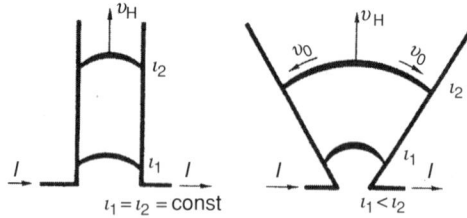

FIGURE 3.43
Arc moving along plates.

$l_1 = l_2 = \text{const}$ $l_1 < l_2$

arc-suppressing device with an arc rotating in a stack of interjacent flat electrodes was patented (patent 576932). In 1951, this idea was further developed (German patent 928655) and quite successfully implemented (Figure 3.46). In this device the arc descends from horns and turns to a grid. Sub arcs between the flat electrodes rotate in annular channels formed by insulating guides (lattices) under a radial magnetic flux, created by permanent magnets.

There are also arc-suppressing devices based on forced "blowing up" of an arc into a slot chamber where the arc cools off and goes out (Figure 3.47). In this device a coil (1) flown over by current creates a magnetic flux (F) that with the help of the magnetic core (2) and ferromagnetic plates (3) is brought to an arc. The interaction of the magnetic flux, Φ, and arc current creates an electromagnetic force driving the arc to the slot, where the arc cools off and is deionized.

There were also some attempts to use insulating and metal elements of different shapes, brought in by a spring between the contacts at the moment of their separation (Figure 3.48), and distorting the trajectory of the plasmic flux of the arc.

In actual constructions of power AC relays (contactors) most arc-suppressing devices in use are based on arc-suppressing lattices (Figure 3.49 and Figure 3.50). Arc-suppressing chambers are large in size and quite heavy. This considerably increases characteristics of mass and size of the switching device itself.

Accordingly, power relays produced by some firms (Figure 3.51), really do look amazing. They switch currents of tens and hundreds of amperes and at the same time are quite compact. Why? Let us recall miniature relays with disproportionately great switched current. The same is true when great currents can be switched on by lower voltages

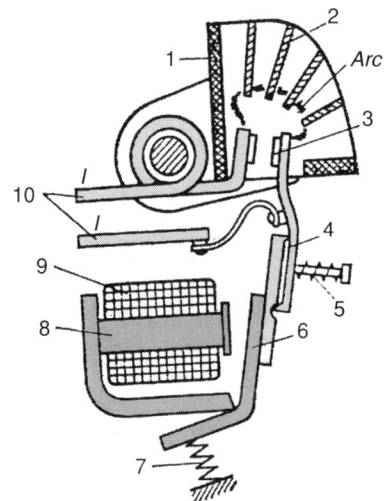

FIGURE 3.44
Principle of an arc-suppressing chamber. 1 — chamber; 3 — main contacts; 2 — lattice; 4 — lever; 5,7, — springs; 6 — armature; 8 — core; 9 — winding; 10 — wireways.

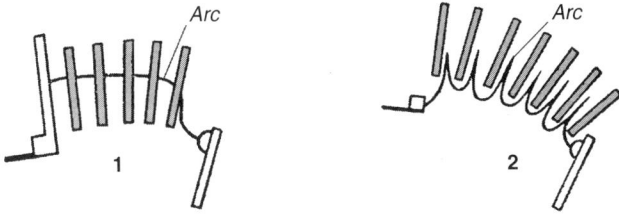

FIGURE 3.45
Principle of arc extinction in a lattice made of metal (1) and of high-temperature insulating material (2).

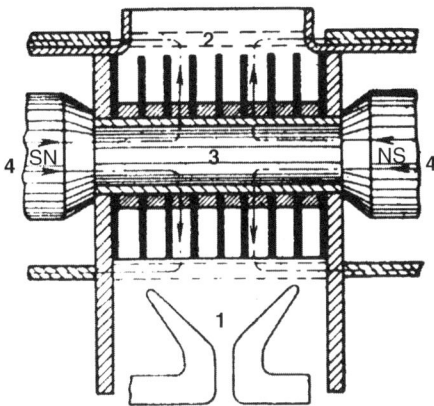

FIGURE 3.46
Arc-suppressing device with a rotating arc (German patent 928655, 1951); 1 — contacts with plates in the form of horns; 2 — lattice formed by round insulating plates; 3 — iron core; 4 — permanent magnets.

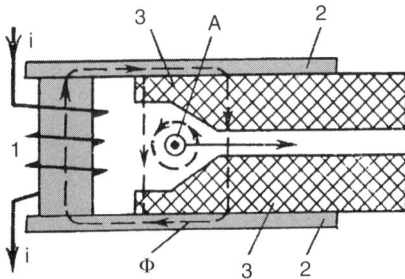

FIGURE 3.47
Slot chamber with magnetic blow-out. 1 — coil; 2 — magnetic core; 3 — ferromagnetic plates; A — arc.

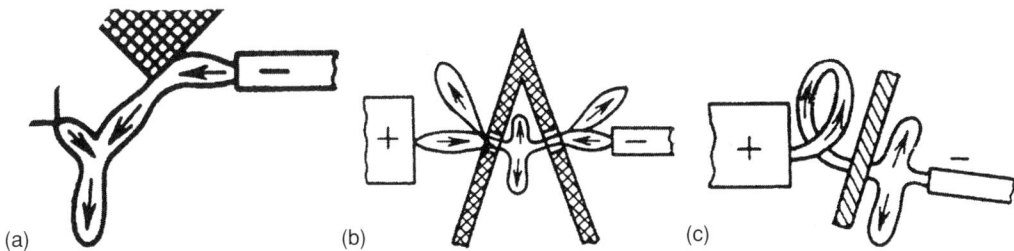

FIGURE 3.48
(a), (b), and (c) Implementation of insulating and metal elements brought in between electrodes to affect the plasmic flux.

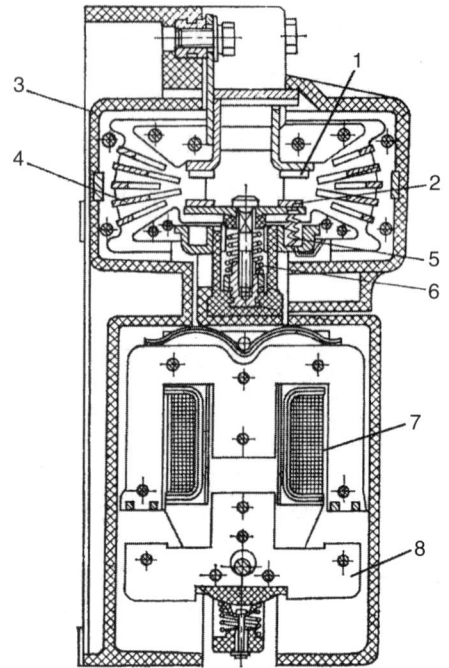

FIGURE 3.49
KTU-2E type power relay (contactor) produced in Russia. Rated current is 63 A (for the AC-4 mode), rated voltage is 1140 V (50 Hz). 1 — stationary contact; 2 — movable bridge contact; 3 — arc-suppressing chamber; 4 — arc-suppressing lattice; 5,6 — springs; 7 — winding; 8 — movable part of the core.

with the help of a simplified contact system. In this case we are speaking about a class of power relays with switched voltages of 12 to 14 and 28 V of continuous current.

The first voltage level is used in cars (Figure 3.51a), and the second is employed in aircraft power systems, tanks, other types of military, and some types of civil equipment (Figure 3.51b and c). Simple construction of a contact system is typical not only of automobile relays with switched current of tens of amperes, but also of military relays with switched currents of hundreds of amperes (Figure 3.52). These systems do not contain any special arc-suppressing elements. It is only natural that in power relays the contacts are more massive and larger in size (as a rule they are made of silver) and contact pressure is harder. These are the main characteristics of such contact systems.

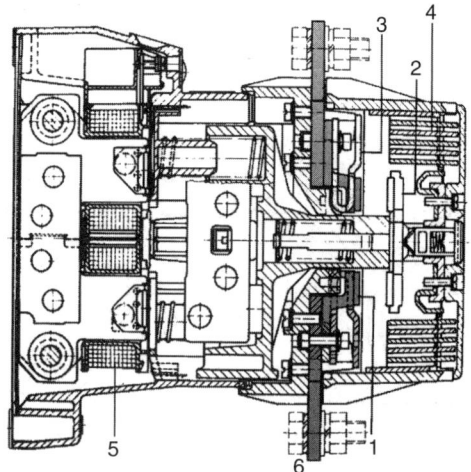

FIGURE 3.50
3TF56-type power relay produced by Siemens. Rated current is 80 A (for the AC-4 mode), rated voltage is 1000 V. 1 — stationary contact; 2 — movable bridge contact; 3 — ferromagnetic plate for blowing up an arc into the lattice; 4 — arc-suppressing lattice; 5 — one of the two coils; 6 — current leads.

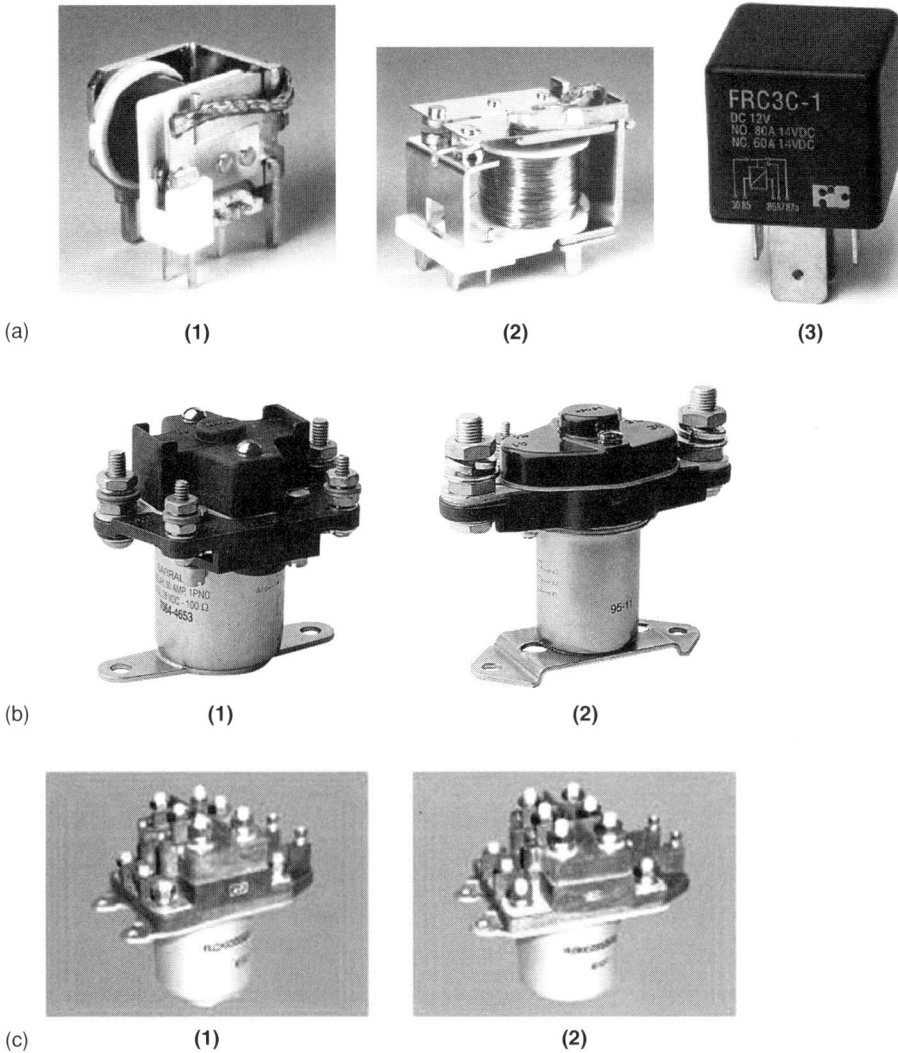

FIGURE 3.51
(a) Miniature relays of car type with simplest contact system, allowing for switch current of tens of amperes. (b) Relay designed according to the military standard for DC voltage of 28 V. Produced by LEACH International. 1 — Relay of 7064-4653 type with switching current of 50 A, size: 70 × 53.4 × 65 mm, weight: 267 g; 2 — relay of 7401-4658 type with switching current of 400 A, size: 139.7 × 62 × 44.3 mm, weight: 1180 g. (c) Power relays designed according to the military standard for DC voltage of 28 V, produced by the Kirov Electric Machine-Building Enterprise "LEPSE" (Russia). 1 — Relay of PDKS133DOD type with 100 A switching current, size: 117 × 68 × 119 mm, weight: 1.35 kg; 2 — relay of PDKS 233 DOD with 200 A switching current, size: 147 × 92 × 148 mm. weight: 2.8 kg.

Low voltage on contacts (12 to 28 V) is not enough to maintain the arc with quite great contact gaps. For example, at a voltage of 28 V for switching currents of 1 A the gap between silver contacts should be 0.8 mm, for currents of 5 A — 0.25 mm, for currents of 15 A — 0.6 mm, etc. In addition, massive silver contacts (Figure 3.52) cause intensive cooling of an arc and it goes out quickly as the contacts separate.

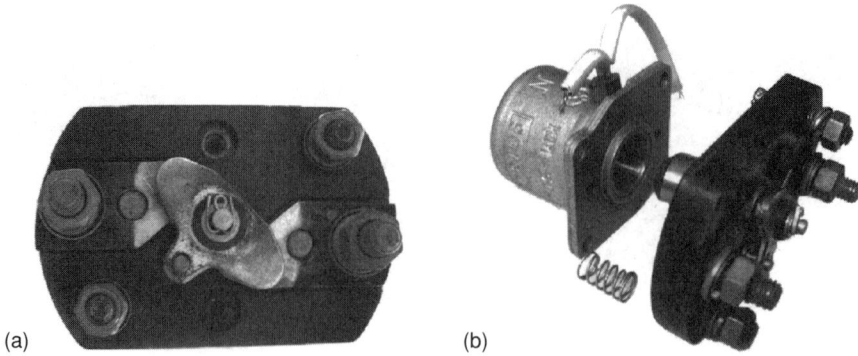

(a)   (b)

FIGURE 3.52
(a) Contact system (of bridge type) and (b) magnetic system (of solenoid type) of a high-current relay for military equipment for a voltage of 28 V.

In practice it is sometimes necessary to switch not only average and above-average currents but also low currents and voltages (micro-volts and micro-amperes), or for example, in circuits of some electronic devices, measuring sensors. The process of switching then differs considerably from that of switching average and greater than average currents (voltages): there are practically no electro-spark processes destroying oxide films on the contacts. Circuits with such small current and voltage values, which cause no electric erosion of contacts in the process of switching, are called "dry circuits." It is obvious that such contacts switching such low currents and voltages are less reliable than contacts switching normal currents and voltages. To enhance reliability of contacting, usually materials with minimum chemical activity, oxidability, and small thermal-EMF etc., are used. Experimental research of different contact material has shown that contacts from platinum and palladium have the worst characteristics. While reasons for failures of such contacts have been studied, a new method of removal of dirt from the contact surface has been suggested. A preheated contact is pressed on a clean sheet of transparent plastic. After cooling of the contact and its extraction from plastic all unnecessary films appearing during contact operation remained in the hole formed in the plastic, while the contact remained absolutely clean.

The imprint of contact dirt was examined with the help of micro-chemical analysis. Thus the source of contact contamination was determined: quite unexpectedly it turned out to be amorphous organic brown powder. Further research proved that no arc or spark will cause this powder to appear. It is formed by slippage with friction of contacts from metals of the platinum group (platinum, palladium, rhodium, iridium, osmium, ruthenium) and also from molybdenum, tantalum and chrome, when there is organic vapor in the air. When this polymeric powder is formed during friction activation, it is called "friction polymer." When switching with arcing takes place, organic films and powders are quickly destroyed. That is why this effect is produced only during "dry circuit" switching.

Such powder does not appear on silver contacts, but silver is not used for the switching of "dry circuits" as they contain sulfur films, which are not prone to destruction during switching of such circuits. Such circuits are well switched by contacts of pure gold. But pure gold is not so hard and in the course of exploitation contacts are often grinded to each other. As there are no oxide films on gold, due to certain contact pressure inter-diffusion of gold atoms of contact surfaces may cause cold welding of the contacts, which is why 8% of silver is usually added to increase the hardness of the contacts. Use of split contacts (see above) also increases reliability of a relay.

Even circuits with a weak spark (currents of more than 10 mA with voltages more than 0.1 V) cannot be switched even once by contacts designed for switching of "dry circuits." That is why for checking such relays one must use not only incandescent lamps, but also light-emitting diodes.

In construction of a contact (containing a contact spring and a multi-layer contact attached to this spring) different metals and alloys are used. For switching of low voltages one should take into account that a pair of contacts from different metals in conditions of elevated humidity forms an electrochemical element (galvanic couple) with an e.m.f. of 0.05 to 0.25 V. In addition, the thermo-electromotive of a pair of contacts ($1 \mu V/°$) is also quite great. For example, the thermo-electromotive of contacts of an RES-22-type relay (Russia) is $100 \mu V$ and of an RES-10-type relay (Russia) is even $250 \mu V$. To avoid distortion of a switched signal the thermo-electromotive value should not exceed more than a few percentage points of working voltage. That is why for RES-22- and RES-10-type relay minimum switching voltages are 1.2 and 5 mV, respectively.

Another problem is commensurability of switched current with current leakage through relay insulation, especially in conditions of elevated humidity. To remove the impact of current leakage on switched current, current leakage should not exceed than a few percentage points of switched current. For example, to switch current of $0.01 \mu A$ with a voltage of 1 V insulation resistance should be not less than 2 GΩ.

This is difficult to implement in open relays, which is why relays for micro-signal switching are usually hermetic. However, this does not solve all the problems, since with low levels of switched signals contacts are affected by a microscopical amount of organic and water vapors released from the winding as it is heated. That is why in most reliable relays double encapsulation is used, with the contacts placed in a separate hermetic shell inside a common hermetic shell. Some other technical solutions may also be applied, for instance removal of winding extruded by plastic from a hermetic shell where only the contacts remain (see "Reprocon" relay below, designed by the author).

3.12 Mercury Displacement Relays

Liquid metal contacts also play quite an important role in the history of the development of contact systems. Liquid metal contacts are contacts in which current is switched not by hard metallic contact surface but with the help of liquid metal. Since the only metal that remains liquid at low temperatures is mercury, that is the metal used

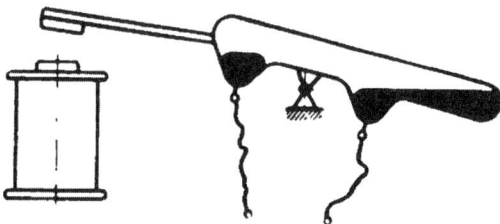

FIGURE 3.53
Principle of construction of a relay with a mercury contact.

in contacts of electric relays, which is why liquid metal contacts are also called mercury contacts.

Why is a mercury contact better than hard metal one? Because when arcing occurs during switching, the contact material does not evaporate irretrievably leaving a crater on the surface; it condenses on the walls of a shell (which usually contains a mercury contact) and drains back into a reservoir. In addition, mercury contacts have quite low and stable transient contact resistance, and do not require contact pressure with great currents.

Vibration of contacts wetted by mercury does not lead to load circuit breakdown and contact welding. Mercury contacts are effective both with increased gas pressures in the shell and in a vacuum. As the good conductive qualities of mercury were known to the first physicists, at the very beginning of the development of electrical engineering, they started using such contacts in relays as well, a long time ago (Figure 3.53). In such relays when power is applied to a coil, the armature is attracted to a core carrying an ampoule with mercury. The ampoule spins about its axis and the mercury flows from the right of the ampoule into the left closing the contacts soldered into the walls of the ampoule. This was the basic principle of construction of industrial relays in the 20 and 30s of the previous century (Figure 3.54).

In these mercury contacts, the mercury is moved (with closing and opening contacts) through displacement by means of a plunger immersed into the mercury (Figure 3.55). In this power relay with a mercury contact system hollow ferromagnetic floats (3) drift in the mercury until the control winding is switched on. When the operating coil is switched on, the floats drawn by the magnetic field of the coil sink into the mercury. The mercury level in both compartments of the body increases and upon reaching the hole (4) in the dielectric barrier, both portions of mercury merge, closing the circuit of the outlet contacts. As the coil is switched off, the floats emerge, the mercury-level increases, and the outlet circuit is opened.

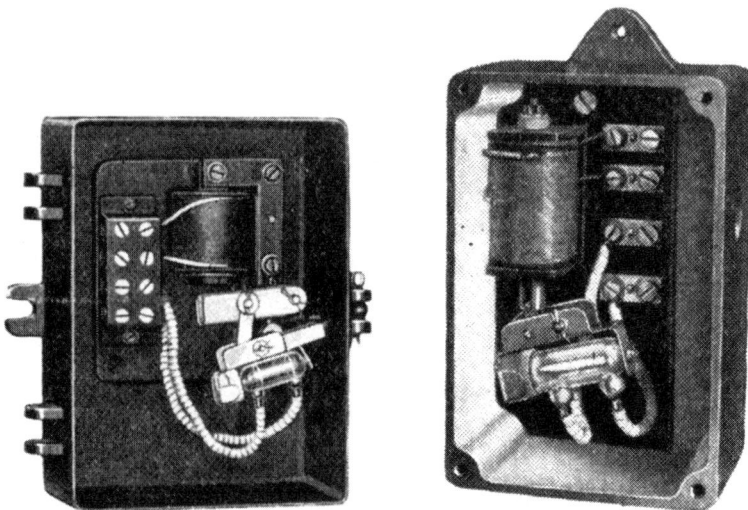

FIGURE 3.54
An industrial relays with turning mercury contacts. Switching current up to 6 A with voltage of 250 V (General Electric, 1935).

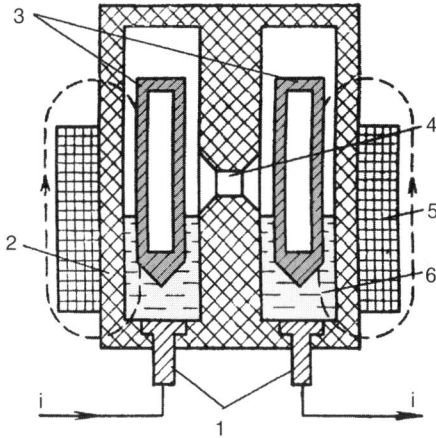

FIGURE 3.55
Power relay with a mercury contact system. 1 — outlet electrodes for attaching of external circuit; 2 — insulating case; 3 — hollow ferromagnetic floats; 4 — hole in barrier between two compartments of the body; 5 — control coil; 6 — mercury. Outside view of contact node.

A similar principle works in power mercury displacement relays with switched currents of 30 to 100 A, produced by a surprisingly large number of firms, but there are still some differences in the construction. The principle of operation of all types industrial mercury displacement relays can be understood by visualizing a nonmagnetic stainless steel tube containing a pool of mercury upon which floats a metallic plunger (Figure 3.56). An electrode is suspended some distance above the surface of the mercury. When the coil power is off, the mercury level is below the electrode tip. No current path exists between the insulated center electrode and the mercury pool. When coil power is applied the plunger is drawn down into the mercury pool by the pull of the magnetic field and the plunger centers itself within the coil. This action raises the mercury level until the mercury covers the end of the electrode and thus completes the current path.

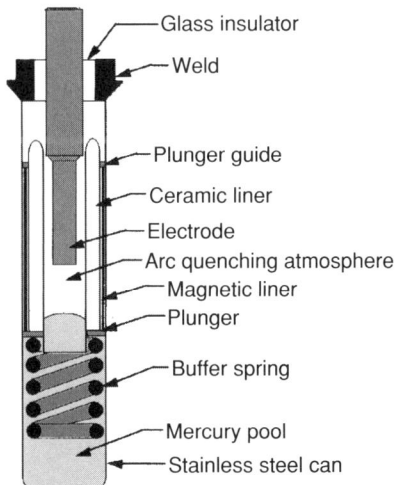

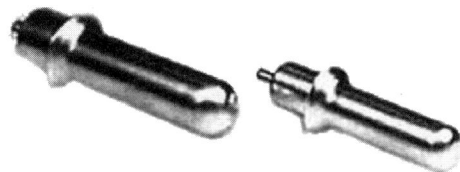

FIGURE 3.56
Construction of a contact node of mercury displacement relays.

When the coil power is turned off, the buoyancy force of the mercury causes the plunger assembly to again rise to the starting position. This drops the level of the mercury and breaks the current path through the center electrode and the mercury pool. A disadvantage of this type of control is that the relay must be mounted upright. The non-magnetic stainless steel can (contact tube) is filled with pressurized gas to minimize arc erosion.

Every contact system is hermetically sealed in a steel tube to provide maximum life, protection to the user from arcing, and from the hazards of switching heavy loads with exposed contacts. Liquid mercury means a new contact surface after every operation. The mercury is self-renewing; it cannot pit, weld, disintegrate or oxidize. The internal resist-

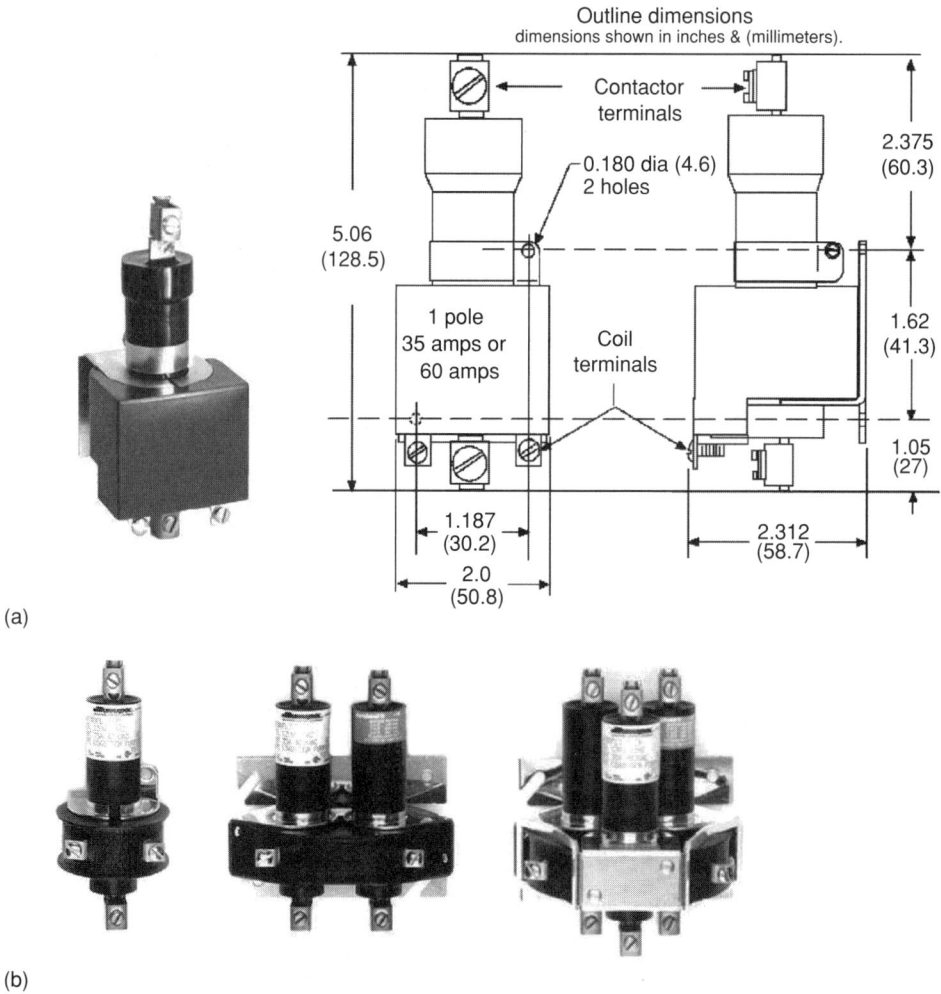

(a)

(b)

FIGURE 3.57
(a) External design and size of 1 pole mercury displacement relay WM60A (Magnecraft and Struthers-Dunn); (b) external design of 1 to 3 pole mercury displacement relays BFL35 for currents of 30 to 100A produced by International Sensors and Controls Co. (USA).

(*Continues*)

(c)

FIGURE 3.57 (*Continued*)
(c) external design of 1 to 3 pole mercury displacement relays HG series for currents of 30 to 100A produced by Watlow (USA).

ances of the contact surfaces typically measure only a few milliohms and are ideal for switching large loads safely.

Due to significant advantages in comparison with common relays with hard contacts, mercury displacement relays for currents of 30 to 100 A are widely used in industry and produced by a large number of companies. However, they are all based on the same principle and look quite alike (Figure 3.57).

4

External Design of Relays

4.1 Environmental Impact on Relays

In practical service conditions, relays are affected by negative environmental factors, which can considerably change their characteristics. Changes in the temperature of the environment may cause changes in the linear dimension of a core, an armature, a case, and other significant elements of a relay. As a result, distortion of movable parts and even jamming may occur. Changes in the resistance of relay windings and the modulus of elasticity of a restorable spring can lead to considerable changes in relay pickup (operate) and dropout (release).

As the temperature increases from $+20$ to $+100°$, the insulation resistance of a relay decreases practically by ten times. Strange as it might seem, even increased air humidity can lead to changes in pickup and release currents of a relay. Oxide films and corrosion in joints of the movable parts of a relay may cause pickup current multiplication of 10 to 15%. Temperature fluctuations varying from 0 to -20-$60°$ may result in malfunctions of a relay caused by freezing of contacts. As air-pressure goes down, its electric strength considerably decreases, in accordance with Paschen's curve (Figure 4.1). As it can be seen from the curve, minimum electric air strength is about 320 V/mm at a pressure of 4 to 5 mm Hg, corresponding to 42 km height. This fact must be taken into account for relays designed for aircraft and rockets.

When relays are used in movable units or stationary equipment affected by vibration, they are prone to external mechanical vibration loads of different frequencies and amplitude. Under vibration the pickup current of a relay usually decreases by 5 to 25% because of recurrent reduction of the magnetic circuit gap, easier relay operation at the moment, and also because of a decrease in the constant of friction between movable and stationary elements. In addition, the pressing strength of the closed contacts may also change occasionally. A weakening of this effort may lead to contact welding. If this frequency of external oscillations is in accordance with the frequency of natural oscillations, resonance may occur, causing a sharp increase in the amplitude of oscillations, bringing about opening of closed contacts or closing of open contacts, breaking winding outlets, and eventually causing mechanical collapse of the relay. Relay specifications usually indicate frequency band and amplitude (acceleration) range, at which no spontaneous closing or opening of the contacts will occur, and at which the pressing effort (strength) remains great enough for a continuation of reliable work.

Apart from vibration, a relay installed on movable units is also prone to linear acceleration. Most relays are affected by strong acceleration during aircraft take-off, in-flight maneuvers of military aircraft, and missile take-off. In these cases if special precautions are not taken, relays may be picked-up spontaneously. Relay with so-called "balanced

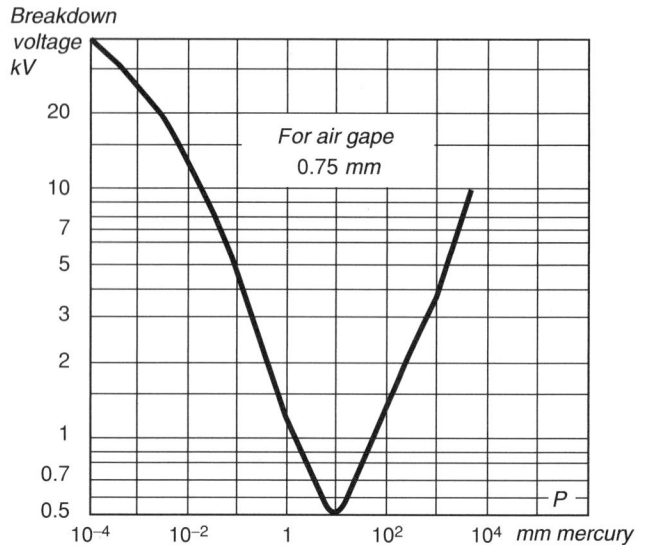

FIGURE 4.1
Paschen's curves for air.

armature'' (see above) have the highest resistance to linear accelerations. This type of relay is the most widespread today. Some years ago in some cases relays with an attracted-armature magnetic system (see Section 2.3.1) were used in operations with linear accelerations. Such relays were designed in such a way that spontaneous energizing in the direction of acceleration could be avoided.

Relays are also very sensitive to current-conducting dust or gases causing metallic corrosion. For example, unprotected relays installed in automation systems of big poultry-yards or stock farms will go out of action quickly due to the corrosive impact of ammonia.

4.2   Wood and Cardboard: First Protection Shield For Relays

Of course the necessity of relay protection was not realized at once. It took decades to form a system of knowledge about negative impacts on relays and means of protection. However, due to intuitive realization of the fact that relays are sensitive and precision devices that can be easily damaged, the first relays were also supplied with some elements of protection.

At first, these were simple frames (Figure 4.2), and then later wooden boxes (Figure 4.3). Simple cardboard boxes were used as shipping containers (Figure 4.4). Wooden boxes for relays had been used for a long time (at that time wooden constructions were widely used in engineering). For example, in the 1935 directory of the biggest English electrotechnical company, General Electric Co. (later GEC Measurements, now Alstom), there are quite a lot of relays in wooden cases for different purposes (Figure 4.5).

By the end of the 19th century, some companies began producing relays in round metal cases with wooden heelpieces (Figure 4.6), supplied with convenient clamps for switching of external circuits. Later on, durable metallic cases were also widely used for other types of relays, in particular for relays used in military equipment (Figure 4.7).

FIGURE 4.2
First half-covered relay (on the left).

FIGURE 4.3
One of the first covered relays.

As production of plastic became widespread in the middle of the 20th century, protection shields for relays gained their modern design (Figure 4.8). According to the modern classification there are open relays and dust-proof and sealed relays. Due to the quite strong environmental impact on relays, open relays are very rarely used now. In constructions of the second type, either plastic cases snapping shut at the heelpiece of the relay, or expanded aluminum cases, are most often used (Figure 4.9). There are a few types of plastic cases (Figure 4.10 and Figure 4.11), that provide different levels of protection.

To protect the winding from mechanical damage and humidity it is often impregnated with epoxide resin.

FIGURE 4.4
Cases for packing of early relays.

FIGURE 4.5
Relays for automation systems in wooden cases produced by General Electric Co. (From the G.E.C. Catalog of Electrical Installation material, 1935.)

FIGURE 4.6
Telegraph relays of the 19th century in round metal cases.

FIGURE 4.7
Relay in heavy metallic case for military equipment produced by General Electric Co. on request of the Mine Department in the 1930s. (From the G.E.C. Catalog of Electrical Installation material, 1935.)

(a)

(b)

(c)

FIGURE 4.8
Relays with plastic protection shields produced in the middle of the 20th century. (a) Ericson relay; (b) Omron relay; (c) Siemens relay of RH-25 type (Russian relay of MKU-48 type has a similar design).

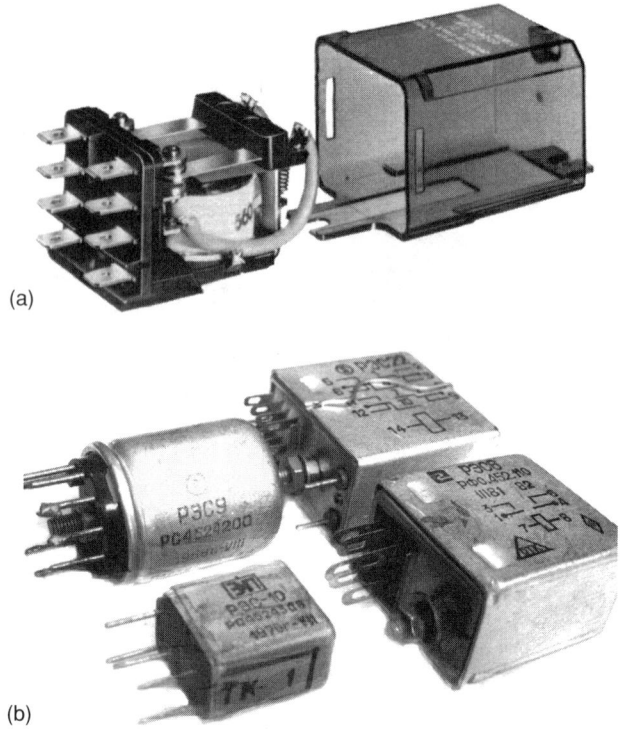

FIGURE 4.9
Modern dust-proof relays in (a) plastic
and (b) aluminum cases.

4.3   Is a Sealed Relay Always Better than an Open One?

On the face of it, all dust-proof relays may seem to be less prone to environmental impact than open ones, but paradoxical as it might seem this is not true.

At increased humidity of the environment, moisture gradually penetrates the relay through the uptight joint of the heelpiece and protective case, and remains for a long period of time because of the lack of ventilation. Also, when there are fluctuations of temperature and air-pressure, humid air may be sucked into the relay. After switching ON and warming of winding, moisture may be condensed on contacts and can sometimes lead to water bridges between the opening contacts. Moisture, together with the difference of potential on the contacts, may cause more intensive processes of decomposition of the contact material. Long-term moisture leads to a sharp decrease in insulation resistance and a further lowering of temperatures — even to icing over of contacts and windings.

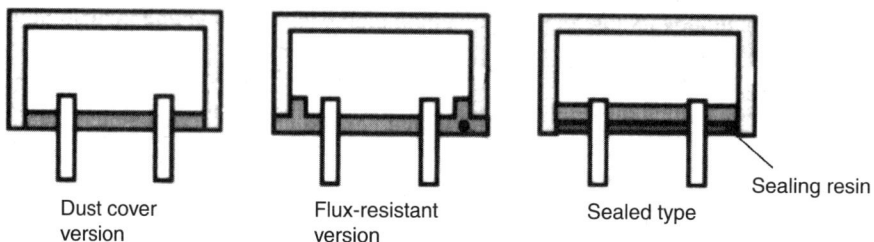

Dust cover version   Flux-resistant version   Sealed type   Sealing resin

FIGURE 4.10
Schemes of plastic cases of different types.

FIGURE 4.11
Relays in plastic cases sealed by epoxide resin (sealed-type relays).

For these reasons dust-proof relays are frequently more prone to high humidity impact than open ones; nevertheless, such relays are still better protected from dust and external mechanical effects.

To enhance reliability of dust-proof relays under high humidity conditions, some models are produced with vent doors in cases of approximately 1 cm^2 in area, covered on the inside with a few layers of poultry netting (8 to 10 thousand of meshes per square centimeter). Such netting protects the relay against dust, but allows air circulation inside the relay.

Sealed relays with welded metal cases have a more perfect design allowing protection of all internal elements of the relay from environmental impact (Figure 4.12). However, special materials and technologies are required to produce them mostly, because even small amounts of substances which usually do not affect open relay operation can have a strong negative impact if found in small stoppered cases of sealed relays. As metal vapors or flux arise during soldering or welding of contacts, or sealing of the case, such substances

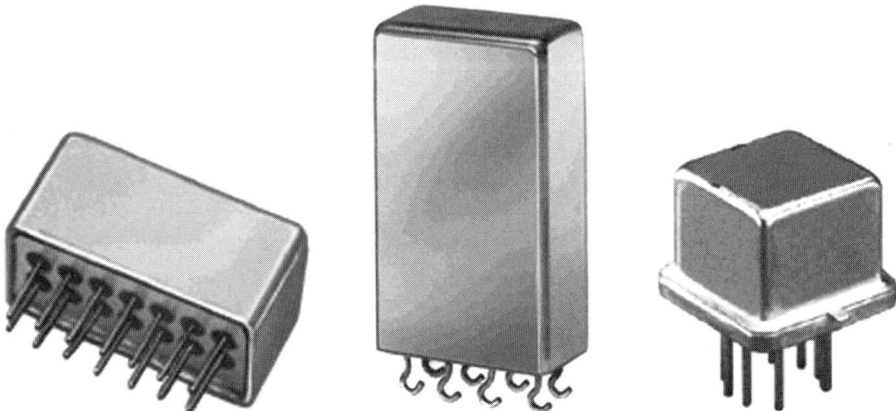

FIGURE 4.12
Relays sealed into metal cases.

can penetrate or pass under the relay shell as it is being produced. Hazardous substances can also be released in the course of relay operation when elements such as plastic coil bobbin or enameled wire are heated, because of which special materials are required for sealed relays, and production must be carried out in absolute cleanliness. This is similar to the process required for production of vacuum electronic devices. In some cases, double sealing is required to avoid gas contamination from the winding: the winding is hermetically insulated from the contact system inside an external sealed case. Before sealing-in, the relay is degassed in a vacuum thermostat under pressure of not more than 10^{-4} mm Hg and a temperature of about 170 °C, and is filled with a dehydrated mixture of nitrogen (90%) and helium (10%). Vacuum-tight sealing of relay outlets is carried out by lead-out pins made from Kovar® (Kovar® is a registered trademark of Carpenter Technology Corporation). Kovar alloy is a vacuum melted, iron–nickel–cobalt low expansion alloy whose chemical composition is controlled within narrow limits to assure precise uniform thermal expansion properties. Outlets, insulated with glass insulators, soldered inside relay with argon-arc welding or in a hydrogen environment (Figure 4.13).

You may think that such relays are perfectly protected against moisture and Paschen's curves, but in fact it is not so. Of course there will be no more problems with internal elements, but how are we to deal with external elements such as outlets, for example?

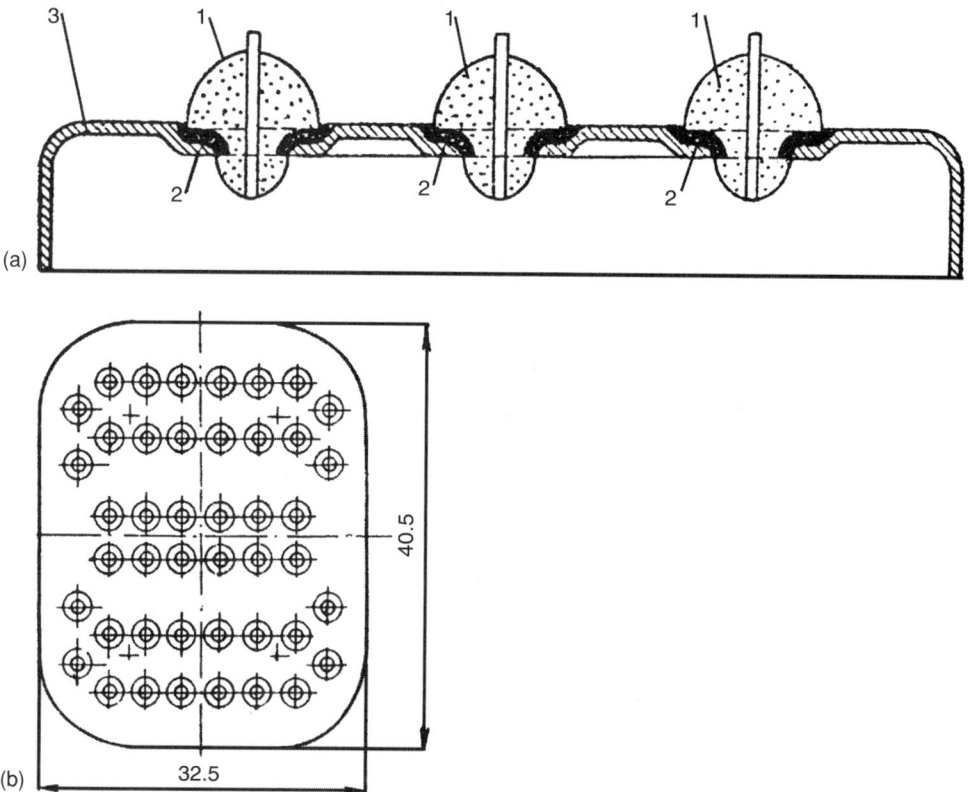

FIGURE 4.13
(a) Sealed relay outlets. 1 — glass insulators, 2 — layer of silver baking into glass; 3 — metal heelpiece. (b) Heelpiece of a DP-12 relay with 12 switching contact groups.

4.4   Outlets, Terminal Sockets and "Containers" for Relays

Multi-contact relays have a number of such elements (Figure 4.14). Since the heelpiece area is quite small, glass insulators are also very small in size (about 3 mm in diameter with about 1 mm of surface leakage path).

Here problems begin to arise. If under normal air-pressure such glass insulators can withstand AC voltage with an active value of more than 2000 V (which is more than enough for miniature sealed relays), at 15 km height (when air-pressure is about 70 mm Hg) breaking-down voltage reduces to 700 V and at 42 km height already to 200 V. Moreover, at working voltage subnormal discharge may arise between the contacts.

External contamination of a relay with dust, and increased humidity may lead to an increase in leakage current on the glass insulator surface, even under normal air-pressure. So safe protection of just the internal elements is not enough for reliable relay operation. It is also essential to protect the outlets. Usually such protection is carried out in the equipment itself after the relay has been installed. In case of point-to-point wiring, the heelpiece of the relay is covered with foamed sealant or silicon. If a relay is installed on a printed circuit board, the whole board and the relay together are covered with several layers of high quality waterproof varnish.

Modern relays can have outlets of different types. Sealed relays have direct outlets for soldering into a printed circuit board or hooks for point-to-point wiring (Figure 4.14). As has been mentioned above, outlets of sealed relays are made of a special alloy called Kovar in order to obtain a coefficient of expansion similar to that of glass, but this material is not the best conductor of electric current, which is why the outlets are sometimes made bimetallic to increase their carrying capacity with the internal copper core pressed into an external tube made of Kovar.

The diversity of outlets for industrial relays is even greater. It used to be quite easy to install such relays in equipment (in control cabinets, for instance) when they were large in size and had their own terminal sockets for connection of external wire (Figure 4.15). The solution was to provide small relays with pin-like (or flat) outlets (Figure 4.16). They were designed to be inserted into terminal sockets (Figure 4.17), and supplied with screw clamps for connection of external wire. Many of them look like sockets of old radio valves and similar size spectrum — heelpiece with eight-pinned radio valves (Figure 4.17) — used for relay outlets. Such terminal sockets can be called an interface-providing joint of

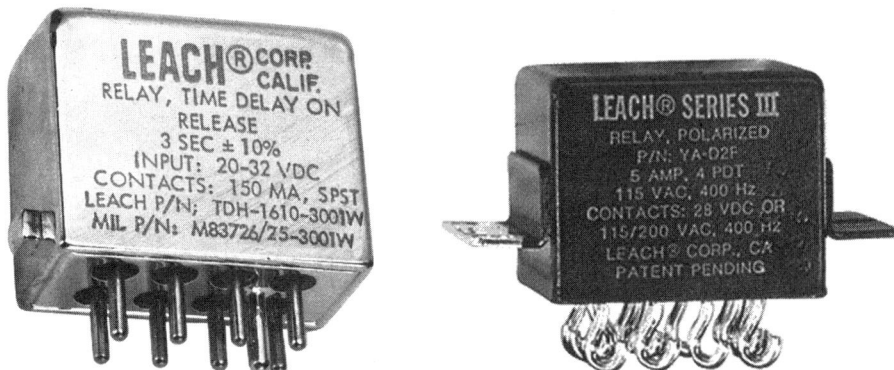

FIGURE 4.14
Various types of outlets of sealed relays.

FIGURE 4.15
A large industrial relays with its own sockets, with screw clamps for connection of external wire.

small relays to external circuits. These sockets may be two and even three-steps in height, thus saving space in the cabinet if the relay has many contacts. Due to such terminal sockets relays can be assembled and dismantled easily, quickly, and conveniently (Figure 4.18), and sockets can be also easily installed on a standard DIN Rail (Figure 4.19). Due to a great number of spring-loaded pins in a terminal socket a relay

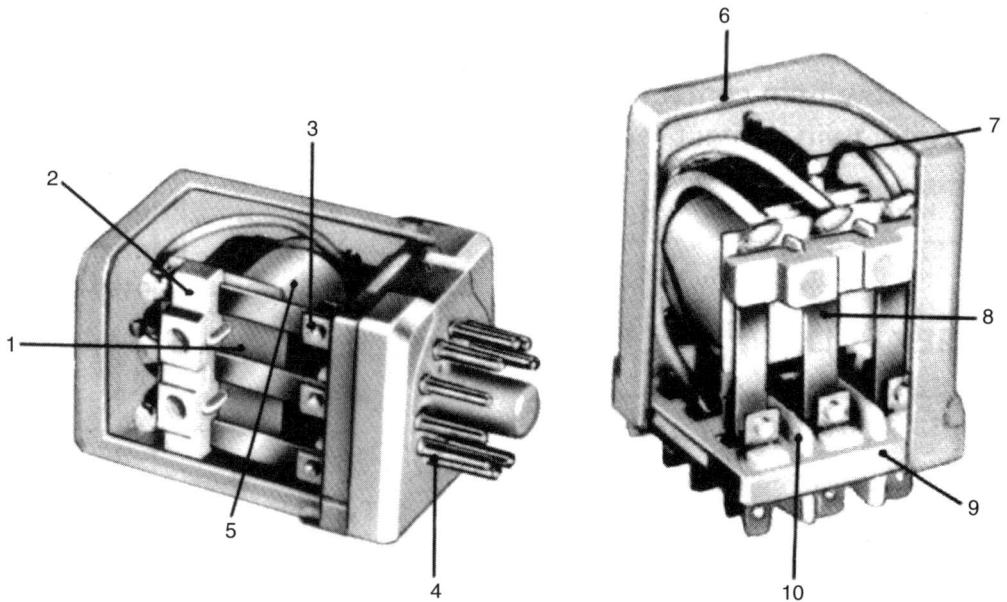

FIGURE 4.16
Industrial relay with pin-like outlets (terminals). 1 — insulation sheet; 2 — movable contact; 3 — fixed contact; 4 — terminals; 5 — coil; 6 — transparent dust-proof case; 7 — releasing spring; 8 — movable contact spring; 9 — base; 10 — insulating barrier.

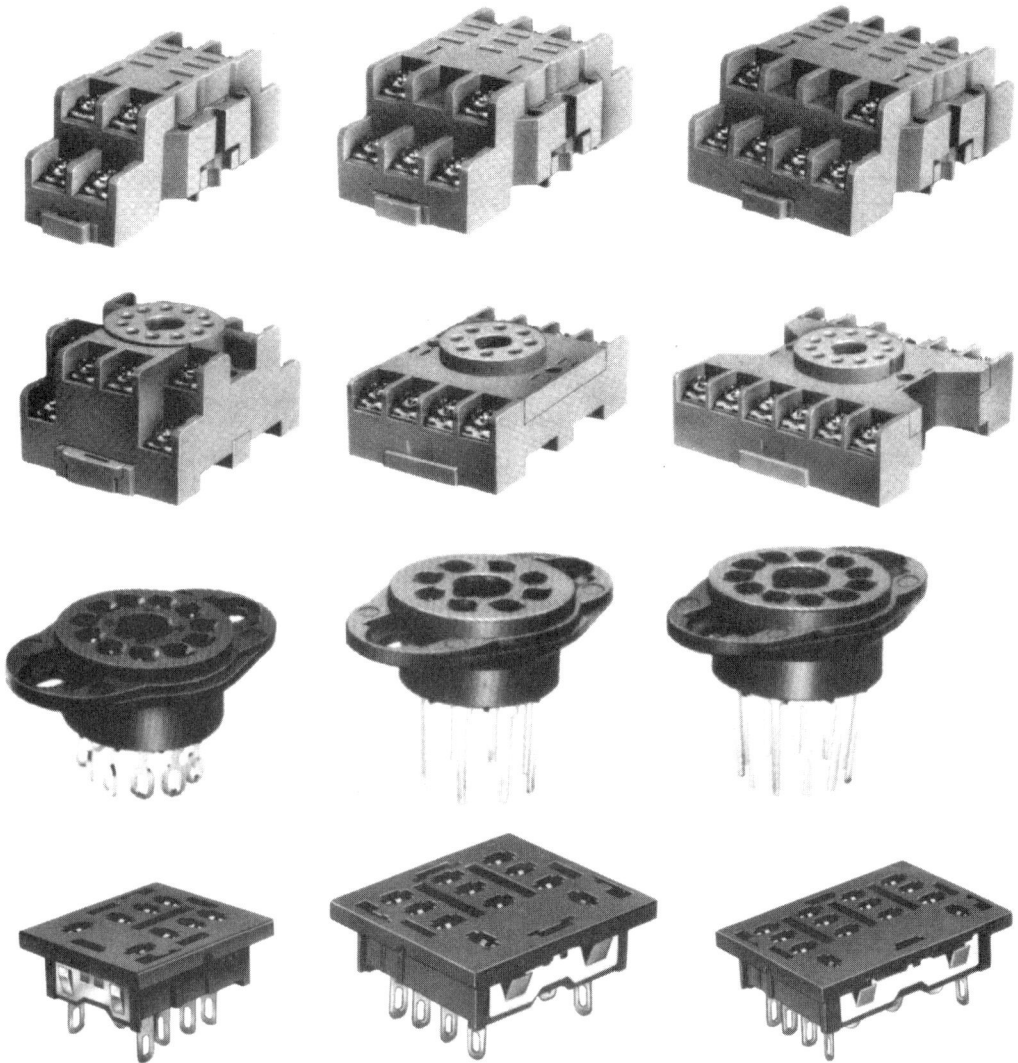

FIGURE 4.17
Terminal sockets for industrial relays.

is well held in a vertical and horizontal position, however, it is not recommended to install relays vertically downwards on such terminal sockets, because long-term vibration can cause the falling out of a relay from a terminal socket.

Many firms produce terminal sockets supplied with a special lock holding a relay in any position in space (Figure 4.20). Sockets of this type allow assembly of relays in electric cabinets with very dense mountings, with the help of a DIN Rail.

The designers of Phoenix Contact created a whole world of original constructions based on the use of terminal sockets holding relays on standard DIN Rails in electric cabinets (Figure 4.21). In particular they designed original "containers" where ready-for-service relays produced by other firms were placed (Figure 4.22). Pin-like outlets of a container allow direct installation of relays on standard terminal-socket connectors used for connections of wire in electric cabinets.

FIGURE 4.18
Installation of a relay on terminal sockets.

The Phoenix Contact Company uses some other original principles of positioning of a relay on standard DIN Rail (Figure 4.23).

Despite the modern tendency to produce miniature relays, some companies still manufacture equipment exceeding modern industrial miniature relays in size. Such equipment is similar (by size, by contact, and by magnetic systems) to relays produced at the beginning of the 20th century. World leaders in the electric-power industry, such as concern ABB, produce such relays, designed for use in electrical power systems.

Production of such large electromagnetic relays is connected with a tendency to unify sizes for protective relays used in the electric-power industry. Protective relays are complex devices containing elaborate magnetic and electronic systems and are quite large in size. Perhaps large electromagnetic relays are better arranged with protective relays in relay protection cabinets, however, in some industrial automation systems there are so many auxiliary relays that several cabinets are required to install and to mount them all (Figure 4.24). In this case it is the consumer who is to pay with his working areas for producers' adherence to their own standards.

Of course such relays do have modern designs (Figure 4.25) and features such as original construction of outlets. By external design as well as by construction, it appears to be a multi-contact connector (Figure 4.26).

The General Electric Company has been producing protective relays in big and heavy metal cases with detachable glass doors for many years. For example, differential relays of PVD, BDD type or others, have size of $380 \times 168 \times 160\,mm$ and a weight of 8 to $10\,kg$

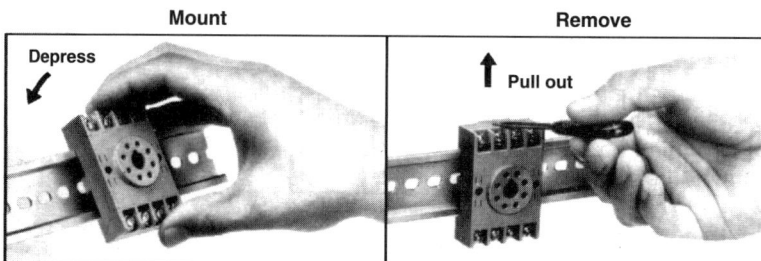

FIGURE 4.19
Installation of a terminal socket on a standard DIN Rail.

FIGURE 4.20
Rail mount terminal sockets are supplied with a special lock (produced by Omron, Shcrack, Idec).

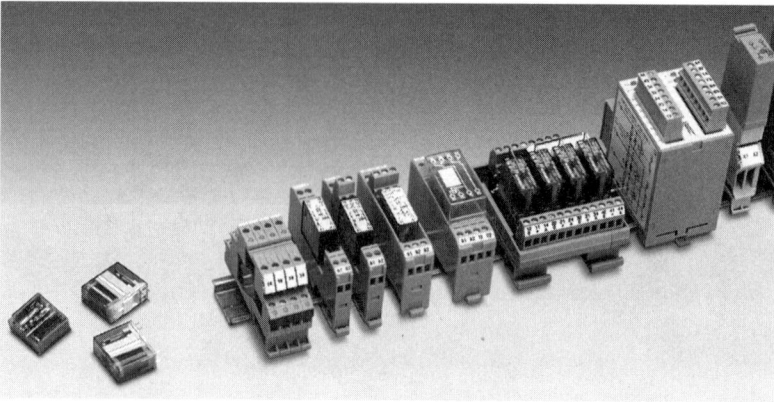

FIGURE 4.21
Terminal sockets produced by Phoenix Contact GmbH & Co KG for mounting of different types of relays on standard rails in electric cabinets. (Picture Courtesy Phoenix Contact GmbH & Co., 2004.)

(Figure 4.27). In spite of the fact that modern relays contain electric and microprocessor-based systems, cases remain the same. This probably happens because such cases are popular around the world since they provide safe protection of the relays against external stress, dust, and magnetic fields. Special plug-in connectors (plug-ins) can be attached to them, allowing connection of external devices to test the relay without having to dismantle it from the circuit. Simple electromagnetic relays have also been produced in similar standard cases (Figure 4.28).

According to different types of mountings (Figure 4.29), modern industrial miniature relays may have different outlets (Figure 4.30). It is obvious that small low-power relays designed for installation on a printed circuit board will have relatively thin straight (or curved, for surface mounting) outlets placed at a standard distance to each other. More powerful relays are supplied with larger outlets designed for soldering of external wire, insertion of them into a terminal socket, or switching of the conductors with the help of a special connector, the so-called "faston" (Figure 4.31).

How are we to deal with a power relay that is to be installed on a printed circuit board?

For this purpose firms have designed and constructed relays with TMP-type outlets containing both thin straight outlets of winding for printed circuit wiring (below) and powerful contact outlets for switching massive external conductors with the help of fastons (above — Figure 4.32).

More powerful relays are supplied with screw clamps for external connection of massive wire. As a rule, in order to save space needed for mounting of the relay these clamps are placed on the top of the relay case (Figure 4.33). There is a connection between modern tendencies of development of electromagnetic relays with an increase of switching capacities, on the on hand, and with the process of micro-miniaturization on the other hand (Figure 4.34).

It took many engineering efforts to develop constructions of cases, fastening elements and outlets, but for micro-miniature relays — a masterpiece of engineering — designers did not have to reinvent these elements: micro-miniature relays were placed into standard cases of transistors or chips (Figure 4.34), which were simply soldered into a printed circuit board without any additional fastening elements.

(a)

(b)

(c)

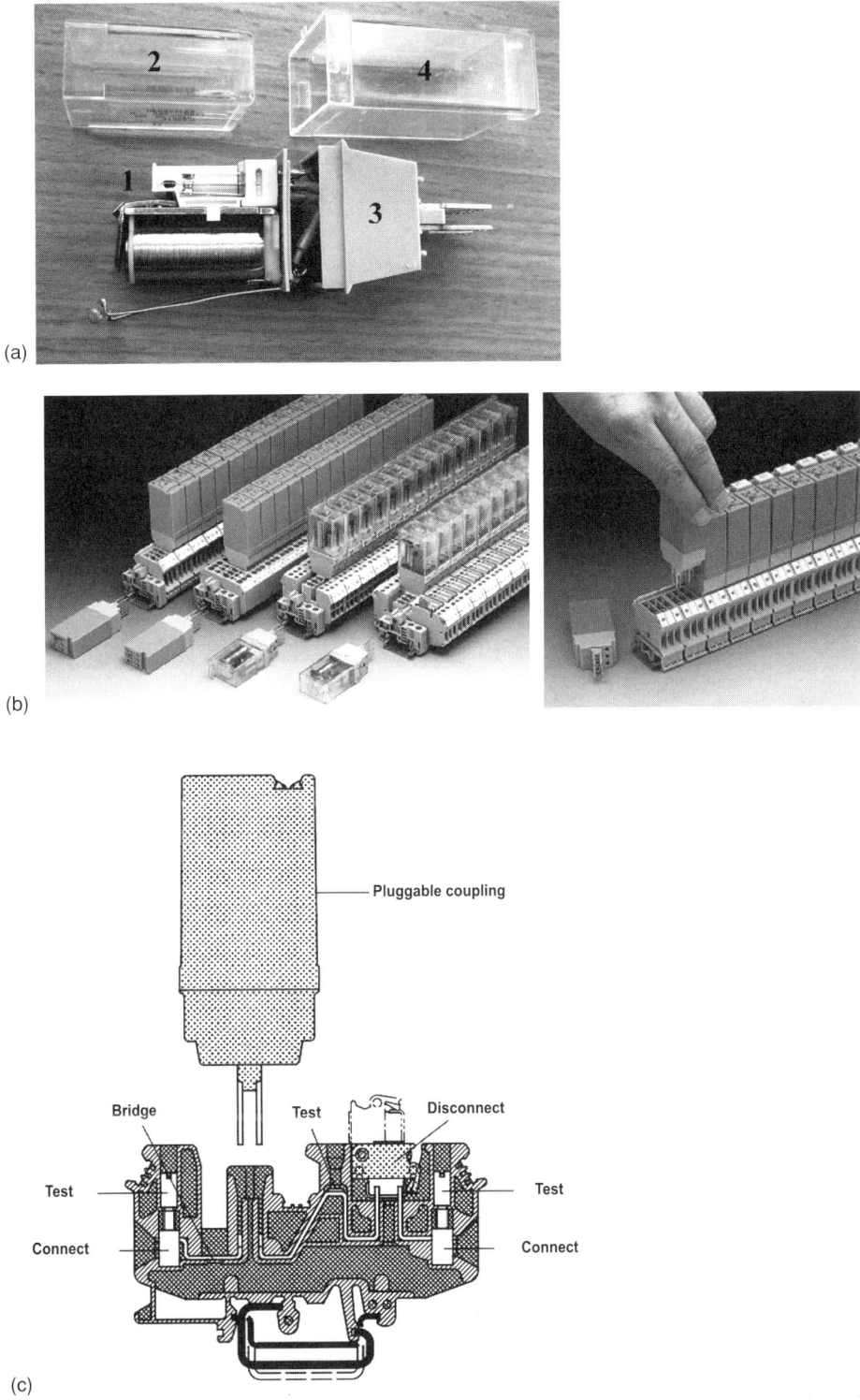

FIGURE 4.22
(a) Construction of a container with a relay installed inside. Produced by Phoenix Contact GmbH & Co KG. 1 — ready-assembled electric relay; 2 — plastic case of the relay; 3 — container with pin-like outlets produced by Phoenix Contact; 4 — plastic case of the container. (b) Installation of containers with electromagnetic relays on mounting terminal sockets (Phoenix Contact). (c) Construction of a cell of a mounting terminal socket produced by Phoenix Contact. (Picture Courtesy Phoenix Contact GmbH & Co., 2005.)

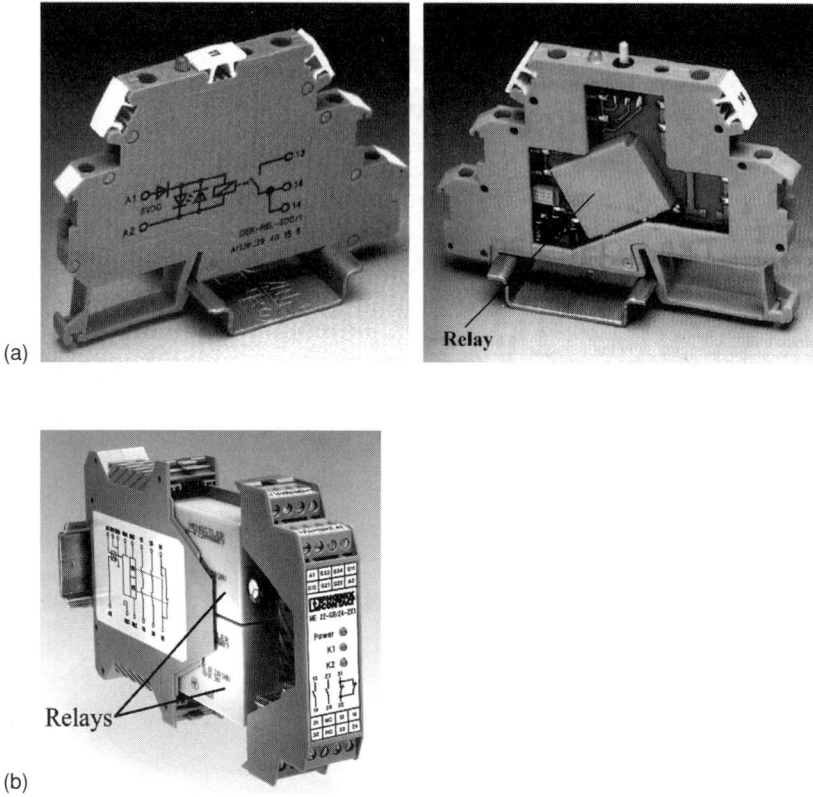

(a)

Relay

Relays

(b)

FIGURE 4.23

(a) Mounting of a ready-assembled miniature relay supplied with a plastic case on a standard DIN Rail (Phoenix Contact). (b) Another method of mounting an industrial relay in a special collapsible plastic case designed for installation on a standard DIN Rail (Picture Courtesy Phoenix Contact GmbH & Co., 2006.)

FIGURE 4.24

Cabinets with auxiliary electromagnetic relays, produced by ABB, installed in one of the industrial enterprises.

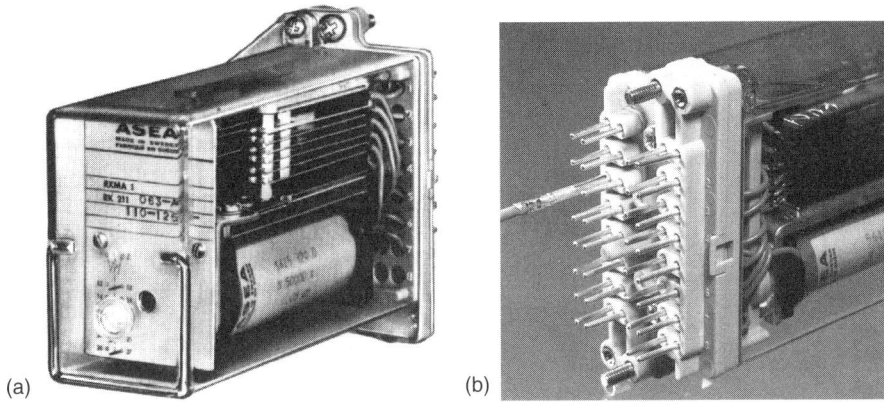

FIGURE 4.25
(a) Modern electro-magnetic relay of RXMA-1 type used in cabinets of relay protection as an auxiliary relay (produced by ABB). (b) Terminal socket of RXMA-1 relay. (From ASEA Relays, Buyers Guide B03-00011E.)

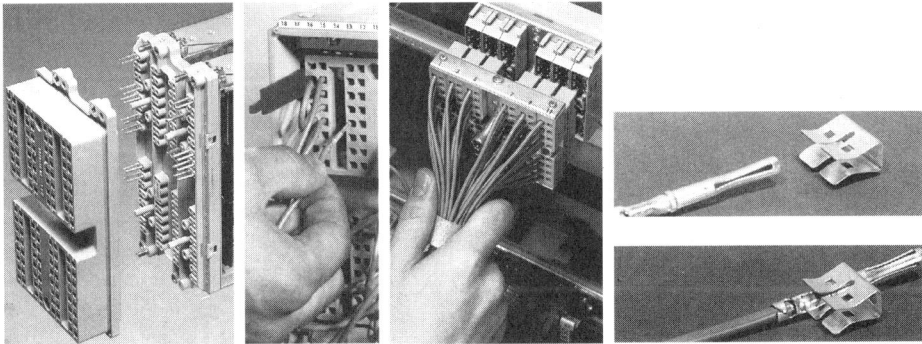

FIGURE 4.26
Peculiarities of mounting and elements of construction of a relay socket produced by ABB. (From ASEA Relays, Buyers Guide B03-00011E.)

FIGURE 4.27
Relays of differential protection produced by General Electric

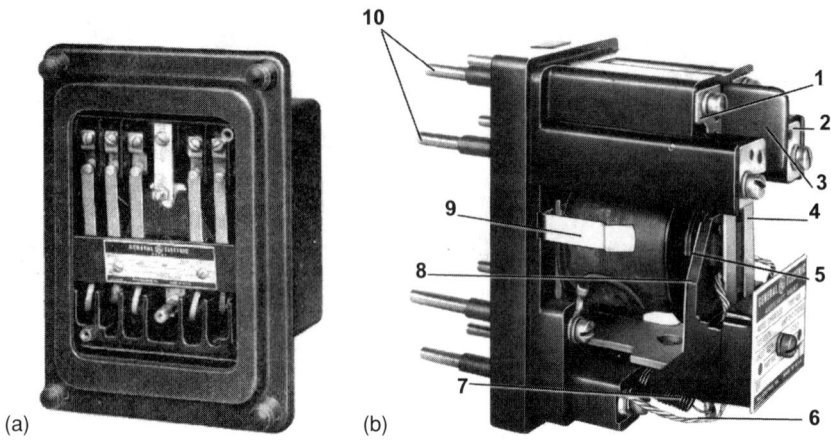

FIGURE 4.28
Electromagnetic auxiliary relays produced by General Electric: (a) — a multi-contact relay in a standard case. (b)— a relay with two switching contacts extracted from a case; 1 — normally open contact; 2 — normally closed contact; 3 — voltage barrier; 4 — spring leaves for moving contacts; 5 — pole piece; 6 — flexible lead; 7 — control spring; 8 — hinged armature; 9 — cover spring clip; 10 — connection studs. (From GE. catalog With permission.)

FIGURE 4.29
Means of mounting of modern relays.

FIGURE 4.30
Types of outlets of modern relays.

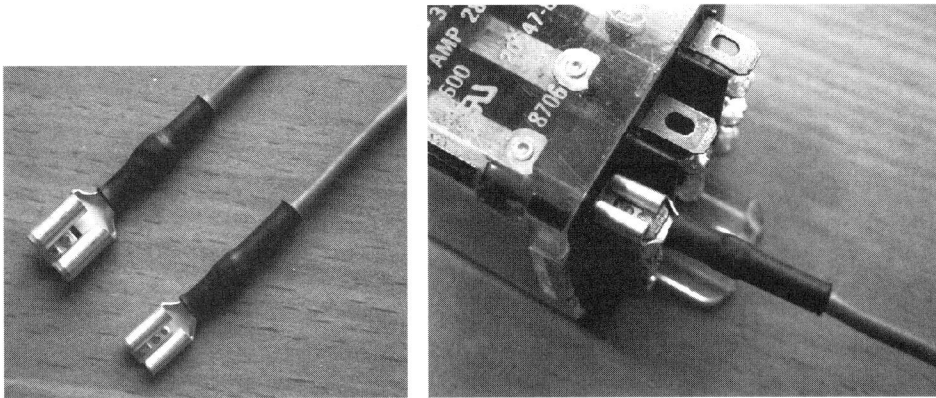

FIGURE 4.31
Faston-type connectors and their use for relay connection

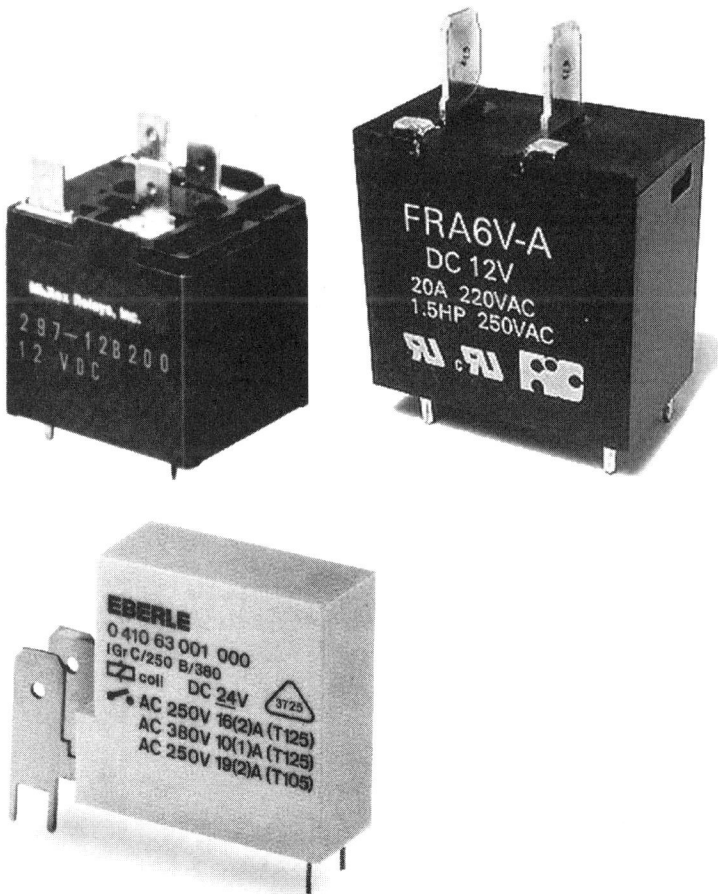

FIGURE 4.32
Relays with hybrid outlets.

FIGURE 4.33
Power relay for switching of high currents with screw clamps for external mountings.

4.5   Indicators of Operation and Test Buttons

Apart from the described peculiarities of constructions of cases and relay outlets, recently additional elements have been placed in cases of industrial relays, making their exploitation more convenient. These include, first of all, light-emitting elements displaying the "on" state of the relay (either miniature baseless neon lamps or light-emitting diodes [LED] are usually used as light-emitting elements). These elements are connected through current-limiting resistance parallel to the winding of the relay (LEDs are sometimes connected in series), usually placed under a transparent case of the relay.

RS-series modules (Figure 4.35a) are compact, space-saving relay terminal modules containing six so-called "card relays" — small flat relays with one normally open contact each. All relays contain built-in coil surge-suppression diodes and operation indicators (LEDs) that simplify circuit design and maintenance. The module is easily-mounted on a 35 mm DIN rail.

The firm Shcrack produces indicator elements with LED in the form of separate modules, which can be attached to a relay by placing it next to the relay into a terminal socket (Figure 4.35b). As similar modules the firm produces additional diodes designed for lowering of switching over-voltages, with a relay controlled by a DC coil.

Another additional element is a so-called test button (Figure 4.36) This, in fact, is the same as a plastic pusher, one end of which sticks out through a hole in the relay case, with the other end touching the armature of the relay. As one presses the pusher with the relay in a deenergized state, the armature moves as if affected by the magnetic field of the coil, and switches the contact. This relay is very convenient for setting up of automation devices and for fault tracing in circuits.

4.6   Relays Which Do Not Look Like Relays At All

Power relays for 16 to 25 As currents are more often supplied with an additional unit of contacts, which can be easily attached to the upper part of a relay with a help of plastic catches (Figure 4.37). In this case a plastic pusher of additional unit of contacts meshes with the pusher of the main relay providing synchronous reliable operation of all contacts. A basic relay of this type usually has room for up to four convertible contact units

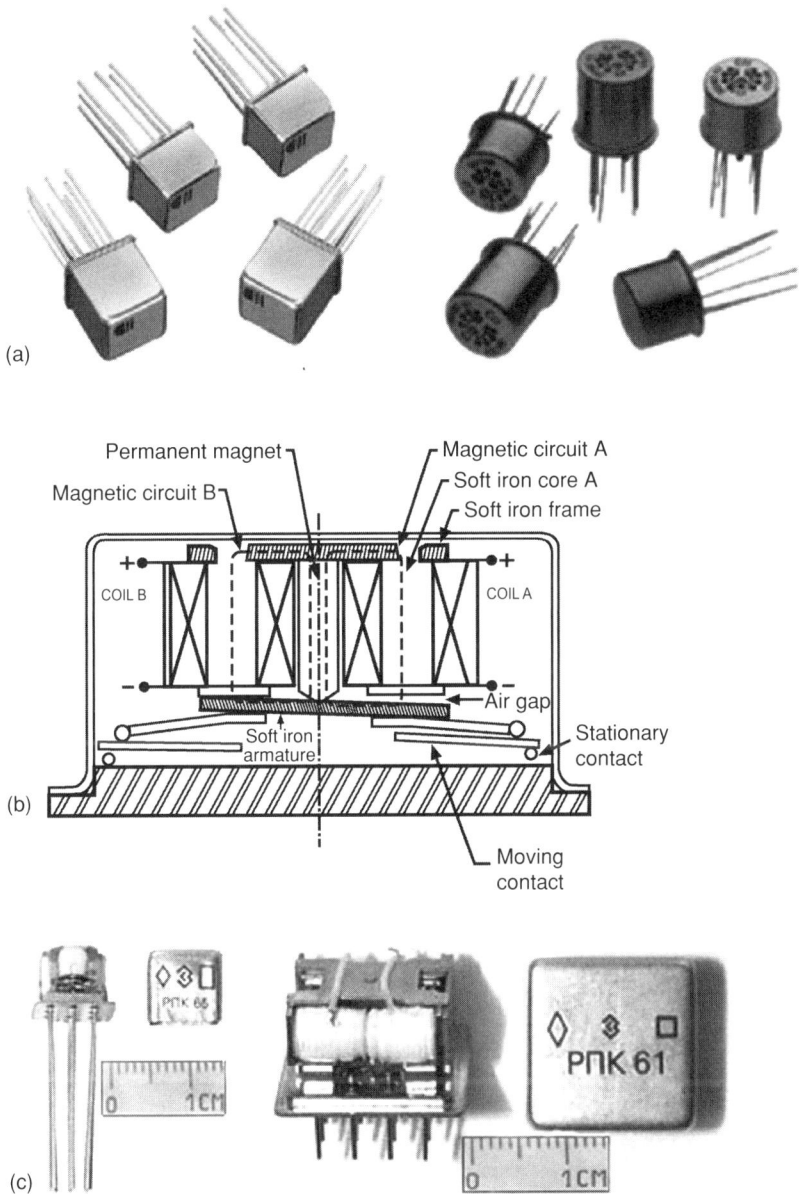

(a)

(b)

Permanent magnet
Magnetic circuit B
Magnetic circuit A
Soft iron core A
Soft iron frame

COIL B

COIL A

Air gap
Soft iron armature
Stationary contact
Moving contact

(c)

FIGURE 4.34
(a) Micro-miniature relays in standard cases used for transistors and integral circuits (IC), produced by many companies: Teledyne Technologies, CII, Nuova HI-G Italia, Guardian Controls, and others. (b) Construction of micro-miniature relays, produced by Teledyne. (c) External design and construction of new micro-miniature relays, developed by the Russian company "Severnaya Zarya."

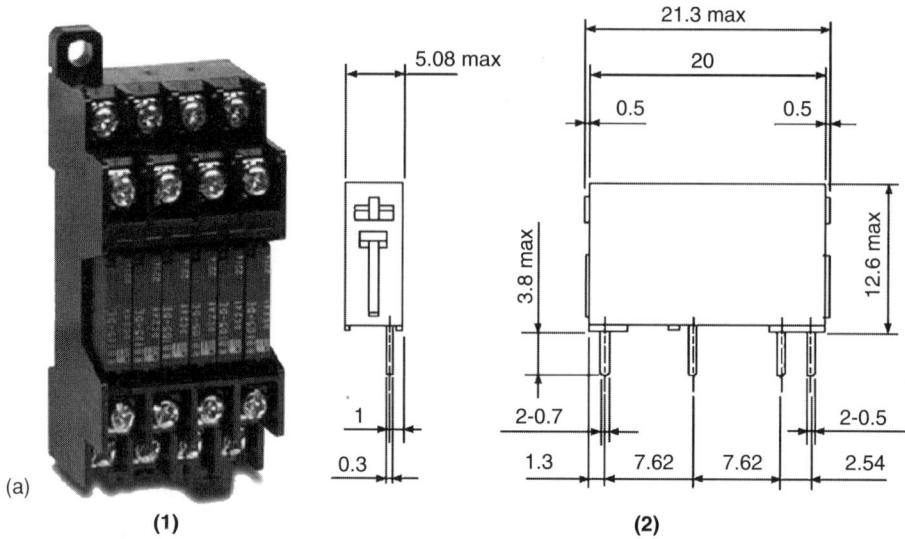

FIGURE 4.35
(a) Module of RS-series card relays and dimensions of separate relay with built-in coil surge-suppression diodes and operation indicators (AutomationDirect Ltd). (b) Modular unit of the firm Shcrack. 1 — indicator units with LED; 2 — relay; 3 — unit with diodes.

(cartridges). It can be expanded to 6 or 8 poles by installing an added deck. A 10 or 12 pole relay can be built by adding a second deck.

The American company Olympic Controls Corp. produced a special so-called T-Bar multi-contact relay (Figure 4.38). T-Bar multi-contact relays are designed for use in controlled environments, such as test areas, computer control rooms, broadcast studios and network management centers. The heart of T-Bar components is the 12 pole-switching wafers. Relay assemblies are available in switching configurations of 4, 8, 12, 24, 36, 48 and 60 form "A" contacts (normally open) and 52 form "C" contacts (changeover). The 900 series, designed for dry circuits of a maximum 1 A switching for use in data, thermocouple and instrumentation circuits, and the 800 series, used for control interlock and for indicator circuits.

Many companies produce power relays consisting of a main unit which can be used separately, and a set of many auxiliary units and elements attaching to the main unit

FIGURE 4.36
Relays with test buttons.

(a)

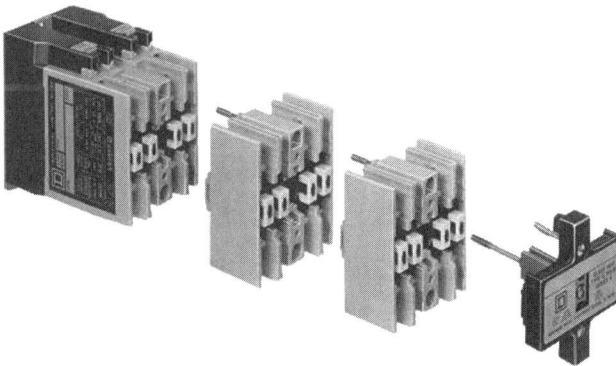

(b)

FIGURE 4.37
Power relays modular construction with attached units of additional contacts. (a — Telemechanique; b — Square-D).

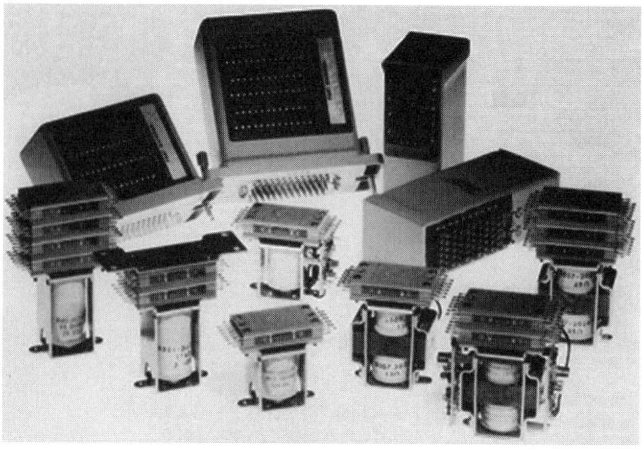

FIGURE 4.38
T-Bar type multi-contact relays, produced by Olympic Controls Corp. (U.S.A.). (From Olympic Control Corp., Online Internet Catalogue, 2004)

(indicator unit, diode unit, additional contacts, time delay unit, overload protection unit, and others) on all sides and considerably broadening the capabilities of the device. Such power relays, designed for reversing start-up of electric motors, are often produced as twin-units supplied with mechanical and electronic blocking prevention from the synchronous switching of both relays, and are supplied with all of the necessary bonds (Figure 4.39).

Especially amazing are the so-called explosion-proof electromagnetic contactors and relays. No, these devices have nothing to do with military equipment. They are used in coal mines. All devices of this type are placed in massive steel cases with hermetic hatches, which can withstand explosion both inside and outside of the shell. They look really original (Figure 4.40). This kind of product (Figure 4.41) is suitable for mining wells with their explosive mixed gas and coal dirt, and for direct starts, stops, and turnovers in the three phase squirrel cage different step electrical motor with AC 50 Hz, rated voltage up to 660 V, rated current up to 120 A.

Relays designed for switching of strong signals of high frequency also look quite unusual (Figure 4.42). The major applications of such relays are:

- High power transmitter switching
- Radar pulse forming networks
- Phased array antenna systems

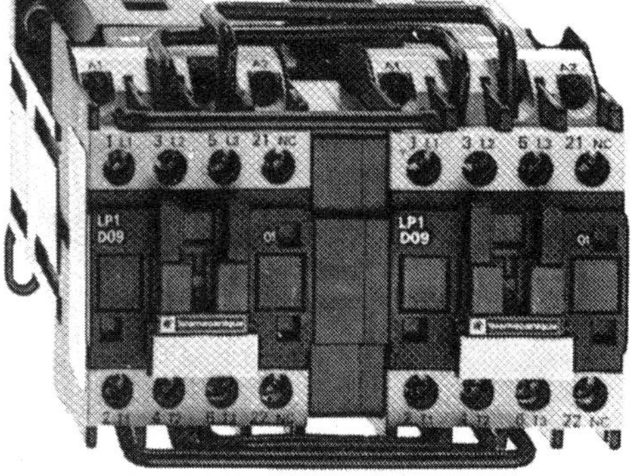

FIGURE 4.39
A twin (so-called "reversing") contactor produced by Telemechanique.

FIGURE 4.40
Explosion-proof electromagnetic contactor of PB-1140 type for 250 A current and 1140 V voltage (Russia). Size: 870 × 850 × 980 mm; weight: 410 kg

- UHF/VHF communications systems
- Magnetic resonance imaging systems

Such relays have high power handling capabilities in a small package. The ability to often handle up to 3 to 5 kW (up to 90 kW for some models of Jennings relays, Figure 4.43) at frequencies of up to 30 MHz is achieved with vacuum-enclosed contacts, minimizing noise and losses. Such rugged switches are capable of "hot" switching kilowatts of 30 MHz with optional special tungsten–molybdenum contacts to avoid pitting when switched with RF power applied. Even with heavy-duty construction, hot-switching will reduce the typical operational life of 1,000,000 cycles significantly — to approximately 10,000 cycles.

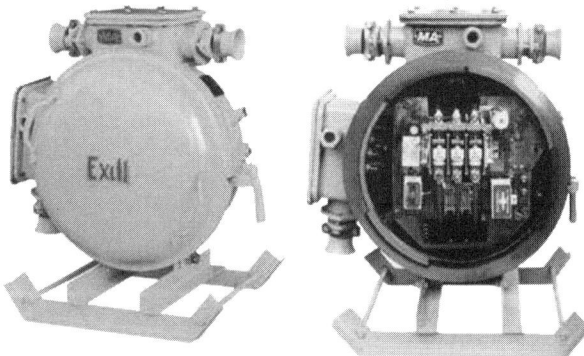

FIGURE 4.41
120ND series mining explosion-proof vacuum changeover electromagnetic starter for 120 A current and 660 V voltage (Yueqing Bada Vacuum Electric Appliances Switches Factory, China).

FIGURE 4.42
Powerful high-frequency relay of 310 series (DowKey Microwave).

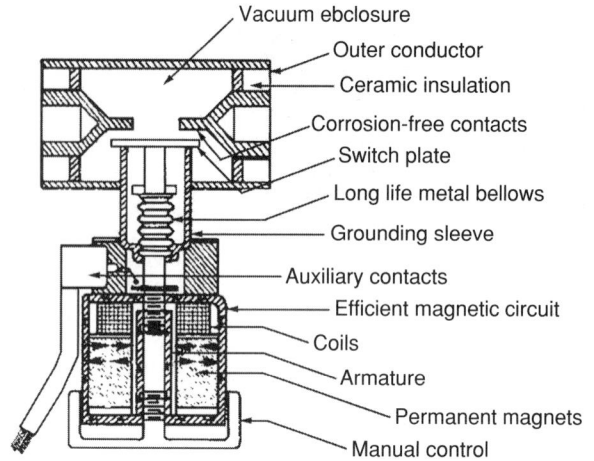

FIGURE 4.43

High Frequency Vacuum Coaxial Relays, produced by Jennings Technology (U.S.A.). (From High Voltage Vaccuum and Gas-Filled Relays, Josyln-Jennings Corp. Catalog. REL-103, 1993.)

TABLE 4.1

Main Parameters of DowKey 310 Series Relay

Frequency (MHz)	VSWR (max)	Isolation dB (min)	Ins. Loss dB (max)	RF Power Watts (CW)
30	1.05	35	0.07	3000
50	1.06	30	0.08	2300
100	1.08	25	0.09	2000
400	1.10	17	0.1	850

Switching capacity of such relays is characterized by a number of specific parameters ("standing wave ratio," "crosstalk attenuation," etc.) denoting distortion and waste in the radio-frequency circuit in the closed state, and by a capability to insulate radio-frequency circuits in the open state. We will not go into details of terminology for high and ultrahigh frequencies as this issue lies outside the scope of this book. Here we will give an example of technical specifications for a DowKey 310 Series relay (Table 4.1).

5

Reed Switches and Reed Relays

5.1 Who Invented a "Reed Switch"?

Many engineers have come across original contact elements contained in a glass shell (Figure 5.1) However, not everyone knows that reed relays differ from ordinary ones not because of the germetic shell (sealed relays are not necessarily reed ones), but because of the fact that in a reed relay, a thin plate made of magnetic material functions as contacts, magnetic system, and springs at the same time. One end of this plate is fixed, while the other end is covered with some electroconductive material and can move freely under the effect of an external magnetic field. The free ends of these two plates, directed towards each other, are overlaped for from 0.2 to 2 mm and form a basis for a new type of a switching device — a "germetic magnetically controlled contact" (in Russian) or "reed switch" (in English). Such a contact is called a "magnetically controlled contact" because it closes under the influence of an external magnetic field, unlike contacts of ordinary relays which are switched with the help of mechanical force applied directly to them. The original idea of such a function mix, which was in fact the invention of the reed switch, was proposed in 1922 by a professor from Leningrad Electrotechnical University, V. Kovalenkov, who lectured on "magnetic circuits" from 1920 until 1930. Kovalenkov received a U.S.S.R. inventor's certificate registered under No. 466 (Figure 5.2).

In 1936, the American company "Bell Telephone Laboratories" launched research work on reed switches. Already in 1938 an experimental model of a reed switch was used to switch the central coaxial cable conductor in a high-frequency telecommunication system, and in 1940 the first production lot of these devices, called "Reed Switches," was released (Figure 5.3). Reed switch relays (that is a reed switch supplied with a coil setting up a magnetic field — Figure 5.4a), compared with electromagnetic armature relays which are similar in size, have higher operation speeds and durability, a higher stability of transient resistance, and a higher capability to withstand impacts of destabilizing factors (mechanical, climatic, specific), in spite of their relatively low switching power.

At the end of the 1950s some western countries launched construction of quasi-electronic exchanges with a speech channel (which occupied over 50% of the entire equipment of an exchange) based on reed switches and control circuits on semiconductors. In 1963, the Bell Company created the first quasi-electronic exchange of ESS-1 type designed for an intercity exchange. In a speech channel of such an exchange more than 690,000 reed switches were used. In the ensuing years the Western Electric Company arranged a lot production of telephone exchanges based on reed switches with a capacity from 10 up to 65,000 numbers. By 1977, about 1,000 electronic exchanges of this type had been put into operation in the U.S.A.

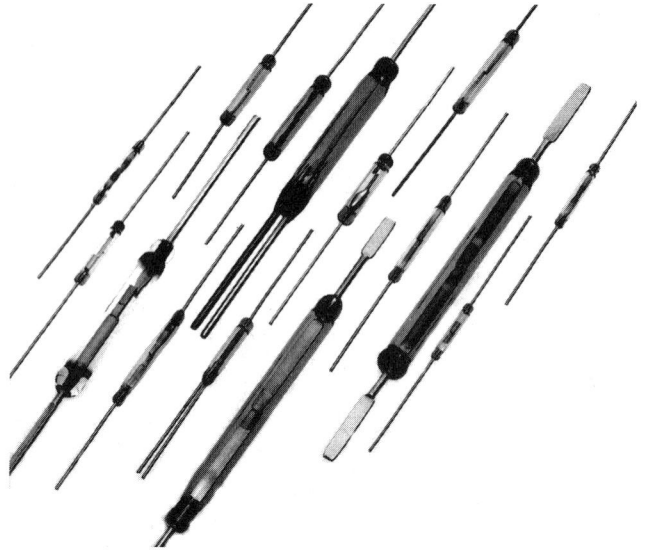

FIGURE 5.1
Different types of modern sealed reed switches.

In, Japan the first exchange of ESS-type was put into service in 1971. By 1977, the number of such exchanges in Japan was estimated in hundreds. In 1956, the Hamlin Co. launched lot production of reed switches and soon became the major producer and provider of reed switches for many relay firms. Within a few years this company built plants producing reed switches and relays based on them in France, Hong Kong, Taiwan, and South Korea. Under its licensed plants in Great Britain and Germany, it also started to produce reed switches in those countries. By 1977, Hamlin produced about 25 million reed switches, which was more than a half of all its production in the U.S.A. Reed switches produced by this firm were widely used in space-qualified hardware, including man's first flight to the Moon (the Apollo program). The cost of each reed switch thoroughly selected and checked for this purpose reached $200 a piece.

In the former Soviet Union lot production of reed switches was launched in 1966 by the Ryazan Ceramic–Metal Plant (RCMP). Plants of the former Ministry of Tele-communication Industry (its 9th Central Directorate in particular) were also involved in production of weak-current relays based on reed switches. At the end of the 1980s there were 60 types of reed relays produced in the U.S.S.R. The total amount of such relays reached 60 to 70 million a year. Economic crises in Russia led to a steep decline in the production of both reed switches and reed relays. In 2001, plants producing relays (those which were still working in Russia) ordered only about 0.4 million reed switches for relay production.

Depending on the size of a reed switch, the working gap between contact-elements may vary between 0.05 and 0.8 mm (and more for high-voltage types) and the overlap of ends of contact-elements between 0.2 and 2 mm. Due to the small gaps between contacts and a

FIGURE 5.2
Kovalenkov's relay. 1 and 2 — contact-elements (springs) made of magnetic material; 3 — external magnetic core (the core of the relay); 4 — control winding (external magnetic filed source); 5 — dielectric spacers; 6 — ends of contact-elements; 7 — working gap in magnetic system and between contacts; 8 — contact outlets for connection of external circuit.

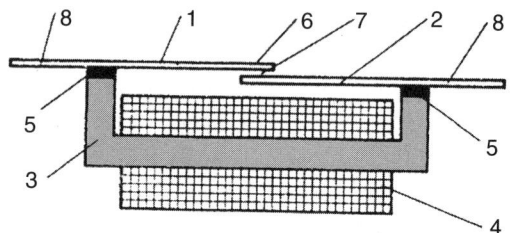

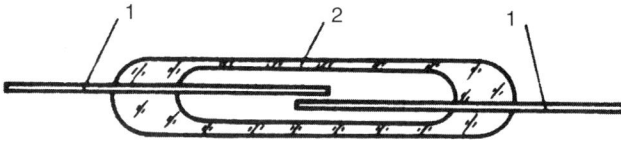

FIGURE 5.3
Construction of a modern reed switch. 1 — contact elements (springs) from permalloy; 2 — glass hermetic shell.

small total weight of movable parts, reed switches can be considered to be the most high speed type of electromagnetic switching equipment with a delay of 0.5 to 2 ms capable of switching electric circuits with frequencies of up to 200 Hz. The smallest in the world reed switches produce Hermetic Switch, Inc. (U.S.A.) (Figure 5.4b). The tiny oval shaped glass balloon of the HSR-0025 reed switch measures a mere 4.06 mm long, 1.22 mm wide, and 0.89 mm high. Maximum switching rating of HSR-0025: 30 V, 0.01 A, 0.25 W. Sensitivity ranges from 2 to 15 A-turns.

Bigger relays can switch higher switching current, as the contacting area of the contact-elements, their section, contact pressure, and thermal conductivity, increase. Most reed switches have round-shape shells (balloons), because they are cut from a tube (usually a glass one), the ends of which are sealed after installation of contact-points. Glass for tubes should be fusible with softening temperatures and coefficients of linear expansion similar to that of the material of contact-elements.

Contact-elements of reed switches are made of ferromagnetic materials with similar coefficients of linear expansion as for glass. Most often it is Permalloy, an iron–nickel alloy (usually 25% nickel in alloy). Sometimes Kovar, a more high-temperature alloy is used. It allows application of more refractory glass for tubes (560 to 600 °C) and as a result, more heat-resistant reed switches are obtained. To provide a better joint with the glass, contact-points are sometimes covered with materials providing better joints with glass than Permalloy. Sometimes contact-elements have more a complex cover consisting of sections with different properties. Contact-elements may also contain two parts, one of which joins well with the tube glass and has the required flexibility, and the other has the necessary magnetic properties. The contacting surfaces of contact-elements of average power reed switches are usually covered with rhodium or ruthenium; low-power reed switches designed for switching of dry circuits are covered with gold and high-power and high-

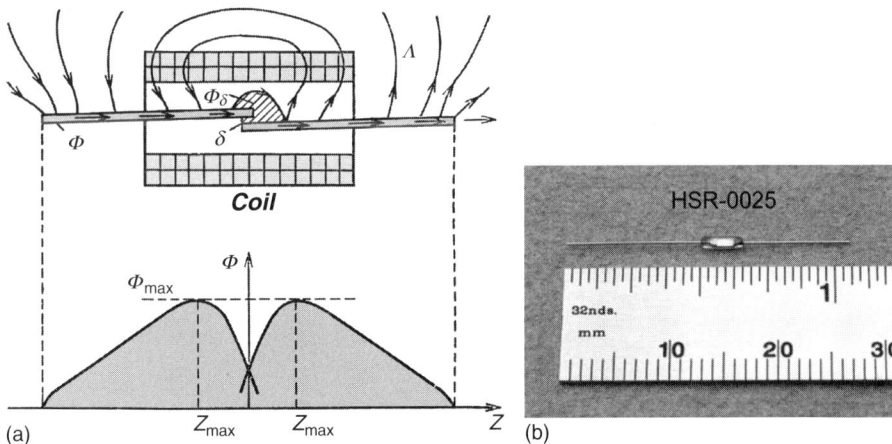

FIGURE 5.4
(a) Magnetic field in a sealed reed relay. δ — working (magnetic and contact) gap. (b) The smallest in the world reed switch, produced by Hermetic Switch, Inc.(U.S.)

voltage reed switches with tungsten or molybdenum. Covering is usually made by galvanization with further heat treatment to provide diffusion of atoms of a cover to the material. It can also be carried out by vacuum evaporation or other modern methods. Contact-elements of high-frequency reed switches are fully covered with copper or silver to avoid loss of or attenuation of high-frequency signals, and after that the contacting surfaces are also covered with gold.

The tube of a medium or low-power reed switch is usually filled with dry air or a mixture of 97% nitrogen and 3% hydrogen. A 50% helium–nitrogen mixture, carbonic acid, and other mixtures of carbon dioxide and carbonic acid can also be used. Carefully selected gas environments effectively protect contact-elements from oxidation and provide better quenching of spark as low powers are being switched. Reed switches designed for switching of voltages from 600 up to 1000 V have a higher gas pressure in the tube, which may reach several atmospheres. High-voltage reed switches (more than 1000 V) are usually vacuumized.

The fact that there are no rubbing elements, full protection of contact-elements from environmental impact, and the possibility to create a favorable environment in the contact area, provides switching and mechanical wear resistance of reed switches estimated in the millions and even billions. Reed switches which are in mass production and which are widely used in practice can be classified by the following characteristics:

1. Size
   - Normal or standard reed switches with a tube about 50 mm in length and about 5 mm in diameter
   - Subminiature reed switches with a tube 25 to 35 mm in length and about 4 mm in diameter
   - Miniature reed switches with a tube 13 to 20 mm in length and 2–3 mm in diameter
   - Micro-miniature reed switches with a tube 4 to 9 mm in length and 1.5–2 mm in diameter

2. Type of a magnetic system
   - Neutral
   - Polarized

3. Type of switching of electric circuit
   - Closing or normally open — A type
   - Opening or normally closed — B type
   - Changeover — C type

4. Switched voltage level
   - Low-voltage (up to 1000 V)
   - High-voltage (more than 1000 V)

5. Switched power
   - Low-power (up to 60 W)
   - Power (100 to 1000 W)
   - High-power (more than 1000 W)

6. Types of electric contacts
   - Dry (the tube is filed with dry air, gas mixture, or vacuumized)
   - Wetted (in the tube there is mercury wetting the surface of contact-elements)

7. Construction of contact-elements
   - Console type (symmetrical or asymmetrical) with equal hardness of the movable unit (Figure 5.3 — main type of reed switches)
   - With a stiff movable unit
   - Ball type
   - Powder type
   - Membrane type, etc.

5.2 Coruscation of Ideas and Constructions

The classification given above is relative and is true for classical constructions of reed switches produced on mass production lines. One should bear in mind, however, that there are so many patents for very original and sometimes even exotic constructions of reed switches that it is almost impossible to include them all in the given classification. Some reed switches are produced in limited numbers for specific purposes, others will always remain examples of engineers' inventiveness. However, even a brief description of their constructions can reveal major problems for designers on the one hand and on the other hand provide some solutions to these problems. For example, the most popular construction of a reed switch with console contact-elements with equal hardness along the length (Figure 5.3) turns out not to be so optimal, since it requires great efforts for curving of the contact-elements. The slightest deformation or inaccuracy of soldering in glass considerably reduces the contacting area and impairs its switching properties. Requirements for contact-elements are quite inconsistent; on the one hand, the bigger the section of the contact-elements, the better their magnetic conductivity and the greater the contact pressure that can be applied with specified magnetic flux of a control coil. On the other hand, the hardness also increases considerably and a greater mechanical effort is required to curve them to the closing point. Is it possible to increase magnetic conductivity of a reed switch without an increase of hardness of the contact-points?

Yes, it is! Moreover, there are several ways of doing it!

In fact, the tube of a reed switch must not necessarily be made of glass. It is true that glass provides a hermetic soldered joint with metal, but the whole tube must not necessarily be of glass. It can also be metallic with glass insulators, as in sealed relays in metal cases. Actually it is not even necessary to put insulators on both sides. One can connect electrically one of the contact-elements to a metal case, and one glass insulator will be enough. A more rational solution, though, would be perhaps to remove this contact-element and use a case instead. This idea was implemented by the inventor of the reed switch, as shown in Figure 5.5.

The use of a steel case instead of one of the contact-elements allows us to increase considerably the magnetic conductivity of the construction, and also contact pressure. Apart from other obvious advantages of replacement of glass with metal, one can also include greater strength of the reed switch, better heat abstraction from contacting area, etc.

The American inventor R. Alley chose another way (patent No. 2987593). He made contact-elements with unequal hardness along the length. A part with a small section resembling a flexible spring provided the required flexibility and stiffer parts with a greater section ending with massive contacts provided high magnetic conductivity (Figure 5.6). Another solution is to use three absolutely stiff ferromagnetic elements

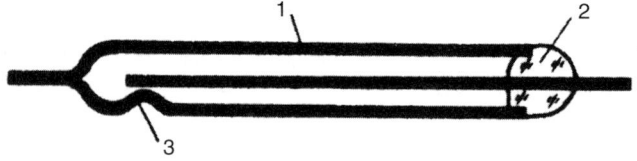

with the central one swinging on hinges or moving linearly (Figure 5.7), linking the two other stiff stationary elements.

Some types of relays require reed switches with one-way outlets (Figure 5.8). In the place of the curve the movable contact-point of the closed reed switch has a smaller section, which also reduces its hardness.

In reed switches with reverse outlets there is an attracting force between differently magnetized contact-elements, while in reed switches with one-way outlets (Figure 5.8), there is a repulsive force between the equally magnetized parts of the contact-elements.

Both of these principles are used in changeover reed switches (Figure 5.9). When an external magnetic field affects such a reed switch, contact-elements 1 and 2 are magnetized differently creating an attracting force, and contact-elements 2 and 3 are magnetized equally, creating a repulsive force. As a result, the movable contact-point 2 curves switching the external circuit.

Another type of switched reed switch was invented by W. Eitel, an employee of the Penta Laboratories firm (U.S.A. patent No. 2360941, Figure 5.10). Reed switches produced by this firm are usually used for switching of high-frequency circuits when minimal capacity between contact-points is required. Such construction also has good magnetic and switching characteristics because the stationary contact-point is supplied with a ferromagnetic packing, which considerably increases permeance of the system and provides reliable contact pressure. A movable contact-point is constituent, as it contains a part of increased flexibility and a stiff part of quite a large section made of ferromagnetic material with a powerful contact at the end. Such a reed switch with a well vacuumized shell can withstand voltage between contacts of up to 20 kV and skip short current pulses with an amplitude of up to 100 A in the closed position.

The T-shape form of a reed switch with two symmetric magnetic cores (Figure 5.11), and two control windings on these cores, allows creation of a relay with a differential function, that is, a relay, the mode of which depends on a difference of currents in the control windings, on current direction in windings and different logical combinations.

In standard electromagnetic relays the magnetic circuit and the contact system are two independent systems connected to each other only by an insulated pusher, therefore only

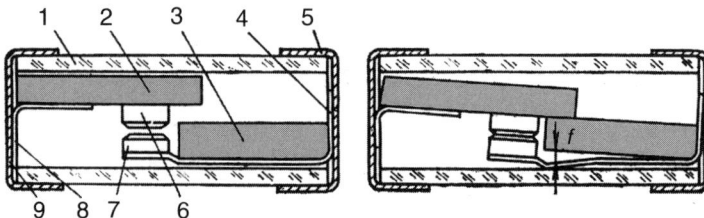

FIGURE 5.6
Power reed switch with contact-elements of unequal hardness. 1 — glass shell; 2, 3 — stiff parts of contact-elements; 4, 8 — flexible parts of contact-elements; 5, 9 — outlet contacts for external connection; 6, 7 — contact straps.

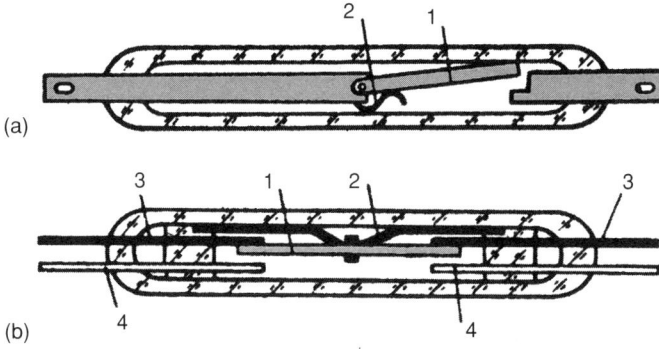

FIGURE 5.7
Reed switches with absolutely stiff contact-elements and an internal movable unit. 1 — movable unit; 2 — spring; 3 — passive (nonmagnetic) contact-points forming an opening contact; 4 — ferromagnetic contact-points forming a closing contact.

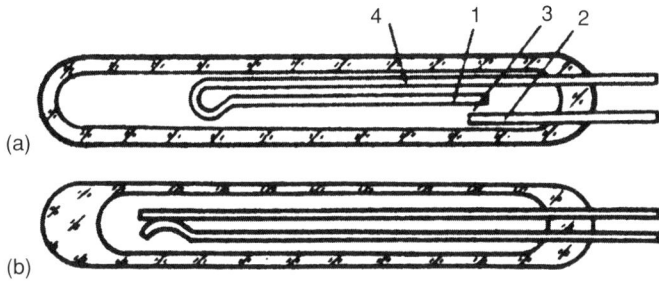

FIGURE 5.8
Closing (above) and opening reed switches with one-way outlets. 1 and 4 — forward and reverse parts of movable contact-elements; 2 — stationary contact-elements; 3 — gap between contacts.

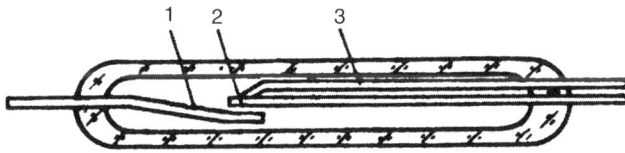

FIGURE 5.9
Changeover reed switch. 1 — stationary contact-element of the closed contact; 2 — flexible movable contact-element; 3 — stationary contact-element of the opening contact.

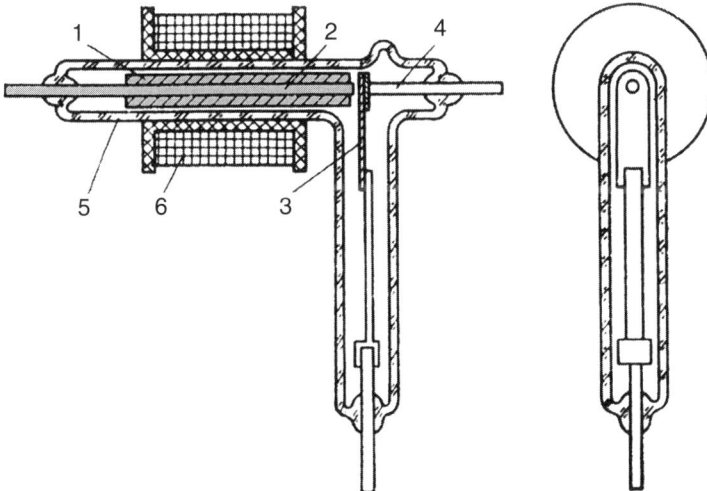

FIGURE 5.10
Switched T-shape vacuum reed switch. 1 — ferromagnetic packing for stationary contact-elements; 2 and 4 — stationary nonmagnetic contact-elements; 3 — stiff ferromagnetic part of the movable contact-point; 5 — glass tube; 6 — control coil.

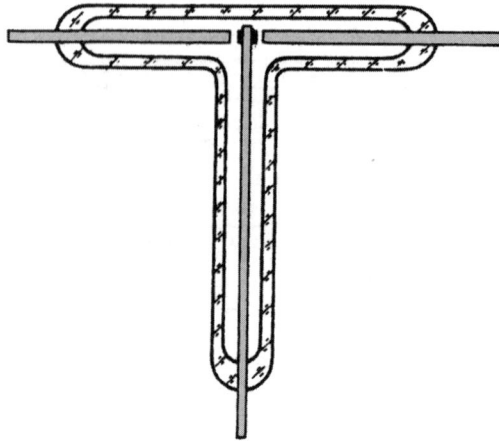

FIGURE 5.11
Symmetric T-shape reed switch.

the contact spring force determines contact pressure. In a reed relay, however, these systems are interrelated and the contact pressure directly depends on the magneto-motive force of control winding. As the reed switch becomes energized, the gap between the contacts is reduced, as well as the nonmagnetic gap in the magnetic circuit. This leads to an increase in electromagnetic tractive effort affecting the contact-elements and to higher moving speed of the contact-elements. As a result, the contact-points collide with more energy, rebound, collide again … causing the switching process of reed switches with dry contacts to be accompanied with considerable vibration of the colliding contact-elements (Figure 5.12). As can be seen from the oscillogram, the dynamic resistance of the reed switch varies from a minimal value tending to zero (this is the static transient resistance of closed contacts of a reed switch) up to infinity, that is up to a full break in the switched circuit.

Parameters of the transient switching process depend mostly on the size of the reed switch, the weight of the contact-elements, their elasticity, etc. It is obvious that vibration undermines wear resistance of contact-elements, which is why it is only natural that designers do their best to remove or at least to reduce vibration. In the German patent No. 1110308 a changeover reed switch with a split movable contact-elements 1 is described (Figure 5.13). Both parts of the split movable contact-point work as independent contact-points. Due to the difference in width their hardness is different and different tractive efforts are required to curve them. As a result, when a reed switch is energized these parts nonsimultaneously collide with the stationary contact-point and the oscillating process of

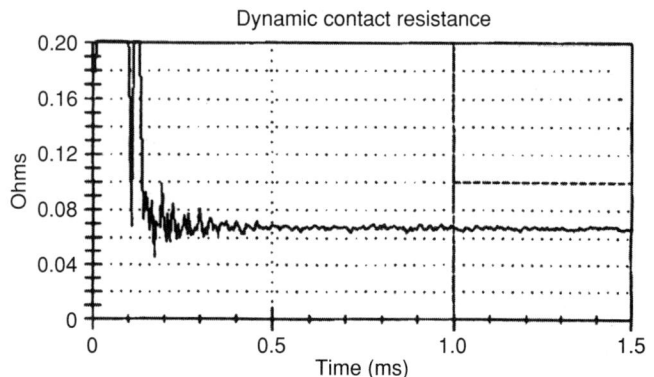

FIGURE 5.12
Oscillogram of the switching process of a closed miniature reed switch.

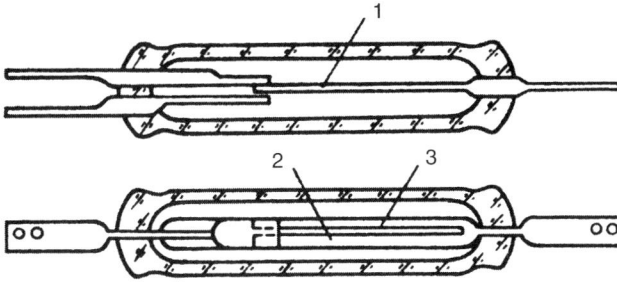

FIGURE 5.13
Bounce-free reed switch with split contact-elements. 1 — split movable contact-element; 2, 3 — parts of different width of the movable contact-element.

one of these parts is practically in antiphase with the other, which is why it is practically immediately dampened.

In another German patent (No. 1117761), there is a description of a reed switch with a movable contact-element supplied with a spring (2, Figure 5.14). As the contact-elements collide and start to vibrate, turns of the spring (2) move with regard to each other, with the considerable friction absorbing kinetic energy of the oscillating contact-elements.

The technical solutions described above were aimed at reduction of vibration, or at removal of the consequences of specific build-up of the magnetic field in a reed switch, while the solution suggested in the patent No. 1146738 registered in the former Soviet Union was aimed at removal of the original source of vibration, that is the increase in tractive effort affecting contact-elements when they are closing. In this construction there are special notches and lugs that fit each other without touching at the last stage of closing-in of contact-points. The configuration of the magnetic field in the gap between the contact-points leads to a weakening of tractive effort and the contact-points touch each other without any vibration (Figure 5.15).

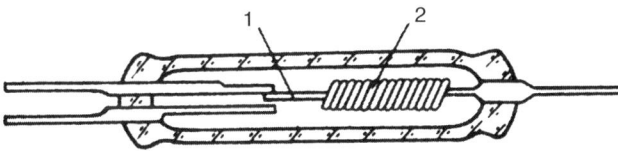

FIGURE 5.14
A bounce-free reed switch with a spring 2 on a movable contact-element 1.

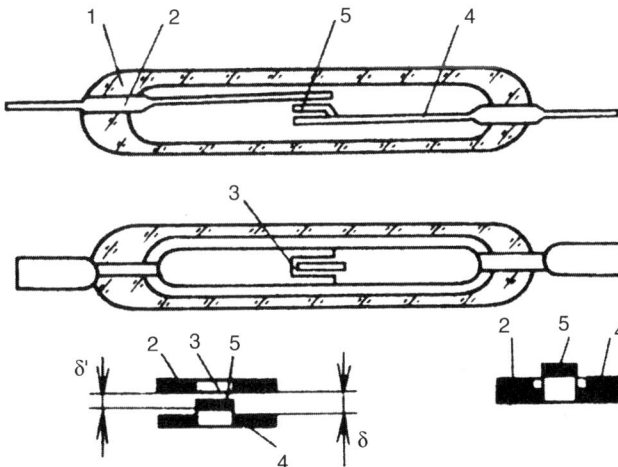

FIGURE 5.15
A bounce-free reed switch with special notches in contact-elements delaying the process of their closing in the last stage of the switching process. 1 — glass tube; 2 and 4 — contact elements; 3 — notch on the end of the contact-elements 2; 5 — deflected part on the end of the contact-elements 4.

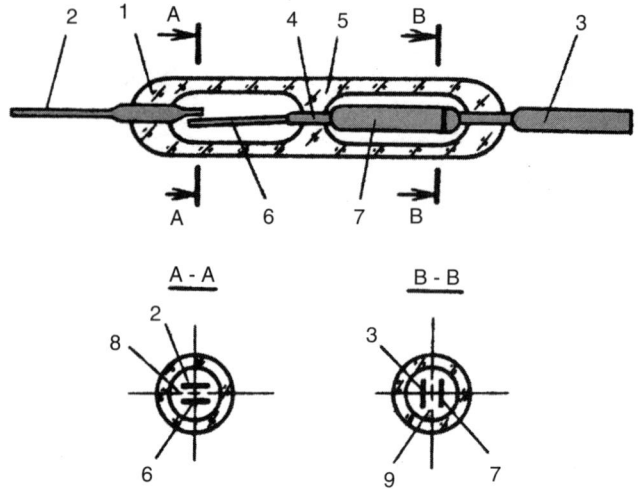

FIGURE 5.16
Vibration-resistant reed switch with three contact-points in mutually perpendicular planes. 1 — glass tube; 2 and 3 — stationary contact-points revolved through 90°; 4 — central (stationary) part of the movable contact-point; 5 — place of fixation of the movable contact in the tube; 6 and 7 — movable parts of the contact-point revolved through 90°; 8 and 9 — working gaps.

Apart from the problem of vibration of contact-elements in the process of operation (closing) of the reed switch, there is also a problem of resistance of the reed switch (and consequently reed relays) to external mechanical effects: vibration, accelerations, shocks. There are a number of technical solutions allowing an increase of resistance of the reed switch to spontaneous closing under mechanical impact.

In the construction described in the patent of the former Soviet Union No. 528624 (Figure 5.16), there are three contact-elements: two movable contact-elements (2 and 3) fused into the glass of the tube and revolved through 90°, and a stationary one fixed in the central part (4) of the tube and having the free ends (6 and 7) also revolve through 90°. The reed switch closes only when both ends (6 and 7) of the movable contact-elements simultaneously close, but since they can move only in perpendicular planes, their closing under external shocks or vibration is practically out of question.

Quite an original solution is described in the patent of the former Soviet Union No. 576618 (Figure 5.17). In this reed under no external magnetic field switch, the movable parts (4 and 5) of the central contact-elements are pressed to a flat stop (2). Under an external magnetic field of the coil both movable parts (4 and 5) are repelled and the stationary contact-elements (6 and 7) close. A flat stop disables simultaneous closing of the stationary part with parts 4 and 5, in case acceleration or shock occurs. It should be noted, however, that fast rotation of such a reed switch might cause separation of parts 4 and 5 under centrifugal force, leading to a closing of the stationary contact-points 6 and 7. For example, in airborne instruments and equipment of noncontrolled missiles, stabilization during the flight is carried out with the help of a special motor spinning the missile along its axis, and with stabilizers placed at certain angles to maintain the rotation of the missile during flight. For such cases, the author of this book has suggested (the patent of the former Soviet Union No. 1387069) filling the internal volume of a reed switch with

FIGURE 5.17
Vibration-proof reed switch with repelled contact-elements. 1 — glass tube; 2 — flat stop fixed on the butt-ends of the tube; 3 — place of fixation of a movable contact-elements in the glass of the tube; 4 and 5 — ends of the movable contact-elements; 6 and 7 — stationary contact-elements.

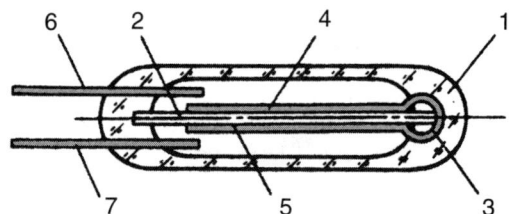

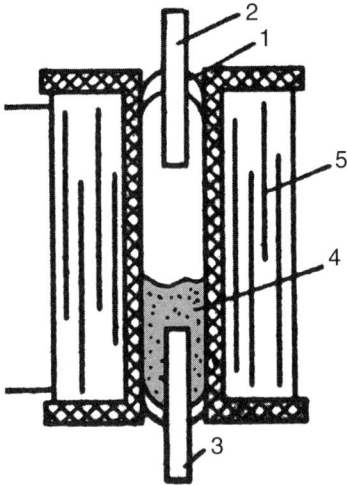

FIGURE 5.18

Reed switch with fusible movable contact-elements. Described in the patent of the former Soviet Union No. 1387069. 1 — glass tube; 2 and 3 — stationary contact-elements; 4 — fusible electro-conductive material with ferromagnetic filling; 5 — control coil.

some fusible material similar to paraffin which can be warmed by the current of the control winding and which melts before turning ON and OFF of the reed switch. When the material inside the tube becomes cool and hard, the reed switch is resistant to any type of external mechanical effects (Figure 5.18).

Moreover, it is possible even to reject the conventional construction of a movable contact-element by replacing it with electro-conductive material with ferromagnetic filling. In this case, two different coils or two different sources should be used to heat the material to form an electrode closing stationary contact-points 2 and 3.

In fact, there are a lot of different and exotic constructions with originally shaped movable contact-elements (Figure 5.19c, d):

- Contact-elements in the form of a ring fixed on one of stationary contacts shrinking under magnetic field of the coil, when it stretches, it closes the circuit between the stationary contact-elements (Figure 5.19a);
- Contact-elements in the form of a ball rolling back and forth under the magnetic field of the coil and closing the proper pair of stationary contact-elements (Figure 5.19b);
- Contact-elements in the form of ferromagnetic electro-conductive powder. Particles of such powder are aligned under the magnetic field of the coil and close stationary contact-elements (Figure 5.19c, d).

If we divide, in the last construction, the volume of the glass tube into several separate parts with the help of electro-conductive but nonmagnetic partitions (1a and 1b), we will have a multi-circuit reed switch capable of simultaneously switching several circuits.

Problems of construction of multi-circuit reed switches have puzzled and occupied designers for a long time. Over the years engineers and designers have found many original solutions. Take the construction shown on Figure 5.20, for example. In such a reed switch the movable contact (3) unwinds (untwists) under the influence of an external magnetic field, closing the stationary contact-elements 4 and 5.

In another construction of a multi-circuit reed switch (a patent of the former Soviet Union No. 595801) the movable contact-element has a more traditional form, while the stationary contacts are transversally located. To provide reliable contact of the movable

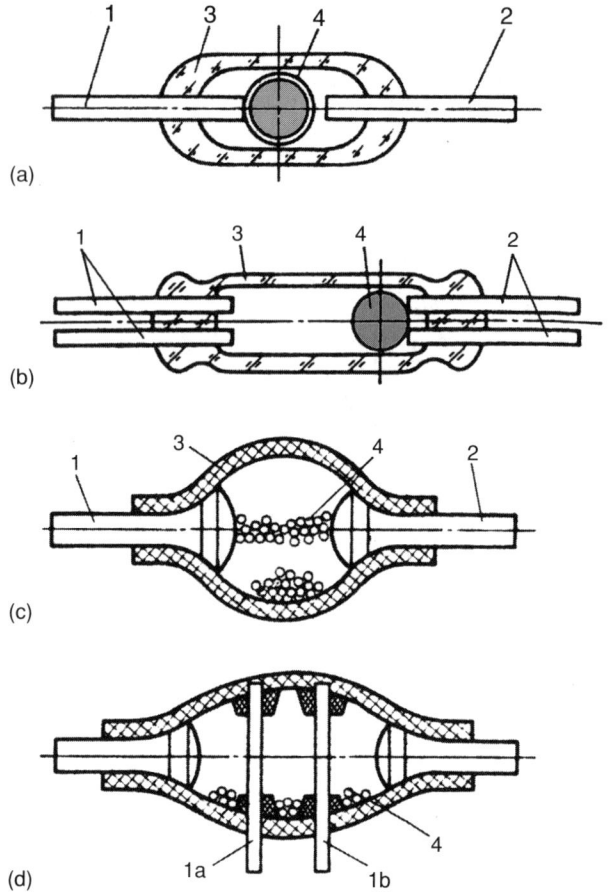

FIGURE 5.19
Reed switches with originally shaped movable contact-elements. 1 and 2 — stationary contact-elements; 3 — glass tube; 4 — movable contact-element.

contact-point with the transverse stationary ones, the latter must be sufficiently flexible (Figure 5.21).

More complex constructions of multi-circuit reed switches were proposed in Germany (Figure 5.22a), and in the U.S.A. (Figure 5.22b). Actually these are several pairs of contact-elements of the traditional form placed in the same glass tube, be it flat (Figure 5.22a) or round (Figure 5.22b). The author has no knowledge of whether such reed switches were actually produced, but a few scientific articles were published in the 1970s in Germany, which dealt with issues concerning the use of such four-polar reed switches in electronic exchanges.

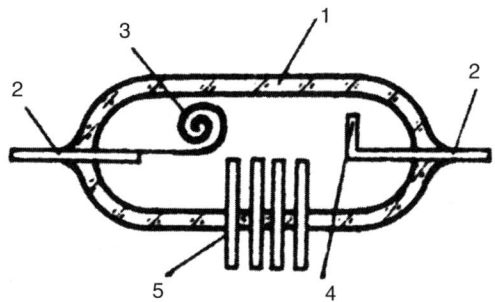

FIGURE 5.20
A reed switch with an unwinding (untwisting) movable contact-point.

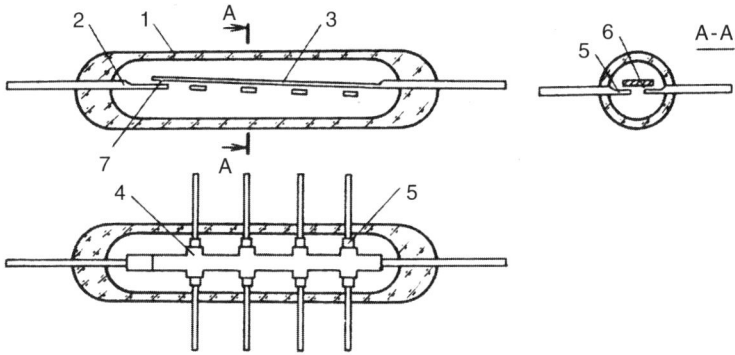

FIGURE 5.21
Multi-circuit reed switch with a movable contact-element of the traditional form. 1 — glass tube; 2 — main stationary contact-element; 3 — movable contact-element; 4 — lugs on the movable contact-element; 5 — additional pairs of stationary contact-elements soldered into glass across the longitudinal axis of the reed switch; 6 and 7 — working gaps between contacts.

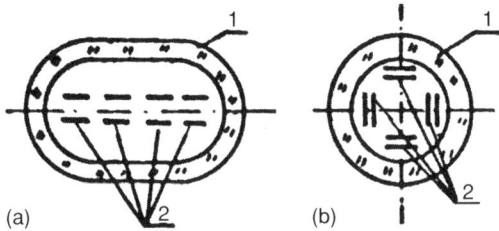

FIGURE 5.22
Four-polar reed switches for electronic exchanges suggested in Germany (a) and in the U.S.A. (b). 1 — flat or round glass tube; 2 — contact-elements of the traditional form.

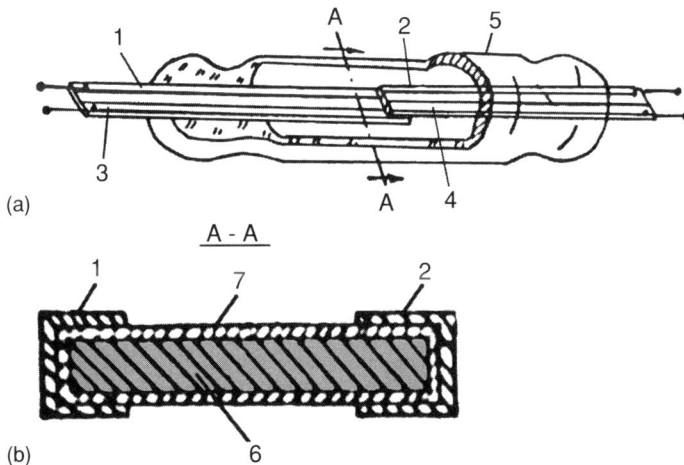

FIGURE 5.23
Double-circuit reed switched designed by the Bell Telephone Laboratories. 1 and 2 — the first pair of closing contacts; 3 and 4 — the second pair of contacts; 5 — glass tube; 6 — Remendur (semi-hard and very ductile magnetic alloy of 49% Co — 48% Fe — 3% V); 7 — glass.

Another complex construction of a multi-circuit, or double-circuit reed switch to be exact, was designed (also for the use in quasi-electronic exchanges) by the Bell Telephone Laboratories (Figure 5.23).

5.3   High-Power Reed Switches

The various constructions of reed switches described above were designed for switching of power up to 60 W, and all of them are considered "low-power reed switches." Apparently, this usually does not have enough power for the reed switches used in industry relays, which is why it has been a long time since research work (and industrial production as well) has been carried out for high-power reed switches.

The English company Brookhirst Igranic Ltd is a pioneer in this field. At the end of 1960s it produced a reed switch 82400 H3100, which is more famous under the "Powered" brand name (Figure 5.24). In this construction, it implemented the already existing idea of division of functions of current switching and current carrying after switching, with the help of two pairs of contacts. One pair should be resistant to electric erosion, and the other pair should be able to conduct high current well in the closed (ON) position. In this reed switch the additional contacting pair (4, 5) from tungsten closes first and opens last. The main contacting pair (7, 8) shunts the additional contacting pair and unshunts it without arcing. The tube is filled with a nitrogen–helium–oxygen mixture. This reed switch was capable of switching ON up to 15 A and switching OFF up to 3 A at 125 V AC voltage, with an inductive load of power factor 0.35. Unfortunately, the complexity and high cost led to a halt of production of such reed switches already in the middle of the 1970s.

Using profound knowledge of physical processes proceeding on contacts when electric current is switched, American inventor J. Santi from the Briggs and Statton Co. patented in the 1970s a complex construction of a reed switch for DC switching (the most difficult mode for reed switches). In such a reed switch the contact having positive potential is made of molybdenum and the contact with negative potential of tungsten, a metal with a higher melting temperature (Figure 5.25).

At the initial stage of the opening process (at low voltages on contacts) molybdenum is transferred from the positive contact to the negative one. At the following stage, when voltage on the contacts increases, the direction of metal transfer changes to the opposite one and the molybdenum is transferred back to the positive contact. As the contact-elements close, under the external magnetic field of control coil, first the molybdenum, then component 2 closes with a part of tungsten strap (1'). Afterwards the tungsten strap (1) reaches the free part of the tungsten strap (1'). The opening process is carried out on the opposite order. Tests showed that this reed switch, supplied with a spark suppressing RC-circuit, can withstand 2×10^9 switches of DC at a voltage of 3000 V.

In 1977, the American A. Beavitt from the Square D Company patented a reed switch with an iron armature capable of switching power of more than 1 kW without spark

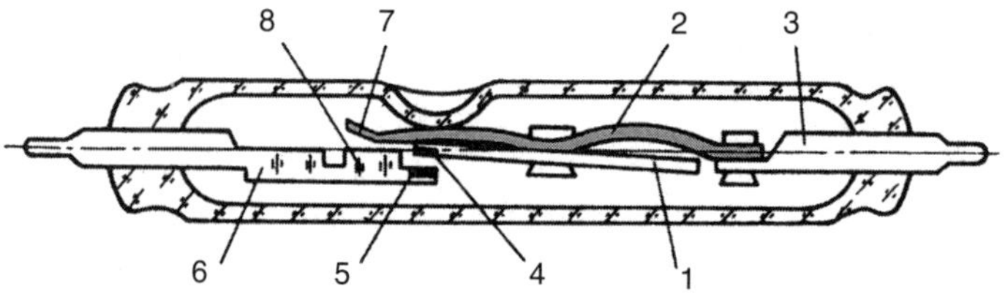

FIGURE 5.24
Powerful 82400 type reed switch ("Powered"). 1 — ferromagnetic armature; 2 — spring; 3 — outlet of the movable contact; 4 and 5 — additional contact pair from tungsten carbide; 6 — outlet of the stationary contact; 7 and 8 — main contacting pair from a silver alloy.

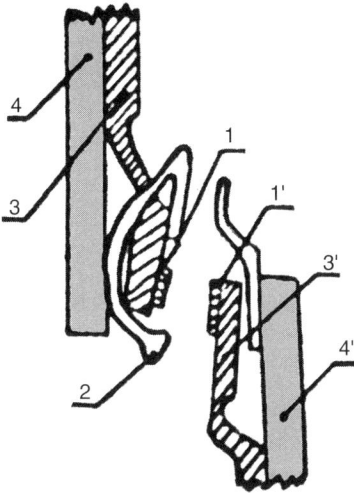

FIGURE 5.25
Power reed switch produced by Briggs and Statton Co. 1 and
1′ — tungsten contact straps; 2 — contact-element made of molybdenum; 3 and 3′ — ferromagnetic springs; 4 and 4′ — supports.

suppressing RC-circuits (Figure 5.26). Such a reed switch can switch currents up to 5 A at
an AC voltage of 220 V.

In the Russian Scientific Research Institute of Relay Production a simplified (by elements, but not by the technical ideas implemented in it) construction of a power reed switch
was designed on the basis of the "Powered" reed switch (Figure 5.27). Unlike the British
reed switch, the Russian variant turned out to be quite easy to produce, which is why it
continues to be produced by the Orel Electronic Instrument-Making Plant until now. In
this reed switch, after closing of the tungsten contacts (1) the armature (2) continues
moving, causing deflection of the spring (3) until it is connected to the stationary contact-
element (4). The opening process has the opposite order. Part 5, with a smaller section (in
later variants there is a special through hole at this place), is quickly saturated as the
magnetic field of the control winding affects the reed switch. As a result, there are two
poles under the lugs of the ferromagnetic armature to which it is attracted.

As it can be seen in this construction, which appears simple at first sight, not only
technical ideas of the "Powered" reed switch, but also of the reed switch produced by
Square D Company are applied (Figure 5.26). Currents switched by this reed switch vary
within 0.001 to 5.0 A at DC voltages of 6 to 380 V, with power not more than 250 W. The
maximum making and breaking current is 20 A (at voltage of 48 V) on AC and 4 A (at
voltage of 24 V) on DC. The ON-time delay of such a reed switch is not more than 7 ms,
and the release time is 5 ms. The tube of the reed switch is 7 mm in diameter and 52 mm in
length, and its weight is 4 g.

When quite high currents (5 to 10 A) flow through closed contact-elements, those
elements warm up considerably, sometimes even up to the Curie Point (magnetic transition temperature). At that point they are no longer held in the closed position by the

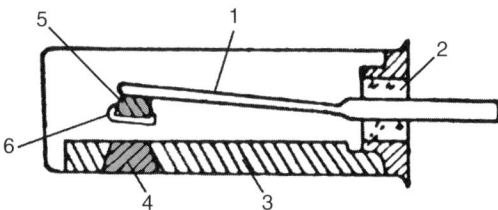

FIGURE 5.26
A powerful reed switch with an iron armature,
patented by Square D Company. 1 — movable contact-element (a spring) from Permalloy; 2 — glass
insulator; 3 — ferromagnetic stationary contact-element; 4 — copper insert as a contact strap; 5 — iron
armature; 6 — tungsten cover.

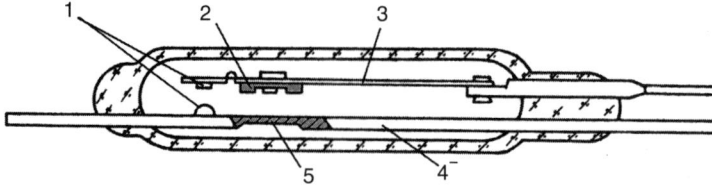

FIGURE 5.27

MKA-52202 type power reed switch (made in Russia). 1 — contact straps from tungsten; 2 — ferromagnetic armature covered with silver; 3 — flexible movable contact-element (a spring); 4 — hard stationary ferromagnetic contact-element covered with silver; 5 — part of the stationary contact-element with smaller section.

magnetic field of the control coil and separate, causing intensive arcing. If the contact-elements are not welded under the impact of arc, they close within a few seconds (after they cool and their magnetic properties are restored), and the process will start all over again. Obviously, after such spontaneous switching with intense arc, there is nothing left for the reed switch to do but to throw itself out.

The solution to this problem is given in the patent of the former Soviet Union No. 440709, which describes a high-power reed switch supplied with a special heat sink on the stationary contact-element (Figure 5.28), providing intensive heat abstraction to the environment. In this construction, plates of the movable contact-element (5) affected by the external magnetic field, open like petals, closing with the stationary contact-element (2).

In one of the industry relays, the author came across an interesting construction of a high-power reed switch with a magnetic core plugged into the glass tube (Figure 5.29). The core of the external control coil (or a permanent control magnet) touches this magnetic core. This allows considerable reduction of the magnetic resistance of the system, and an increase of contact pressure. Such a reed switch has an important distinction from all the other constructions described above: its magnetic system is partially detached from the contact system.

This principle is also applied in some other constructions of reed switches, for example, in the construction shown in Figure 5.30. Such separation of the magnetic system from current-carrying components of a reed switch, even partial, allows avoidance of a negative peculiarity of reed switches: dependence of contact pressure (as well as drop-out values and reset ratio) on current flowing through the reed switch. Such dependence is obvious for reed switches of standard constructions: current carrying through contact-elements of the reed switch creates its own magnetic field, interacting with the magnetic field of the control coil. We have already described above the source of forces tending to open contacts in standard electromagnetic relays as very high currents (usually short-circuit currents) pass through them.

In reed switches this phenomenon is intensified because of a merging of the magnetic and contact systems. As a result, affected by even smaller currents than in a standard

FIGURE 5.28

Reed switch with a heat sink on the stationary contact-element. 1 — glass tube; 2 — stationary ferromagnetic contact-element in the form of cup; 3 — heat sink; 4 — outlet of the stationary contact-element; 5 — movable contact-element consisting of two plates welded at the point of the lead-in into glass; 6 — working gap between contacts.

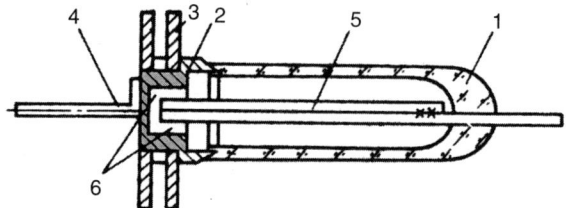

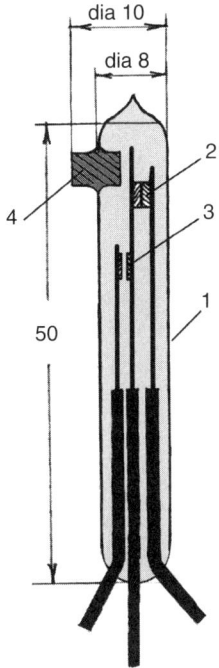

FIGURE 5.29

Power reed switch with a magnetic core plugged into the glass tube. 1 — glass tube; 2 — normally closed contact pair; 3 — normally open contact pair; 4 — magnetic core plugged into the glass tube.

electromagnetic relay, contact-elements may open. In constructions of reed switches shown in Figure 5.29 and Figure 5.30, such danger can be avoided.

The American branch office of the Yaskawa Company advertises its R14U and R15U type power reed switches, produced under the brand name "Bestact" (Figure 5.31), which belong to the same class of reed switches with partially detached magnetic and contact systems. This reed switch has current-carrying capacity up to 30 A. It can break similar current (as emergency one) 25 times in an AC circuit with a power factor of 0.7. Switched power in the AC circuit is 360 VA (inductive load), maximum switched current is 5 A, maximum switched voltage is 240 V. The electric strength of the gap between contacts is 800 V AC, mechanical life 100.000 operation for R15U and 50.000 for R14U. Operating and release time is 3 ms.

On the basis of such reed switches, the Yaskawa Company produces a great number of different types of switching devices: relays, starters, push-buttons, etc. The principle of separation of magnetic and contact systems, and inserting an additional magnetic core with a larger section inside the tube, was the basis of constructions of high-power reed switches designed by M. Koblenz in the All Union Scientific Research Institute of Electrical Apparatus (now a part of the Ukraine, Kharkov city). The external design of such

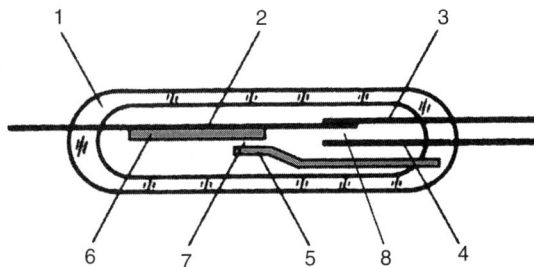

FIGURE 5.30

Reed switch with partially detached magnetic (elements 5 to 7) and contact (elements 2 to 4, 8) systems. 1 — glass tube; 2 — nonmagnetic movable contact-element; 3 and 4 — nonmagnetic stationary contact-element; 5 — stationary ferromagnetic component; 6 — movable ferromagnetic component fixed on the movable nonmagnetic contact-element 2; 7 — magnetic gap; 8 — gap between contacts.

FIGURE 5.31

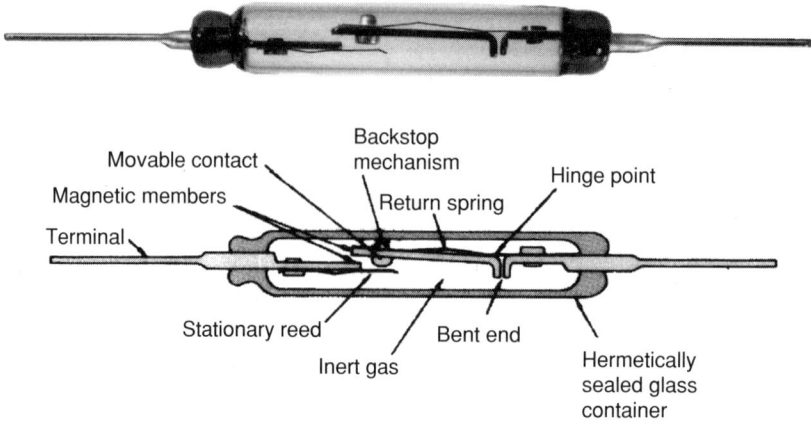

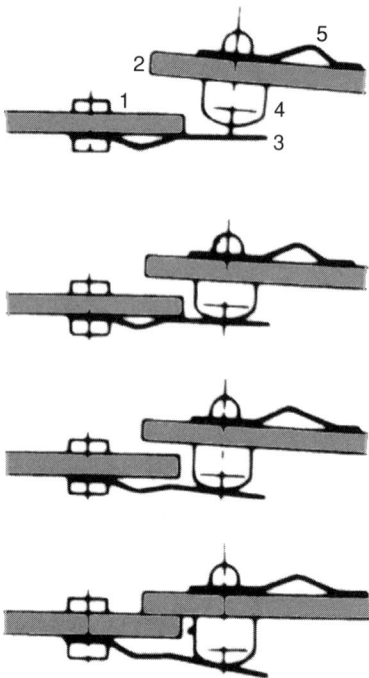

Power reed switch Bestact™ produced by Yaskawa Electric America and its closing/opening process. 1 and 2 — main silver coated contacts; 3 and 4 — auxiliary contacts from tungsten; 5 — spring. Dimensions of glass tube: diameter — 6 mm, length — 37 mm.

devices (called "hersicon," from Russian words: "Hermetical Power Contact") differs greatly from that of traditional reed switches (Figure 5.32–Figure 5.34).

Through the hermetic shell of a high-power reed switch made of ceramics two ends (instead of as in the power reed switches described above), the massive magnetic core passes. This helps to reduce losses in magnetic circuit so that they do not exceed the losses in standard electromagnetic relays. The construction of the armature (6) allows us to combine to some extent some contradictory requirements for a movable armature (the term "contact-elements" is not appropriate for such a construction): the largest sectional area possible should be combined with the greatest flexibility possible. These

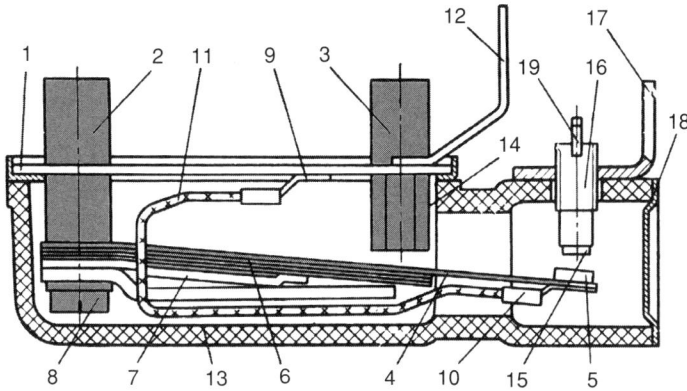

FIGURE 5.32
High-power reed switch (hersicon) of KMG12-19 type for 6.3 A nominal current and 440 V AC voltage (switching power up to 3 kW). 1 — board; 2 and 3 — poles of the magnetic system; 4 — spring-armature; 5 and 15 — contact straps; 6 — ferromagnetic element of the armature consisting of a package of thin flexible plates; 7 — limiter; 8 — screw; 9 and 19 — tips of flexible copper wire (11) used as a shunt; 12 and 17 — wireways; 13 — ceramic case; 14 — insulation layer unavailable to metallization; 16 — adjusting screw; 18 — cover; 19 — nipple.

requirements can be partially met by using a package of thin flexible ferromagnetic plates, the number of which is reduced as the moving end of the armature approaches. Of course, it should be noted that such a relatively successful solution only partially meets the requirements, because the section of the armature is not as big as it should be and its over-stiffness does not allow an increase in the gap between the contacts of more than 1.5 mm.

Flexible copper wire (shunting such a package of ferromagnetic plates with not very good electro-conductivity) provides almost a full separation of magnetic and electric circuits in the high-power reed switch. This is a key fact, because full separation of electric and magnetic circuits is typical for standard electromagnetic relays. Reed switches are distinguished from standard relays mostly by the fact that magnetic and electric circuits in them are combined in the same elements. Therefore a hersicon is just a hermetic contact node of the console type, with adjoining parts of the magnetic core sticking out of

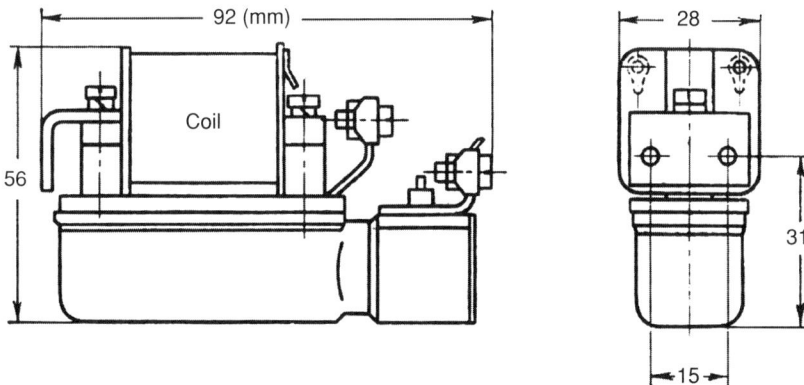

FIGURE 5.33
High-power reed switch (hersicon) of KMG12-19 type with a control coil fixed on poles of the magnetic system.

FIGURE 5.34
External view of single-pole hersicons for currents from 6.3 to 63 A produced by Electroceramics Plant (the Ukraine).

the hermetic shell, and with separate contact outlets. Being connected to the control coil such a contact node makes up a standard electromagnetic relay.

But if that is really so, the following questions arise: Why is it necessary to use an armature of a console construction as in standard reed switches, in such a relay, and then to do one's best to increase its flexibility and at the same time through a big section by introducing an additional wire shunt, further complicating the construction? Why not use a conventional turning stiff armature of the required section, supplied with a restorable spring with the required stiffness, as in a standard electromagnetic relay?

Such arguments brought the author of the book to the idea of a creation of a new type of switching device called REPROCON (RElay with PROtective CONtacts) (Figure 5.35). The description of such a device was published in 1994.

Having analyzed the history and tendencies of development of high-power reed switches, one may conclude that it is possible to enhance reed switches only up to certain limits of powers (up to 500 VA). An attempt to design much more powerful devices will lead to significant changes in the construction, so that the designed device can no longer be considered to belong to the category of "reed switches." In such cases, the application of basic principles of construction of reed switches to such devices becomes unjustified.

5.4 Membrane Reed Switches

In membrane reed switches (including "petal" reed switches) the movable contact-element is made in the form of a membrane (petal) from ferromagnetic material supplied with slots (Figure 5.36). Under the effect of the magnetic field of the core (5), the leaf (4) sags and closes the circuit. There is, of course, no need to say that both the ferromagnetic core (5) and the leaf (4) must have a good electro-conductive cover.

Elements 1 to 5 are ferromagnetic with an electro-conductive cover. The cover (3) and the heelpiece (2) are welded along the contour, forming an internal hermetic volume of the reed switch filled with gas.

When the stationary contact-element (1) is magnetized, the central part of the membrane (8) sags, closing the circuit between the outlets joined to the heelpiece (2) and the stationary contact-element (1) (Figure 5.37).

Due to the complex configuration of notches, the central part of the membrane has several degrees of freedom and when moved fits cleanly with the butt-end of the stationary contact-element (1) even in the case of inaccurate assembly. This also reduces

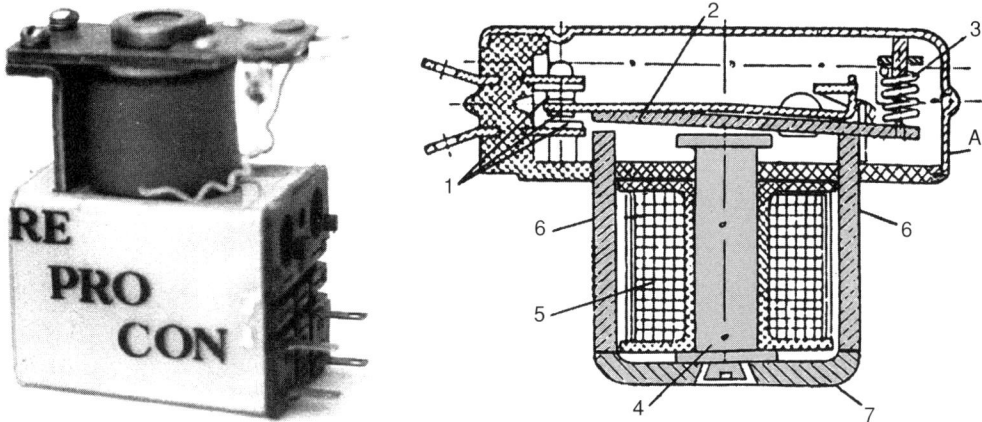

FIGURE 5.35
Reprocon designed by the author. 1 — contact straps; 2 — stiff ferromagnetic armature; 3 — restoring spring; 4 — core sticking out of the hermetic shell A; 5 — control coil; 6 — parts of the magnetic core sticking put of the hermetic shell A; 7 — removable part of the magnetic core.

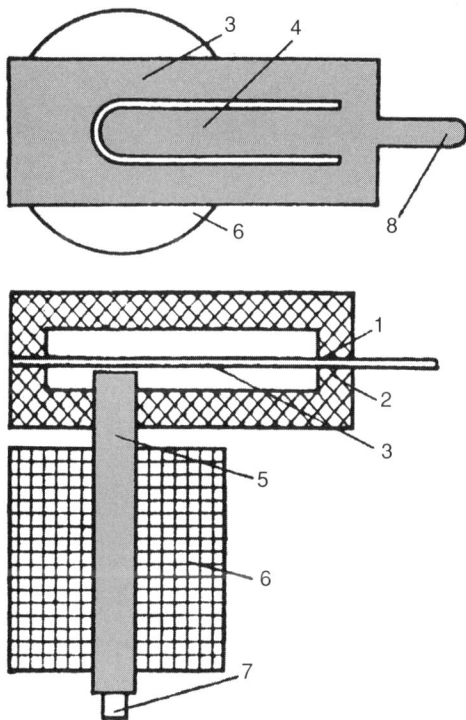

FIGURE 5.36
Principle of construction of a petal reed switch. 1 — cover; 2 — heelpiece; 3 — plate with a figured slop in the form of a peal; 4 — peal (membrane); 5 — ferromagnetic core; 6 — control coil; 7 and 8 — wireways.

bounce of the contact surfaces when they collide. It is much easier to choose the material for such a construction because the membrane is not welded to anything (while in standard reed switches the material of contact-element should be welded well into the glass, have a good adhesion to it and a coefficient of linear expansion similar to glass).

The patent of the former Soviet Union No. 750591 deals with a membrane reed switch with increased resistance to vibration, having two membranes instead of one, each

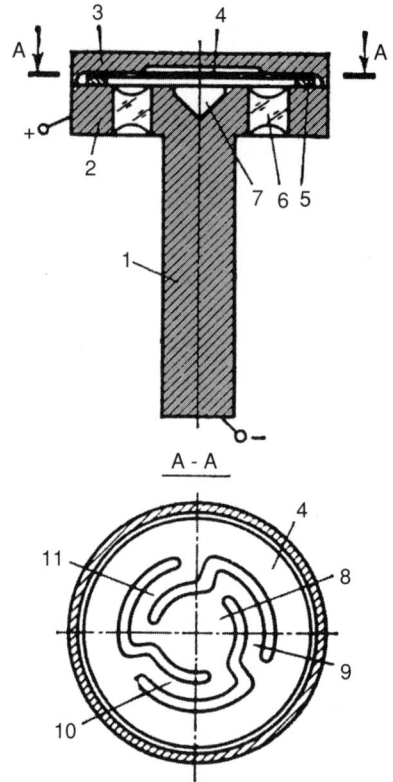

FIGURE 5.37
Membrane reed switch. 1 — stationary ferromagnetic contact-element; 2 — ring-shaped heelpiece; 3 — cover; 4 — membrane; 5 — gage block; 6 — glass seal; 7 — recess; 8 — central part of the membrane; 9 to 11 — membrane leaves.

moving towards the other when the device is energized (Figure 5.38). The cavities (4 and 5) are hermetically separated from each other and are filled with gasses of different pressures, to increase resistance of the membrane to vibration.

As is the case with standard reed switches, designers try to increase as much as possible the power switched by membrane reed switches by following the same way (Figure 5.39).

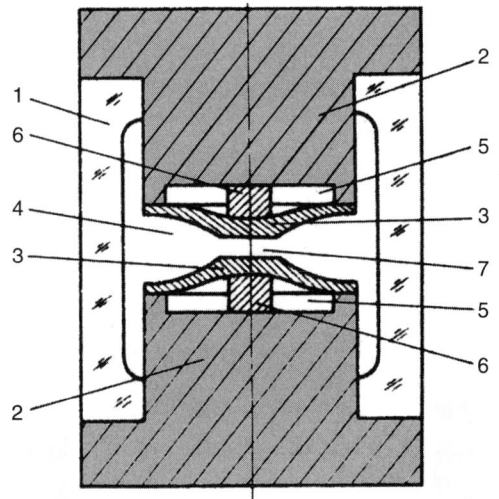

FIGURE 5.38
Vibration-resistant membrane reed switch with two membranes. 1 — glass tube; 2 — ferromagnetic cores; 3 — crimped membranes; 4 — main cavity filled with gas; 6 — nonmagnetic stops; 7 — working gap.

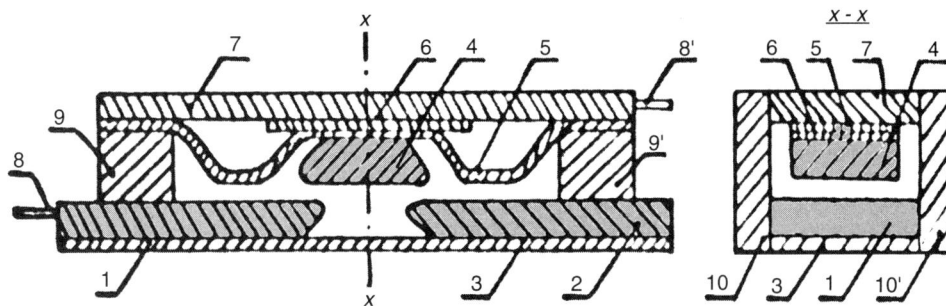

FIGURE 5.39
High-power membrane reed switch produced by Square D Company. 1 and 2 — stationary contact-elements; 3 — heelpiece; 4 — ferromagnetic armature (a movable contact-element); 5 — elastic diaphragm; 6 — damping plate; 7 — cover of the reed switch; 8 and 8′ — outlets of the reed switch; 9 and 9′ — end walls of the reed switch (plastic); 10 and 10′ — sidewalls (plastic).

In this construction the ends of the contact-elements (1 and 2) are splayed in the contacting area to reduce the stray flux. The contacting surfaces of the contact-elements are covered with tungsten.

Under the effect of the longitudinal magnetic field, the armature (4) is attracted to the stationary contacts (1 and 2), causing the membrane to sag. When the electric circuit between the outlets 8 and 8′ is closed, the electric load current flows through the armature and the membrane. Such a reed switch can switch currents up to 5 A with voltages up to 250 V AC. If the heelpiece (3) is made of dielectric material and the outlet (8′) is connected to the element (2) the switched load current will not pass through the thin membrane and it increases by several times.

5.5 Mercury Reed Switches

Mercury reed switches belong to the class of liquid–metal switching devices. These are devices in which current conducting elements are wetted with liquid metal fully or partially.

As mercury is the only pure metal capable of being in a liquid state at room temperature, when we say "liquid–metal devices," we usually mean only mercury devices. The tube in such reed switches is filled with mercury of 0.1 to 0.15 volume (Figure 5.40), so that it does not overflow onto the contacts.

That is why most mercury reed switches can operate only in a vertical position (the maximum admissible deviation from the vertical is 30°). The mercury in such reed switches is necessary only for wetting of the contacting surfaces of the contact-elements. Wetting of contact-elements is carried out by pumping of the mercury from a reservoir, through capillaries or semicapillaries placed on the surface of the contact-elements.

The first constructions of mercury reed switches and relays already appeared in the latter half of the 1940s and by the 1960s they already looked liked modern ones. In earlier constructions the capillaries were made from two parallel pieces of wire while in modern ones they are made from several longitudinal cuts on the flat surface of the movable contact-elements. Surfaces that come into contact with the mercury are covered with a special amalgam providing for good wetting with mercury.

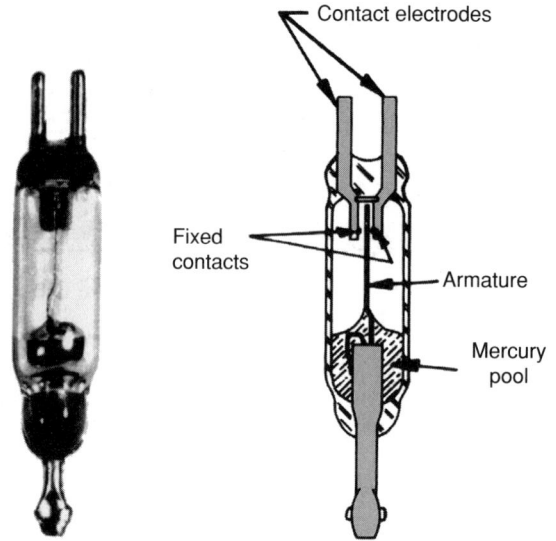

FIGURE 5.40
External view and construction of the most popular type of mercury reed switches.

When the contacts switching OFF, a mercury bridge is formed between them. As the contacts move apart, this bridge becomes thin and snaps (Figure 5.41).

Electric spark and even arc (in case it occurs) lead to vaporization of the mercury drop which then condenses on the sidewalls of the tube and flows down back to the reservoir. The surface of the contact-elements remains clean and undamaged. Apart from considerable multiplication of switching cycles, in mercury reed switches vibration of contact-elements during closing does not lead to a break of the bridge, which is why it does not affect the reed switch and external circuits as well. For safety purposes the tube in mercury reed switches is usually made of thicker glass than in dry reed switches and has quite a high strength. This allows the tube to be filled with hydrogen under pressure of up to 2000 kPa, which considerably increases switched current (up to 5 A), voltage (up to 600 to 800 V) and power (250 W) (Figure 5.42a).

In certain types of reed switches it is possible to considerably raise the switched voltage by increasing the gap between the contacts. For example, the Russian reed switch MKAP-58241 can switch voltage of up to 4500 V with insulation strength of up to 8000 V. However, enlargement of the gap between contacts in such a reed switch may lead to a prolongation of make delay, and a release time of up to 10 ms, increasing the magnetomotive force of operation up to 500 to 700 A, and reducing the switching frequency to 25 Hz. The tube of such a reed switch is 58 mm in length and 14.5 mm in diameter. Another interesting peculiarity of such a reed switch is a protective plastic shell

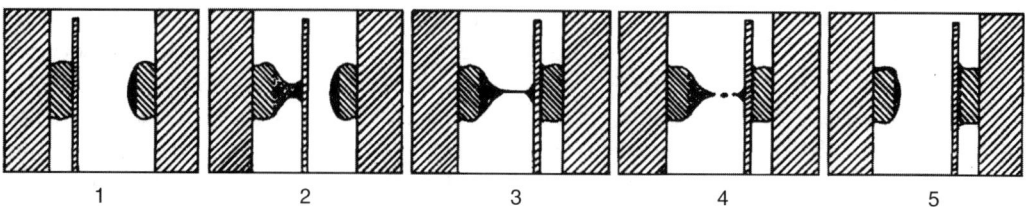

FIGURE 5.41
Stages of the process of switching of electric current by a switched contact wetted with mercury.

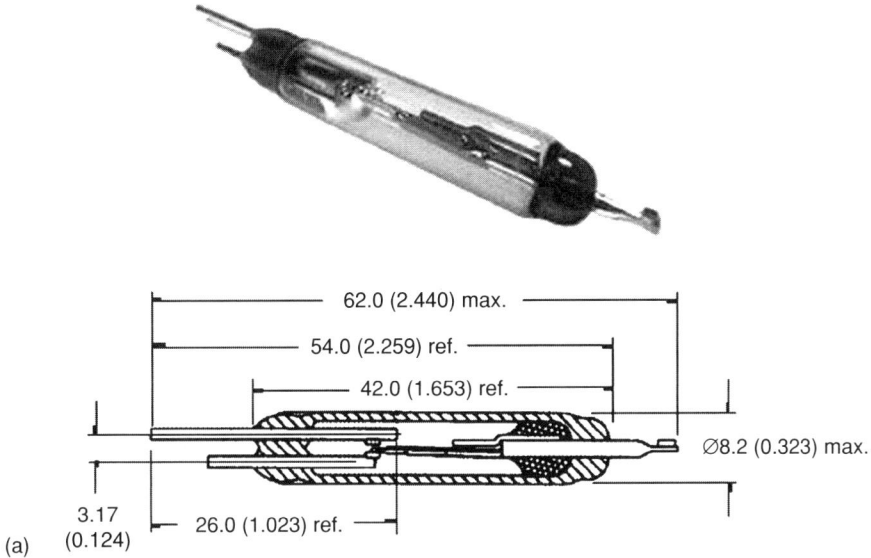

FIGURE 5.42
(a) Mercury wetted power reed switch HG 07 120 180 type, produced by Gunther Belgium. (Comus Group Companies) (Vertical mounted required)

preventing ingress of mercury in the ambient space in case of rapid depressurization of the glass tube.

The necessity of vertical installation of standard mercury reed switches with a mercury reservoir led to the invention of reservoir-free mercury reed switches in which the mercury is in the capillaries only. This allows us to exploit such reed switches in any position, but restricts switching capacity because of the small amount of mercury. For example, the Russian reservoir-free mercury reed switch MKAR-15102 has switched power up to 30 W, while practically the same mercury reed switch MKAR-15101 with a mercury reservoir up to 50 W.

The two shining parts of the tube, from the two sides of overlap of the contact-points, are made of a layer of amalgam covering the internal surface of the glass tube. To provide a reliable supply of mercury for contacting surfaces of contact-elements, most of the internal surface of the glass tube in such reed switches produced by CP Clare Corp. is covered with an amalgam based on nickel, thus two parts with a gap are formed in the overlap of contact-elements (Figure 5.42b). The mercury in this part is sufficient to provide a reliable contact.

One of disadvantages of mercury reed switches is their limited range of working temperatures: $-38 + 125°C$ when properties of mercury remain valid. Unfortunately, in the course of exploitation of a reed switch sometimes even within this interval of temperatures one can observe an increase in the tenacity and surface tension of the mercury. This may lead to maintenance of a contact bridge in the extreme position of separated contact-elements, that is to unopening of the reed switch. In addition, increased

(b)

FIGURE 5.42
(b) Mercury reed switch insensitive to the position in space. (Produced by CP Clare Corp.)

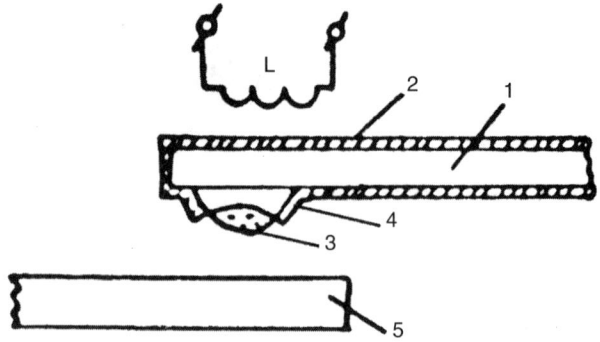

FIGURE 5.43
Original construction of contact-elements for a mercury reed switch. 1 and 5 — contact-elements; 2 — cover unwettable by mercury (tantalum, niobium); 3 — drop of mercury; 4 — cup-shaped element the internal surface of which is wettable by mercury.

temperature can sometimes cause inter-diffusion of metal of the contact-elements through a thin mercury film. To prevent this phenomenon S. Bitko (U.S.A. patent No. 3644603) suggested an original construction of the contact-elements (Figure 5.43). In this construction the surface tension makes a drop of mercury penetrate the cup-shaped element (4). Mercury leavings are absorbed into the internal cavity of this element from its surface, thus preventing unopening of the reed switch.

Deviation from use of traditional console construction in mercury reed switches allowed production of a miniature device with good switching properties (Figure 5.44). The stationary contact-elements (1 and 7) are welded into two separate glass tubes (2 and 6), which are linked by a metal bush (3), thus forming a common tube. The bush serves as a guide for the plunger (4) wetted by mercury. A disk outlet (5), together with outlets 1 and 7, are the stationary elements of the switching contact, which changes its positions as the plunger (4), moves under the effect of the external magnetic field.

The "Logsell-1" reed switch is very small in size: its tube is 7 mm in length and 1.2 mm in diameter, however, its switching characteristics are quite high for miniature reed switches: switching power is up to 15 W, current is up to 1 A and voltage up to 200 V. Minimum life is $5 \times 10^7 - 10^9$ switching cycles (depending on the parameters of the switched circuit). This reed switch will function in any position in space because there is no danger of overflowing of mercury in it. In addition, the Logsell-1 switch has quite a high console construction switching frequency, even for dry reed switches, of up to 200 Hz.

After removal of the external magnetic field the plunger in such a reed switch remains in one of its extreme positions by the forces of surface tension of the mercury film. Unfortunately, such a unique reed switch did not become a popular element due to its complexity and the high cost of its production.

Attempts to create new constructions of liquid–metal reed switches still continue. In patent descriptions it is possible to find a lot of original constructions, which have not succeeded in becoming commercial devices, like a hybrid of a ball and mercury reed switch for instance (Figure 5.45), which is controlled by a permanent magnet or a reed

FIGURE 5.44
Mercury reed switch "Logsell-1" of the plunger type. 1 and 7 — stationary contact-elements with internal butt-ends wetted with mercury; 2 and 6 — separate parts of the glass tube; 3 — guide bush; 4 — ferromagnetic plunger wetted with mercury; 5 — disk outlet.

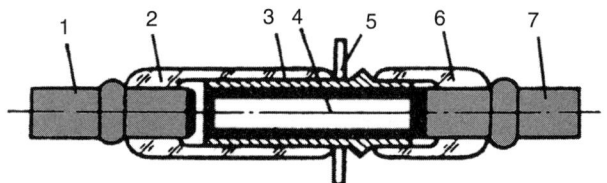

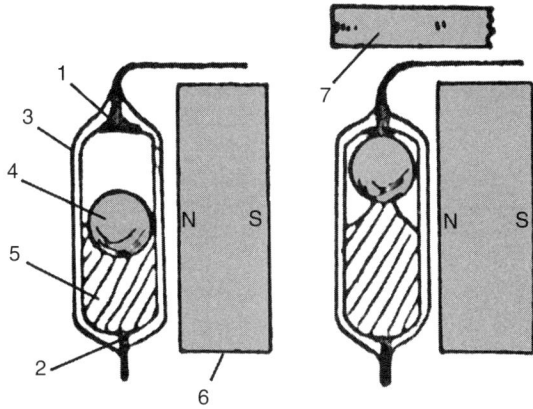

FIGURE 5.45
Ball mercury reed switch (Germany patent No. 1515775). 1 and 2 — contacts; 3 — glass tube; 4 — steel ball; 5 — mercury; 6 — permanent magnet; 7 — ferromagnetic control plate.

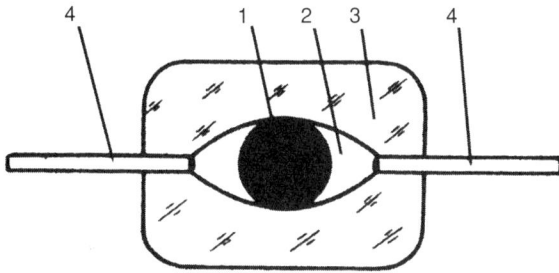

FIGURE 5.46
Reed switch with a movable contact-element in the form of a drop of electro-conductive ferromagnetic liquid (patent of the former Soviet Union No. 851522) 1 — a drop of electro-conductive ferromagnetic liquid; 2 — internal cavity of an original form; 3 — glass tube; 4 — ferromagnetic outlets.

switch with a drop of electro-conductive ferromagnetic liquid as a movable contact-element (Figure 5.46), etc.

In such a reed switch the movable contact-element is made in the form of a drop of a suspension of finely dispersed magnet-soft powder in an electro-conductive liquid. Its internal cavity (2) has such a form that under no external magnetic field will the drop conform to the shape of a ball. Only when the external longitudinal magnetic field affects it, will the ball-shaped drop stretch and close the outlets (4).

5.6   High-Voltage Reed Switches

In fact, classification of reed switches into low- and high-voltage ones is quite relative. In electrical engineering of strong currents devices designed for operation with voltages of more than 1000 V are considered to be high-voltage switches, while in electronic devices with voltages of more than a few hundred volts are also called high-voltage. According to such a division it appears that almost all gas-filled mercury reed switches are high-voltage ones. Unlike them, high voltages in dry reed switches are reached not through gas filling under high pressure but through deep vacuumization of the tube. There is practically no construction difference between reed switches for working voltages of up to 5–10 kV, and low-voltage reed switches, although the former ones have characteristic features — a remaining stem through which air is evacuated, and tungsten covering of the contact-elements.

FIGURE 5.47
High-voltage vacuum reed switch for
20 kV DC. 1, 9 — outlets of contact-elements; 2 — restorable spring; 3 — ferromagnetic armature; 4 — ceramic stop;
5 and 7 — contact-elements; 6 — stem;
8 — tube.

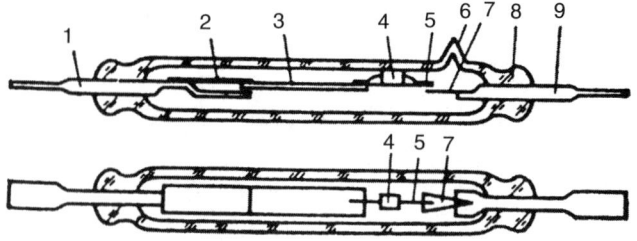

Of course such reed switches have a bigger gap between contacts (and therefore
increased magneto-motive operation force) and the proper size (as a rule, the tube is 50
to 55 mm in length and 5–6 mm in diameter). Reed switches for voltages of 15 to 20 kV
have some additional constructive elements (Figure 5.47). The power of such reed
switches with switched voltages of up to 1000 V does not usually exceed 50 W and with
voltages of more than 1000 V up to 10–20 W. Switched current is only a fraction of a
milliampere (current-carrying 3 A).

In the 1970s the main producers of high-voltage vacuum reed switches were the
English firm FR Electronics and the American one Hamlin, Inc. In the period 1973–1974
the latter produced vacuum reed switches of the DRVT-30 type with switched voltage of
up to 27.5 kV! The tube of such a reed switch was 58 mm in length and 7.4 mm in
diameter, and was capable of switching currents up to 1 mA under working voltage.
Operation time of such a reed switch was 20 ms and its magneto-motive operation force
was 500 A.

In the former U.S.S.R. (Lvov firm "Polyaron") original vacuum reed switches of the
BB-20 type for working voltages of up to 10 kV in the no-current switching mode and up
to 5 kV in the current switching mode (up to 2 A) were produced at the end of the 1970s. It
was possible to let current pass up to 20 A through the closed contacts. In the butt-ends of
the tube there were deeply recessed pin-like outlets soldered into the tube. They were
designed to be connected to the receptacle of the high-voltage sockets, which provided
very high resistance of leakage along the surface and allowed insulation of the high-
voltage outlets from the control coil. Ceramic bushes with sockets for standard banana-
like connectors were used for experiments. These were quite large devices, with a tube
150 mm in length and 30 mm in diameter and with a movable contact-element of the
plunger-type. The tube moved forward with typical noise from the end stationary part of
one contact-element towards the stationary part of the second contact element, touching it
with its butt-end. Operating time corresponded to the size of the construction and made
up 15 ms. This was perhaps the biggest reed switch in the history.

In high-voltage reed switches, there is a problem alien to low-voltage constructions:
electrostatic attractive force between the contact-elements. Such force is approximately
proportional to squared voltage and the area of the overlap of contact-elements, and is in
inverse proportion to the distance between them. With voltages of 15–20 kV the force may
become so strong that under certain circumstances it can cause spontaneous closing-in of
the contact-elements, to the distance that could allow a breakdown of the gap between
contacts. On the other hand, those same forces prevent the contact-elements from opening
after removal of the control magnetic field, which is why the area of the overlap of the
contact-elements with such high voltages should be as small as possible, with switched
current not exceeding a few milliamperes, anyway.

Because of the increased gap between the contacts the operating time of such reed
switches is usually more than in low voltage reed switches, and is usually 3 to 5 ms,
however, measurements carried out by the author show that the closing time of the

external circuit with a voltage of 5 to 10 kV with the help of such a reed switch is considerably less than the closing time of a low-voltage circuit with the help of the same reed switch. This can be explained by the occurrence of high-voltage breakdown between the contact-elements as they close in, and holdover commensurable with small working currents, long before the contact-elements touch.

5.7 Reed Switches With Liquid Filling

In patent descriptions one can often come across a number of constructions of reed switches with a tube filled with chemically inert insulating liquids such as silicone oil, instead of gas (Japanese patent No. 4814590; U.S.A. patent No. 2547003; East Germany patent No. 53152; U.S.S.R. patent No. 477478; England patent No. 1520080; Germany patent No. 2512151, and many others).

Reed switches filled with such liquids have higher insulating characteristics and breaking-down voltages, and better dynamic and thermal properties. Trough-shaped contact-elements closing in such liquids do not rebound after closing because their movement is slowed down by the liquid. Such slowing down of movement does not change total operating time of a reed switch considerably, since due to high dielectric properties of dielectric liquids the gap between the contacts in such reed switches can be reduced to 0.025–0.07 mm. Moreover, there is even a saving of operating time with such reed switches.

The use of hollow movable contact-elements (in the form of a flat tube — Figure 5.48a), also allows it to obtain its "neutral buoyancy" and to increase its resistance to external mechanical shocks and vibrations. As it turns out, it is possible to fill not only dry reed switches with dielectric liquid, but also reservoir-free mercury ones (Figure 5.48b — England patent No. 1520080).

The appropriate selection of liquid and nonmagnetic material for contact-points wetted with mercury allows avoidance of vibration of the contact-points during closing and the so-called dynamic noise caused by magneto-strictive effects, after closing of standard reed switches. Moreover, it turns out that it is also possible to fill reed switches with dielectric liquid containing much mercury in the reservoir (Figure 5.49). In such a reed switch, under the effect of the magnetic field the ferromagnetic liquid (3) moves down, forcing

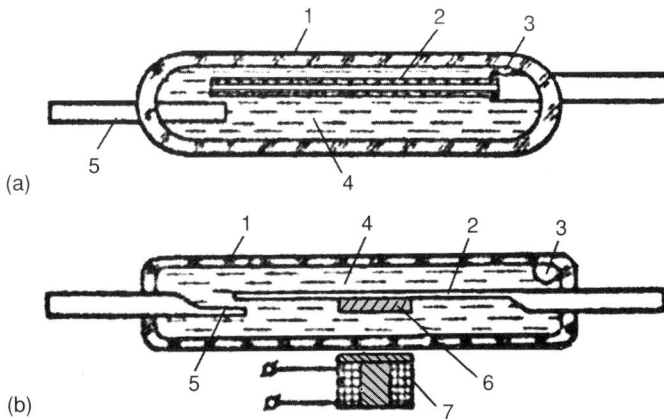

(a)

(b)

FIGURE 5.48
Reed switches with a liquid-filled tube. 1 — glass tube; 2 — movable contact-point; 3 — damping gas bubble; 4 — electro-insulating liquid; 5 — stationary contact-point; 6 — disk-shaped ferromagnetic armature fixed on the movable contact-point; 7 — control coil.

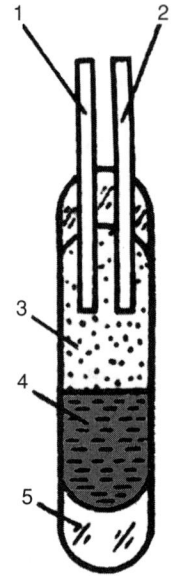

FIGURE 5.49
Mercury reed switch with ferromagnetic liquid. 1 and 2 — outlets of the reed switch;
3 — ferromagnetic insulating liquid; 4 — mercury; 5 — glass tube.

the mercury (4) upward, where it closes outlets 1 and 2. Ferromagnetic liquids are usually based on organic or inorganic (silicon, for instance) oils with finely dispersed ferromagnetic powder with a particle size of under 100 A (1 A $= 10^{-10}$ m).

Experiments carried out in the 1970s in the U.S.S.R which dealt with filling of tubes of standard dry reed switches with transformer oil showed electrical noise reduction by 5 to 6 times. Filling of the tubes with a more heavy substance caused reduction by more than 30 times. The American Magnavox Company used such reed switches in high-speed multi-way switches, carried out a number of successful experiments in this field in the 1960s.

Liquids filling the tube must retain their properties in the closed and small volume of the tube, without replacement during the entire service life of the reed switch. This is a difficult task because high-temperature electric spark and even the slightest arcing on opening contact-elements can cause destruction of most organic liquids, with a release of solid particles of carbon. There is a tendency to apply flurohydrocarbon liquids or Freon $C_3Cl_4F_4$, $C_5F_{10}HF$ $(C_3F_6O_2)O$; ether $(C_8F_{16}O)$; polyorganosiloxane liquids. The high costs of such liquids, and the considerable complexity of production of such reed switches, restrain their mass fabrication for the time being.

5.8 Polarized and Memory Reed Switches

Polarized reed switches are those reed switches that are sensitive to the polarity of the control signal applied to the control coil, in other words to the vector direction of the magnetic field F (Figure 5.50). Such sensitivity is caused by an additional static magnetic field produced by a permanent magnet placed nearby (or by an additional polarized winding, which is rarer). The external magnetic field of the control signal may have the same direction as the magnetic field of the permanent magnet. In this case as their fluxes are summed up, causing operation of the reed switch, the sensitivity of the reed switch to the control signal increases considerably.

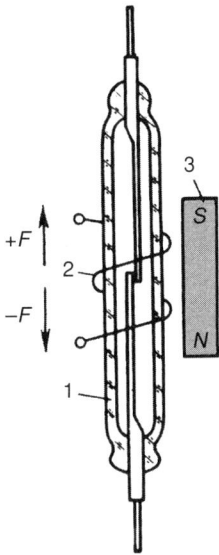

FIGURE 5.50
Polarized reed switch. 1 — neutral reed switch; 2 — control coil; 4 — polarized permanent magnet.

If vectors of the magnetic fluxes have opposite directions, the resultant magnetic flux is so small that the reed switch cannot be energized. One of the most important applications of such polarized reed switches is obtaining an opening (normally closed) contact out of a standard normally open one. In this case the magnet is selected in such a way that its magnetic field is enough to energize and hold a standard normally open reed switch in such a state. If the direction of the control magnetic field of the coil is opposite to the direction of the magnetic field of the permanent field, the total value of the magnetizing force affecting contact-elements will be less than their elastic force, and they will open, affected by these forces.

As far as construction is concerned a magnet can be placed not only along the tube or outside it, as it is shown in Figure 5.50. There are a number of different constructions with original combinations of control coils and permanent magnets, some of which are shown in Figure 5.51. With the help of permanent magnets it is possible to produce a three-position reed switch with a neutral mid-position, which would switch this way or that under effect of the magnetic field of the control coil of this or that polarity (Figure 5.52). Using several control coils placed in different parts of a reed switch, instead of one, it is possible to produce reed switches capable of carrying out standard logical operations AND, OR, NOT, EXCLUSION, NOR (OR-NOT), NAND (AND-NOT), XOR (EXCLU-SIVE-OR), etc. (Figure 5.53). If such multi-wound reed switches are combined with permanent magnets (Figure 5.54), it is possible to obtain quite complex functional elements with adjustable operation thresholds, and remote switching of certain options. The number of such combinations is practically endless. This allows designers to implement almost fantastic projects.

Taking into account that the reset ratio of reed switches is less than 1 (that is, for operation a stronger magnetizing force is needed than for release) one may try to choose a magnet of such a strength which is sufficient for operation of a reed switch, and at the same time capable of holding closed the contact-elements which have already been closed by the control coil field. In this case the reed switch is switched ON by a short current pulse in the control coil and remains in this state even after the control impulse stops affecting it (that is, it "memorizes" its state). The reed switch can be switched OFF by applying a control pulse of the opposite polarity to the coil. Such a switching device, though in fact capable of operating, is not used in practice. There are several reasons for

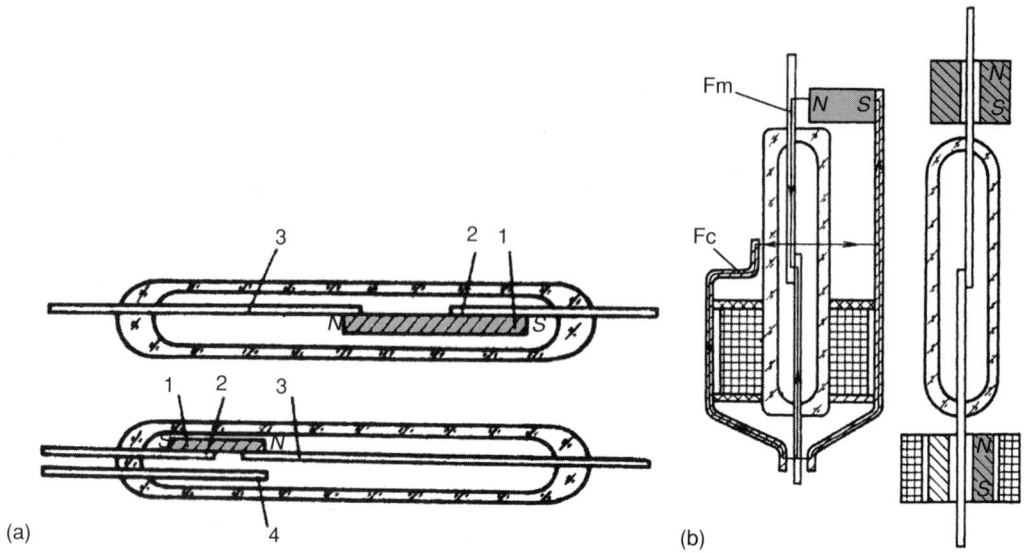

FIGURE 5.51
(a) Polarized reed switches with internal arrangement of magnets. 1 — permanent magnet with electroconductive covering; 2 — stationary contact-element to which a magnet is welded; 3 — movable contact-element; 4 — second stationary contact-element. (b) Polarized reed switches with external arrangement of magnets. Fm — magnetic flux of the permanent magnet; Fc — control magnetic flux.

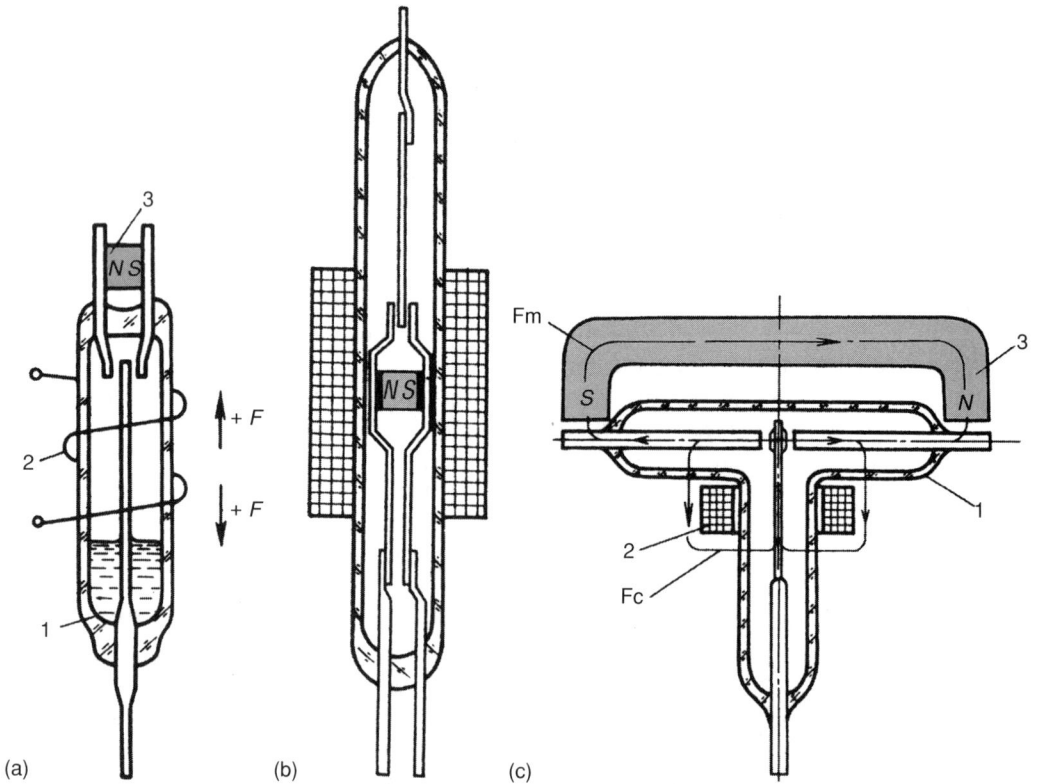

FIGURE 5.52
Three-position polarized reed switches: (a) mercury reed switch with an external magnet; (b) dry reed switch with an internal magnet; (c) high-frequency reed switch. 1 — glass tube; 2 — control coil; 3 — permanent magnet with external insulating covering which can also be made of ferrite.

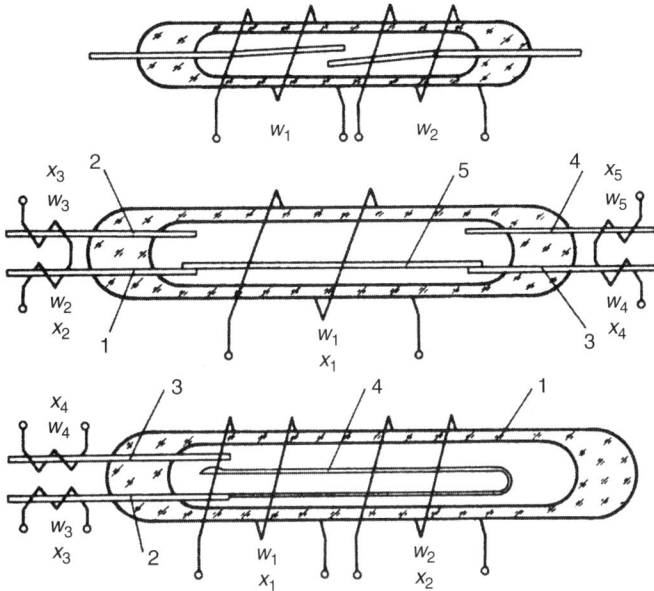

FIGURE 5.53
Multi-wound reed switches designed to carry out standard logical operations.

this. Firstly, such a device requires very accurate adjusting because the slightest excess of magnetizing force of the permanent magnet will cause spontaneous closing of a reed switch. If the magnetizing force is not strong enough, the reed switch will not remain closed after the control impulse stops affecting it. Taking into account great technologic differences between parameters of reed switches, magnets and control coils, each device will require individual adjusting, which is impossible for mass production. That is why even a device adjusted beforehand to a certain temperature may malfunction at other temperatures.

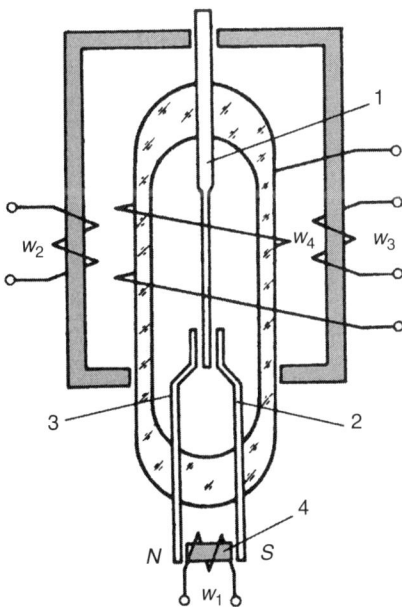

FIGURE 5.54
Combined switching logical device on a reed switch. 1 to 3 — contact-elements; 4 — permanent magnet.

In 1960 A. Feiner and his colleagues from Bell Laboratories published in the *"Bell System Technical Journal"* an article entitled "Ferreed — A new switching device," in which they suggested that the way to overcome these difficulties was by creating memory reed switches. The idea was that the permanent magnet should become a magnet only when it and the reed switch are affected by the control pulse of the coil. Other details were technical. They chose magnetic material with a medium coercive force, which could be magnetized during the time of affecting of the control pulse, and remain magnetized for a long time, until the pulse of the magnetic field of the opposite polarity affects it (such material is called "remanent"). This device, consisting of a reed switch and a ferrite element, was called "ferreed" (by the first letters of the words "ferrite" and "reed switch"). Later, for advertising purposes, some firms began to name devices operating on the same principle in a different way: "remreed," "memoreed," etc.

It turns out that ferrite can be remagnetized for 10 to 50 μs while closing of contact-elements requires a time of 500 to 800 μs. This allows us to use very short pulses for ferreed control (in practice pulses with a reserve of up to 100 to 200 μs for closing are used). This means that contact-elements are not only held after closing of the magnetized ferrite, but also continue of closing process using the magnetic flux of the ferrite after the short control impulse stops affecting it.

It is obvious that ferreed with one control winding will be critical to the amplitude of the control pulse. When the amplitude of the switching current pulse (for switching OFF) is not high enough, the core in the control coil is not fully magnetized and the contacts will remain in a closed position. If the control signal is quite strong the core can be reversely remagnetized and obtain the opposite polarity. In that case the contact-elements will remain in a closed position also. To avoid this two control windings are applied (Figure 5.55).

The magneto-motive force of each winding is not enough to magnetize the core to the degree necessary for closing of the contact-elements. Only when switching-on current pulses of opposite polarity are applied to both windings is the total magnetizing force enough to magnetize the core, closing the contact-elements.

To open the contact-elements, switching-on current pulses of the same polarity are simultaneously applied to both windings. The polarity of magnetization of the halves of the ferrite core will be opposite, and both contact-elements are likely to be magnetized, therefore a repulsive force causing the contacts to open will arise between them. An additional shunt (4) made from soft magnetic material enhances configuration of the magnetic field in the overlap of the contact-points, and reliability of operation of the device.

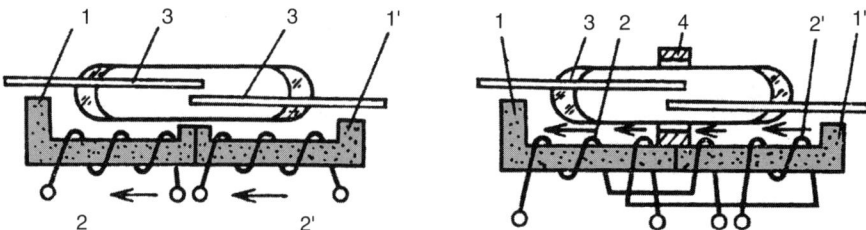

FIGURE 5.55
Ferreeds with two control windings. 1 — core from remanent material; 2 and 2' — control windings; 3 — contact-elements; 4 — additional magnetic shunt.

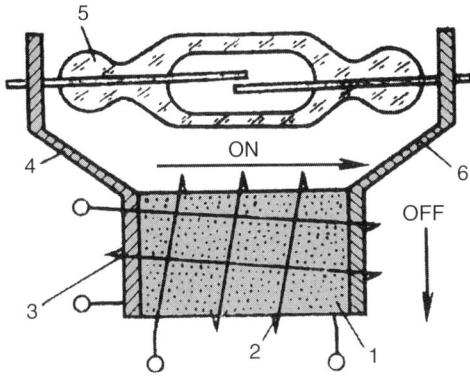

FIGURE 5.56
Ferreed with orthogonal control. 1 — core from remanent material; 2 — switching-on winding; 3 — switching-off winding; 5 — reed switch; 6 — magnetic core.

In a ferreed with a so-called orthogonal control (Figure 5.56), in order to change the state of the magnetization the vector turns by an angle of 90°, instead of 180° as in the previous case. The first such solution was patented by A. Feiner, from Bell Laboratories (U.S.A. patent No. 2992306). In his construction the magnetic flux of the winding (2) for switching ON passes through a magnetic gap between the contact-elements, and the magnetic flux of the winding (3) for switching OFF does not pass through the gap between the contact-elements, providing reliable switching OFF of the reed switch.

As in multi-wound reed switches, on ferreeds it is quite easy to implement single- or multi-circuit automation logic elements (Figure 5.57). For example, in multi-circuit relays with a cross-shaped core (Figure 5.57), there are 16 possible combinations of closed and opened reed switches, depending on what windings are switched ON.

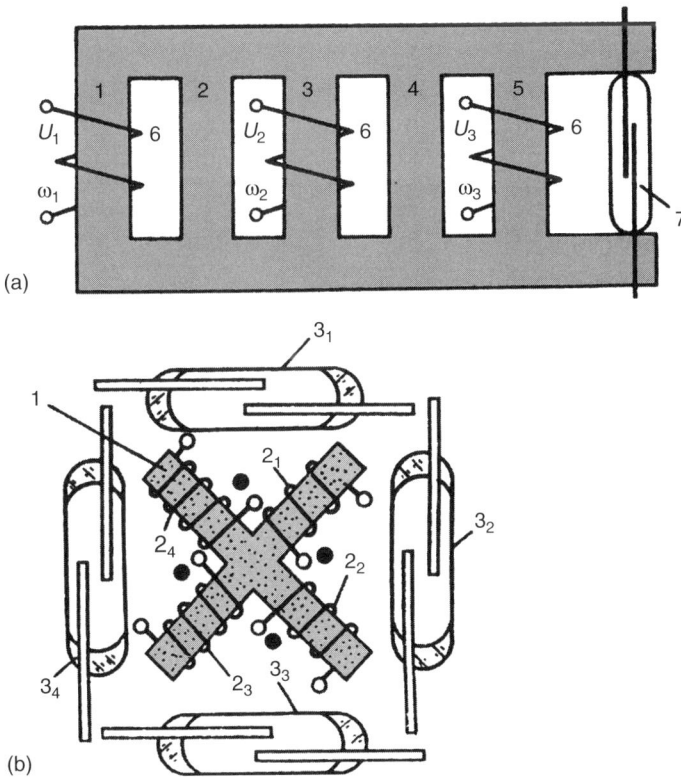

FIGURE 5.57
Automation logic elements on ferreeds.

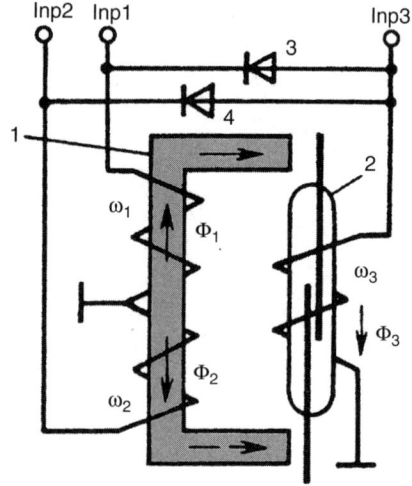

FIGURE 5.58
Device in which memorization option can be disabled.

In some constructions one can enable or disable memorization options with the help of additional control signals (Figure 5.58). In the construction described above a memory element of the external type is used. Since the 70s one can observe rapid development of ferreeds with internal memory, which were produced by Hamlin, FR Electronics, and Fujitsu. Their external design was practically identical to that of dry reed switches, but their contact-elements were made of special alloys, providing sealing of the reed switch after it was affected by the pulsed magnetic field. Thus no external elements are needed for such ferreeds. Originally, contact-elements of such ferreeds consisted of two parts: an elastic one and a hard magnetic one (from remanent material), but there were also excess joints with increased magnetic resistance. Later on, hard magnetic alloys were invented and used, so that the contact-elements could be made more flexible and elastic enough. Such an alloy consists of 49% cobalt, 3% vanadium, and 48% iron, or of 30% cobalt, 15% chromium, 0.03% carbon, and the rest iron. There are also bimetallic contact-elements (U.S.A. patent No. 3828828), the internal rod of which consists of an alloy of 81.7% iron, 14.5% nickel, 2.4% aluminum, 1% titanium, and 0.4% manganese, with the shell of the section made of an alloy containing 42% iron, 49% cobalt, and 9% vanadium.

5.9 Reed Switch Relays

Unlike electro-mechanic relays with many interacting elements, the simplest reed relay contains nothing but a reed switch and a winding (Figure 5.59).

In some cases the relay is supplied with a ferromagnetic shield protecting the reed switch from magnetic field impact, and that is it! The simplicity of construction and low

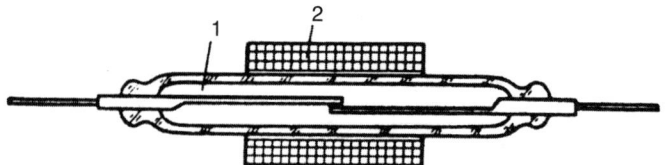

FIGURE 5.59
Simplest Reed Relay. 1 — reed switch; 2 — winding.

cost in mass production are the main advantages of reed relays, however, despite such simplicity there are several schemes for arranging magnetic circuits of reed relays by using additional ferromagnetic elements forming magnetic fields of the best configurations, and reducing flux leakage (Figure 5.60). There are also magnetic systems with internal (Figure 5.60a and b) and external (Figure 5.60c) reed switches. The choice of this or that variant is determined by many factors, including the size of the reed switch, the required sensitivity of the relay, size limitations, etc. In mass-produced relays all of the above variants are used.

In the 60s and 70s the most popular type in mass production was a relay with or without a ferromagnetic shield, in rectangular plastic cases covered with epoxide resin or silicon rubber (Figure 5.61). Later on, a ferromagnetic tube was used as a relay case. It allowed miniaturization and simplification of the construction. Typical examples of this are relays of the RES-55 and RES-64 type, which were produced in the U.S.S.R. in the 1970s (Figure 5.62). They were the smallest reed relays ever produced in the U.S.S.R. and belonged to the class of subminiature relays. The RES-55 type relay is based on a changeover reed switch (KEM-3), capable of switching voltages up to 127 V and currents up to 1 A (for voltage up to 36 V) with an active load and switched power of up to 30 W. The weight of such a relay is not more than 6 g. A RES-64 relay contains a normally open reed switch (KEM-2) switching under active load voltages of up to 130 V, currents up to 0.25 A (for voltage up to 30 V) with switched power of up to 9 W. The weight of this relay is also not more than 6 g.

Traditionally industrial automation relays, based on larger and more powerful reed switches, are considerably larger in size (Figure 5.63). In addition, relays designed for relay protection systems were produced in standard cases of electro-mechanic relays, whose volume exceeded that occupied by a reed relay (Figure 5.64). The main advantage of such a relay, compared with similar electro-mechanic ones, is its high speed of

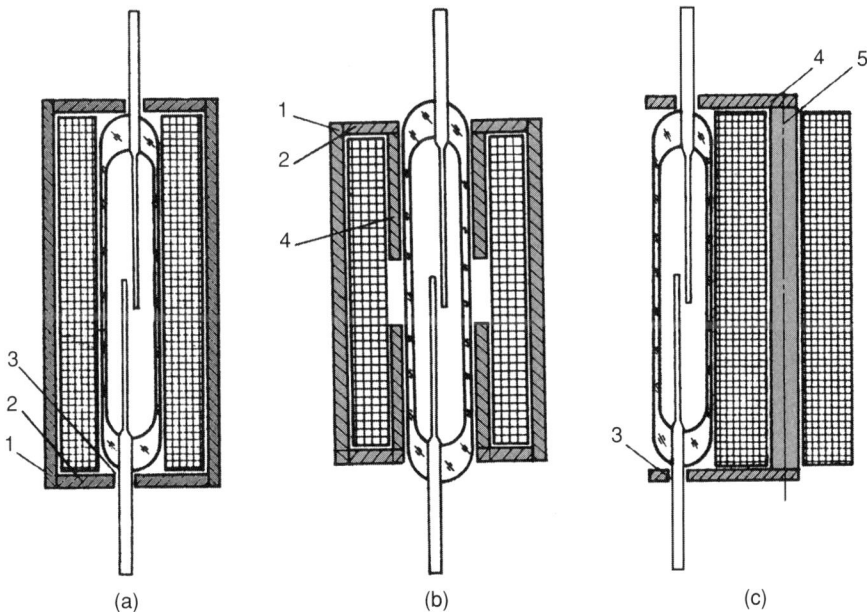

FIGURE 5.60
Construction schemes of magnetic circuits of reed relays. 1 — ferromagnetic shield (the same as part of a magnetic core); 2 — magnetic terminals; 3 — insulating gap; 4 — additional terminals; 5 — core.

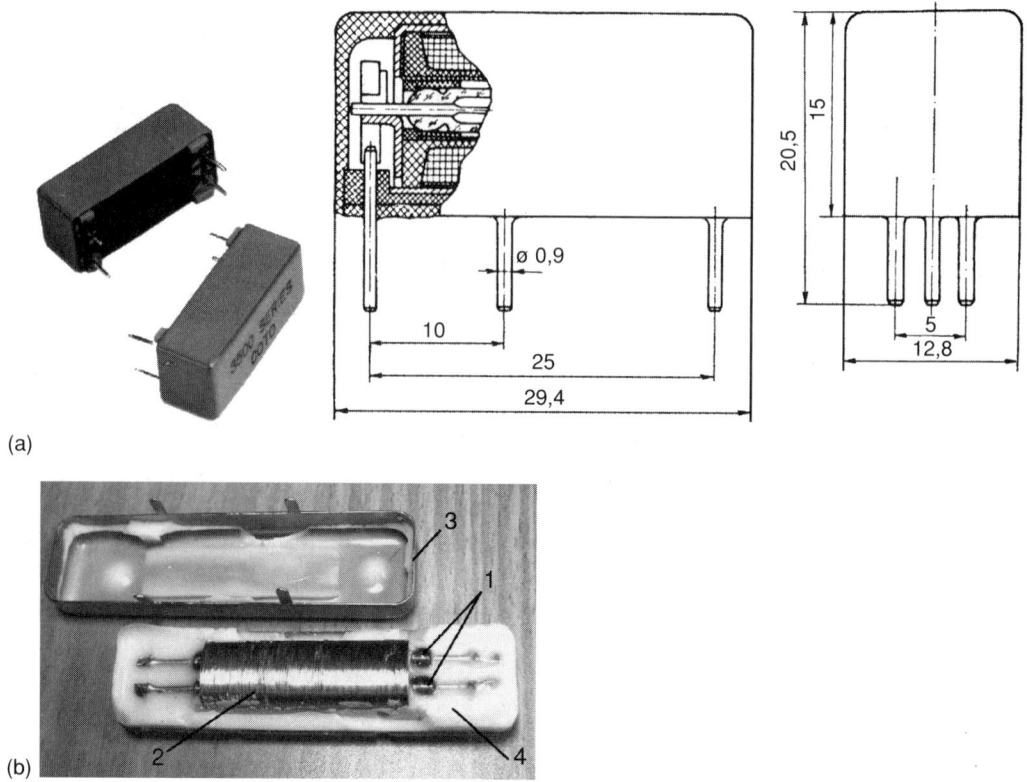

(a)

(b)

FIGURE 5.61
(a) External design and construction of typical low-power reed relays in plastic cases produced by many companies in the 70s. (b) Reed relay of ARID-B-2A2 type with a ferromagnetic cover coated with soft silicon rubber (ERNI). 1 — reed switches; 2 — free-bobbin coil; 3 — ferromagnetic cover-shield; 4 — silicon rubber.

operation. In reference sources of the firm ASEA, this relay is specified as an especially high speed one.

As chips appeared and came into wide use in electronic devices and other electronic components installed on printed-circuit boards became smaller, the size of reed switches was no longer satisfactory for designers of electric equipment. In addition, enhancement of standard electro-mechanic relays led to considerable miniaturization of relays having the same switching capacities compared with reed switches (Figure 5.65).

Because of the necessity of miniaturizing electric devices, engineers started designing reed relays in so-called dial in-line package (DIP) and single in-line package (SIP) cases of a size similar to that of chips, and with distances between outlets similar to a standard grid with a pitch (1.0″ × 0.1″ or 1.0″ × 0.15″ — Figure 5.66). Enhancement of the construction of reed relays and reed switches allowed considerable reduction of their weight and size (Figure 5.67), and restored competitiveness compared with miniature electro-mechanic relays.

In practice, construction of multi-contact relays is similar to that of single-contact ones, the only difference being that instead of one reed switch, an unit of a few reed switches was inserted into the coil (which was, of course, large in size (Figure 5.68). Relays in such cases were produced in the U.S.S.R. in the 1960–70's. Their external design and size were very much like diode-transistor logic elements produced at that time from the series "Logika-T" (designed for construction of automation control systems and control

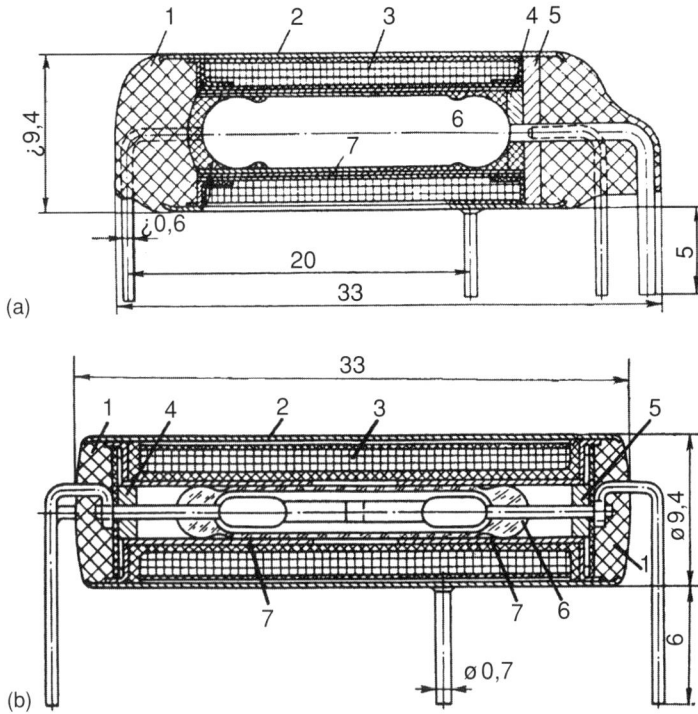

FIGURE 5.62
Reed Relays (RES-55 (a) and RES-64 (b)) with a ferromagnetic shield as a case, produced by plants of the former U.S.S.R. in the 70s. 1 — epoxide resin; 2 — case (a steel tube with a sidewall 0.2 mm thick); 3 — winding; 4 and 5 — ferroelastic disks; 6 — reed switch; 7 — electrostatic shield from brass 0.1 mm thick.

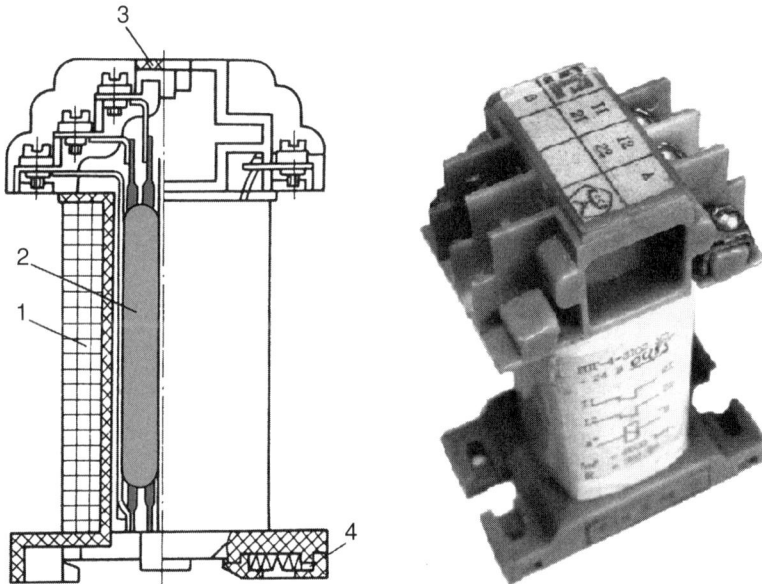

FIGURE 5.63
Industrial automation relay of the RPG-4 type, on the basis of power reed switches MKA-52202, for switching currents up to 4 A, voltages up to 380 V, with power up to 280 W (Russia). 1 — winding; 2 — reed switch; 3 — cover with a terminal socket; 4 — fixation elements of the relay on a standard DIN rail.

FIGURE 5.64
Reed relay of RXMT-1 type for devices of relay protection of power systems (produced by ASEA).

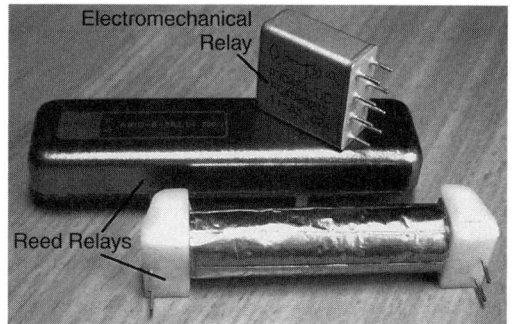

FIGURE 5.65
Multi-contact electro-mechanic relays with similar switching characteristics are smaller in size than reed ones.

(a) (b) (c)

1 FORM A, 1 FORM C

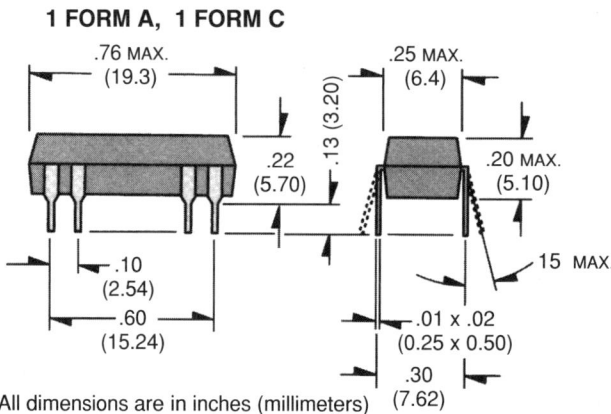

(d) All dimensions are in inches (millimeters)

FIGURE 5.66
Modern miniature reed relays (produced by ALEPH) in (a) DIP and (b, c) SIP cases.

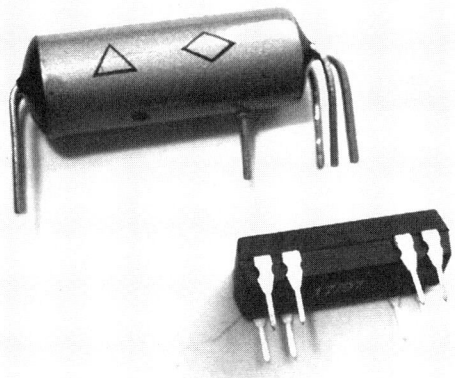

FIGURE 5.67
The smallest reed relays with a changeover reed switch produced in the 1970s — upper (In Russia it is still being produced) and modern ones in a DIP case — bottom.

of industrial processes), because they were used as outlet nodes of these logic elements. At that time such relays corresponded to the technical level. They switched currents up to 1 A and voltages up to 250 V with power of 50 W. When at the beginning of the 1980s production of logic elements of the "Logika-I" series started, based on antinoise integrated circuits, the external design of the industrial reed relays was also changed, causing them to very much resemble these logic elements (Figure 5.69), and a new type of installation — on standard DIN rail used in western countries — was applied.

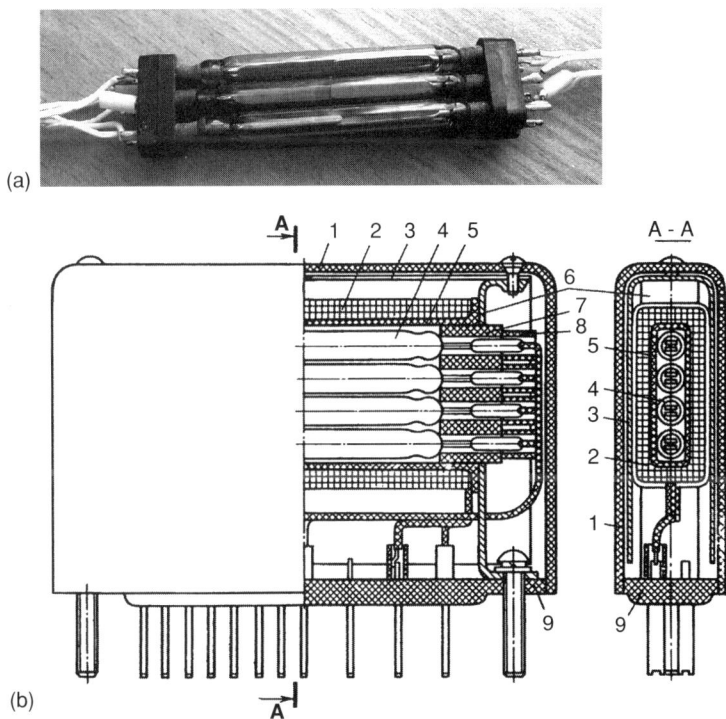

FIGURE 5.68
(a) An unit of six reed switches of standard "A" form size (three reed switches in two lines) for installation in the internal cavity of a coil of a multi-contact relay. (b) Construction of multi-contact relay of RPG type based on reed switches of standard size for industrial automation systems. 1 — plastic case; 2 — winding; 3 — ferromagnetic shield; 4 — reed switches; 5 — coil bobbin; 6 — cramp; plastic sockets; 8 — rubber tubes; 9 — heelpiece.

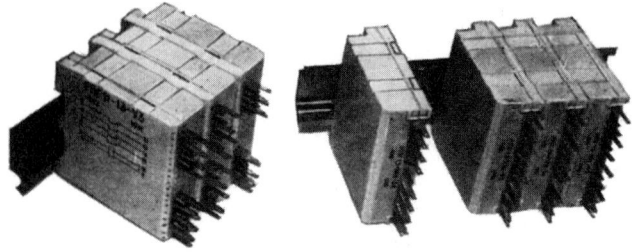

FIGURE 5.69
Reed relays RPG-11, RPG-13 for industrial automation systems in cases of logic elements of "Logika-I" (the U.S.S.R, Russia).

In the magnetic systems of multi-contact relays with external reed switches, the latter are placed outside the coil circle-wise (Figure 5.70). According to such construction schemes reed relays of the RPG-14 type, based on high power reed switches (4 A, 380 V, 250 W) were produced (Figure 5.71). Lately the production of low-power (50 W) and higher-power (up to 250 W) reed switches of normal size, and reed relays based on them (Figure 5.66–Figure 5.69) has been considerably reduced. This becomes clear when one compares multi-contact reed relays of these types with electro-mechanic relays with similar characteristics (Figure 5.72). In comparison, the reed relays were seen to be inferior, and it was obvious that large reed relays based on reed switches of normal size, designed for industrial automation systems, have no future (if we do not take into account some exceptional and unusual cases). At the same time miniature reed relays

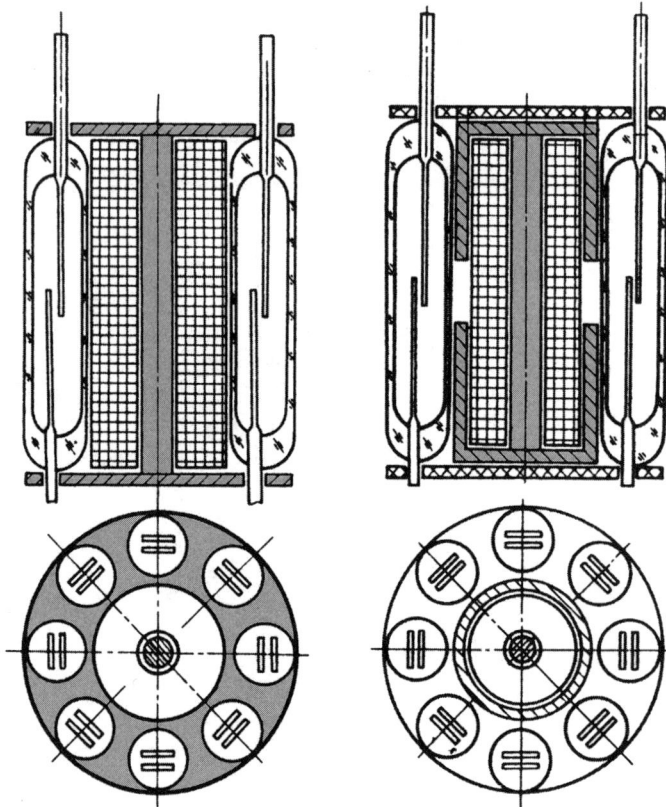

FIGURE 5.70
Magnetic systems of multi-contact relays with external reed switches.

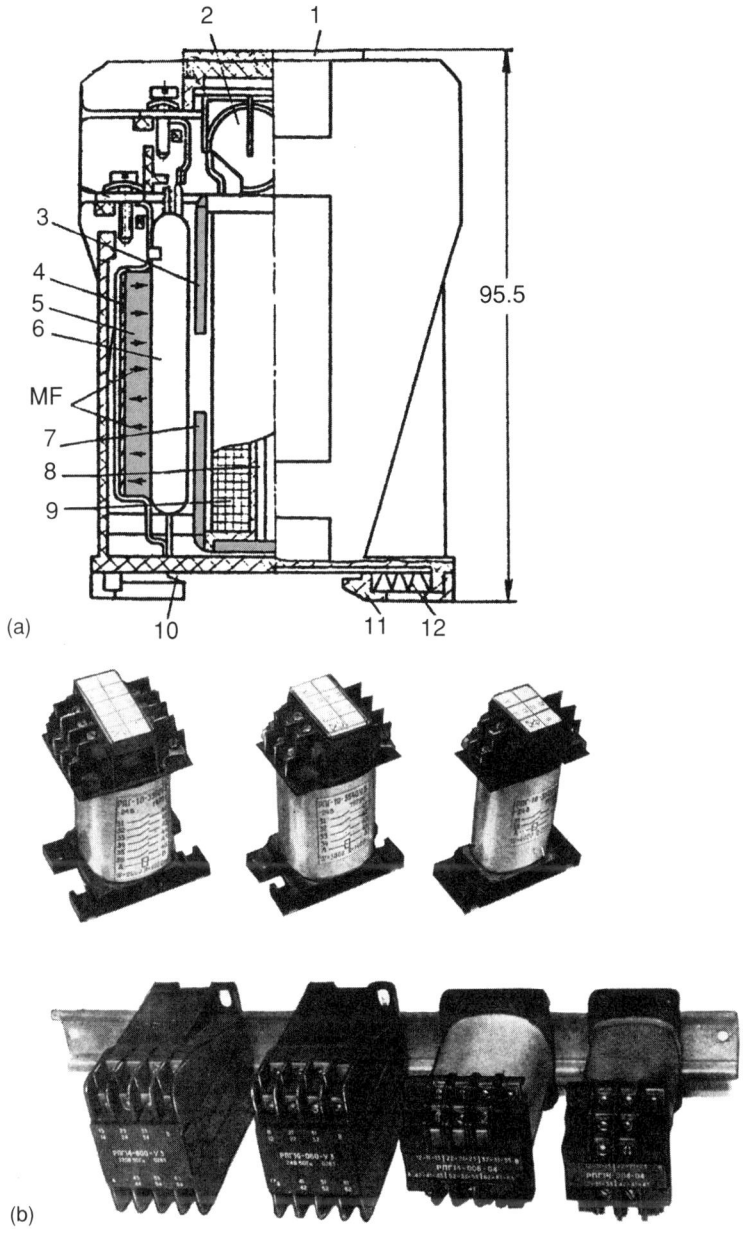

FIGURE 5.71

(a) Construction of a mass produced multi-contact relay RPG-14 based on power reed switches placed outside the coil. 1 — cover; 2 — varistor connected in parallel to the winding; 3 and 7 — magnetic cores; 4 — ferromagnetic plates; 5 — permanent magnet; 6 — reed switch; 8 — ferromagnetic core; 9 — winding; 10 to 12 — elements of relay fixation on a standard DIN rail. (b) Multi-contact power reed relays of the RPG-10 type (above) and RPG-14 (below), produced in the U.S.S.R. (Russia).

with switching reed switches in DIP and SIP cases continue to be in demand on the relay market, and their production has been scaled up.

It is not quite clear yet whether contactors based on high-power reed switches (hersicons) of the KMG type (see above), constructed in the former U.S.S.R. (Figure 5.73), will be used, as there is not yet enough international field experience. Apparently such

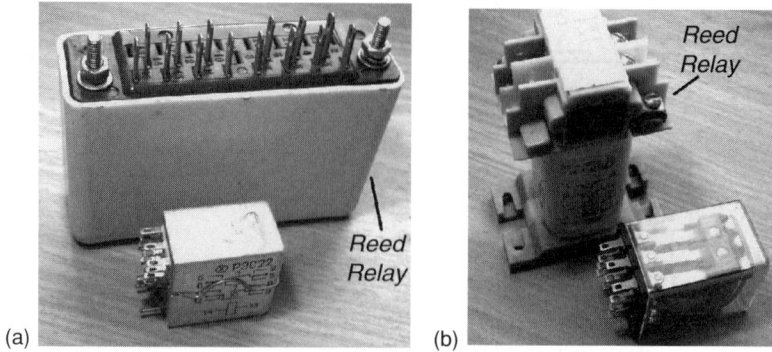

FIGURE 5.72
Comparison of sizes of relays based on (a) low-power and (b) higher-power reed switches of normal size with electro-mechanic relays with similar characteristics.

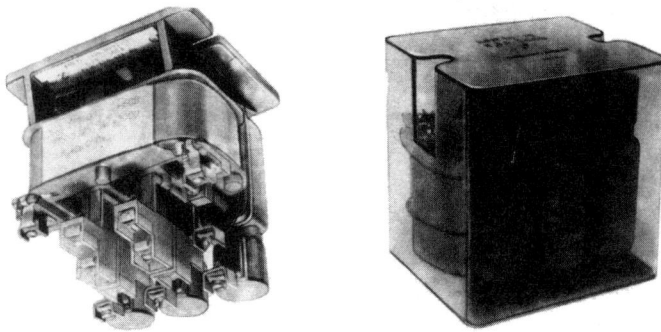

FIGURE 5.73
Three-phase contactors based on high-power reed switches.

constructions may be quite useful under special conditions, such as dust-laden and explosive atmospheres, intensive ammoniac vapors (in agriculture), salt spray atmosphere impact, etc.

5.10   Mercury Reed Relays

Relays based on mercury reed switches have been produced serially for a long time, perhaps since such reed switches were invented. The first constructions were produced in cases unusual for relays, similar to those for vacuum radio valves (Figure 5.74). Such cases provided good protection of a mercury reed switch from mechanical damages and had well-developed production techniques. When the air had been pumped out, the tube of the reed switch of this relay was filled with hydrogen under about 15 atm pressure, providing break-down voltage between contacts up to 8500 V, and preventing oxidation of the mercury. The relay was 28.1 mm in diameter and 81.2 mm in length. The weight was 113 g.

Because of mercury and high pressure, such relays are labeled with one or more of the following warnings (Figure 5.74b):

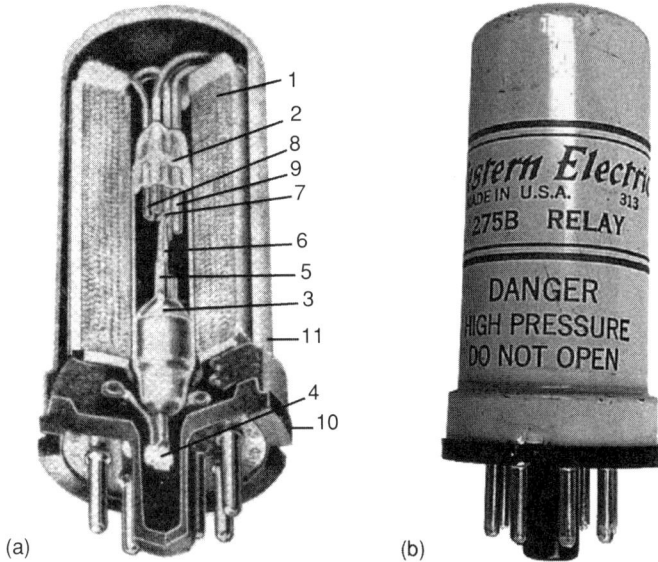

FIGURE 5.74
(a) Relays of HG type based on mercury reed switch in a case of a metallic radio tube (CP Clare Co., 1947). 1 — coil; 2 — glass tube of the reed switch; 3 — mercury; 4 — stem; 5 — lower pole terminal; 6 — armature; 7 — movable contact fixed on the armature; 8 — stationary contacts; 9 — terminals with stationary contacts; 10 — octal heelpiece; 11 — metallic case. (b) Mercury relay 275B type, produced by Western Electric Co. in the 1950–1960s years.

- "UP"
- "Danger"
- "High Pressure, Do Not Open"

At present, a wide range of relays based on mass-produced mercury reed switches is manufactured. One can come across both relays in standard round cases, and in more common for relays rectangular (metal) cases (Figure 5.75). A typical feature of such relays is a large pointer on the relay case, which indicates its working position.

5.11 Winding-Free Relays

The winding in a reed relay is used to produce the magnetic field needed for reed switch operation, however, such a magnetic field can also be created by other sources, for instance by a permanent magnet or copper wire with a large current passing through it. In fact such sources of magnetic fields are widely used to control reed switches.

Different firms produce a great number of position pickups, liquid level detectors, pressure sensors, etc., based on reed switches controlled by moving permanent magnets. Such devices belong to detectors rather than to relays, and it is practically impossible to describe all of their peculiarities in the book. A reed switch placed at some distance from the conductor line, with currents of hundreds of amperes operating at a certain value of the current, is a current relay (Figure 5.76).

The operation threshold (pickup current) of such a relay (that is, its sensitivity) with invariable current value in the line, depends on the distance X between the line and the

FIGURE 5.75
Modern relays based on mercury reed switches produced by Midtex company.

reed switch, and the angle α between the longitudinal axis of the reed switch and the longitudinal axis of the line. Apparently, the maximum sensitivity of the device will be at minimum distance, and the angle $\alpha = 90°$. One can adjust the pickup current of a relay by changing these parameters.

To obtain a nonlinear (sharper) change of the magnetic flux affecting a reed switch, a ferromagnetic shunt with a smaller section at the overlap of contact-elements is used, when current value in the line approximates the pickup current of the relay (Figure 5.77). At low values of current in the bus, far from the pickup current of the relay, the magnetic flux Φ in the upper part of the line closes through the shunt (3) and does not affect the reed switch. As current is increased to a certain value, the reduced part of the shunt is quickly saturated and the magnetic flux bulges saltatory at this part of the shunt. The reed switch is energized, affected by the magnetic flux.

Taking into account the sensitivity of real reed switches, and the necessity of holding the insulating distance X between the line and the reed switch, it is possible to provide a minimal pickup current starting from 50 to 100 A. In cases when that is not enough, an additional magnetic core concentrating the stray flux of the conductor line and directing it to the reed switch area is used (Figure 5.78). One can increase the sensitivity of a relay with an additional magnetic core by several times.

For current control in three-phase circuits, relays with three reed switches and a magnetic core of original construction are used (Figure 5.79). Principles of construction of winding-free reed relays described above can be applied to both DC and AC. In the latter case, relay "operation" means vibration of contact-points of the reed switch with doubled circuit frequency.

A vibrating reed switch can be included in the simplest electronic circuit, transforming a variable signal into a standard continuous one. Sometimes that is not very convenient, and sometimes it is just impossible, for instance, in case when a relay must be in the

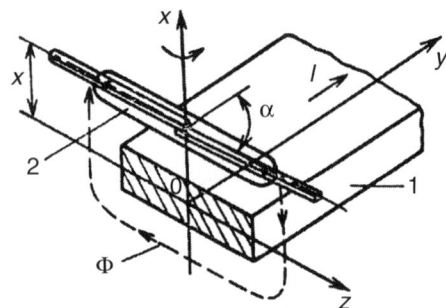

FIGURE 5.76
Winding-free current relay based on reed switches.
1 — conductor line; 2 — reed switch.

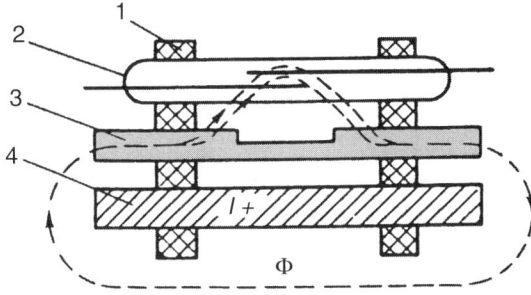

FIGURE 5.77
Winding-free reed current relays with a magnetic shunt. 1 — insulating fastening elements; 2 — reed switch; 3 — magnetic shunt; 4 — current-carrying bus.

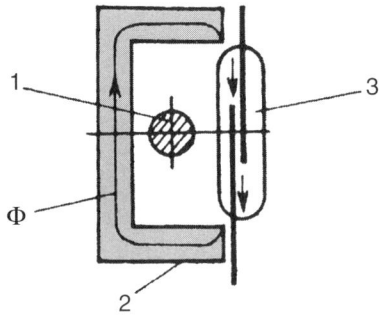

FIGURE 5.78
Winding-free reed relay with an additional magnetic core. 1 — current-carrying bus; 2 — external magnetic core; 3 — reed switch.

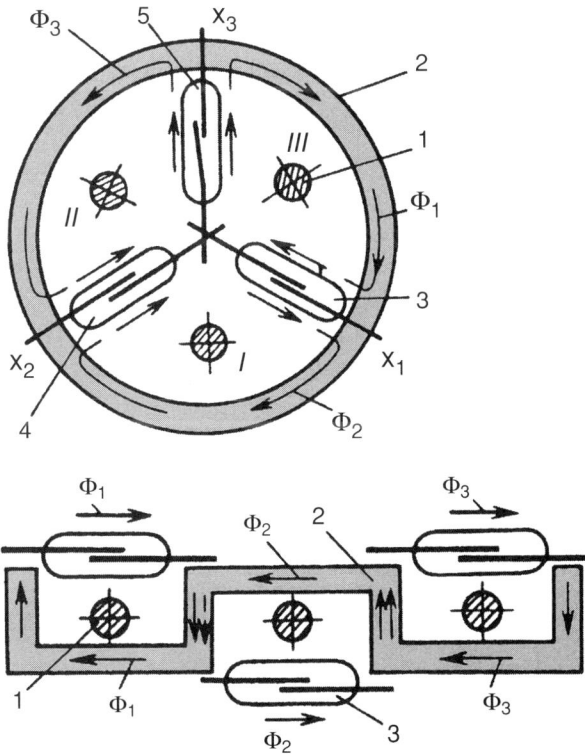

FIGURE 5.79
Winding-free reed relays for three-phase circuit. 1 — current-carrying bus; 2 — magnetic core; 3 to 5 — reed switches.

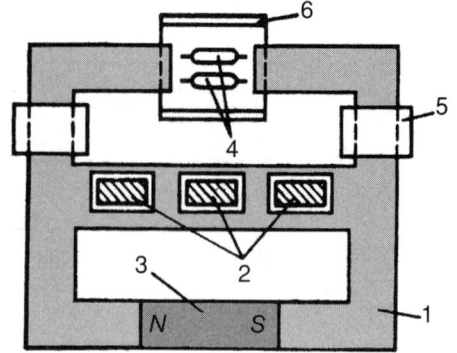

FIGURE 5.80
Three-phase winding-free reed AC relay, operating without vibration of the reed switch. 1 — magnetic core; 2 — current-carrying buses of the three-phase system; 3 — permanent magnet; 4 — reed switches; 5 — short-circuit windings; 6 — protective shield.

operated state for quite a long time. In these cases, certain trick techniques are used (Figure 5.80). In such a relay with no current (or of very small value) in buses (2) the ferromagnetic jumper strap is not saturated and the magnetic flux of the permanent magnet (3) is shunted by this strap, and does not affect the reed switches (4). As current in the lines increases, the strap becomes saturated. It stops shunting the magnetic flux of the permanent magnet (3) and the reed switches (4) are energized, under effect of the magnetic field of this permanent magnet. In the three-phase system of this relay, pulses of the resultant magnetic flux of the three phases affecting the reed switch are insignificant and, of course, do not cause any vibration of the reed switches.

The same solution can be applied in a single-phase relay. In order to smooth pulses of the magnetic flux one can use additional short-circuit windings (5). The patent No. 10003169 (U.S.S.R.) describes a single-phase winding-free reed AC relay where the alternating magnetic flux in the magnetic core, affecting the reed switch, seems to be "rectified" with the help of additional windings shunted by diodes (Figure 5.81).

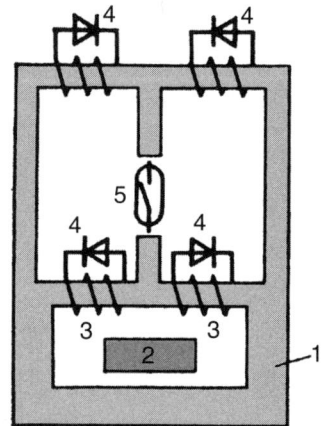

FIGURE 5.81
Winding-free reed relay with "rectifying" of the alternating magnetic flux. 1 — magnetic core; 2 — AC current-carrying bus; 3 — additional windings on the magnetic core; 4 — rectifier diodes; 5 — reed switch.

6

High-Voltage Relays

6.1 What is a "High-Voltage Relay?"

Rapid development of electric technologies applying high voltages (power lasers, industrial accelerators, high-frequency metal and dielectric heating, etc.), use of powerful electronic equipment operating under high voltages (radar, TV, and radio transmitters) and also the need for systems for testing insulation of electrical devices of different voltage levels, were the causes that bought about the spread of high-voltage (HV) relays, operating under voltages from 5 to 300 kV and higher. Such relays can be divided into two groups: relays with HV insulation for all current-carrying elements switching high voltages, and relays with low-voltage (LV) contacts and high voltage insulation between the input elements (control coil) and the output ones (contacts). The second group marks a new direction in construction of relays, founded by the author in the 1970–80's. In fact, the founding of new directions in engineering is quite a rare phenomenon. Tens of patents and articles in technical–scientific journals published in the Ukraine and Russia and translated in the U.S.A. provide additional evidence of the author's precedence in founding this new direction in relay construction. His latest research in this field is published in his book, *Protection Devices and Systems for High-Voltage Applications* (Gurevich, V. 2003). The relays of the first group are applied in equipment similar to the relays described above (only under higher voltages), while relays of the second group have a more specific field of application: these are insulating interfaces designed for transmission of control instructions, alarm signaling and protection (over-current) of components of equipment operating under high potential differences.

Relays switching high voltages can be divided into contact, solid-state (semiconductor), and cathode-ray ones. Contact HV relays may be open or sealed (gas-filled or vacuum), and also reed ones.

6.2 Open Relays for High-Voltage Switching

Open HV relays for working voltages up to 5 kV alternating current (AC) and direct current (DC) are quite simple and cheap devices. The main differences between them and LV relays are an increased gap between contacts and some additional plastic components increasing electric strength between elements with opposite potentials (Figure 6.1). The increased gap between contacts in such relays requires greater armature travel and

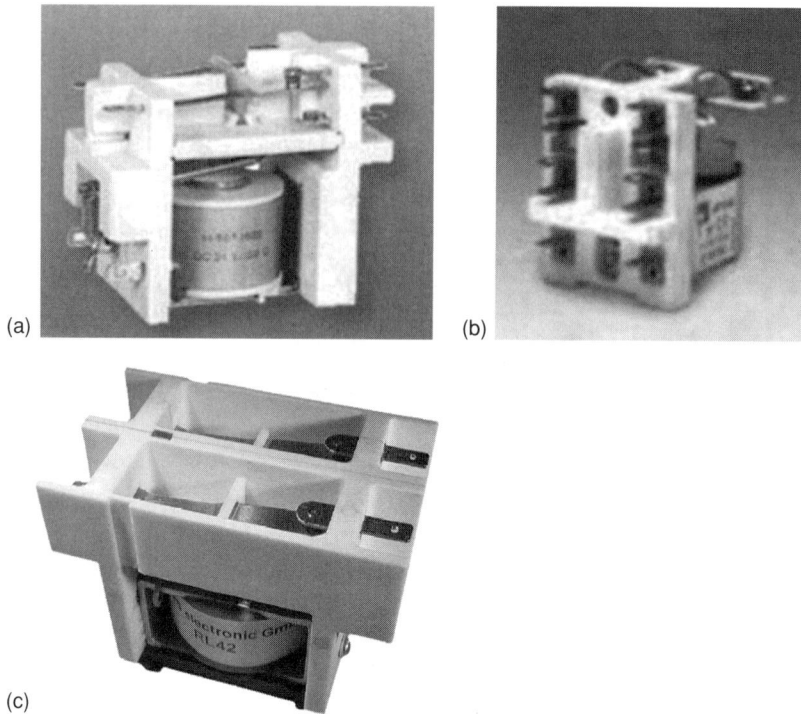

FIGURE 6.1
HV relays (with maximum switched voltage 5 kV) of the open type produced by different firms: (a) Hengsler-Ka Co; (b) Italiana Rele; (c) SPS Electronic GmbH.

therefore a bigger initial gap in the magnetic circuit of the relay. This leads to a considerable increase of power consumed by the coil of the relay, and overheating.

In the W158HVX relay produced by the German company Magnecraft, this problem is solved by introduction of an additional corner toggle between the armature and the movable contact, at small travel of the armature (Figure 6.2). The power consumed by the coil of this relay is only 5 W. Further growth of working voltage necessitates a sharp increase in the relay size, and manufacture of contacts in the form of cylinders with hemispherical ends, and additional insulating rods linking the armature with the movable contact (Figure 6.3).

The Ross Engineering Company produces a great variety of such relays for voltages of 12 to 300 kV. It is necessary to take into account that Ross Engineering has specified in their catalogs that their relays withstand peak values of test voltages. Here is Ross Engineering's explanation for this:

> The peak test rating of high voltage relays should be 1.2 to 5 times the normal high voltage circuit operating voltage, depending upon the application. For lower power systems, where transients are unlikely or intermittent flashover is of no consequence, a safety factor of 1.2 to 1.5 may be suitable. For medium power systems, or where moderate transients are likely, a safety factor of 1.5 to 3 is desirable and 2 to 3 recommended. For higher power systems, or where transient over-voltages are expected, a safety factor of 2.5 to 5 should be considered and the factor should be

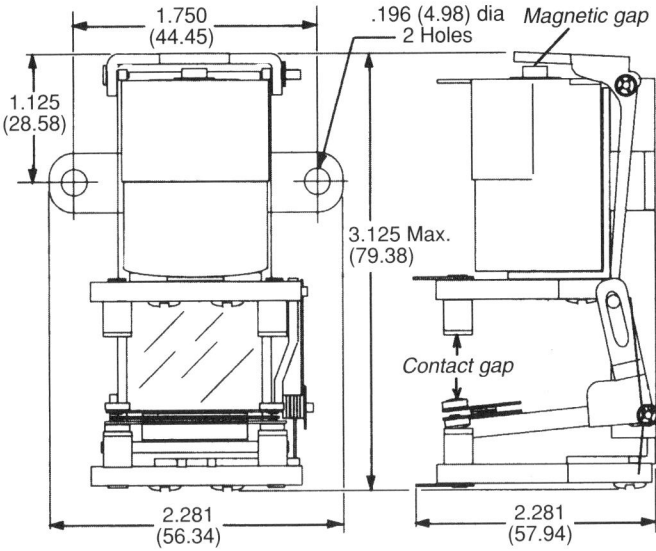

FIGURE 6.2
Open HV relay of W158HVX type with a large gap between contacts, and a small magnetic gap. Produced by Magnecraft.

based on the maximum probable transient. The peak test voltage to ground rating should be selected in the same way as is the contact to contact value.

From this explanation it becomes clear why it is impossible to specify the certain value of rated voltage that can be applied to all consumers. For one consumer the relay at 300 kV (catalog value) can be used in installations with a rated voltage of

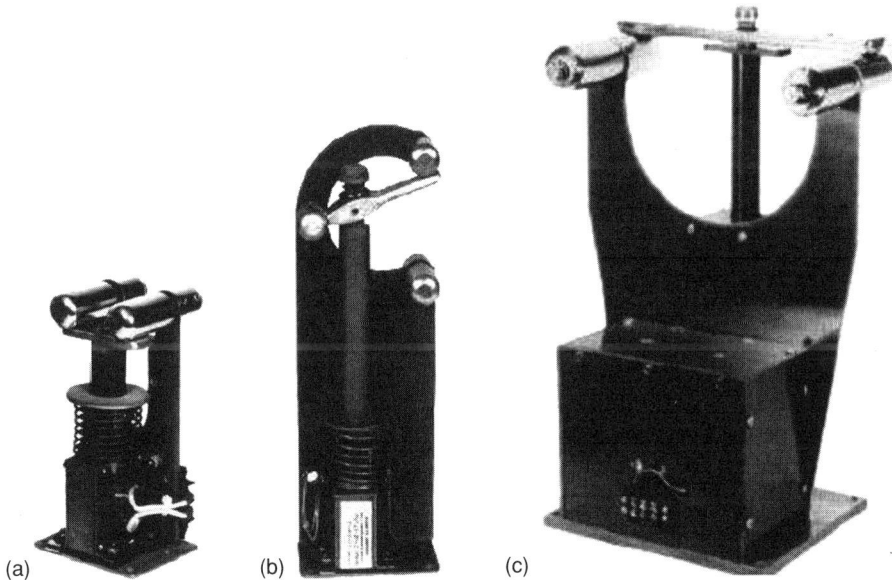

FIGURE 6.3
Open HV relays for voltages of 12, 60, and 300 kV (values indicated in the catalog) produced by Ross Engineering.

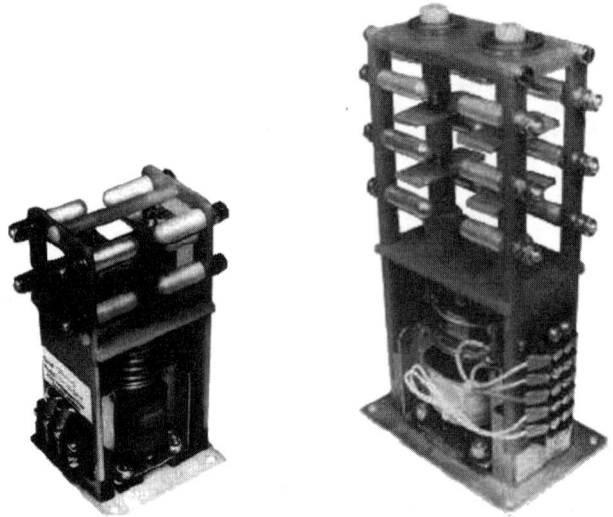

FIGURE 6.4
Multi-contact HV open relays produced by
Ross Engineering.

200 kV, whereas for others, the same exact relay can be used only upto 50 kV. This is one of problems that we already noted above when discussing the level of isolation of the relays intended for use in LV electronic equipment and in powerful systems of industrial automatics. In our opinion, the Ross Engineering approach to this problem is fair and correct. It is also necessary to take into account that the value is given for relays in air, and if they are put into a tank with oil or into a hermetic reservoir filled with SF_6 gas under certain pressures, the working voltage can be increased twofold.

The size of a relay with 300 kV voltage indicated in the catalog is $650 \times 914 \times 1725$ mm. The company also produces multi-contact relays operating according a similar principle (Figure 6.4). The Ross Engineering Company also produces HV relays with a pneumatic drive (Figure 6.5). Open relays produced by Ross Engineering are designed for shorting HV circuits for safety's sake, condenser discharge, etc., and are not designed for current breaks in the HV circuit. When charged, the HV capacitors are abridged, and the relay contacts are capable of withstanding closing pulse currents of up to tens of kilo-amperes (during the first 20 μs), and continuous currents in a closed position of up to 50–200 A.

FIGURE 6.5
HV open relay for voltages of 12 to 40 kV with a pneumatic drive, produced by Ross Engineering.

6.3 Vacuum and Gas-Filled High-Voltage Low Power Relays

Use of vacuum as a dielectric environment allows enhancement of the switching characteristics of a relay. Depending on the type of switching, low-power relays can be divided into two categories:

- Cold-switching of high voltages 12–70 kV
- Hot-switching with voltages up to 3–10 kV

There are two types of relays for hot-switching:

- Make-only relays, with pulse currents of up to a few kilo-amperes, with a duration of a fraction or a few milliseconds
- Power-switching relays, for currents of a few amperes

Industrial vacuum low-power relays (from the first category) were produced already in the 1950s by the General Electric Co. (Figure 6.6). Some of them had a small internal vacuum chamber with contacts and a flexible bellow (a crimped membrane providing moving of the movable contact by a certain value through a hermetic shell) and a winding placed outside the vacuum area (Figure 6.6a). Others had a bigger vacuum chamber containing all the elements of the relay including a winding (Figure 6.6b). Each of these constructions has advantages and disadvantages. The flexible bellow complicates a relay and makes it more expensive, and a coil placed in vacuum should have a ceramic bobbin

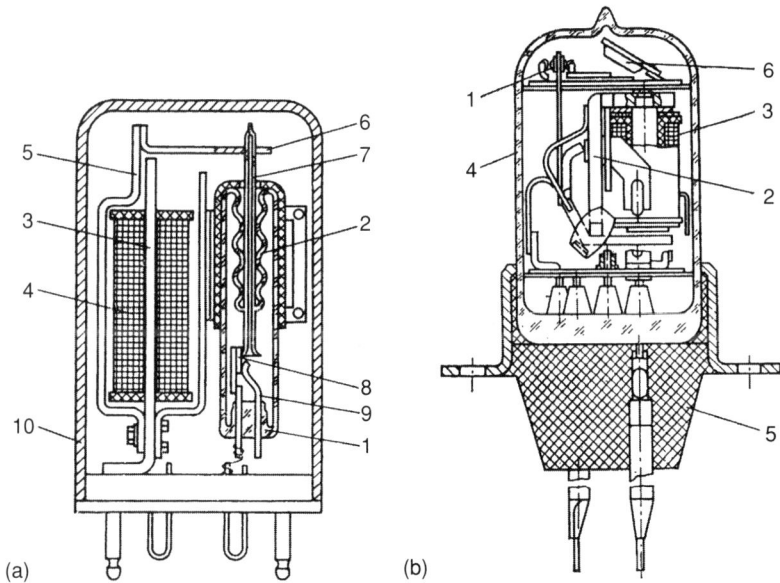

(a)   (b)

FIGURE 6.6
(a) HV vacuum relay with a flexible bellow and a coil placed outside the vacuum area. 1 — Glass shell of the vacuum chamber; 2 — flexible bellow (flexible membrane); 3 — core; 4 — winding; 5 — armature; 6 — pusher of the armature (steel); 7 — pusher of the contact system (glass); 8 and 9 — contacts; 10 — metal case. (b) HV vacuum relay with all elements placed in vacuum. 1 — Changeover contact; 2 — magnetic core with an attracted armature; 3 — coil; 4 — glass shell; 5 — heelpiece; 6 — baric getter.

and should be wound by a special wire in glass insulation, which does not evolve gas as it is being heated.

Vacuum is almost a perfect environment for relay contacts with very high dielectric strength (in relays vacuum with residual pressure of not more than 10^{-6} mm Hg is usually used. This provides a breaking-down voltage of up to 100 kV/mm), keeping the contacts absolutely clean, maintaining low resistance between the contacts, and also allowing use of magnetic systems with small armature travel and small weight, and therefore quick-operating.

The process of contact opening begins with a gradual reduction of contact pressure and an increase of resistance between the contacts, from a very small value up to infinity (when the contacts are opened). At this moment even at small currents the contacting points become intensely heated, up to melting temperature of metals, and a molten bridge appears on the separating contacts. As this bridge is broken, the arc begins to burn in the metal vapors of the contacts. On AC this arc goes out as soon as the current sinusoid passes through zero. There is no new arcing because the speed of restoration of electric strength in the vacuum is very high. The electric strength of the gap between the contacts is already restored entirely within 50 to 10 µs after current zero. A DC arc in vacuum does not decay of its own accord if the current value is enough for melting and evaporation of the contact material (for standard contact materials currents are a few amperes) and the source voltage is higher than arc-drop (about 20 V for tungsten contacts). It is interesting to note that even in very powerful vacuum contactors (which will be described below), switching OFF AC with amplitudes of tens of kilo-amperes are capable of switching OFF DC of only a few amperes. Here are specifications of the HB-204 vacuum contactor of Ross Engineering Corp., for example:

Voltage: 200 kV RMS (single phase operation)
Current: 50 to 1200 Amps continuous; 2,000 to 28,000 A AC, 1/2 cycle interrupt; 10 A
          DC interrupt; 5,000 to 80,000 A peak momentary.

That is why special schemes containing an LC oscillatory circuit forming a false zero current during switching are used to quench the arc.

Even microscopic doses of gases starting to evolve from metals and insulating materials at high vacuum are capable of impairing the dielectric properties of the vacuum, and can lead to malfunctioning of the relay. It is quite difficult to maintain a high vacuum during the service life in the complex construction consisting of tens of different elements with seals of glass and metal, which must remain absolutely hermetic for quite a long period of time and under considerable changes of temperature. In addition, any small-size construction containing two metallic electrodes in vacuum, between which a voltage of 10 to 20 kV is applied, can become a source of X-radiation, and HV vacuum relays are not an exception. As in a vacuum, the speed of restoring voltage on separating contacts is very high (10 to 20 kV/µs). Circumstances for arc decaying arise on AC of sinusoidal form before the current passes the zero value. A sharp break of current circuit occurs (so called "current chopping") if there is inductance in the load. Quite considerable spikes, capable of damaging the insulation of electric equipment, accompany such a sharp break of current.

Sulfur-hexafluoride (SF_6) or a mixture of it and helium filling the tube under pressure of few atmospheres, are another alternative to vacuum for use in HV relays. Such gas has an electric strength exceeding air strength by 2.5 times, and under increased pressure in a closed shell its insulating properties are almost similar to those of technical vacuum used in relays.

Compared with air sulfur-hexafluoride has a twofold energy-to-volume thermal capacity. That is why cooling capacity of this gas is noticeably higher than that of air, which is very important in small-size relays with heavily loaded current-carrying parts. Affected by high temperature of an arc, this gas decomposes and turns into a monatomic mixture of sulfur and fluorine. As soon as the arc decays, the mixture recombines, turning to the original gas again.

However, if there are admixtures of water and oxygen even in microscopic quantities, recombination is not full and properties of the switching device worsen sharply. In addition, remains of free sulfur in the closed shell have a negative impact on the contact surface as they form films that conduct current badly. As a result, transient contact resistance increases by a factor of 10^2. In real constructions of relays, it can reach 500 to 1500 mΩ, which restricts the wide use of sulfur-hexafluoride in relays. The peculiarity of this gas is that as the strength of the electric field between contacts of the relay increases (and this happens when the contacts approach each other during the process of closing), gas ionization occurs and becomes conductive, completing the circuit before the contacts of the relay close, and keeping the circuit closed as the contacts rebound when they collide. This prevents erosion of the contacts caused by vibration, especially when circuits with great pulse currents (caused by shorting of a charged HV capacitor) are switched. That is why in such cases it is almost always recommended to use a gas-filled relay; however, long ionization of this gas caused by the corona in nonhomogeneous electric fields, may lead to decomposition of the gas.

Decomposition products have strong toxic and corrosive properties, which is why relay construction must exclude the possibility of corona occurrence. International production of modern HV vacuum and gas-filled relays for switching voltages from 4 to 70 kV is almost entirely done by the American companies Kilovac, Jennings Technologies, and Gigavac, who produce relays with similar construction, similar external design, and similar characteristics (Figure 6.7).

The magnetic system of a HV relay in a glass case (Figure 6.7e), is a traditional one, with an attracted armature moving contacts with the help of an insulating rod. The upper end of the ferromagnetic core is hermetically embedded in a vacuum chamber with contacts. The rest of it is outside the vacuum chamber. A coil molded with plastic is installed onto this external part of the core. Such construction prevents contamination of the vacuum chamber by gases evolving from the materials of the coil, and in addition one can always replace the coil with another one, with the required characteristics.

In a relay of the diaphragm type (Figure 6.7g), the coil is inside the case and cannot be replaced, but it is also separated from the hermetic contact area with the help of a flexible diaphragm. The magnetic system is of the same type, with an attracted armature. Transmission of the effort on the movable contact, as in the previous case, is carried out with the help of an insulating rod.

In Russia, relays of P1D, V1V, V2V-1V, and other types, which are based on similar principles and have similar fields of application, have been constructed and produced by the Penza Scientific Research Institute of Electromechanical Devices for a long time already (Figure 6.8). It is worth mentioning that only a few types of HV vacuum relays produced by Jennings are designed for operation in hot switching modes. Such relays are capable of switching currents not exceeding 3 A, with voltage up to 2.5 kV. The firm Kilovac indicates in its catalog the possibility of using some of these types of relays, "for power switching low current loads," without mentioning particular values of switched currents and voltages. Each specification of switching devices as a rule contains so called "life curves," illustrating dependence of a number of operating cycles on switched power for several voltage levels. Such curves are never given for HV vacuum relays by firms producing them, which is also evident that the main purpose of such relays is not current switching.

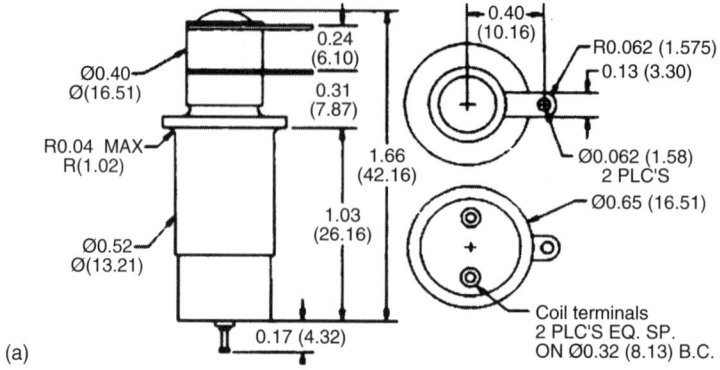

(a)

(b)

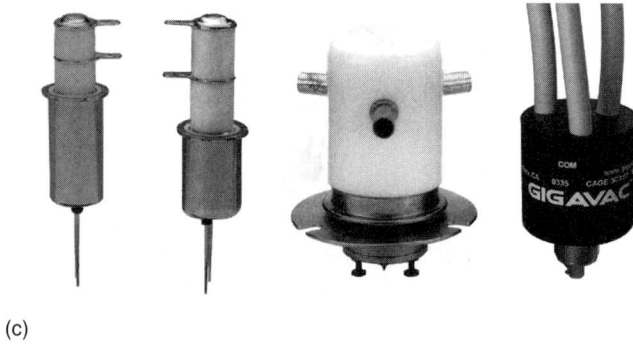

(c)

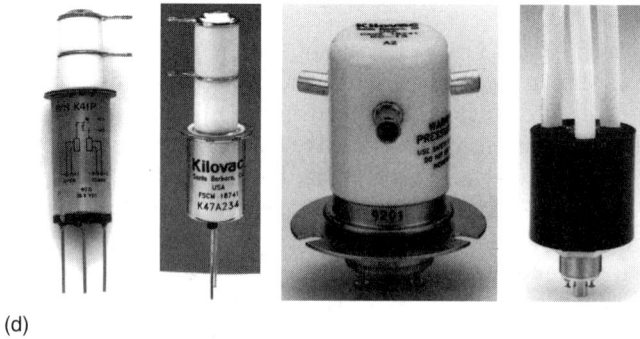

(d)

(e)

Glass envelope
Terminal lug
Armature assy
Hinge pin
Spring
Coil
Molded bobbin
Terminal plate
Coil terminal

(f)

(g)

Normally open
fixed contact

Movable contact
Diaphragm
Insulator

Normally open contact
Normally closed
fixed contact
Normally closed contact
Common contact
Ceramic insulators
Return spring
Armature
Coil assembly

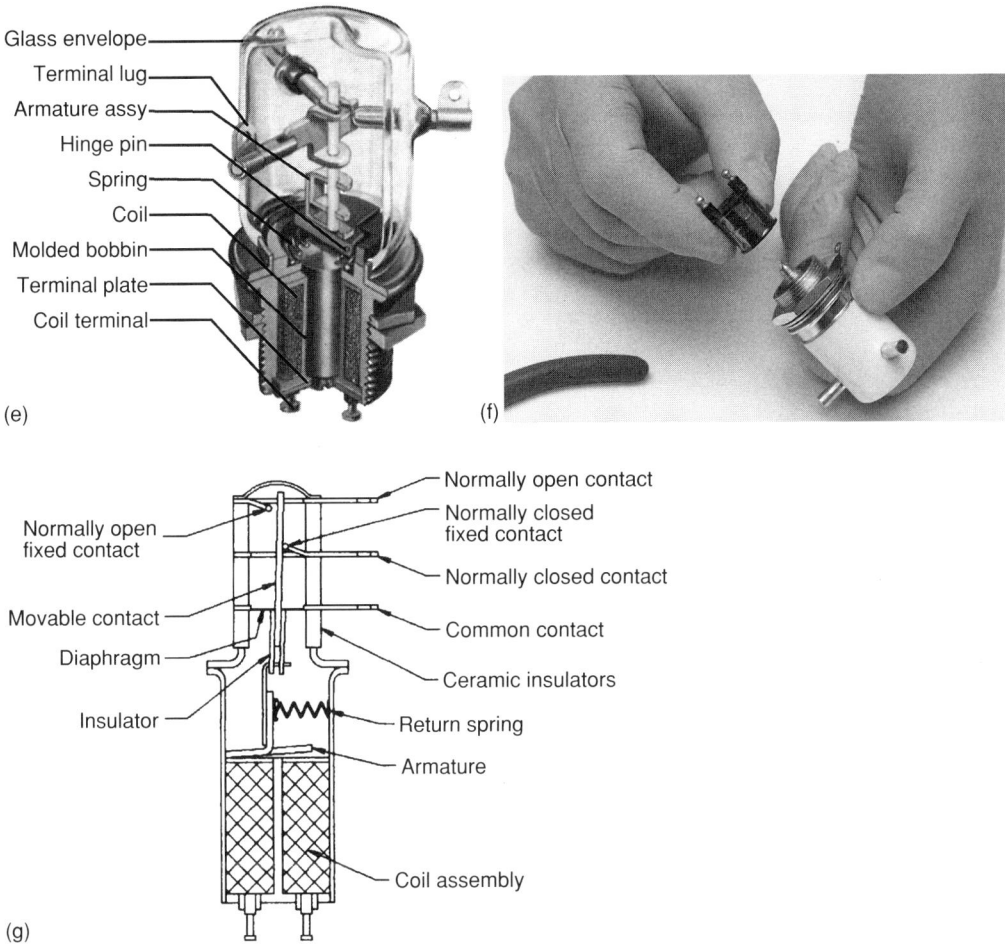

FIGURE 6.7 (*Continued*)
(a) Sizes of the HV vacuum relay of RF41–26S type (4 kV, 12 A) produced by Kilovac. (b) HV relay produced by Jennings (U.S.A.). (c) HV relay in metal–ceramic cases produced by GIGAVAC (Santa Barbara, CA). (d) HV relays in metal–ceramic cases produced by Kilovac (Santa Barbara, CA). (e) Construction of a HV relay in a glass case produced by Kilovac. (f) Process of assembly of the HV relay on the Gigavac company. The coil mounting in the relay after hermetic sealing and vacuuming of a shell with contacts. (g) Construction of a HV relay of the diaphragm type, in a metal–ceramic case.

The only difference in the construction of relays operating in hot switching modes is tungsten contacts, due to the fact that contact resistance of such relays is three times as much as that of relays with standard contacts. Arcing in the vacuum proceeds as long as the energy emerging on the contacts is enough to maintain the concentration of metal vapors and arcing. If tungsten, which does not vaporize intensely under the effects of electric arcing (the boiling-point is about 6000°C), is used, such relays can switch quite small direct currents (a few amperes) with voltages of only a few kilovolts. As vacuum is a very poor heat conductor, heat abstraction from contacts in vacuum relays is difficult, which is why in some types of small-size or miniature relays, as continuous currents of over 10 to 15 A pass through the contacts, radiators are applied to prevent overheating of such small-size relays.

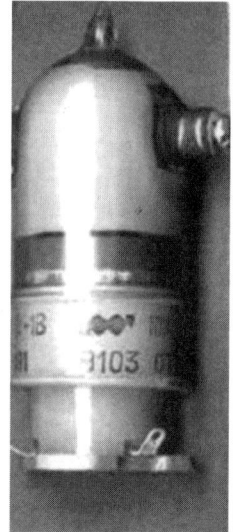

FIGURE 6.8
Russian vacuum relay of V2V-1V type (15 A cold-switching, 4 kV).

6.4   High Power Vacuum Relays and Contactors

High power relays and contactors designed for switching currents of hundreds of amperes are also produced with vacuum insulation. According to data we have, the pioneer in this field was the firm "Motor and Control Gear Division," which declared creation of the first high power vacuum contactor for voltage of 3.3 kV in 1965. Each contact in such devices was made as a separate item: the so-called vacuum interrupter (Figure 6.9). A solid case, from vacuum-tight ceramics and metal flanges, provides maintenance of pressure inside the chamber on the level of 10^{-5} Pa during the whole service life. In their initial state the contacts of the vacuum chamber, affected by the pressure difference inside and outside the chamber, are always closed. To open the contacts it is necessary to tug the external outlet of the current-carrying core 2. Freedom of movement of this core in the closed volume is provided by metal flexible membrane 5 (the so-called flexible bellow). The metal shield (3) protects the internal surface of the chamber from small parts of molten metal, from the electrodes of the chamber formed under arc effect when strong currents are switched off, and also equalizes the electric field strength in the contact area. Before a vacuum chamber is assembled, its elements are heated for a few hours at temperatures of over 400 °C, for degassing.

Contacts of the vacuum chamber are shaped in such a way that working current forms a magnetic field, forcing out the arc. A number of projects and patents are known on this subject. Contacts with spiral leaves are widely used (Figure 6.10). Contacts with spiral leaves have the form of disks, with peripheral parts cut by a spiral slot into sections which are joined in the central part. With strong current, when such contacts open under electro-dynamic forces, the arc moves to the periphery of the disks in the direction of the bend of the spiral notches, and then starts to rotate on the surface of the electrodes. This prevents overheating and intensive melting of contacts in some places. As the contacts of the vacuum chamber (usually copper ones) are closed for a long period of time and are pressed to each other with a considerable effort, and also have clean unoxidized surfaces, they may be prone to the so-called *cold welding*, caused by inter-diffusion of atoms of the metal of the contacting surfaces. Welding of contacts can also be caused by spark breakdown when contacts approach each other while closing. Such problems are solved

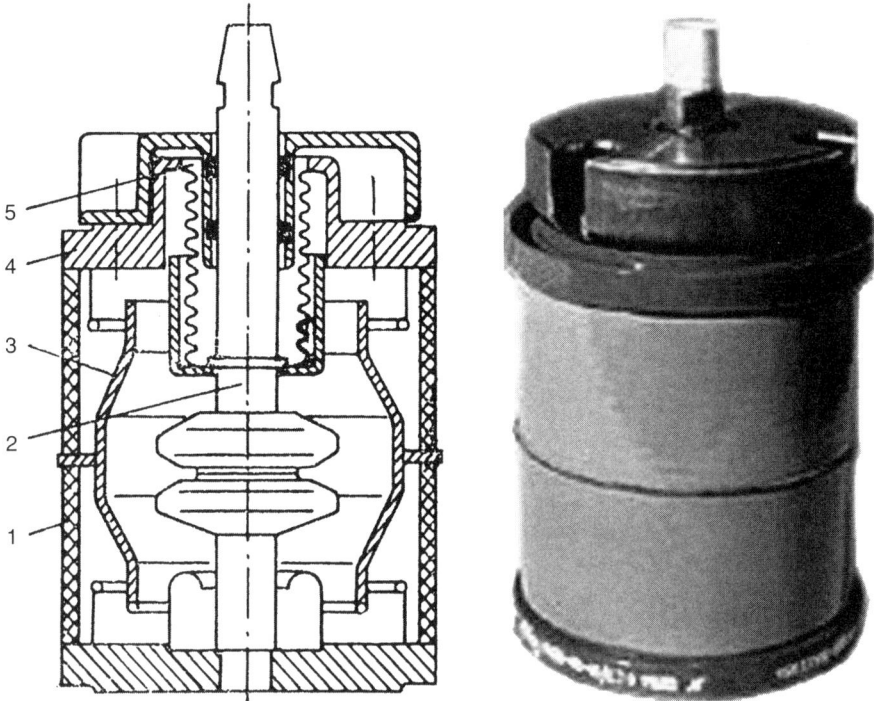

FIGURE 6.9
Construction and external design of vacuum interrupter. 1 — Ceramic chamber; 2 — current-carrying core with a movable contact at the end; 3 — metal shield; 4 — flange; 5 — metal bellows.

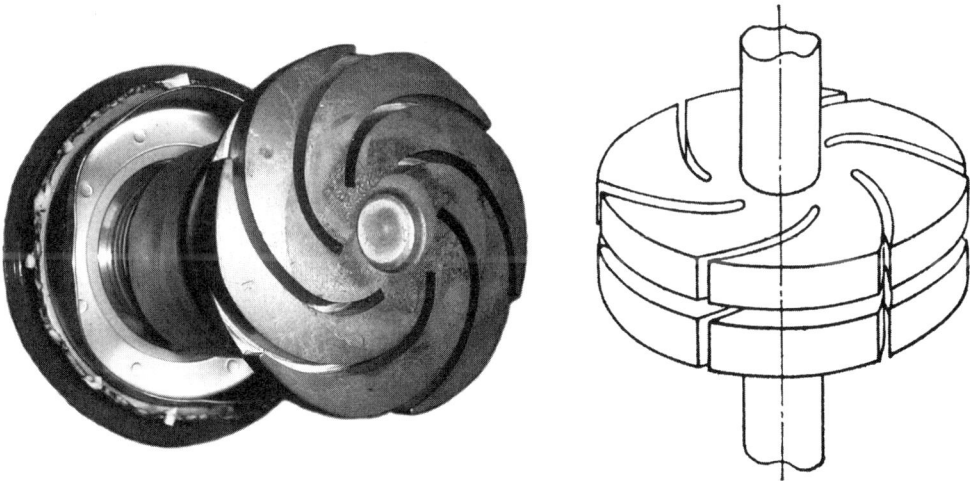

FIGURE 6.10
Contacts of a vacuum chamber with arc-suppressing leaves.

by using small amounts of bismuth, chrome, or beryllium. Such admixtures allow the relay to maintain arcing of alternating current in the vacuum almost until the sinusoids pass through the zero value, thus preventing current-chopping (that is arc extinction before crossing the zero value, caused by overheating).

In addition, the vacuum chamber contactor also contains a powerful electromagnet joined through a mechanical system, with part of the movable contact sticking out (Figure 6.11). Three-phase constructions contain three vacuum chambers installed on the same plate and equipped with three electromagnet coils, connected in parallel. HV contactors are equipped with insulating elements and current-carrying buses placed at certain distances. First constructions were quite big (Figure 6.12), but modern HV contactors are more compact (Figure 6.13). To increase switched voltage, the vacuum chambers are joined in series, and equipped with a common operating mechanism.

A typical example of such a construction is a HV single-pole contactor produced by Ross Engineering (Figure 6.14). It consists of four vacuum chambers connected in series. Toroidal shields are used for equalization of electric field strength in the construction, and for prevention from corona. Ross Engineering indicates a withstanding test voltage of AC (200 kV of peak) for this contactor. The consumer can choose maximum value of switched voltage on grounds of particular technical requirements and necessary voltage reserve.

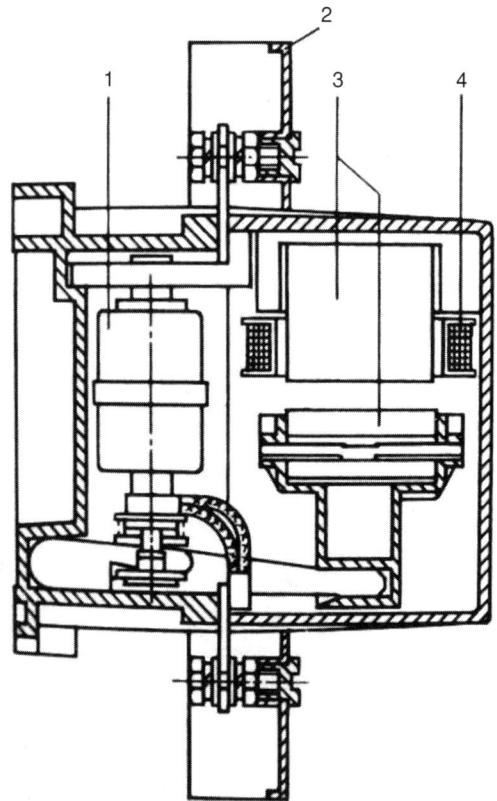

FIGURE 6.11
Construction of power single-pole vacuum contactor of 3TF68AC type for current of 630 A and voltage of 690 V, produced by Siemens. 1 — Vacuum chamber; 2 — terminal cover; 3 — ferro-magnetic core; 4 — coil.

FIGURE 6.12
One of earlier constructions of HV vacuum contactors for 10 kV voltage. VC — vacuum chambers.

(a)

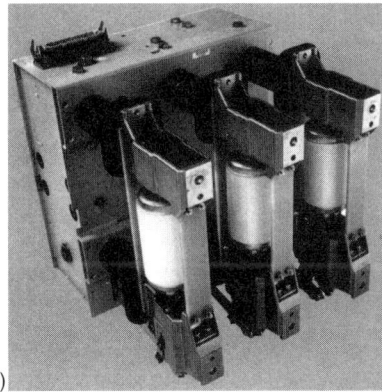

(b)

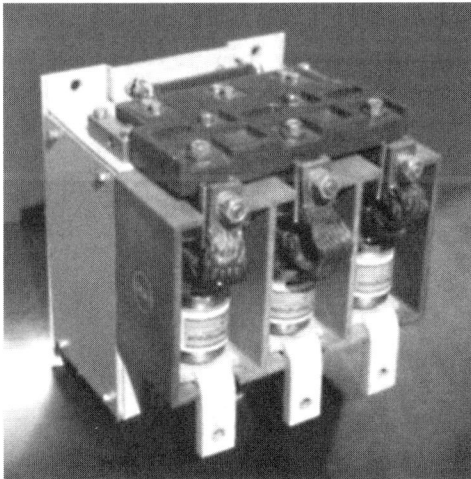

(c)

FIGURE 6.13
Modern HV three-phase vacuum contactors. (a) VD4 type, 12 kV, 1250 A (ABB); (b) CV-6 KA type, 7.2 kV, 720A (Toshiba); (c) CCV-JT type, 1.14 kV, 400 A (ICPE S.A., Romania).

FIGURE 6.14
HV single-pole contactor of HB-204 type (test voltage is 200 kV of peak), based on four vacuum chambers connected in series. Produced by Ross Engineering.

6.5 High-Voltage Reed Relays

HV reed relays differ from LV relays by the use of HV vacuum reed switches and increased insulation of the control coil from the reed switch. HV switched vacuum reeds (considered above) are standard items produced in mass lines by different firms, and are designed for switching of small currents with voltage of 5 to 10 kV DC (switched power is up to 50 W). Maximum switched current (with similar switched power, of course) can reach 3 A. The insulation of the coil from the reed switch is made for the same level of voltage as the insulation of the reed switch, which is for 5 to 10 kV of working voltage DC.

HV reed relays have some particular construction forms:

- *An insulated coil and an open reed switch* (a small coil entirely molded with plastic and placed in the central part of the reed switch; the HV outlets of the reed switch are open and distant from the coil; the reed switch can be easily put in and taken out of the coil — Figure 6.15).

- *An insulated reed switch and an open coil* (the open coil is wound on an insulating bobbin supplied with "wings" in the form of tubes, entirely covering the reed switch with its outlets — Figure 6.16);

- A reed switch with a coil entirely molded with plastic in the form of a solid construction (Figure 6.17).

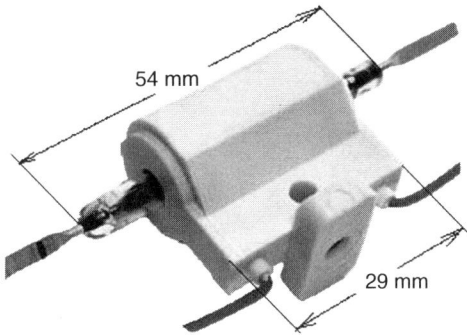

FIGURE 6.15
HV reed relay with an insulated coil and an open reed switch.

A switching device from the "Goliath" series, designed by the author in 1991 (Israeli Patent 130440), can also be referred to as a HV reed relay (Figure 6.18). Such a relay is constructed on the basis of one of the largest reed switches in the world, where transformer oil or sulfur-hexafluoride is used as a dielectric environment. A movable contact-element of this reed switch is made in the form of a turned T-shaped element, with a massive bridge contact and cylindrical ferromagnetic core fixed in the center of the bridge. This contact-element can freely move inside the dielectric shell within 60 mm distance on special guides along the standing axis of the device. Fixation of the movable contact-element in extreme positions is carried out with the help of permanent magnets (Figure 6.19).

"Goliath" is a new type of high performance latching commutation device with the following unique characteristics: high voltage, low cost and relative small size. The "Goliath" design is based on reed switch technology and includes a minimal number of components with no need for vacuum technology, while assuring reliable operation at low cost. "Goliath" can be switched ON or OFF (fixed position) by control signals and consumes energy only during transition time. The actual position of the apparatus may be indicated by a light-emitting element (LED) on the operator's control console.

The "Goliath" consists of a dielectric body (1) formed with two isolated compartments: a large compartment for the contact system, a small compartment for magnetic systems and one more insulated open compartment which is concentric with the small compartment. In the large compartment are two fixed contacts, symmetrically positioned (2), a movable bridge-type contact (3), a fixed magnet (9) and a movable magnet (8), fixed in a housing that is connected to the central part of the bridge-type contact (3). This housing enters special guides (11) that limit the degree of freedom of the housing. From the other side of the central area of the bridge-type contact (3), a dielectric rod is attached (having a certain clearance) with a ferromagnetic core (6) fixed to its other end.

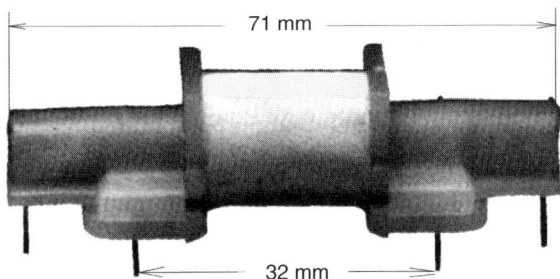

FIGURE 6.16
HV reed relay with an insulated reed switch and an open coil.

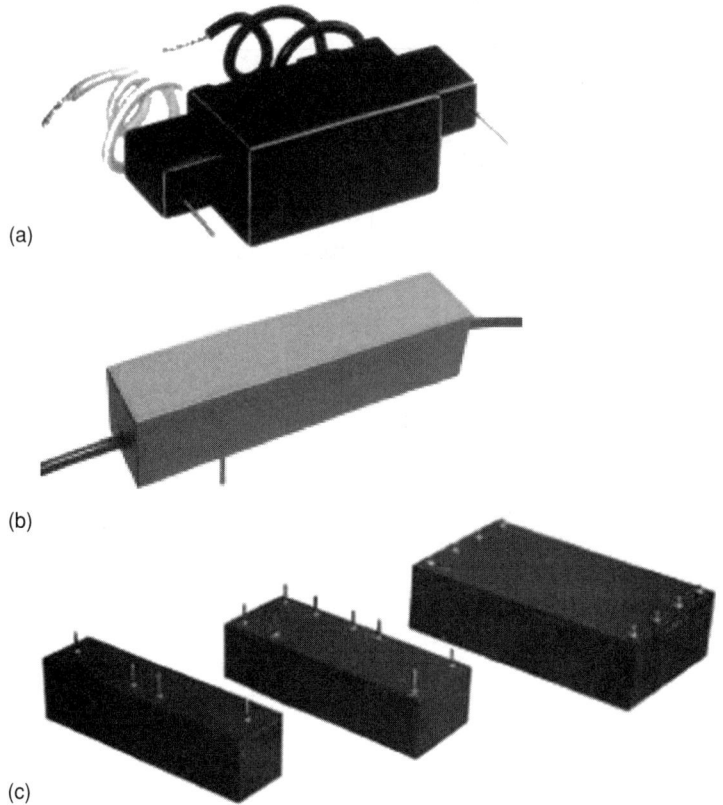

FIGURE 6.17
HV reed relays in the form of a solid construction, with a reed switch and a coil which are entirely molded with plastic. (a) — Relay with reed switch outlets for printed circuit and coil outlets — flexible wire; (b) relay with coil outlets for printed circuit and reed switch outlets from HV wire; (c) relay with reed switch and coil outlets for mounting on printed circuit board.

The permanent magnet (7) with steel bushings is located in the small compartment, and is free to move in the upper part of the compartment along its longitudinal axis by 1 to 1.5 cm. Two control coils (5 — the lower) and (4 — the upper), mounted in the small compartment and provided with magnetic circuits, are located in an open concentric cavity. After the control coils are mounted in the small compartment, it is filled with an

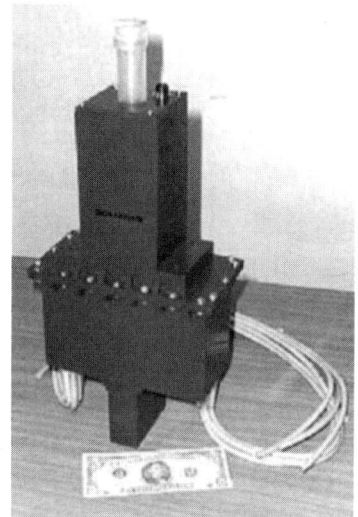

FIGURE 6.18
The biggest HV reed relay in the world, "Goliath," designed by the author.

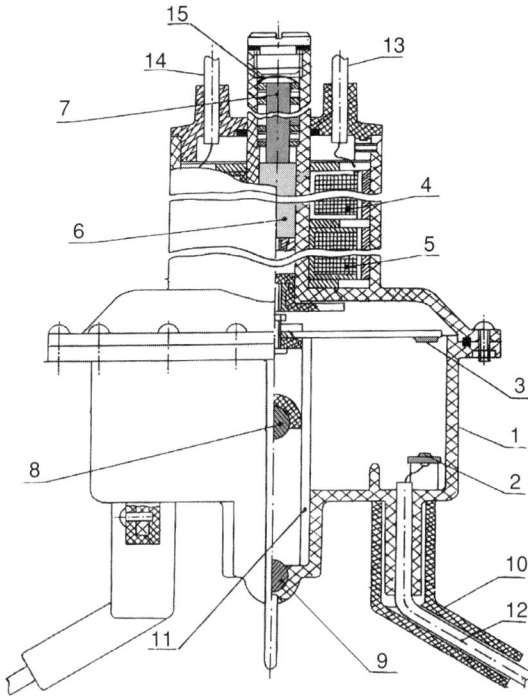

FIGURE 6.19
Construction of a HV reed relay of the "Goliath" type.

epoxy compound and covered with a cover having bushings with outputs (13, 14) formed as wires with HV insulation drawn through them. The ferromagnetic core (6) is 20 to 30% longer than the control coils. The outputs of the fixed contacts are formed as high voltage wires (12) drawn through the bushings (10).

With the external power supply cut OFF, the relay can be in one of its extreme positions: (a) either engaged, when core 6 with contacts 3 are in their lower position and magnet 8, being attracted to magnet 9, fixes this position and generates the required contact pressure, or (b) disengaged when core 6, with contacts 3 and magnet 8, are in their upper position, which is fixed owing to core 6 attraction to permanent magnet 7.

When lower coil 5 is connected to a DC power supply (rectifier), the magnetic field generated by this coil detaches core 6 from permanent magnet 7 and imparts a motion pulse to the core. As a result, the core quickly moves downward, carrying away contacts 3 and magnet 8 coupled with it. As the fixed contacts 2 are attained and the velocity is reduced by the elasticity of the spring located at the center of the bridge (that interconnects contacts 3, core 6, and all), the mobile elements connected to it are stopped in the lower position so that a few millimeters gap is left between magnets 8 and 9. Mutual attraction between magnets 8 and 9 prevents the springing back of contacts 3 from contacts 2 at their initial touch, providing the required contact pressure and fixation of the movable device elements in the lower position.

When upper coil 4 is connected to a DC power supply, the magnetic field generated in this coil acts on core 6 and imparts a motion pulse, as a result of which magnet 8 is detached from magnet 9 and, together with contacts 3, quickly moves upward until core 6 reaches the permanent magnet 7. Since the magnet is not fixed permanently and can move along its axis, inelastic impact of core 6 on magnet 7 is prevented, so that after they come into touch, their joint movement is preceded and damped by spring elasticity (15).

Here is "Goliath" main parameters:

Maximum switched voltage (kV) AC (rms)	60
Dielectric strength (kV) AC (rms), 1 min	120
Continuous current through closed contacts (A) (rms)	100
Current spike (for 20 ms duration) (A)	1500
Control voltage (V) DC	12, 24, 110
Control power (W)	5
Minimum control pulse width (duration) (ms)	300
Operating time (ms)	50
Temperature range (°C)	$-10 + 55$
Dimensions (mm)	$235 \times 95 \times 435$
Weight (kg)	5.2

6.6   High-Voltage Interface Relays

HV equipment (10 to 100 kV) has become very popular over the last few years. It is utilized in military and civil radar stations, powerful signal transmitters for communication, broadcasting and TV systems, technological lasers, X-ray devices, powerful electronic and ion devices, devices for inductive heating and melting of metals, technological electron accelerators for material irradiation, electro-physical and medical equipment, and industrial microwave ovens, among others.

Technical difficulties caused by the presence of functional components isolated from each other, not permitting direct connection owing to a high difference of potentials, are encountered when designing systems for control and protection against emergency conditions (over-current, sparks) in modern power HV equipment. To guarantee information and electrical compatibility, as well as to implement the required algorithms for interaction of functional components of equipment, special control instruments are required that have been called "interface relays," or "insulating interfaces" (in technical literature).

Apart from problems connected with transmission of commands between parts with opposite potentials of HV equipment, there are also problems of current overload protection (level current trip) of such devices, caused by HV circuit insulation breakdowns or breakdowns in the high voltage devices. These problems still remain acute. The first is related to unfavorable conditions that cause moisture and dust to penetrate the equipment, and the second to unpredictable internal breakdowns in high voltage vacuum electronic elements (klystrons, tetrodes, etc.) or semiconductor elements (HV rectifier).

Current overload protection in such devices is usually resolved by inclusion of current sensors and electronic relays into the LV or grounded circuits. However, such protection is not necessarily efficient and in itself can cause many problems, which is why high-performance systems of protection of HV equipment from current overloading are based on interface relays.

The general principle of design of interface relays is the presence of a special galvanic decoupling unit between the receiving and final controlling systems of the relay. Interface relays with a working voltage of more than 10 kV have the greatest interest for these areas of engineering, to which the present study is devoted. In the design of devices classified as interface relays, some of the widely used physical principles may not be used in electrical relays of other types.

It is well known that any electromagnetic relay has a specific level of isolation of the output circuits from the input circuits, that is, it functions secondarily as an interface relay. However, in ordinary relays, this function is not decisive and is not at all considered in the existing system of classification. In the interface relay, the property of galvanic decoupling of the circuits has been repeatedly intensified, and the parameters of the galvanic decoupling unit are decisive from the standpoint of the function performed by this relay. On the other hand, the parameters associated with switching capacity are secondary and, significantly, there can be interface relays with the same level of galvanic decoupling. In this regard, an artificial assignment of interface relays to existing classes does not seem to be expedient. It seems more appropriate, rather, to classify them as a separate type of electrical equipment, having an intrinsic structure based mostly on a classification according to the characteristics of the galvanic decoupling unit. For example, according to the decoupling voltage level:

- Low level (to 10 kV)
- Medium level (10 to 100 kV)
- High level (above 100 kV)

According to principle of action:

- Opto-electronic
- Pneumatic
- Radio-frequency
- Electrohydraulic
- Transformer
- Ultrasonic
- Electromagnetic, and with mechanical transmission

According to speed:

- Super fast (up to 100 μs)
- Fast (100 μs to 2 ms)
- Inertial (above 2 ms)

Although such classifications may seem arbitrary, they fully reflect the most important properties of interface relays that have a decisive effect on the functions performed by them.

The simplest interface relays of the optoelectronic type typically consist of an LED built into the semiconductor structure (power SCR, triac) or LED and matching low power photothyristor or phototransistor in a switching mode, mounted close together and optically coupled within a light-excluding package having a galvanic decoupling voltage up to 4 kV.

Some companies (see above) produce high voltage reed relays for commutation voltage of up to 10 to 12 kV DC, and therefore have a galvanic decoupling voltage on the same level. All of these relays are intended for use only in DC circuits under normal climatic conditions and have no reserves for withstanding voltage required for high-power equipment.

In order to significantly increase the galvanic decoupling level of interface relays of the optoelectronic type, a fiber optic cable of appropriate length is installed between the LED and photo-receiving elements. These relays are also equipped with an electronic pulse shaper and an electronic amplifier. At a length of 5 to 20 mm of fiber optic channel connecting the transmitting and receiving units, the galvanic decoupling voltage ensured by the interface relay can reach 15 to 50 kV DC (for low power electronic equipment only!) (Figure 6.20a).

The input of the OPI1268 (Figure 6.20a) device consists of a high efficiency GaAlAs (Gallium Aluminum Arsenide) LED; a photodiode in the output integral circuitry (IC) detects incoming modulated light and converts it to a proportionate current. The current is then fed into a linear amplifier that is temperature, current, and voltage-compensated. The OPI125, OPI126, OPI127, and OPI128 (Figure 6.20b) each contain gallium arsenide infrared emitting diode coupled to a monolithic integrated circuit which incorporates a photodiode, an optic insulator, a linear amplifier, and a Schmitt trigger on a single silicon chip. The devices feature TTL compatible logic level output.

It is necessary to note, that it is as much as possible allowable levels of a voltage which should not be exceeded in any modes. In practice, certainly, working voltage should be chosen in 1.5 to 2 times less for the electronic equipment. For the industrial and power equipment this voltage should be even less. Additional essential reduction in a working voltage is required at work on an alternating current. In result, we shall have working voltage about 2.5 to 3.5 kV instead original 15 to 16 kV.

Interface relays of the optoelectronic type have also found application in electrical power configurations in which the transmitting and receiving units are connected by hollow porcelain insulators of fairly large dimensions, equipped with a built-in optical system. Such interfaces are used in 110 to 400 kV power networks to control the drives of HV circuit breakers as a device for protecting shunt capacitor batteries, etc. (Figure 6.20).

The developmental trends of interface relay technology suggest the use of optoelectronic systems as the prevailing design principle for galvanic decoupling units. It is agreed that the most important characteristic feature of optoelectronic systems is their noise immunity and insensitivity to electromagnetic fields; however, what is not considered here is that in addition to the fiber optic line itself and the output actuator, such a system includes a shaper of light pulses on the transmitting and electronic amplifier, with triggering units on the receiving end that are generally based on IC. It is precisely these elements, which have low activation levels, that are damaged by pulse noise on the side of high voltage power equipment (interferences, spikes, and high voltage discharges), which negates the main advantage of optoelectronic systems. Moreover, the optical fibers themselves are subject to a severe negative effect of ionizing radiation and external mechanical influence (very important for military applications). The arrangement of input and output circuits of such systems should be widely spaced (optical fiber length is 0.5 to 1 m for voltage 40 to 150 kV for power equipment), and it is this factor that determines the overall dimensions of the interface unit.

All of this indicates that the preferred use of an optoelectronic galvanic decoupling unit in interface relays is not always warranted, and it sometimes is merely the consequence of stereotypical thinking of developers, or a peculiar technical style. A new type of HV interface relays, based on the reed switch, were proposed by the author for the first time in 1977 (U.S.S.R. Patent 758462). Analysis of the characteristics of this type of reed switch-based HV interface relays (Relays of Gurevich — "RG-relays") developed by the author, as well as experience in creating and using them (see Gurevich V. *Protection Devices and Systems for HV Applications.* Marcel Dekker, New York, 2003) shows that they have a definite area of use within which they enjoy distinct advantages over other types of interface relays. These parameters include transmission of discrete control commands,

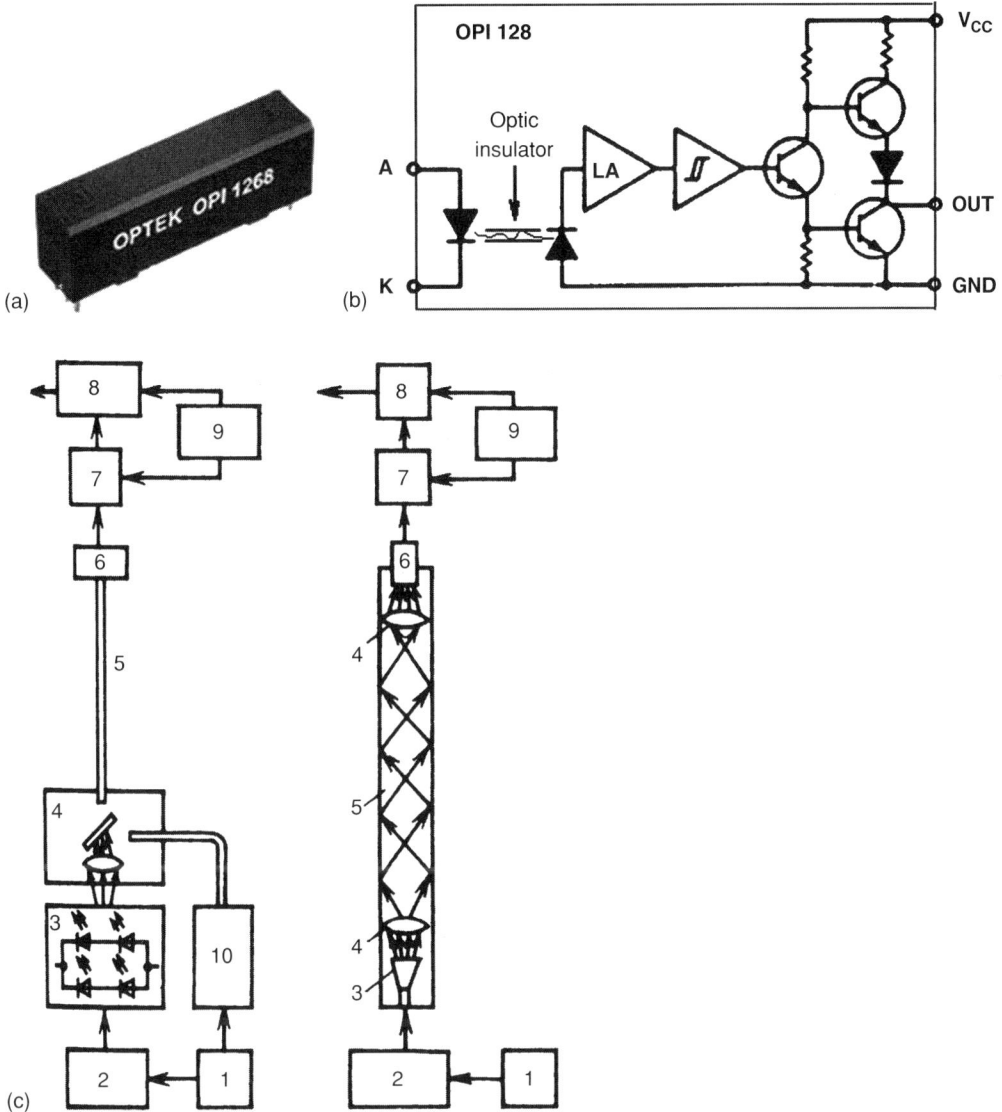

FIGURE 6.20

(a) Optical-insulating interface OPI 1268 type for maximum voltage input—output 16 kV DC. Dimensions: 9.0 × 6.5 × 28.0 mm (Optek Technology). (b) Function diagram of optical-isolated interface OPI 128 type (Optek Technology) for voltage level 15 kV DC (3750 V AC). (c) Interface relays of the optoelectronic type based on a fiber optic cable (on the left) and a hollow porcelain insulator (on the right). 1 — Power source on ground potential; 2 — electronic pulse shaper; 3 — optical emitter (transmitter); 4 — optical system; 5 — optical channel; 6 — phototransistor; 7 — electronic amplifier on the high potential; 8 — output final control element; 9 — power source on the HV potential; 10 — transmitter monitoring unit.

protection, and binary warning transferred by a frequency of up to 50 to 100 Hz, and an admissible speed of 0.8 to 1.5 ms, between parts of equipment under a potential difference of up to 100 kV. Within these parameter values, RG-relays are characterized by the highest degree of simplicity and reliability, and possess broad functional capabilities. Particularly attractive are such interface relay properties as a large overload capacity of the control circuit, a large power output circuit, insensibility to pulse noise, mechanical strength of

FIGURE 6.21
HV interface RG-series relays for industrial and
military applications.

the design, and preservation of serviceability over a wide range of temperatures, pressure
and humidity, suitable for military standard MIL-ST-202 requirements.

The relatively low cost of interface relays is also of no small importance in a number of
cases. These properties of RG-relays are responsible for their widespread use for industrial and military applications in on-board, mobile and stationary powerful radio-electronic equipment, in relay protection and automation systems of electrical networks of the
6 to 24 kV, in electro-physical installation, in power converter technology, etc.

RG interface relays are a new type of HV device designed for automation systems for
overload protection, fault indicating, interlocking of HV equipment, as well as for transfer
of control signals from ground potential to HV potential (reverse connection). The series
consists of the following devices: RG-15, RG-25, RG-50, RG-75, which are designed to
operate under voltages of 15, 25, 50, and 75 kV DC, respectively (refer to Figure 6.21 and
Table 6.1).

The operation of these devices is based on separation of the electric and magnetic
electromagnetic field components. Each device is based on a magnetic field source
(coil), connected to a high potential current circuit, a reed switch and a layer of high
voltage insulation, transparent for the magnetic component of the field, and completely
insulating the reed switch from the electric field component (Figure 6.22).

TABLE 6.1

Main Parameters of the RG Devices

RG-Relay Type	RG-15	RG-25	RG-50	RG-75
Nominal voltage (kV) DC	15	25	50	75
Test DC voltage 1 min (kV)	25	35	70	90
Control signal power (W)	0.2…0.4	0.2…0.5	0.5	0.9
Maximal switching voltage in the output circuit (V)				
DC	600			
AC	400			
Maximal switching output circuit current (A)	0.5			
Maximal response frequency (Hz)	100			
Maximal response time (ms)	0.5,…,0.8			
Maximal dimensions (mm)	Ø26 × 47	56 × 27 × 70	Ø75 × 150	Ø75 × 190
Weight (g)	45	130	370	620

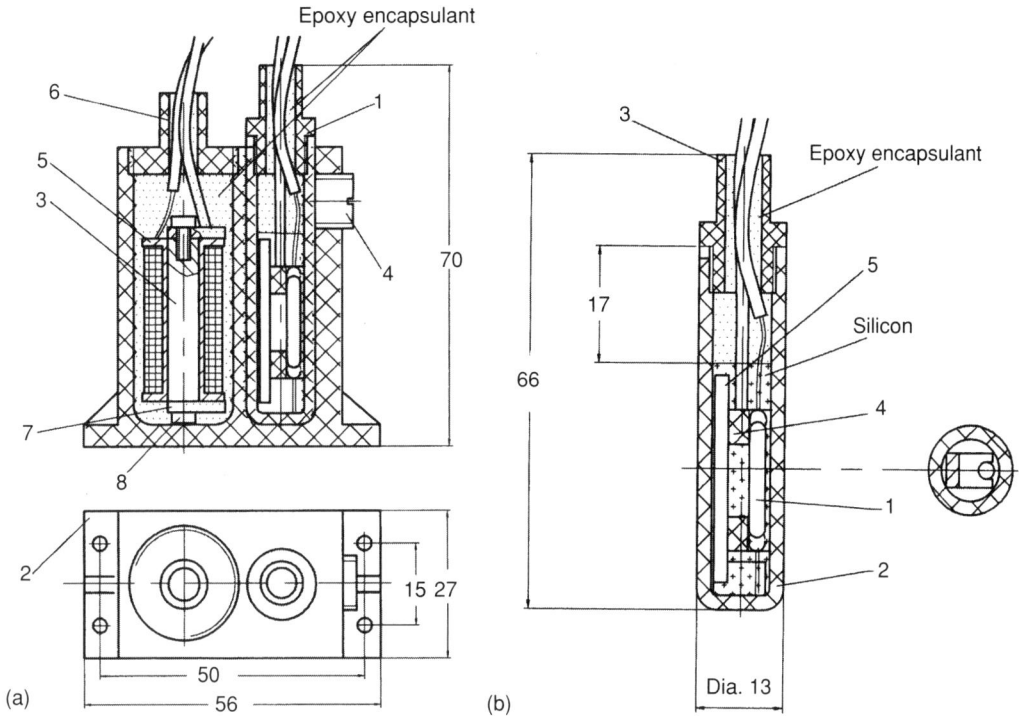

FIGURE 6.22

(a) Insulation interface RG-25 series intended for power lasers, industrial microwave ovens, medium power radar. 1, 6 — Bushings; 2 — main insulator; 3 — ferromagnetic core; 4 — plastic screw; 5 — coil, 7 — pole. (b) Revolving assembly part of RG-25. 1 — Reed switch; 2 — insulator; 3 — bushing; 4 — support; 5 — ferromagnetic plate.

The current pickup levels can be adjusted up to 50% (for each subtype). The option of operation threshold adjusting is an important peculiarity of interface relays when they are used as current relays in systems of protection from current overloading. Such adjusting is necessary to compensate parameter spread of the elements and accurate relay adjustment for any given operating current. Basically, there are a lot of ways of adjusting operation currents for reed relays. For HV interface relays only those methods are adequate which allow us to avoid, in the course of adjusting, parasitic air gaps in the HV construction, because in such a construction corona charge arises, which can destroy insulation. In an interface relay of the RG-25 type, when a movable insulator with a reed switch turns around its longitudinal axis the latter moves away from the terminals of the magnetic core, and a magnetic shunt takes its place (Figure 6.22b). Contacting surfaces of the movable insulator (with a reed switch) and the stationary insulator (with a winding) have conductive covering and are smeared with conductive grease, shunting entrapped air.

The RG-75 (and RG-50) relay (Figure 6.23) comprises the main insulator (1) formed as a dielectric glass, whose cylindrical part is extended beyond flange 2. The flat external surface of the bottom (3) of this glass smoothly mates with the extended cylindrical part (4), having threaded internal (5) and external (6) surfaces. The relay also includes control coil (7), with a Π-shaped ferromagnetic core (8) located inside the main insulator and a reed switch (9) located in an element for reed switch rotation through 90° (10). This element (10) is formed as an additional thin-walled dielectric glass with walls grading

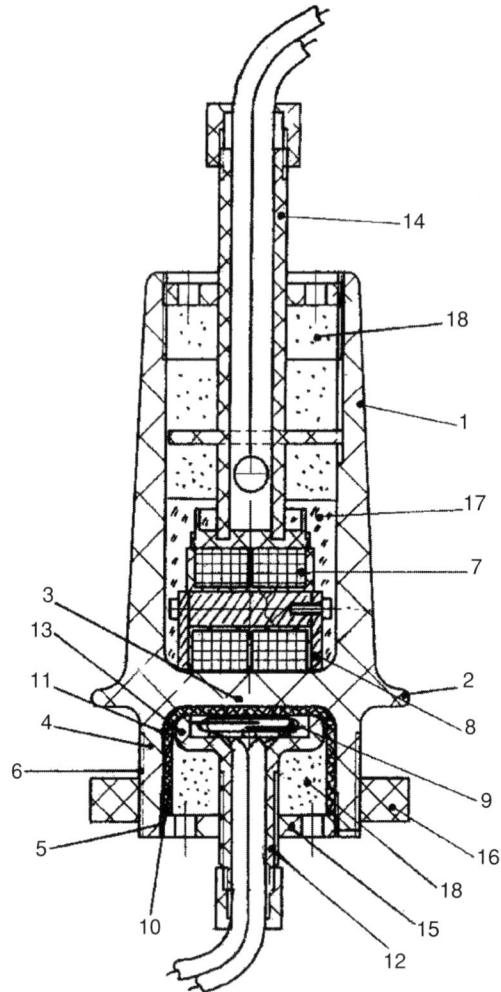

FIGURE 6.23
RG-75 and RG-50 series relay design.

into the bottom and mating with the inner surface of cylindrical part (4). These mated surfaces are coated with conductive material (11). Reed switch outputs (9) are conveyed through an additional insulator (12), formed as a tube extending beyond the reed rotation element body (10). The lower end of this tube is graded into oval plate (13), covering the reed formed with the conducting external coating. The control coil (7) outputs are also conveyed through the tube-shaped insulator (14), extending beyond the main insulator. The reed switch position fixation element is formed as disk (15) with a threaded side surface and a central hole, with insulator (12) conveyed through it. External attachment of the device is effected with dielectric nut (16). The lower layer of epoxy compound (17), filling the main insulator to the control winding, performs conduction by the addition of copper powder (60 to 70% of the volume). The rest of the filling compound (18) has been made dielectric. The element space (10) is filled with the same dielectric epoxy compound.

The shapes of the main insulator and the reed switch rotation element are chosen so that their mating surfaces, which contact with the conducting coating, do not form sharp edges emerging on the main insulator surface and at the same time provide for safe shunting of the air layer between them, removing the thin conducting sharp-edged layer from the design.

Significant reduction of the field intensity generated by the sharp outputs of the reed switch is achieved by adding one more tube-shaped insulator, extending beyond the main insulator used to convey the reed switch outputs, and causing the inner end of this tube to function as a plate with conducting coatings covering the reed switch. Applying the lower layer of epoxy compound, which fills the main insulator conducting space (holding the control coil with the ferromagnetic core), thus reduces the intensity of the field generated by the winding outlets and neutralizes the action of the air bubbles remaining between the coil windings.

Implementing the reed fixation element as a simple threaded disk, threaded into the respective part of the main insulator, forces the reed rotation element. Use is made of an additional dielectric nut threaded on the appropriate part of the main insulator as an element of the relay external attachment assembly, and the main insulator flange is used as a stop for this attachment assembly.

Device operation is based on the action of the magnetic field of the control coil (penetrating through bottom (3) of high voltage insulator (1) to reed switch (9).) When the reed switch threshold magnetic flux value is attained, it becomes engaged and appropriately switches the external circuits of the installation. The reed switch engagement threshold value is adjusted by changing its position relative to the magnetic field source. This change is effected by rotation of element (10) with reed switch (9) by an angle of 90° relative to the poles of Π-shaped ferromagnetic core (8). The position of element (10) with the reed is fixed by forcing element (10) as disk 15 is screwed in.

Each device from this series functions as four separate devices:

- Current level meter in an HV circuit
- Trip level adjustment unit
- Galvanic isolation assembly between the HV and LV circuits
- Fast response output relay in LV circuit

In current overload protection systems, the RG-Relays are usually connected to the open circuit of the HV power supply between the rectifier bridge and filter capacitor, when the acting current does not exceed 10 A (pulsating current amplitude up to 30 A), however, when the current is above 10 A, they are connected to the shunt. The RG-Relay is triggered when the current in the HV circuit exceeds the pickup level.

The RG-24-bus device (Figure 6.24), is designed to be used in overload protection units for 3 to 24 kV AC power networks, powerful electric motors, etc. The device output is a 100 Hz signal with 100 to 150 V DC or a standard "ON–OFF" type relay protection signal. The device design envisages its installation directly on a high voltage current-carrying bus or cable, as well as allowing for the possibility of wide range variations of the operation threshold (5 to 5000 A). Operating time is 1 ms.

The main advantage of these devices, as compared to those available on the market, is their possible direct installation on HV buses and output connection to LV automatic circuits. Medium voltage compact switchboard and switchgear cubicle systems (including SF$_6$ filled) can be significantly improved by using these devices. Built-in fault detectors and other automatic systems can now be produced as factory-standard equipment, and at affordable prices, and obtained without any alteration whatsoever of the HV equipment design.

The author designed a whole range of HV interface relays with specific properties for operation in strong magnetic fields, for instance (Figure 6.25). An interface relay with increased insulation level and effective protection from external magnetic fields was designed especially for use in electro-physical equipment (Figure 6.26).

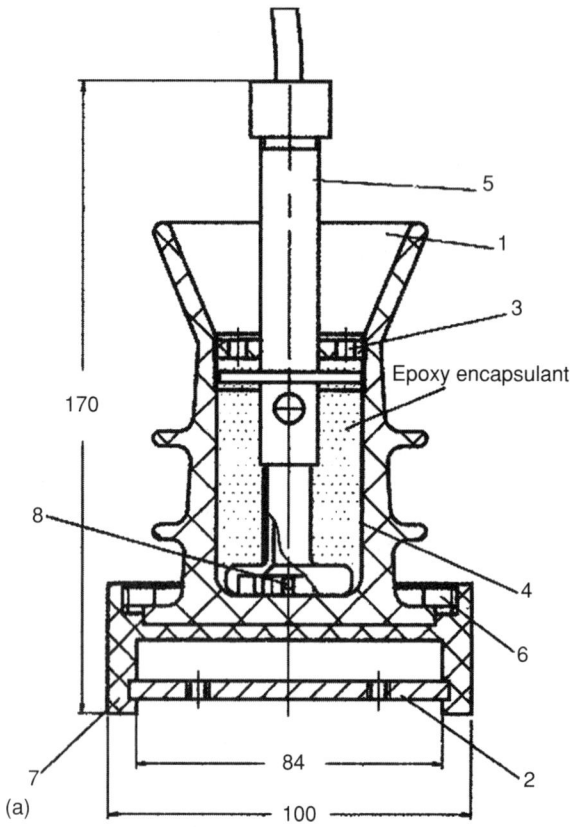

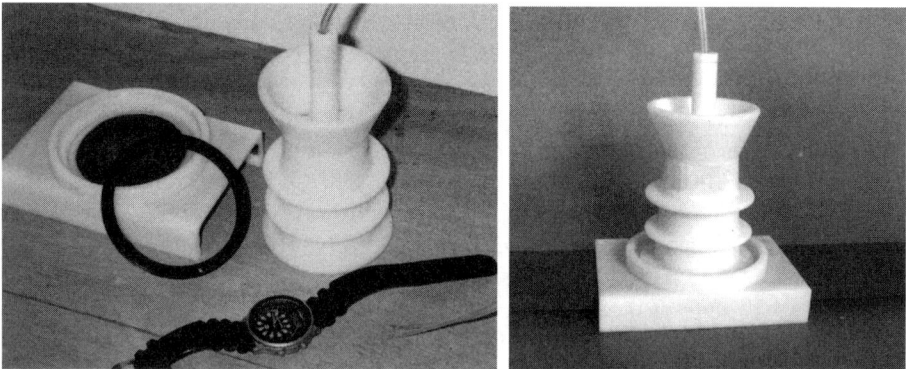

FIGURE 6.24

(a) RG-24-bus device. 1 — Main insulator; 2 — fixative plate; 3 — inside nut; 4 — semiconductive cover; 5 — bushing; 6 — fixative nut; 7 — fastener; 8 — reed switch. (b) Installation of RG-24-bus device on a high voltage bus bar.

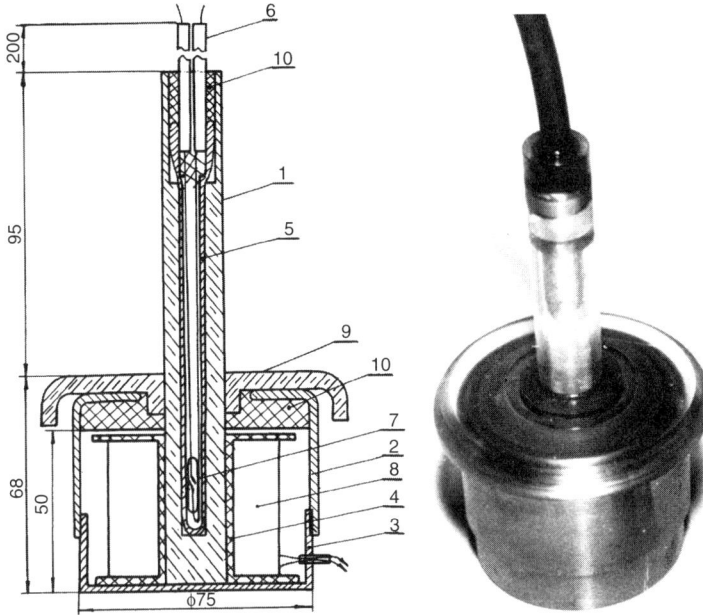

FIGURE 6.25
RG-relay with a high level protection from an external magnetic field. 1 — Main insulator (made as one unit with element 9); 2, 3 — thick wall ferromagnetic shield; 4 — bobbin; 5 — electrostatic shield; 6 — HV wires (reed switch's leads); 7 — reed switch; 8 — operate winding; 10 — epoxy encapsulant.

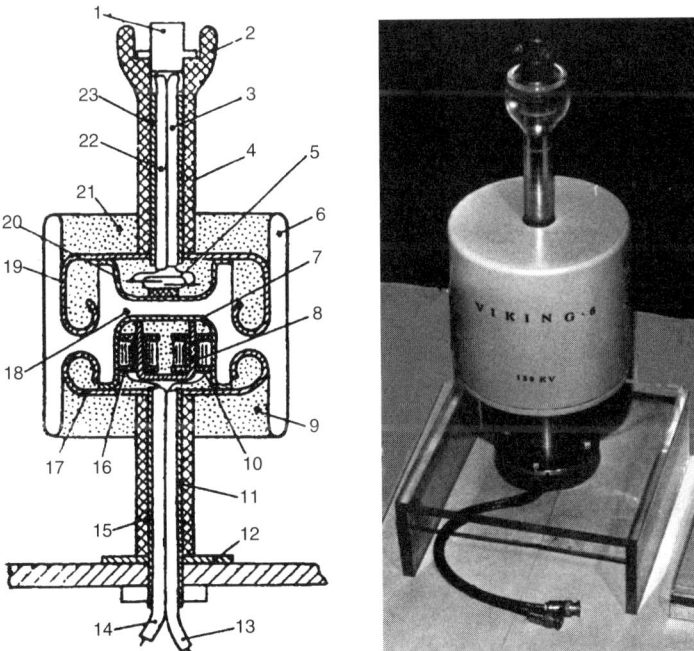

FIGURE 6.26
Ultra high voltage RG-relay. 1 — Connector; 2 — insulator for connector; 3, 22 — HV cables (reed switch leads); 4 — HV bushing; 5 — reed switch; 6 — main insulator; 7, 20 — aluminum shields; 8 — magnetic core; 9, 21 — epoxy encapsulant; 10, 16 — operating coils; 11, 15, 23 — lead shields; 12 — fastening element; 13, 14 — HV cables (operating coil leads); 17, 19 — ferromagnetic shields.

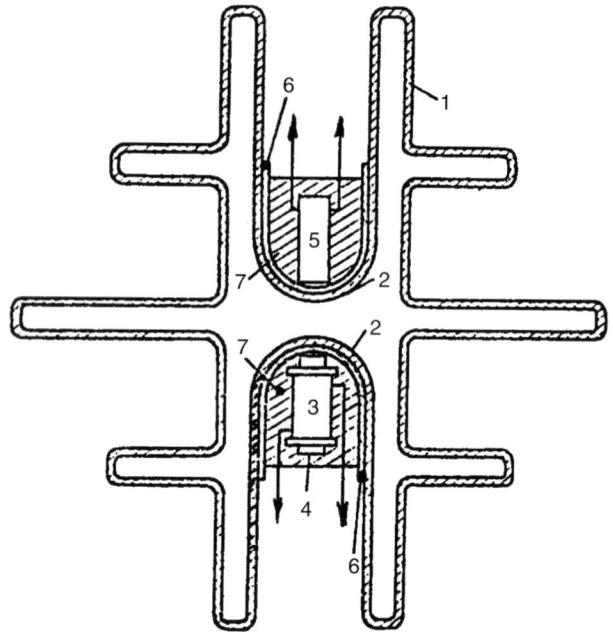

FIGURE 6.27
RG-relay with vacuum insulator. 1 — Main insulator (vacuum chamber); 3 — operate coil; 4 — ferromagnetic core; 5 — reed switch; 6 — electrostatic shield or conductive cover; 7 — epoxy encapsulant.

An original technical solution was found for HV interface relays with a vacuum chamber as a main insulator: to avoid the risk of vacuum failure caused by gases evolving from elements of the construction, they are all removed from the vacuum area and placed on the external surface of the vacuum chamber (the U.S.S.R. patent 836704, 1979 — Figure 6.27).

Over many years of work in this field, the author has designed many original constructions of HV interface relays. If you are interested in these constructions, you can find descriptions of them in the following books:

- Gurevich V., *High-Voltage Automatic Devices with Reed Switches*. Haifa, 2000, 367 p. (in Russian);

- Gurevich V., *Protection Devices and Systems for High-Voltage Applications*. Marcel Dekker, New York, 2003, 292 p.

7

Electronic Relays

7.1   Was It Thomas A. Edison who Invented a Vacuum Light Lamp?

The history of electronic relays, like of all relays in general, begins with basic components from which relays are constructed. These were radio lamps (or "electron tubes" or "vacuum tubes") first, and then semiconductor devices.The basis for an amplifying radio lamp was a standard illuminating lamp. Thomas Alva Edison is usually considered to be the inventor of the vacuum light lamp. In fact he was not the first pioneer, but just developed an experimental illuminating vacuum lamp designed by the English physicist Joseph Wilson Swan (Figure 7.1a).

In 1860, Swan used carbonized paper as a filament in his lamp; however, low vacuum level and source power prevented Swan from success. It was only 15 years later that Swan resumed his experiments and then, due to the use of better vacuum and carbonized thread, he managed to demonstrate an operating incandescent lamp (Figure 7.1b). Moreover, in 1880 Swan arranged a first international trade fair of electric lamps in Newcastle, England, but still the lamp invented by Swan was imperfect; he lost interest in his invention soon afterwards and devoted himself to other problems.

Edison performed thousands of experiments, selecting appropriate materials for filaments and developing the lamp's construction. Unlike Swan, he was more purposeful and persistent in the achievement of his goal, and sought to turn a commercial profit from the implementation of his lamps. The first lamps coming into the market were rightfully called "Edison–Swan Lamps" (sometimes simply: "EdiSwan" — Figure 7.2). Later, for different reasons, the name Swan was gradually forgotten and now Edison is "known" to be the inventor of the electric light lamp. During his numerous experiments in 1883, Edison came across an unknown (at that time) effect which was later called "Edison's effect" and it became the basis of the whole of modern radio engineering.

Edison discovered that if a metal plate is placed near the filament and is connected to the positive battery terminal (Figure 7.3), electric current will pass between the filament and the plate. At that time the reason was inexplicable, as there was no electric current conductor. The fact that electric current stops when battery polarity is changed remained even more unclear. Despite his instinct for profit and his intuition, Edison failed to apply the effect invented by him. It was only implemented in engineering after more than 20 years.

In 1904, on the basis of this effect, the English physicist John Ambrose Fleming designed and patented the first radio lamp in the world, called a "radio valve" or "thermionic diode," designed for the conversion of alternating current to direct current and for radio signal detection (Figure 7.4). Many inventors tried to design a more perfect Fleming diode for higher-quality detection of radio signals and wireless telegraphy, and

(a)

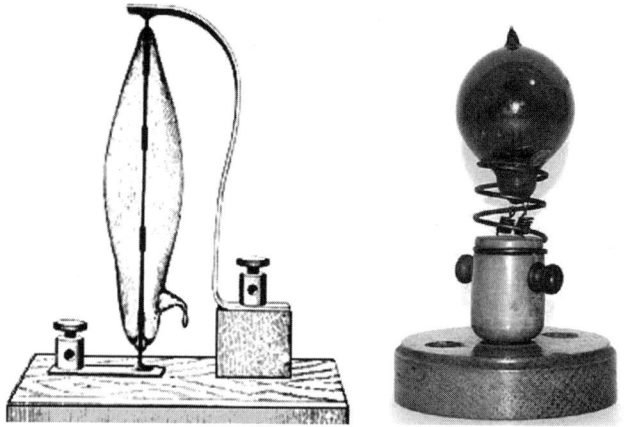

FIGURE 7.1
(a) Joseph Wilson Swan; (b) Swan's
first lamps.

(b)

FIGURE 7.2
First commercial light lamps bearing the
name of Edison and Swan ("EdiSwan
Lamps").

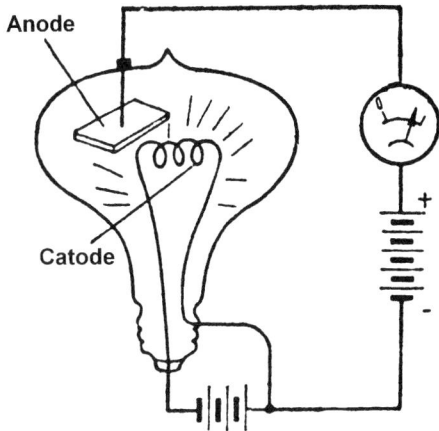

FIGURE 7.3
Edison's effect.

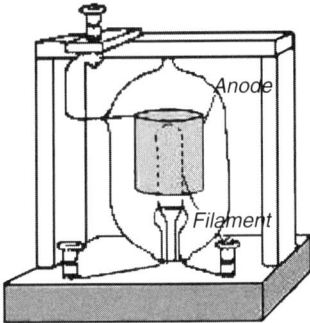

FIGURE 7.4
Fleming's vacuum diode with an additional electrode "anode" in the form of a cylinder encircling the filament.

FIGURE 7.5
One of the first constructions of vacuum diodes of increased power (for current up to 6 A) based on the incandescent lamp.

later sought to apply it to rectification of alternating current of increased power (Figure 7.5). It can be well seen in Figure 7.5 that this diode differs from an incandescent lamp with a spiral filament (a later construction of an incandescent lamp) by an additional plate only (an "anode").

7.2   Lee De Forest Radio Valve: From its First Appearance Until Today

The American engineer Lee De Forest achieved more success in this field when he placed an additional curved electrode between the cathode and anode (Figure 7.6).

In 1907, Lee De Forest patented a new construction of its vacuum tube, naming it "Audion" (now such type of vacuum tubes with three-electrodes called "triodes") (Figure 7.7) . The anode of this device was constructed in the form of a split cylinder, with the first electrode placed between the filament (the spiral), and the external anode made in the form of a large spiral with a large lead covering the internal spiral (Figure 7.8).

Later De Forest founded a company, "De Forest Wireless Telegraph Co" (according to other sources it was called "DeForest Radio Telephone & Telegraph Company"), where he launched production of vacuum tubes and radio receivers (Figure 7.9).

De Forest created the "Audion Piano," the first vacuum tube instrument in 1915. The Audion Piano was a simple keyboard instrument but was the first to use a beat-frequency (heterodyning) oscillator system and body capacitance to control pitch and timbre. The heterodyning effect was later exploited by the Leon Termen (Russian radio-engineer Lev

FIGURE 7.6
Lee De Forest.

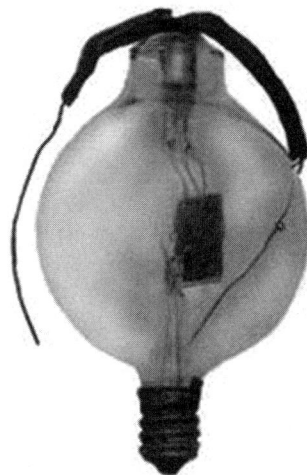

FIGURE 7.7
The first three-electrode vacuum tube constructed by Lee De Forest.

FIGURE 7.8
One of Lee De Forest's constructions of "Audion."

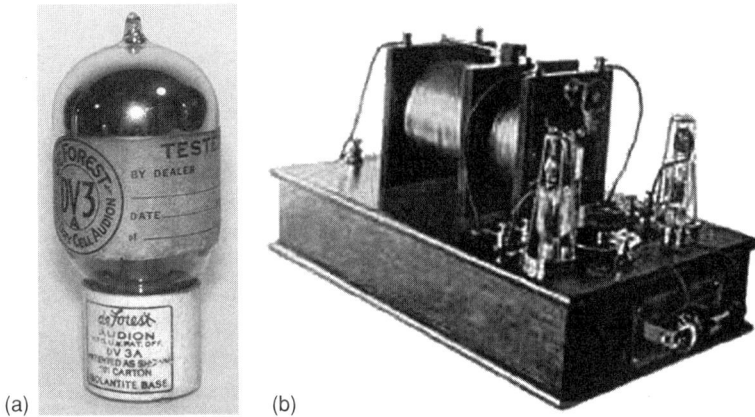

(a)   (b)

FIGURE 7.9
(a) Vacuum tube produced by "De Forest Company," and (b) the first radio set constructed by Lee De Forest in 1907.

Sergeivitch Termin) with his "Theremin" (also "Termenvox," "Aetherophone") — series of unique electronic musical instruments. The Audio Piano used a single triode per octave, which was controlled by a set of keys allowing one note to be played per octave. The output of the instrument was sent to a set of speakers that could be placed around a room, giving the sound, a dimensional effect.

De Forest later planned a version of the instrument that would have separate triode per key allowing full polyphony. It is not known if this instrument was ever constructed. De Forest described the Audio Piano as capable of producing: "Sounds resembling a violin, Cello, Woodwind, muted brass, and other sounds resembling nothing ever heard from an orchestra or by the human ear up to that time — of the sort now often heard in nerve racking maniacal cacophonies of a lunatic swing band. Such tones led me to dub my new instrument the "Squawk-a-phone." (Lee de Forest Autobiography, "The Father of Radio," 1915, pp. 331–332.)

During the 1930s de Forest developed Audion-diathermy machines for medical applications and, during World War II, conducted military research for Bell Telephone Laboratories. Although bitter over the financial exploitation of his inventions by others, he was widely honored as the "father of radio" and the "grandfather of television." He was supported strongly but unsuccessfully for the Nobel Prize for Physics.

The first commercial vacuum tubes produced in different countries had quite different external designs, but construction was practically similar to De Forest's Audion (Figure 7.10).

In Russia first vacuum tubes were named "cathode relay" and "pustotnoye relay" ("pustota" is vacuum in Russian)." The first serial vacuum tubes produced in Russia was constructed in the Nizhni Novgorod Radio Laboratory under the supervision of M. Bonch-Bruevich and was called PR-1 (Pustotnoye Relay, model No. 1). The name of a receiving vacuum tube of the R-5 type produced in 1922 by Petrograd Electrovacuum Plant stands for "Relay, Development No. 5." A new tube produced in 1923 with a thoriated-tungsten cathode, consuming ten times less filament current than an R-5 relay, was called "Micro" valve.

A space-charge tetrode with a "cathode grid," which was also quite economical by incandescence, was called MDS standing for "micro-two-grid." By 1920–30's, vacuum tubes had similar external designs to the modern ones now (Figure 7.11). Despite

FIGURE 7.10
One of the first commercial three-electrode vacuum tubes (triodes) replicating the internal construction of "Audion" of De Forest, produced by different firms.

FIGURE 7.11
"Radiotrones" produced by the American firm RCA in 1930: 6A7, 75, 80, 6D6.

amazing progress in semiconductor technology, vacuum tubes are still produced and used in high-end audio equipment of different types and specific electronics. A vacuum tube, operating in the mode of generation of a powerful ultra-high-frequency signal, with voltages on the electrodes of up 30–45 kV, is indispensable in powerful broadcasting transmitters and in radar. One can come across very powerful vacuum tube of this type (Figure 7.12). Modern vacuum tubes (Figure 7.13) are products of high-tech industry and cost a lot.

FIGURE 7.12
High-power triode.

FIGURE 7.13
Modern vacuum radio valves in glass and metallic tubes, produced up to now in many countries.

7.3   How a Vacuum Tube Works

When a tube cathode is incandesced, it seems to be surrounded by a cloud of electrons flying out from it. Affected by the electric field of the positive anode, electrons start to move towards the anode, creating anode current of the tube. The higher the voltage on the anode, that is the stronger its electric current, the more the current will be. If a metal grid is placed between the cathode and the anode and is not under voltage, the situation will not change.

Electrons will freely pass through the grid and will rush to the anode, because the holes of the smallest grid are enormous in comparison with the size of the electrons, but as soon as an electric charge is applied to the grid, an electric field occurs around it and the grid starts to influence the process of passing of electrons towards the anode.

If the grid is positively charged with respect to the cathode, it will help the anode attract electrons, thus increasing the anode current. If it is negatively charged, it will repel electrons, preventing them from passing through the grid and thus reducing the anode current (Figure 7.14). Thus, the grid controls the anode current of the vacuum tube and it is worth mentioning that the slightest changes of voltage on it will considerably change the anode current. Such property allows a vacuum tube to amplify electric oscillation.

Thus, alternating voltage applied to the controlling grid of a vacuum tube is transmitted to the anode circuit with amplification, but there is capacitive coupling between the anode and the control grid. Due to such coupling voltage changes are transmitted back to the circuit of the control grid and two variants are then possible; first of all, feedback voltage may increase the total voltage on the control grid, if positive and negative half-cycles of the feedback voltage and the voltage applied to the control grid concur. An increase of voltage amplitude on the control grid will cause an increase of voltage amplitude on the anode, and this in its turn will lead to higher feedback voltage and therefore to an increase of voltage on the control grid, etc. Such feedback is called "positive." If it is strong enough, progressive increase of voltage amplitude on the control grid and on the anode of the valve will lead to spurious oscillation of the amplifier circuit, but the feedback voltage may also bring about a decrease in total voltage on the control grid. Such feedback is called "negative" because it causes a decrease of amplification of the vacuum tube, which is why feedback — especially considerable feedback — through the capacity "anode-grid" is undesirable. Of course feedback in electric circuits arises not only due to capacity between the anode and the grid. Capacities between circuits of the electrodes, between components and wire, etc., also can play an important part in feedback occurrence; however, well-thought-out arrangements of the components and wire, and accurate assembling will make such capacities negligible. It may seem that it is possible to decrease capacity between the control grid and the anode by just miniaturizing

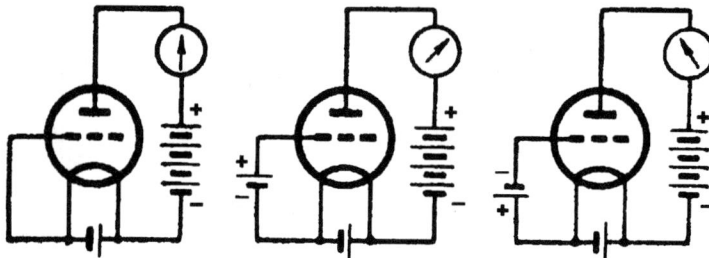

FIGURE 7.14
Principle of functioning of a three-electrode vacuum tube–triode.

the anode and the grid, and by moving them away from each other, but this leads to a sharp deterioration of the amplifier properties of the valve, lowering its power and capacity to operate under very high frequencies.

That is why some other methods of reduction of capacity between the anode and the control grid of the valve were needed. It turned out that it was possible to considerably reduce such capacity by introducing a shield in the form of an additional grid between the electrodes. Positive voltage could be applied to it, but lower in value than the anode one. Such a new grid did not prevent electrons moving to the anode, because of its positive potential. On the contrary, it even helps to "attract" electrons. It is called a "screening grid," and a vacuum tube with two grids is called a "tetrode" (the total number of electrodes is four).

The third grid, which in a vacuum tube is called a "pentode," can also protect the valve from the so-called "dynatron effect," when electrons hitting the anode at a high speed knock out secondary electrons, which when rebounding from the anode are attracted by the positive shield grid, thus causing reverse electric flux deteriorating the valve operation. To remove such a negative phenomenon, another additional grid had to be placed between the anode and the positive shield grid, but it should be negatively charged with respect to the anode. Such a grid repels electrons flying out from the anode back to the anode and is called "protective."

In most cases automation devices, radio sets, and other electronic devices are fed from standard AC network (through a rectifier, of course). Incandescence of the filament is fed with similar AC supply voltage, which is lowered up to several volts with the help of a transformer, but as the current is alternating, the temperature of the incandescent filament changes in accordance with its changes. If the filament serves as a cathode (as is shown on the simplified scheme, Figure 7.15), number of electrons flying out will change simultaneously with the filament temperature, which is why the anode current of the vacuum tube will also vary together with the AC frequency. To avoid this, the filament should be insulated from the cathode. The filament only heats up the massive cathode, and due to its considerable thermal inertia, the variation of filament temperature does not affect the number of electrons flying out. This is a so-called "cathode with indirect heating." Sometimes the number code of the tube indicates the heater voltage: for example, 12BY7A is a tube that has a heater operated at 12.6 V (not 12 V), and 7 means 6.3 V.

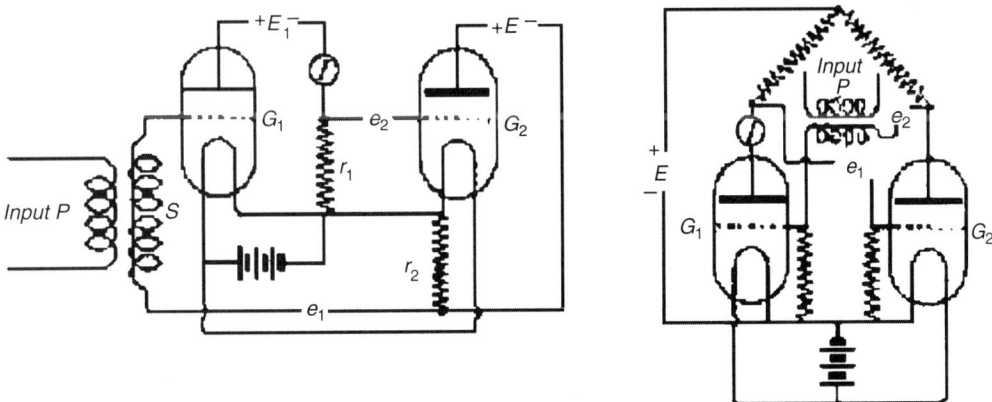

FIGURE 7.15
Circuit diagram of the first electronic relay described by W.H. Eccles and F.W. Jordan in 1919. (From W. Eccles and F. Jordan, A Trigger Relay Utilizing Three-Electrode Thermionic Vaccuum Tubes, Vol. 1, No. 3, 1919. With permission)

These voltages originally came from the days of using lead acid batteries as a power source, since many people did not have any other source of electricity. The voltage of a fully charged "6 V" lead acid battery (2.1 V per cell) is 6.3 V and the voltage for a six cell battery such as those used now in all cars (three cell batteries were used until the late 1950s in the U.S.) is 12.6 V. This is the reason why we have seen power transformers in the old days with 6.3 V on the secondary windings as a supply of heater voltage.

7.4 Relays with Vacuum Tubes

Very often devices consisting of a sensor of some physical quantity (pressure, temperature, light, etc.), an electronic amplifier and an electric relay at the output, are called electric relays. Such devices are really energized at a certain threshold value of the input quantity (pickup) and operate like any other relay, however that fact can be explained only by an electromagnetic relay at the output in such a device. In fact, we have in this case an electromagnetic relay with a preamplifier of the input signal. This is another modification of a relay that will be considered below. In this section we are considering only electronic devices with relay characteristics.

The first electronic device based on vacuum tubes with relay properties (Figure 7.15), was described in an article by W.H. Eccles and F.W. Jordan, "A trigger relay utilising three-electrode thermionic vacuum tubes." *Radio Review*, Vol. 1, No. 3 (December 1919): pp 143–146.

In order to understand how an electronic relay operates, it is necessary to consider a schematic of the simplest two-stage resistance-coupled amplifier on triodes (Figure 7.16 and Figure 7.17 — filaments are not usually shown on such schemes).

When negative voltage u_{in} (section AB) on Figure 7.17 is high, the triode VT1 is blocked and its voltage on the anode equals the voltage of the power source U_a. Voltage on the grid of the triode VT2 exceeds the negative offset voltage applied from the source of the offset, and the triode becomes unblocked, that is maximum anode current i_{a2} passes through it. Voltage on its anode is on the contrary minimal, and is determined by the difference between the power source voltage and the voltage drop on the anode resistor R_{a2}, that is $U_a - i_{a2}R_{a2}$. When the growing voltage u_{in} reaches pickup potential of the triode VT1

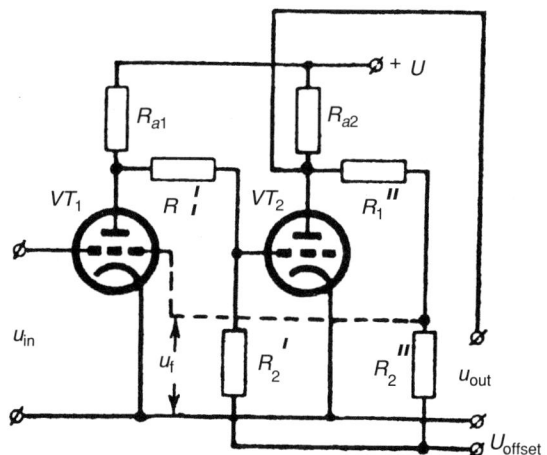

FIGURE 7.16
Two-stage electronic amplifier on triodes. Dependence of output voltage u_{out} on input voltage u_{in} of an amplifier is shown in Figure 7.17.

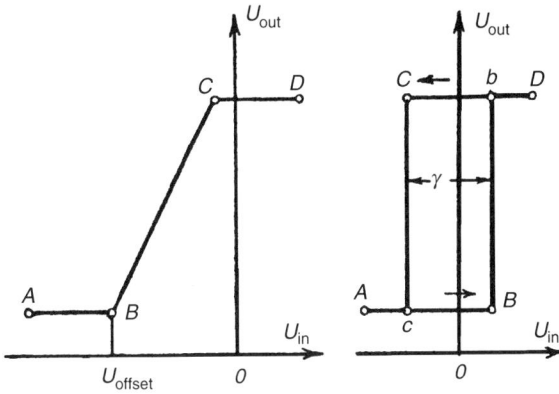

FIGURE 7.17

Transient characteristics (dependence of an output signal on an input one) of the electronic amplifier: on the left — without feedback; on the right — with positive feedback.

(point B), the triode is opened. Voltage on its anode will decrease (i.e., it is in the similar mode as triode VT2 has been before). This will lead to a decrease in the positive voltage on the grid of the triode VT2 and therefore an increase of voltage on its anode. Output voltage u_{out} will also increase (section BC). Finally, voltage on the anode of the triode VT1, and therefore on the grid of the triode VT2, will decrease to such an extent that the triode VT2 will be blocked (C is the blocking point of the triode VT2). Then an increase of the input voltage u_{in} on the grid of the triode VT1 will not affect the voltage of the anode of the triode VT2, and therefore will not affect the output voltage u_{out} (section CD).

If positive feedback is introduced into this scheme (as shown in the pointed line in Figure 7.16), the process of opening of the triode VT1 and closing of the triode VT2 at increased input voltage, will be very short. Indeed, increase of input voltage (after the point B of the characteristic) will lead to a decrease of voltage on the anode of the triode VT1 and therefore to an increase of voltage on the anode of the triode VT2, that is to additional decrease of negative voltage on the grid of the triode VT1 (because of the feedback). In its turn, this will cause further voltage decrease on the anode of the triode VT1 and a voltage increase on the anode of the triode VT2, etc. A chain reaction seems to arise and as a result the circuit is rapidly, turned to a new state when the triode VT1 is opened and the triode VT2 is closed.

Thus the amplifier with positive feedback can function on three states. The first one implies that the triode VT1 is blocked and the triode VT2 is enabled. In this state, the device is in the initial position selected by us. It will be stable because without input voltage the device can function in such a state as long as possible. In order to switch the device to another stable state, one should apply to the grid of the triode VT1 a positive firing pulse of such a value that the voltage on the triode grid becomes lower than the blocking potential. When it reaches such a position (point B in the right part of Figure 7.17) the device will be turned to another stable state (point b), characterized by the upper branch of the CD characteristic. The device may operate in the stable state when the triode VT1 is enabled and the valve VT2 is disabled, as long as possible.

In order to turn the device back to the initial stable state characterized by the branch AB, theoretically it is necessary to apply a negative blocking pulse on the grid of the triode VT1, and after that the device will be turned stepwise to point c of the characteristic, however in practice, the negative blocking pulse may not necessarily have the amplitude equal to the potential of triode blocking, because even the slightest voltage reduction on the grid of this triode (similar to a voltage decrease on the coil of an electromagnetic relay) will lead to a considerable voltage increase on its anode, and therefore to a voltage increase on the grid of the triode VT2. As a result, the triode will be enabled, a chain reaction will arise and the device will be "overturned" to the initial stable state.

Finally, a third state is possible when the device is at the moment of switching from one state to another one. In such a state both triodes are enabled and the state is unstable. The circuit turns saltatory from this position to one of the stable ones. In the right part (Figure 7.17), it can be seen that at continuous input voltage change, output voltage is extremely volatile (the same as the state of an electromagnetic relay). Upsurges occur when input voltage passes threshold values determined by the extreme points B and C of characteristic. The distance between these points is marked by γ and is called hysteresis voltage, as it is in electromagnetic relays. Hysteresis voltage depends on the circuit amplification factor, and the more it is, the more the amplification factor.

7.5 Gas-Tubes with Relay Characteristics

A triode gains a completely new quality when it is filled with some rare gas. In a gas-filled appliance, electrons that have flown out from the heated cathode are accelerated by the positive field of the anode. They collide with gas atoms and ionize them, and as a result the number of electrical current carriers in the tube goes up sharply.

Charged particles with both signs, which are in great concentration between the cathode and the anode, form electronic-ionic plasma. The process of current, passing through the gas interval followed by a formation of plasma, is called an electric discharge. In gas-filled triodes, the effect from the grid differs from that in vacuum triodes. In vacuum triodes considered above, the grid allows fluent changes of the value of the electric flux passing through it, while in gas-filled triodes the grid just controls the moment of occurrence of an electric discharge between the cathode and the anode, and is unable to quench the discharge after it has been initiated.

Fixation of the moment of discharge initiation is carried out by applying a considerable negative potential with respect to the cathode, to the grid, and by replacing it with a more negative (or even a positive) one when it is initiated (Figure 7.18). While the negative field predominates in the area between the grid and the cathode and also in the grid holes, electrons going out from the cathode are slowed down by the field, which is why the number of electrons reaching the area between the grid and the anode is very small.

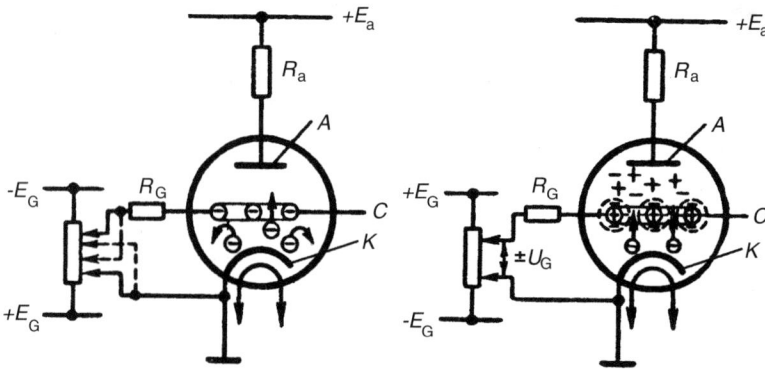

FIGURE 7.18
Scheme of the grid functioning in a gas-filled triode: on the left — before the discharge is initiated; on the right — after its development.

When a considerable negative potential on the grid is replaced with a less negative (or more positive) grid potential, the slowing down effect of the grid is considerably weakened. That is why a great number of electrons reach the area between the grid and the cathode. In that area electrons are accelerated by the positive field of the anode and receive enough energy for ionization of gas atoms. This is the starting point for discharge development. Further stages of development are concerned with avalanche-like multiplication of carriers, which leads to occurrence of arc discharge within a short period of time.

Such a gas-filled triode changing its state spasmodically (that is having a relay characteristic) is called a thyratron. A thyratron is in fact a real electronic relay. The discharge in the thyratron belongs to the nonindependent type, because primary agents (electrons) providing the ground for the discharge, are emitted by an incandescent cathode to which power is applied from the outside. Like vacuum tetrodes and pentodes, gas-discharge thyratrones can be with one or two additional grids. Thyratrones with an arc discharge can be both low power (Figure 7.19), and power capable of working under high voltages (tens of kilovolts) and strong currents (tens and hundreds of amperes in the continuous mode, and tens of thousands of amperes in the mode of short pulses switching — Figure 7.20).

FIGURE 7.19
Low-power thyratrones with an incandescent cathode, produced in Russia.

FIGURE 7.20
Modern high-voltage power pulse thyratrons in metal cases (produced by EG & G).

There are thyratrones not only with an incandescent cathode, but also with a cold one (Figure 7.21). The cathode of such a thyratron is made in the form of a metal cylinder activated by cesium. The anode is a molybdenum core placed in the glass tube with a free end sticking out of the glass. The starting electrode (a grid) has the form of a disk with a central hole and is placed between the anode and the cathode. The glass tube where the electrodes are placed is filled with neon with a small admixture of argon under a total pressure of 20 to 30 mmHg. The distance between the electrodes and the gas pressure is chosen in such a way that a discharge between the controlling electrode and the cathode occurs under lower voltage than a discharge between the anode and the cathode.

Along with single-grid thyratrones (triodes), double-grid thyratrones (gas-filled tetrodes) are widely used. When a voltage pulse of positive polarity is applied to the circuit of the control grid (control electrode) the potential of the grid increases and the field strength in the area between the control grid and the cathode is enough for gas ionization. An additional discharge occurs between the control grid and the cathode. Then it moves to the anode, and the thyratron is started.

After the thyratron is started, the grid loses its controlling properties. It is impossible to change the value of the anode current or to extinguish the thyratron by changing the potential of the grid of a started thyratron. In order to extinguish the thyratron, it is necessary to switch OFF the anode supply, or to decrease the anode voltage to a value that is lower than the voltage of combustion. A scheme providing a cut-off of the thyratron is shown in Figure 7.22. The capacitor in this scheme is charged through the open thyratron, which is open at that moment, and is discharged the next moment when the second thyratron is enabled. Thus, if the thyratron T_1 is enabled, the capacitor C is charged through this thyratron and the resistor R_{a2}, with the polarity indicated on top of the capacitor. When the positive pulse enables the thyratron T_2, the capacitor C is discharged through both thyratrones, creating direct current in thyratron T_2 and back current in the thyratron T_1. After thyratron T_1 closing, capacitor is charged through the resistor R_{a1} and then the open thyratron T_2. In such schemes load (resistors R_{a1}, R_{a2}) is switched ON and OFF just as in electromechanical relay. The value of the

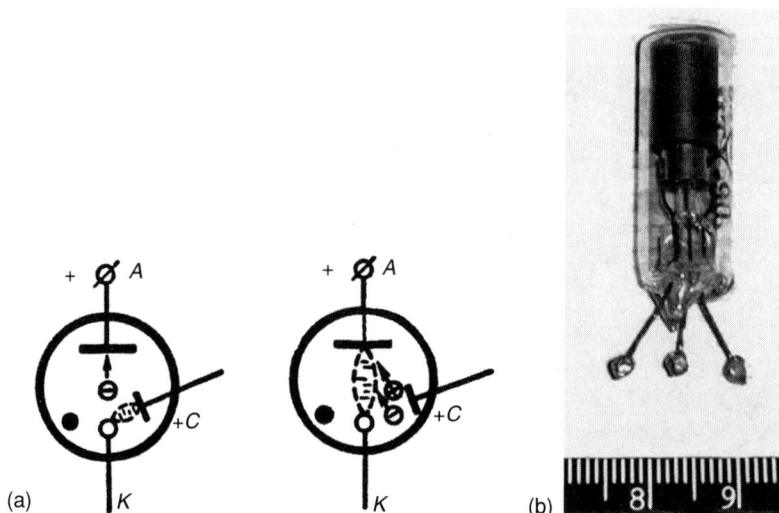

FIGURE 7.21
(a) Principle of functioning and (b) external design of the thyratron with a cold cathode.

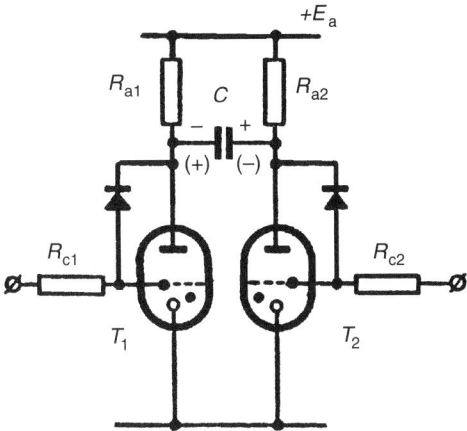

FIGURE 7.22
Scheme of the so called "forced commutation" of thyratrons.

capacitor *C* must be enough so that after switching the current from one thyratron to the other, the former has negative voltage for some time in order to restore its blocking properties.

7.6 Power Mercury Valves

Power mercury valves in which a liquid mercury cathode and graphite anodes are used should be considered separately. Such devices were widely used in industry in 1940–60's. According to the construction, they were split into glass and metal tubes, single-anode and multi-anode (2-, 3-, 6-, 12- and 24-anode) ones, with natural air or forced-water cooling.

The simplest glass mercury valve (Figure 7.23), consists of a vacuum glass tube (1) with two anode sleeves containing graphite anodes 2 and 3. The tube which is relatively big in size is used for better cooling of the device and creates better conditions for condensation

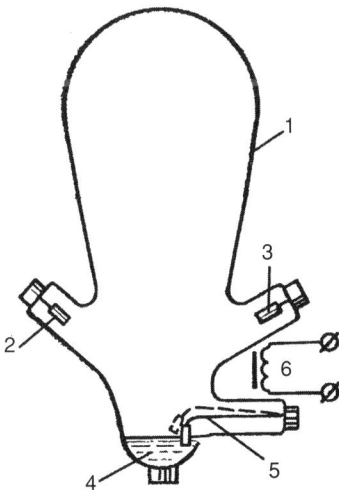

FIGURE 7.23
Power double-anode mercury valve.

of mercury vapors. In the lower part of the tube, there is a mercury cathode (4) and a movable igniting anode (5) moved by the electro-magnet. Alternating voltages in anti-phase are applied to the anodes 2 and 3 from the secondary winding of the transformer. To initiate the valve, a short-term control signal is applied to the electromagnet (6), which sinks the igniting anode (5) into the mercury and closes the circuit of the igniting anode. When the igniting anode is lifted, the mercury "bridge" between the cathode and the igniting anode breaks and at the break point an electric arc occurs. The arc occurrence is accompanied by a liberation of free electrons.

Affected by the electric field of anodes, free electrons move to one of the anodes, which has the positive potential with respect to the cathode at that moment, and ionizes mercury vapors in the area between the cathode and the anode. Electric ion plasma, through which load current passes, is formed between the cathode and the anode. The discharge is maintained by electrons coming from the so called "cathode spot" formed on the surface of the cathode. To maintain such a cathode spot and to provide the needed number of electrons, current not less than 3 to 5 A is required.

Unlike in the mercury valves considered above, in which igniting anodes are used for maintenance of arcing, in devices called ignitrons arc initiation occurs during each positive half-period of anode voltage, with the help of an additional electrode called igniter. An ignitron is a glass or metal vacuumized tube (Figure 7.24 and Figure 7.25), containing mercury cathode, anode, and igniter.

An igniter is the most important element of the ignitron. It has the form of a conic rod made of a mercury-unwettable semiconducting material as carborundum, or boron

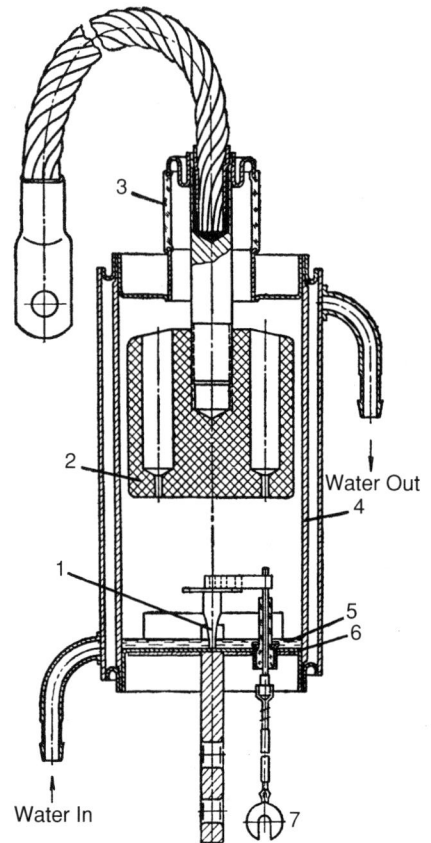

FIGURE 7.24
Construction of an ignitron with water cooling in metallic case.
1 — igniter; 2 — graphite anode; 3 — insulator from quartz glass; 4 — metal case with double-layer walls; 5 — mercury (cathode); 6 — metal bowl for mercury; 7 — ignitor's outlet

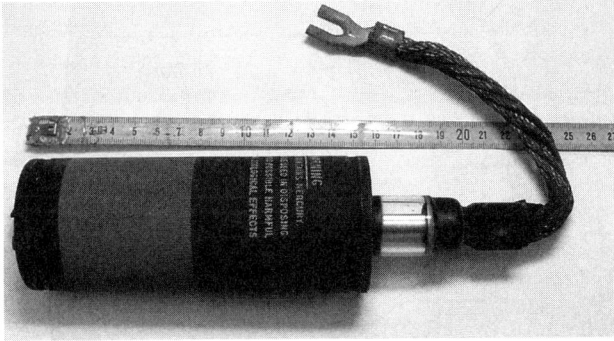

FIGURE 7.25
External design of an ignitron of the GL-5550-1 type, without additional cooling, in a metal case (General Electric Co.).

carbide for instance, submerged 3 to 5 mm into the mercury cathode. An insulation microfilm is formed between the igniter and the cathode. Voltage pulses of up to 170–200 V with a current upto 30 A are applied to the igniter.

If an igniting pulse is applied with positive voltage on the anode, arcing arises and plasma occurs. A cathode spot emerges on the surface of the mercury, that surface being the source of electrons maintaining the discharge. If the voltage half-period is negative, there will be deionization of the mercury vapors, and arc decaying, which is why during each positive half-period of the anode voltage, it is necessary to apply an igniting pulse on the igniter. Apparently, igniting pulses must be applied synchronously with the anode voltage. The function of an igniter in an ignitron is similar to that of a control grid in a thyratron.

A glass ignitron of the I-100/1000 type (Figure 7.26), designed for a rectified current of 100 A, with permissible reverse voltage of 1000 V, is made in a welded construction containing a copper cylinder (4) cooled by water which is the outlet of the cathode, and a glass cylinder (molybdenum glass) (2) — an anode tube. The graphite anode (3) has the

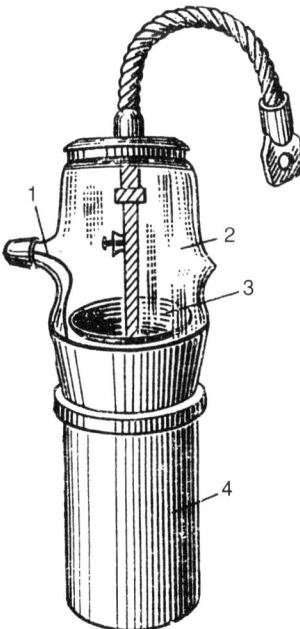

FIGURE 7.26
External design of a glass ignitron of the I-100/1000 type (U.S.S.R.).

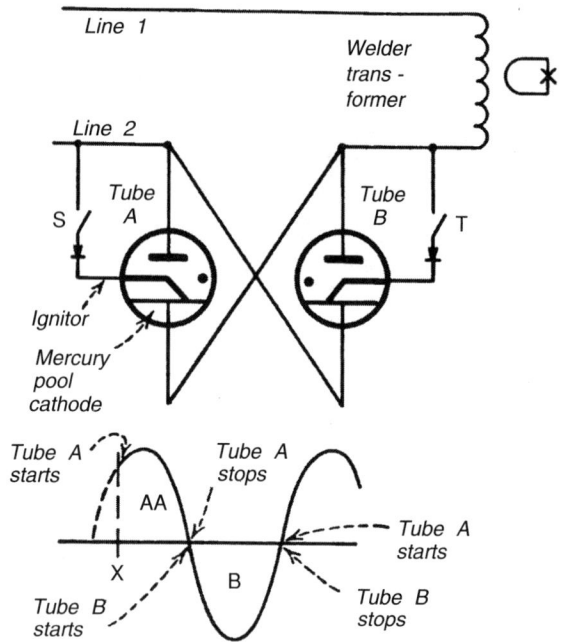

FIGURE 7.27
AC contactor on ignitrons for control of a power welding transformer (1940s—1950s).

form of a cylinder or a hemisphere. The outlet of the igniter (1) is made in the sidewall of the glass cylinder.

As a separate ignitron can conduct electric current only within a half-period of AC voltage, two parallel-opposition connected devices are used for switching both of the half-waves (Figure 7.27).

7.7   Electron-Beam Switching Tubes

An electron-beam switching tube can be considered a hybrid of a multi-anode valve and an electron-beam tube (that is, a television tube used in TV sets and computers). It is a device in which quick switching of circuits is carried out by moving an electron beam, by an electric field, created by deflected electrodes. The difference between an electron-beam switching tube and a standard television tube is that instead of a luminophore it contains a system of electrodes shorted by an electron beam (Figure 7.28). Modern electron-

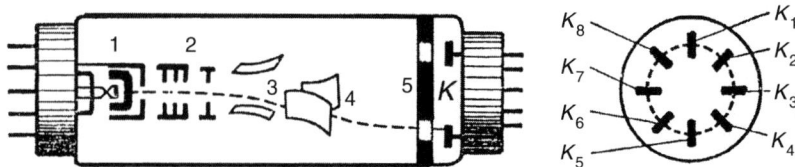

FIGURE 7.28
Principle of functioning of electron-beam switching tubes. 1 and 2 (cathode and anode) — the so called "electron gun" or "electron injector," creating an electron beam; 3 and 4 — vertically and horizontally deflecting electrodes; 5 — protective disk with holes ("mask"), K — contact electrodes.

FIGURE 7.29
Power high-voltage electron-beam switching tubes, produced in Russia.

beam switching tubes are capable of switching circuits with currents of tens and hundreds of amperes, and with voltages of hundreds of thousands of volts (Figure 7.29).

7.8 Semiconductor Relays

7.8.1 First Experiments and First Semiconductor Devices

The history of semiconductor relays starts with the history of the application of the first crystals for radio-signal detection. As far as I know, the earliest paper on asymmetrical conduction was written by Karl Ferdinand Braun in 1874: ''Ueber die Stromleitung durch Schwefelmetalle'' (''On current flow through metallic sulfides''), Poggendorf's Annalen. He discovered that galena (lead sulfide) and copper pyrites, among others, could rectify. Of course, these experiments predated radio, so application of Braun's discoveries to wireless communication took a couple more decades. Perhaps the first to apply semiconductor diodes to the radio art was the remarkable Jagadis Chandra Bose, who applied for a patent on a galena detector in 1901 (it was finally awarded in 1904). This was used in his research work in the area of millimeter-wave RF!

The Peroxide of Lead detector devised by S.G. Brown of England is illustrated in Figure 7.30. This detector has proven quite successful and is used with a pair of sensitive telephone receivers and a critically adjusted battery current. The instrument comprises a

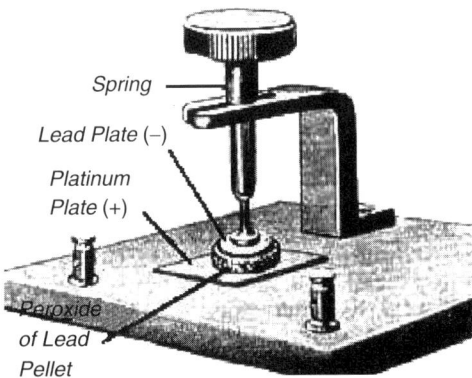

Spring

Lead Plate (−)

Platinum
Plate (+)

Peroxide
of Lead
Pellet

FIGURE 7.30
The peroxide of lead detector.

peroxide of a lead pellet, mounted between an upper lead disk and a lower platinum one; the pressure on the peroxide of the lead pellet being adjustable by means of a thumb screw and spring in the usual fashion. This detector has been termed, more or less correctly, the dry electrolytic detector and its action is supposed to depend upon the fact that an incoming oscillation intensifies the counter-electromotive force set up by the cell (an electro-chemical reaction due to the lead peroxide–lead platinum couple) and which is opposite to the applied battery current (about 1.5 V), thus causing the detector to increase its effective resistance. This results in a drop of current in the phone circuit; as soon as the oscillation ceases the phone current increases.

Henry Harrison Chase Dunwoody (a Brigadier-General of U.S. Army and later a vice-president of "DeForest Company") patented a silicon carbide (carborundum) detector (U.S. Patent 837616) in 1906 and that device also worked quite well, although a bias was essential for proper operation. In all of these point contact structures, the ohmic contact was made either by immersing the crystal specimen in a low melting-point alloy (Wood's Metal), or simply with a clamp of some sort (Figure 7.31). The carborundum detector in its usual form comprises two rather stiff springs, adjustable as to pressure, between which the carborundum (carbide of silicon) crystal (preferably an extremely jagged, greenish specimen) is placed. A pair of high resistance telephones are shunted across the detector and the incoming Hertzian wave oscillations, representing the points and dashes of the telegraphic code, are manifested as short and long signals in the phones, owing to the fact that the carborundum crystal will pass currents several hundred times better in one direction than it will in the reverse direction.

This action is enhanced by mounting the crystal in a cup or clamp of a large section, and making the second electrode of very small contact area. A steel needle has been used effectually as the small electrode and in one commercial instrument an even smaller electrode has been made. Greenleaf Whittier Pickard worked harder than anyone else to develop point-contact detectors. He tested over 31,000 combinations of minerals and wires in a search for the "best" detector. He patented a silicon-based point-contact detector in early 1907, and it worked exceedingly well.

The silicon detector (Figure 7.32) employs a piece of the mineral silicon embedded firmly in a brass cap. A solder or low heat alloy such as Hugonium metal is best used in mounting such minerals, so as not to injure their radio detecting properties or sensitivity. The Silicon detector is generally used without any battery and acts as a rectifier, similar to the carborundum detector. A pair of 2000 Ω phones or higher resistance ones are usually shunted across the detector, and owing to the rectifying action already described, the incoming Hertzian wave currents are manifested as short and long sounds in the phones. As it turned out, the quality of signal detection and sensitivity of the detectors depended much on he properties of the joint point of the metal needle and the crystal.

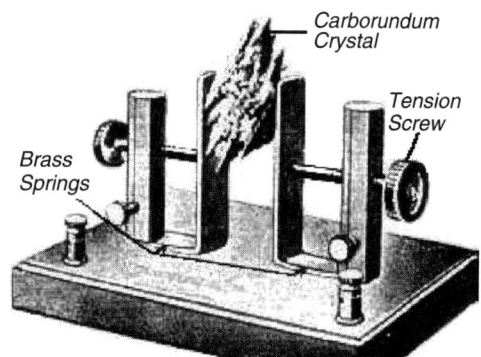

FIGURE 7.31
The carborundum detector, discovered by H.H.C. Dunwoody.

Spring

Metal
needle

Silicon

Adjustable
cup

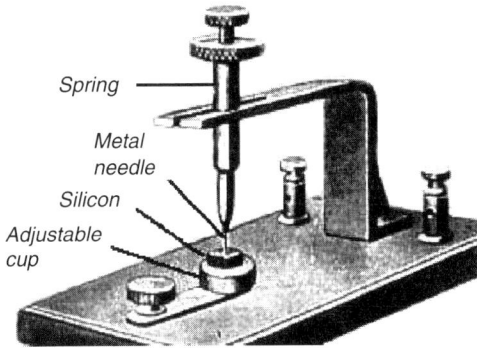

FIGURE 7.32
Silicon detector.

Such properties were very unstable, which is why in the course of work one had to search for an optimal contact point. That's why all earlier constructions of detectors were equipped with simple mechanisms for moving, changing pressure degree and fixation of the working point, and also for replacement of crystals (Figure 7.33). Due to the rapid development of industrial electronics and automation, in the middle of the 20th century demand for relatively powerful (for currents of a few and even tens of amperes) and cheap rectifier diodes arose. Detector diodes used in radio engineering were not quite adequate for these purposes. In 1940–50's copper-oxide diodes were very popular (Figure 7.34). The n-type electroconductivity area is formed inside the copper-oxide area with oxygen deficiency. The p-type area (with oxygen excess) is formed on the surface of the layer. Very thin soft leaden disks are used as packing, providing good electric contact of current conducting lamellas with a valve unit (the disks — 3). One of

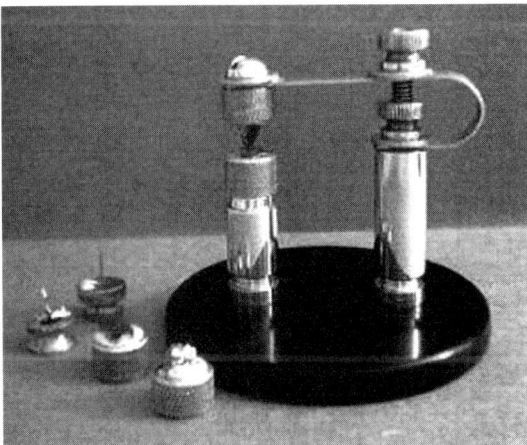

FIGURE 7.33
Crystal radio detectors, with replaceable "cups" so you can change to different minerals and wires.

FIGURE 7.34
External design (a) of the column of four copper-oxide diodes and elements (b) of its construction (produced by Westinghouse). 1 and 5 — current conducting lamellas; 2 and 4 — lead disks; 3 — copper disk with copper oxide (Cu_2O) layer applied to one of its sidewalls and silver plates in the form of semirings.

such elements rectifies alternating current with voltage of not more than 8 to 10 V. Permissible current density is 40 to 60 mA/s m^2.

To increase working voltage, such elements are connected in series. Copper-oxide valves had not very high, but stable characteristics and that's why they were used even in electrical measurement equipment. These were such simple and reliable elements that they had been applied in engineering for quite a long period of time. It is worth mentioning that cheap copper-oxide diodes for low voltages were produced until the 1970s, along with more complex and at the same time more expensive silicon diodes.

Selenium diodes can be referred to as a whole epoch (Figure 7.35). All electrical engineers who started to work in the 1950–60's remember well such ribbed small and large (with the ribs of up to 15 cm) devices. There still exist electric installations with selenium rectifiers. The author happened to come across large selenium valves in powerful battery charger produced in the 1960s, which were still functioning in 2003! Selenium rectifiers, like copper-oxide ones, consist of separate valve units in the form of round or square disks beaded on an insulated stud. A selenium valve unit consists of an

FIGURE 7.35
Showing how average power rectifiers for currents of a few amperes, assembled from selenium valves, look.

aluminum disk with a polished area in the center, with a layer of crystalline selenium obtained from the amorphous state with the help of thermal treatment.

For better contact of selenium and aluminum, a thin bismuth layer is sometimes sprayed between them. Such crystalline selenium has conductivity of the p-type (see below). On the surface of selenium, a tin–cadmium alloy in the molten state is applied. The selenium layer with a cadmium admixture forms an n-type layer (see below). Permissible current density for such valves was 0.8 to 1.0 A/s m^2. Big aluminum disks were used as heat sinks cooling the valve unit and were always a component of power and high-power diodes. Low current high-voltage selenium diodes were made in the form of a dielectric tube 6 to 8 mm in diameter and 10 to 16 cm in length. Tens of thin selenium disks, pressed by electrodes twisted in from the ends, were attached to it.

The rectifying unit consisted of tens of separate disks (diodes) connected in series (it should be noted that modern silicon diodes do not allow such connection in series without additional elements equalizing voltage distribution on the diodes that are connected in series). Selenium diodes also had another interesting property not common in modern diodes: localization of the breakdown spot in the crystal and removal of the damaged area from work, that is restoration of the working capacity after partial crystal disruption. Modern crystal diodes are elements designed for detection of radio signals and rectification of alternating current in power electronics and automation. They resemble the first detectors a lot: a similar crystal; similar metal needle (Figure 7.36). Of

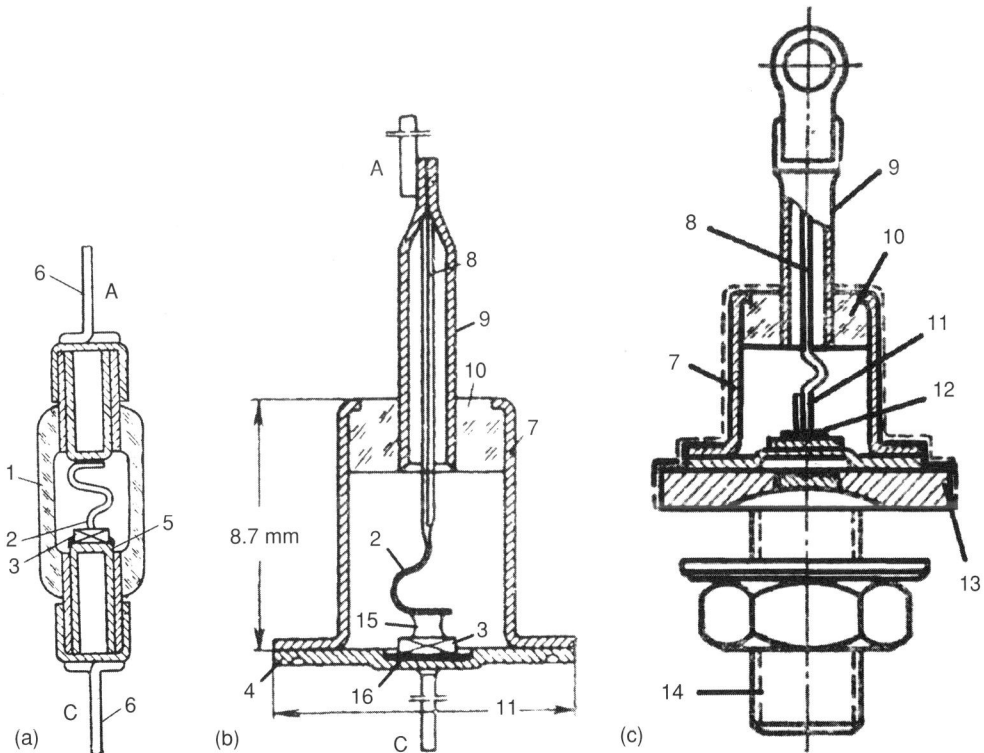

FIGURE 7.36
Construction of modern diodes: a, b — germanium low power and power diodes; c — silicon high-power diode. 1 — ceramic or glass case; 2 — upper tungsten electrode; 3 — crystal; 4 — metal flanges; 5 — crystal enclosure; 6 — wire terminals; 7 — metal case; 8 — flexible anode outlet (copper multiple stranded conductor); 9 — steal tube; 10 — glass insulator; 11 — contact outlet from the p-layer crystal; 12 — silicon crystal with a p—n barrier; 13 — copper heel piece; 14 — cathode outlet in the form of a screw-bolt; 15 — indium; 16 — outlet from the n-layer.

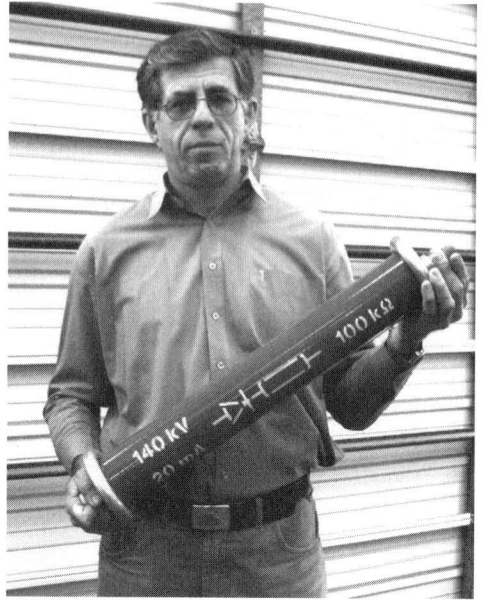

FIGURE 7.37
Semiconductor diode for voltage of 140.000 V in a joint
case with a resistor for 100 kΩ, connected in series.

course now applied materials are modern and technologies of mass production are
different, and the needle is not simply pressed to the crystal but fused into it, thus
avoiding instability of contact properties. Modern technologies allow production of
semiconductor diodes for such voltages that were unimaginable for designers of the
first diodes (Figure 7.37).

Design and application of the first detectors (rectifiers) did not require an understanding of physics of the processes taking place in them. Until the 1940s they were based only
on numerous experiments with different materials. Later, it turned out that the physics of
these processes is very complex and that their full understanding is possible only on the
basis of knowledge of modern physical theories of substance structure. This book is not
aimed at considering fundamental theories of solid-state physics, which is why we will
give only a simplified explanation of the processes taking place, in those elements which
concern us.

7.8.2 Semiconducting Materials and P-N-Junction

As is known, all substances depending on their electro-conductivity are divided into
three groups: conductors (usually metals) with a resistance of 10^{-6}–10^{-3} Ω·cm, dielectrics
with a resistance of 10^{9}–10^{20} Ω·cm, and semiconductors (many native-grown and artificial
crystals) covering an enormous intermediate range of values of specific electrical
resistance.

The main peculiarity of crystal substances is typical, well-ordered atomic packing into
peculiar blocks — crystals. Each crystal has several flat symmetric surfaces and its
internal structure is determined by the regular positional relationship of its atoms,
which is called the lattice. Both in appearance and in structure, any crystal is like any
other crystal of the same given substance. Crystals of various substances are different. For
example, a crystal of table salt has the form of a cube. A single crystal may be quite large
or so small that it can only be seen with the help of a microscope. Substances having no
crystal structure are called amorphous. For example, glass is amorphous, in contrast to
quartz which has a crystal structure.

Among the semiconductors which are now used in electronics, one should point out germanium, silicon, selenium, copper-oxide, copper sulfide, cadmium sulfide, gallium arsenide, and carborundum. To produce semiconductors two elements are mostly used: germanium and silicon.

In order to understand the processes taking place in semiconductors, it is necessary to consider phenomena in the crystal structure of semiconductor materials which occur when their atoms are held in a strictly determined relative position to each other due to weakly bound electrons on their external shells. Such electrons, together with electrons of neighboring atoms, form valence bonds between the atoms. Electrons taking part in such bonds are called valence electrons. In absolutely pure germanium or silicon at very low temperatures, there are no free electrons capable of creating electric current, because under such circumstances all four valence electrons of the external shells of each atom that can take part in the process of charge transfer, are too strongly held by the valence bounds. That is why that substance is an insulator (dielectric) in the full sense of the word — it does not let electric current pass at all.

When the temperature is increased, due to the thermal motion some valence electrons detach from their bonds and can move along the crystal lattice. Such electrons are called free electrons. The valence bond from which the electron is detached is called a hole. It possesses properties of a positive electric charge, in contrast to the electron, which has a negative electric charge. The more the temperature is, the more the number of free electrons capable of moving along the lattice, and the higher the conductivity of the substance is. Moving along the crystal lattice, free electrons may run across holes — valence bonds missing some electrons — and fill up these bonds. Such a phenomenon is called recombination. At normal temperatures in the semiconductor material, free electrons occur constantly, and recombination of electrons and holes takes place.

If a piece of semiconductor material is put into an electric field by applying a positive or negative terminal to its ends, for instance, electrons will move through the lattice towards the positive electrode and holes — to the negative one. The conductivity of a semiconductor can be enhanced considerably by putting specially selected admixtures to it — metal or nonmetal ones. In the lattice the atoms of these admixtures will replace some of the atoms of the semiconductors. Let us remind ourselves that external shells of atoms of germanium and silicon contain four valence electrons, and that electrons can only be taken from the external shell of the atom. In their turn, the electrons can be added only to the external shell, and the maximum number of electrons on the external shell is eight.

When an atom of the admixture that has more valence electrons than required for valence, bonds with neighboring atoms of the semiconductor, additional free electrons capable of moving along the lattice occur on it. As a result, the electro-conductivity of the semiconductor increases. As germanium and silicon belong to the fourth group of the periodic table of chemical elements, donors for them may be elements of the fifth group, which have five electrons on the external shell of atoms. Phosphorus, arsenic, and stibium belong to such donors (donor admixture).

If admixture atoms have fewer electrons than needed for valence bonds with surrounding semiconductor atoms, some of these bonds turn out to be vacant and holes will occur in them. Admixtures of such a kind are called p-type ones because they absorb (accept) free electrons. For germanium and silicon, p-type admixtures are elements from the third group of the periodic table of chemical elements, external shells of atoms of which contain three valence electrons. Boron, aluminum, gallium, and indium can be considered p-type admixtures (accepter admixture).

In the crystal structure of a pure semiconductor, all valence bonds of neighboring atoms turn out to be fully filled, and occurrence of free electrons and holes can be caused only by deformation of lattice, arising from thermal or other radiation. Because of this, conductivity of a pure semiconductor is quite low under normal conditions. If some donor admixture is injected, the four electrons of the admixture, together with the same number in the filled valence, bond with the latter. The fifth electron of each admixture atom appears to be "excessive" or "redundant," and therefore can freely move along the lattice.

When an accepter admixture is injected, only three filled valence bonds are formed between each atom of the admixture and neighboring atoms of the semiconductor. To fill up the fourth, one electron is lacking. This valence bond appears to be vacant. As a result, a hole occurs. Holes can move along the lattice like positive charges, but instead of an admixture atom, which has a fixed and permanent position in the crystal structure, the vacant valence bond moves. It goes like this. An electron is known to be an elementary carrier of an electric charge. Affected by different causes, the electron can escape from the filled valence bond, having left a hole which is a vacant valence bond and which behaves like a positive charge equaling numerically the negative charge of the electron. Affected by the attracting force of its positive charge, the electron of another atom near the hole may "jump" into the hole. At that point recombination of the hole and the electron occurs, their charges are mutually neutralized and the valence bond is filled. The hole in this place of the lattice of the semiconductor disappears. In its turn a new hole, which has arisen in the valence bond from which the electron has escaped, may be filled with some other electron which has left a hole. Thus, moving of electrons in the lattice of the semiconductor with a p-type admixture and recombination of them with holes can be regarded as moving of holes. For better understanding one may imagine a concert hall in which for some reason some seats in the first row turn out to be vacant. As spectators from the second row move to the vacant seats in the first row, their seats are taken by spectators of the third row, and so on. One can say that in some sense vacant seats "move" to the last rows of the concert halls, although in fact all the stalls remain screwed to the floor. "Moving" of holes in the crystal is very much like "moving" of such vacant seats.

Semiconductors with electro-conductivity enhanced due to an excess of free electrons caused by admixture injection, are called semiconductors with electron-conductivity or in short, n-type semiconductors. Semiconductors with electro-conductivity influenced mostly by moving of holes are called semiconductors with p-type conductivity or just p-type semiconductors.

There are practically no semiconductors with only electronic or only p-type conductivity. In a semiconductor of n-type, electric current is partially caused by moving of holes arising in its lattice because of an escaping of electrons from some valence bonds, and in semiconductors of p-type current is partially created by the moving of electrons. Because of this, it is better to define semiconductors of the n-type as semiconductors in which the main current carriers are electrons and semiconductors of the p-type as semiconductors in which holes are the main current carriers. Thus, a semiconductor belongs to this or that type depending on what type of current carrier predominates in it. According to this, the other opposite charge carrier for any semiconductor of a given type is a minor carrier.

One should take into account that any semiconductor can be made a semiconductor of n- or p-type by putting certain admixtures into it. In order to obtain the required conductivity, it is enough to put in a very small amount of the admixture, about one atom of the admixture for 10 millions of atoms of the semiconductor. All of this imposes special requirements for the purification of the original semiconductor material, and accuracy in dosage of admixture injection. One should also take into consideration that the speed of current carriers in a semiconductor is lower than in a metal conductor or in a vacuum. Moving of electrons is slowed down by obstacles on their way in the form of inhomogeneities in the crystal. Moving of holes is half as slow because they move due to jumping of electrons to vacant valence bounds. Mobility of electrons and holes in a semiconductor is increased when the temperature goes up. This leads to an increase of conductivity of the semiconductor.

The functioning of most semiconductors is based on the processes taking place in an intermediate layer formed in the semiconductor, at the boundary of the two zones with the conductivities of the two different types: "p" and "n". The boundary is usually called the p–n junction or the electron–hole junction, in accordance with the main characteristics of the type of main charge carriers in the two adjoining zones of the semiconductor.

There are two types of p–n junctions: planar and point junctions, which are illustrated schematically in Figure 7.38. A planar junction is formed by moving a piece of the admixture — for instance indium, to the surface of the germanium — of n-type, and further heating until the admixture is melted. When a certain temperature is maintained for a certain period of time, there is diffusion of some admixture atoms to the plate of the semiconductor, to a small depth, and a zone with conductivity opposite to that of the original semiconductor is formed. In the above case, it is p-type, for n-germanium.

Point junction results from tight electric contact of the thin metal conductor (wire), which is known to have electric conductivity, with the surface of the p-type semiconductor. This was the basic principle on which the first crystal detectors operated. To decrease dependence of diode properties on the position of the pointed end of the wire on the surface of the semiconductor, and the clearance of its momentary surface point, junctions are formed by fusing the end of the thin metal wire to the surface of a semiconductor of the n-type. Fusion is carried the moment a short-term powerful pulse of electric current is applied. Affected by the heat formed for this short period of time, some electrons escape from atoms of the semiconductor, which are near the contact point, and leave holes. As a result of this some small part of the n-type semiconductor in the immediate vicinity of the contact turns into a semiconductor of the p-type (area 3 on Figure 7.38a).

Each part of semiconductor material, taken separately (that is before contacting), was neutral, since there was a balance of free and bound charges (Figure 7.39a). In the n-type area, concentration of free electrons is quite high and that of holes quite low. In the p-type area on the contrary, concentration of holes is high, and that of electrons low. Joining of semiconductors with different concentrations of main current carriers, causes diffusion of these carriers through the junction layer of these materials: the main carriers of the p-type semiconductor — holes — diffuse to the n-type area because the concentration of holes in it is very low. And vice versa, electrons from the n-type semiconductor,

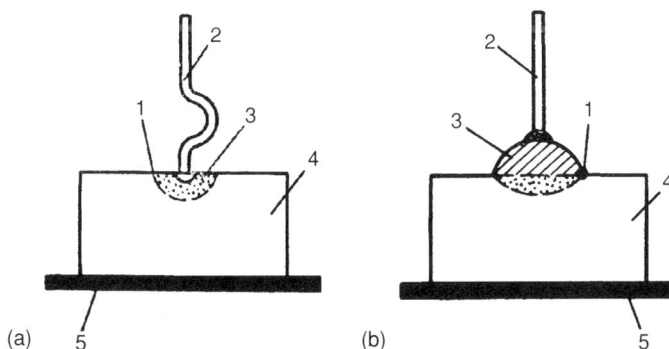

(a)   (b)

FIGURE 7.38
Construction of point (a) and planar (b) p—n junctions of the diode. 1 — p—n junction; 2 — wire terminal; 3 — p-area; 4 — crystal of n-type; 5 — metal heel piece.

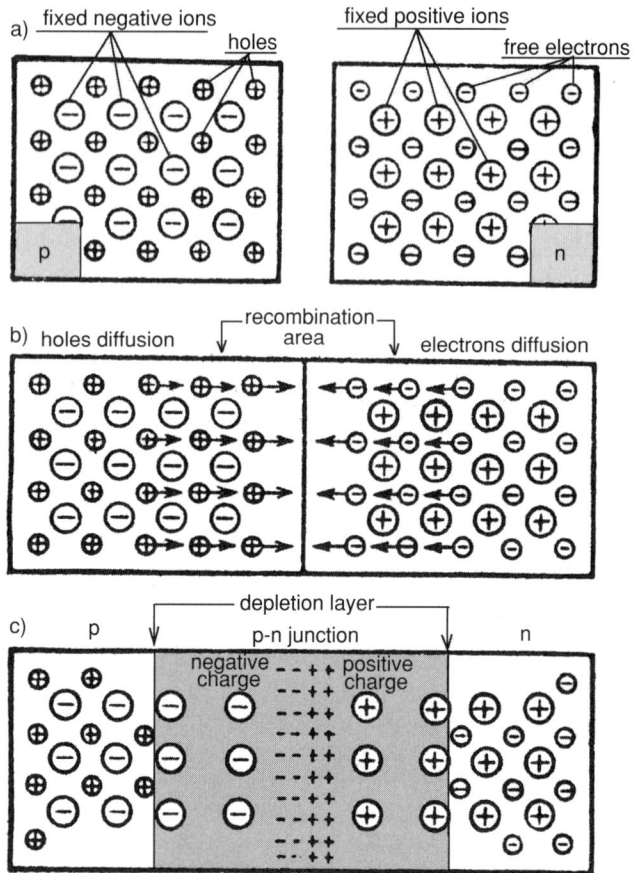

FIGURE 7.39
Formation of a blocking layer when semiconductors of different conductivity are connected.

with a high concentration of them, diffuse to the n-type area, where there are few of them (Figure 7.39b).

On the boundary of the division of the two semiconductors, from each side a thin zone with conductivity opposite to that of the original semiconductor is formed. As a result, on the boundary (which is called a p-n junction) a space charge arises (the so-called potential barrier), which creates a diffusive electric field and prevents the main current carriers from flowing after balance has been achieved (Figure 7.39c).

Strongly pronounced dependence of electric conductivity of a p–n junction from polarity of external voltage applied to it, is typical of the p–n junction. This can never be noticed in a semiconductor with the same conductivity. If voltage applied from the outside creates an electric field coinciding with a diffusive electric field, the junction will be blocked and current will not pass through it (Figure 7.40). Moreover, moving of minor carriers becomes more intense, which causes enlargement of the blocking layer and lifting of the barrier for main carriers. In this case it is usually said that the junction is reversely biased. Moving of minor carriers causes a small current to pass through the blocked junction. This is the so-called reverse current of the diode, or leakage current. The smaller it is, the better the diode is.

When the polarity of the voltage applied to the junction is changed, the number of main charge carriers in the junction zone increases. They neutralize the space charge of the blocking layer by reducing its width and lowering the potential barrier that

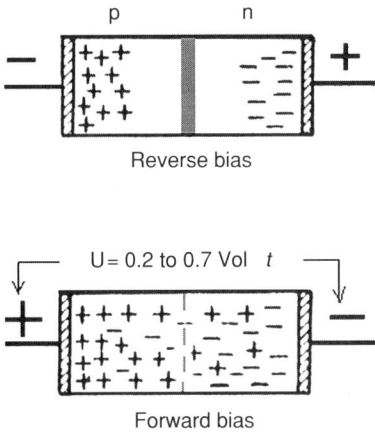

Reverse bias

Forward bias

FIGURE 7.40
p—n-junction with reverse and forward bias.

U = 0.2 to 0.7 Vol *t*

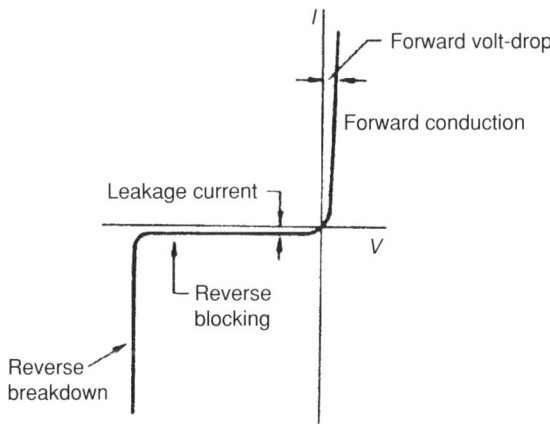

Forward volt-drop

Forward conduction

Leakage current

V

Reverse
blocking

Reverse
breakdown

FIGURE 7.41
Volt–ampere characteristic of a single p–n junction (diode).

prevented the main carriers from mobbing through the junction. It is usually said that the junction is forward biased. The voltage required for overcoming of the potential barrier in the forward direction is about 0.2 V for germanium diodes, and 0.6 to 0.7 V for silicon ones.

To overcome the potential barrier in the reverse direction, tens and sometimes even thousands of volts are required. If the barrier is overpassed, irreversible destruction of the junction and its breakdown takes place, which is why threshold values of reverse voltage and forward current are indicated for junctions of different appliances. Figure 7.41 illustrates an approximate volt–ampere characteristic of a single junction, which is dependence of current passing through it on the polarity and external voltage applied to the junction. Currents of forward and reverse directions (up to the breakdown area) may differ by tens and hundreds of times. As a rule, planar junctions withstand higher voltages and currents than point ones but not work properly with high frequency currents.

7.8.3 Diode Switch of Electric Circuits

A diode can be used for switching of electric signals (Figure 7.42), like relays. But in this device control voltage u_{contr} must be higher than the voltage of the power source (and of course higher than the working voltage of the load). Depending on the polarity of the control voltage, the working point of the diode may be biased to the direct and reverse branch of the volt–ampere characteristic.

The diode operates either in the mode of conductivity — the switch is closed (point *A* in Figure 7.42) or in the cut-off mode — the switch is open (point *B*). To make a long story short, if the polarity of the control voltage coincides with the polarity of the power source (E), the diode is open and current flows to the load from the power source. If it does not, the diode is closed and the load is without current (it is implied that the voltage of the control signal is high by the absolute value than the voltage of the power source). If changes in the value control voltage of the blocking polarity are applied to the diode, the diode will be automatically enabled and let current flow to the load, the moment the control voltage is lower than the voltage of the power source.

There are also types of diodes with characteristics specially selected for operation in the switch mode. For example, "tunnel diodes" (Figure 7.43). The tunnel diode was invented by Leo Izaki in 1958. It was named "tunnel" after the effect on the basis of which it operates. This is a very complex physical effect, which can be described in simple terms as original behavior of electrons which cannot pass through the potential barrier of the blocking layer in the usual way, and therefore pass under the barrier, as if through a "tunnel." Such a tunnel is formed when there is a high concentration of admixtures (semiconductor degenerates to "semimetal") and junction depletion region so narrow

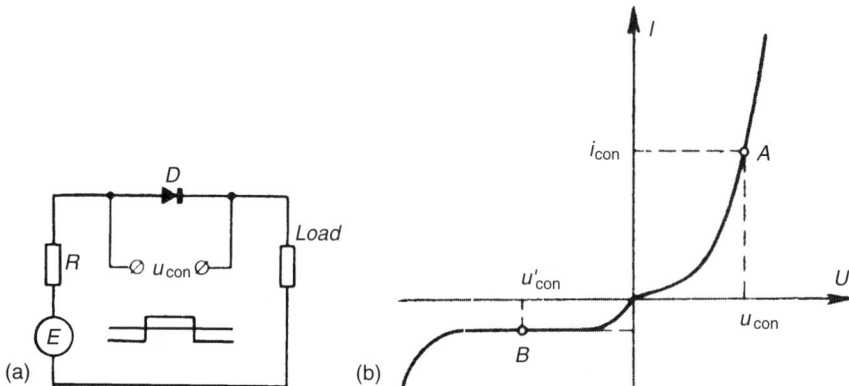

FIGURE 7.42
Diode signal switch and changes of the working point of the diode on the static volt–ampere characteristic during work.

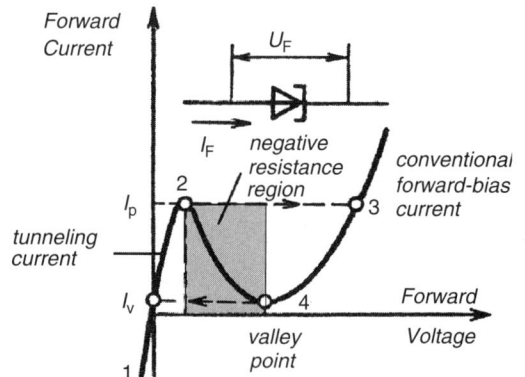

FIGURE 7.43
Notation and volt–ampere characteristic of a tunnel diode.

that both holes and electrons can transfer across the p–n junction by a quantum mechanical action called "tunneling." The conductivity of such material becomes close to the conductivity of the metal when forward or reverse voltage is increased (the "tunnel effect"). However, further increase of forward voltage causes weakening of the effect and the behavior of the diode becomes regular, which is why after a certain sharp curve (the area with negative resistance (NR) where the positive voltage increment corresponds to the negative current increment) the volt–ampere characteristic of the tunnel diode becomes regular. In other words, the resistance to current flow through the tunnel diode increases as the applied voltage increases over a portion of its region of operation. Outside the NR region, the tunnel diode functions essentially the same as a normal diode.

The three most important aspects of the characteristic curve are (1) the forward current increase to a peak (I_p) with a small applied forward bias, (2) the decreasing forward current with an increasing forward bias to a minimum valley current (I_v), and (3) the normal increasing forward current with further increases in the bias voltage. The portion of the characteristic curve between I_p and I_v is the region of NR.

Tunneling causes the NR action and is so fast that no transit-time effects occur even at microwave frequencies. The lack of a transit-time effect permits the use of tunnel diodes as super high frequency switch in a wide variety of microwave applications. Unfortunately, the tunnel diode is difficult to use in power applications due to low working voltages and strong temperature dependence.

7.8.4   The Transistor: A Piece of Silicon with Three Wires that Has Changed the World

Despite the wide use of semiconductor diodes, vacuum tubes have continued to be used for amplification, generation of signals, and high-speed switching of electric current for a long period of time. Over the decades, vacuum tubes were improved and used in more and more complicated circuitry. At the 1939 World Fair, for example, vacuum tubes were showcased in fully electronic television, and by 1945 the high-speed computer ENIAC was built, containing more than 17,000 tubes. Although successful, ENIAC and its offspring showed the real limitations of vacuum tubes: to make more powerful computers more tubes were needed, but at some point available space and energy prevented further growth. Vacuum tubes were bulky, used a lot of energy, were somewhat fragile and easily overheated. Engineers knew that they needed to find something better. The telephone companies had problems with vacuum tubes, too, and hoped to find something better for switching telephone calls. The idea of somehow using semiconductors (solid materials such as silicon that conduct electricity, but not as well as a conductor such as copper) had been tossed about before World War II, but the knowledge about how they worked was scant, and manufacturing semiconductors was difficult.

In 1945, however, the vice president for research at Bell Laboratories established a research group to look into the problem. The group was led by William Shockley and included Walter Brattain, John Bardeen, and others, physicists who had worked with quantum theory, especially in solids. The team was talented and worked well together. In 1947, John Bardeen and Walter Brattain, with colleagues, created the first successful amplifying semiconductor device (Figure 7.44). They called it as a transistor (from "transfer" and "resistor"). In 1950, Shockley made improvements to it that made it easier to manufacture. His original idea eventually led to the development of the silicon chip. Shockley, Bardeen, and Brattain won the 1956 Nobel Prize for the development of the transistor. It allowed electronic devices to be built smaller, lighter, and even cheaper.

Bell Laboratories began to license the use of transistors (for a royalty) and offered courses on transistor technology, helping spread the word throughout the industry.

FIGURE 7.44
(a) Inventors of the transistor, the Nobel Prize laureates: William Bradford Shockley, John Bardeen, and Walter Houser Brattain. (b) This is how the first laboratory sample of a transistor looked.

W. Shockley left Bell Laboratories in 1955 and served as visiting professor and consultant at various universities and corporations. Bardeen and Brattain continued in research (Bardeen later won another Nobel prize).

W. Shockley started his own semiconductor company in Palo Alto to develop transistors and other devices. The business changed hands a few times and finally folded up in 1968, but its staff went on to invent the integrated circuit (the "chip") and to found the Intel Corporation. In 1963, Shockley was appointed professor of engineering at Stanford University, where he taught until 1975.

It can be seen in Figure 7.45 that a transistor contains two semiconductor diodes connected together, and has a common area. Two utmost layers of the semiconductor (one of them is called an "emitter" and the other a "collector") have p-type conductivity with a high concentration of holes, and the intermediate layer (called a "base") has n-type

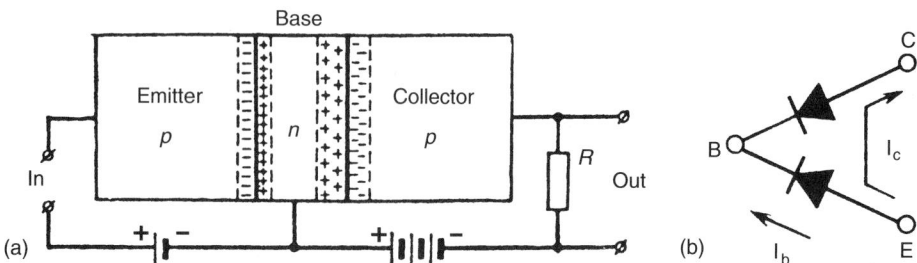

FIGURE 7.45
Circuit (a) and the principle(b) of operation of a transistor.

conductivity with a low concentration of electrons. In electric circuits, low voltage is applied to the first (the emitter) p–n junction because the junction is connected in the forward (carrying) direction, and much higher voltage is applied to the second (the collector) junction, in the reverse (cut-off) direction. In other words, emitter junction is a forward biased and collector junction is a reverse biased. The collector junction remains blocked until there is no current in the emitter-base circuit. The resistance of the whole crystal (from the emitter to the collector) is very high. As soon as the input circuit (Figure 7.45) is closed, holes from the emitter seem to be injected (emitted) to the base and quickly saturate it (including the area adjacent to the collector). As the concentration of holes in the emitter is much higher than the concentration of electrons in the base, after recombination there are still many vacant holes in the base area, which is affected by the high voltage (a few or tens of volts) applied between the base and the collector, easily overpassing the barrier layer between the base and the collector.

Increased concentration of holes in the cutoff collector junction causes the resistance of this junction to fall rapidly, and it begins to conduct current in the reverse direction. The high strength of the electric field in the "base-collector" junction results in a very high sensitivity of the resistance of this junction in the reverse (cutoff) state to a concentration of the holes in it. That is why, even a small number of holes injected from the emitter under the effect of weak input current can lead to sharp changes of conductivity of the whole structure, and considerable current in the collector circuit.

The ratio of collector current to base current is called the "current amplification factor." In low-power transistors, this amplification factor has values of tens and hundreds, and in power transistors — tens. The transistor, named M1752 (Figure 7.46a) uses a very small

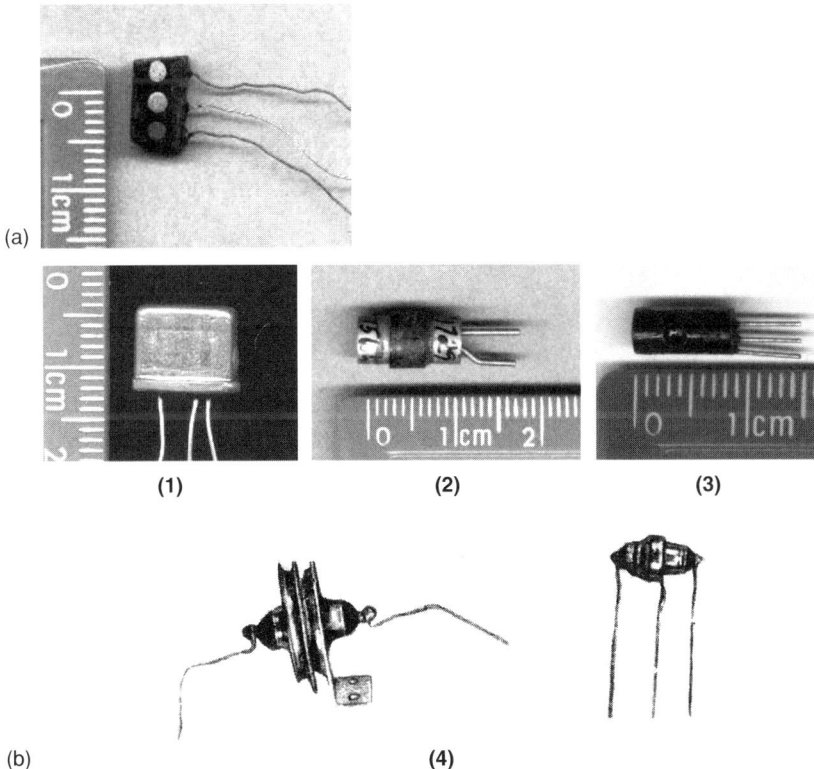

(a)

(1)　　　　　　(2)　　　　　　(3)

(b)　　　　　　　　(4)

FIGURE 7.46
(a) First industrial sample of a transistor produced by Bell in 1951. (b) First industrial transistors produced by Motorola (1), Western Electric (2), General Electric (3), and by the former U.S.S.R. (4) in the 1950s.

FIGURE 7.47
A transistor in a glass case, found by the author in an old electric device produced by AEG in 1961.

plastic package. It is identified, like most early Bell types, by a four-digit number, coded in colored paint dots on the package. The colors are purple–green–red, namely 7–5–2, to which 1000 must be added. The lead wires are made of steel, not copper, because the thermal expansion coefficient of steel matches that of glass and a small glass block was used to anchor the wires.

Some companies began to produce transistors using their electro-vacuum technology for the production of vacuum tubes (Figure 7.47).

7.8.5 Bipolar ... Unijunction ... Field ...

In the 1970s transistor engineering developed very rapidly. Hundreds of types of transistors and new variants of them appeared (Figure 7.48). Among them appeared transistors with reverse conductivity or n–p–n transistors, and also unijunction transistors (as it contains only one junction such a transistor is sometimes called a two-base diode) (Figure 7.49). This transistor contains one junction formed by welding a core made from p-material to a single-crystal wafer made from n-type material (silicon). The two outlets, serving as bases, are attached to the wafer. The core, placed asymmetrically with regard to the base, is called an emitter. Resistance between the bases is about a few thousand ohms. Usually the base B_2 is biased in a positive direction from the base B_1. Application of positive voltage to the emitter causes strong current of the emitter (with insignificant voltage drop between the emitter E and the base B_1). One can observe the area of NR (see Figure 7.49) on the emitter characteristic of the transistor where the transistor is very rapidly enabled, operating like a relay.

In fact, modern transistors (Figure 7.50), are characterized by such a diversity of types that it is simply impossible to describe all of them in this book devoted to relays, therefore only a brief description of the most popular types of modern semiconductor devices, and the relays based on them, are presented here.

Besides the transistors described above, which are called bipolar junction transistors or just "bipolar transistors" (Figure 7.51), the so-called field effect transistors (FET — Figure 7.52) have become very popular recently. The first person to attempt to construct a field effect transistor in 1948 was again William Shockley. But it took many years of additional experiments to create a working FET with a control p–n junction called a "unitron" (unipolar transistor), in 1952. Such a transistor was a semiconductor three-electrode device, in which control of the current caused by the ordered motion of charge carriers of the same sign between two electrodes, was carried out with a help of an electric field (that is why it is called "field") applied to the ford electrode.

Electrodes between which working currents pass are called source and drain electrodes. The source electrode is the one through which carriers flow into the device. The third electrode is called a "gate". Change of value of the working current in a unipolar transistor is carried out by changing the effective resistance of the current conducting area, the semiconductor material between the source and the drain called the "channel".

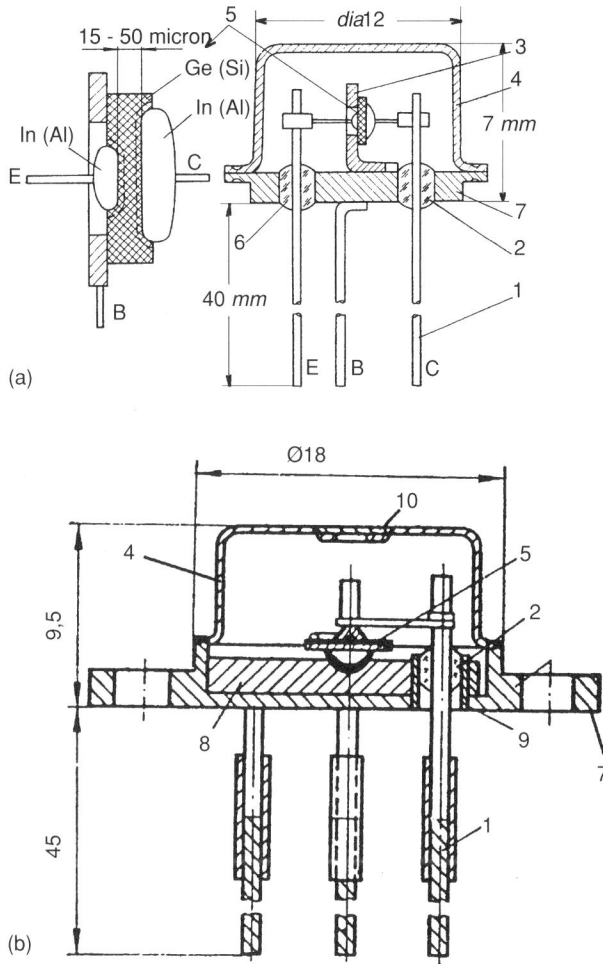

FIGURE 7.48
Transistors produced in the 1970s: (a) low-power transistor; (b) power transistor. 1 — outlets; 2 and 6 — glass insulators; 3 — crystal holder; 4 — protection cover; 5 — silicon (germanium) crystal; 7 — flange; 8 — copper heat sink; 9 — Kovar bushing; 10 — hole for gas removal after case welding and disk for sealing-in.

That change is made by increasing or decreasing area 5 (Figure 7.52). Increase of voltage of the initial junction bias leads to expansion of the depletion layer. As a result, the rest area of the section of the conductive channel in the silicon decreases and the transistor is blocked, and vice versa, when the value of the blocking voltage on the gate decreases, the area (5) depleted by current carriers contracts and turns into a pointed wedge. At the same time, the section of the conductive channel increases and the transistor is enabled.

Depending on the type of the conductivity of semiconductor material of the channels, there are unipolar transistors with p and n channels. Because of the fact that control of the working current of unipolar transistors is carried out with the help of a channel, they are also called "channel transistors". The third name of the same semiconductor device — a "field transistor" or FET (field effect transistor) points out that working current control is carried out by an electric field (voltage) instead of electric current as in a bipolar transistor. The latter peculiarity of unipolar transistors, which allows them to obtain very

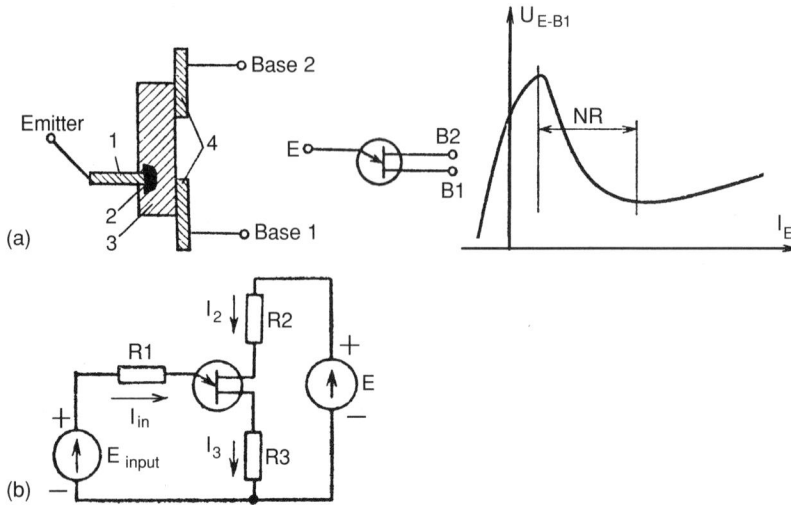

FIGURE 7.49
(a) A unijunction transistor (or two-base diode) and (b) its circuit. 1 — p-type core; 2 — p—n-junction; 3 — n-type plate; 4 — ohmic contacts; (c) NR — negative resistance area.

high input resistances, estimated in tens and hundreds of meg-ohms, determined their most popular name: field transistors (Figure 7.53).

It should be noted that apart from field transistors with p–n junctions between the gate and the channel (FET), there are also field transistors with an insulated gate: metal-oxide-semiconductor FET (MOSFET) (Figure 7.54). The latter were suggested by S. Hofstein and F. Heiman in 1963. Field transistors with an insulated gate appeared as a result of searching for methods to further increase input resistance and frequency range extensions of field transistors with p–n junctions. The distinguishing feature of such field transistors is that the junction biased in a reverse direction is replaced with a control structure "metal-oxide-semiconductor," or a MOSFET-structure in abbreviated form. As shown in Figure 7.52 this device is based on a silicon mono crystal, in this case of p-type. The source and drain areas have conductivity opposite to the rest of the crystal, that is of the n-type.

The distance between the source and the drain is very small, usually about 1 μm. The semiconductor area between the source and the drain, which is capable of conducting current under certain conditions, is called a channel, as in the previous case. In fact the channel is an n-type area formed by diffusion of a small amount of the donor admixture to the crystal with p-type conductivity. The gate is a metal plate covering source and drain zones. It is isolated from the mono crystal by a dielectric layer only 0.1 μm thick. The film of silicon dioxide formed at this high temperature is used as a dielectric. Such film allows us to adjust the concentration of the main carriers in the channel area by changing both value and polarity of the gate voltage. This is the major difference of MOSFET, as opposed to field ones with p–n junctions, which can only operate well with blocking voltage of the gate. The change of polarity of the bias voltage leads to junction unblocking and to a sharp reduction of the input resistance of the transistor.

The basic advantages of MOSFET are as follows: first there is an insulated gate allowing an increase in input resistance by at least 1000 times in comparison with the input resistance of a field transistor with a p–n junction. In fact it can reach a billion megohms. Second, gate and drain capacities become considerably lower and usually do not exceed

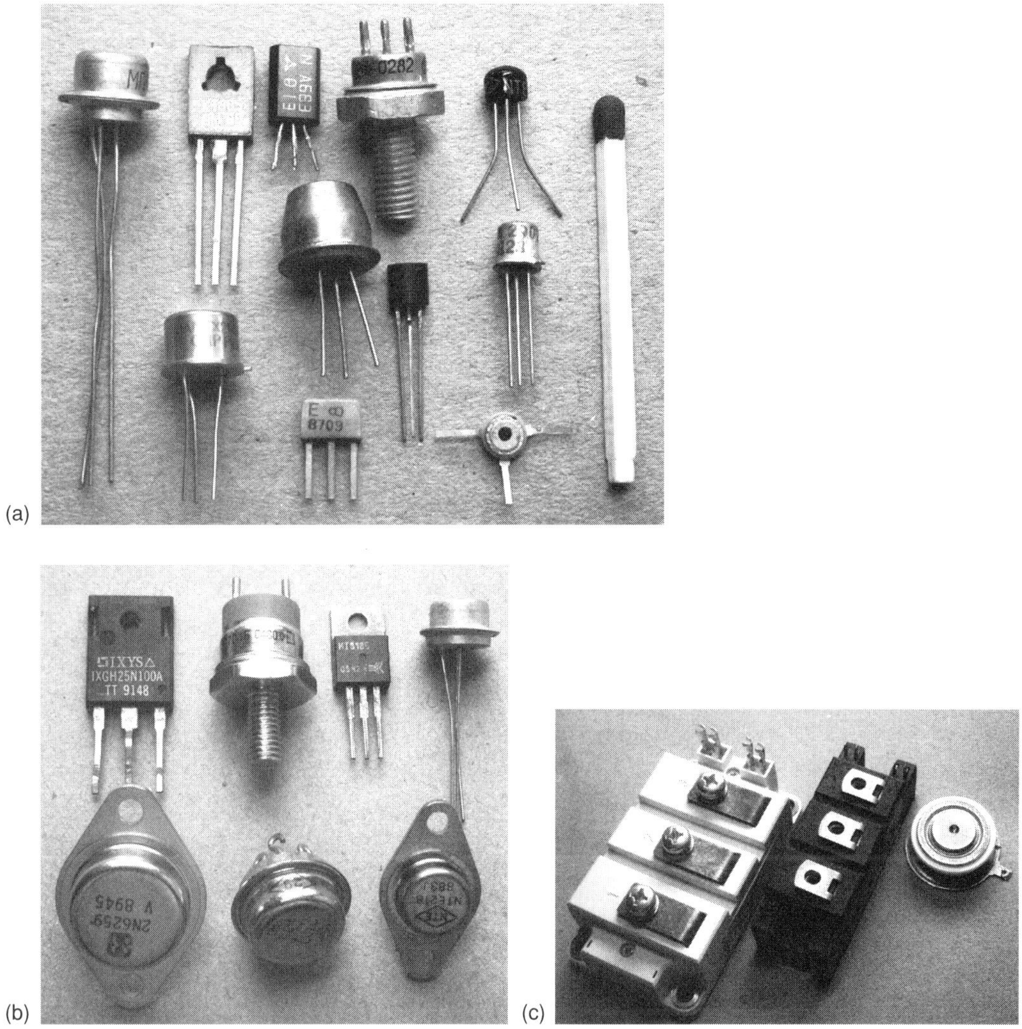

FIGURE 7.50
This is how modern (a) low-power transistors, (b) power transistors and (c) high power transistors look.

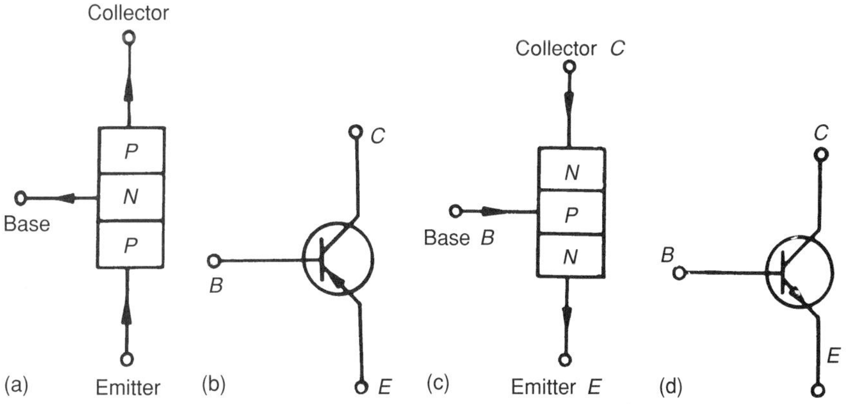

FIGURE 7.51
Structure and symbolic notation on the schemes of bipolar transistors of (a, b) p—n—p and (c, d) n—p—n types.

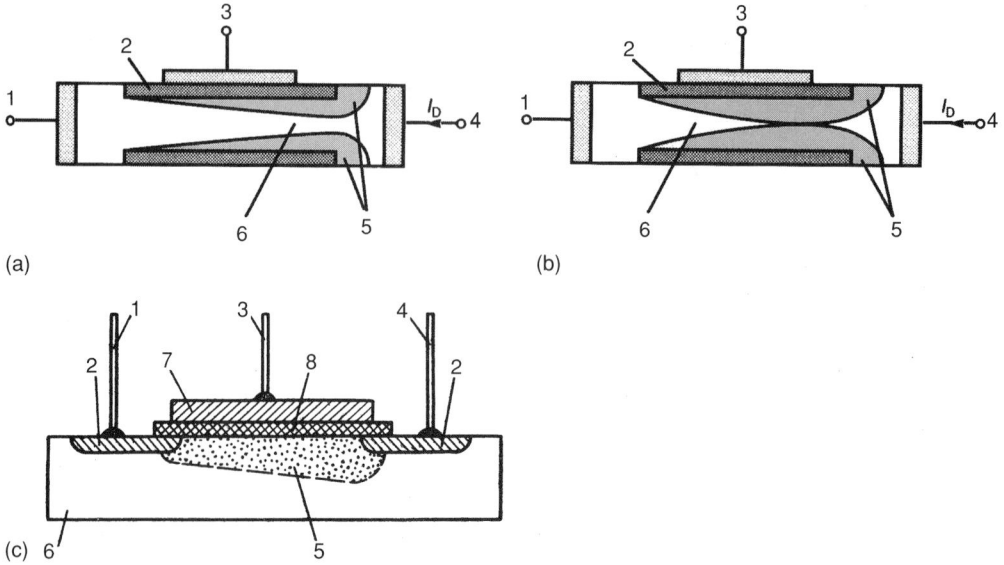

FIGURE 7.52

(a, b) Simplified structure of FET and (c) MOSFET. 1 — source; 2 — n-type admixture; 3 — gate; 4 — drain; 5 — area consolidated by current carriers (depletion layer); 6 — conductive channel in silicon of p-type; 7 — metal; 8 — silicon dioxide.

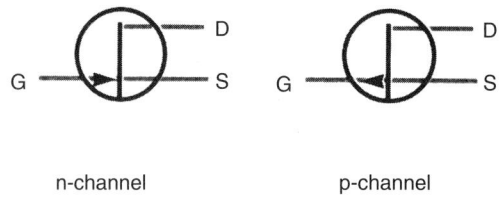

FIGURE 7.53

Symbolic notation of FET with n- and p-channels.
G — gate; S — source; D — drain.

n-channel   p-channel

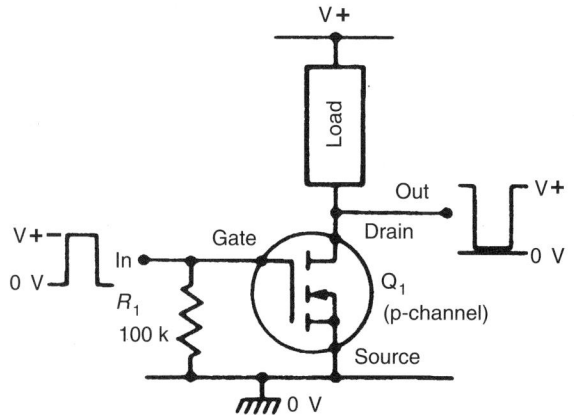

FIGURE 7.54

Symbolic notation and the circuit of a MOSFET.

1 to 2 pF. Third, the limiting frequency of MOSFET can reach 700 to 1000 MHz, which is at least ten times higher than that of standard field transistors.

Attempts to combine in one switching device the advantages of bipolar and field transistors led to the invention of a compound structure in 1978, which was called a "pobistor" (Figure 7.55). The idea of a modular junction of crystals of bipolar and field transistors in the same case was employed by "Mitsubishi Electric" to create a powerful switching semiconductor module (Figure 7.56).

Further development of production technology of semiconductor devices allowed development of a single-crystal device with a complex structure with properties of a "pobistor": an insulated gate bipolar transistor (IGBT) transistor. The IGBT is a device which combines the fast-acting features and high power capabilities of the bipolar transistor with the voltage control features of the MOSFET gate. In simple terms, the collector–emitter characteristics are similar to those of the bipolar transistor, but the control features are those of the MOSFET. The equivalent circuit and the circuit symbol are illustrated in Figure 7.57.

Such a transistor (Figure 7.58) has a higher switching power than FET and bipolar transistors and its operation speed is between that of FET and bipolar transistors. Unlike bipolar transistors, the IGBT does not operate well in the amplification mode and is designed for use in the switching (relay) mode as a powerful high-speed switch.

The IGBT is enabled by a signal of positive (with regard to the emitter) polarity, with voltage not more than 20 V. It can be blocked with zero potential on the gate, however, with some types of loads a signal of negative polarity on the gate may be required for reliable blocking (Figure 7.59). Many companies produce special devices for IGBT control. They are made as separate integrated circuits or ready-to-use printed circuit cards, so-called drivers (Figure 7.60). Such drivers are universal as a rule and can be

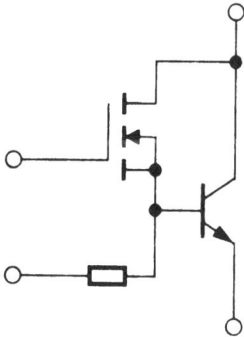

FIGURE 7.55
Compound structure — "Pobistor."

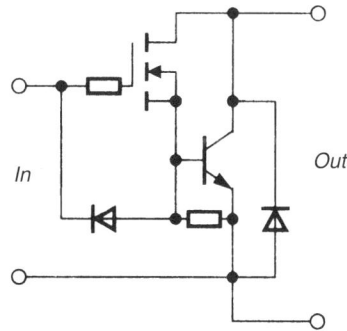

FIGURE 7.56
Scheme of a power switching module CASCADE-CD, with a working voltage of 1000 V and currents more than 100 A ("Mitsubishi Electric").

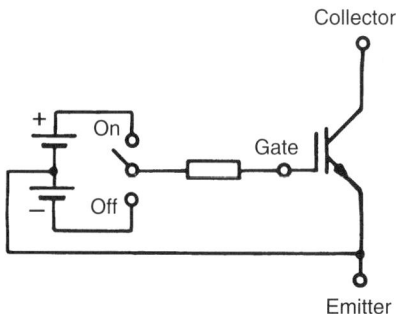

FIGURE 7.57
Insulated gate bipolar transistor (IGBT).

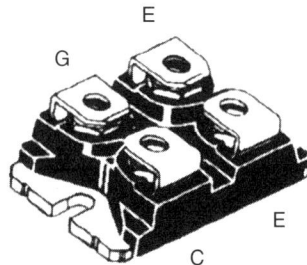

FIGURE 7.58
An IGBT IXDN75N120A produced by IXYS with a switched current up to 120 A, and maximum voltage of up to 1200 V (dissipated power is 630W). With such high parameters the device is quite small in size: 38 × 25 × 12 mm.

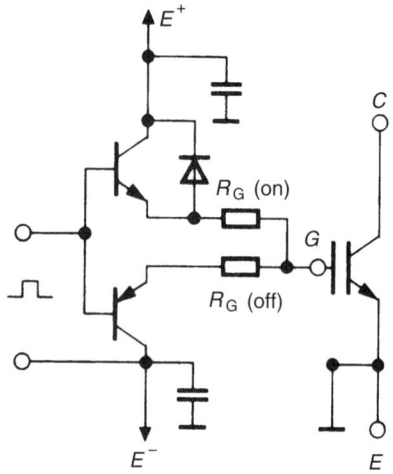

FIGURE 7.59
Model scheme of the IGBT control, providing pulses of opposite
polarity on the gate required for reliable blocking and unblocking
of the transistor.

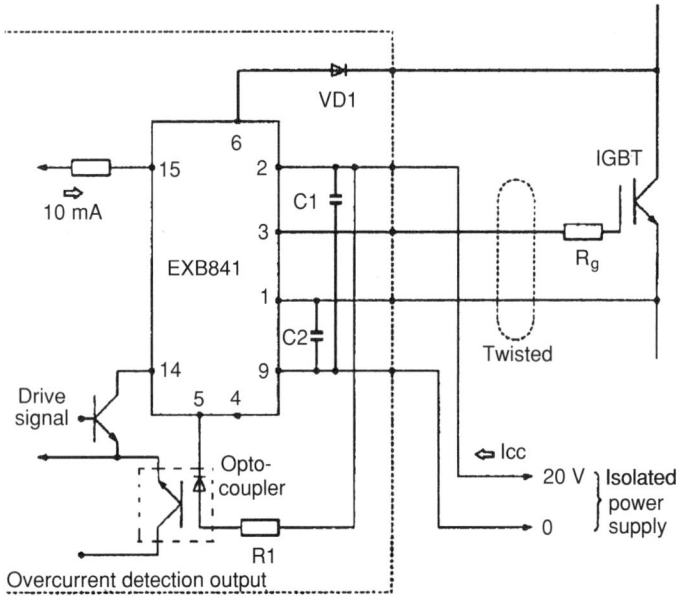

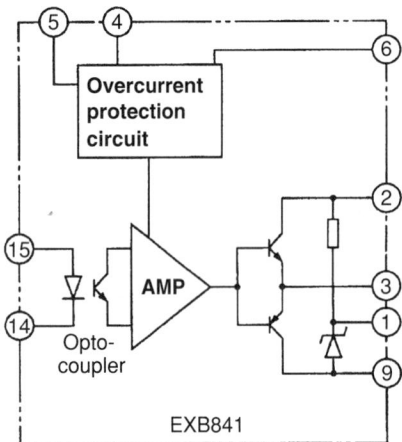

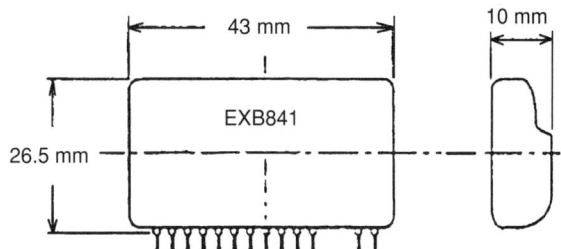

FIGURE 7.60
IGBT-driving hybrid integral circuit EXB841 type (Fuji Electric).

applied to any type of power IGBT. Apart from forming control signals of the required level and form, such devices often protect the IGBT from short circuits.

In spite of progress in IGBT development, different firms continue producing standard high-power bipolar transistors in capsule packages (Figure 7.50c). In power devices such transistors, equipped with big aluminum heat sinks and fans, are united to power blocks (Figure 7.61), which can weigh tens of kilograms. Heat sinks for such transistors are made in the form of two separate halves pulled together with special screw-bolts, with insulation covering between them, inside of which there is a transistor.

To provide a good thermal contact between the transistor and the heat sink the hold-down pressure must be strong enough, but must not exceed the threshold value of the transistor. Special torque spanners or spring disks with a scale are used (Figure 7.62). To increase switched current transistors are connected in parallel (Figure 7.63). Current grading through appliances connected in parallel is carried out with the help of low-value resistors cut in to a circuit of emitters of the transistors. When there is a great number of parallel connected transistors (Figure 7.63b), the total current of all base electrodes (control current) becomes commensurable with the working (collector) current, which is why in this case an additional transistor is used on the input side (Figure 7.63b).

FIGURE 7.61
Unit of bipolar transistors in a capsule package equipped with aluminum heat sink.

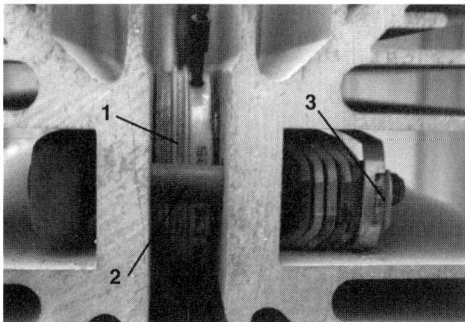

FIGURE 7.62
Attachment point of power wafer transistor in a heat sink. 1 — transistor; 2 — insulated screw bolt; 3 — torque measuring disk with a scale.

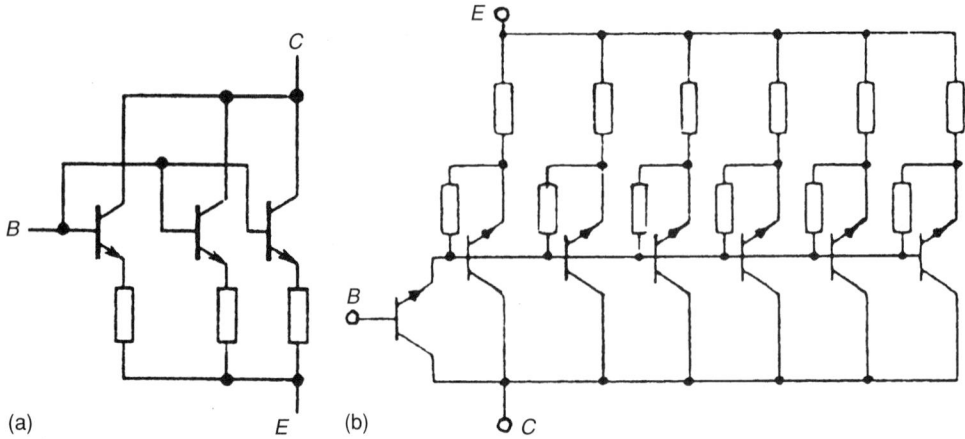

FIGURE 7.63
Parallel connection of bipolar transistors.

7.8.6   From Micromodules to Microchips

As engineering tends to develop in different, often opposite directions, micro-modular electronics, along with power transistor modules, also began to develop in the 1950–60's (Figure 7.64a). The first compact modules were produced from standard elements placed on printed circuit cards, assembled to a solid pack or a stack, and then from special elements including case-free transistors in the form of a ball 1 to 1.5 mm in diameter, with very thin outlets from golden wire.

Such micro-modules (so-called "multi-chip circuit") were mounted on small ceramic plates, with high packaging density (up to 30 elements per 1 s m^3) (Figure 7.64b). Some plates were linked with the help of welding or soldering. The ready micro-module was covered with epoxy resin. The use of this new (for that time) technology allowed miniaturization of feeble current equipment by almost 20 times.

In those days, electrical engineers were aware of the potential of digital electronics, however, they faced a big limitation known as the "Tyranny of Numbers." This was the metaphor that described the exponentially increasing number of components required to design improved circuits, against the physical limitations derived from the number of components that could be assembled together. Both Jack St Clair Kilby (born in 1923) at Texas Instruments, and Robert Norton Noyce (1927–1990) at Fairchild Semiconductor were working independently on a solution to this problem during 1958 and 1959 (Figure 7.64c). The solution was found in the monolithic (meaning formed from a single crystal) integrated circuit (Figure 7.64d). Instead of designing smaller components, they found the way to fabricate entire networks of discrete components in a single sequence by laying them into a single crystal (chip) of semiconductor material. Kilby used germanium and Noyce used silicon.

Kilby wrote in a 1976 article titled "Invention of the IC,"

> Further thought led me to the conclusion that semiconductors were all that were really required — that resistors and capacitors (passive devices), in particular, could be made from the same material as the active devices (transistors). I also realized that, since all of the components could be made of a single material, they could also be made *in situ* interconnected to form a complete circuit.

Two electrical engineers, working separately, each filed for patents for an invention. Texas Instruments filed for a patent in February 1959. Fairchild Semiconductor did the

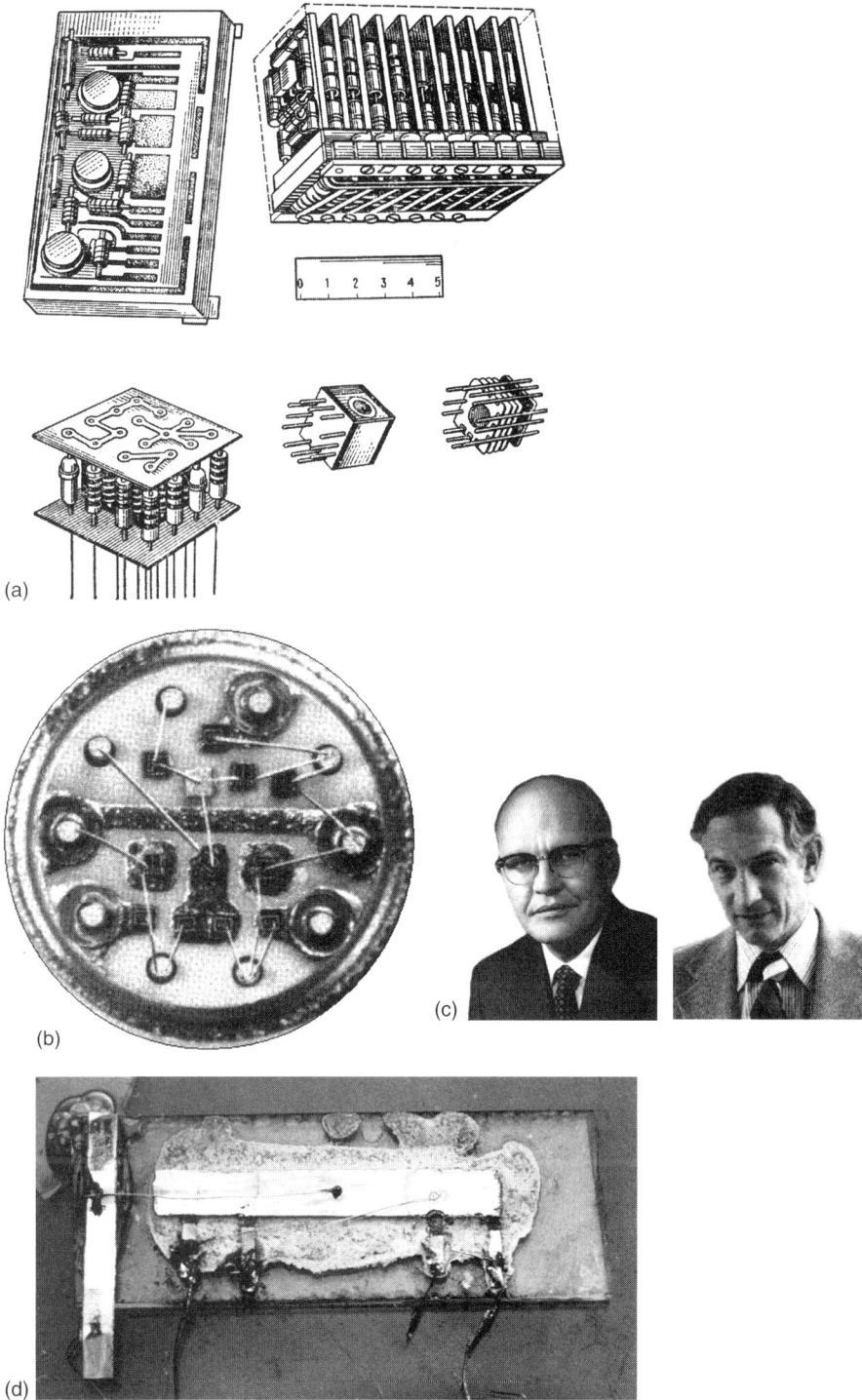

FIGURE 7.64

(a) Electronic micro-modules on discrete elements in the 1950–60's; (b) Multi-chip circuit placed on a standard transistor metal case. (c) Jack Kilby (left) and Robert Noyce (right) in 1958, the year they invented the world's first integral circuit IC. (d) The first Jack Kilby's 7/16 × 1/16-in. IC comprised of only a transistor, three resistors, and a capacitor on a slice of germanium. (From the Texas Instruments Website: www.ti.com)

same in July 1959. Naturally, both firms engaged in a legal battle that lasted through the decade of the 60s until they decided to cross-license their technologies. In the end, the patent No. 3,138,743 ("Miniaturized Electronic Circuits") was issued to Jack S. Kilby and Texas Instruments in 1964, and the patent No. 2,981,877 ("Planar Integrated Circuit") was granted to Robert Noyce.

Jack Kilby was named, along with three Russian scientists, as winners of the 2000 Nobel Prize in physics for their work in laying the foundations of information technology. Zhores Alferov and Herbert Kroemer of Russia, with Kilby from the U.S. share one half of the $1 million prize for work on developing semiconductors. Kilby, of Texas Instruments won the award for his part in the invention of the integrated circuit and as a co-inventor of the pocket calculator. The JK flip-flop (one of variants of bistable SR-multivibrator) is named after him, as is the The Kilby Center, TI's research center for

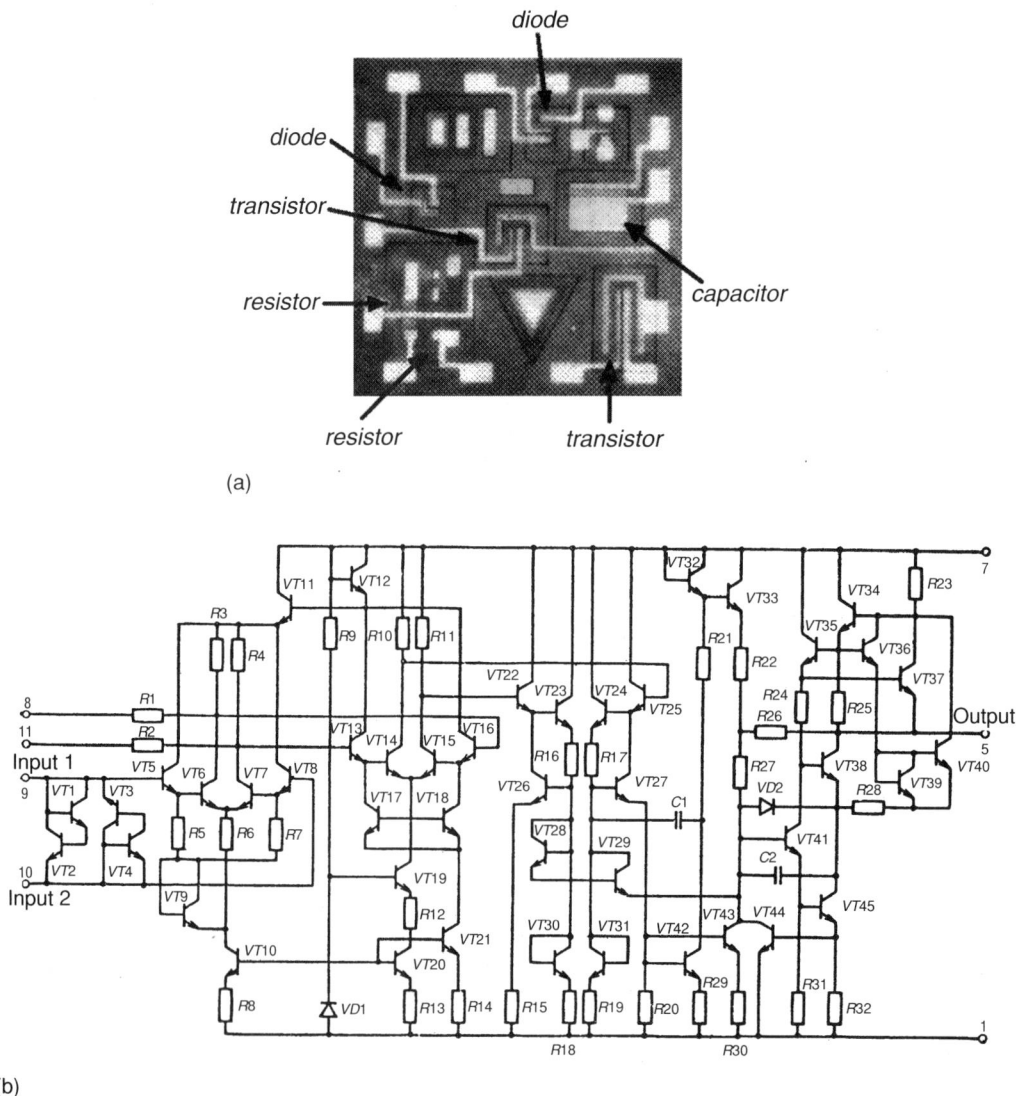

(a)

(b)

FIGURE 7.65
Electronic amplifier produced by integrated technology (45 transistors, 32 resistors, 2 capacitors, and a diode) and placed in the small case: structure fragment (a) and circuit diagram (b).

silicon manufacturing. Kilby officially retired from TI in the 1980s, but he has maintained a significant involvement with the company that continues to this day. In addition, he still consults, travels, and serves as a director on a few boards.

Kilby's and Noyce's inventions changed the world. They virtually created the modern computer industry — the basis of our future. During the 1970s the modern technology of integrated circuit production was developed. Packaging density was brought up to 2000 elements per 1 s m^3. Such devices (Figure 7.65) are made on the basis of an artificial crystal with the p–n junctions of transistors and diodes inside. Capacitors and resistors also form electron-hole junctions biased in a nonconducting direction.

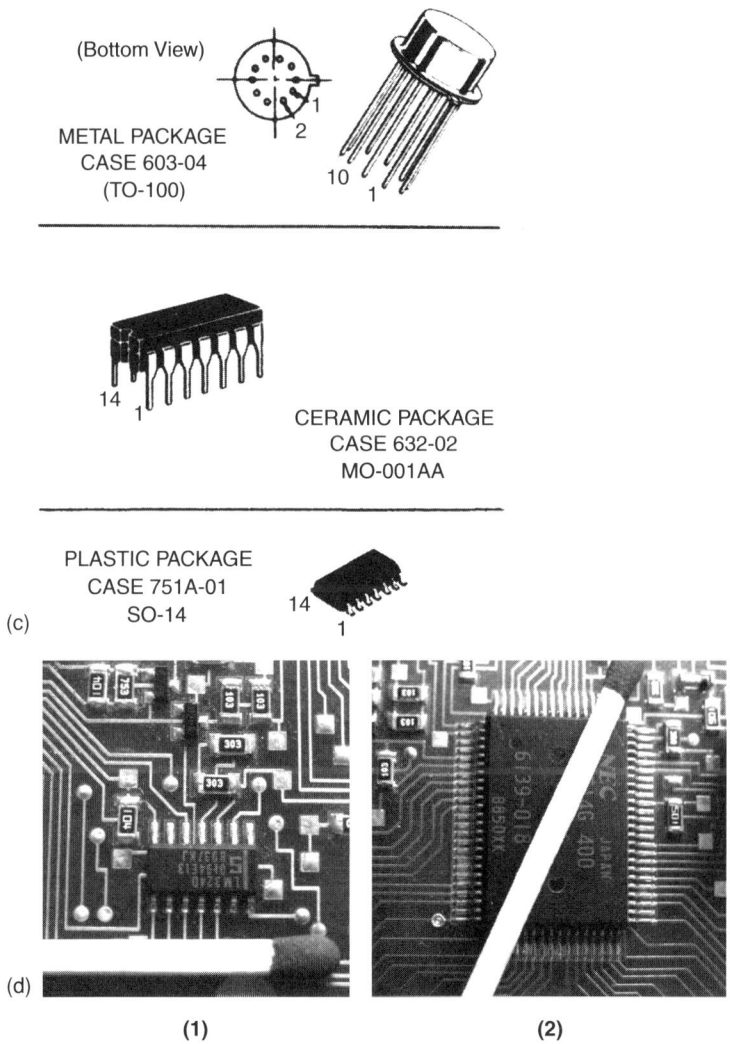

FIGURE 7.65 (*Continued*)
External design (c) of modern integrated circuits with 10 and 14 outlets for surface mounting on the printed circuit board and for ordinary montage. Fragments of the printed circuit (d) board with surface mounting, compared with a match of the standard size. (1) an integrated circuit LM3240; (2) an integrated circuit produced by NEC.

With the help of the same technology numerous variants of triggers (even several items in a case), logical elements, and other circuits with relay characteristics, are made. When a very weak signal is applied to the base of the transistor, a small current I_b begins to flow through the base circuit and is enabled, letting strong current I_c pass through the load. In this case it carries out another relay function — signal amplification.

7.8.7   Transistor Devices with Relay Characteristics

One of the often-used operation modes of a transistor is the switching (that is relay) mode; even a single transistor can work as a high-speed switch (Figure 7.66). For switching current from one circuit to another one, a two-transistor circuit (Figure 7.67) is used. In this circuit stable offset voltage is applied to the base of the transistor (T_2) and control voltage to base T_1.

When $u_{inp} = u_{offset}$, the currents and voltages in the arms of the circuit are the same. If the input voltage (u_{inp}) exceeds the offset voltage (u_{offset}), transistor T_2 is gradually blocked and the whole current flows only through transistor T_1 and load resistor R_{C1}, and vice versa. When input voltage decreases below the level of the offset voltage ($u_{inp} < u_{offset}$), transistor T_1 is blocked and T_2 is unblocked, switching the sole current to the circuit of the resistor R_{C2}.

As is known, contacts of several electromagnetic relays, connected with each other in a certain way, are widely used in automation systems for carrying out the simplest logical operations with electric signals (Figure 7.68). For example, the logical operation AND is implemented with the help of several contacts connected in series, switched to the load circuit (Figure 7.68a). The signal y will be the output of this circuit (that is the bulb will be alight) only if signals on the first input x_1 and on the second input x_2, operate simultaneously (that is when both contacts are closed). Another simple logical operation OR (Figure 7.68b), is implemented with the help of several contacts connected in parallel. In this circuit in order to obtain the signal y on output (that is for switching-on of the bulb), input of signal or on the first input (x_1), or on the second input (x_2), or on both of the inputs simultaneously, is required. Implementation of logical operations with electric circuits is one of the most important functions of relays. Transistor circuits successfully carry out this task. For example, the function NOT can be implemented on any type of single transistor (Figure 7.69).

In the circuit in Figure 7.69, when an input signal is missing, the transistor is blocked, that is the whole voltage of the power source E is applied between the emitter and the

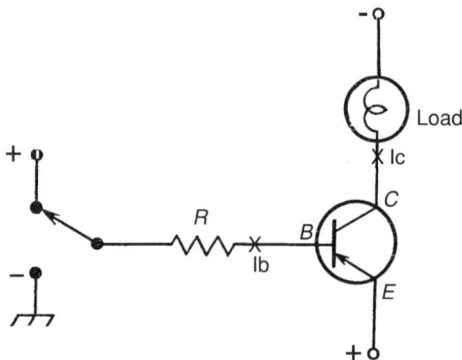

FIGURE 7.66
Electronic switch on a single transistor.

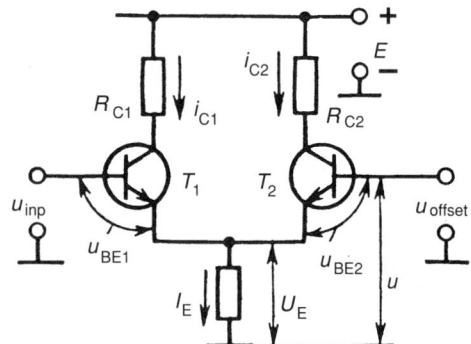

FIGURE 7.67
Transistor switch for two circuits.

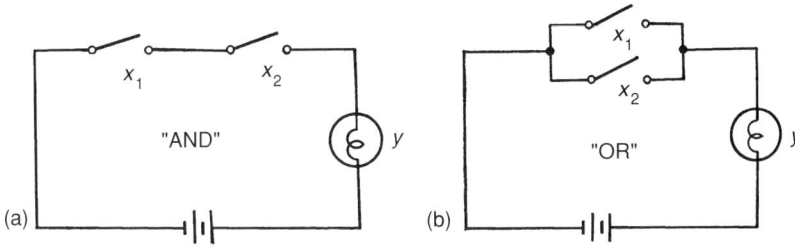

FIGURE 7.68
(a) and (b) Implementation of the simplest logical operations, with the help of electromagnetic relay contacts.

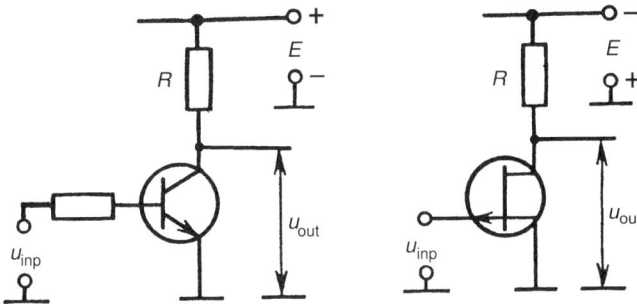

FIGURE 7.69
Logical element NOT implemented on a bipolar and field transistors.

collector (the drain and the source) of the transistor, and since the output signal is voltage on the collector (the source) of the transistor, that means that if there is no signal at the input, there will be a signal at the output of this circuit. And vice versa, when the signal is applied at the input, the transistor is unblocked and voltage drops to a very small value (fractions of a volt) and therefore the signal disappears at the output.

The logical element AND-NOT can be implemented by different circuit methods. In the simplest case, this is a circuit from transistors connected in series (Figure 7.70a). When control signals are applied to both inputs x_1 and x_2 simultaneously, both transistors will be enabled and the voltage drop in the circuit with two transistors connected in series will decrease to a very small value. This means no output signal Y. In the second circuit diagram (Figure 7.70b) even one signal on any input (x_1 or x_2) is enough for the voltage on output Y to disappear.

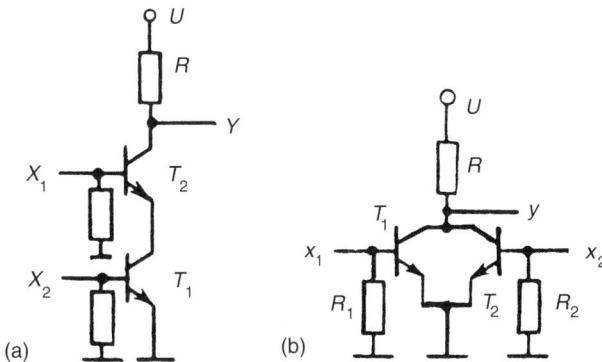

FIGURE 7.70
Transistor logical elements AND-NOT (a) and OR-NOT (b).

Self-contained logical elements are indicated on circuit diagrams as a special signs, Table 7.1. A signal strong enough for transition of a logical element from one state to another is usually marked as "1." No signal (or a very weak signal incapable of affecting the system state) is usually marked as "0." The same signs are used for indication of the state of the circuit elements: "1" — switched-on; "0" — switched-off. Such bi-stable (that is having two stable states) devices are called triggers.

When supply voltage is applied to such a device (Figure 7.71) one of the transistors will be immediately enabled and the other one will remain in a blocked state. The process is avalanche-like and is called regenerative. It is impossible to predict which transistor will be enabled because the circuit is absolutely symmetrical and the likelihood of unblocking of both transistors is the same.

This state of the device remains stable just the same. Repeated switching ON and OFF of voltage will cause the circuit to pass into this or that stable state. The essential disadvantage of such a trigger is no control circuit, which would enable us to control its

TABLE 7.1

Basic logical elements (according to certain standards logical elements are also indicated as rectangles)

Logical Function	Conventional Symbols	Boolean Identities	Truth Table		
			Inputs		Output
			B	A	Y
AND	A, B, Y	$A \cdot B = Y$	0	0	0
			0	1	0
			1	0	0
			1	1	1
OR	A, B, Y	$A + B = Y$	0	0	0
			0	1	1
			1	0	1
			1	1	1
NOT	A, $\overline{A}$	$A = \overline{A}$		0	1
				1	0
AND-NOT (NAND)	A, B, Y	$\overline{A \cdot B} = Y$	0	0	1
			0	1	1
			1	0	1
			1	1	0
OR-NOT (NOR)	A, B, Y	$\overline{A + B} = Y$	0	0	1
			0	1	0
			1	0	0
			1	1	0

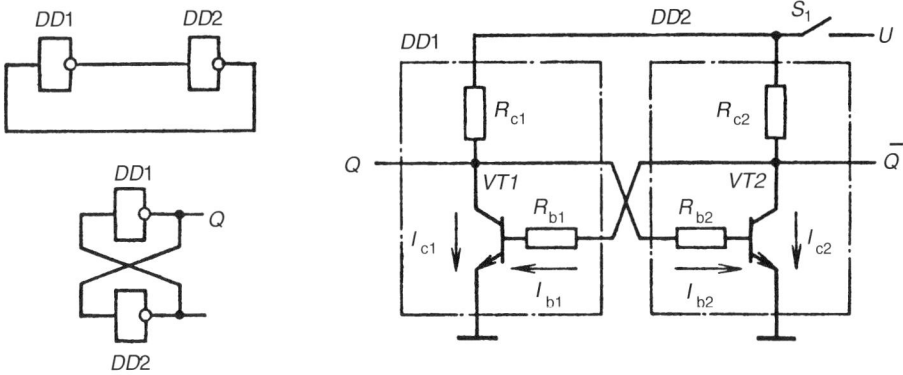

FIGURE 7.71
Bistable relay circuit with two logical elements NOT.

state at permanent supply voltage. In practice, the so-called Schmitt-Trigger are often used as electronic circuits with relay characteristics. There are a lot of variants of such triggers, possessing special qualities. In the simplest variant, such a trigger is a symmetrical structure formed by two logical elements connected in a cycle of the type AND-NOT or OR-NOT, (Figure 7.72); it is called an asynchronous RS-trigger.

One of the trigger outlets is named direct (any outlet can be named so as the circuit is symmetrical) and is marked by the letter Q, and the other one is called inverse and is marked by the letter $\bar{Q}$ ("Q" under the dash), to signify that in logical sense, the signal at this output is opposite to the signal at the direct output. The trigger state is usually identified with the state of the direct output, that is to say that the trigger is in the single (that is switched-on) state when $Q = 1$, $\bar{Q} = 0$, and visa versa.

Trigger state transition has a lot of synonyms: "switching," "change-over," "over-throw," "recording," and is carried out with the help of control signals applied at the inputs R and S. The input by which the trigger is set up in the single state is called the S input (from "set") and the output by which the trigger turns back to the zero position — the R input (from "reset"). Four combinations of signals are possible at the inputs, each of them corresponding to a certain trigger position (Table 7.2).

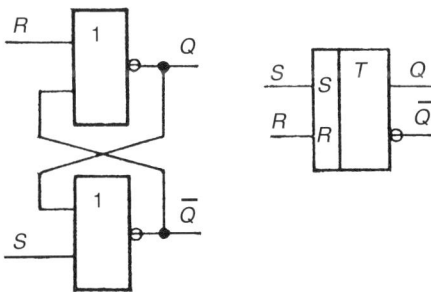

FIGURE 7.72
Asynchronous RS-trigger formed by two logical elements NOR.

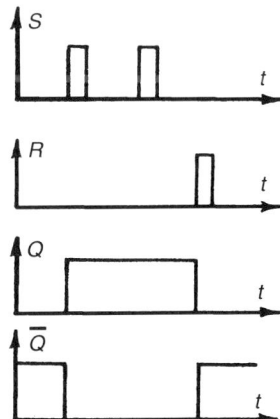

FIGURE 7.73
Time diagram of an asynchronous RS-trigger.

TABLE 7.2

Combinations of Signals at the Inputs and the RS-Trigger Position

Input			Output for Logical Element Type			
			AND-NOT		OR-NOT	
S (Set)	R (Reset)	Notes	Q	Q	Q	Q
0	0	Forbidden mode for *AND-NOT*	Uncertainty		Without changes	
1	0		1	0	1	0
0	1		0	1	0	1
1	1	Forbidden mode for *NOR*	Without changes		Uncertainty	

As can be seen from the table, when there are no signals on both of the trigger inputs on the elements AND-NOT, or when there are signals on both of the trigger inputs the elements OR-NOT (NOR), the trigger state will be indefinite which is why such combinations of signals are prohibited for RS-triggers.

From the time diagram of the asynchronous RS-trigger (Figure 7.73), it can be seen that after transfer of the trigger to the single state no repeated signals on the triggering input S are capable of changing its state. The return of the trigger to the initial position is possible only after a signal is applied to its "erasing" R input. The disadvantage of the asynchronous trigger is its incapacity to distinguish the useful signal of starting from noise occurring in the starting input by chance. Therefore, in practice so-called synchronous or D-triggers, distinguished by an additional so-called synchronizing input, are frequently used.

Switching of the synchronous trigger to the single state (ON) is carried only with a both signals: starting signal at the S input and also with a simultaneous signal on the synchronizing input. Synchronizing (timing) signals can be applied to the trigger (C input, Figure 7.74) with certain frequencies from an external generator.

Apart from increase of resistance to noises, synchronization provides time registration of signals and unites the operation of many units of the equipment. Different types of triggers are produced by many forms; in standard cases in the form of integrated circuits (Figure 7.75).

Simple relay devices are constructed also on the basis of so-called operational amplifiers (OA). OA are complex many-stage transistor circuits (see Figure 7.65b, for example) with a very high coefficient of amplification made by the integrated technology in

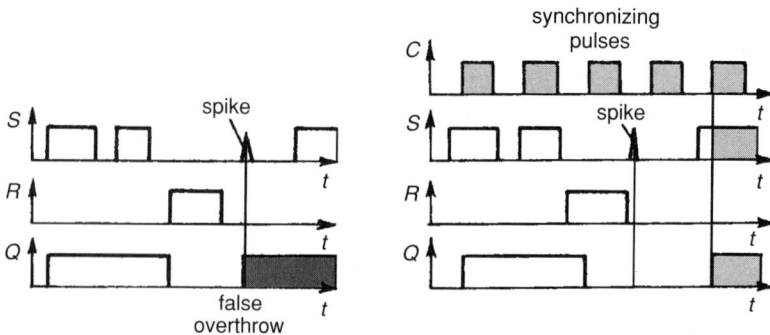

FIGURE 7.74

Time diagrams of operation of an asynchronous trigger (left) and synchronous trigger (right) when there is noise.

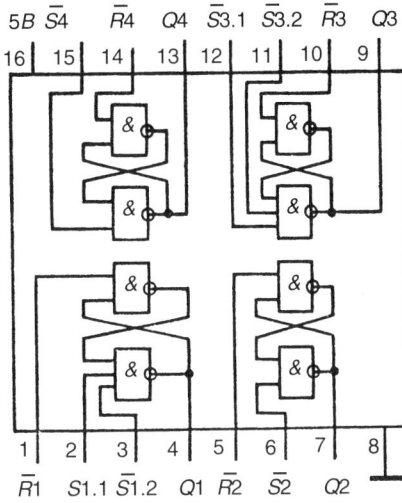

FIGURE 7.75
Structure of an integrated circuit of the 74LS279 type, containing four RS-triggers on logical elements AND-NOT.

standard cases of chips. When they are used directly for that purpose (as amplifier), that is for signal amplification, such OA are supplied, as a rule, with negative feedback (a signal from the output of the amplifier is applied to the input but with the opposite polarity), which slightly reduces amplification but increases operational stability and the quality of amplification considerably (Figure 7.76).

It should be noted that "+" and "−" on the scheme indicate the direct and inverse inputs of the amplifier, and not the polarity of the supply (supply circuits of OA are usually not indicated in order not to complicate the scheme). If an OA is supplied with positive feedback instead of negative (Figure 7.77), such an amplifier will start to work as a trigger (or a relay), being energized when input voltage exceeds a certain level, and turning to the initial position when the level of the input signal decreases.

It is very convenient to use a miniature OA as electronic relay, but not obligatory. A simple amplifier with two transistors, with positive feedback, also has similar properties (Figure 7.78). In initial position, when there is no voltage (or when voltage is very low) at the input of the circuit, transistor VT1 is closed (locked up). There is voltage on VT1 collector, which opening transistor VT2. The emitter current of transistor VT2 causes a voltage drop on the resistor R3, which blocks transistor VT1 and holds it in closed position. If input voltage exceeds the voltage in the emitter on VT1 transistor, it will be opened and will become saturated with very small collector–emitter junction resistance. As a result, the

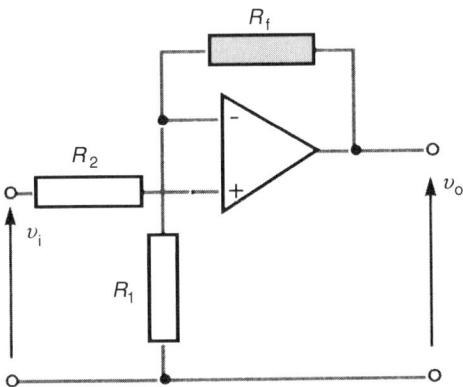

FIGURE 7.76
Operational amplifier with negative feedback carried out through a resistor R_f. v_i — input signal; v_o — output signal.

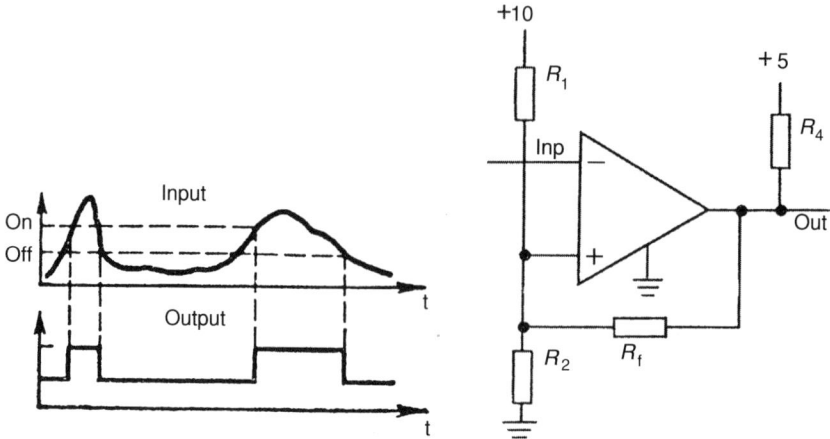

FIGURE 7.77
Operation amplifier with positive feedback working as a trigger.

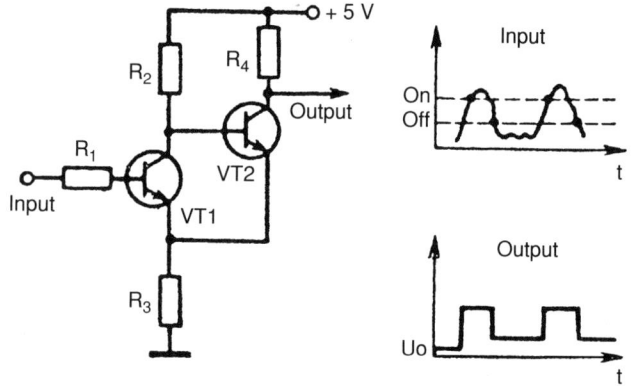

FIGURE 7.78
A simple two-transistor trigger.

potentials of the base and the emitter of transistor VT2 will be equal. Transistor VT2 will be blocked. At the output there will be voltage equal to the supply voltage. When input voltage decreases, transistor VT1 leaves the saturation mode, and an avalanche-like process occurs. Emitter current of transistor VT2, causing blocking voltage on resistor R_3, accelerates closing of the transistor VT1. As a result, the trigger returns to its initial position.

7.8.8 Thyristors

The history of development of another remarkable semiconductor device with relay characteristics begins with the conception of a "collector with a trap," formulated at the beginning of 1950s by William Shockley, familiar to us from his research on p–n junctions. Following Shockley, J. Ebers invented the two-transistor analogy (inter-bounded n–p–n and p–n–p transistors) of a p–n–p–n switch, which became the model of such a device (Figure 7.79).

In 1954–1955 John Moll estimated the performance capabilities of a p–n–p–n switch and a group of scientists from "Bell Telephone Laboratories" under his direction constructed the first working silicon p–n–p–n devices. The work of this group and the principles of operation of such devices were described in 1956 by Moll, Tanenbaum, Goldey and

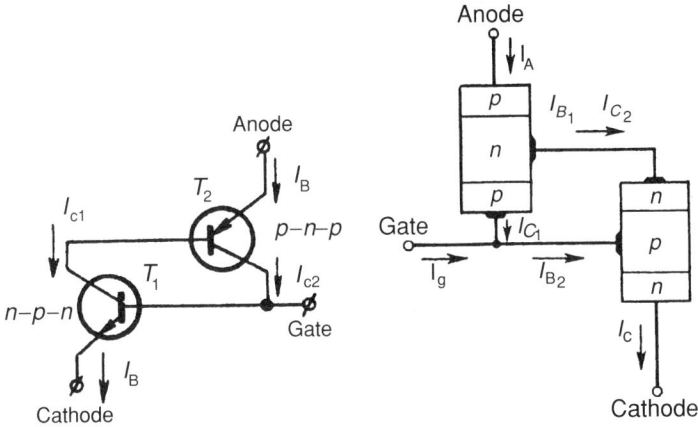

FIGURE 7.79
Two-transistor model of a thyristor.

Holonyak, in a scientific article which became the basis for further research carried out in this field.

The p–n–p–n switch had a destiny similar to some other devices in 1956–1957: not many people understood the principles of its operation and in practice it was not used very much, but R.A. Iork (from the General Electric Co.) realized the importance of the researched carried out in Bell Laboratories, became interested in the semiconducting "thyratron," and initiated a successful project of production of a silicon switch for strong currents (Figure 7.80).

The working element of this new semiconductor device with relay characteristics was a four-layer silicon crystal with alternating p- and n-layers (Figure 7.80). Such a structure is made by diffusion into the original monocrystal of n_1-silicon (which is a disk 20 to 45 mm in diameter and 0.4 to 0.8 mm thick, or more for high-voltage devices) admixture atoms of

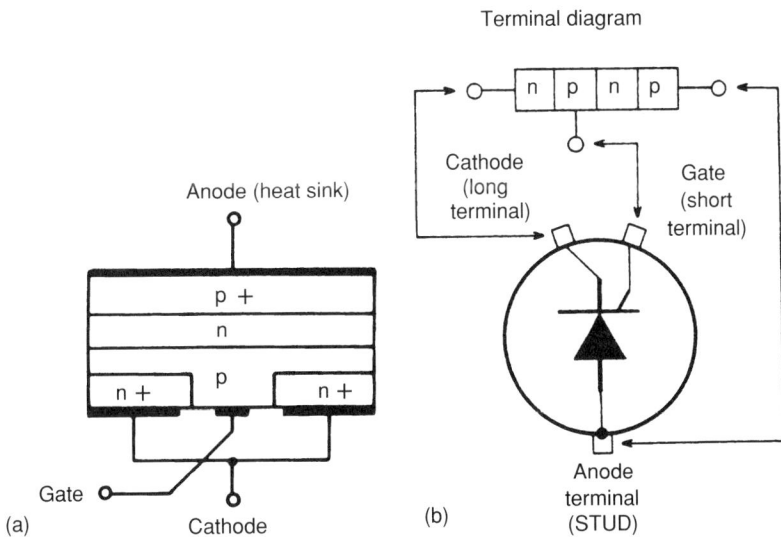

FIGURE 7.80
(a) and (b) Structure and symbolic notation of a semiconducting thyratron — "thyristor."

aluminum and boron from the direction of its two bases to a depth of about 50 to 80 μm. Injected admixtures form p_1 and p_2 layers in the structure.

The fourth (thinner) layer n_2 (its thickness is about 10 to 15 μm) is formed by further diffusion of atoms of phosphorus to the layer p_2. The upper layer p_1 is used as an anode in the thyristor, and the lower layer p_2 — as a cathode. The power circuit is connected to the main electrodes of the thyristor: the anode and the cathode. The positive terminal of the control circuit is connected through the external electrode to layer p_2, and the negative one — to the cathode terminal.

The volt–ampere characteristic (VAC) of a device with such a structure (Figure 7.81), much resembles the VAC of a diode by form. As in a diode, the VAC of a thyristor has forward and reverse areas. Like a diode, the thyristor is blocked when reverse voltage is applied to it (minus on the anode, plus on the cathode) and when the maximum permissible level of voltage (U_{Rmax}) is exceeded there is a breakdown, causing strong current and irreversible destruction of the structure of the device.

The forward area of the VAC of the thyristor does not remain permanent, as does that of a diode, and can change, being affected by current of the control electrode, called the "Gate." When there is no current in the circuit of this electrode, the thyristor remains blocked not only in reverse but also in the forward direction, that is, it does not conduct current at all (except small leakage current, of course). When the voltage applied in the forward direction between the anode and the cathode is increased to a certain value, the thyristor is quickly (stepwise) enabled and only a small voltage drop (fractions of a volt) caused by irregularity of the crystal structure remains on it.

If low current is applied to the circuit of the gate, the thyristor will be switched ON to much lower voltage between the anode and the cathode. The more such current, the lower the voltage that is required for unblocking of the thyristor. At a certain current value (from a few milli-amperes for low-power thyristors up to hundreds of milli-amperes for power ones) the forward branch of the VAC is almost fully rectified and becomes similar

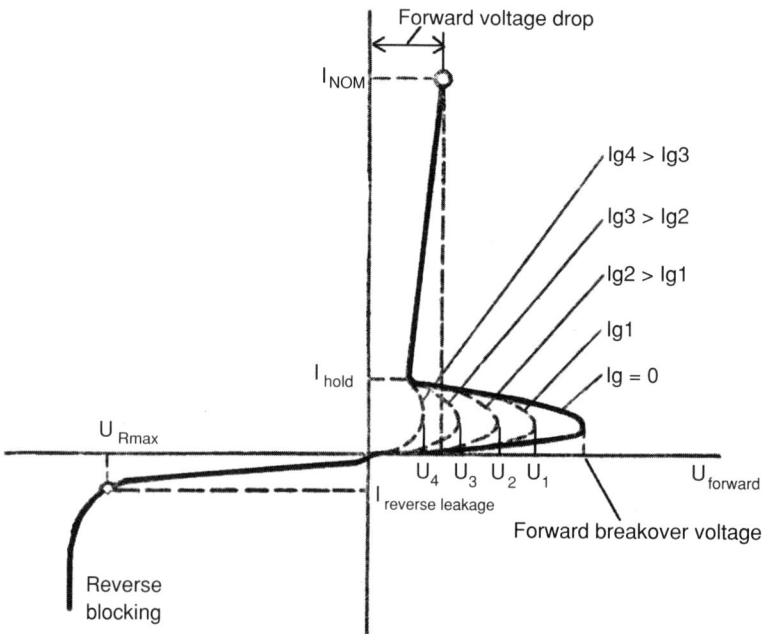

FIGURE 7.81
Volt–ampere characteristic (VAC) of a thyristor.

to the VAC of a diode. In this mode (that is, when control current constantly flows in the gate circuit) the behavior of the thyristor is similar to that of a diode that is fully enabled in the forward direction and fully blocked in the reverse direction. However, it is senseless to use thyristors in this mode: there are simpler and cheaper diodes for this purpose.

In fact, thyristors are used in mode when working voltage applied between the anode and the cathode does not exceed 50 to 70% of the voltage, causing spontaneous switching ON of the thyristor (when there is no control signal, the thyristor always remains blocked) and control current is applied to the gate circuit only when the thyristor should be unblocked and of such a value that would enable reliable unblocking. In this mode, the thyristor functions as a very high-speed relay (unblocking time is a few or tens of microseconds).

Perhaps many have heard that thyristors are used as basic elements for smooth current and voltage adjusting, but if a thyristor is only an electric relay having two stable states like any other relay: a switched ON state and a switched OFF one, how can a thyristor smoothly adjust voltage? The point is that if nonconstant alternating sinusoidal voltage is applied, it is possible to adjust unblocking moment of the thyristor by changing the moment of applying a pulse of control current on the gate with regard to the phase of the applied forward sinusoidal voltage. That is, it is as though a part of the sinusoidal current flowing to the load were cut off (Figure 7.82). The moment of applying a pulse of unblocking control current (such pulses are also called "igniting" by analogy with the control pulses of the thyratron) is usually characterized by the angle α.

Taking into account that average current value in the load is defined as an integral (that is the area of the rest part of the sinusoid) the principle of operation of a thyristor regulator becomes clear. After unblocking, the thyristor remains in the opened state, even after completion of the control current pulse. It can be switched OFF only by reducing forward current in the anode–cathode circuit to the value less than hold current value. In AC circuits the condition for thyristor blocking is created automatically when the sinusoid crosses the zero value. To unblock the thyristor in the second half-wave of the voltage it is necessary to apply a short control pulse through the gate of the thyristor. To control both half-waves of alternating current two thyristors connected antiparallel are used. Then one of them works on the positive half-wave and the other one — on the negative one.

At present such devices are produced for currents of a few milli-amperes to a few thousands amperes, and for blocking voltages up to a few thousands volts. The first industrial examples of power and high-power thyristors produced in different countries had the so-called "pin-like" (stud and flat base types) construction (Figure 7.83). As it can

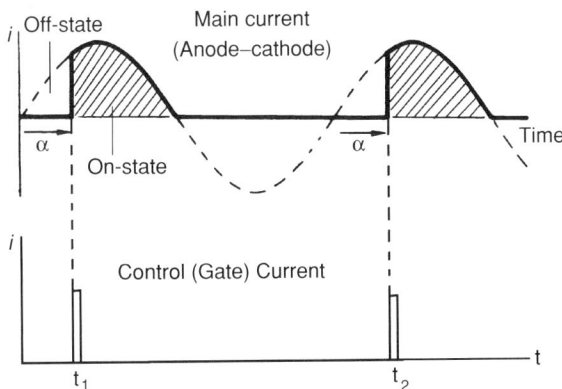

FIGURE 7.82
Principle of operation of a thyristor regulator.

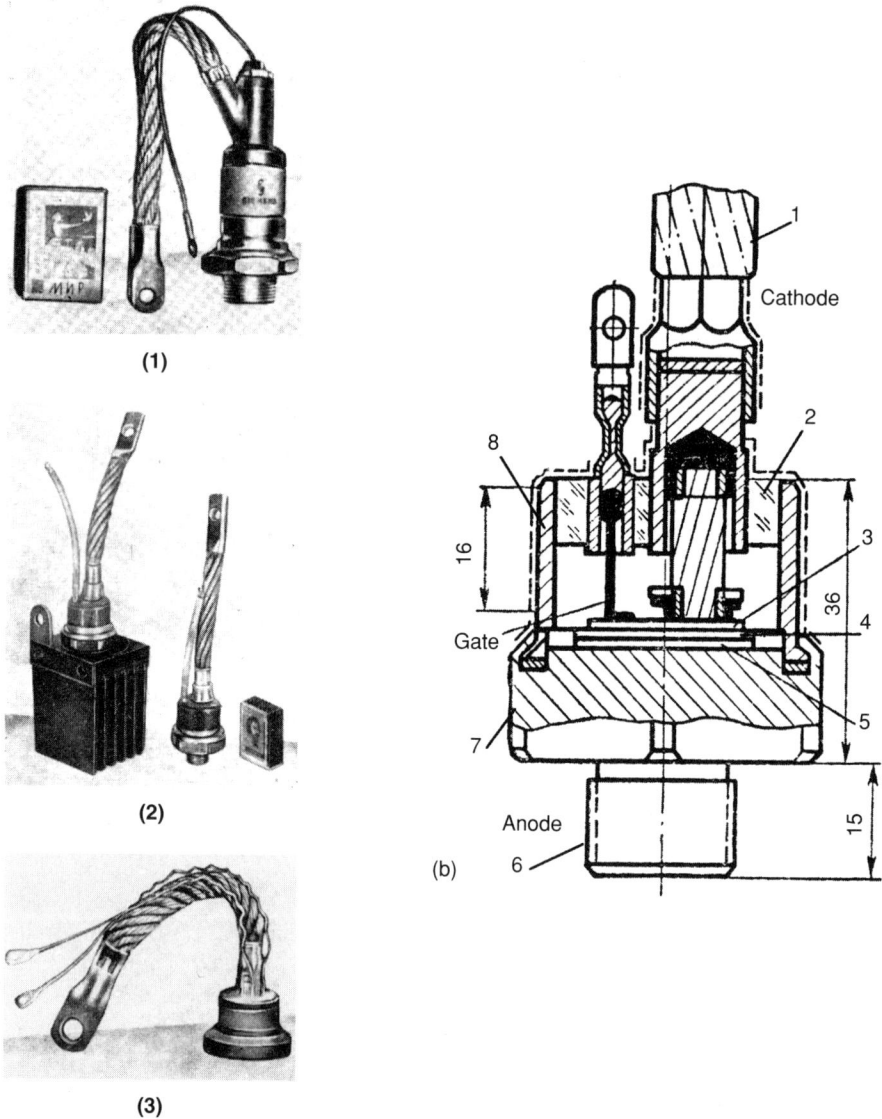

FIGURE 7.83

(a) Industrial samples of the first pin-like (1, 2 — stud; 3 — flat base) thyristors for currents up to 100–150 A, produced in 1960s. 1 — Bst L02 (Siemens); 2 — BTY-16 (AEG); 3 — BKY-100 (Russia). (b) Construction of power stud type thyristor. 1 — multiple-strand flexible copper braid with a point at the end; 2 — glass or ceramic insulator; 3 — layer n_2 of the semiconductor structure; 4 — silicon monocrystal (layer n_2); 5 — layer p_1 of the semiconductor structure; 6 — anode outlet made in the form of a screw-bolt; 7 — copper heel piece; 8 — steel cylindrical case.

be seen from the VAC, a certain voltage drop takes place even on a fully open thyristor because of imperfections in its crystal structure. This voltage is very small in comparison with working voltage. It totals only fractions of a volt; however, when strong working currents pass through the thyristor, such a voltage drop may lead to considerable power dissipation of the thyristor. For example, with voltage on an open thyristor of 1.5 V and current of 200 A, thermal power equal to 300 W is constantly being released. This is very high power and if certain measures for cooling the thyristor are not taken, its temperature will quickly exceed 150 to 160 °C and voltage will cause a breakdown of the crystal structure.

That is why all high power thyristors are always equipped with heat sinks. These are big ribbed constructions from aluminum alloy for air cooling or more compact for water cooling (Figure 7.84). Another problem with heating of thyristors was destruction of the joint points of the silicon crystal with a copper heelpiece and a cathode outlet, which were made with the help of standard tin solder. In the first power thyristors, already after several tens of thousands of switched OFF and ON cycles (when the thyristor was heated up to 100 to 120 °C and then to be cooled up to 20 to 30 °C) there was cracking of solder caused by the difference of linear expansion coefficients of the various materials.

Later on, they managed to cope with this disadvantage by introducing special temperature compensators and using pressure contacts instead of soldered ones (Figure 7.85).

(a) (b)

FIGURE 7.84
Air (a) and water (b) heat sinks for thyristors.

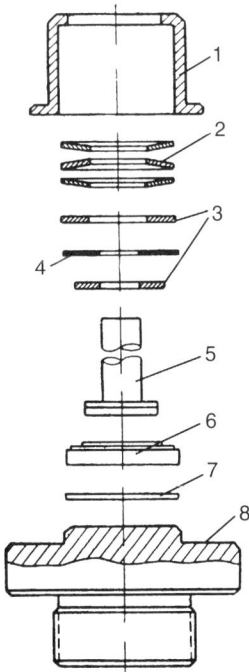

FIGURE 7.85
Construction of a stud-like thyristor with pressure contacts and a temperature compensator. 1 — pressure cup; 2 — disk spring; 3 — metal disk; 4 — mica disk; 5 — contact stamp with a temperature compensating plate; 6 — semiconductor crystal structure on a temperature compensating plate; 7 — silver contact spacer; 8 — copper piece heel of the case.

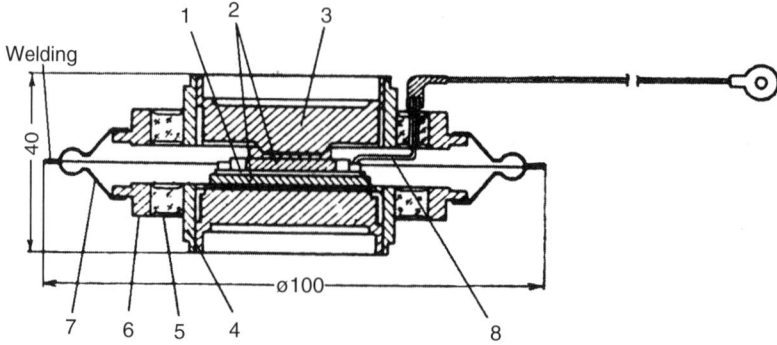

FIGURE 7.86
Earlier construction of a capsule type thyristor with pressure contacts. 1 — semiconductor crystal structure; 2 — pressure tungsten disks; 3 — copper contact elements (anode and cathode); 4 and 6 — metal rings; 5 — glass insulator; 7 — spring goffered disk; 8 — gate.

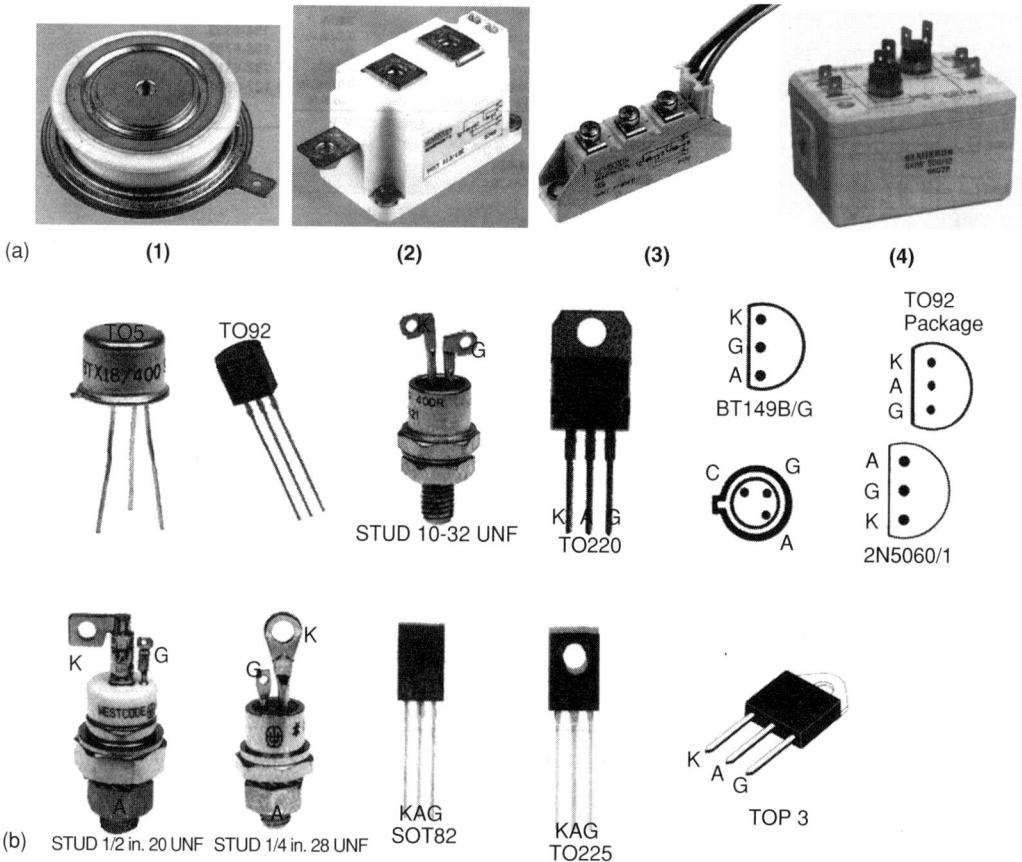

FIGURE 7.87
(a) Modern high-power thyristors. 1 — single thyristor; 2 and 3 — dual thyristor module with common anode or common cathode; 4 — high current (900 A) antiparallel thyristors with isolated (via aluminum oxide — AlO_2) water flow. (b) Modern low power and power thyristors.

Later, it turned out that it is more convenient to produce and exploit tablet construction (called capsule types) in the form of a disk, as in the construction of pressure contacts (Figure 7.86). When such constructions began to be produced, high-power pin-like thyristors were almost entirely forced out of production. The stud-like construction remained only for low power and power thyristors (for currents up to a few tens of amperes).

Modern thyristors have a distinguishing diversity of forms and sizes (Figure 7.87). Current progress in solid-state physics and new technological advances in semi-conductor production have allowed development of mass production of thyristors with characteristics that the inventors of thyristors could not even imagine (Figure 7.88 and Figure 7.89).

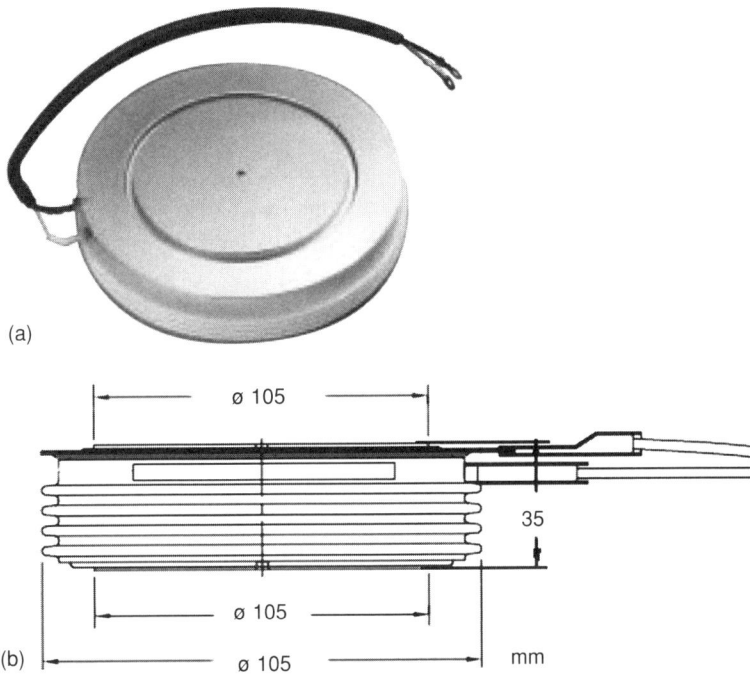

(a)

(b)

ø 105

ø 105

ø 105

35

mm

FIGURE 7.88
Ultra high-voltage thyristor FT1500AU-240 type (12 kv; 1500 A) produced by Powerex and Mitsubishi Electric.

FIGURE 7.89
Commercial thyristor of the SF3000GX21 type (Toshiba) with switched current up to 3000 A (surge current is 60.000 A) and switched voltage up to 4000 V. (From Toshiba Commercial Thyrstor online catalog 2004.)

7.8.8.1 *Control of Thyristors on Direct Current*

As has been mentioned above, the thyristor in the initial position is blocked in both current directions and for correct (nonemergent) unblocking, it is necessary to create certain conditions for current and voltage:

- Forward voltage not exceeding forward breakover voltage should be applied to the thyristor (Figure 7.81)
- In the "Gate–Cathode" circuit there should be current of positive direction enough for thyristor unblocking both by value (0.05 to 0.2 A for power thyristors) and by duration (tens and hundreds of microseconds)

Under such conditions, the thyristor will be switched ON and current will flow through its main "anode–cathode" junction. The control junction (gate-cathode) will be shunted by forward current, and further operation of the thyristor will not depend on current in the gate circuit. The thyristor state after it has been enabled will be fully determined by the forward current value in the "anode–cathode" circuit which is by load resistance. If this current exceeds the hold current (I_{HOLD} in Figure 7.81), the thyristor will be conductive. If it is less than the hold current, the thyristor will be immediately switched OFF.

In the scheme shown in Figure 7.90a, the thyristor VS1 is switched ON at the moment when the resistance R_1 decreases to a value sufficient for the gate current in the circuit, corresponding to the unblocking control current of the given thyristor. When the thyristor is enabled, the resistance R_1 is shunted by low resistance of the main opened junction (anode–cathode) and does not affect the state of the thyristor.

In the circuit shown in Figure 7.90b, current in the gate circuit arises only at the moment of closing of the control contact (S_1). The resistor R_2 is almost always used in such circuits to prevent the pulse noise from getting to the gate circuit, and spontaneous thyristor opening. In the circuit shown in Figure 7.90c, the control junction of the thyristor (gate-cathode) is constantly shunted by the contact (S_1). When the contact opens, current of the resistor R_1 changes its direction, passing to the circuit of the gate and opening the thyristor.

And how can an opened thyristor be switched OFF? It is not that easy to do on direct current (Figure 7.91). Methods applied in practice usually come to break circuits of anode current (a); shunting of the thyristor with an additional contact (b) or a transistor (c); reduction of anode current to a value less than the hold current (d); use of the charged capacitor C, which is connected parallel to the thyristor, at the moment when the thyristor

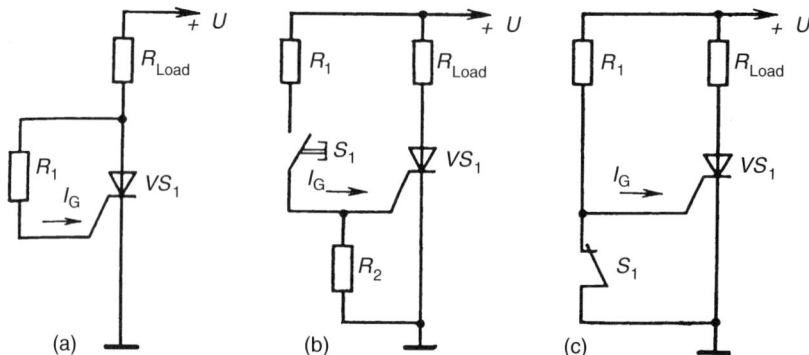

FIGURE 7.90
Connection and control of thyristors on DC.

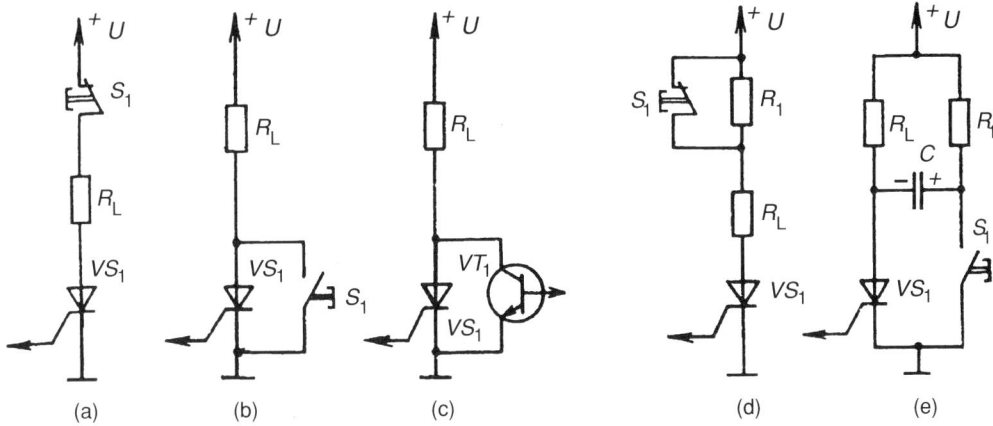

FIGURE 7.91
Principles of switching OFF of thyristors on direct current.

must be switched OFF and runs down on it, creating current of the opposite polarity blocking the thyristor (e). All of these methods of forced closing of thyristors are called "forced commutation" (in contrast to "natural" commutation on AC). The method of closing of thyristors with the help of capacitors, like similar methods used for switching OFF of thyratrones (Figure 7.22) was the most popular. In the circuit in Figure 7.91e the resistance of the resistor R_1 is much less than the load resistance R_L, which is why at the first moment after thyristor switching ON its node, the current passes not through the load but through the resistor R_1 charging the capacitor C. After that, the current ceases passing through the capacitor, which is why the anode current of the thyristor passes to the parallel branch with the load R_L. When the contact S_1 is closed (an additional thyristor VS2 can be used instead of it, see Figure 7.92), the voltage of the charged capacitor is applied to the thyristor with the opposite polarity ("plus" to the cathode, "minus" to the anode), causing blocking of the thyristor.

Pulse circuits of thyristor control, with a transformer in the gate circuit, were very popular due to the fact that such small transformers allowed application of control pulses to the gate circuit of the power thyristor, which is under the full potential of the power source (which can be hundreds and even thousands of volts), directly from low voltage

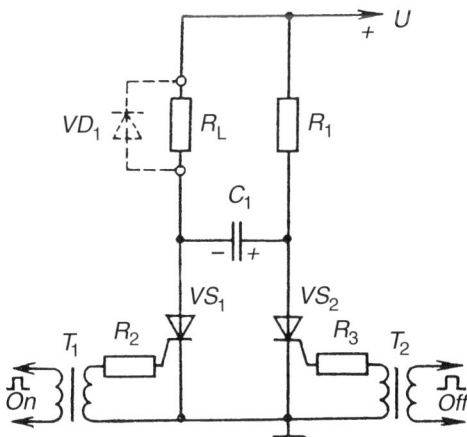

FIGURE 7.92
Pulse control circuit providing "forced commutation" of the main thyristor (VS1) on direct current.

FIGURE 7.94
Parallel connection of thyristors with balancing reactors.

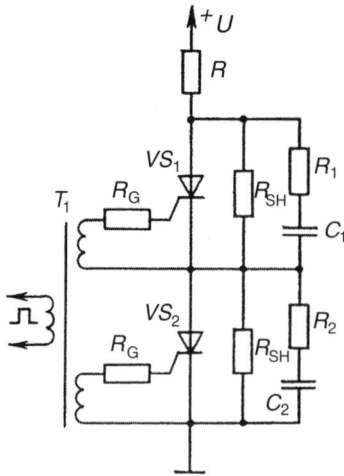

FIGURE 7.93
Series connection of thyristors with pulse control. R_{SH} — shunting resistors equalizing voltage distribution between thyristors connected in series; R_1C_1 and R_2C_2 —circuits protecting thyristors from spikes during switching processes.

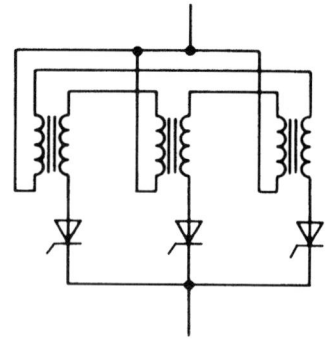

microelectronic control devices, and also to control a group of thyristors connected in series and designed for work under high voltages (Figure 7.93).

Sometimes it is necessary to connect thyristors in parallel in order to increase switched current. Like in the case of the series connection, one has to balance the work conditions of all the thyristors connected to the group because of the natural parameters of dispersion of the thyristors, but instead of balancing of voltage it is essential to balance currents passing through thyristors connected parallel, which is much more difficult. In such cases, more ponderous and expensive inductive reactors must be applied (Figure 7.94).

7.8.8.2 Control of Thyristors on Alternating Current

In AC circuits, the thyristor can be used without a forced cut-off, because every half-period the sinusoidal current passes through the zero value and at that moment conditions for thyristor cut-off are created, however, for switching of both half-waves of current, two inverse-parallel connected thyristors (Figure 7.95a), or a thyristor connected to the diagonal of the rectifier bridge (Figure 7.95b), are required.

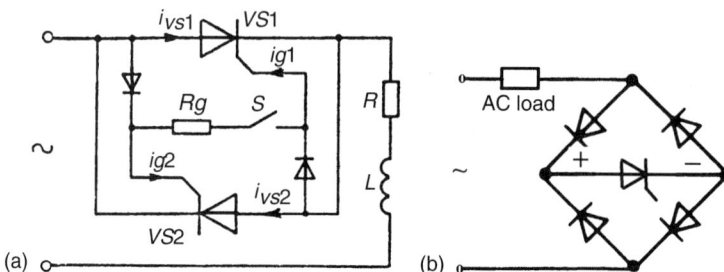

FIGURE 7.95
Thyristor AC switches.

In circuits of AC switches controlled by an additional contact, (Figure 7.95a — and this can be a reed switch) in the closed position of the contact in the gate circuit of thyristors, quite short control pulses are automatically formed from the anode voltage (Figure 7.96). In order to switch three-phase loads, a three-phase switch constructed on the same principle is used (Figure 7.97). Why is forced commutation of thyristors (Figure 7.98) needed if they are cut off while crossing the zero value of the current sine?

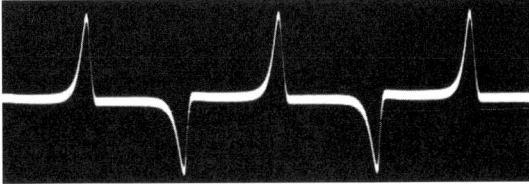

FIGURE 7.96
Oscillogram of pulses of the gate current automatically formed in the gate circuit of a thyristor AC switch (Figure 9.95a).

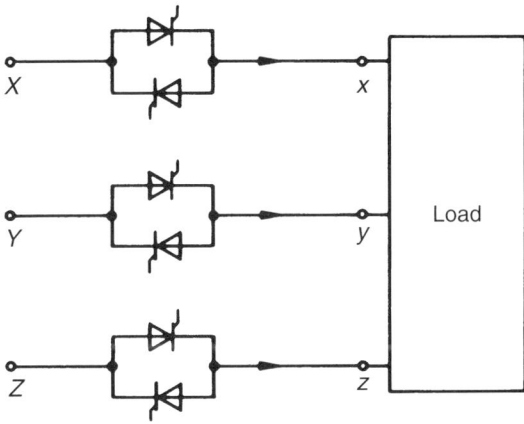

FIGURE 7.97
Three-phase thyristor switch based on inverse-parallel connected thyristors.

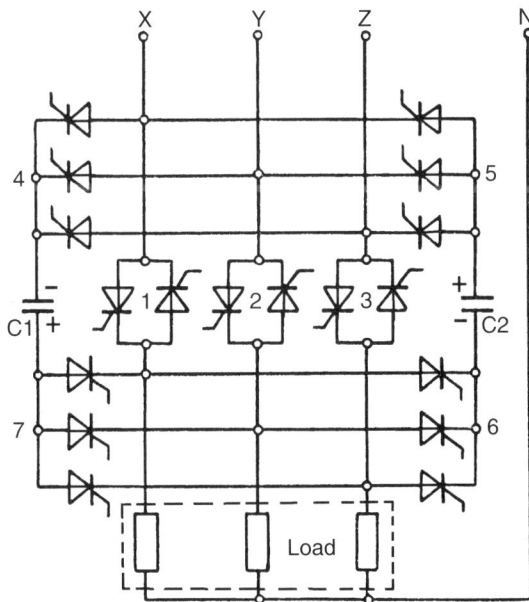

FIGURE 7.98
Three-phase contactor with forced commutation. 1—3 — Main groups of thyristors switching the load; 4—7 — additional groups of thyristors controlling capacitors C1 and C2.

Forced commutation is used on alternating current when one needs to speed up the cut-off of thyristors without waiting for current to cross the zero value. Such necessity arises in high-speed switching devices. The principle of accelerated switching by a thyristor on alternating current is similar to that on direct current; the use of previously charged capacitors connected to the thyristor by the opposite polarity; some solutions concerning circuits may differ, though (Figure 7.98).

7.8.8.3 Diac, Triac, Quadrac

As in the case with transistors, there are several types of thyristors differing by their properties and characteristics. First of all, this is the so-called symmetrical thyristor — "triac" (the last two letters stand for "alternating current"). The symmetrical thyristor as it follows from its name has a symmetrical VAC (Figure 7.99), that is, when there is a control signal, it applies current in both directions and can replace two standard thyristors connected inverse-parallel (Figure 7.100). It is obvious that a triac has a more complex structure than a standard thyristor. It is no longer a four-layer device, like a thyristor. It is five-layer device, with a thyristor only as a part of a more complex structure.

In theory, the triac can be enabled at any combination of voltage polarities on main electrodes and on the gate. That is why it is quite senseless to indicate the main electrodes as an "anode" and a "cathode" and they are marked simply as M1 and M2. But it is correct for so-called "four-quadrant" (or 4Q) triac only.

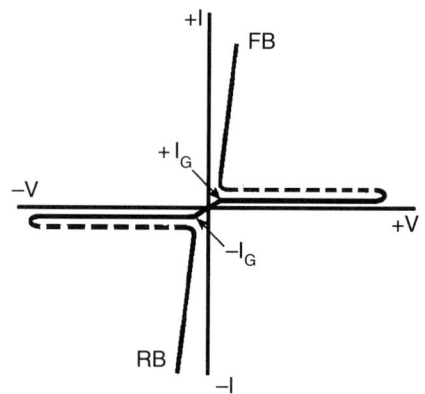

FIGURE 7.99
VAC of a symmetrical thyristor "triac" (thyristor for alternating current). FB — forward branch; RB — reverse branch; I_G — gate current.

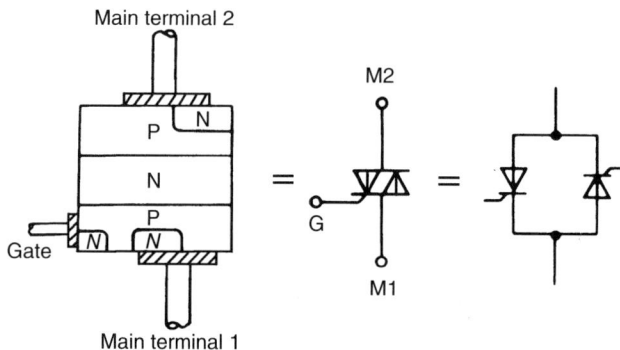

FIGURE 7.100
Triac structure, symbolic notation and thyristor equivalent.

Such triac can be triggered in all four quadrants, Fig. 7.101. Three-quadrant triacs (3Q triac) allow triggering in quadrants I, II and III only. 3Q triacs are more efficient in applications that have non-resistive loads, such as motor control applications, transformer loads, ignition circuits, etc.

For these types of applications 4Q triacs must include additional protection components to minimize the effects of false triggering (uncontrolled triac conduction). These include RC snubbers across the main terminals of the triac and an inductor in series with the triac. 3Q triacs have eliminated or reduced the need for protection components, making system design for non-resistive loads more reliable, cheaper, and smaller. At more prevalent and more reliable 3Q triac VAC does not look as nice as in Fig. 7.99, and the symmetrical thyristor is not in fact all that symmetrical: the turn-on gate current with certain (reverse) voltage polarity on the main electrodes, appears to be 3–5 times as much as with other (forward) polarity. Of course one can construct a control system capable of generating more powerful control pulses compensating for this difference in sensitivity.

As it can be seen, rejection of the names of the main electrodes "anode" and "cathode" is not quite defensible, since despite triac "symmetry" it is essential to designate from which electrode the control signal will be applied to the gate.

Like a standard thyristor, a triac can be controlled in different ways in real constructions of switching devices (Figure 7.102). It should be taken into account that physically the triac does not comprise two thyristors connected in parallel, as it is shown in Figure 7.100. It only functions as two inverse-parallel connected thyristors on alternating

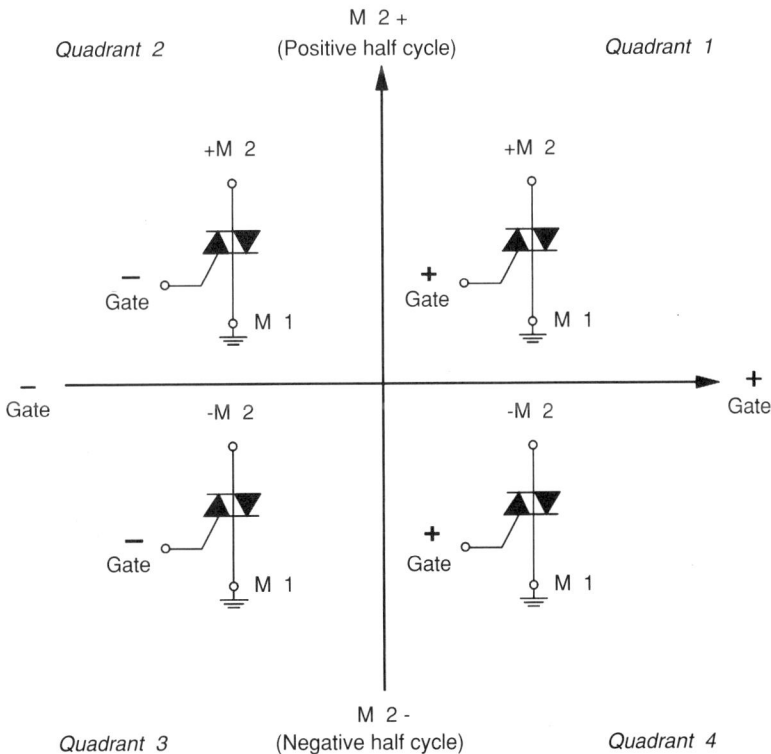

FIGURE 7.101
Preferred combinations of polarities of the signal of control and voltage on the main outlets for triac.

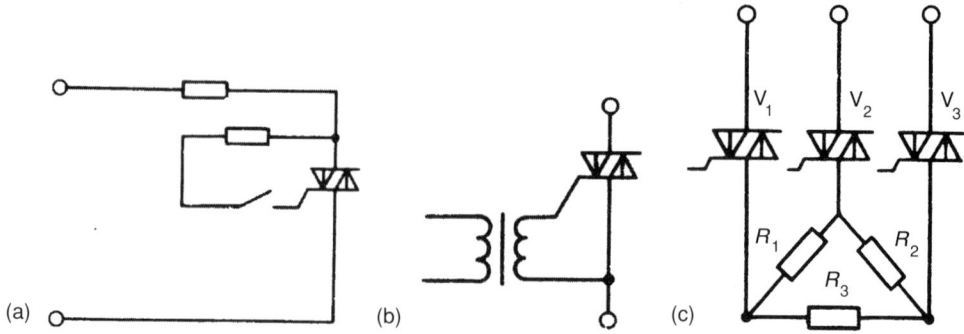

FIGURE 7.102
Some methods of triac control and its connection to a three-phase circuit.

current. (*Note*: Alternating current; a triac is not designed for work on DC and unlike a pair of inverse-parallel connected thyristors, does not operate very stably on DC.)

Besides a triac there are some other variants of thyristors, like a dinistor, for instance (Figure 7.103a), which in fact is a standard thyristor without a gate outlet, being enabled when the voltage applied to it (between the anode and the cathode) is increased to the level of forward breakover voltage (Figure 7.81). Such devices are produced in Russia, but are not well known in the West. More popular are devices controlled by voltage without a gate based on a triacs, not thyristors. They are called a "diac" (Figure 7.103b). Some firms produce semiconductor devices with a triac and a diac combined in its structure (Figure 7.103c). Such devices are called a "quadrac."

The so-called "gate turn-off thyristors" ("GTO-thyristors,") are of special interest. As one can conclude from its name these are thyristors which can be not only turned ON but also turned OFF by a signal received on the gate (Figure 7.104). You probably remember how the control circuit should be complicated in order to disable the thyristor at the required moment. The GTO-thyristor allows simplification of the solution to the problem (Figure 7.105).

The GTO-thyristor is enabled like a standard thyristor, only it requires a longer gate current pulse for reliable enabling, and has a quite high hold-on current (I_{HOLD}), that is it requires a higher value of direct anode current in order to remain in the open state after completion of the gate current pulse. The thyristor is disabled by the pulse of current in the gate of the opposite polarity, with greater amplitude than the pulse of enabling current has (which can reach one fifth up to one third of anode current!). That's why circuits of the GTO-thyristor usually contain storage reactive elements (capacitors,

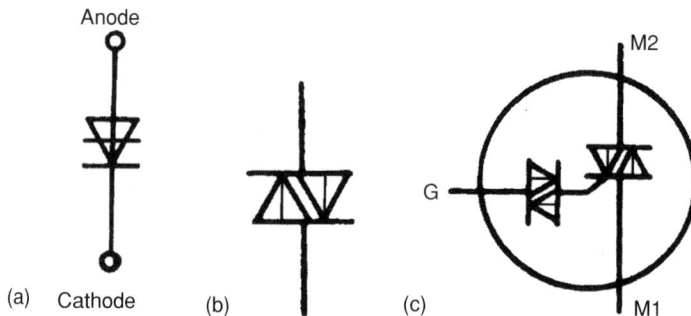

FIGURE 7.103
Different types of thyristors: dinistor (a), diac (b), quadrac (c).

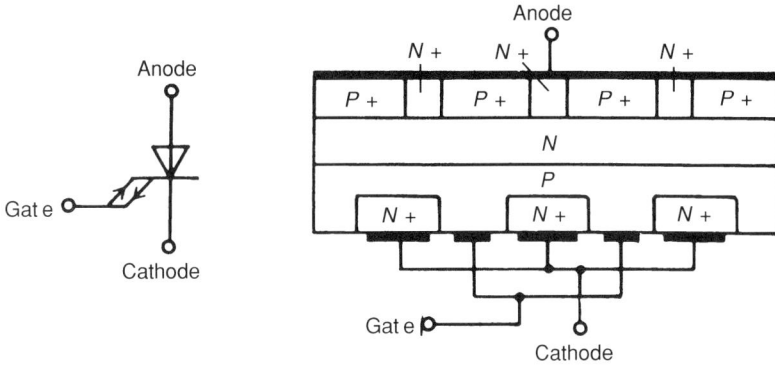

FIGURE 7.104
Notation used in schemes and structure of A GTO-thyristor.

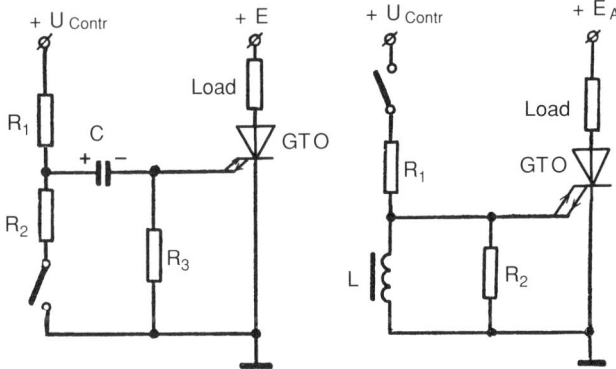

FIGURE 7.105
Control circuits of a GTO-thyristor.

chocks), creating strong current pulses needed for thyristor closing. For example, in the circuit on the left (Figure 7.105), the thyristor is enabled by the pulse of a charge of the capacitor (C) through the resistor (R_1) when control voltage (U_{contr}) is applied, and is disabled by the current of a discharge of this capacitor through the resistor R_2 when the control contact is closed.

In very high power applications, which require operating at high voltage, high current, and high temperatures, the use of silicon (Si) devices is restricted. Silicon carbide (SiC) has superior physical properties like a wider band gap, higher thermal conductivity, higher breakdown voltage, and higher temperature handling capability, which make it potential material to overcome the limitations of Si. For the last few years various SiC devices have been developed to be used in power applications, including SiC GTO thyristors for voltages above 6 kV and an operating temperature of 250°C.

In the last few years, many new super-powerful semiconductor switching devices have appeared in the market, so many new names and abbreviations, that even for the expert it is difficult to understand all of them. Here for example, are only some of the new thyristor types:

- MCT (MOS-Controlled Thyristor), Figure 7.106a
- FCTh (Field-Controlled Thyristor)
- SITh (Static Induction Thyristor)

- MTO (MOS Turn-Off Thyristor), Figure 7.106b
- IGTT (Insulated Gate Turn-Off Thyristor)
- IGT (Insulated Gate Thyristor)
- IGCT (Integrated Gate-Commutated Thyristor)
- ETO (Emitter Turn-Off Thyristor)

Within the framework of our book we cannot consider in detail all of the newest types of power electronic switches, but two recent ones should be mentioned nevertheless.

ABB semiconductors recently developed a new type of power switch with an architecture combining the best features of an IGBT and a GTO thyristor. Called the Integrated Gate Commutated Thyristor (IGCT), the new solid-state switch is for high-voltage applications of up to 6.9 kV, with maximum ratings of 4000 A (Figure 7.107). As it is possible to see from Figure 7.107, an ignition or control circuit of thyristors (driver) is made on a printed-circuit-board and integrated into the common module together with the power structure. It relieves users of "headaches" caused by complex and difficult systems based on such thyristors, and simplifies their use greatly.

The method to achieve unity gain used in the ETO thyristor is to insert an additional switch in series with the cathode of the GTO. The cathode of the GTO is the emitter of the internal n–p–n transistor, so the series switch is referred to as the emitter switch (Figure 7.108). The ETO thyristor was developed at Virginia Tech under a program directed by Sandia National Labs. The ETO thyristor was initially developed as an extremely high-power switching device to be used in power conversion systems within electric utility grids. However, the ETO thyristor has properties which also make it an attractive option for other high-power applications, such as large multi-megawatt electric motor drive controllers. ETO thyristors are capable of switching current up to 4 kA and voltage up to 6 kV. Although the ETO has the highest power handling capabilities of all solid-state switches, its greatest benefits may well be its low cost and reliability. Other competing power switching technologies for high-power GTO and IGCT — are complex and bulky with either less than half the switching speed (GTO) or costly (IGCT). The commercialized ETO switch is projected to cost less than $1000, compared with $1900 for a typical high frequency IGCT switch.

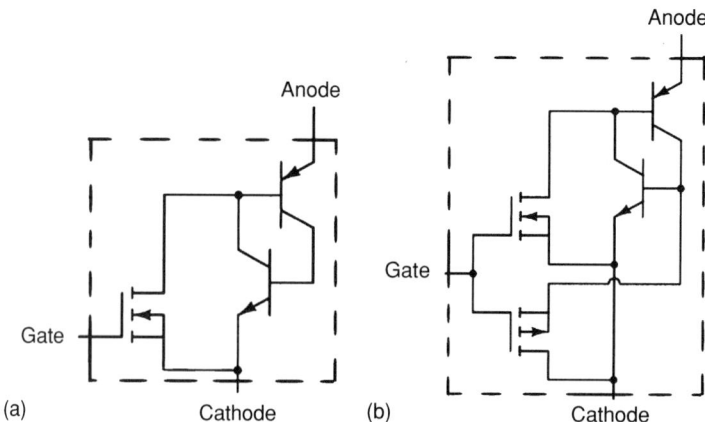

FIGURE 7.106
Equivalent circuit diagram for MCT (a) and MTO-thyristors (b).

FIGURE 7.107
Integrated gate-commutated thyristors (IGCT), developed by ABB semiconductor. (Odegard B. ABB 12/2002 Applying IGCT Gate Units.)

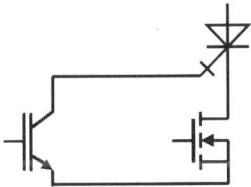

FIGURE 7.108
Equivalent circuit diagram for ETO-thyristor.

7.9 Optoelectronic Relays

The blocked n–p junction in semiconductor devices (diodes, transistors, thyristors) may begin to allow electric current to pass under the effect of energy of photons (light). When the n–p-junction is illuminated, additional vapors of charge carriers — electrons and holes causing electric current in the junction — are generated within it. The higher the intensity of the luminous flux on the n–p junction is the stronger the current is. Optoelectronic relays (Figure 7.109) comprise a light-emitting element which is usually made on the basis of a special diode (light emission diode [LED]), an n–p junction emitted by photons when current passes through it, and a receiver of the luminous flux (a photodiode, a phototransistor, a photothyristor). Usage of the two series connected photo-MOS transistors ("A" connection in Figure 7.110) as output elements allows the optoelectronic relay to switch either AC or DC loads with nominal output current rating. Connection "B" with the polarity and pin configuration as indicated in the schematic, allows the relay to switch DC load only, but with current capability increases by a factor 2.

The photo-emitting and the photo-detecting elements are almost entirely isolated from one another, can be placed in the same case, or may be separated by flexible glass fiber of 5 to 10 m or more in length (Figure 7.111). There is a great diversity of

Package dimensions

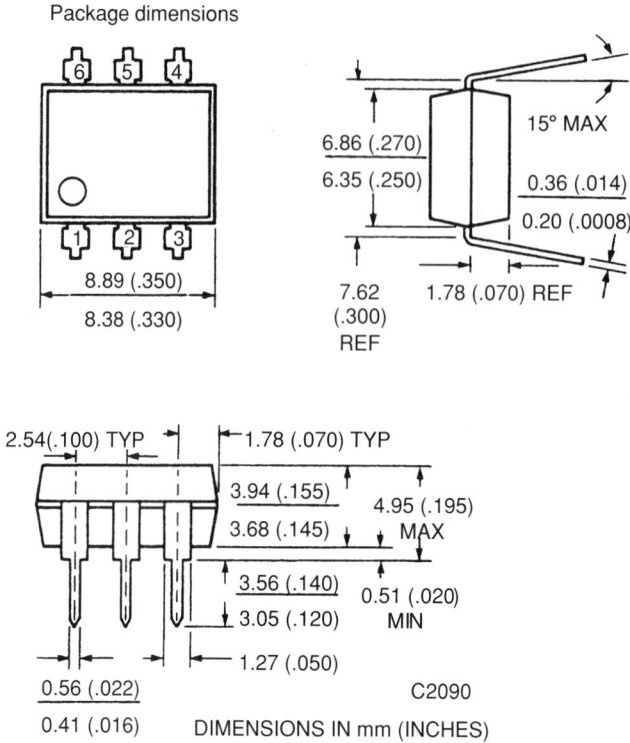

6.86 (.270)
6.35 (.250)

15° MAX

0.36 (.014)
0.20 (.0008)

8.89 (.350)
8.38 (.330)

7.62
(.300)
REF

1.78 (.070) REF

2.54(.100) TYP

1.78 (.070) TYP

3.94 (.155)
3.68 (.145)

4.95 (.195)
MAX

3.56 (.140)
3.05 (.120)

0.51 (.020)
MIN

1.27 (.050)

0.56 (.022)
0.41 (.016)

C2090

DIMENSIONS IN mm (INCHES)

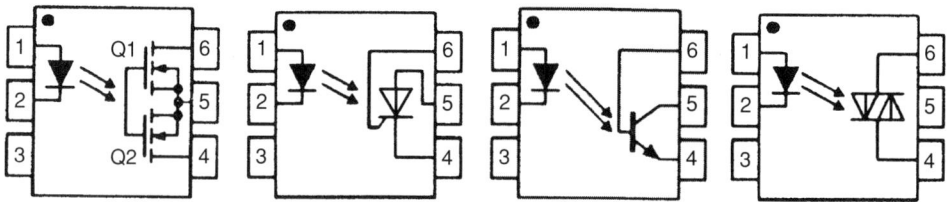

FIGURE 7.109
Optoelectronic relays designed in a standard DIP case.

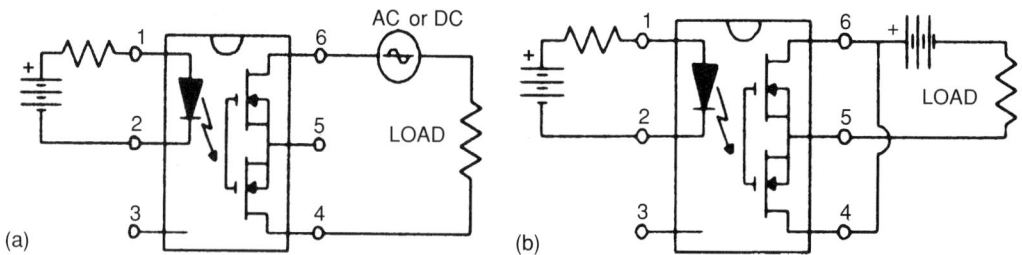

AC or DC

LOAD

LOAD

(a)   (b)

FIGURE 7.110
Connection of bidirectional switched photo-MOS relay.

circuits and constructions of optoelectronic relays, including those containing built-in invertors or amplifiers (Figure 7.112). A similar principle serves as a basis not only for miniature devices in chip cases, but also for practically all power semiconductor relays and contactors.

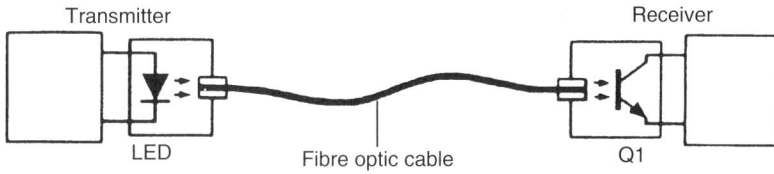

FIGURE 7.111
Optoelectronic relay with light-transmitting fiber.

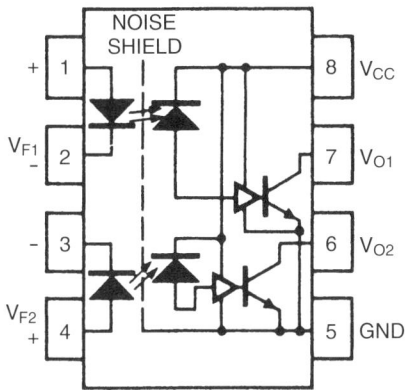

FIGURE 7.112
Twin optoelectronic relay with built-in amplifiers of power.

FIGURE 7.113
Modern semiconductor optoelectronic relays for currents of 3 to 5 A, produced by different companies.

It should be noted that the external designs of not only miniature optoelectronic relays in chip cases, but also of more powerful semiconductor relays of various firms, are very much alike (Figure 7.113). Such relays are usually constructed according to a similar scheme (Figure 7.114), with only some slight variations. As a rule they comprise an RC-circuit (the so-called "snubber"), and a varistor protection outlet, protecting the thyristors from overvoltages. They often contain a special unit (a zero voltage detector) controlling the moment when the voltage sinusoid passes through the zero value and allowing it to

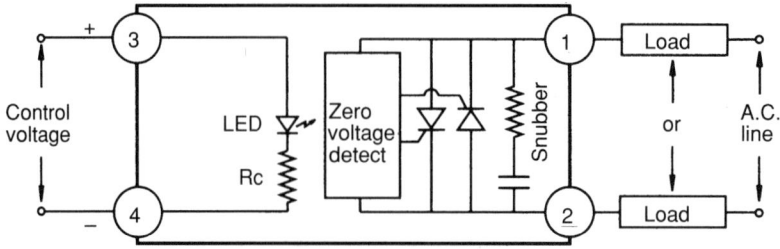

FIGURE 7.114
Standard scheme of a power single-phase optoelectronic AC relay.

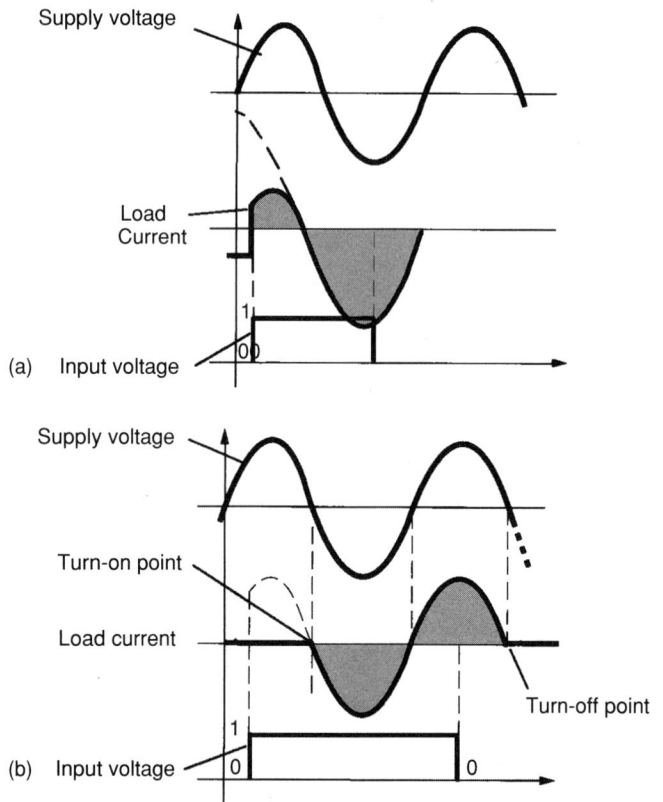

FIGURE 7.115
Oscillograms of switching of a semi-conductor AC relay. (a) Asynchronous switching without zero voltage detector; (b) synchronous switching with a zero voltage detector.

enable (and sometimes to disable) the thyristor at the zero value of voltage (so-called "synchronous switching" — Figure 7.115). Synchronous switching (especially at its high frequency) allows a considerable reduction of both the number and the amplitude of spikes, and high-frequency harmonics arising at transient switching processes in the load circuit.

More powerful single-phase and three-phase contactors for currents of 10 to 150 A, produced by different companies, also have a similar construction (Figure 7.116). Like the monophase variant, three-phase contactors can comprise built-in RC-circuits and varistors, and a zero voltage detector. It is only natural that when current of tens of amperes must be switched, a compact contactor requires the use of quite a large and

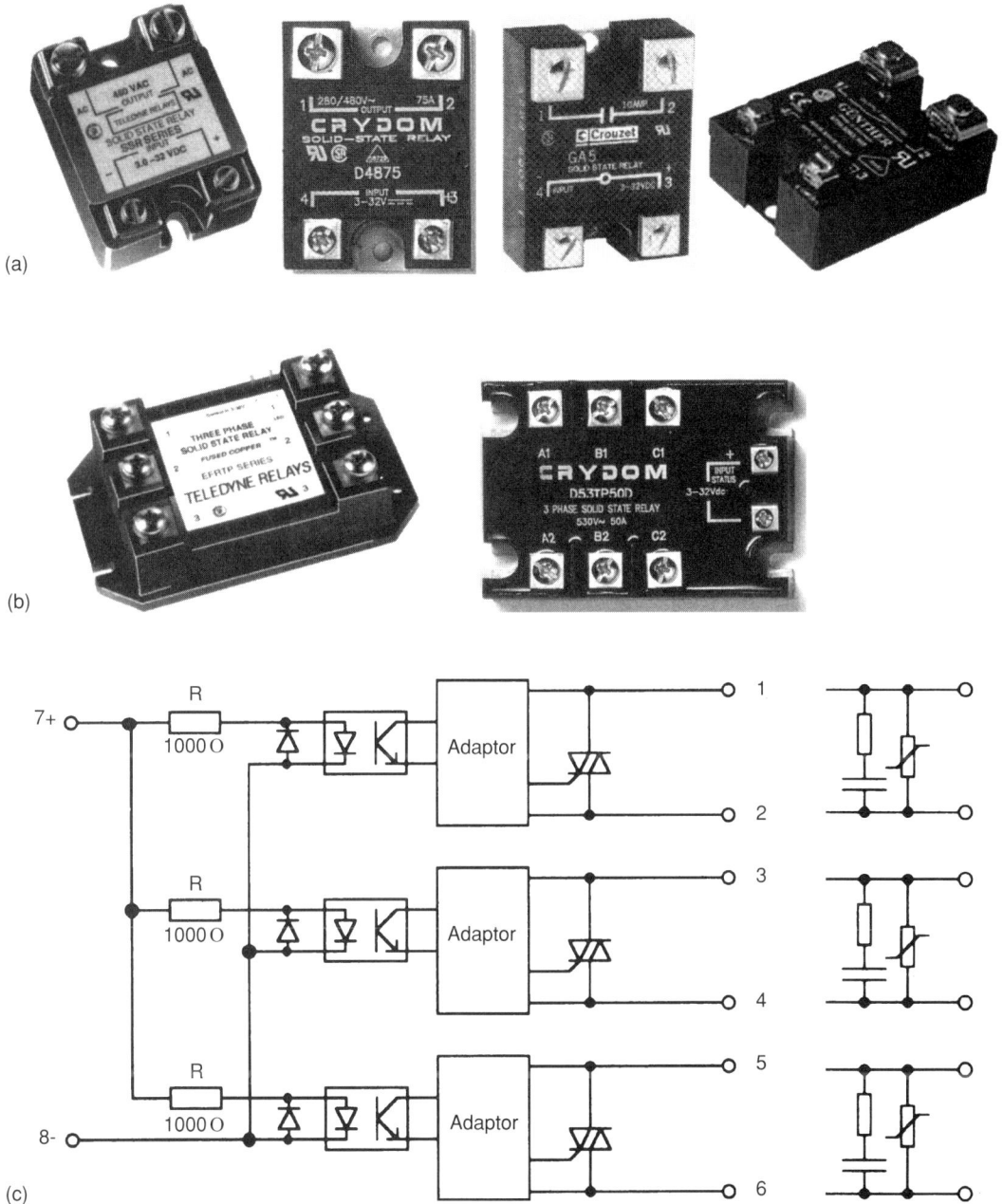

(b)

(c)

FIGURE 7.116
(a) Single-phase semiconductor AC contactors for currents of 10 to 75 A, produced by the companies: Teledyne, Crydom, Crouzet, Gunther. (b) Three-phase semiconductor contactors for currents of 50 to 150 A and voltage of 630 V, produced by different companies. (c) Standard scheme of a three-phase optoelectronic AC contactor.

heavy heat sink. (Figure 7.117). As with standard electromagnetic relays, Phoenix also produces optoelectronic relays in cases of peculiar shapes, designed for installation on a standard DIN rail (Figure 7.118).

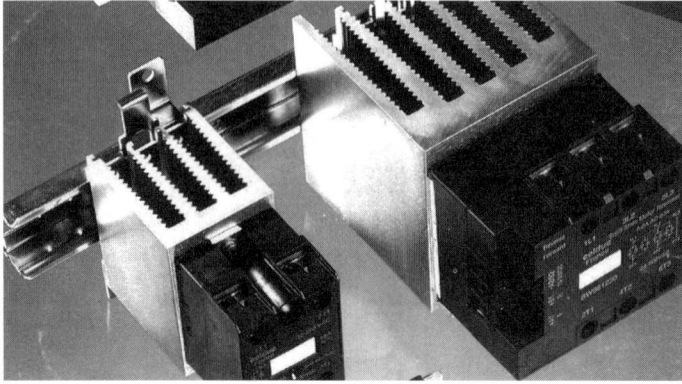

FIGURE 7.117
Thyristor contactors with heat sinks.

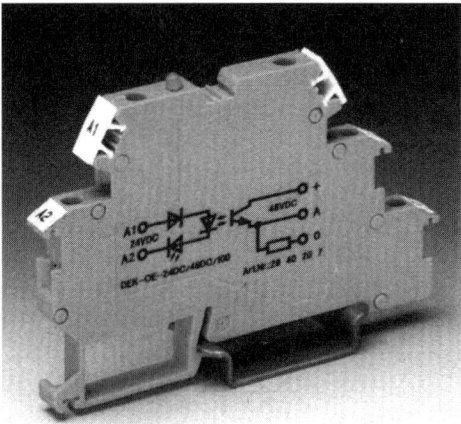

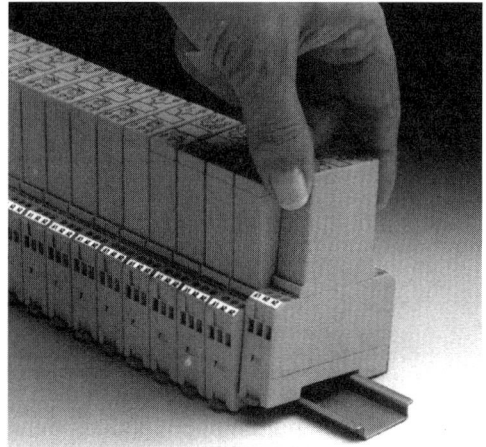

FIGURE 7.118
Optoelectronic relays produced by Phoenix. (Phoenix Contact GMbH & Co. catalog.)

7.10 Super-Power Electronic Relays

Absolutely unique high-speed solid switches (relays) for voltages of tens of kilovolts and currents of tens and hundreds of amperes (though in the form of very short pulses), are produced by the German company Behlke (Figure 7.119). The solid structure of such relays consists of a great number (up to a few hundreds) of series connected layers of MOSFET or IGBT elements (transistors) placed on the same ceramic plate. Such relays are capable of switching voltages up to 65 kV and pulse currents with amplitudes of up to 10 kA (with a pulse width up to 100 μsec).

More powerful switching devices for working voltages of up to hundreds of kilovolts, capable of allowing currents of up to thousands amperes pass for long periods of time, are constructed on the basis of opto-thyristors (Figure 7.120) connected in series. Such

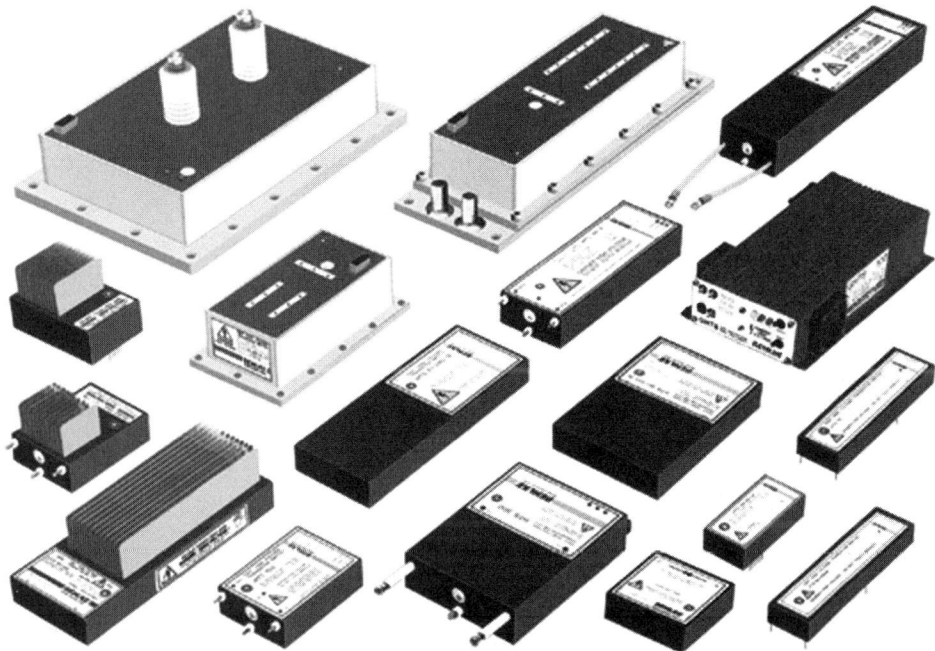

FIGURE 7.119
High-speed high-voltage pulse transistor relays produced by Behlke. (Behlke-Electronic catalog. Fast High Voltage Solid State Switches 1998.)

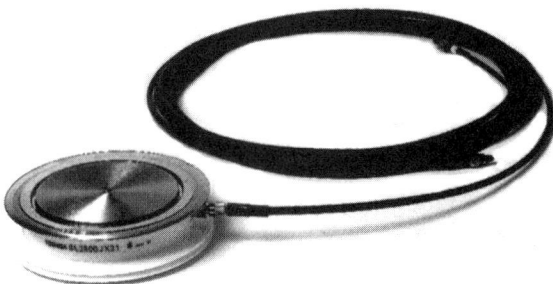

FIGURE 7.120
Light-triggering thyristor of the SL2500JX21 type (Toshiba), for switching voltage 6 kV and a nominal current 2500 A. (Toshiba Corp. catalog. High Power Semiconductors.)

thyristors contain built-in optical fiber several meters in length, directing the luminous flux to the area of the semiconductor structure, which is responsible for enabling of the thyristor (Figure 7.121). The dielectric properties of optical fiber allow entire insulation of the thyristor connected to the high-potential circuit from the earthed control system. This allows the manufacture of unique constructions of such thyristors connected in series. First of all, the thyristors are mounted in modules (Figure 7.122) containing transient elements of the control system, and elements of the thyristor are protected from over-voltages.

These modules are then used to construct huge thyristor units (Figure 7.123). Such units serve as a basic component of so-called High-Voltage Direct Current Links, for high-voltage power transmission lines, which have been quite popular all over the world lately. High-voltage direct current links for AC power lines allow linkage of power-supply systems of different countries, having different voltage levels and different requirements for characteristics of electric energy. Such links also allow a considerable increase in the robustness of the power-supply system. Of course such thyristor units are

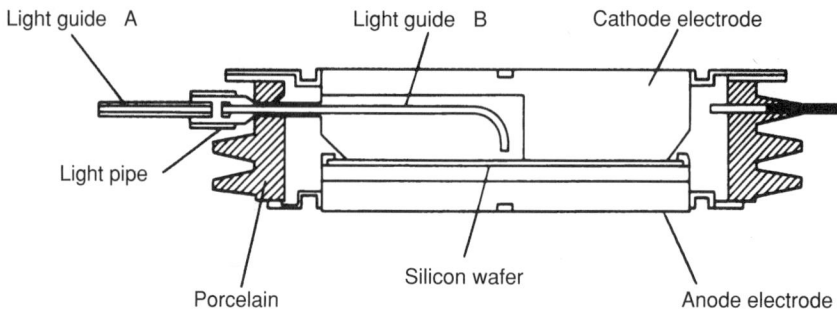

FIGURE 7.121
Construction of a light-triggered thyristor. (Toshiba Corp. catalog. High Power Semiconductors.)

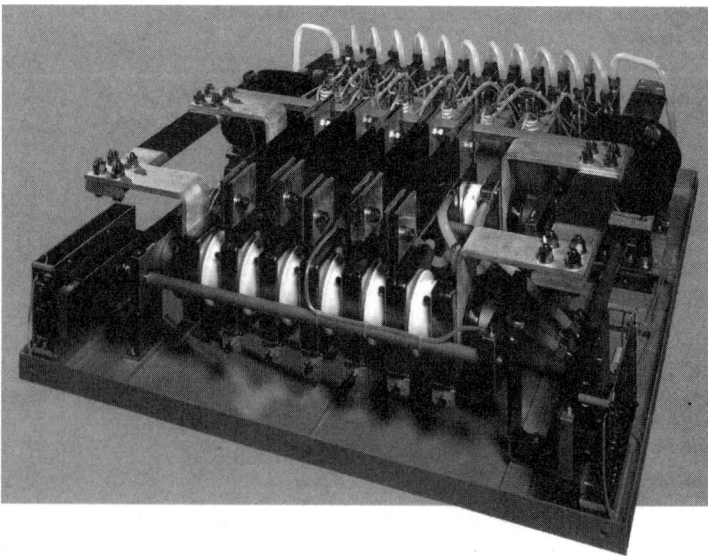

FIGURE 7.122
Light-triggered thyristors module (Toshiba). (Toshiba catalog. Directly Light-triggered Thyristor Valve.)

FIGURE 7.123
Thyristor unit for voltage of 125 kV and nominal current of 2400 A, produced by Toshiba Corp. (Toshiba catalog. Directly Light-triggered Thyristor Valve.)

only part of the most complicated complexes, comprising computer systems for control and protection in emergency modes, and a special system for cooling of the thyristors, with the help of deionized (that is dielectric) water, supplied directly to the high potential, and to many other sophisticated systems.

7.11 Hybrid Relays

After we have considered some specific devices along with many of the problems frequently occurring in them — arcing on contacts, and alternatives of noncontact devices which are capable not only of switching circuits but also of synchronous switching with circuit voltage, the following question can arise: If noncontact relays are so good, why are contact ones still in use along with them?

As the reader has probably guessed, there is no perfect device (like there is no perfect friend, wife, car, etc.), which can meet all the requirements. One of the most essential disadvantages of semiconductor relays is increased (in comparison with the closed contact) resistance in the open state, which accounts for great heating of semiconductor elements when nominal current passes through them, and the necessity for big and heavy heat sinks with forced air or water cooling. Apart from additional costs for electric energy, there are also problems in compact portable and airborne equipment concerning utilization of unnecessary heat released by powerful semiconductor relays. In many types of such equipment one has to fight practically for every watt of heat. In addition, semiconductor-switching devices are more sensitive to overloads in emergency modes, and to overvoltages, than contacts of electromechanical relays.

As soon as designers had realized these problems of semiconductor switching devices, they began to attempt to solve them. They were not very original in their solutions: suggesting combining useful properties of each device, having taken into account advantages and disadvantages of each device. This was the usual solution in engineering, biology, and chemistry.

The first attempts of this kind were aimed at increasing the effectiveness of arc extinction, with the help of a powerful diode (Figure 7.124). Shunting of the bridge contact of the power contactor, of the KTU-4A type (Russia) by a diode, as shown in Figure 7.124, enhanced conditions of arc extinguishing a little bit, and this made possible the use of a contactor with a nominal voltage of 630 V, for a voltage of 1140 V. In standard (not bridge type) contact systems with an arc chute, the diode is connected to an additional electrode fixed between the arc-suppressing horns on the contacts (in this case both circuits are equivalent during arc extinguishing). During one of the voltage half-periods of power supply the diode is opened, and shunts a part of the arc (A2, or an arc on one of the contacts of the bridge) through the additional electrode placed in the arcing zone. This part of the arc is extinguished, and the gap between the contacts where the arcing occurred, is deionized. By the next voltage half-period of the power supply, when the diode is blocked, the arc will not be restored. Total time of arcing in such devices is reduced by 2 to 4 times.

In the device shown in Figure 7.125, the main contacts MC are shunted by two parallel circuits, each of which contains a diode, auxiliary contacts and a current coil. Auxiliary contacts 1 and 2 are linked with a drive of the main contact, and are closed simultaneously with it. Opening of the main contacts occurs without arcing, as load current turns to one of parallel branches (depending on the voltage polarity). The drive of the main contact unblocks (relieves) auxiliary contacts 1 and 2. One of them opens immediately and the second remains closed under the retaining effect of the electromagnet (3 or 4) with the

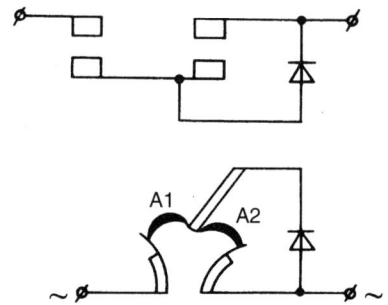

FIGURE 7.124
Simplest hybrid-switching devices, with a diode facilitating extinguishing of an electric arc.

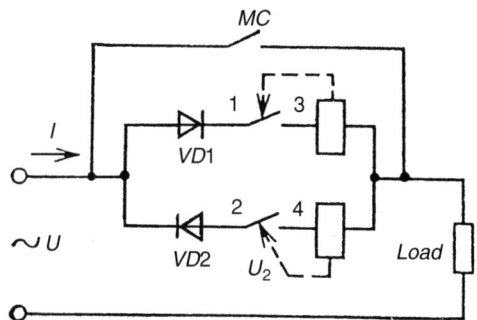

FIGURE 7.125
A hybrid switching device with diodes providing switching of the load at the zero value of the current.

corresponding polarity. When the current sinusoid approaches the zero value, the force of this electromagnet becomes so weak that the corresponding auxiliary contact opens (without arcing and without current) breaking the load circuit once and for all.

Thus there is no arcing on the main contact as it opens because it is shunted with an auxiliary contact, and arcing does not occur on the auxiliary contact because it opens at the zero value of the current. A contactor of the KVK type for a nominal current of 250 A (maximum making current 6900 A, breaking current — 3250 A) and voltage of 1140 V, sized $400 \times 354 \times 190$ mm, was designed on the basis of a similar principle in the former U.S.S.R. in the 1970s. Testing of this device proved that after 440 commutation cycles with a current of 1000 A, wear of the contacts deviated only 10% from the wear that was deemed permissible.

When high-power thyristors appeared, engineers focused on the use of them instead of diodes in hybrid devices, which allowed considerable simplification (and sometimes even leaving out) of the most unreliable parts of these devices — the mechanical blockings (Figure 7.126). In such a circuit, when the main contact MC is closed, the thyristors are

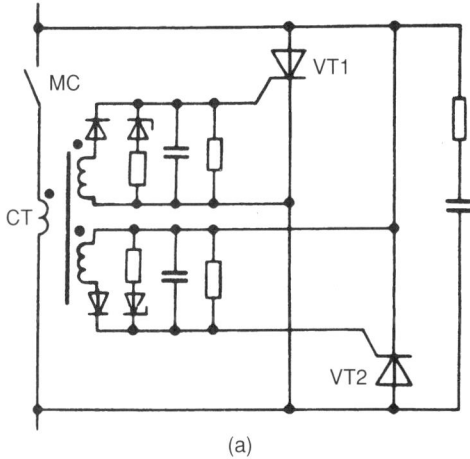

(a)

(b)

FIGURE 7.126
Scheme and construction of a one pole of a thyristor unit of three-phase hybrid contactors KT64 and KT65 (Russia). 1 — case; 2, 6, 8 — outlet wire with the main current; 3 — thyristors VT1, VT2; 4 and 5 — insulated flexible outlets; 7 — magnetic core of the current transformer; 9—11 — elements of thyristor protection.

FIGURE 7.127
External design of hybrid thyristor contactors
KT64 and KT65 (Russia).

enabled and allow the current to pass if the anode–cathode voltage reaches 6 to 10 V. Such conditions are created at the beginning of the process of opening of the main contact, when there is a short arcing with voltage of over 10 V. This voltage applied to the thyristors is enough to enable them. The current flows to the thyristor circuit with the corresponding voltage polarity at the given moment, and the arc on the main contacts decays. The control signal in the gate circuits of the thyristors disappears, but the thyristor, enabled before that, remains conductive until the current sinusoid passes through the zero value, when it is fully blocked, disabling the load.

In 1970s and 80s in the former U.S.S.R., a whole family of contactors were produced on the basis of such circuits, for nominal currents from 160 up to 630 A and voltages of 380 to 660 V (Figure 7.127). Similar devices were produced by the some Western companies. Some disadvantages of such devices can be explained by the powerful current transformer, which is not always adequate in different modes. That is why there were some attempts to get rid of it due to low-power miniature auxiliary contacts of such construction that would not complicate the mechanical parts of the contactor. It was suggested that a reed switch placed near the control coil of the main contact (Figure 7.128), could be used as such an auxiliary contact. Such a contact operates simultaneously with the main contact, but it is not connected to it mechanically.

In high-voltage hybrid-switching devices current transformers are usually used (Figure 7.129). The drawbacks of the devices mentioned above are that they only weakened the impact of arcing, but were not capable of eliminating it. There are a great

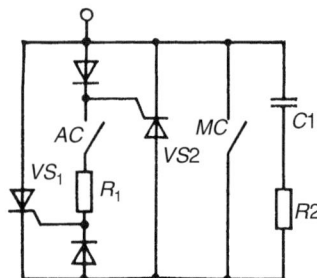

FIGURE 7.128
Hybrid thyristor contactor with a
control reed switch (AC).

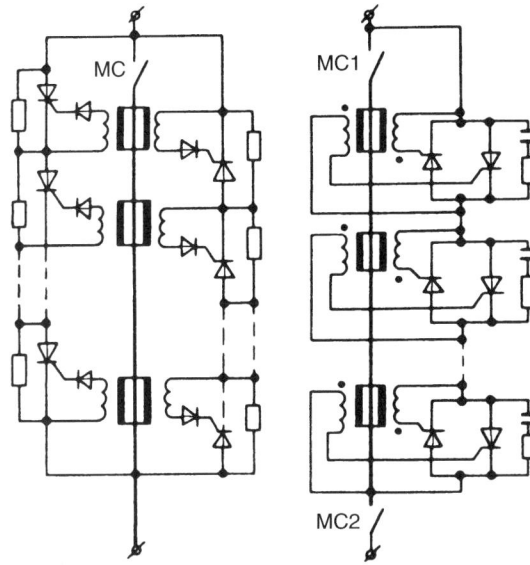

FIGURE 7.129
High-voltage hybrid switching device on thyristors.

number of patents describing hybrid relays with a complex control system, based on complex integral circuits (IC) which analyze load voltages and current curves, and give commands for enabling and disabling of high-power electronic elements in such a way

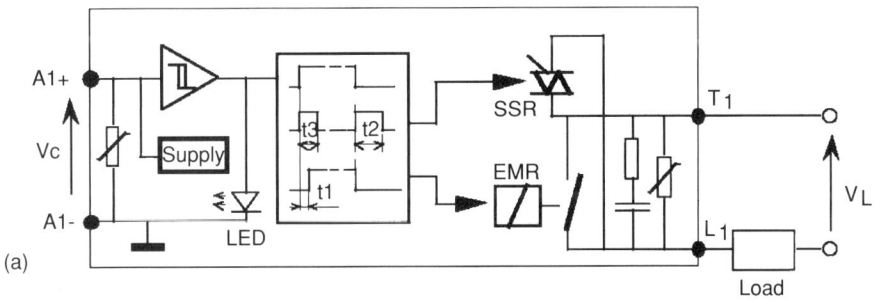

(a)

(b)

FIGURE 7.130
Functional diagram (a) and external design (b) of a hybrid relay XV series (Teledyne Relays). Nominal current 30 A, voltage 420 V AC, dimensions: 61.3 × 44.5 × 45 mm. (Teledyne Relays online catalog 2004.)

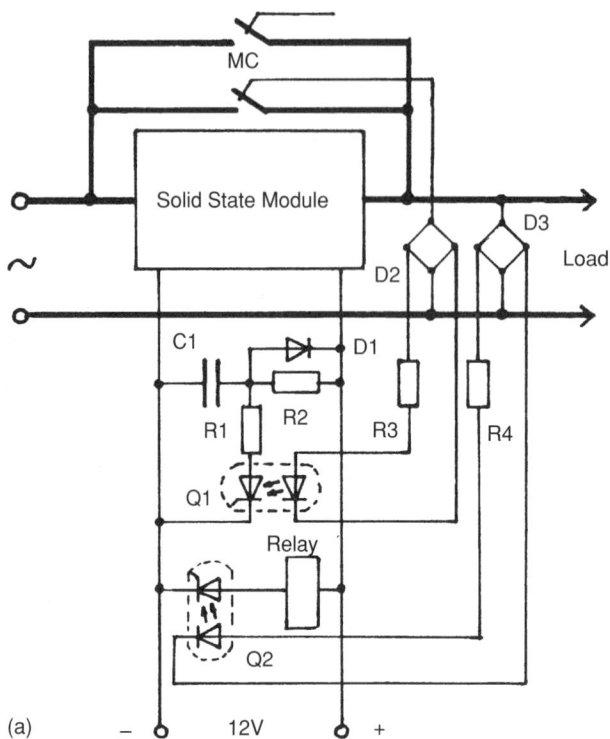

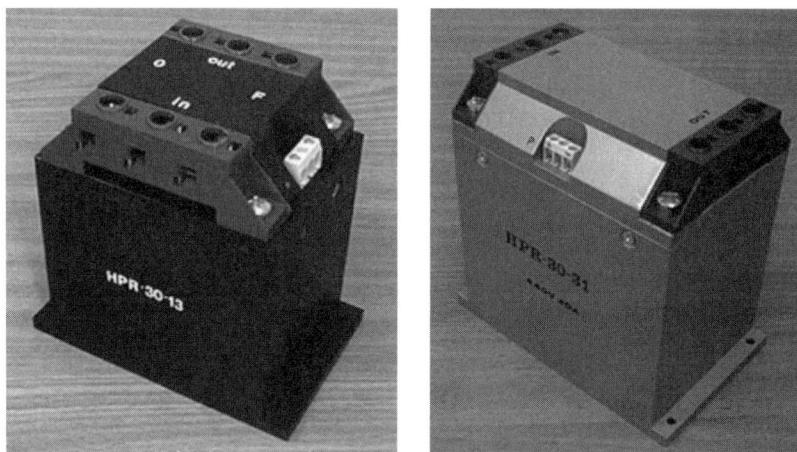

FIGURE 7.131
(a) Scheme of a hybrid relay based on the principle of event-tracing. (b) External design of preproduction models of hybrid relays in one and three-phase construction for current of 50 A and voltage of 440 V, designed by the author.

that the possibility of arcing on the main contact is entirely ruled out. Such circuits require current sensors (though miniature ones), voltage, and cope well with the problem of synchronization of operation of semiconductor and electromagnetic elements in normal modes of exploitation, though they may be inadequate in different emergency and transient modes or when there are higher harmonics or overvoltages in the current. Besides, they are too complex and expensive. Another type of electronic control circuit is based on the commands of enabling and disabling of electronic elements, and on an internal electromagnetic relay with certain time delays. Such devices do not require control of current and voltage phases, and they are much simpler than the former ones, but for reliable operation there should be reserves of time intervals because small time delays may lead to nonsynchronous operation of high-power elements (internal electromagnetic relays have quite large dispersals of make delays). Such devices are less "intelligent," even less so than earlier models with current transformers, because the operation is based on a fixed internal algorithm, independent from the real modes of operation of the switching device. Nevertheless, this very principle (as the simplest one) was the basis for some production-run models (Figure 7.130). The author of this book tried to contribute to the invention of hybrid relays and designed a construction without sensors of current phase and sensitive electronic amplifiers. It did not have a fixed internal algorithm based on fixed time delays as well. The principle of such a solution was based on tracing of events which is actuation of a certain element of the circuit on the signal from another element, providing such actuation in a certain operation mode (Figure 7.131). This provided a minimal number of simple elements, reliability of functioning, and independence of the circuit operation from changes of parameters of certain elements in time (or in temperature), minimal make delay and dropout time of the device.

8

Time Relays

Let us remember that relays are usually defined as devices that can only be in extreme stable states and can switch from one state to another, stepwise, even when the input actuating quantity varies smoothly. In that definition, there is not a single word about the character of the actuating action. Most often, such actuating action is current, which is why relays that are energized by electrical current (voltage) are the most widespread. The bulk of this book is devoted to this type of relays, but this is not the only type of relay that exists. There are relays responsive to light, temperature, location in space, air or liquid pressure, air or liquid speed, etc. Obviously, it is impossible to consider all known types of relays in detail within just one book, but in order to get a complete picture one should be familiar with at least some of them.

One of the most widespread types of relays (after electrical relays) is "time relays." Usually these are relays operating with a certain delay with regard to the signal applied to the relay input, which is why frequently the term "time-delay relay" is used. As the change of state of a relay is accompanied by a certain delay with regard to the signal applied to its input, one can say for sure that, apart from its other functions, every relay also functions as a time relay. Sometimes standard electromechanical relays are used to enhance stability of complex automatic control systems. Their only function is to provide a certain signal delay, the value of which equals its own make delay. In terms of engineering, "time relays" or "time-delay relays" are usually defined as relays in which the time-delay function dominates, and in which the characteristics of that function are enhanced, by one means or another.

8.1 Electromagnetic Time Relays

Let us remember that the pick-up (and drop-out) time of a standard electromagnetic relay includes two major composites: time of increasing (or decreasing) of current while winding up to the operating (releasing) current value, and armature traveling time. The simplest way to increase the pick-up (drop-out) time of a standard electromechanical relay is to increase the first composite. For this purpose an additional short-circuited winding is placed on the relay core (Figure 8.1), with resistance R_2, number of turns W_2 and inductance L_2.

When working voltage is applied to the main relay winding, the current in it builds up from zero to its steady-state value. According to the law of electromagnetic induction, current variation in the main winding (and therefore also the magnetic flux $[\Phi_1]$ in the core with the additional winding) causes current of the opposite direction in the

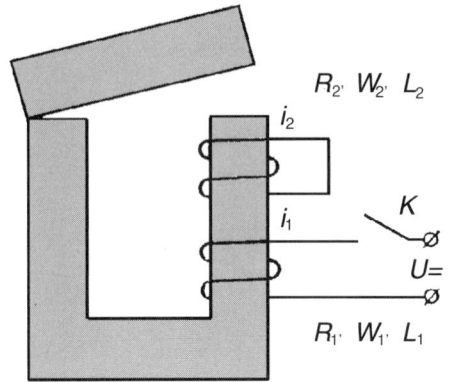

FIGURE 8.1
Electromagnetic time delay relay with an additional short-circuited winding (R_2, W_2, L_2) on the core.

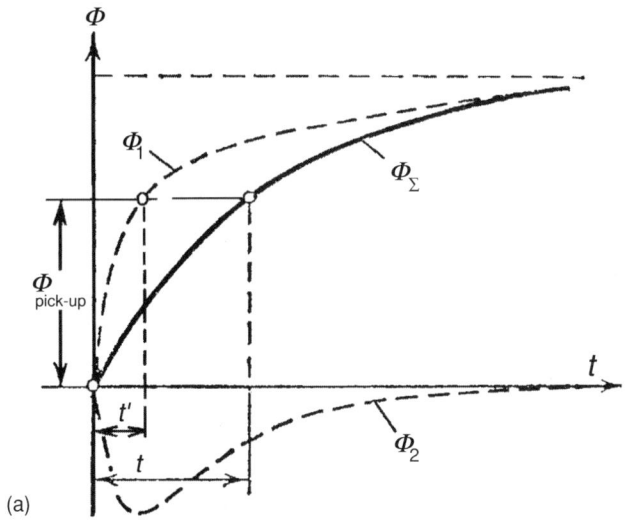

(a)

FIGURE 8.2
Building up of the magnetic flux when the relay with the short-circuited winding is switched ON (a) and fading of the magnetic flux as the relay is switched OFF (b). Φ_1 — Magnetic flux of the main winding; Φ_2 — magnetic flux of the short-circuited winding; Φ_Σ — compound flux; t' — pick-up (drop-out) time of the relay without short-circuited winding; t — pick-up (drop-out) time of the relay with short-circuited winding.

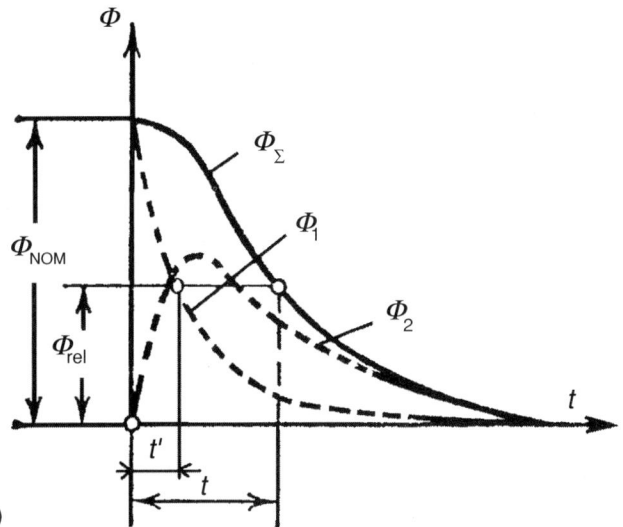

(b)

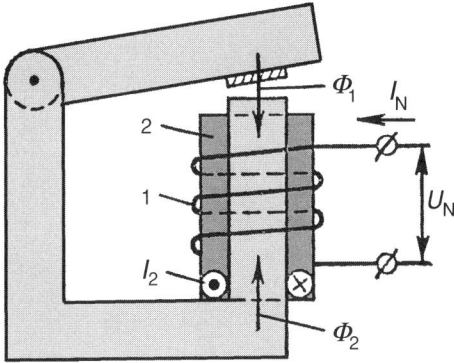

FIGURE 8.3
Principle of construction of a time relay with an electro-magnetic delay. 1 — winding; 2 — copper cylinder.

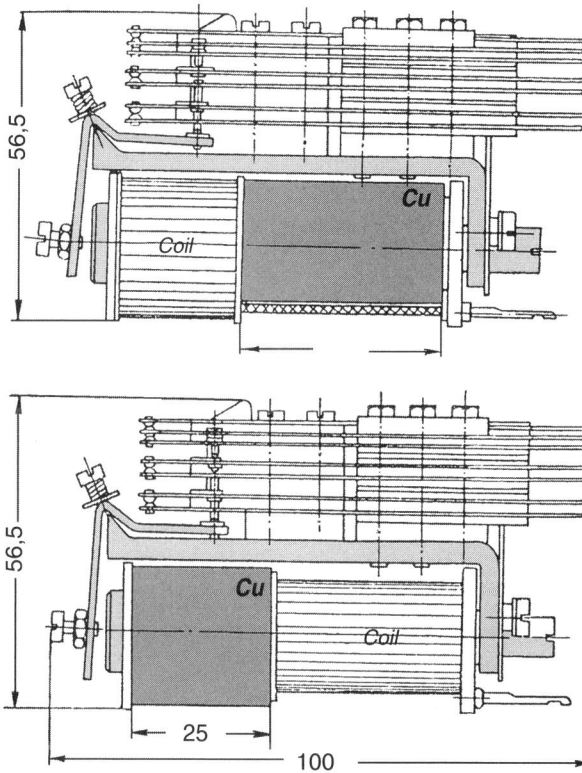

FIGURE 8.4
Relay with prevalence of delay for drop-out (above) and prevalence of delay for pick-up (below).

additional short-circuited winding. This additional current creates an additional magnetic flux [Φ_2] whose direction is opposite to the main magnetic flux and therefore weakens it thus delaying the relay pick-up (Figure 8.2a), and vice versa, when the voltage of the power supply is switched OFF, the main magnetic flux falls to the zero value and the magnetic flux of the short-circuited winding prevents this fall by delaying the drop-out time of the relay (Figure 8.2b).

The less the resistance of the short-circuited winding R_2, the more it affects the pick-up and drop-out times of the relay. That is why in practice thick copper slats (2) or disks, which are put directly onto the core under the main winding 1 (Figure 8.3), are used instead of short-circuited winding.

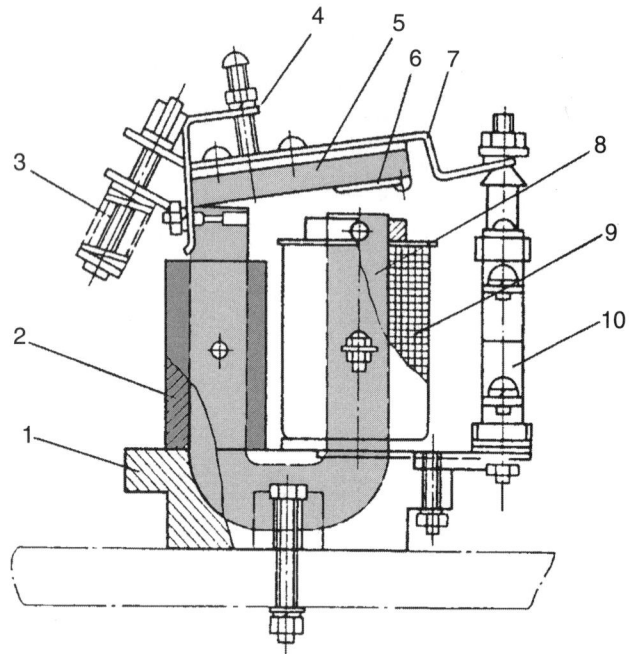

FIGURE 8.5
High-power relay with electromagnetic delay. 1 — aluminum heelpiece; 2 — copper slate; 3 — return spring; 4 — thrust cramp with a screw; 5 — armature; 6 — nonmagnetic layer; 7 — control-rod; 8 — U-shaped core; 9 — coil; 10 — main movable-bridge-type contact assembly.

It appears that if this copper slat is shorter than the core and is placed not in the center but along the edges of the core (Figure 8.4), then in one case, we will have a relay with a prevalence of the delay for pick-up, and in the other case, we will have a relay with a prevalence of the delay for drop-out. When the size of the copper slat and of the winding

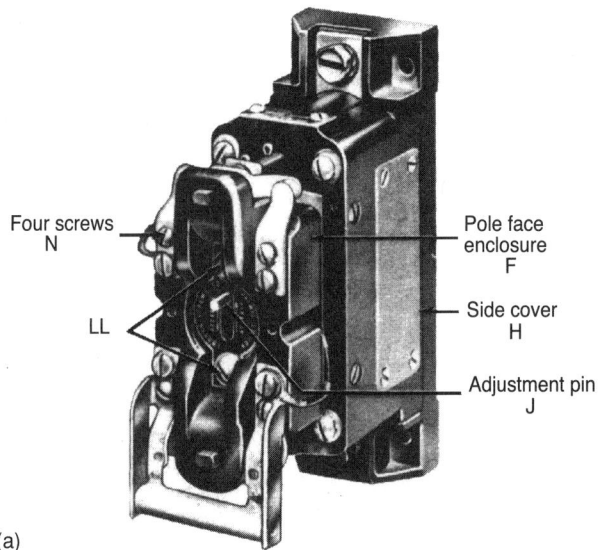

(a)

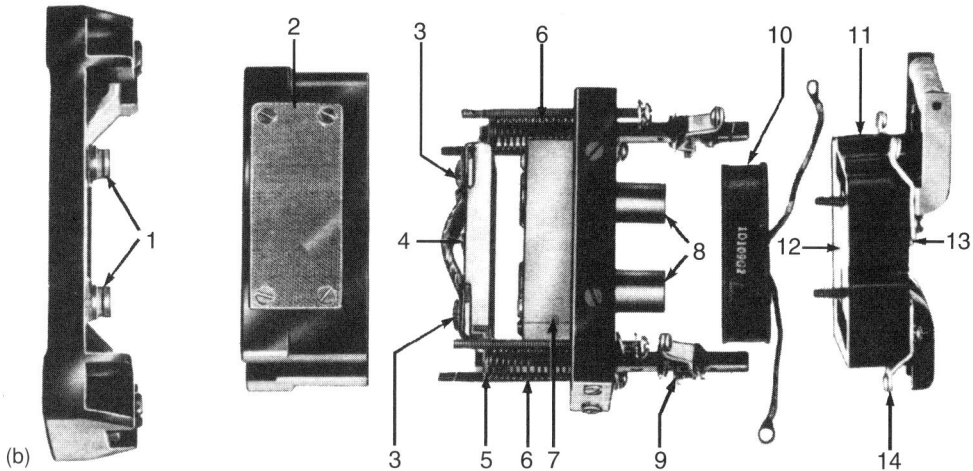

FIGURE 8.6 (Continued)
Time-delay drop-out contactor CR5481-33 type (General Electric). 1 — main stationary contact; 2 — side cover; 3 — main movable-bridge-type contact assembly; 4 — holding screws; 5 — slot; 6 — return spring; 7 — copper block; 8 — magnetic poles; 9 — movable interlock contact spring; 10 — coil; 11 — pole-face assembly; 12 — slotted steel block; 13 — pin; 14 — stationary interlock contacts.

of the relay is the same, the delay for the dropout is almost two times as long as that of the pick-up, which is that in order to equalize these values the copper slats and coils must have different sizes for relays with a delay for pick-up, and for relays with a delay for drop-out (Figure 8.4).

Though the time delay in this case is not very long and usually does not exceed 0.5 to 3 sec, relays with electromagnetic delays have been used for years due to their simplicity and reliability, and not only in miniature relays but also in high-power relays (Figure 8.5). In some old constructions (Figure 8.6), there was even continuous adjustment of the time delay within 0.75 to 3 sec by moving a slotted steel block (12) in the "increase" or "decrease" direction by means of a pin (13).

8.2   Capacitor Time Relays

It is possible to hold back building up of the current in the winding of a DC relay by quite a large value with the help of a capacitor shunting the winding of the relay (Figure 8.7). In the first case the capacitor (C) practically does not affect the pick-up time of the relay, because its charging current is limited by the resistor (R). In the steady-state mode, the capacitor is charged up to the value of the power supply voltage. When the supply circuit of the winding of the relay is broken, the charged capacitor starts to discharge, through the resistor (R). The value of the discharge current of the capacitor is limited by the same resistor and also by the resistance of the winding of the relay, as in order to hold the relay in a closed position much weaker holding current (the current of the capacitor discharge) is required than pick-up current (the current absorbed in the charging capacitor in order to delay the pick-up of the relay). Such circuit in most cases is able to cope with its task quite successfully, but when the natural resistance of the winding is quite high (tens of $k\Omega$, for example), the resistance of the limiting resistor R must also be high, otherwise the capacitor C will affect the pick-up time of the relay. But in this case total resistance of

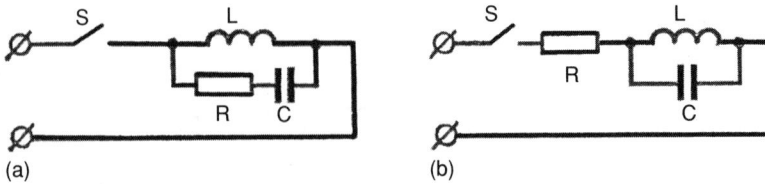

FIGURE 8.7
Use of a capacitor for increased time delay of dropout (a) and pickup (b).

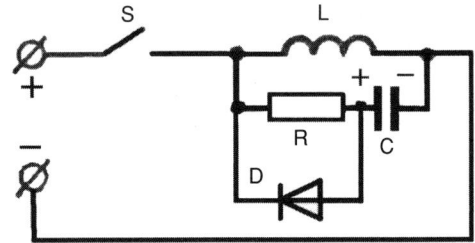

FIGURE 8.8
Circuit of a capacitor time relay with an additional diode.

the winding and of the resistor R appears to be so high that discharge current of the capacitor with switched OFF power supply of the relay is not enough to hold the relay and obtain the required time delay. In such cases one may use an additional diode D, which allows the capacitor to discharge through it directly to the relay winding, leaving out the limiting resistor R (Figure 8.8). But the charged capacitor C (Figure 8.7b) has internal resistance close to zero, and at the first moment when the relay is switched ON will short-circuit the winding of the relay.

When the capacitance is quite high, the voltage on the winding is completely determined by the charge level. In its turn the charging speed of the capacitor with a permanent voltage of the supply source is determined by the so-called "time constant" (τ) of the circuit: $\tau = RC$.

By changing the resistance of the resistor R and the value of the capacitor C in this circuit, it is possible to change the delay time of such a relay. Usually simple time relays with a time delay of up to 10 sec are based on this principle.

8.3   Relays with Clockwork

The integration of the electromagnetic relay with clockwork provides practically unlimited time delay. The clockwork may have a spring drive or an electric drive. Earlier time relays with clockwork look very much like ordinary clocks (Figure 8.9). Their construction does not differ much from that of a standard clock: similar clock spring, similar mechanism. The only difference is that electric contacts are attached to the clockwork. Later on, electric drives appeared: solenoid, inductive, motorized ones. In some constructions a spring clockwork was applied and an electric motor used for automatic winding-up of the spring. In the 1960s time relays with clockwork launching with the help of a solenoid were widely used (Figure 8.10). The construction of this device consists of a separate clockwork in a disk-shaped steel case, with a scale, contacts, and a solenoid. The solenoid is linked with the clockwork with the help of a pin (4) from the clockwork case. The

FIGURE 8.9
Time relay produced in 1935 (General Electric Co.); (a) with a spring and manual winding up; (b) with an electric motor. (General Electric G.E.C. catalog of electrical installation material 1935.)

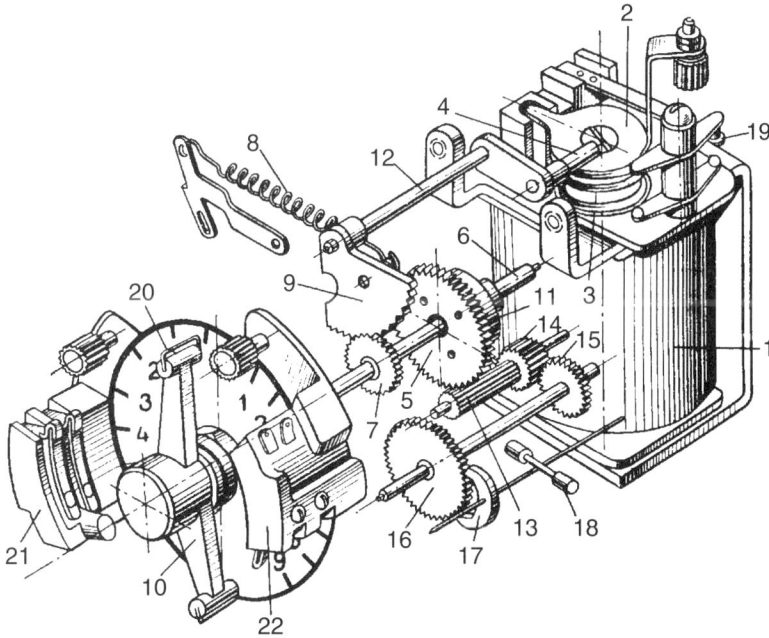

FIGURE 8.10
A time relay with a spring clockwork and a starting solenoid. 1 — solenoid winding; 2 — upper part of the solenoid armature; 3 — return spring; 4 — pin; 5–7 — elements of mechanical gear; 8 — drive spring of the clockwork; 9 — ratchet latch; 10 — contact traverse; 11 — friction device; 12 — axis; 13–18 — elements of the clockwork; 19 — snap-action contacts; 20 — movable contact; 21 — stationary contact; 22 — slipped contact.

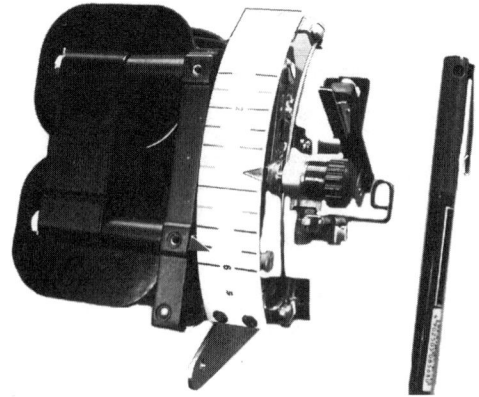

FIGURE 8.11
Internal section of a time relay with a spring clockwork
and a powerful two-coil starting solenoid of the RZf type.
(AEG)

clockwork is fixed on the solenoid in such a way that the pin in the initial position is lifted
up and the spring (8) is stretched. When the solenoid is energized, its armature (2) is
retracted and releases the pin (4). Affected by the spring (8) the clockwork begins to turn
the traverse (10) with a movable contact (20). The time delay is determined by the distance
between the movable contact (20) and the stationary one (21) and is adjusted by reposi-
tioning of the stationary contact (21). When the power supply of the solenoid is switched
OFF, the strong restoring spring (3) returns the drive spring of the clockwork next to the
pin (4) to its initial position. Relays of such type were quite massive (Figure 8.11) and
heavy (more than 1.5 kg).

In the time relay with a motor drive, the clockwork was set in motion by a small
synchronous electric motor with a reduction gear and a solenoid was used to return it
to its initial position (Figure 8.12). In such a relay the required time delay is displayed on
the scale. When the relay is switched ON (when the power supply is applied to it) the DC
motor with a reduction gear and the solenoid are both powered at the same time. The
motor rotates the mechanism until the time set on the scale. At that point the output

FIGURE 8.12
A time relay of the MC-13 type (General Electric),
with a motor drive and a solenoid, produced in the
1950s. 1 — oil bearings; 2 — contact D; 3 — contact
arm H; 4 — contact arm G; 5 — contacts C; 6 —
contact arm A; 7 — contact B; 8 — latch; 9 — solenoid's armature; 10 — solenoid's coils; 11 — pointer.
(General Electric catalog. Definite time-control relay
type MC-13.)

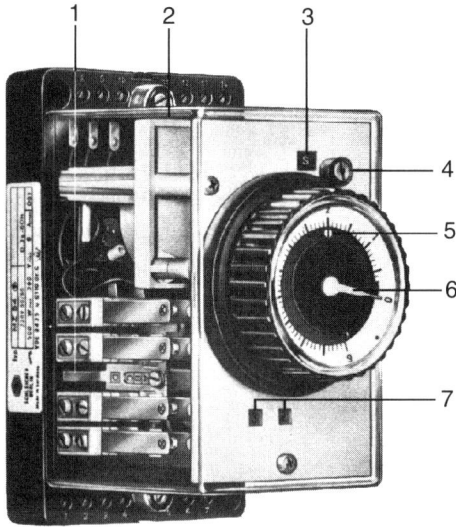

FIGURE 8.13
High-precision time-delay relay series MZ-54 with synchronous motor and solenoid clutch (Schleicher). 1 — Function selector; 2 — internal frequency selector; 3 — timing range indicator; 4 — timing range selector; 5 — setting mark; 6 — elapsed-time indicator; 7 — contact position indicator.

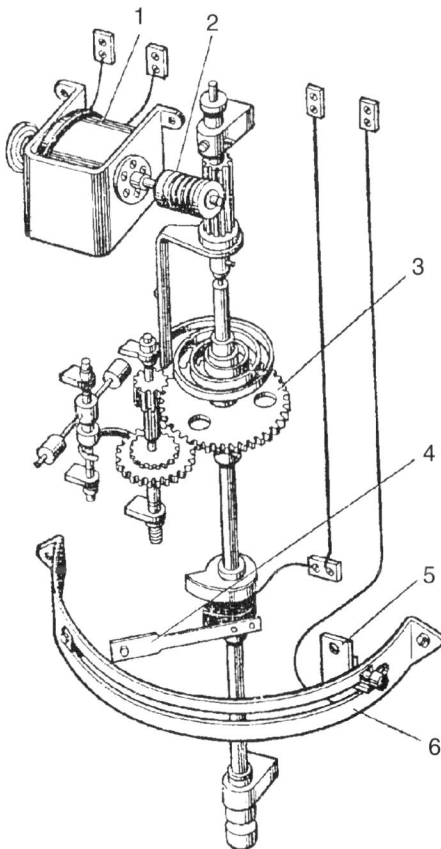

FIGURE 8.14
Construction of the time relay with a solenoid drive. 1 — solenoid; 2 — worm-gearing; 3 — clockwork; 4 — movable contact; 5 — stationary contact; 6 — scale of the time delay.

contact closes, the additional contact breaks the supply circuit of the motor, and the solenoid opens. Switching-off of the solenoid causes the mechanism affected by the restoring spring to return to its initial position.

As the voltage frequency in the AC circuit is quite stable, many companies at the same time, produced relays based on an AC synchronous electric motor, the motor speed of which depended on the frequency of the supply main. The company Schleicher produced a whole series of time relays based on this principle (Figure 8.13). Upon energization, the solenoid couples the swing-out gear train axle to the timing mechanism, actuates the instantaneous contacts, and cocks a switch lever by spring tension. Simultaneously, the synchronous motor starts to rotate and the timing period commences. Upon the expiry of the preset time, the timing mechanism releases the cocked switch lever, which snaps the delay contacts into the operating position and de-couples the gear train axle. The timing mechanism reverts immediately to the "before-start" condition. Upon de-energization, the solenoid and all contacts revert to the "before-start" position.

There are constructions in which a solenoid rather than a motor is used in order to make it cheaper (Figure 8.14). The armature (1) of such a solenoid is connected to the worm gear (2) transforming the linear displacement of the armature into the rotating

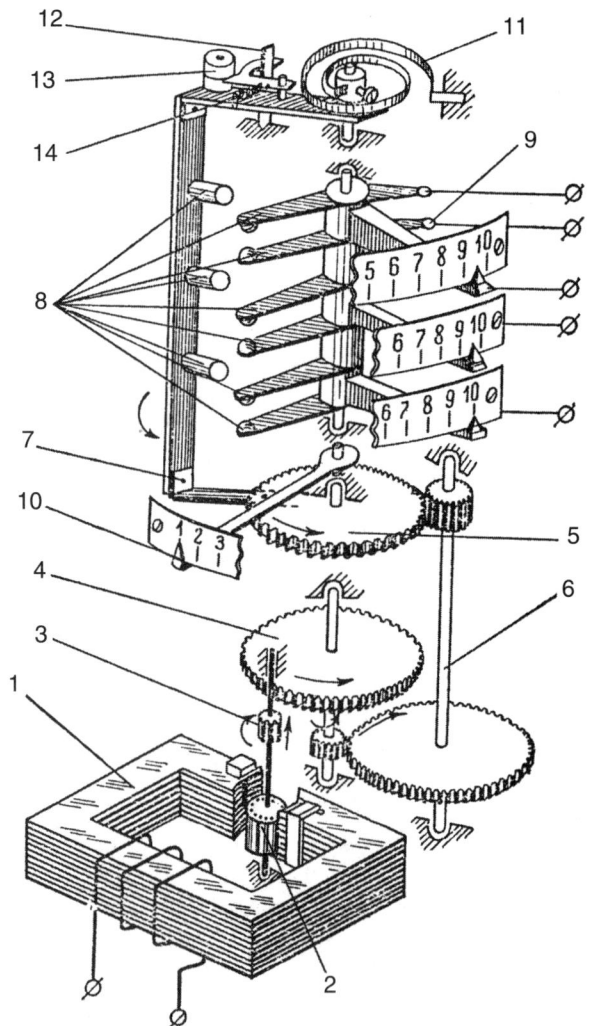

FIGURE 8.15
Construction of the time relay of the RBM-12 type with a synchronous motor and a retractable (like a solenoid) rotor. 1 — motor starter; 2 — rotor; 3, 4, 5 — gear wheels; 6 — reduction unit; 7 — thumb frame with contacts; 8 — contacts; 9 — drag-bar for adjustment of stationary contacts (time delay); 10 — pointer; 11 — return spring; 12 – stop; 13 — lock; 14 — spring.

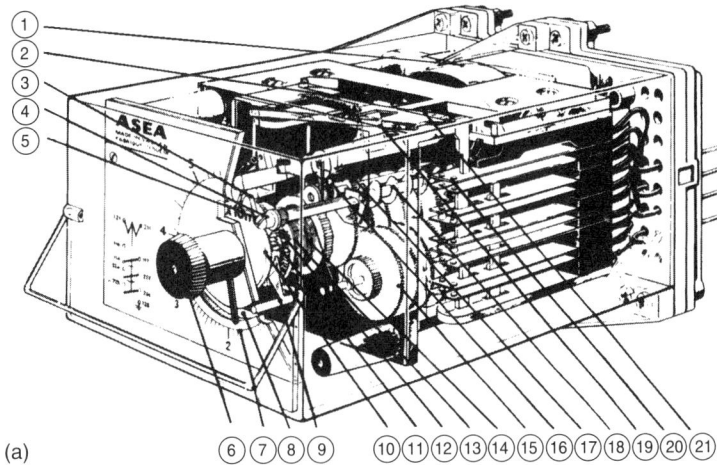

FIGURE 8.16
(a) Construction of a motor time relay of the RXKP-2 (ASEA-ABB) type with delays from 30 sec to 60 h. 1 — synchronous motor; 2 — spring; 3 — projection on intermediate disc 11; 4 — setting screw for scale range; 5 — scale range index; 6 — setting knob for operating time; 7 — operating time scale; 8 — running time scale; 9 — running time recorder; 10, 13 — crown wheels; 11 — intermediate disc; 12 — planet wheel; 14 — return springs; 15 — gear train; 16 — clutch; 17 — pinion; 18 — worm; 19 — stop; 20 — instantaneous contact; 21 — armature. (b) External design of a motor-time relay of the RXKP-2 type without cover (ASEA-ABB). 1 — Synchronous motor; 2 — contact system; 3 — mechanical transmission. (c) Fragment of mechanical transmission of a motor-time relay of the RXKP-2. (ABB)

FIGURE 8.17
Modern time relays with a motor drive.

motion of the elements winding up the spring. The moment of rotation of the spring through the clockwork makes the movable contact (4) move evenly until it closes with the stationary contact (5).

In the time relay shown in Figure 8.15, the rotor (2) is retracted to the starter (1) (it goes up), when the current in the winding reaches a certain value and its gear wheel (3)

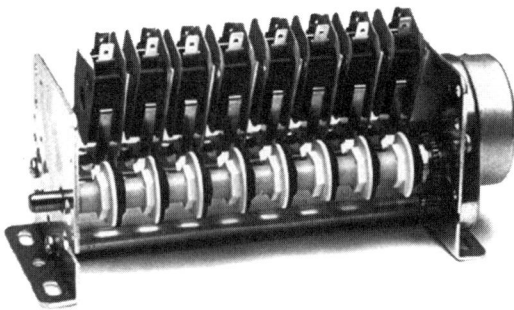

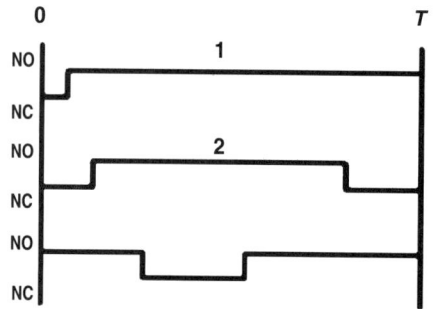

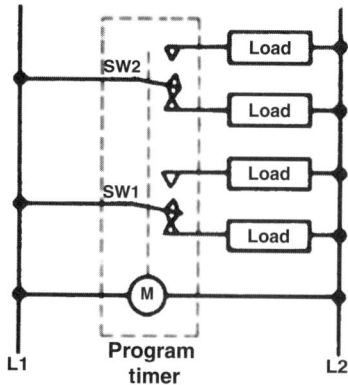

FIGURE 8.18
Synchronous motor driven cam program timer of the 150 type, timing diagram and typical wiring. (American Control Products, a division of Precision Timers Inc online catalog 2004.)

engages the gear wheels of the reduction unit. At that point, the wheel (5) with the frame (7), with contacts fixed onto it, begins to rotate. As the frame (7) rotates, the contacts fixed onto it touch the corresponding stationary contacts, according to the time defined for the position of the contacts. Such relays were produced in the 1960s by the Cheboksar Electrical Equipment Plant (the former U.S.S.R.).

Similar relays were produced by many companies until the 1970–80's, and some of them are still used today (Figure 8.16). Of course modern relays, based on the same principle (a small synchronous electric motor with a reduction unit that sets the clockwork in motion) look more modern (Figure 8.17), but do not differ much from constructions designed years ago, in the 1970–80's.

One version of a motor-driven time relay is a Cam Program Timer (Multi-cam Timer, Re-cycle Timer, Repeat Cycle Timer, etc.), (Figure 8.18). When energized, a program timer continuously and synchronously repeats a preestablished sequence of ON–OFF switching events. The program timer may be set for frequency and duration of timed events. The frequency (interval) is the time between the "On" and "Off" cycles. The frequency may be set from intervals of several minutes to several hours. The duration (cycle) is one revolution of the output camshaft (camshaft speed). Duration may be set from 1 sec to 30 h. An automatic reset feature (in some models of such devices) allows function synchronizing with other system controls. This feature provides an interesting variety of timing combinations.

8.4 Pneumatic and Hydraulic Time-Delay Relays

Time relays with a clockwork allow one to obtain very long delays, measured in tens of hours, but such delays are not always necessary and it is not always financially favorable to use expensive relays in order to obtain delays from just a few seconds up to one minute (the most frequent range of delays), which is why along with complex mechanisms based on clockwork or an exact electric motor, there are also simpler devices, consisting of a solenoid and an air or liquid damper delaying retraction and return of the relay armature (core) to its initial position.

The Allen West & Co. produced relays with time delays, provided by deceleration of the solenoid core with the help of a viscous fluid (Figure 8.19). In this construction an additional rod with a plate and holes was fixed on the end of the core. It was placed in a vessel with silicone oil. When current was applied to the winding of the solenoid, the core was slowly retracted into it, as the oil flowing through the small holes in the plate prevented its movement. Of course, such a relay could not provide delays as long as motor relays, but it was quite suitable for delays of a few seconds.

The essential disadvantage of such relays was a strong dependence of the time delay on the ambient temperature (Figure 8.19b), caused by changes of oil viscosity due to changes in its temperature. In addition to that, the time delay also was very dependent on the voltage applied to solenoid.

Pneumatic time relays in which air was used instead of viscous fluid lacked this disadvantage (Figure 8.20). In the relay in Figure 8.19, the block (2) links three elements of the relay: the solenoid core (1), the micro switch (4), and the rubber diaphragm (5) of the pneumatic decelerator. When the solenoid (1) is switched ON, its core is immediately retracted into the coil (in contrast to relays with a liquid damper), the pusher (8) goes down and releases the block (2). Affected by the spring (3), the block (2) begins descending,

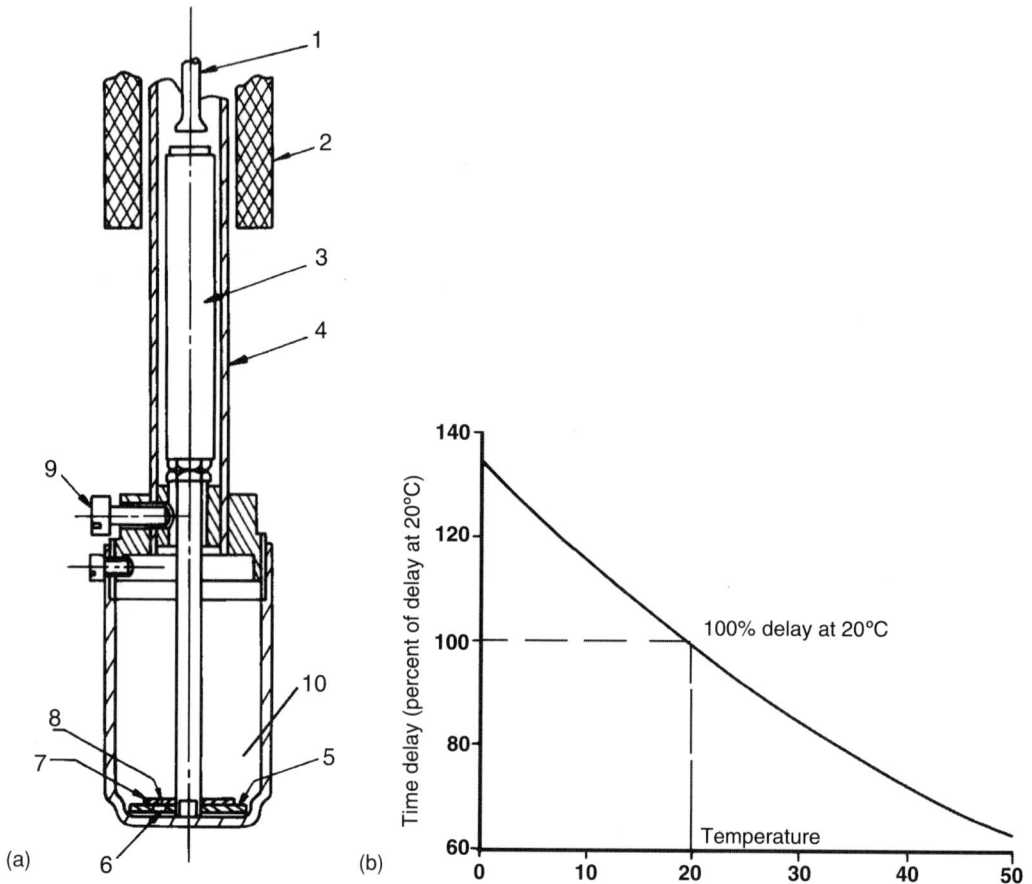

FIGURE 8.19
Element of a relay with delay provided by deceleration of the solenoid core with the help of liquid damper. (a) —
Construction; (b) — temperature dependence of the delay. 1 — Trip pin; 2 — coil; 3 — plunger; 4 — plunger
casing; 5 — piston; 6 — hole; 7 — plate; 8 — hole; 9 — setting screw; 10 — silicon oil.

following the pusher (8), although slowed by the diaphragm (5), which straightens gradually up when the upper part of the decelerator is filled with air. The air is drawn into this
part though a small hole, the section (and therefore the time delay) of which is adjusted by
the needle (6). The micro switch (4) picks up when the upper part of the decelerator has
already been filled with air (Figure 8.21).

Such relays provided accuracy of the time delay within 10 to 12% in a wide range of
temperatures. The coil of the solenoid could be switched both to AC and DC circuit and
the time delay did not depend on the voltage in the circuit. The simplicity of obtaining
time delays from a few seconds up to tens of seconds, and even as much as a few minutes,
as well as the relative stability of the time delays, made this type of relay very popular on
the market. Such relays were produced by many firms and had various external designs,
but the principle of operation and basic elements remained the same (Figure 8.20).

Pneumatic time relays are still used nowadays. These are usually small light add-on
devices to relays and contactors, adjoined to electromechanical relays like an additional
unit of contacts (Figure 8.22).

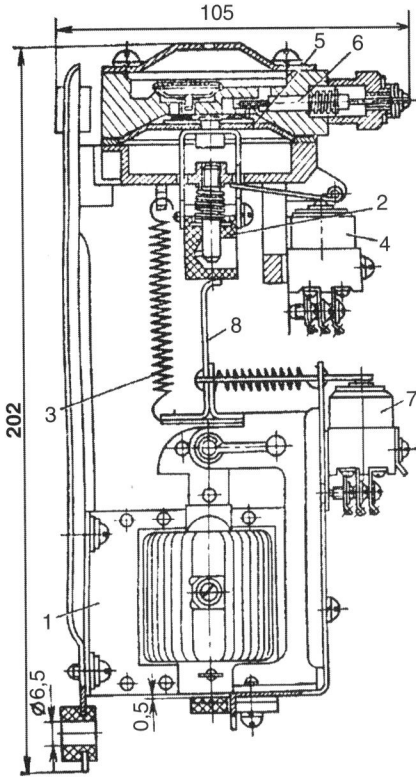

FIGURE 8.20

Pneumatic relay with delay 0.4 to 180 sec long, produced in the U.S.S.R. in the 1950–70's. 1 — Electric magnet; 2 – block; 3 — spring; 4 — micro switch operating with delay; 5 — rubber diaphragm; 6 — needle; 7 — snap-action micro switch; 8 — pusher.

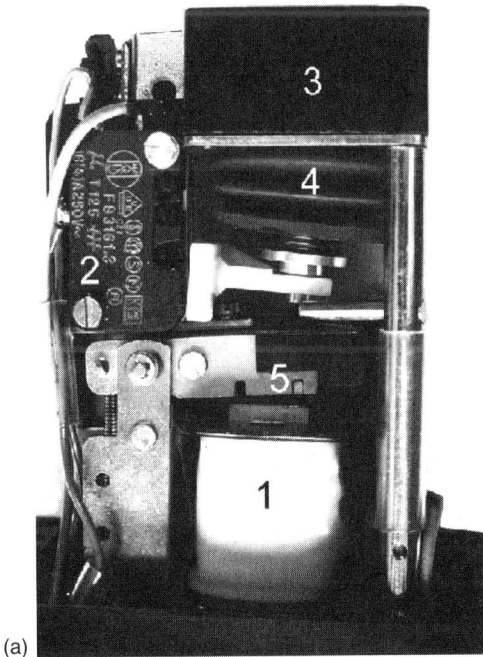

(a)

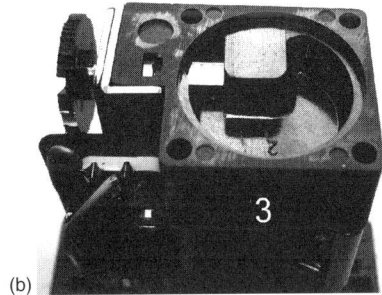

(b)

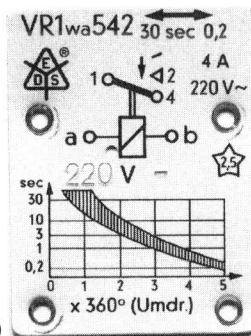

(c)

FIGURE 8.21

Pneumatic time relay of the VR1wa542 type with a time delay within 0.2 to 30 sec. 1 — Coil of the electromagnetic drive; 2 — standard micro switch; 3 — air-chamber; 4 — rubber siphon; 5 — armature of the electromagnetic drive.

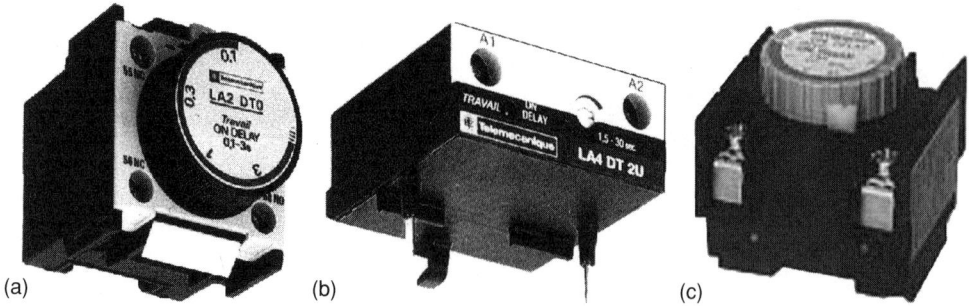

FIGURE 8.22

Miniature pneumatic time delay modules for standard contactors. (a) — LA2 DTO for time delay 0.1 to 3 sec (Telemecanique); (b) — LA4 DT 2U for time delay 1.5 to 30 sec (Telemecanique); (c) — UN-TR4AN (Mitsubishi).

8.5   Electronic Time-Delay Relays

One may say that electronic relays are perhaps the best relays in this class of devices. They provide more stable time delays throughout a very wide range and may be adjusted with very high accuracy; however, it is not correct to say that electronic time relays have superseded motor driven or pneumatic relays, since electronic time relays were produced when both transistors and thyristors were unknown, and developed along with motor, hydraulic, and pneumatic relays.

The similar RC-circuit was the basis for an electronic relay. It was used to obtain time delays in capacitor time relays (see above). The general idea behind electronic time relays was that an electronic amplifier was inserted between the timing capacitor and the output relay (Figure 8.23). This allowed a considerable reduction in the current used by the relay from the capacitor. On the one hand it helped to increase time delays to tens and hundreds of seconds, and on the other hand it allowed reduction of the volume of the timing capacitor and enhanced its stability.

In 1950–70's electronic time relays based on gas-discharge thyratrons were especially popular (see above). Many variants of such relays were produced by the AEG Company (Figure 8.24), and some others. Depending on the value of the installed capacitor, time delays of such relays ranged as 0.1 to 5 sec, 1 to 10 sec; 5 to 50 sec, etc. With the help of an additional external capacitor, it was possible to increase the time delay up to 3 min. Accurate adjusting of time within the range was carried out with the help of a potentiometer (P) (Figure 8.24). In this device, the pick-up of the thyratron (Tr) takes place at a certain voltage value ($U_{pick-up}$) on the grid (gate) of the thyratron. At the first moment after applying voltage of power supply to device, when the capacitor (C) is discharged, the voltage on the grid equals zero. As the capacitor (C) is charged through the divider (R_3) and potentiometer (P), the voltage on it (and therefore on the gate of the thyratron)

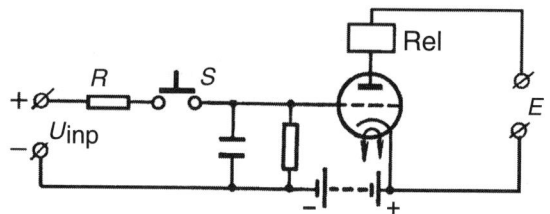

FIGURE 8.23

Electronic time relays with amplifier on vacuum triode.

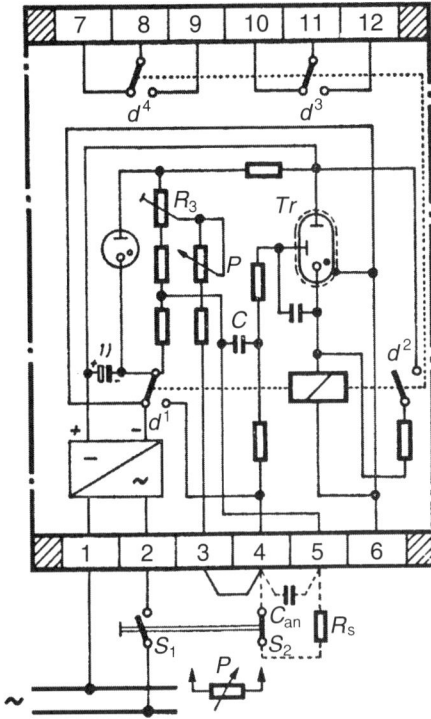

FIGURE 8.24
Circuit diagram of electronic relay of time of RZSg series (AEG) on gas-discharge thyratron — Tr. (AEG, Elektronische Zeitschalter Typ ZSg and RZSg, 1966.)

gradually increases up to the pick-up voltage of the thyratron, which energize of the output relay (D). The position of the curve characterizing the speed of charging depends on the charging resistance ($R_3 + P$).

The more this resistance is, the more mildly the curve goes up (Figure 8.25): $P_3 > P_2 > P_1$. The time constant is τ is defined as the multiplication of RC, and graphically can be illustrated as the intersection of the tangent passing through the initial point of the curve of the charge, with a horizontal right line E corresponding to 0.63 of the voltage on the capacitor established at the end charge. This section is chosen to be a working one because the linearity of the characteristic remains there. The disadvantages of such devices are a small working current of the gas-discharge thyratron in the open state, which requires

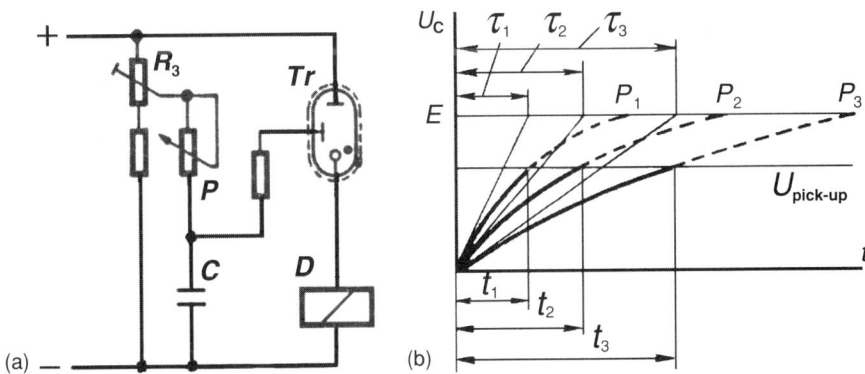

FIGURE 8.25
Simplified circuit diagram of thyratron relay and its time delay (*t*) characteristics, depending on values of position (resistance) of the potentiometer *P*. (AEG, Elektronische Zeitschalter Typ ZSg and RZSg, 1966.)

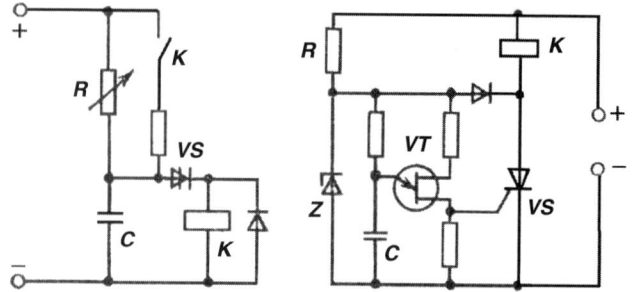

FIGURE 8.26
Electronic time relays on solid-state elements. (a) — With a dinistor VS; (b) — with unijunction transistor VT and on a thyristor VS.

application of a high-sensitivity output relay, and also quite a strong dependence on the opening of the thyratron from ambient temperature.

When such solid-state devices as transistors and thyristors appeared, the production of relays gradually started to redirect to them. Time relays on solid-state devices have turned out to be very simple and reliable (Figure 8.26). In the device based on a dinistor (Figure 8.26a), the latter remains closed and the output electromagnetic relay without current until the capacitor charges to the voltage of breakdown of the dinistor. After that the capacitor is discharged through the open dinistor to the relay winding, causing its energization. The energized relay then starts to self-feed through its own contact K.

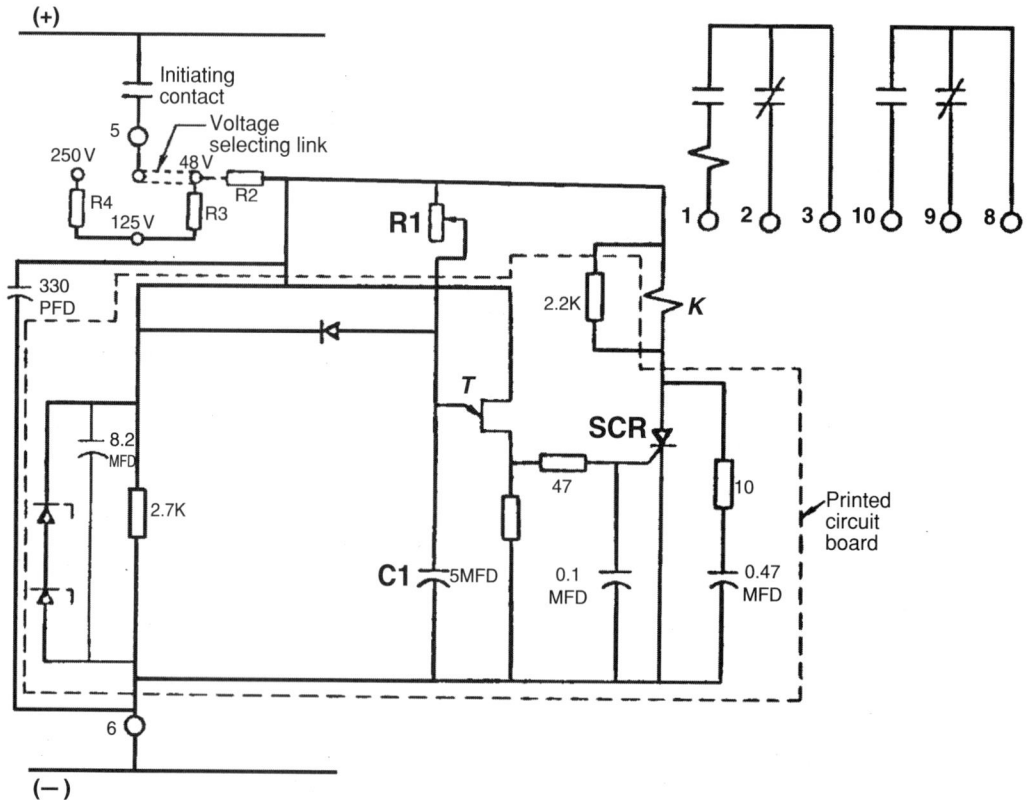

FIGURE 8.27
Circuit diagram for time-delay relay SAM-11 type. (General Electric. Timing Relays. Instruction GEC-7393D.)

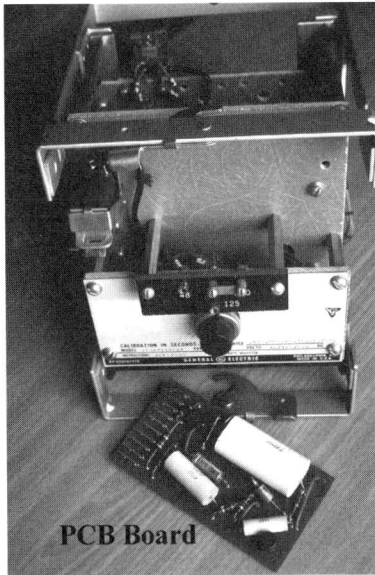

PCB Board

FIGURE 8.28
Time-delay relay of the SAM-11 type in the knocked-down form.

In the second device after the capacitor C is charged to the voltage on the Zener diode Z, the unijunction transistor VT is enabled and the capacitor C is discharged through the control gate of the thyristor VS. The latter is switched ON and energized the output relay K. This scheme is very reliable and has been popular for decades. Suffice it to say that the General Electric Co., has produced time relays until recently for important systems of relay protection based on this scheme (Figure 8.27). In the foreground one can see a small printed-circuit board with electronic components.

As one can see, the electronic components of the relay, which look huge and heavy for such a simple circuit, occupy a very small part of the construction, and this is not the only peculiarity of the construction. Note how slantwise and awry the elements are placed on the printed-circuit board (Figure 8.28). This is because a work piece with previously applied conducting cooper lines is used as a printed-circuit board, since it does not require etching.

Such work pieces are usually used by young radio-amateurs who do not have enough money to buy required materials. Another peculiarity of this construction is an "original" scale of time delays made in the form of dents (in some constructions in the form of hairlines) extruded by a pointed instrument on an aluminum plate. You may ask how is such a scale used? Here is the answer: you should open the instruction manual and find the matching explanation for each point on the scale!. Why is it all so difficult? Because the great dispersion of parameters of electronic components prevents us from using a previously graduated scale in this circuit. Of course one may have provided for additional foot-note elements to compensate the range of values, but GE found another solution: they just fix hairlines or points on each relay unit according to the results of testing.

Despite these peculiarities this construction is still reliable and well fits the standard range of protective relays produced by the GE. On the basis of this scheme GE also produced twin time relays (Figure 8.29), with two independent time delays for use in relays of remote control of high-voltage lines. These relays provided accuracy of maintenance of time intervals of $\pm 4\%$ within temperatures of -20 to $+60°C$.

Relays of the 7PS10 type produced by Siemens (Figure 8.30) are based on the same principle of operation. However, unlike relays of the SAM-11 type, this relay is compact

Zone
targets

D.C voltage
selection
like

Calibration
knob zone 2
timing citcuit
(Rheostat R1)

Calibration
knob zone 3
timing circuit
(Rheostat R2)

(a)

Zone 3
printed
citcuit
card

Zone 2
printed
citcuit
card

Timing
capacitor

Locking
nut

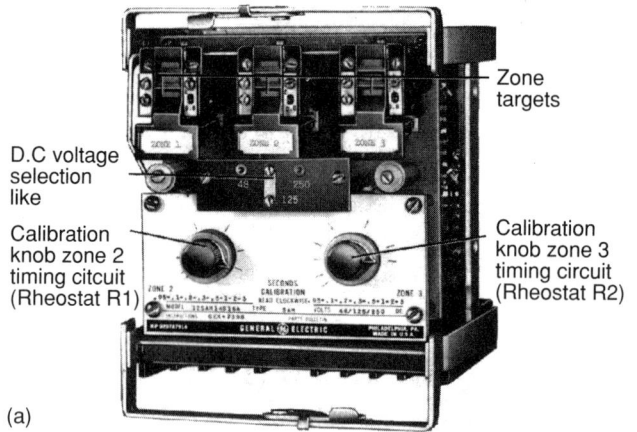

FIGURE 8.29
Twin time relays designed for two stages
of time delay in zones of remote protections of high-voltage lines. (General Electric. Timing Relays. Instruction GEC-
7393D.)

(b)

FIGURE 8.30
7PS10-type time relay (without a protective case) based on an RC-
circuit with a unijunction transistor and thyristor. (Siemens. Protective Devices Catalog NS 1-89.)

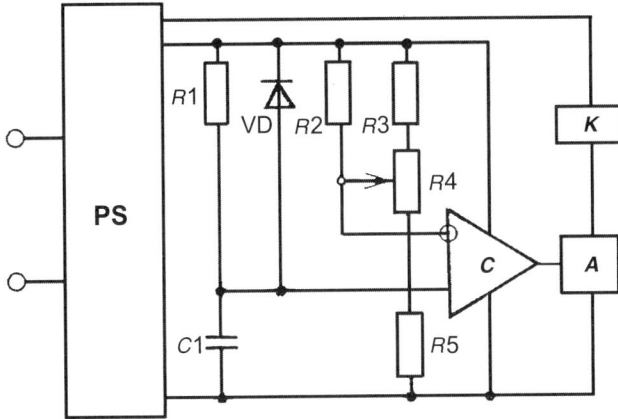

FIGURE 8.31
Principle of construction of an electronic time relay with chips, based on a standard RC-circuit. PS — Power supply; C — comparator; A — amplifier; K — output electromagnetic relay.

and light because an especially designed plastic case is used for its production. In addition, the relay has a normal scale of time delays, which are observed with quite high accuracy. For example, the inaccuracy of a relay with a maximum time delay up to 10 sec is only $\pm 0.5\%$. Naturally, an increase of the time delay causes a lowering of accuracy. For relays with a working range of 10 to 100 sec, inaccuracy is already $\pm 8\%$. These are quite good figures for such simple relays. They are obtained due to the individual adjusting of each relay during the process of production.

When integrated circuits appeared, characteristics of electronic relays were enhanced considerably. The use of a high-quality operational amplifier, made in the form of chips, allowed production of high-accuracy time relays with very long time delays (tens and hundreds of seconds) on the basis of the RC-circuit (Figure 8.31).

In this device the first operational amplifier (C) works in the comparator mode (that is the mode of the comparison circuit) and the second one (A) — in the mode of amplification. Increasing voltage from the charging capacitor (C1) is applied to the direct input of

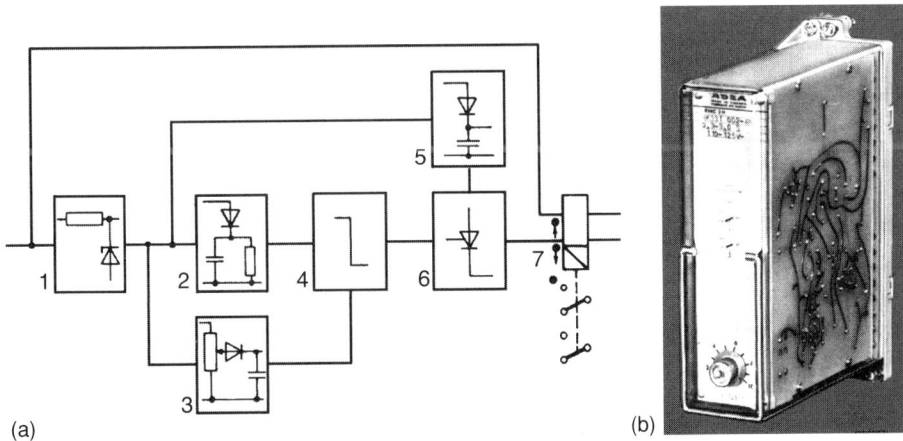

(a)   (b)

FIGURE 8.32
Function diagram (a) and external design (b) of a time relay of the RXKC-2H type (ASEA-ABB). 1 — stabilizing circuit; 2 — timing circuit; 3 — reference-voltage circuit; 4 — voltage level detector (comparator); 5 — buffer capacitor; 6 — thyristor; 7 — output relay. (ABB Relay Units and Components Buyer's Guide, 1990)

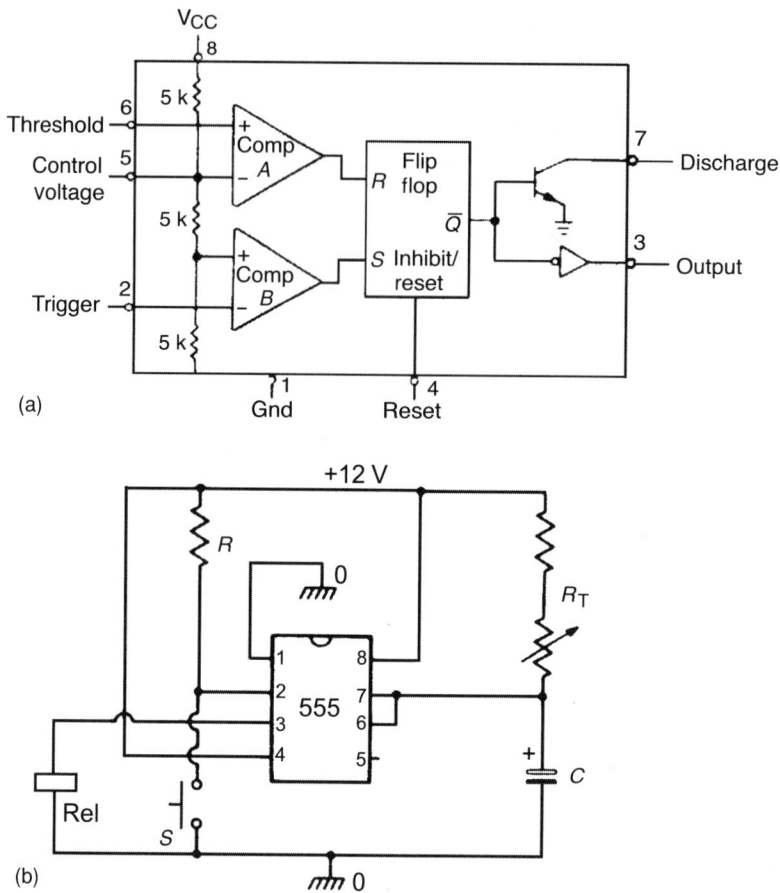

FIGURE 8.33
Structure (a) and circuit (b) of external adjunctions of the integral chip of the 555 series. R_T and C — Timing elements; Rel — output relay.

the comparator, and stable reference voltage from the divider ($R2 \ldots R5$) is applied to the second (inverse) one. When voltage on the charging capacitor ($C1$) reaches the reference one, the comparator (C) pick-up and its output signal will be amplified by the operational amplifier and applied to the winding of the output electromagnetic relay (K). The time delay is adjusted by changing the reference voltage with the help of the resistor ($R4$). A great number of time relays for industry and power engineering were produced on the basis of this principle by various companies in the 1970–80's (Figure 8.32). Moreover, several series of special-purpose IC chips containing most of the necessary elements for production of time relays based on this principle were produced. The most popular one was the chip of the so-called 555 series.

With this chip all that has to be done to obtain a high-quality time relay is to connect several resistors, a timing capacitor and an output relay to the chip (Figure 8.33). Several timers can be connected to drive each other for sequential timing (Figure 8.34). The sequence is started by triggering the first timer, which runs for 10 msec with R1C1. The output then switches low momentarily, and starts the second timer, which runs for 50 msec with $R2C2$, and so forth. It is clear that an increase of values of R and C will cause considerable increase in the time delay.

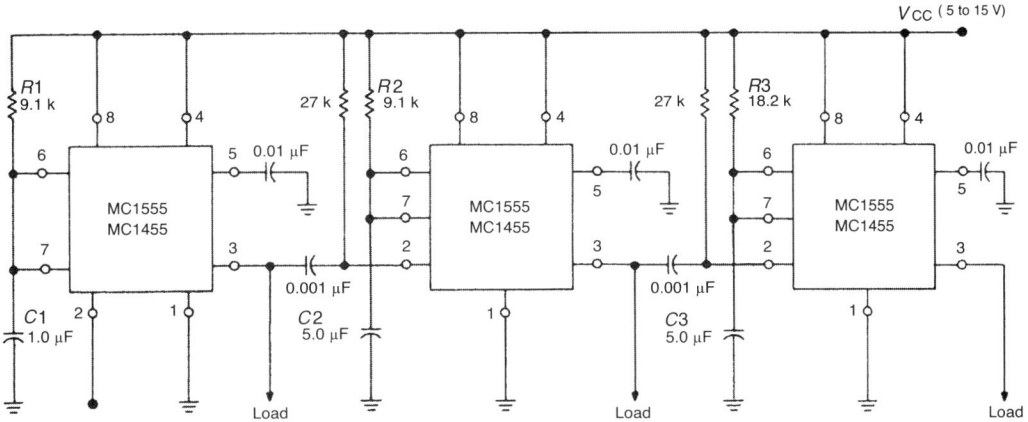

FIGURE 8.34
Circuit diagram for sequential timing with IC 555 series.

In order to increase time delay it was suggested to use a pulsing charge of the timing capacitor (*C*), with short rectangular-shaped pulses. When the pulse applied to the capacitor from the special generator is short, it does not have enough time to be charged fully and the voltage on it slightly increases. At that point there is a pause, during which the capacitor is not charged, and then again a period of charging (Figure 8.35).

The most perfect type of electronic time relay is the digital relay, which operates without charging (or discharging) the CR-circuit (Figure 8.36). Basic elements of such a time relay are a highly stable pulsed oscillator (*G*) and a pulse counter (CR).

The counter starts counting pulses until the number reaches a given value (with the help of the set-circuit) corresponding to the required time delay. At that point that a signal appears at the input, which is amplified and proceeds to the output relay. Simultaneously, the pulse counting stops and as the input voltage is switched OFF, the device returns to its initial state.

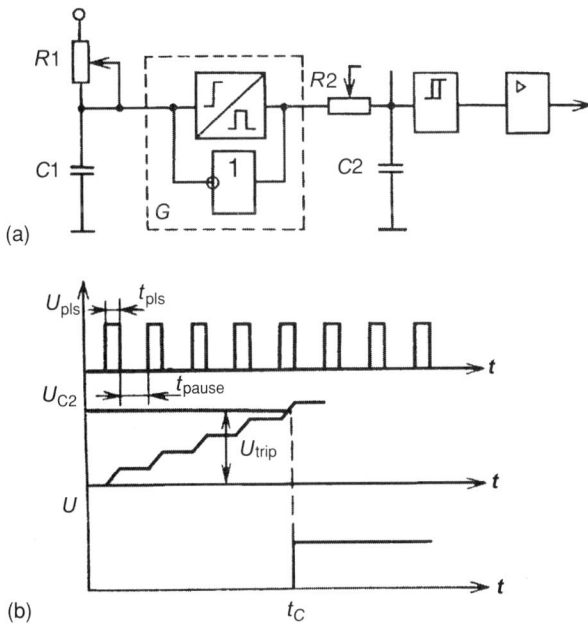

FIGURE 8.35
Structure (a) and diagram (b) of operation of a semiconductor time relay with pulsing charge of the capacitor.

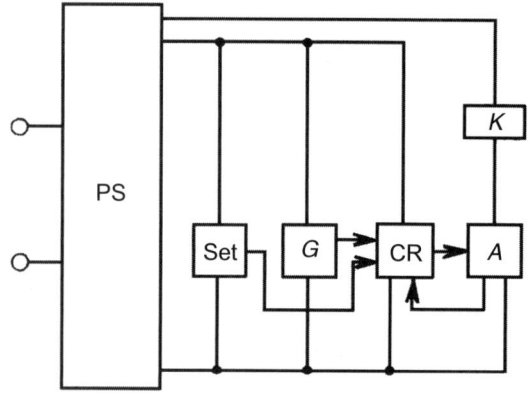

FIGURE 8.36

Simplified block diagram of a digital time relay. PS — stabilized power source; Set — installation circuit of the initial state; G — pulse generator; CR — pulse counter; A — amplifier; K — output electromagnetic relay.

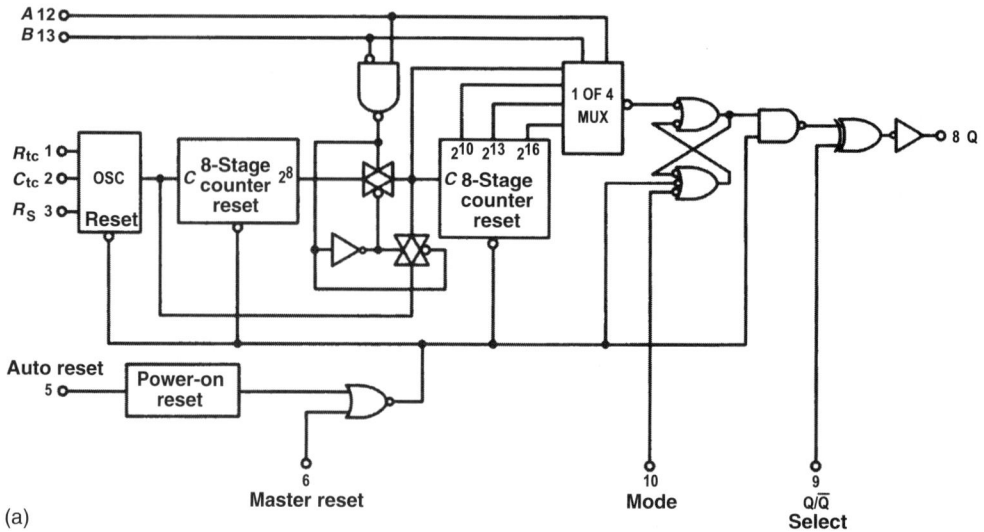

(a)

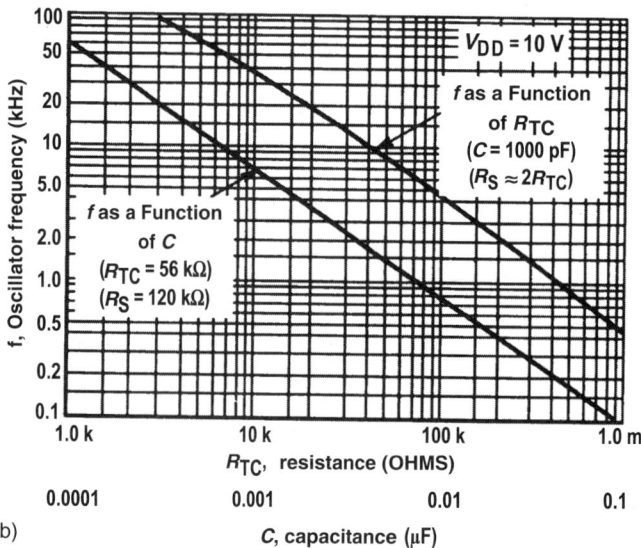

(b)

FIGURE 8.37

(a) Structure of a programmable timer of the MC14541B type (Motorola). (b) Oscillator frequency as a function of R_{tc} and C_{tc}.

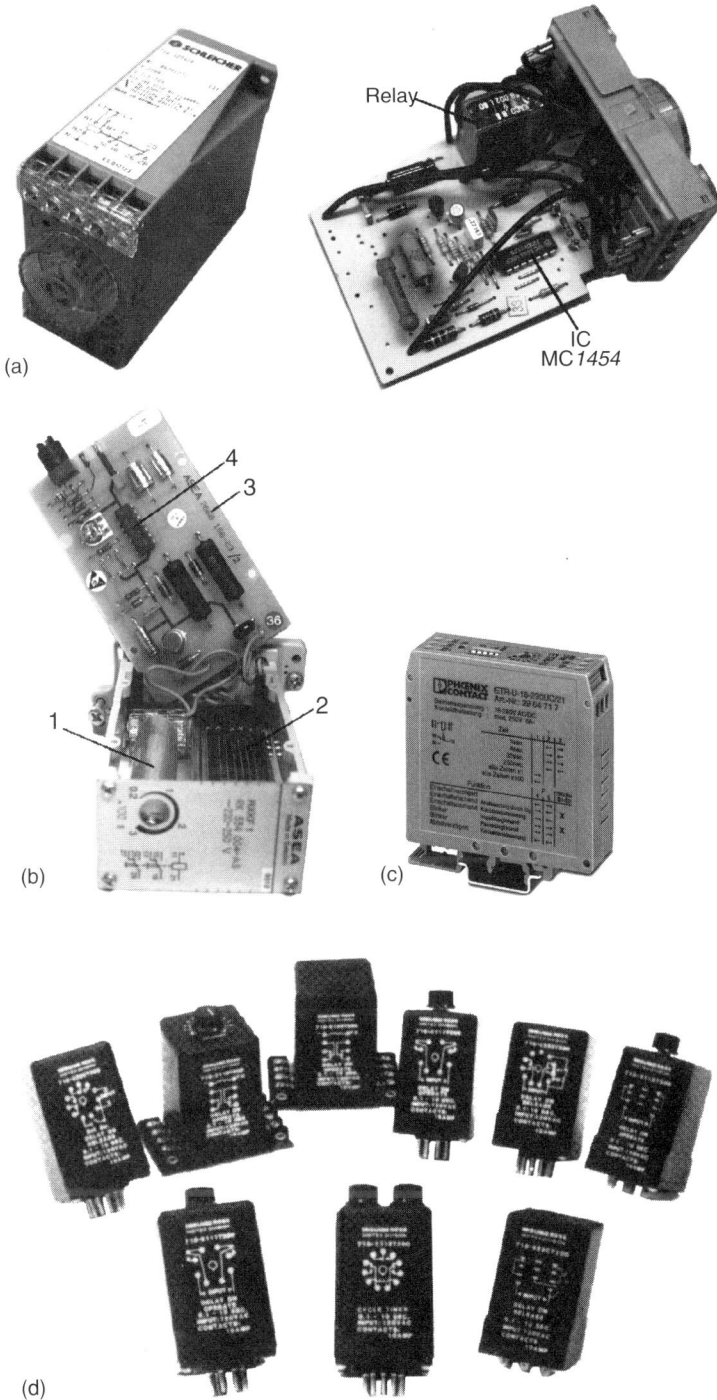

FIGURE 8.38

(a) Electronic time relays of the SZT 420 type with a time delay adjusted within 1.5 to 30 sec by a potentiometer (Schleicher). (b) Electronic time-delay relay of the RXKF-1 type (ASEA), 80 msec to 300 sec. 1 — output relay coil; 2 — contacts; 3 — printed circuit board; 4 — programmable timer MC14541B type (see Figure 8.37). (c) Electronic relay of the ETR-U type with a time delay of 1 to 250 sec, fixed with the help of a set of micro switches. The case is designed for installation on standard DIN rails (Phoenix Contact). (d) Electronic time-delay relays of the 715 type (Midtex). The enclosure of these relays (d) and plugs are industry standard 8 and 11 pin octal, with a side terminal strip for flush mounting, and an 11 pin square base. . . . . (*Continues*)

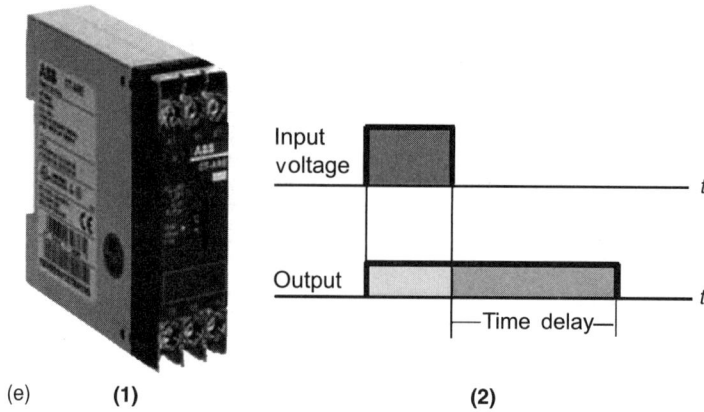

(e)   **(1)**   **(2)**

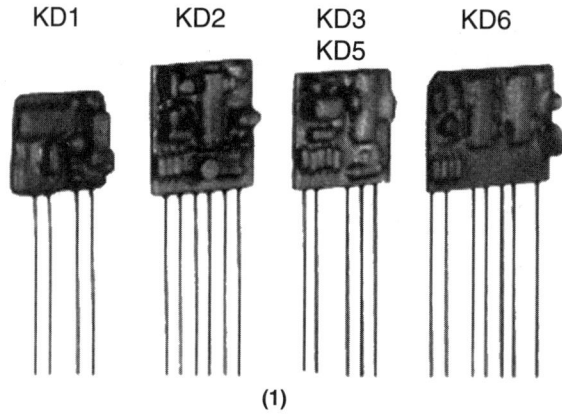

(1)

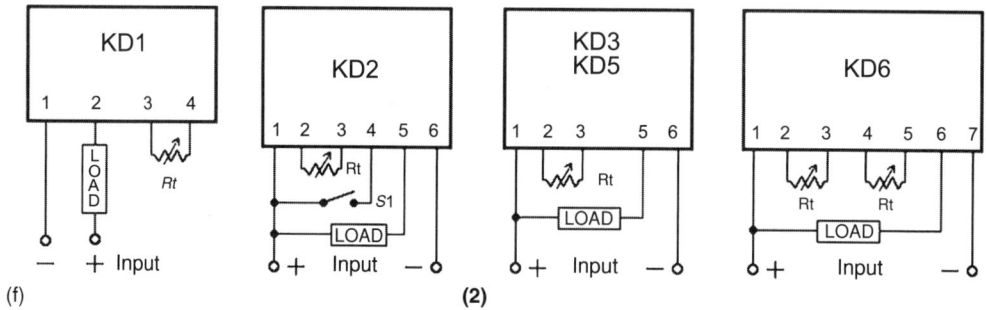

(f)   **(2)**

FIGURE 8.38 (*Continued*)
(e) True-OFF delay relay CT-ARE type without auxiliary supply, DIN-rail mounted (ABB). (f) Timer modules KD-series with thick conformal coating. Changing the value of the external resistor in range 0 to 1 MΩ gives minimum and maximum time delay. (Instrumentation and Control Systems Inc., 2004.)

In order to simplify development and production of digital time relays, producers of chips put on the market special-purpose chips containing inside a case, a oscillator, a pulse counter and timing RC elements providing a specific frequency of the oscillator, depending on the required time delay (Figure 8.37). A wide range of time relays containing a special integrated circuit, an electromagnetic output relay, and special auxiliary elements, are produced in different cases (Figure 8.38) by different companies.

One of versions of the time-delay relays is the so-called "True-OFF Delay Relay," Figure 8.38e. When voltage is applied to the input terminals, the relay energizes. Timing

FIGURE 8.39
Electronic time relays for military and aerospace application (LEACH, 2004).

starts when power is removed from input terminals (OFF delay begins without auxiliary voltage). At the completion of the delay cycle, the relay is de-energized and the output contact transfer is made. If voltage is reapplied during the delay period, the relay remains picked up and the timer resets to zero. Voltage must be applied for a minimum of 0.5 sec to assure proper operation. Energy in delay cycle for such relays is provided by internal capacitor and time delay generally does not exceed 300 sec.

Instrumentation and Control Systems, Inc. is known for unique line of timer modules designed for printed circuit mounting. These timers are thick film hybrid analog circuits or digital countdown circuits with thick conformal coating. Conformal coating is the process of spraying a dielectric material onto a device component (on printed circuit board, for example) to protect it from moisture, fungus, dust, corrosion, and thermal shock. M.G. Chemicals is the largest manufacturer of such cold sprays and protective coatings (urethane, silicone, and acrylic). With such coatings, the timers (small, light, inexpensive) have proved reliable year after year. Such timers are very convenient for using as elements at designing complex electronic devices. Numerous time ranges (0.1 sec to 10 h), functions, and voltages (12 to 240 V AC, DC) are available.

The LEACH Company produces a whole series of electronic time relays with a powerful integral output relay, with a time delay ranging from 0.1 to 600 sec (by changing the value of the external resistor) (Figure 8.39). LEACH time-delay relays are designed with thick film hybrid microelectronic timing circuits, are packaged in a hermetically sealed military style enclosure and designed to withstand severe environmental conditions encountered in military and aerospace applications. These relays are suitable for use in power control, communication circuits, and many other applications where power switching and high reliability are required over a wide temperature range (−55 to 125°C).

8.6 Attachments to Standard Electromagnetic Relays

Lately, universal attachments to standard electromagnetic relays containing electronic elements of time delay, made in the form of separate modules connected to standard electromagnetic relays, have become especially popular (Figure 8.40). Type 618 is an electronic time-delay module that provides a 0.5 or 2 A SCR output to drive relays. By wiring in series with multiple power relays, a versatile timing function is attained. The module is available with 3/16 in. quick-connect, solid axial, or flying leads. Fixed and externally

Applications and Use for Delay-on-Operate

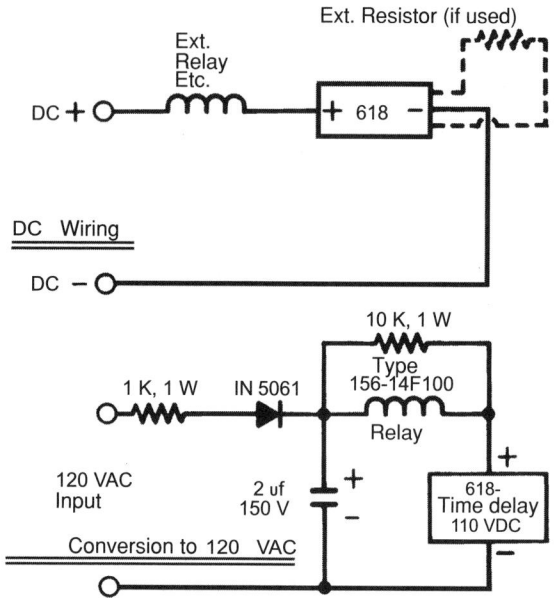

FIGURE 8.40

Type 618 universal time delay electronic module and connection diagram, for DC and AC applications (Midtex). Time delay with external resistor: 0.1 to 300 sec.

FIGURE 8.41

Universal electronic module of the UTC type. (Artisan Controls Corporation.)

adjusted resistor units are available. The units are fully encapsulated and are versatile and economic, especially when used in conjunction with power relays in the 30-A rating category.

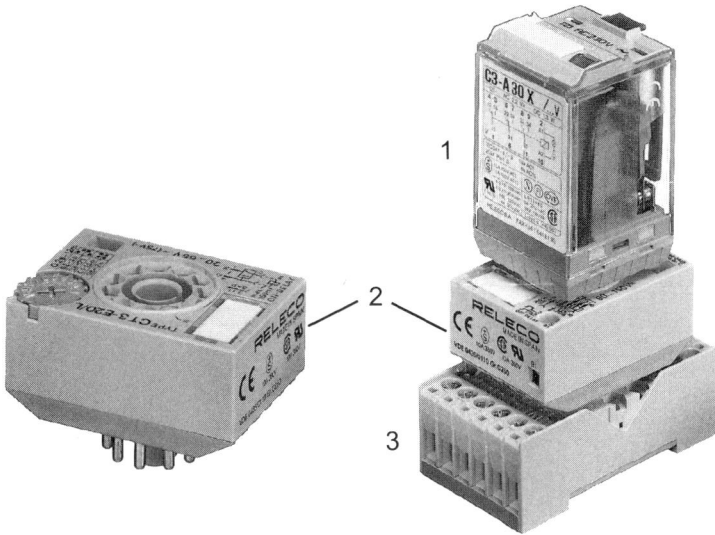

FIGURE 8.42
Time cube (2) with elements of time-delay (from 0.2 sec to 30 min) installed between the standard relay (1) and the terminal block (3). (produced by RELECO Company.)

Similar modules are produced by many other companies (Figure 8.41). Module UTC is designed for a series connection with a load of any type with nominal voltage of 24 to 240 V of alternating or direct current, up to 1 A. When an external variable resistor with 10 kΩ resistance is connected, the time delay may be adjusted within 1 to 1000 s.

Internal elements of the construction are covered with epoxy resin. The case is equipped with a universal fastening. Such modules containing a thyristor or a triac as an output switching element may be used not only for delayed switching-ON of an electromagnetic relay, but also as an independent relay for switching-ON of solenoids, signal lamps and other low-power loads. Especially produced for relays with plugs for industry standard 8 and 11 pin octal, many companies manufacture small "time cubes" inserted between a standard plug-in relay and its socket (Figure 8.42).

Inside such a "time cube," there is a small printed-circuit board with electronic components providing time delay of the external relay.

8.7   Microprocessor-Based Time-Delay Relays

Recently universal microprocessor timers with new functions and interesting features have appeared in the market (Figure 8.43). Some of these timers have unique characteristics, for example, time relays produced by ABB-SSAC (Figure 8.44). The TRDU Series is a versatile universal time-delay relay with 21 selectable single and dual functions. With the progress in microprocessor technique and distribution on the market universal multi-function timers, necessity to classify these functions has appeared. Now the technical

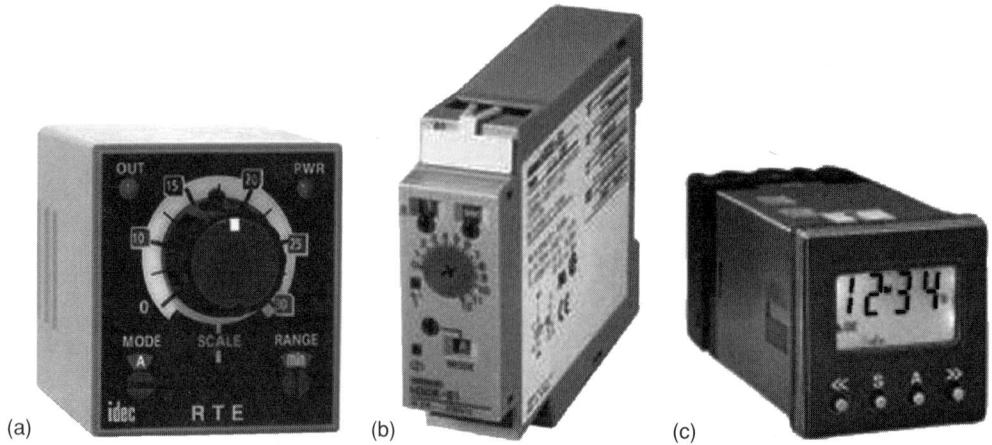

FIGURE 8.43
Multifunction microprocessor timers produced by different companies.

specifications on the timers specify numbers of standard functions, frequently without additional explanations.

Some standard functions of multifunction timers are shown in Table 8.1. The dual functions replace up to three timers required to accomplish the same function. Both the function and the timing range are selectable with switches located on the face of the unit.

The TRDU is the first in market series of time delay relays and includes 21 timing functions. One TRDU replaces hundreds of conventional time delay relays. Any one of the 10 single-functions or 11 dual-functions is easily selected by transferring one or more of the 6 programming switches. Timing functions include: delay-on-make, delay-on-break,

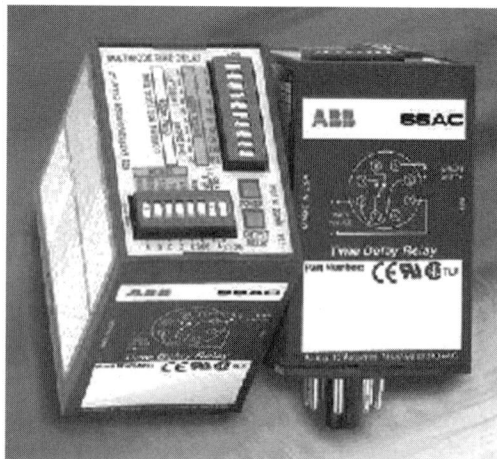

FIGURE 8.44
Microprocessor-based plug-in time-delay relay TRDU type with unique features. Dimensions: 1.78 × 2.39 × 3.1 in. (45 × 61 × 79 mm). Produced by ABB-SSAC.

TABLE 8.1

Some Standard Functions of Microprocessor-Based Timers

Standard Function Number	Flow Diagram	Function Name
11		ON-delayed
12		OFF-delayed
16		ON- and OFF-delayed
21		Fleeting output on making
22		Fleeting output on breaking
42		Flashing (blinked)
81		Pulse generating
82		Pulse forming

interval, single shot, recycling, motion detector, accumulating functions, inverted functions, and combinations of these.

The TRDU's time delay is adjusted by closing a combination of 10 binary DIP switches. Available time delays range from 0.1 sec to over 1705 h in eight ranges. Timing accuracy is $\pm 0.1\%$ and setting accuracy is $\pm 0.1\%$ over the full adjustment range. Industry standard 8 and 11 pin base wiring makes substitution simple and fast, with no rewiring required.

8.8 Accelerated (Forced) Relays

In modern automatic control systems there is a need not only for relays with increased make delay time but also for relays with reduced (in comparison with standard ones) make delay time. The simplest way to accelerate pick-up of the relays is a series connection of an additional resistor to the winding, designed for reduced pick-up voltage (Figure 8.45). At first sight such a technical solution seems quite strange: a lot of energy which may be used for the creation of a more powerful magnetic flux caused by the coil of the relay is dispersed on such additional resistors.

But all this is not so simple, and relays were designed by quite inventive man. The make delay time of the electromagnetic relay that is the time from the beginning of the voltage supply to the winding until the moment when the armature stops:

$$t = t_s + t_m$$

t_s is the starting time, that is the time from the moment of voltage supply to the winding until the beginning of armature motion and

t_m is the time of motion (traveling) of the relay armature from the initial position to the final one.

As the winding of the relay has quite considerable inductance, the current in it does not reach the steady-state value immediately, but increases gradually. Starting time is a period of time necessary for current increase in the winding to the value when the electromagnetic force is enough for overpassing of frictional forces, and the elastic force of the spring, and the armature begins to move. It is obvious that this time will mostly depend on the inductance of the winding. Theoretical conclusions prove that the starting time may be calculated by the formula:

$$t_3 = \frac{L}{R} \ln\left(\frac{1}{1 - \frac{I_s}{I}}\right)$$

where L is the inductance of the winding, R the resistance of the winding, I_s the current when the armature starts to move and I the steady-state current value in the winding.

FIGURE 8.45
Relay of the RXMS-1 type (ABB) with two powerful resistor (the second resistor is not seen on the figure), connected in series with the winding (placed between contacts).

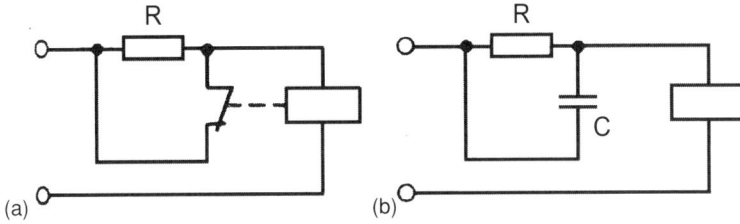

FIGURE 8.46
Principle of forcing of the relay.

The value $L/R = \tau$ is called as the "time constant of the winding."

To reduce the time constant of the relay (in order to enhance its speed of operation), one employs a winding for pick-up voltage two to three times lower than the used supply voltage (apparently such a winding contains two to three times fewer turns) and the voltage excess is reduced by additional resistors connected in series with the winding of the relay. Such a connection causes reduction of the general inductance (L) of the circuit and an increase of the resistance (R). According to the given formula, this leads to a sharp reduction of the time constant of the relay. Thus the total pick-up time of the relay may be reduced by 40 to 50%.

The very same method of enhancement of the speed operation of electromagnetic relays is used in the case considered above. Frequently another method of enhancement of the speed operation of the relay is used. It consists of shunting of this additional resistance for the period of relay pick-up (Figure 8.46). Increased voltage is applied to the winding of the relay (that is, increased current flows through it) during the pick-up. Such method of enhancement of the operation speed of the relay is called "forcing."

The principle of forcing is that voltage from the power supply, exceeding the value permissible by conditions of heating of the winding, is applied to the winding of the relay with the help of a control element. This voltage is applied to the winding for the short period of time, only for the pick-up.

At that point the voltage on the winding is reduced with the help of the control element to the level permissible by the conditions of heating. Thus during the starting period, a large starting current flows through the winding of the relay, which develops a great attractive effort. After the pick-up, the current of the winding and its magneto-moving force (m.m.f.) decrease, but the armature remains in the attracted position because when gaps are small the attractive effort of the electromagnet is great, even with low currents. An additional contact of the same relay shunting the additional resistance (R) during the pickup (Figure 8.46a), or the capacitor (C) (Figure 8.46b), may serve as a control element. In case a capacitor is used, the relay is energized by the large charging current of the

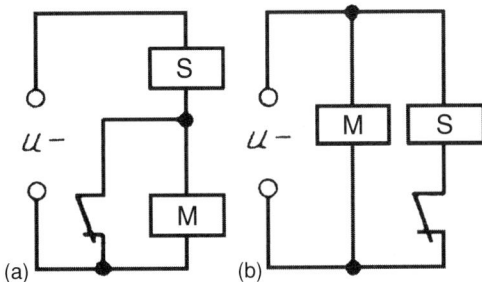

FIGURE 8.47
Circuits of forcing of the relay with two windings. M — main coil; S — start coil.

capacitor. After charging, the current does not flow through the capacitor, whose function is similar to the function of a break contact.

Ratios of fixed values of voltage, current, and m.m.f during starting, corresponding to values during the maintenance period, are called forcing coefficients of voltage, of current, or of m.m.f. accordingly.

Another widespread method of forcing of power relays is the use of two windings: a main one (a holding one) and an auxiliary one (a starting one) (Figure 8.47). During the pick-up of the relay the main winding may be shorted by the additional contact (Figure 8.47a), or may be connected parallel to the start winding. Quick pick-up of the relay is carried out by the powerful start winding, having low resistance and small inductance for short-term passing of large currents. After the relay is energized and the additional contact opens, the start winding is connected parallel to the main one, or is simply disabled.

9

Thermal Relays

Temperature or thermal relays belong to the second (or even probably to the first) most popular type of special-purpose electric relays. There are two basic types of such devices: relays with input energizing quantity in the form of heat, and relays with an input quantity in the form of electric current. Relays of the first type are applied for direct temperature control of different units. Relays of the second type are used as protective relays from current overload, for various electrical customers. In the latter case, electric current is transformed to the heat inside the relay first, and when the temperature of the internal thermal element reaches a certain value (and the relay is energized) — to an output electric signal. As the temperature of the internal heating element of the relay depends both on the value of the current passing through it and on the time of exposure of this current, such a relay turns out to be "intelligent": its speed of operation appears to be dependant on the overload current of the unit being protected. If the overload is small, such a relay allows the protected unit to operate for quite a long time without being disabled, because such overload may be temporary and may be caused by end-around of operation of the unit. If the overload current is quite high, the relay quickly disables the protected unit and does not allow it to break down. The higher the overload current is, the quicker the heating element is heated and the quicker the relay disables the protected unit (Figure 9.1).

In thermal relays of both types, there must necessarily be an additional functional element transforming the heat to an electric signal. In thermal relays, two basic types of transformation are used:

1. With an intermediate transformation of the heat to the mechanically moving internal components and with further exposure to the contact connected to the external electric current.

2. Direct transformation of the heat to an electric signal: variable resistance or voltage, for example. This voltage is amplified by an electronic amplifier with an electromechanical or semiconductor (solid-state) relay on output.

9.1 Relays Based on a Bimetal Thermal Element

In relays with intermediate mechanical moving of elements affected by heat, several basic types of the following elements are used: bimetallic, dilatometer, hydraulic, mercury.

The bimetal element is most frequently used in relays. Such an element is a straight or coiled plate (or more rarely a coiled thin belt) made of two layers of metals with different linear expansion coefficients. If one heats such a plate, it will bend in the direction of the

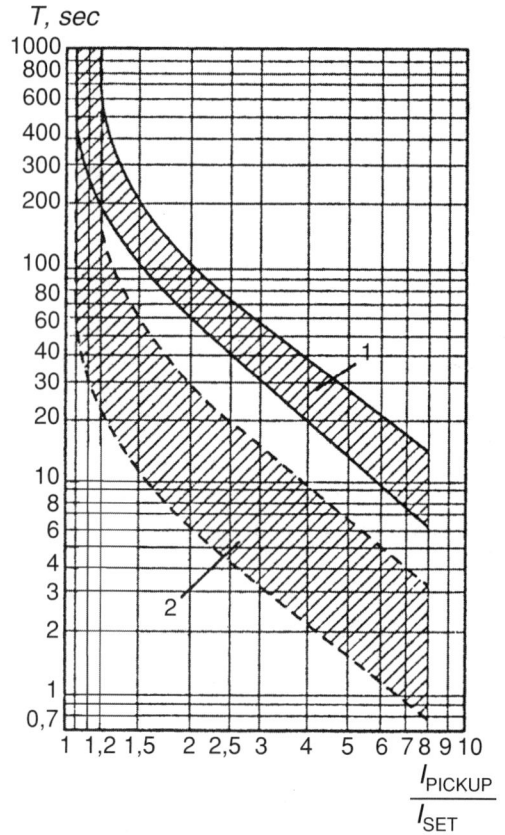

FIGURE 9.1
Typical relationship between the operating time and the overload factor (the ratio of the actual pick-up current to the adjustable preset current) for a thermal relay controlling the current in the circuit of the protected unit. 1 — Zone of protective characteristics when the relay is energized from the cold state; 2 — when the relay is energized from the hot state (after it has been previously heated by the overload current).

metal with the smaller linear expansion coefficient (Figure 9.2). The bimetal strip is heated by the source, the temperature of which is under control. In some cases, it may be heated with the help of a heating spiral, as shown on the Figure 9.2, or by direct current passing through the plate, and in other cases both by the spiral and direct current. When strong currents (50 A and more) are applied, a current shunt is used to unload the bimetal strip from over-current.

The mechanical force developed by such a plate while bent is used to operate the relay contacts. As bending of the bimetal strip (like the moving of mechanical parts of all other types of thermal relays) takes time, there is usually no direct connection of such a plate with the contacts, as shown on the Figure 9.2, otherwise they would be strongly scorched by electrical arc when they slowly separate.

To accelerate disconnection of contacts, a special spring is installed between them and a bimetal strip (Figure 9.3b,c,e). It makes the movable contact "jump," and is made in the form of an "instant-make" spring or snap disk (Figure 9.3d), or the contacts are equipped with some other element providing an instantaneous contact switch (snap-action contact) as the actuating force smoothly increases (Figure 9.3a).

FIGURE 9.2
Principle of operation of a bimetal heating element. 1 — Metal with a small linear expansion coefficient; 2 — metal with a large linear expansion coefficient; 3 — heater.

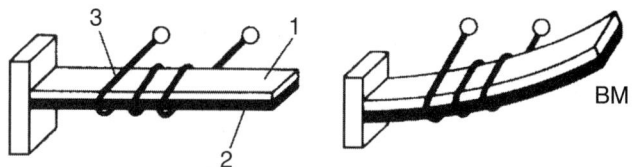

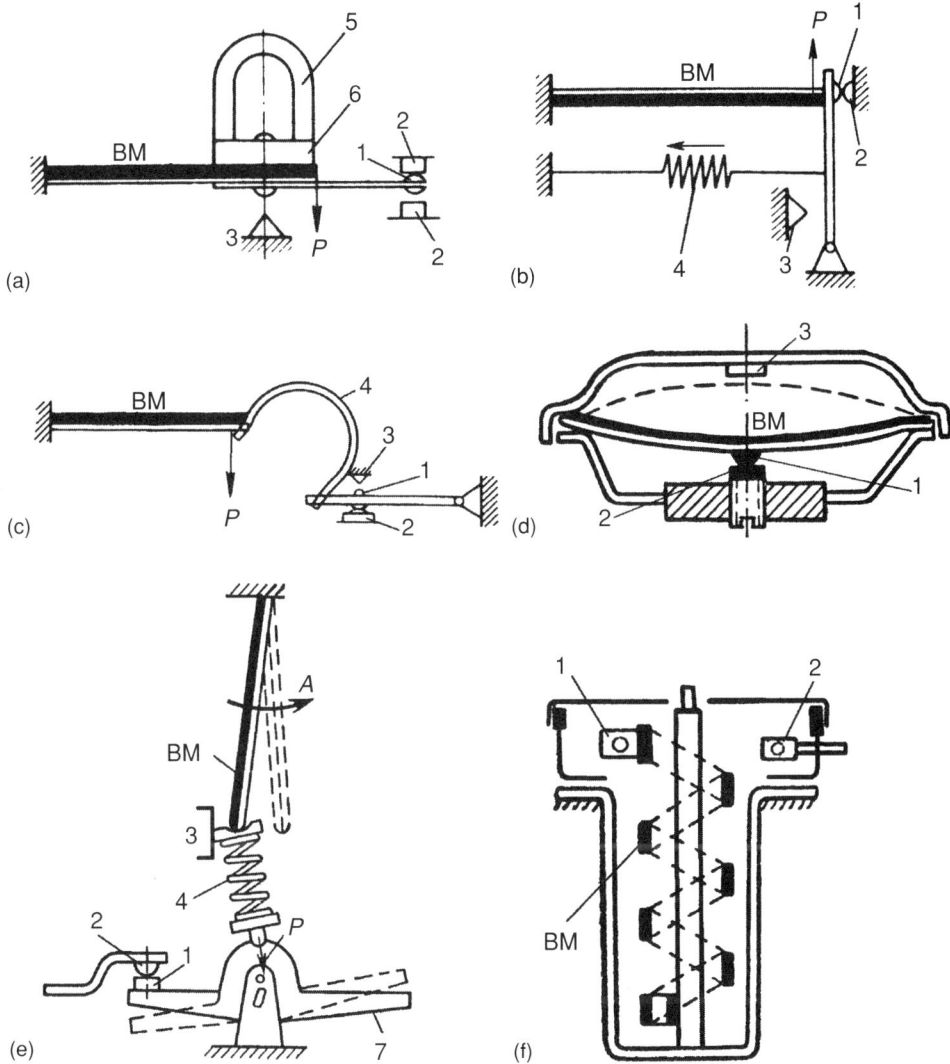

FIGURE 9.3

Principle of construction of thermal relays with a bimetal element (BM). 1 — Movable contact; 2 — stationary contact; 3 — stop; 4 — spring; 5 — permanent magnet; 6 — ferromagnetic armature; 7 — beam.

As the temperature level must be controlled in various different devices (ranging from cases of semiconductor devices and windings of electric motors, to electric kettles and boilers for water heating), constructions of relays may be very different. For example, thermal relays in round copper cases having 10 to 25 mm in diameter (Figure 9.4), based on the construction diagram in Figure 9.3d, are widely used as integral elements molded to the controlled unit or installed on its surface.

Thermal relays of such type are quite reliable and are applied even in military equipment (Figure 9.5). These relays are capable of functioning in a very wide range of temperatures: $-60 + 265°C$ (fixed pick-up temperatures ranging from $+30$ to $+250°C$) and their contacts can switch current up to 25 A at a voltage of 27 V.

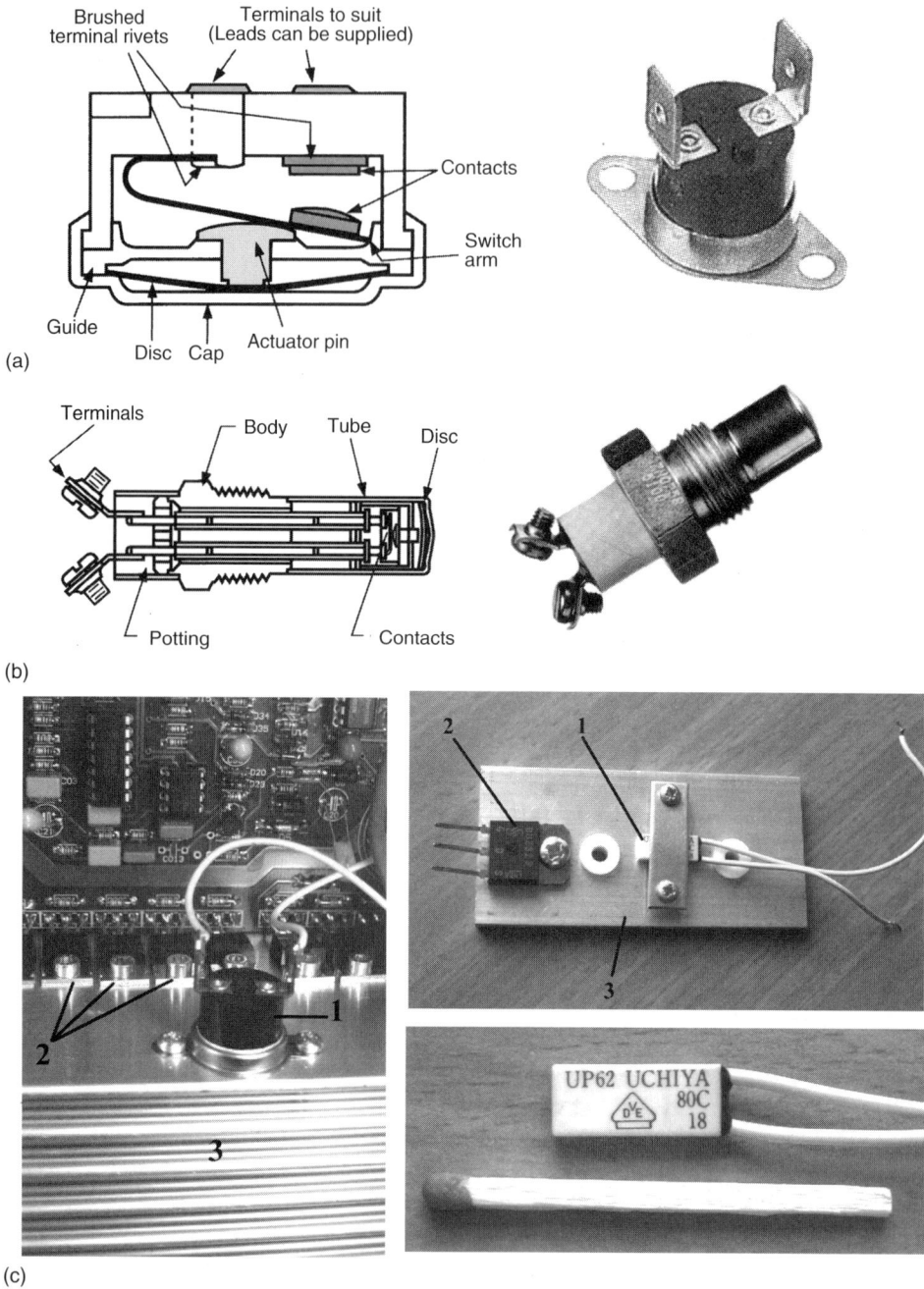

(a)

(b)

(c)

FIGURE 9.4

(a) Thermal relay based on bimetal element in the form of a preliminary bent disk — snap disk (a) Relay in plastic cylindrical case widely used for protection of household and industrial appliances. (b) Series 5100 immersion-type, hermetically sealed relay (Airpax Corp.) used on water-cooled engines, hydraulic systems, degreasers, industrial air compressors, and tanks. (c) Thermal relays installed on a heat sink of powerful semiconductor devices. 1 — thermal relay; 2 — powerful transistors evolving heat; 3 — aluminum heat sink for cooling of transistors.

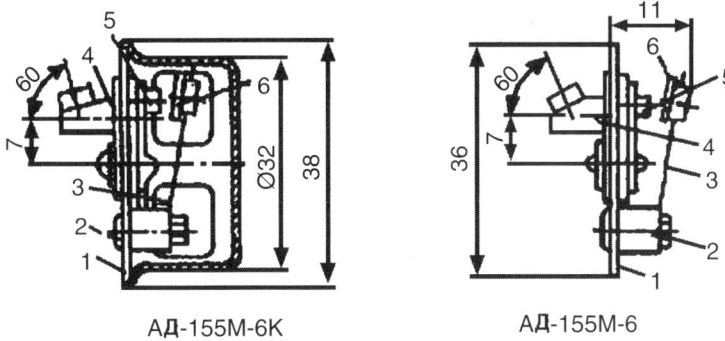

FIGURE 9.5
Thermal relay of the AD-155M type for application in military equipment (Russia). 1 — Board; 2 — column; 3 — sphere-shaped bimetal strip; 4 — terminals for external connections; 5 — stationary contact (silver); 6 — movable contact (silver).

An auto-cutout of an electric kettle, so well known to us, also contains this bimetal strip (Figure 9.6). The bimetal element BM in this switch is made in the form of a disk with two notches and a reed placed in the center. When the element is heated with steam vapors, its reed bends slowly at first and then roughly (like an "instant-make" spring) and affects the contact system with an indexing mechanism through the pusher.

In 1967, in the former U.S.S.R. the differential temperature relay of the DTP-3M type was designed to protect power electric motors from overloads (Figure 9.7). This relay had great selectivity to the overload type of the motor and was more adequate to the real state of the protected unit. This was accomplished by the use of two bimetal strips placed at different distances from the heat-conducting cover of the relay. At long overloads of the motor with a repetition factor of 1.5 to 2 of the nominal value, the speed of winding temperature increase does not exceed $0.5°C/s$. At such low speed of temperature change, both bimetal strips (4 and 5) are heated practically to the same temperature and bend equally.

When the limiting temperature is reached and the bending ends of bimetal strips 4 and 5 touch the stop (10) they cease moving together. It is only the upper plate (4) that continues bending. The pin (2) passes through the hole in the lower (already stationary) plate (5) and unbends the spring (6), with the contact (8) breaking the circuit (Figure 9.7b). In emergency modes, the rate of temperature change of the windings of the motor increases sharply. The upper bimetal strip (4) is quickly heated and begins to bend while the lower plate remains in the initial position. Because of this, the contacts

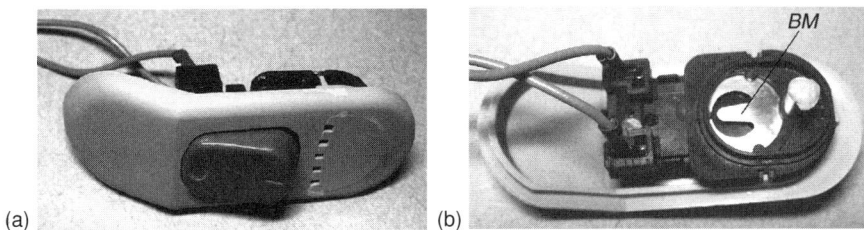

FIGURE 9.6
Construction of an auto-cutout of an electric kettle.

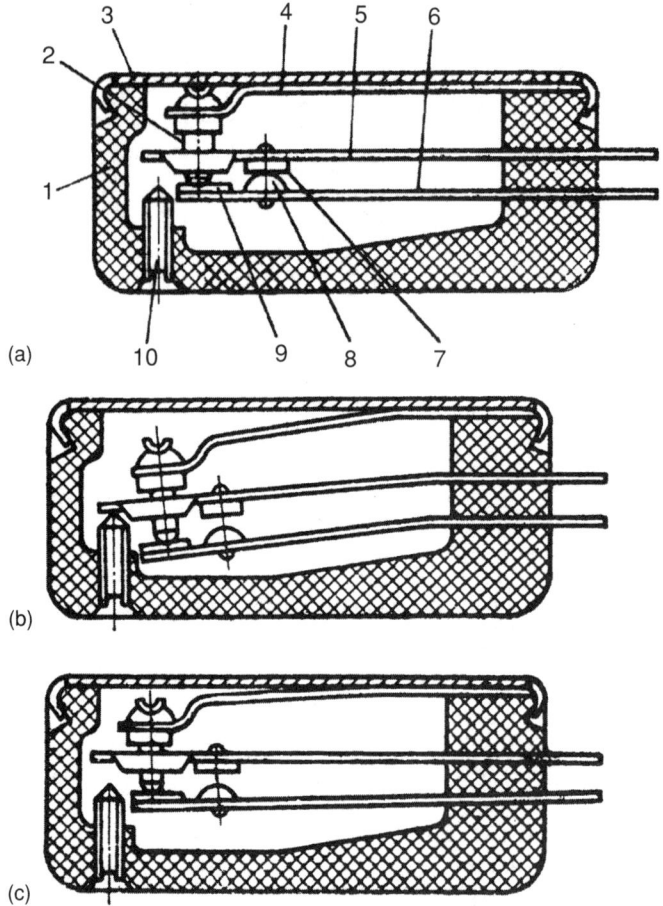

FIGURE 9.7
Differential temperature relay of the DTP-3M type (Russia). 1 — Plastic case; 2 — insulating bolt; 3 — heat-conducting copper cover; 4 — upper bimetal strip; lower bimetal strip; 6 — spring; 7 and 8 — contacts; 9 — insulating layer; 10 — adjusted stop; a — Construction and the initial state; b — pick-up at the long slow heating; c — pick-up at the quick temperature increase.

open at considerably lower temperatures on the cover of the relay and at a slight bend of the bimetal strip 4 (Figure 9.7c). Pick-up of this relay is biased at increase of temperature rate.

9.2   Protective Thermal Relays

In the protective thermal relay for high currents (up to 200 to 300 A — Figure 9.8), a shunt for current unloading of the bimetal element and a complex method of heating of the bimetal element (by the direct current passing through it and an additional heater) are applied. For large currents, an ordinary current transformer is sometimes used, also, the secondary winding of which (as is usual with a current of 5 A) supplies a small heating element.

The contact system is constructed according to Figure 9.3c and is equipped with an instant-return spring, providing stepwise opening of the contacts. Relays of such type are used for protection from overload of electric motors with power up to 100 to 200 kW, DC

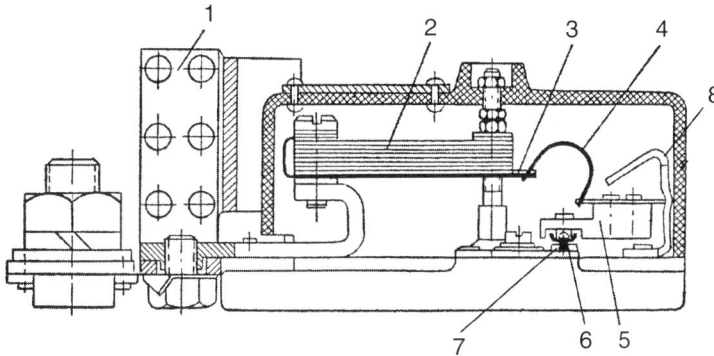

FIGURE 9.8
Protective thermal relay of the TPA and TPB series (Russia) for high currents. 1 — External shunt; 2 — heater; 3 — bimetal strip; 4 — instant-return spring; 5 — intermediate chock; 6 — movable contact; 7 — stationary contact; 8 — stop.

and AC, with difficult starting conditions (that is with great starting duration, and a greater starting current ratio). The external shunt is used with load currents over 50 A.

Contacts of thermal relays designed for protection of powerful consumers cannot switch the full current of the load (100 to 200 A, for instance) and are usually used only for control of a powerful contactor (Figure 9.9). As it can be seen from the circuit shown in Figure 9.9, the shunt, the heating element, and the bimetal element of the thermal relay are connected directly to the circuit of the main current, and its main contacts — to the supply circuit of the coil of the external powerful contactor.

In most cases, the contactor is usually combined with the thermal relay (Figure 9.10), into one construction so that the circuit of the main contacts of the contactor is connected in series with the circuit of heating element, and the output contact of the thermal relay (normally closed) is switched in the supply circuit of the contactor coil.

Due to such connection, as the thermal relay picks up, the load circuit is broken not by its low-power contacts, but by the main contacts of the contactor. The heating elements (Figure 9.10b) are often made as removable ones (like in the example considered above) so that they can be easily and quickly replaced. For each type of heating element, the current and proper pick-up time of the thermal relay are indicated by the manufacturer. Usually heating elements are heated to high temperatures during operation and glow a bright red, which is why cases of thermal relays are made of heat-resistant plastic.

Protective thermal relays for smaller currents (a few amperes protractedly and tens of amperes in case of emergency switching) were produced in the 1940–70's of the last

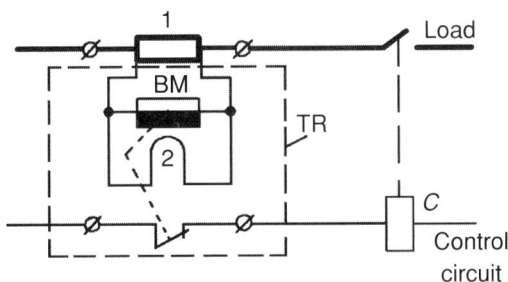

FIGURE 9.9
Circuit diagram for connection of a thermal relay (TR) to an external contactor (C). 1 — Shunt (or current transformer); 2 — heating element; BM — bimetal element.

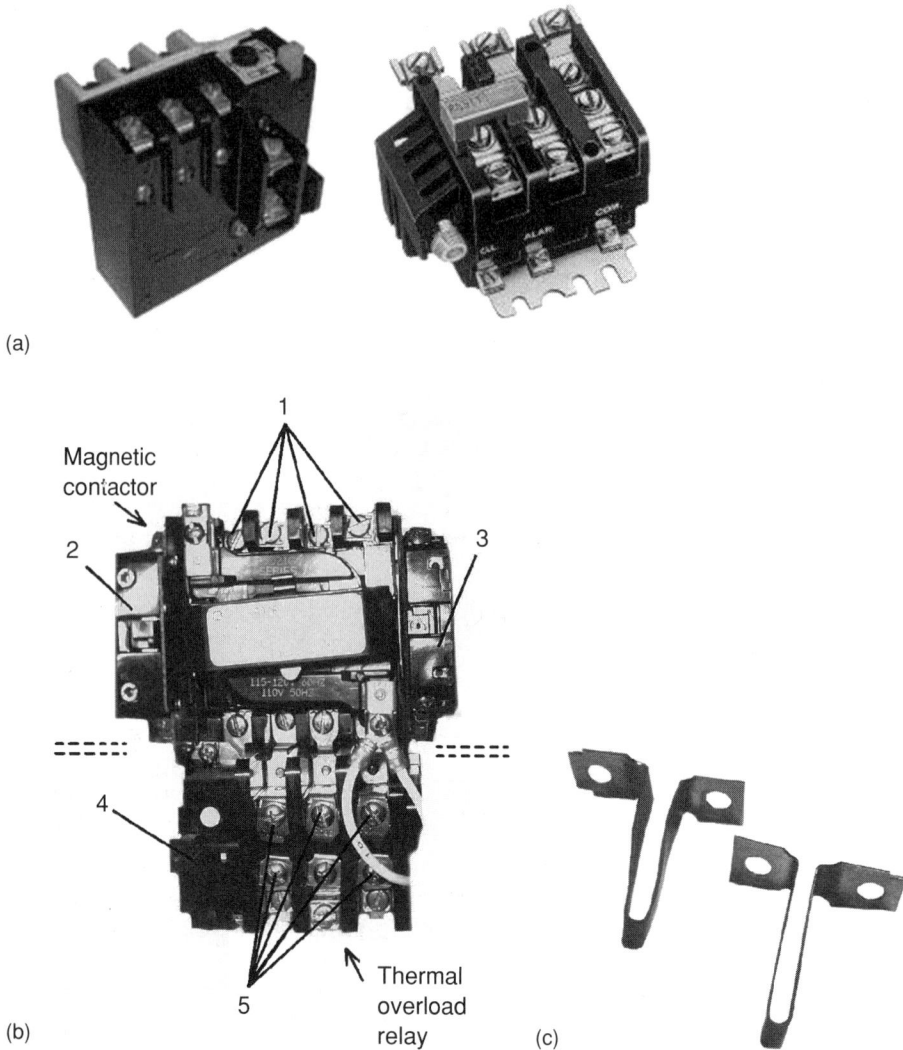

FIGURE 9.10
(a) Modern protective thermal relays designed for mounting directly on the power relay (contactor). (b) Power contactor integrated with the thermal relay assembly. 1 — Main outlets of a contactor; 2 and 3 — units of auxiliary contacts; 4 — main contact unit of the thermal relay assembly; 5 — fastening and electrical bond of the heating elements. (c) Removable heating elements.

century as separate devices (Figure 9.11). Some constructions had a removable heating element for different currents. The thermal unit consists of a current coil (as primary transformer coil) (Figure 9.12) placed over a bimetal helix that acts as a short-circuited secondary transformer coil. The current heats the helix causing it to rotate in a direction that closes the hand reset contacts. Tripping current is adjustable from 90 to 110% of coil rating.

The instantaneous unit is the small electromagnetic relay with hinge-type armature mounted on the right front side of the relay. It operates over a 4 to 1 range and has its calibration stamped on a scale mounted next to the adjustable pole piece. Why is the

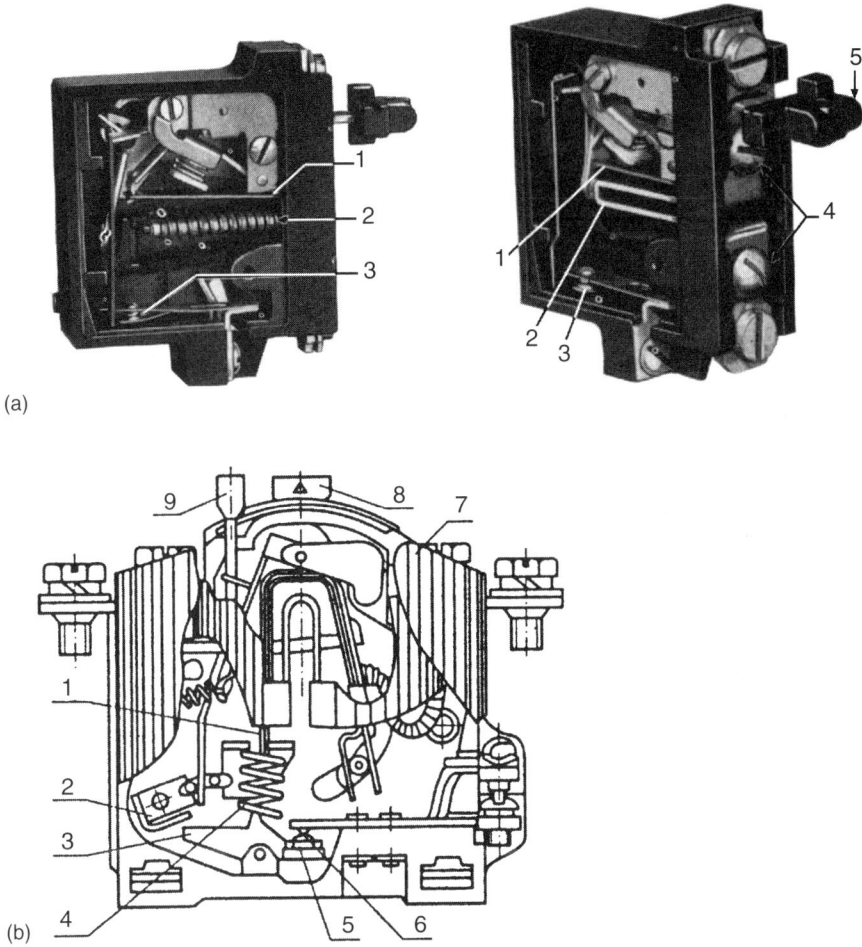

(a)

(b)

FIGURE 9.11
(a) Protective thermal relay of the CR-2824-41 type (General Electric, 1953). 1 — Bimetal strip; 2 — heater; 3 — silver contacts; 4 — heater screws; 5 — reset knob. (b) Protective thermal relay of the TRP type (Russia) made according to the construction diagram (Figure 9.3e). 1 — Bimetal strip; 2 — return stop; 3 — beam; 4 — spring; 5 — movable contact; 6 — stationary contact; 7 — removable heater; 8 — knob of the setting current controller; 9 — reset button.

instantaneous unit needed? The answer is that the thermal process of heating and mechanical moving of the end of the bimetal strip is inertial. Even at a five-to tenfold overload current, a certain time is needed to heat the strip. As a rule, such sharp current rushes in the controlled circuit occur because of a short circuit, when there is no need for the time delay produced by the bimetal strip. On the contrary, the short circuit must be disabled as soon as possible. It is the instantaneous unit that helps to accelerate the pick-up of the relay at high currents.

The protective thermal relay of the TMC11A type, produced by General Electric Co. (Figure 9.12), is notable for its original construction.

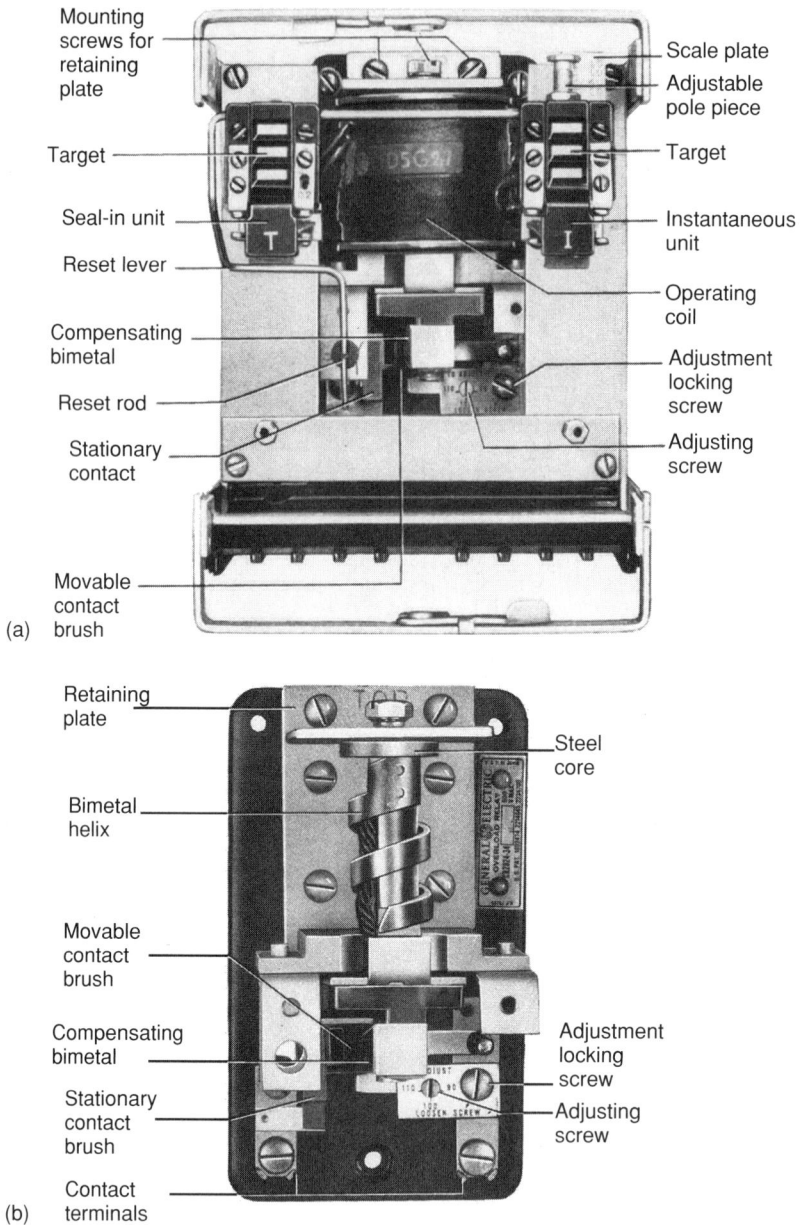

FIGURE 9.12
(a) Type TMC11A thermal protection relay (GE). (b) Thermal unit of TMC11A relay with operating coil removed.

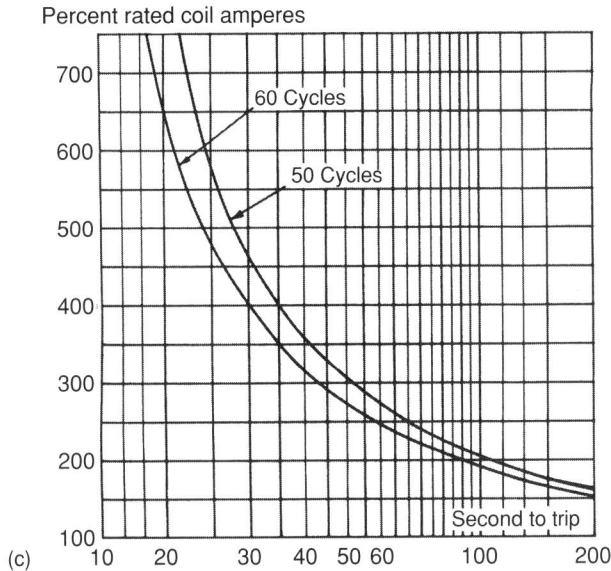

FIGURE 9.12 (*Continued*)
(c) Average time-current characteristic curve for thermal unit of TMC11A relay (General Electric. Instruction Manual GEI-28826A, 1935).

9.3 Automatic Circuit Breakers with Thermal Elements

Relays of the type described above are used for protection of powerful electric motors having high ratios of starting currents. For similar purposes, so-called "automatic circuit breakers" are used (Figure 9.13). These contain a bimetal heating element with an integral electromagnet with a coil for high currents — like the relays considered above. An armature of the clapper-type (or solenoid type) of this electromagnet is mechanically connected to the same contact mechanism as the bimetal strip.

Such complex relays are usually called "protective circuit breakers," "automatic circuit breakers," or just "circuit breakers," although there is no logic in this name. By the principle of operation this device is a protective relay of direct action and is designed for protection from overloading. This is the main function of the device. Due to a special lever connected to a release, the device may also be used for manual switching of the load in normal modes; that is, it may serve as a standard circuit breaker (this probably reflects the origin of the name of the device), however, if one takes into account that some types of thermal and electromagnetic relays (see the section on "Latching Relays" in Chapter 16 of this book) also have a lever for manual closing and opening of contacts, the use of the word "circuit breaker" to denote a protective relay does not seem very logical.

In this book, which is devoted to relays (not to circuit breakers), we will continue to call such complex devices "relays," using the term which correctly reflects the functional characteristics of these devices. Such relays have complex constructions, with a great number of interacting elements. Devices produced by various companies differ considerably from each other by construction; however, they are all based on the same principle, contain similar basic elements (Figure 9.14), and have similar design characteristics (Figure 9.15).

The "e" zone of the characteristic belongs to short-circuit currents exceeding nominal currents of relay adjustment by 50 to 100 times. An additional increase of speed of operation in this zone is provided due to the electrodynamic repulsive force of the contacts. You remember that we considered methods of compensation for this repulsive force (see Chapter 3). In constructions of protective relays with high commutation current, this effect is not compensated, rather, it is additionally accelerated by the same technical methods, only applied vice versa (Figure 9.16).

Thermal and electromagnetic releases in such powerful devices are equipped as a rule with adjusting elements placed on the front panel (Figure 9.17). In power protective relays (Figure 9.19), a microprocessor release is sometimes used. This simulates the time–current characteristic by a purely electronic method, and provides the working order of the device, similar to complex action relays.

Use of a microprocessor not only allows reiteration of the time–current characteristic of the complex relay but also construction of each period of this characteristic with the required parameters (Figure 9.18). According to the type of time–current characteristics and functions, protective relays of the types considered above are divided into the following classes:

In particular devices, especially ones of the microprocessor type, combinations of several classes are possible. For example: LI, LSI or LSIG. According to the standard IEC 60898-1: "Circuit-breakers for over-current protection for household and similar installations," the relays illustrated in Figure 9.13 are classified by the pick-up current as follows:

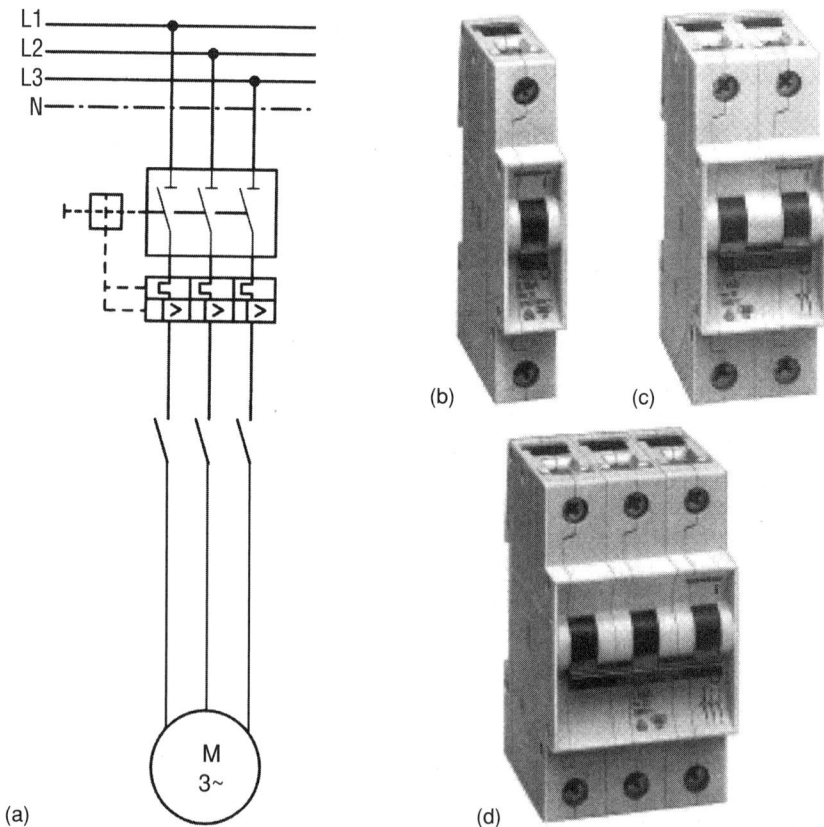

FIGURE 9.13
Modern protective complex action relays: single-, double- and triple-circuit ones and their circuit diagrams (in the three-phase form).

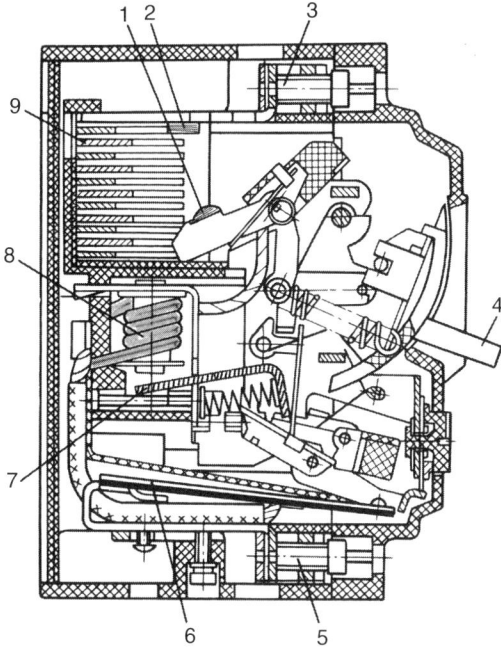

FIGURE 9.14

Typical construction of a protective complex action relay. 1 — Movable contact; 2 — stationary contact; 3 and 5 — clips for attachment of external circuits; 4 — knob for manual control; 6 — bimetal plate; 7 — armature of the electromagnet; 8 — coil of the electromagnet; 9 — arc-suppressing lattice.

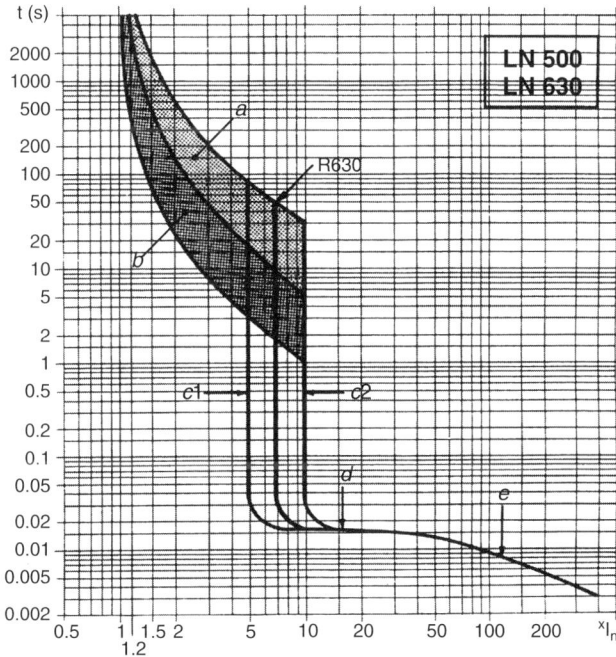

FIGURE 9.15

Time—current characteristic of the LN500 and LN630 type protective complex action relay (ABB SACE), with nominal currents of 500 and 630 A. a — Thermal releases (bimetal element) in cold conditions; b — thermal releases (bimetal element) in service conditions; $c1$ — adjustable electromagnetic releases (minimum setting value); $c2$ — adjustable electromagnetic releases (maximum setting value); d — total break time of electromagnetic releases; e — break times by electrodynamic effect (ABB).

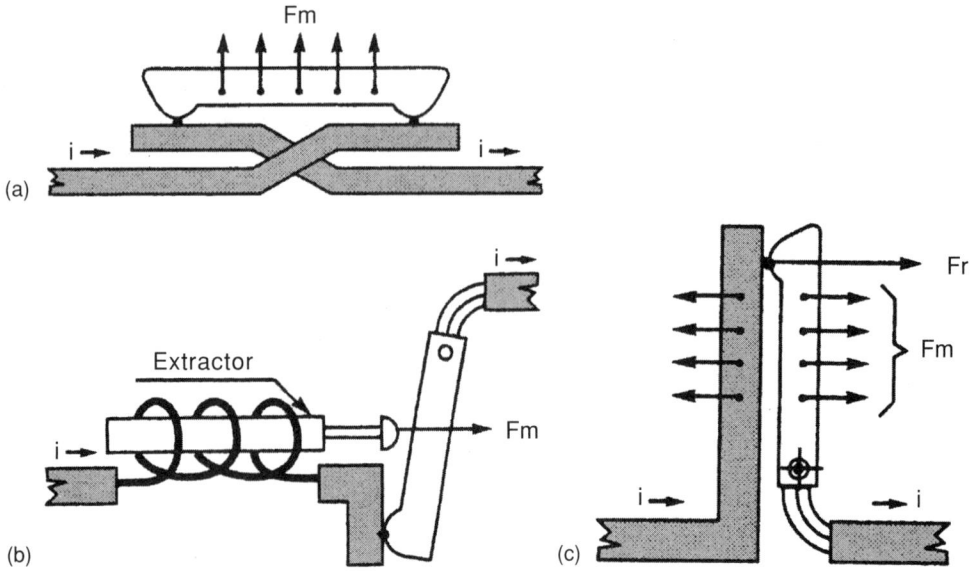

FIGURE 9.16
Contact repulsion principles, used in complex action protective relays. (a) Double contact with double repulsion force; (b) use "extractor" (magnetic core pushes or pulls) for moving contact; (c) simple repulsion loop.

Type	Ranges of instantaneous tripping current
B	Above $3I_N$, up to and including $5I_N$
C	Above $5I_N$, up to and including $10I_N$
D	Above $10I_N$, up to and including $50I_N$

The inscription "C16" on the relay means, for example, that the nominal current (that is the nominal load current continuously admissible without break-up) of this relay equals 16 A. When this current is exceeded, the overload protection, based on a bimetal strip with a time delay inversely proportional to the passing current, begins to work. When currents of short-circuits are above $5 \times 16 = 90$ A, an electromagnetic release (without time delays) immediately responds. Switching capacity of this relay allows switching of short-circuit current up to $10 \times 16 = 160$ A, for many cycles.

As it can be seen from all of the above, there exist a number of classification systems of such relays (automatic circuit breakers). One of the most widespread in European systems

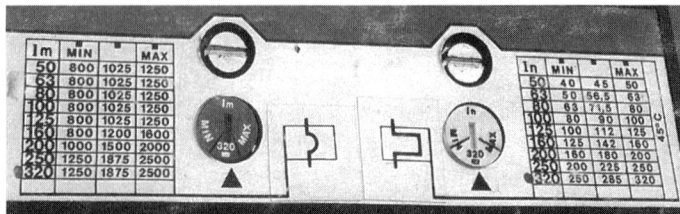

FIGURE 9.17
A fragment of the front panel of a power protective relay, with elements of adjustment of the operation threshold of the electromagnetic (to the left) and thermal (to the right) releases. In tables: nominal pick-up currents I_m and I_n, and corresponding minimum and maximum values.

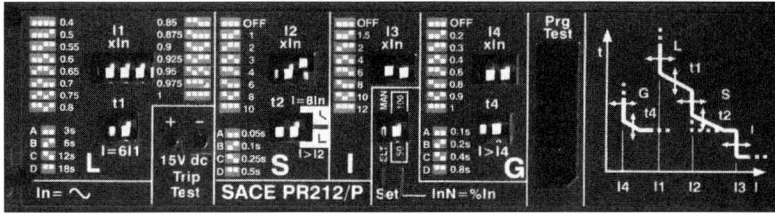

FIGURE 9.18
A fragment of the front panel of a microprocessor protective relay, designed for setting parameters for all intervals of the time–current characteristic.

was a German standard DIN VDE 0641 and 0660. According to these standards, automatic circuit breakers were divided into the four following groups (Table 9.1). Moreover, it turns out that many companies (for example, ABB) simply "invented" their own classifications, having nothing in common with other standards, and applied only for items produced by their company (Table 9.2).

The new type of protection devices, named: "Hydraulic–magnetic circuit breakers" now available in market (see Chapter 16).

TABLE 9.1

Circuit Breaker Classification According to German DIN VDE Standard

Type	Nominal Current, I_N (A)	Trip Level for Thermal Unit	Trip Level for Electromagnetic Unit
L	6 to 10	$1.5 I_N$–$1.9 I_N$	$3.6 I_N$–$5.25 I_N$
	16 to 25	$1.4 I_N$–$1.75 I_N$	$3.36 I_N$–$4.9 I_N$
	32 to 50	$1.3 I_N$–$1.6 I_N$	$3.12 I_N$–$4.55 I_N$
B	6 to 63	$1.13 I_N$–$1.45 I_N$	$3 I_N$–$5 I_N$
C	6 to 63	$1.13 I_N$–$1.45 I_N$	$5 I_N$–$10 I_N$
K	0.2 to 63	$1.05 I_N$–$6 I_N$	$8 I_N$–$14 I_N$

TABLE 9.2

ABB Classification for Circuit Breakers

Type	Nominal Current, I_N (A)	Trip Level for Thermal Unit	Applications
B	6 to 63 (in 10 steps)	$3 I_N$–$5 I_N$	Designed primarily for use in cable protection applications
C	63, 80, 100, 125	$5 I_N$–$10 I_N$	For medium magnetic start-up currents
K	6 to 63	$8 I_N$–$12 I_N$	For high in-rush magnetic start-up
D	6 to 63	$15 I_N$	currents from motors, transformers and other equipment
Z	6 to 63	$2 I_N$–$3 I_N$	A very low short circuit trip setting, in order to protect semiconductor or other sensitive devices

FIGURE 9.19
Power protective relays (automatic circuit breakers) with complex and microprocessor releases (ABB SACE).

9.4   Dilatometer Relays

Dilatometer relays have quite a simple construction (Figure 9.20). As it can be seen from the above principal diagram, the dilatometer thermal relay contains three basic components: elements with a high and low linear expansion coefficient, and electric contacts. Elements with different linear expansion coefficients are rigidly mounted with each other at one point in such a way that as these elements are heated, the free end of one of them (the rod) shifts from the free end of the other one (the tube) and operates the electric contacts. Very often, dilatometer relays are made in the form of a brass (or nickel, for high temperatures) tube 5 to 8 mm in external diameter and 100 to 300 mm in length. Inside this tube there is a rod made from special material (Invar for temperatures up to 200°C, quartz and porcelain for temperatures up to 1000°C). The rod and the tube are rigidly mounted at the end.

On the open end of the tube, there is a plastic case with a contact system and a snap-action contact (Figure 9.21b). The relay of such a type can be found in various kinds of household and industrial appliances. Most likely, your water-heating boiler contains this very type of a thermal relay.

Thermal relays have so many widespread applications that some companies completely specialize in the manufacture of such relays (Control Products, Portage Electric Products, Claus Schafer GmbH). Some of these relays have unusual designs, for example:

- A dual set-point snap-action thermal switch allows control of two circuits at different temperatures

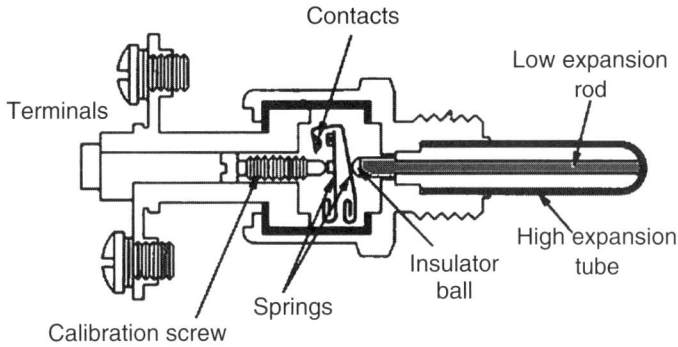

FIGURE 9.20
Dilatometer thermal relay: principal diagram.

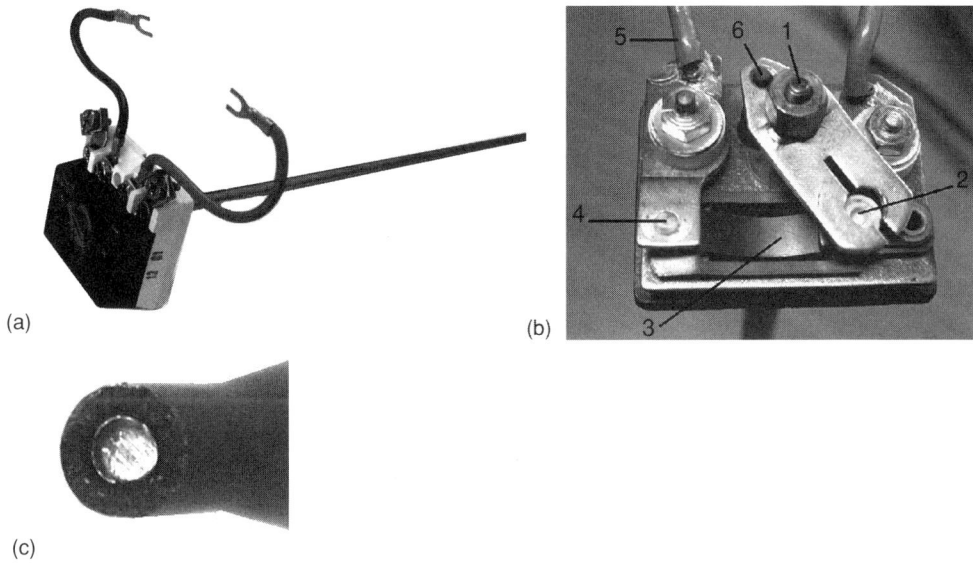

FIGURE 9.21
Dilatometer thermal relays: external design. (a) Ready-assembled relay; (b) construction of the contact system; (c) end of the pin-part of the sensor (in the center one can see the end of the rod with the low linear expansion coefficient). 1 — Output of the internal rod with low linear expansion coefficient; 2 — pusher; 3 — instant-return spring with a movable contact at the end; 4 — stationary contact; 5 — relay outlets; 6 — stop.

- A nonresettable thermal switch; a snap-action single operation switch whose reset temperature is less than −35°C prevents the thermo-switch from resetting under normal operating conditions.

9.5 Manometric Thermal Relays

In large industrial refrigerators and other types of industrial plants, thermal relays of the manometric type were widely adopted (Figure 9.22). Such relays are called "manometric" because they are constructed with a combination of a manometer and a contact.

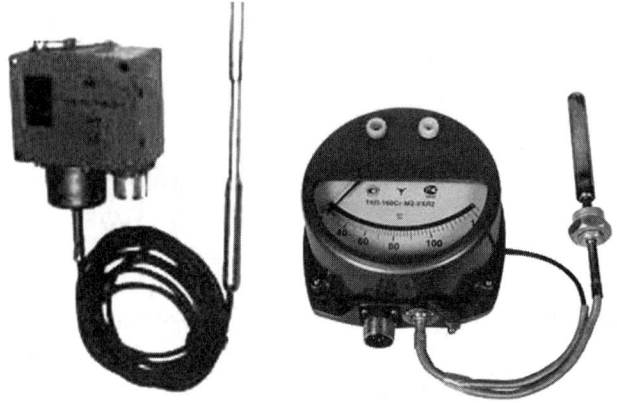

FIGURE 9.22
Manometric temperature relays. To the right there is a so-called "indicating" relay with a scale, which is used as a relay and as a measuring tool at a time.

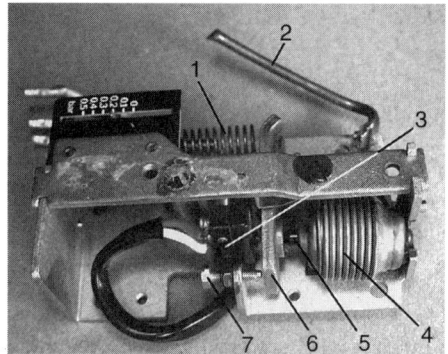

FIGURE 9.23
Construction of a manometric thermal relay. 1 — Spring; 2 — tube connected to the hermetic ampoule with liquid; 3 — micro switch; 4 — manometric coiled metal tube; 5 — pusher; 6 — intermediate strip; 7 — adjusting screw.

But unlike the open system of manometer, connected to a unit with controlled gas or liquid pressure, the thermal relay already contains a hermetically closed vessel filled with liquid or gas. The pressure inside it depends on the temperature. Usually this vessel is made in the form of a small metal ampoule connected to the relay case by a long flexible tube. In the case, there is a standard for the manometer metal tube (coiled around) and a contact system, which usually consists of ready-made micro switches (Figure 9.23). A hermetic metal ampoule with a controlled temperature is placed inside the unit. As the temperature increases, the pressure of the liquid in this ampoule increases as well. It is then transferred to the coiled monometric tube. Affected by the increased pressure, the turns of this coil start to move apart (like in a manometer) and operate the contacts through the pusher. Sometimes they connect a pointer to the coil, thus obtaining a temperature scale, giving us a so-called "indicating" relay that is a relay combined with a measuring tool. Such devices are installed on all power high-voltage transformers for control of the temperatures of the oil filling such transformers.

9.6 Mercury Thermal Relays

Mercury electro-contact thermometers, differing from standard mercury thermometers by a sealed wire leading from the capillary with the mercury, may be also considered to be variants of "indicating relays" (Figure 9.24). Mercury going up the capillary closes these wire leads as it reaches a fixed temperature. The current switched by such relays is not very strong (not more than tens of milliamperes) and requires amplification for

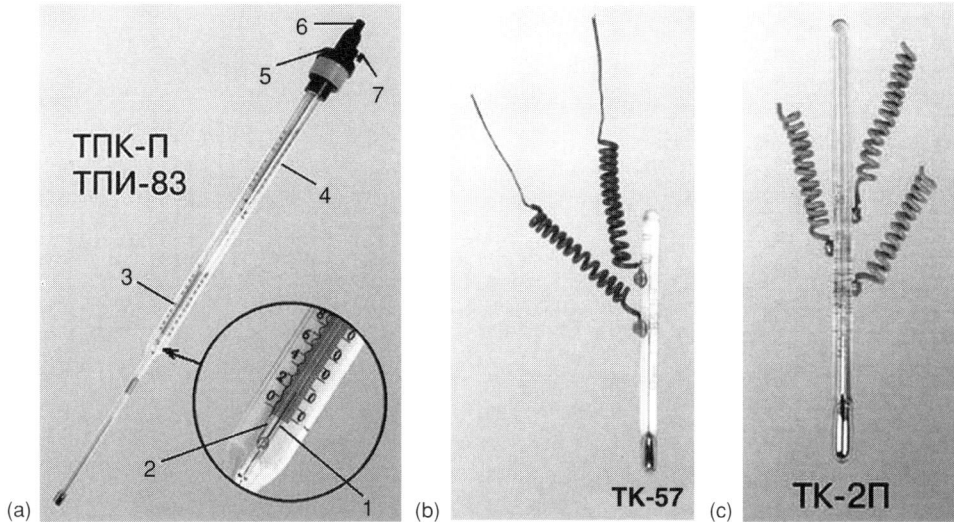

FIGURE 9.24
Electro-contact mercurial thermometers (Russia); (left) ones with adjusting operation threshold; (right) ones with fixed operation threshold, two- and three-position ones. 1 — Mercury; 2 — internal conductor linked with mercury (works as a stationary contact); 3 — lower scale for the capillary with mercury; 4 — upper scale for the capillary without mercury (adjusting one); 5 — permanent magnet; 6 — knob for rotation of the permanent magnet; 7 — holder of the magnet position.

control of the power final control elements. Some variants of such devices have a mechanism for adjusting of the operation setting within the whole temperature range permissible for each given thermometer. Such relays are notable for two scales: an upper one for the capillary without mercury (which serves only for the adjustment of the fixed operation temperature and indication of the cursor position) and a lower one — for the capillary with mercury. The fixed temperature is adjusted with the help of a very thin straight piece of wire, buried in the capillary with mercury to the fixed depth. The wire is moved with the help of the permanent magnet (5), molded with plastic and equipped with a knob for rotation. Inside the thermometer shell there is a second magnet rotating freely under the effect of the magnetic field of the external magnet (5). This internal magnet transmits its rotation to the thin screw where there is a slider with a wire-electrode linked to it. With the help of this simple mechanism, the rotating movement of the magnets is converted (translated) to displacement of the wire-electrode in the capillary the mercury.

9.7 Thermal Relays with Reed Switches

Due to their precisely defined operational threshold (pick-up) to the smooth increase of their actuating value (a very important property of thermal relays), reed switches were applied in thermal relays. But as for reed switches, the actuating quantity is a magnetic flux so in such devices, transformation of heat to a variable magnetic flux is required. Such transformation is carried out in thermal relays with the help of permanent magnets moved by elements sensitive to temperature (Figure 9.25). Usually this is a bimetal strip, made of metals with a memory effect or silphone filled with gas and expanding as the temperature increases.

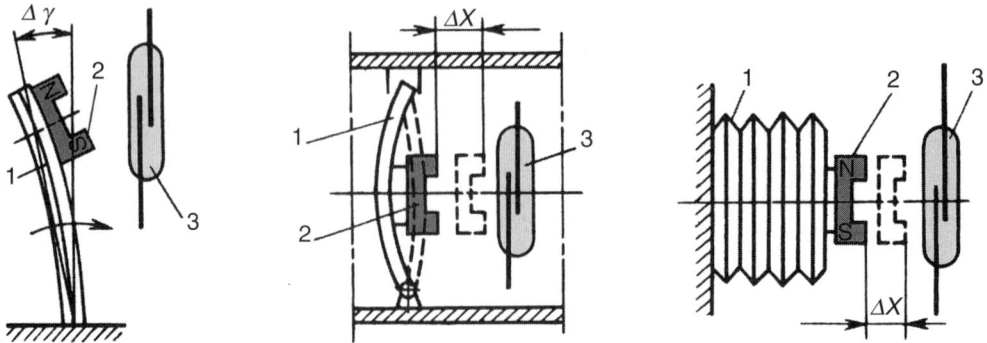

FIGURE 9.25
Thermal relays with reed switches. 1 — temperature sensor; 2 — permanent magnet; 3 — reed switch.

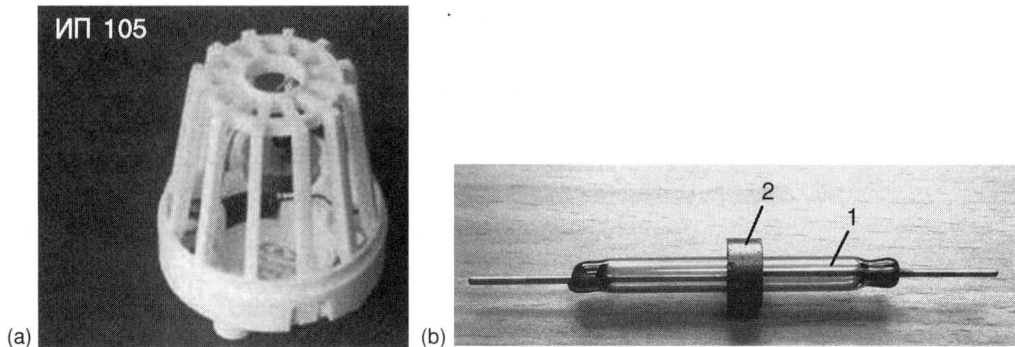

FIGURE 9.26
Thermal relay with reed switch for fire alarm systems (Russia). 1 — Reed switch; 2 — ceramic magnet.

The property of permanent magnets to reduce sharply the magnetic flux as increased temperatures approach the Curie point is applied in the simplest relays of fire alarms responding to increased temperature. The Curie point of standard magnets made from metal alloys is in the high temperature area and inapplicable in fire alarm systems.

However, magnets produced from ferrite powder by pressing, lose all of their magnetic properties at temperatures of about $70°C$. Such a magnet, in the form of a small ring, put on the reed switch, is used in fire alarm relays (Figure 9.26).

9.8   Semiconductor Thermal Elements and Thermal Relays

Various electronic temperature relays are widely used and produced. There are a great number of modifications of such relays, but they are all based on the same principle. Such a relay contains a semiconductor thermo-sensitive element, changing its electric resistance with temperature change, an electric amplifier — usually in the form of a chip, and an output switching element at the output (a standard electromagnetic relay or a power solid-state switch). Thermo-sensitive elements may have different forms and sizes: drop-shaped, in the

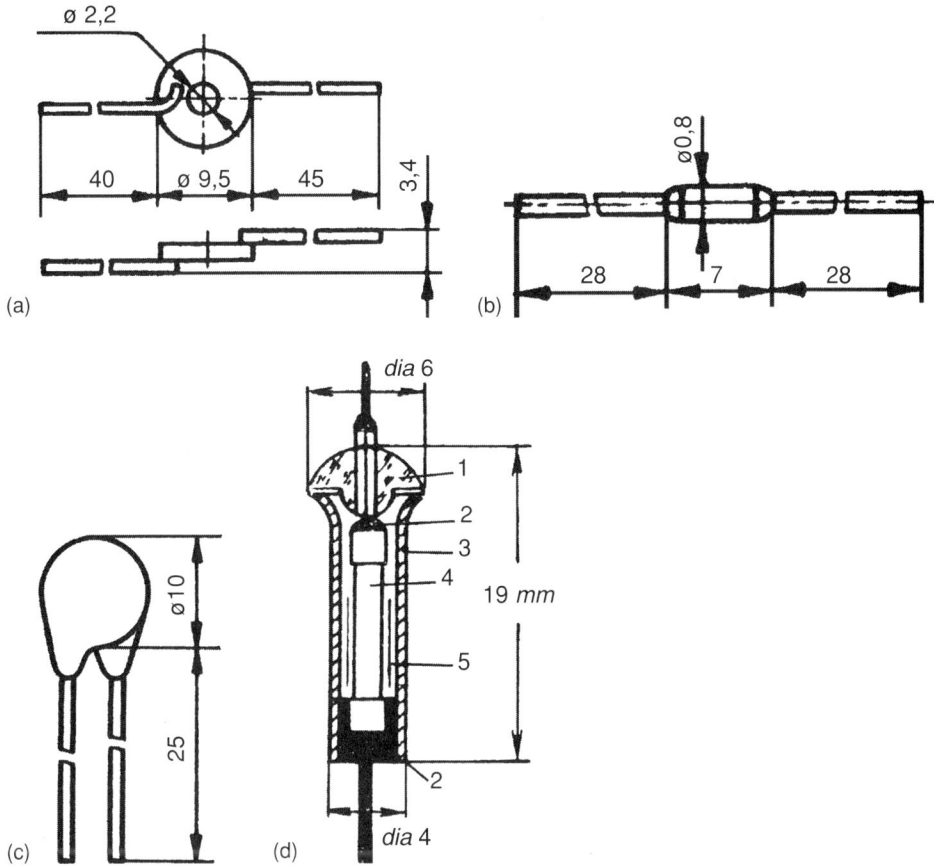

FIGURE 9.27
Thermo-sensitive semiconductor elements of different types (d-in hermetic shell). 1 — Glass insulator; 2 — tin; 3 — metal tube; 4 — body of the thermo-sensitive element; 5 — winding made from a few foil layers.

form of rings, resembling standard resistors, and even sealed ones (Figure 9.27). Depending on the type of the temperature characteristic, there are two classes of thermo-sensitive elements: "thermistors" and "posistors." The resistance of the former decreases as the temperature rises (negative resistance temperature coefficient). The resistance of the latter increases in the working zone (positive resistance temperature coefficient) (Figure 9.28).

These elements are also called "NTC termistor" and "PTC termistor," or "NTC resistor" and "PTC resistor" (negative temperature coefficient [NTC]; positive temperature coefficient [PTC]). Materials for production of thermo-sensitive elements are various different oxides belonging to the class of semiconductors: ZnO, MgO, Mn_3O_4, Fe_2O_3, and some sulfides as Ag_2S and others.

The electronic amplifiers with the output switching elements may also be very different. The external design and constructive diagrams of some of these amplifiers are shown in Figure 9.29 as examples.

One of the applications of this device is protection of power motors from overheating. Thermo-sensitive elements are placed in different parts of the winding of the motor stator and are connected to the electronic block (Figure 9.29b). Cera-Mite Corp. (U.S.A.) produces RTC resistors especially for direct protection of circuits from overcurrents (Figure 9.30).

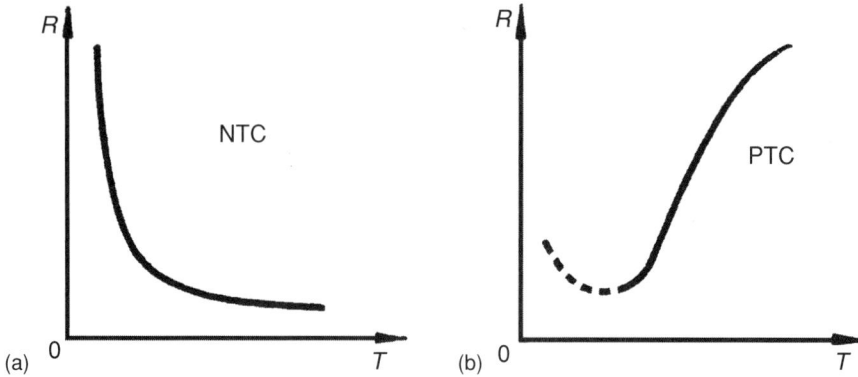

FIGURE 9.28
Temperature characteristics of a thermistor and a posistor.

The PTC resistor's unique property of dramatically increasing its resistance above the Curie temperature makes it an excellent candidate for overcurrent protection applications. Overcurrent situations in electronic devices occur due to voltage fluctuations, changes in load impedance, or problems with the system wiring. PTC thermistors monitor current in series connected loads, trip in the event of excess current, and reset after the overload situation is removed, creating a new dimension of flexibility for designers. Hold currents from 5 mA to 1.5 A are available in PTCR diameters from 4 to 22 mm ("hold current" — I_H — is the maximum continuous current at which a PTCR can be maintained in a low resistance ON state, while operating at rated ambient temperature (typically 25°C). To prevent nuisance tripping, choose the rated hold current to be greater than the normal current expected).

During normal operation, the PTCR remains in a low base resistance state. However, if current in excess of the hold current I_H is conducted, I^2R losses produce internal self-heating. If the magnitude and time of the overcurrent event develops an energy input in excess of the device's ability to dissipate heat, the PTCR temperature will increase, thus reducing the current and protecting the circuit.

The current required to trip (I_T) is typically specified as two times the hold current (2 × I_H). Trip current is defined as the minimum rms conduction current required to guarantee thermistor switching into a high resistance state (Figure 9.31), at a 25°C ambient temperature. PTCRs reset after an overcurrent situation. Protection levels may be set lower than possible with fuses, without having to worry about nuisance trips.

The temperature at which the PTCR changes from the base resistance to the high resistance region is determined by the PTCR ceramic material. The switching temperature (T_{SW}) described by the boundary between regions 1 and 2 (Figure 9.31) is the temperature point at which the PTCR has increased to two times its base resistance at a 25°C ambient ($R_{SW} = 2 \times R_{25}$). Design flexibility is enhanced by Cera-Mite's wide selection of ceramic PTCR materials with different switching temperatures (Figure 9.32). Cera-Mite offers a wide selection of ceramic PTC materials providing flexibility for different ambient temperatures. Close protection levels are possible by designing resistance and physical size to meet specific hold current and trip current requirements.

PTC current limiters are intended for service on telecom systems, automobiles, control transformers (as inrush current limiter), or in similar applications where energy available is limited by source impedance. They are not intended for application on AC line voltages where source energy may be high and source impedance low.

Lately temperature sensors have appeared based on operational amplifiers, releasing current or voltage proportional to the temperature of the case, and having a linear

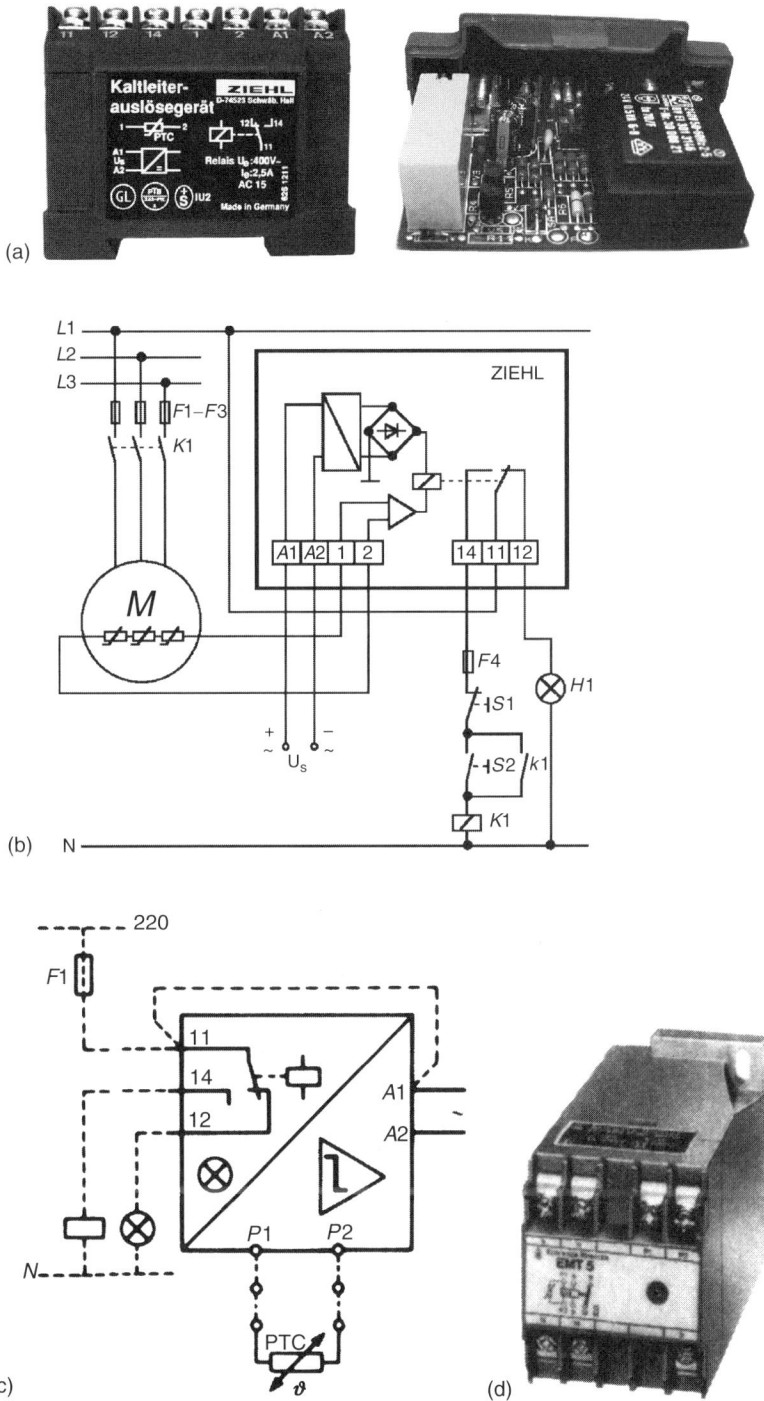

FIGURE 9.29
(a) Electronic amplifier with relay characteristic designed for work with a thermo-sensitive element as a thermal relay (ZIEHL). To the left on the printed circuit board, one can see an output electromagnetic relay in a white plastic case; to the right on the printed circuit board — the transformer of the internal power supply. (b) Electric circuit of an electronic thermal relay (ZIEHL) with thermo-sensitive elements for control of temperature of windings of the electric motor (*M*). (c) External design and circuit diagram of an EMT-5-type termistor overload relay. (Klokner-Moeller, Germany, 2004. Online Internet catalog.)

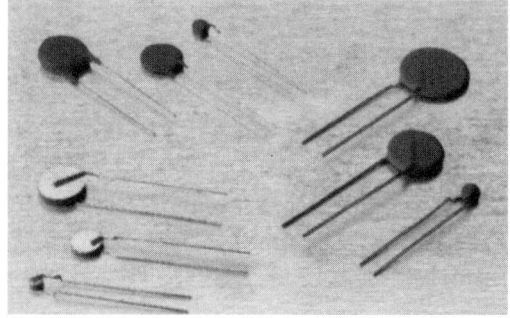

FIGURE 9.30
Resettable PTC resistors for overcurrent protection
(Cera-Mite Corp., U.S.A.).

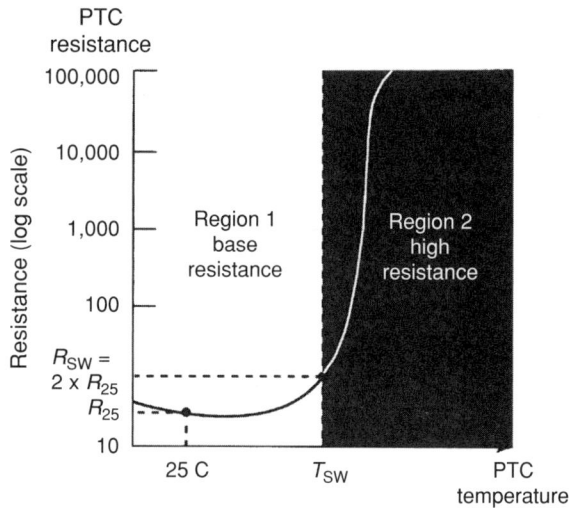

FIGURE 9.31
R vs. T operating characteristics of a PTCR.

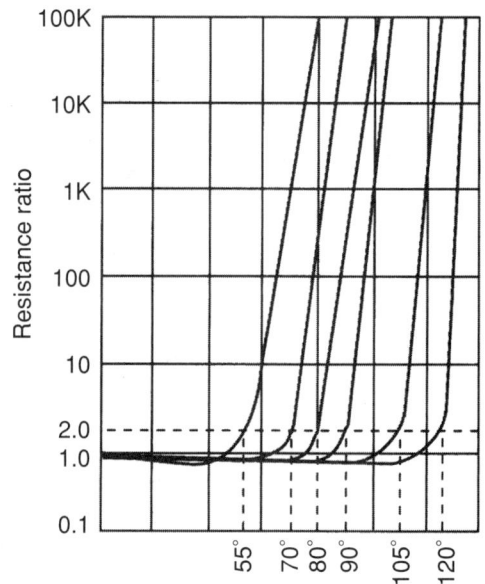

FIGURE 9.32
Characteristics of ceramic PTC materials for different
ambient temperatures.

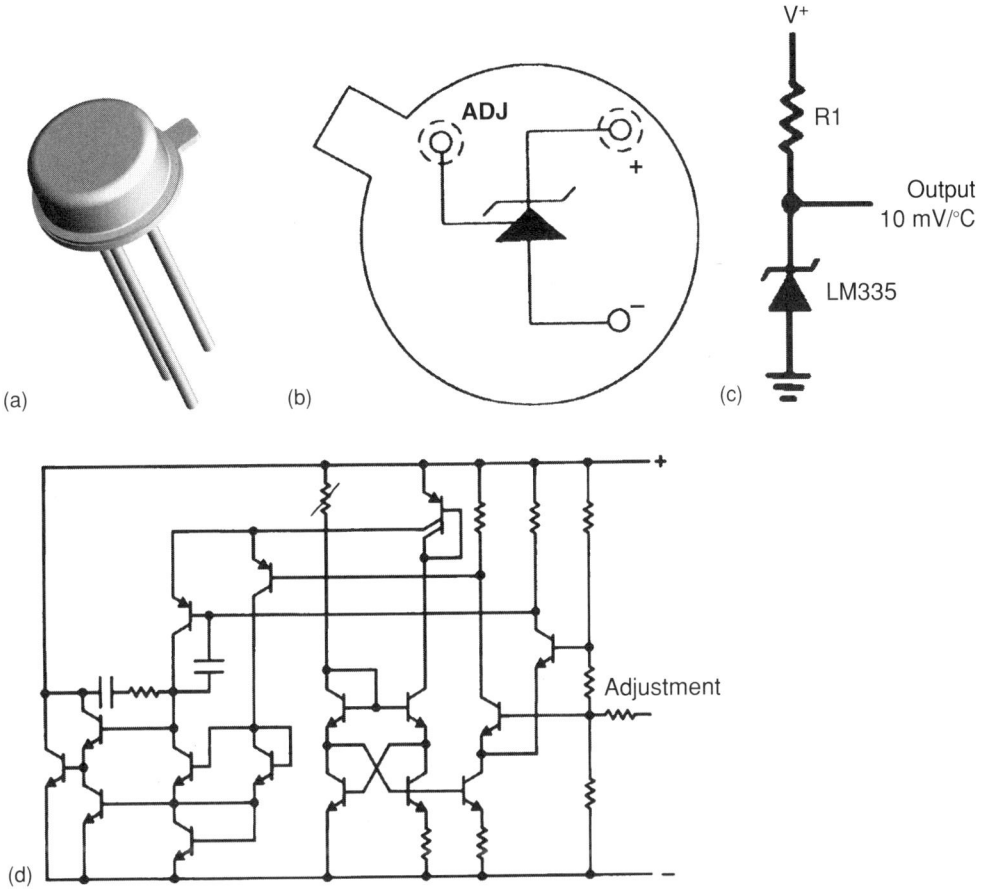

FIGURE 9.33
(a) External view, (b) and (c) connection, (d) internal circuit diagram of LM135H series precision temperature sensor (Motorola, National Semiconductor, etc.).

characteristic at the output, for example, the LM135/235/335 series (Figure 9.33). IC 135/235/335 series is a precision temperature sensor. Operating as a two terminal Zener diode, it has a breakdown voltage directly proportional to absolute temperature at +10 mV/°C. Applications for the device include almost any type of temperature sensing over a −55 to +150°C temperature range, with a 200°C over range also available.

Other examples: LM35CZ-type and 590kH-type integral circuits (Figure 9.34). The LM35 is a three-terminal integrated circuit TO-92 plastic packaged temperature sensor, giving a linear voltage output of 10 mV/°C. Available in two versions, one operating from 0 to +100°C (DZ version), the other from −40 to +110°C (CZ version). Ideally suited for ambient temperature measurements, such as providing cold junction compensation for thermocouples. Accuracy: ± 0.4°C.

The 590 kH is functionally a two-terminal circuit temperature transducer, which produces an output current proportional to absolute temperature. The device acts as a high impedance, constant current regulator, passing 1 μA/°C. Laser trimming of the chip is used to calibrate the device to 298.2 ± 2.5 μA at 298.2 K (+25°C). Since the 590 kH is a current sourcing device, it is ideally suited for remote sensing applications where the output can easily be transmitted over a two-wire twisted pair line, without degradation of

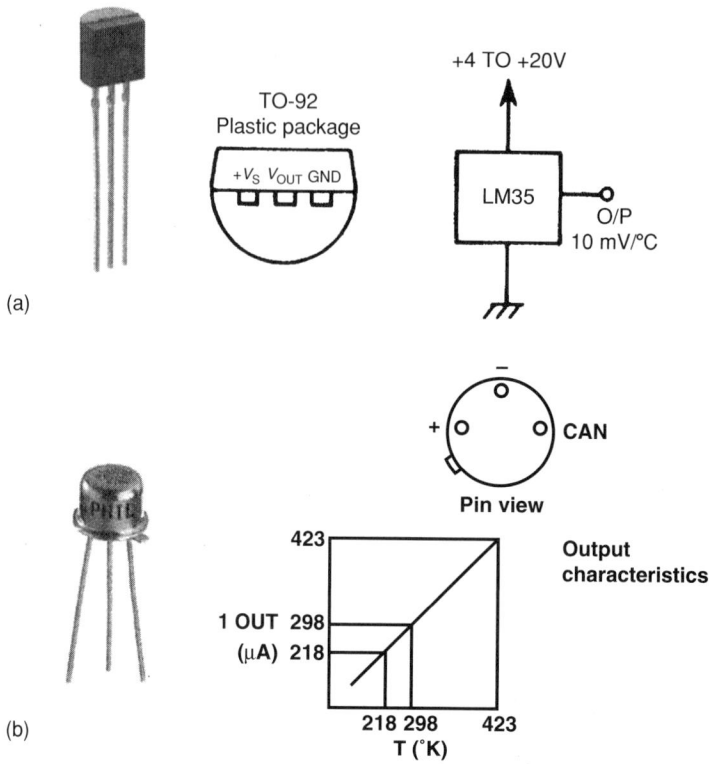

(a)

(b)

FIGURE 9.34
Temperature sensors based on operational amplifiers. (a) LM35CZ type (National Semiconductors); (b) 590 kH type (Phillips).

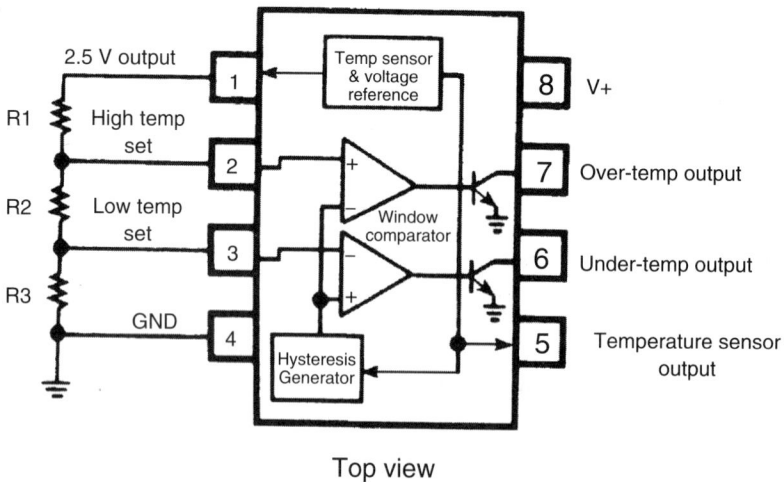

Top view

FIGURE 9.35
Programmable temperature controller TMP01FP type in 8-pin Mini-DIP package (Analog Devices).

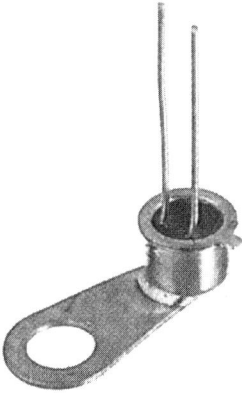

FIGURE 9.36
Small solid-state thermal relay for fixed operation temperatures: $+57$ or $+75°$C.

performance due to line resistance, connector resistance or noise. Operating temperature range: -55 to $+150°$C.

Besides such simple temperature sensors in the form of chips, whole programmable temperature controllers are also produced (Figure 9.35). The TPM01PT is a temperature sensor which generates a voltage output proportional to temperature (from -55 to $+150°$C) and a control signal from one of two outputs when the device is either above or below a specific temperature range. High and low temperature trip points are determined by user-selected external resistors. Typical applications include: over or under temperature sensor or alarm, board level temperature sensing, temperature controllers, electronic thermostats, remote sensors, and process control.

Some semiconductor temperature sensors change resistance stepwise, that is they have relay characteristics and can be realistically called "thermal relays" (it is obvious, that it is not difficult for realizing with the help of electronic schemes, see Chapter 7), for example, a solid semiconductor thermal relay, encapsulated in small T018 type packages, with electrically isolated mounting tabs (Figure 9.36). When heated, the sensor exhibits a high resistance until the transition temperature region is reached, centered around $+57$ or $+75°$C, depending on type. The resistance then changes rapidly from approximately $100\,k\Omega$ down to approximately $100\,\Omega$ for approximately a $10°$C change. The reverse characteristic, with little hysteresis, is followed when cooling.

The Amperite Company produced a TSW series power thermal cut-out solid-state relays for fixed cut-out temperature from 0 to $150°$C, provide a simple means to monitor temperature, using a solid-state sensing device (Figure 9.37). When the temperature exceeds the customer specified value, the output relay change state. Upon sensor cool-down, the contacts return to the de-energized position. Maximum switching voltage is 125 V AC (up to 50 VA) or 60 VDC (up to 30 W).

FIGURE 9.37
TSW series power thermal cut-out solid-state relay in epoxy sealed enclosure with dimensions: $2 \times 2 \times 3/4$ in. (Amperite).

10

Protective Current and Voltage Relays

10.1 What are "Protective Relays"

What are "current relays" and "voltage relays" and what is the difference between them?

These relays are specially designed for current or voltage level control in electric circuits of high and low voltage, and for generation of certain output signals, when the current or voltage level deviates from a preassigned value. Such relays are also called "measuring relays," since in the process of operation they constantly measure the level of the actuating value. Very often the output signal of such relays affects the power shutdown device, de-energizing the load and thus protecting it (or the main supply) from damages in emergency modes, which is why such relays are also called "protective relays."

Some relays of this type have powerful contacts directly switched to the protected circuit, or a powerful electromagnet mechanically linked with the power shutdown device. Such relays are called "direct-action" relays (Figure 10.1a). Low-power relays, which only give control signals to an independent power-switching device (a circuit breaker, for example), are called "indirect-action" relays (Figure 10.1b). In the above examples the relays are connected directly to the controlled current circuit. Such relays are called "primary" ones. As a rule, coils of primary current and voltage relays are designed for currents not exceeding 50 to 200 A, and for voltages of not more than 400 V.

10.2 Current and Voltage Transformers

And what should be done when one needs to control currents of hundreds and thousands of amperes, or very high voltages? In this case the relays are connected not directly to the strong current or high-voltage circuit but through special matching transformers called "current transformers" (CT) and "voltage transformers" (VTs). Relays connected to the controlled circuit through such intermediate transformers are called "secondary" ones (Figure 10.2).

What are CTs and VTs? Like any transformer, a low-voltage CT consists of a primary winding designed for the current in the controlled circuit, a secondary winding to which relays and measuring devices are connected, and also a laminated steel core (Figure 10.3a). Sometimes there is no primary winding in a CT, and the transformer itself looks like a toroid (Figure 10.3b). In this case copper wire or the bus bar is used as a primary winding (it is considered to contain one turn). The primary currents of CT can reach tens of thousands amperes. Standard values for secondary currents are 5 or 1 A. It is for these

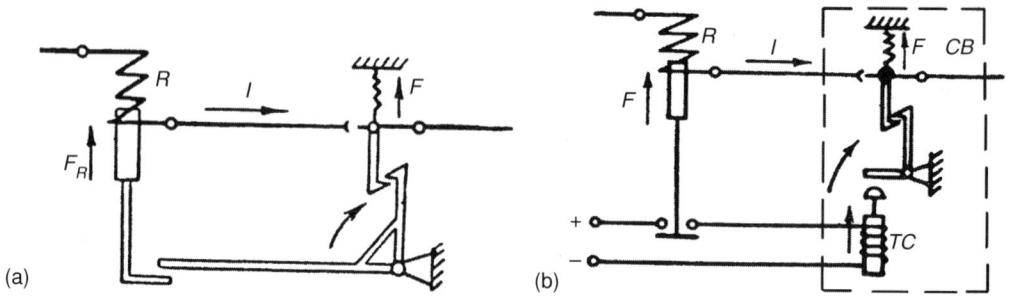

FIGURE 10.1
A schematic plot illustrating the principle of application of primary relays, "direct-acting" (a) and "indirect-acting" (b). R — relay; TC — trip coil; CB — circuit breaker.

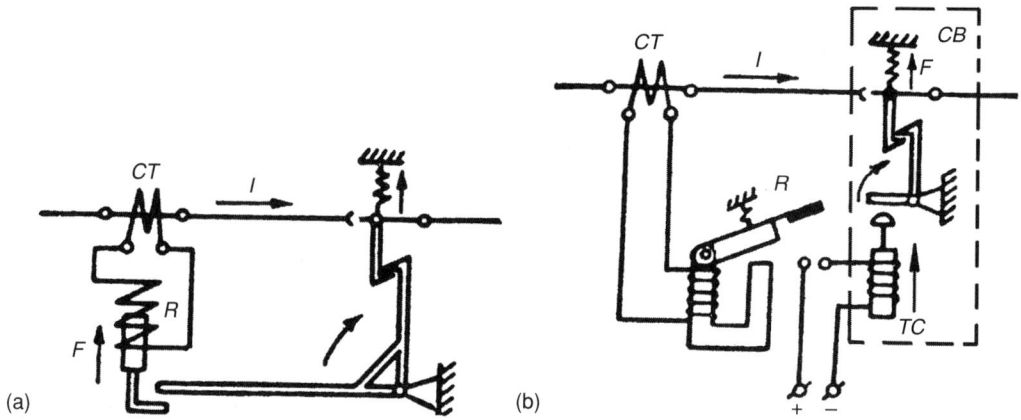

FIGURE 10.2
A schematic plot illustrating the principle of application of secondary relays, "direct-acting" (a) and "indirect-acting" (b). CT — current transformer.

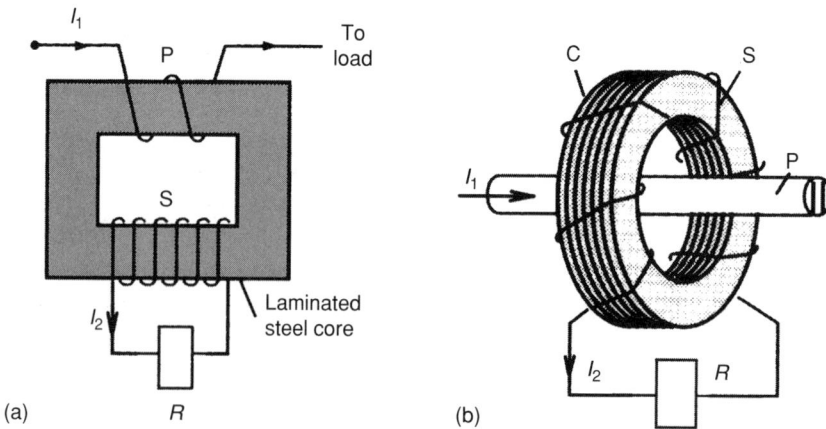

FIGURE 10.3
Principle of construction of CTs. P — primary winding; S — secondary winding; C — laminated steel core; R — relay.

FIGURE 10.4
JS-1 type CTs for internal installation of the 15 kV class, with insulation in the form of cloth tapes impregnated with asphalt-base insulating compound (General Electric Instruction GEC–333B 1955).

current values that most relays in the world are constructed today. The ratio of the primary current (voltage) to the secondary current (voltage) is called the transformer ratio. As a rule, in the CT there is a layer of insulating material between the windings and the core. Transformers designed for operation under high voltages are insulated with particular care. Earlier transformers for middle-class voltages (6 to 36 kV) used to be insulated with the help of special cloth tapes impregnated with a so-called asphalt-base insulating compound, after having been wound around (Figure 10.4).

As it can be seen from Figure 10.4, it was only the piece of bus bar used as a primary winding that was insulated. The magnetic core in such construction was not insulated. After assembly, the whole transformer together with the magnetic core was covered with black varnish (based on a similar asphalt-based insulating compound). Such transformers did not look aesthetically beautiful.

Modern CTs of such class are molded with a special high-quality epoxy resin after assemblage, at the factory. The molding process is carried out with vacuum and increased pressure cycling, thus providing full impregnation of all CT elements and preventing the formation of air bubbles inside both the transformer and the magnetic core. The process results in quite aesthetic solid constructions (Figure 10.5a).

Recently are more and more widely applied CTs with the split core (Figure 10.5b). Such transformers are convenient for mounting and dismounting. After installation of two half of core and connection among themselves, they strongly snuggle up to each other. Magnetic properties of such core do not differ almost from solid core. One kind of such CT is the CT with flexible core (Figure 10.5c).

The constructions of transformers designed for current control in high-voltage transmission lines (160, 400 kV and more) differ considerably from the ones considered above due to special internal and external elements providing the required insulation level between the windings, and also between the windings and the case (Figure 10.6) . The main insulation in such transformers consists of winding around of the primary turn by many layers of specific paper tape alternating with aluminum foil, and further filling of all free space with liquid transformer oil (Figure 10.7).

CTs and VTs are essential parts of relay protection systems in the power industry, where relays simply cannot be used without such transformers. Moreover, there are a number of peculiarities in the operation of such transformers, which require the relays to have specific characteristics. In this connection we have to consider in greater detail CT and VT.

FIGURE 10.5
Modern CTs. (a) bushing and toroidal type; (b) split core; (c) with flexible core

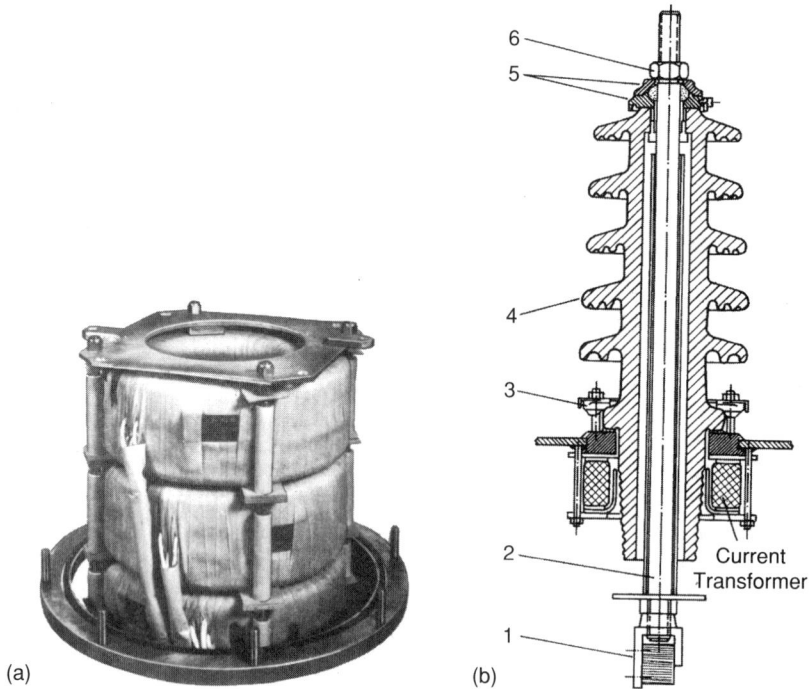

FIGURE 10.6
(a) Bushing CTs assemblies (three separate CTs). (b) Location of a single CT, or CT assemblies, on a power transformer's HV bushing. 1 — Lower connecting piece; 2 — undoing connection bolt; 3 — lower fastening; 4 — replacing porcelain; 5 — head armature; 6 — loosening nut.

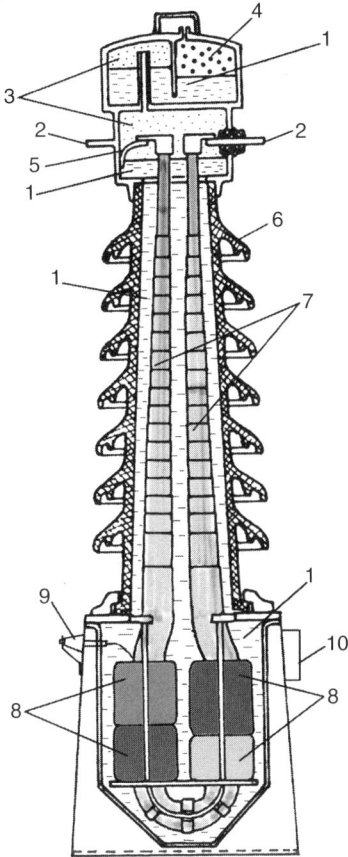

FIGURE 10.7

One of the most widespread constructions of a high-voltage CT (160 kV), with oil-impregnated paper insulation. 1 — Oil; 2 — copper plates for connecting to the controlled current circuit; 3 — volumes filled with nitrogen; 4 — air; 5 — jumper strap linking one of the outlets with the metal case of the upper reservoir; 6 — porcelain insulator; 7 — insulated primary conductor (primary turn); 8 — four independent secondary windings, each with its own magnetic core; 9 — insulator on the outer wall of the tank, with the outlet connected to the tank; 10 — terminal box with outlets of secondary windings.

First of all, the construction diagram in Figure 10.7 is not the only one that exists. Secondary windings with cores can be placed not only on the lower part of the transformer but also in the center, and even on the upper part (Figure 10.8). The traditional scheme, with a lower position of secondary windings, allows us to place the center of gravity of the construction very low, providing high resistance of the upper transformer to turnover. Upper position of the secondary windings, on the contrary, sharply reduces transformer resistance but allows us to simplify and reduce the price of construction of the high-voltage insulation. Availability of windings entirely separated from each other, supplied with separate magnetic cores (these are windings with different parameters and characteristics as a rule) means that in the case of a CT there are several (usually three to four) separate CTs, the primary windings of which are cut into the common controlled current circuit.

The "head" of the transformer, that is an upper metal reservoir, has a complex construction with many barriers, tubes, and valves. There is more than one purpose for this part of the transformer. First, it serves as a shield equalizing the electric field in the area of current outlets connected to the high-voltage line, which is why one of the outlets of the primary winding is connected with the internal (in some construction, external) bridge to the reservoir. Second, this reservoir serves as damper absorbing excess of oil from the main part of the transformer when the temperature increases. Third, this reservoir, with the help of the so-called water seal, prevents transformer oil contact with outdoor air containing moisture.

FIGURE 10.8
(a) Construction diagrams and (b) external design of high-voltage oil-filled CTs. In the photo: to the left — with upper; to the right — with lower position of windings.

The purpose of the aluminum foil layers (or a thin grid of aluminum foil, to be more exact) in the main insulation of the primary turn is even more interesting; at first sight it may seem that inclusion of metal layers to the insulator is unnatural.

You may be even more surprised to learn that the outer layer of this foil is linked with a special outlet placed on the outer surface of the transformer tank, and that this outlet is not a simple one. It passes through a small ceramic insulator (9) and is grounded with the help of a bridge to the tank (Figure 10.9). Why is this outlet insulated from the tank with the help of an additional insulator, and than connected by the bridge to the same tank? What a strange conglomeration of structural components absolutely absurd at first sight? In fact everything has its important purpose. Metal foil, for example, serves as an electrode of the capacitor, forcedly equalizing high-voltage distribution by the thickness of the insulating layer, and also as an electrostatic shield equalizing local electric field strengths in the transformer construction. This allows enhanced reliability and enhanced service life of the high-voltage insulation. The additional insulator on the tank, through which the conductor linking the foil with the grounded tank comes out, is used only for diagnostics of the high-voltage insulation state. As the foil inside the tank is fully insulated both from high potential and from ground potential, if one measures such insulation characteristics as resistance, capacity, polarization index, and the dielectric dissipation factor between the external outlet (9) of the foil, and the outlets connected to the high-voltage circuit, one can make conclusions about the state of some parts of the high-voltage insulation of the transformer (unfortunately, a more thorough consideration of these problems is beyond the limits of this book). The second part of the insulation is

FIGURE 10.9
Test outlet of an oil-filled CT of the 160 kV class, designed for
control of the high-voltage insulation state.

FIGURE 10.10
CT of the 420 kV class, with sulfur-hexafluoride (SF$_6$) insulation
(A Reyrolle & Co. Ltd).

checked between the outlet (9) and the grounded tank of the transformer. As these
measurements take place at relatively high voltages, the outlet (9) must be well insulated
from the tank. With measurements completed, the outlet (9) is again connected to the tank
with the help of the bridge.

In sealed switchgears with gas insulation, filled with sulfur-hexafluoride (SF$_6$), CTs of
appropriate construction, also filled with this gas, are used (Figure 10.10). Once the
necessity arises to use external CTs and VTs for voltages of 400 kV and more, cascade
constructions consisting of two to three units connected in series, are used. Each of these

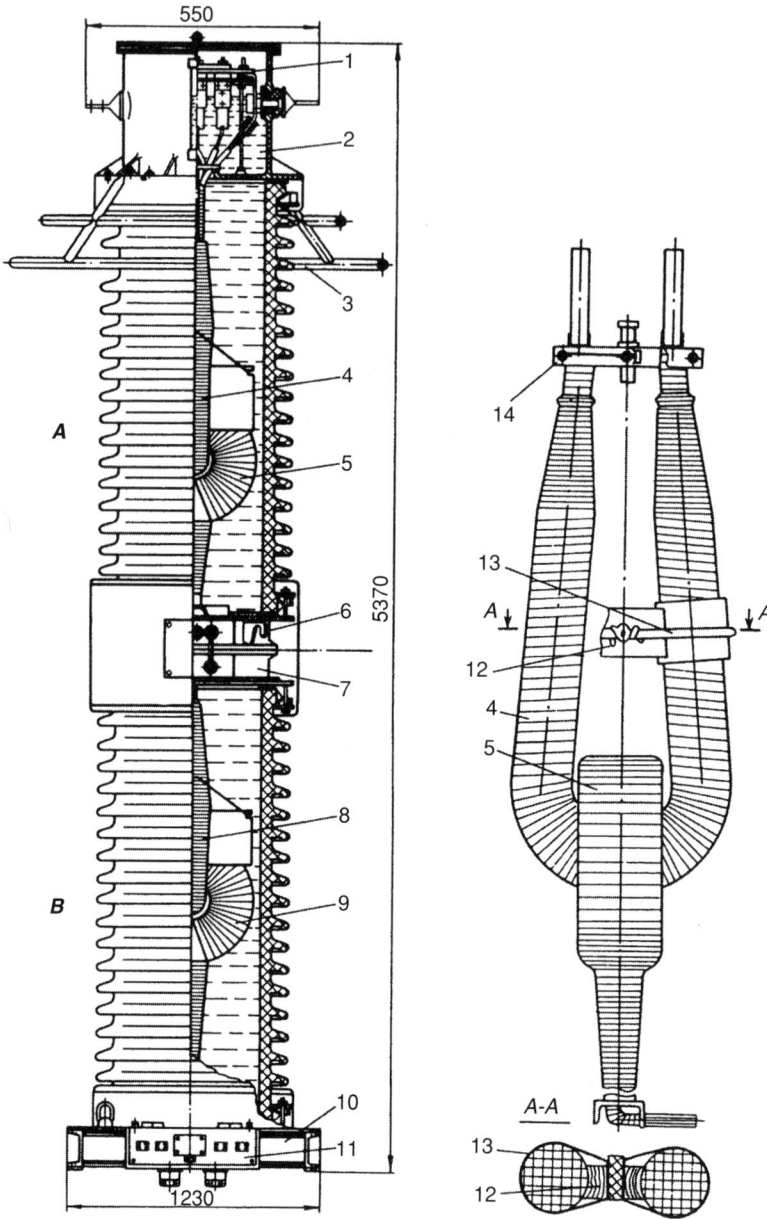

FIGURE 10.11
Cascade two-stage CT of the 500 kV class, of the TF3M500 type (Russia). 1 — Switch of turns number of the primary winding; 2 — oil-conservator; 3 — high-voltage shield; 4 and 8 — primary windings of units 5 and 9 — secondary windings of units; 6 — eye; 7 — base; 10 — heel piece; 11 — terminal box with outlets; 12 — block; 13 — fixing band; 14 — holder.

units is a self-contained transformer (Figure 10.11). An interesting peculiarity of the CT construction, shown in Figure 10.11, is a switch (1) also linking sections of the primary winding (four single-turn U-shaped sections) in different combinations, in series or parallel, providing values of nominal primary currents: 500, 1000, and 2000 A.

In a CT the primary winding consists of the first half-turn, with very low resistance, while in a VT the primary winding has many turns and possesses very high impedance

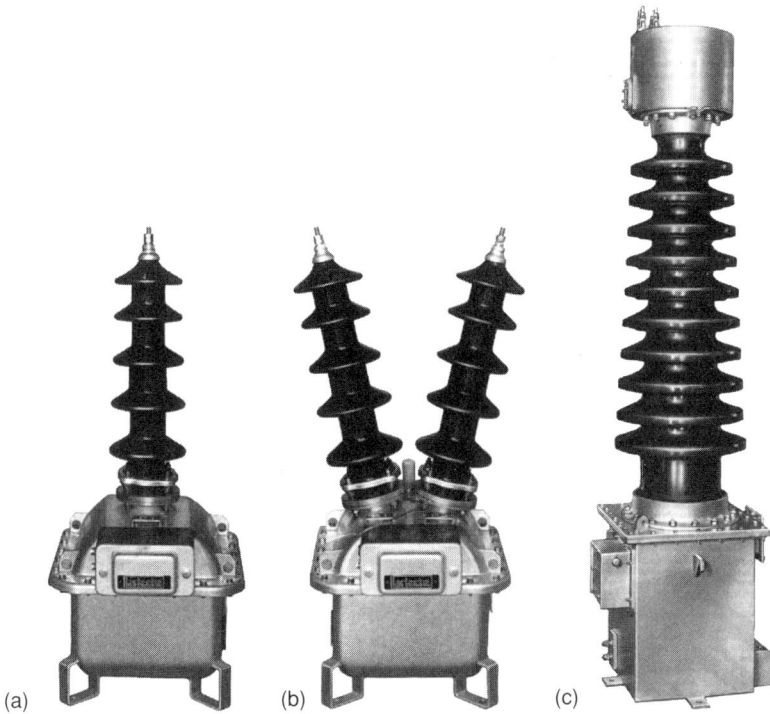

FIGURE 10.12
VTs of the 33 kV (a) and (b) and 145 kV (c) class.

because the total working voltage is applied to this winding. When the primary winding is switched to the so-called "phase" voltage, which is between the phase of high-voltage and the "ground" (the most often used connecting mode of a VT), only one outlet of the winding is supplied with a high-voltage outer insulator. Its second outlet is usually linked with the VT case (as a rule, with the help of an additional small insulator on the tank), which has similar construction and for some constructions (usually at voltages of no more than 36 kV) both transformer outlets may have high-voltage insulation (Figure 10.12b).

For very high voltages this so-called cascade construction is used. It consists of several transformers, connected in series and placed in the same case (Figure 10.13), and also of transformers of the capacitor type (Figure 10.14). The principle of operation of the latter differs greatly from all others. In fact one can hardly call them transformers. According to their principle of operation, these devices might belong to voltage dividers rather than to transformers (Figure 10.14b).

As can be seen in Figure 10.14a, a transformer for a voltage of 800 kV has quite an exotic design, due to two toroids, in the central part and at the upper point. These toroids are made of separate elements — as a rule aluminum ones, with a semicircular polished surface, and serve the purpose of equalizing the electric field and reducing its strength. Recently, capacitor-type transformers have also been produced for lower voltages (Figure 10.15).

Like in any voltage divider, in the capacitor-type VT (Figure 10.14b) there is a high-voltage arm (C_1)-on which the bulk of the high-voltage drop, and a low-voltage arm (C_2), from which the low level voltage appears. In fact, the high-voltage arm is formed not by one, but by several capacitors (a VT of 765 kV class contains 6 such capacitors), connected in series, which are easy to see in constructions of such transformers. In some VTs the

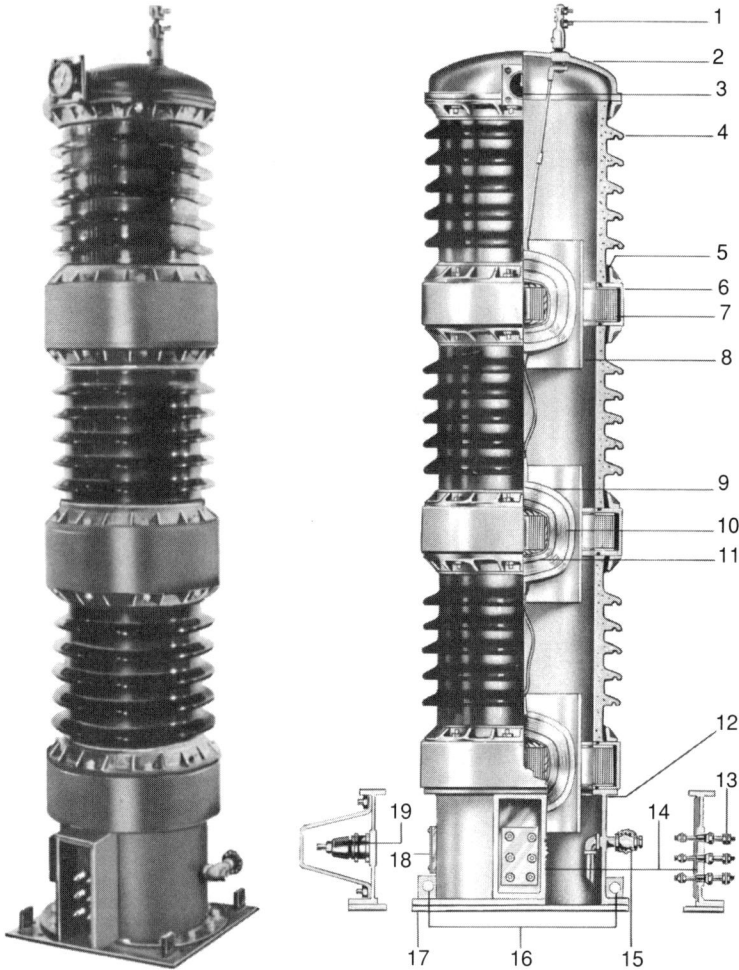

FIGURE 10.13

EU type oil immersed cascade VT (General Electric Co.). 1 — HV terminal; 2 — dome-shaped cover; 3 — liquid-level gage; 4 — porcelain shell; 5 — clamping ring; 6 — cylinder assembly; 7 — formed core; 8 — insulation barriers; 9 — coupling winding; 10 — primary winding; 11 — secondary and tertiary winding; 12 — base tank; 13 — secondary and tertiary terminals; 14 — terminal housing; 15 — drain valve; 16 — eye-bar for lifting a complete VT; 17 — bolter base plate; 18 — nameplate; 19 — neutral (ground) bushing. (General Electric Instruction GEH–1629H.)

low-voltage arm is supplied with a throttle (L) and a low-VT (T) providing the VT with characteristics similar to a standard coil VT. The capacitor-type transformers are much cheaper than the standard ones at high voltage levels, and are not prone to such VT "diseases" as "ferro-resonance," which usually leads to the "lethal outcome" of the VT and severe accidents in the circuit.

CTs and VTs of the opto-electronic type have been developed for a few decades in many countries (Figure 10.16). They are based on the application of Kerr and Pockelce electro-optic effects (for voltage measurements) and the Faraday magneto-optic effect (for current measurement). The Faraday effect can be found in the rotation of the polarization plane of linearly polarized light in optically active material, caused by an external magnetic field.

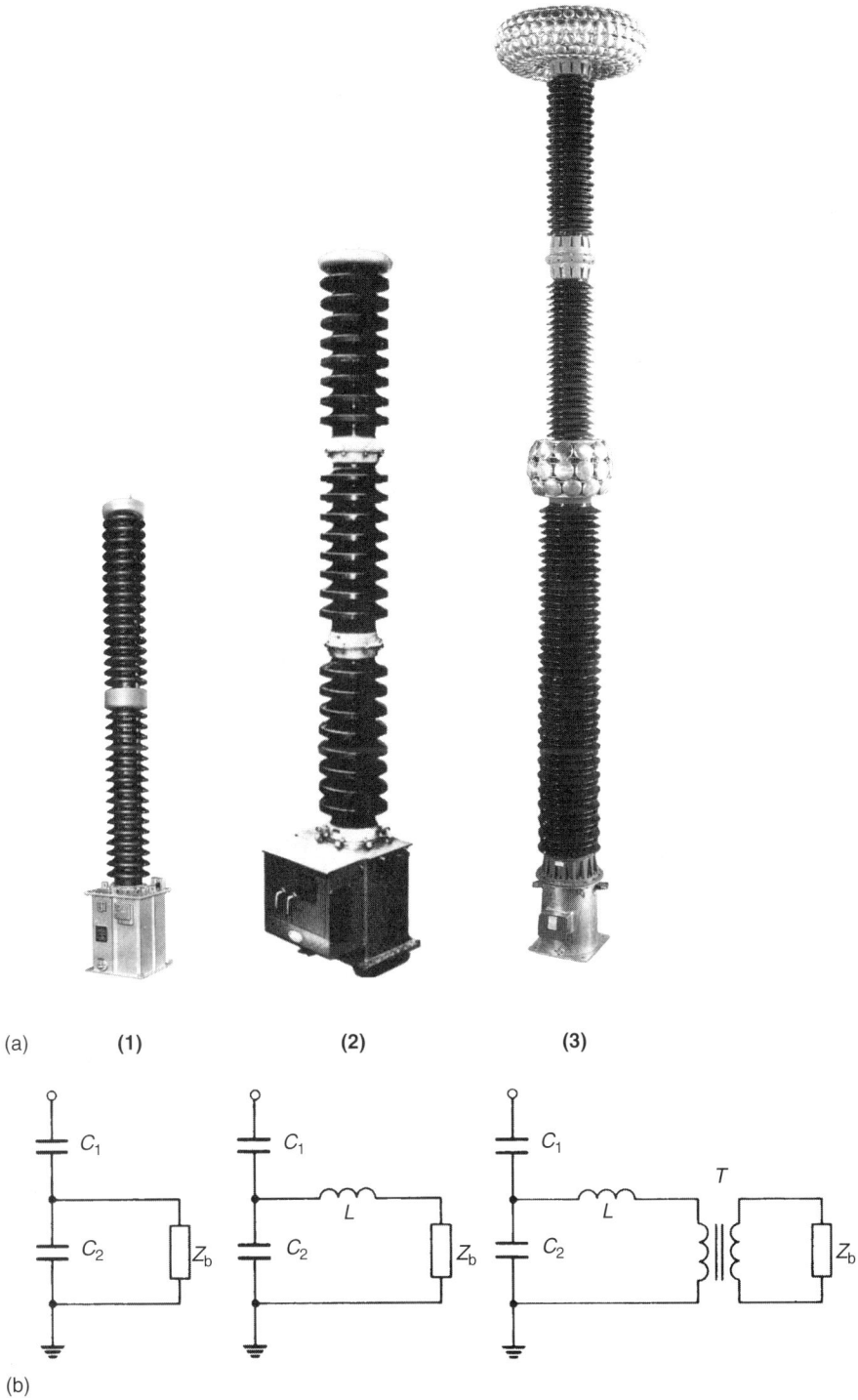

FIGURE 10.14
(a) High-voltage VTs of capacitor type of the 345, 362, and 800 kV class. (b) Principle of construction of VTs of the capacitor type.

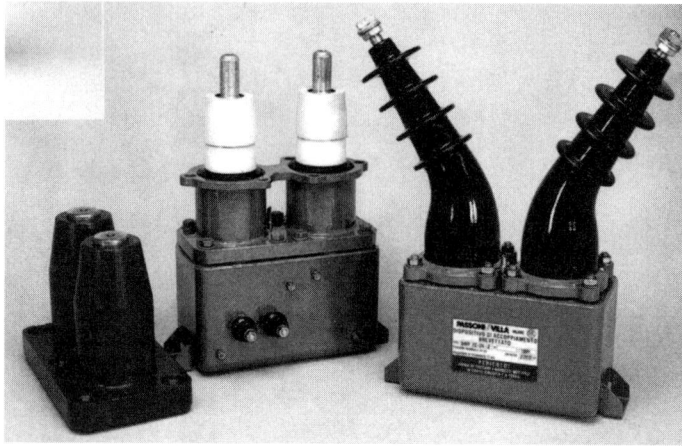

FIGURE 10.15
A capacitor-type VT, for voltage of 24 kV (Passoni Villa, Italia).

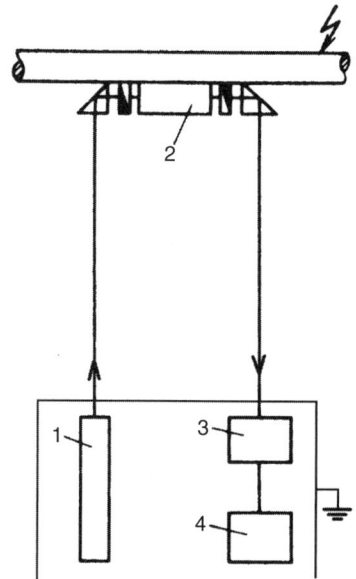

FIGURE 10.16
Principle of construction of an opto-electronic transformer. 1 — Source of polarized light; 2 — electro-optical transducer; 3 — light analyzer; 4 — photo-transducer and amplifier with an outlet final control element.

If the transducer is placed in the magnetic field of the measured current, one can determine the current strength by measuring the angle of rotation of the plane of light polarization. As a working substance in magneto-optical converters, glasses containing lead oxide (so-called flints, crowns) are usually used, and also fused quartz. Iron–garnet films are especially sensitive to magnetic fields. In this device (Figure 10.16) the polarized ray from the grounded source comes through an optical fiber (or through any other type of light-guiding fiber) to the Faraday cell (2), placed on the high potential. In this optical cell the light flux changes its polarization vector, depending on the value of the magnetic flux affecting it. At that point the light ray, modulated in such a way, returns to the ground potential, where it is converted to electric current and is amplified.

In VTs Kerr or Pockelce's cells are used instead of Faraday cells (Figure 10.17). The light flux in them is modulated not by the magnetic field but by the electric field in the active material placed between the electrodes, to which the measured voltage is applied. The

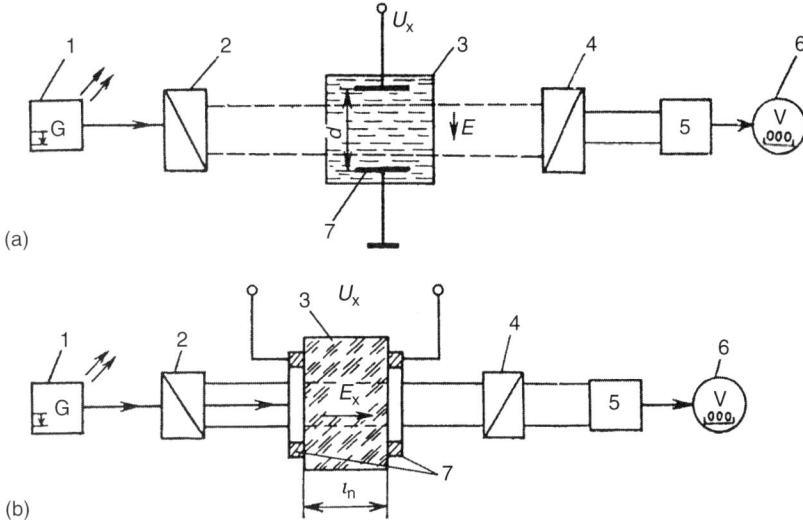

FIGURE 10.17
Electro-optical transducers of Kerr (a) and Pockelce (b). 1 — Light source; 2 — light polarizer; 3 — active material; 4 — polarization analyzer; 5 — photo detector; 6 — output element; 7 — electrodes to which the measured voltage is applied.

Kerr effect occurs in many isotropic substances (benzene, epoxy resin, etc.), very often nitrobenzene, which produces the biggest effect, is used. The Pockelce linear electro-optic effect can be observed in piezoelectric crystals placed into the electric field. This effect is seen most apparently in crystals of ammonium dihydro-phosphate ($NH_4H_2PO_4$) and potassium hydro phosphate (KH_2PO_4), in the longitudinal electric field created by ring electrodes (7), (Figure 10.17b). Pockels cells typically work with five to ten times lower voltages than the equivalent Kerr cell.

Such devices have been designed for the last 30 to 40 years already, but only recently have optical transformers appeared on the scene. The NxtPhase Optical Current Sensor replaces conventional CTs and brings a new level of accuracy to current sensing over a range of 1 A (rms) to 63 kA (rms), from 115 to 500 kV (Figure 10.18). The sensor is based on the Honeywell Fiber Optic Gyro system, which has a trusted reputation for accuracy and reliability in aeronautic and space industries for both commercial and military applications. The sensor can be column mounted on an advanced polymeric insulating column, or bus mounted with a suspension insulator to bring the optical fiber to ground.

The NxtPhase Optical Current Sensor consists of a specialized opto-electronic convert signal (1) from a light emitting diode into two linearly polarized signals, both sent through a polarization maintaining optical fiber to the sensing head. At the top of the column there is a circular polarizer (2) that converts the two linear-polarized light signals into right and left circular polarizations. The light signals (3) travel around the conductor many times. The magnetic field created by the current flowing in the conductor slows one signal and accelerates the other (the Faraday effect). As the circularly polarized signals complete their path around the conductor, they are reflected by a mirror (4) and travel back through the fiber, the direction of their polarizations having been reversed. Along this reverse path, the effect is doubled. Both signals make their way back to the circular polarizer, which converts them back into linearly polarized light beams. Light 6 travels back down the column to the opto-electronics (1). The difference in the speed of propagation has been translated into a phase shift between the linear signals. Since both signals

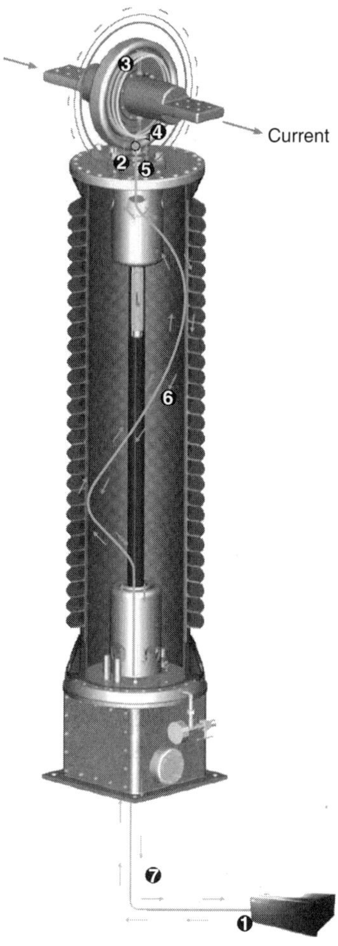

Current

FIGURE 10.18
Optical CT developed by the NxtPhase Corp. (NxtPhase Co. catalog 2005).

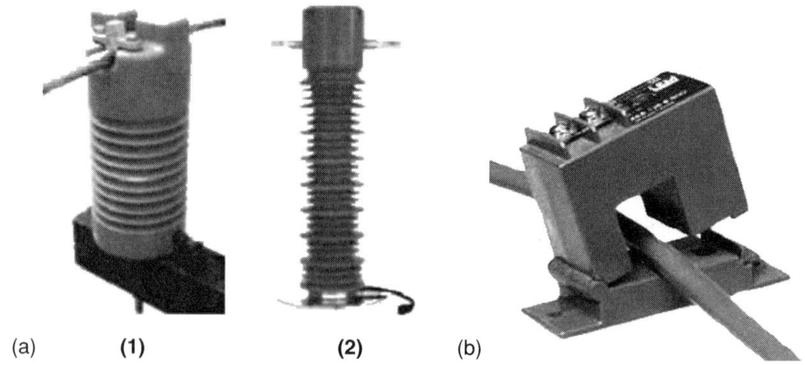

(a)   **(1)**   **(2)**   (b)

FIGURE 10.19
(a) Hybrid current-voltage sensors produced by Lindsey (U.S.A.). Insulating class: 15 to 35 kV (1) and 69 kV (2). Current ratio 600:5 A. (b) AKS series device — combine a current sensor and a relay in a single split-core package (LEM online catalog 2005).

have traveled an identical path, vibration and temperature changes have affected them equally — the highly accurate current measurement remains unaffected.

Hybrids of CTs and VTs are also produced in lots. In fact, CTs or VTs are never used alone in electric power stations or substations. They are always used together. Very often both the CTs and VTs are switched to the same high-voltage line. Each of them is equipped with expensive high-voltage insulation. It occurred to the engineers of the Lindsey Company (U.S.A.) to combine a current and a voltage sensor in one construction, so a hybrid of them appeared (Figure 10.19), which is much cheaper than two separate transformers and which uses less place than is necessary for two separate transformers. For the time being such hybrids do not have very high characteristics for current and voltage, but they are sufficient for many practical applications.

Other interesting hybrids are the combination of current sensor and protective relay in a single package (Figure 10.19b). LEM Components has introduced families of current monitoring relays that uniquely combine a current sensor and a relay in a single split-core package. By requiring fewer connections, the new devices offer increased reliability and up to a 50% reduction in installation time. This makes them particularly suitable for cabinet applications in process automation. The new product families are available for AC current (AKS series) and DC current (DKS series) from 1 to 200 A. They can be powered internally or externally. The output can be normally open or normally closed with a solid state or a dry contact relay compatible with AC or DC secondary circuits. The design's clamp-on function allows mounting on site without splitting conductors, reducing the number of actions (no disconnection of the cable) and cutting installation time still further thereby increasing safety.

The most important characteristic of VTs and CTs is their accuracy, which greatly depends on their load level, and for CTs also on the current ration in the primary circuit. As we know, other conditions being equal, current I in the circuit depends inversely on the load resistance R: $I = U/R$. But this means that current in the input circuit of the current relay will depend on resistance of the input circuit itself, which is on the parameters of the relay. And what shall one do if several relays of different types are connected to the same CT? What accuracy can we speak about in this case? Actually, current in the secondary circuit of the CT (unlike in all other transformers types) does not depend on the resistance of the load. And does the Ohm's law work? It does. Only not the law for a sub-circuit:

$$I = \frac{U}{R}$$

but rather this one, for a whole circuit

$$I = \frac{U}{R + r}$$

where R is the resistance of the load, and r is the internal resistance of the source, that is impedance of the secondary winding of the CT. Under the stipulation that $r \gg R$, the current in the circuit does not depend on the resistance of the load (relay coil).

In standard power transformers which are so-called "voltage sources," the resistance of the load is much higher than the internal resistance of the windings ($R \gg r$), which is why the load current is inversely proportional to its resistance. The CT operates in the mode of "current source" and differs from all other transformers by the fact that the resistance of its secondary circuit is higher than the resistance of the load, and that determines the current in the circuit. The secondary current depends only on the primary current and on the transformation ratio. In order for the CT to operate well in this mode, the load resistance cut in to the circuit of the secondary winding must be very small. The CT operates well even with fully shorted secondary winding, and vice versa; it "feels bad" if the load resistance gets high or the secondary circuit appears to be broken. In the latter case, the CT works as a step-up transformer with a large transformation ratio and the voltage level on terminals of the secondary winding may reach several thousands of volts (Figure 10.20). Such voltage is dangerous, and besides it may lead to damages of insulation of the low-voltage secondary winding. There have been cases of explosion of CTs caused by gases accumulated in the transformer because of long-term exposure to partial discharges at an open-circuit winding, which is why such mode of operation of the CT must be ruled out. In cases where, in multi-winding CTs, some windings are not used, they must be shorted by jumper straps. To protect the CT from spontaneous opening of secondary circuits the author suggested the use of simple electronic devices which short-circuit the secondary winding of the CT in cases of inadmissibly high voltage on it (see V. Gurevich, *Protection Devices and Systems for High-Voltage Applications*, 2003, Marcell Decker, NewYork).

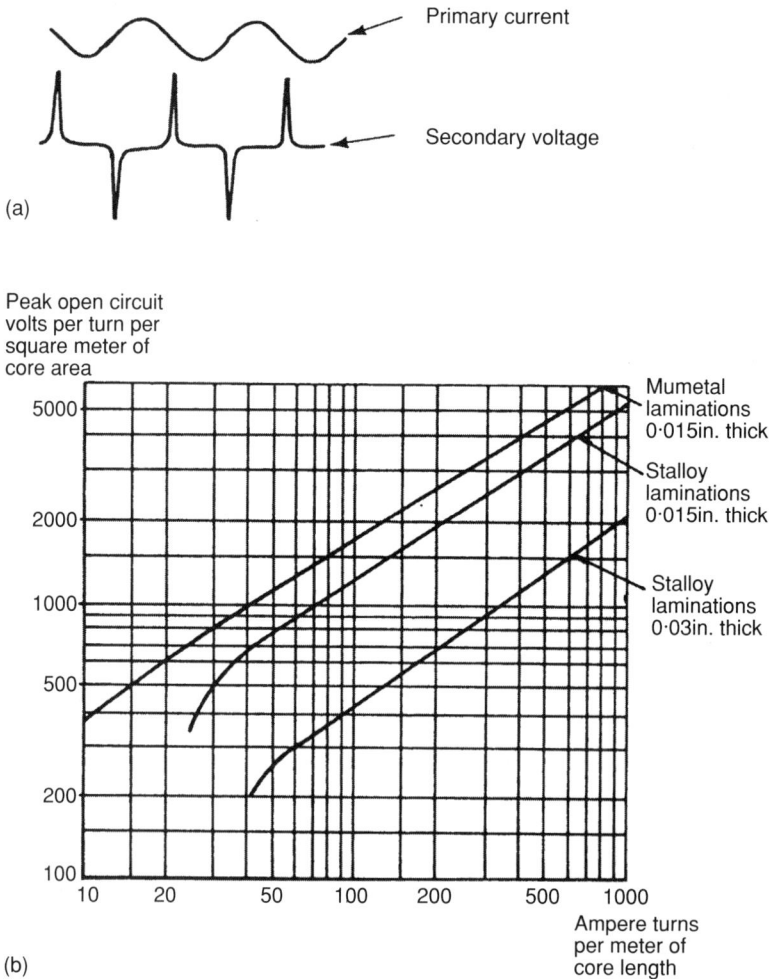

FIGURE 10.20
(a) The voltage shape and (b) level on the terminals of the secondary open-circuit winding of the CT and the voltage amplitude, depend on construction peculiarities and the current in the primary winding of the CT.

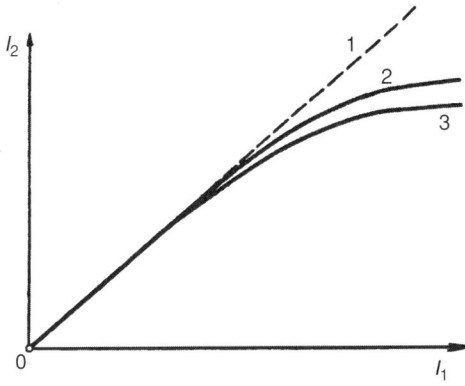

FIGURE 10.21
Dependence of the secondary current (I_2) on the primary one (I_1) in the CT. 1 — Ideal curve; 2 — actual curve with rate burden $Z_{2\ nom}$; 3 — actual curve for large burden $Z_2 > Z_{2\ nom}$.

As in any other technical device, losses also take place in CTs. Because of these losses not all primary current is transformed to the secondary circuit. Such losses may cause current error in the CT. In addition, current in the secondary circuit is slightly displaced in phase with regard to primary current, which causes angle error of the CT. Losses in CTs depend mostly on the state of their magnetic circuits. Until the magnetic core iron is saturated, there remains a directly proportional dependence between primary and secondary currents. When the primary current rises, the degree of saturation of the magnetic core iron increases as well, and the corresponding characteristic begins to turn down (Figure 10.21), and when the CT load increases, the degree of deviation of the characteristic increases as well (as the demagnetizing effect of the secondary current decreases).

To estimate the condition of the iron of the CT, its volt–ampere characteristic is read while gradually increasing alternating current is applied to the secondary winding, and the voltage on the terminals of this winding is measured. At that point it is compared with its manufacturing characteristic (Figure 10.22). One should bear in mind that these characteristics are obtained for simulated conditions of CT testing and do not reflect real correlations between currents and voltages in normal CT operation, but they do allow us to trace some faults in CT, which is why they are always read from the CT when new equipment is put into operation or at periodic tests.

CTs designed for measuring purposes work within the nominal currents on the straight-line portion of the characteristic, which is why they possess fine precision. Measuring CTs are produced for the following accuracy classes: 0.2, 0.5, 1, 3, 5 (with the class number corresponding to the error in percentage).

CTs used with protective relays work in emergency modes with currents considerably exceeding nominal ones, that is the curvilinear portion of the magnetization curve (Figure 10.23). That is why the indication of CT classes for relay protection also contains primary current ratio limits with respect to its nominal value when the indicated error is still to be taken into account. For example, the indication 5P30 means that the error for that given CT does not exceed 5% at primary currents exceeding the nominal value by 30 times.

Other conditions being equal, in order to provide the given error, the load power connected to the secondary circuit of the CT must not exceed the nominal power of the CT. With the nominal current, the load power will be determined by its resistance:

$$P = Z_2 \times I_2^2$$

where Z_2, load resistance; I_2, secondary current.

Voltage, V

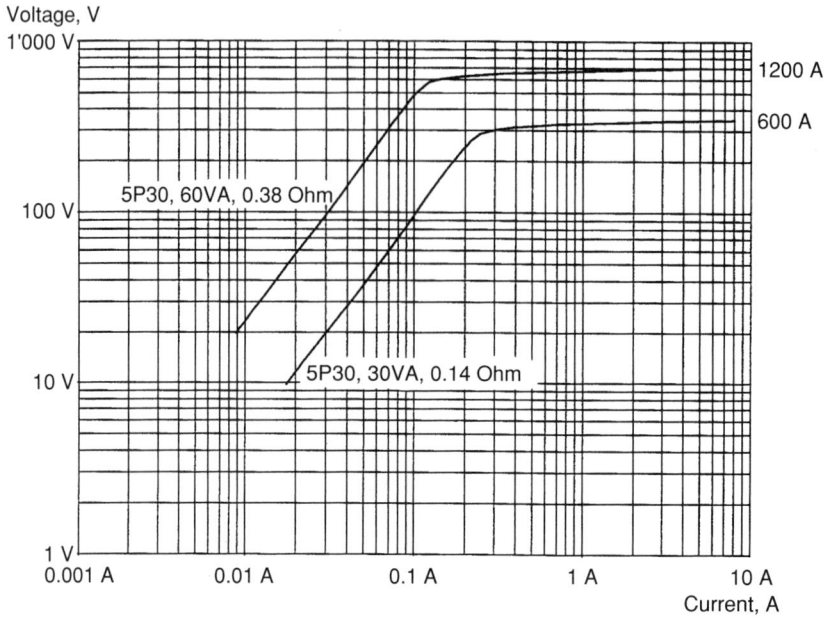

FIGURE 10.22
Actual volt–ampere characteristics of CTs with different transformation ratios are indicated in technical specifications.

That is why one can say that the lower the resistance of the external circuit connected to the CT, the less the loading degree of the CT and the less the error is. The character of the load affects the CT error considerably: an increase of the inductive load component leads to an increase of current error and a decrease of angle error.

There is a great variety of winding connections between the CT and the relay in three-phase networks. Some of them are shown in Figure 10.24. The scheme of the so-called "full star" (Figure 10.24a) responds to all types of short-circuits (between phases and one phase to ground) and is used in circuits with grounded neutral, in which short-circuit currents are possible even only in one phase. Simplified and cheaper circuits of the so-called "unfull star" (Figure 10.24b and c), are often used in electric circuits with an insulated neutral in which considerable short-circuit currents are possible only at a phase-to-phase fault, when considerable currents always flow in two phases. At any combination, fault of phases short-circuit current will flow through at least one CT. One can simplify the scheme even more by connecting the CT to the current difference between the two phases. In this case, it is enough to have just one current relay to protect a three-phase line.

On the scheme in Figure 10.24e, current in the circuit equals the vector sum of the secondary currents of the three phases. In the normal mode, this sum is about zero.

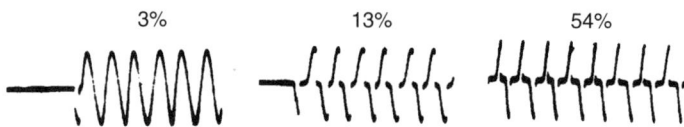

FIGURE 10.23
Forms of secondary currents in CT in emergency modes (overcurrents). The error of the CT is indicated in percentage points.

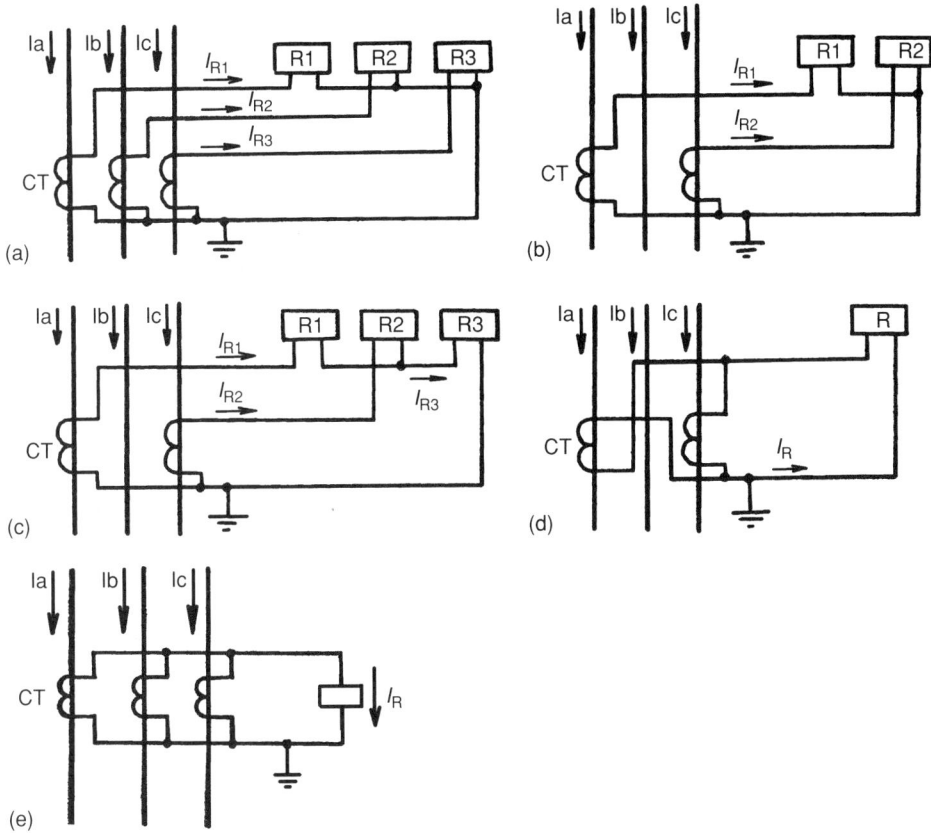

FIGURE 10.24
Some variants of connection of CTs and relays in three-phase circuits. CT — current transformers, R — current relays.

Current in the relay appears only when one or two phases are completed to the ground. Such a connection scheme is also called a "zero-sequence current filter."

In some rare cases, two CTs (with the same transformation coefficient) are used rather than one CT, on the same circuit (in one phase). Their secondary windings are connected with each other in series or parallel (Figure 10.25). In series connections, the current in the load does not change in comparison with the current from one CT (that is, the transformation coefficient does not change either), and in parallel connections it equals the sum of both CTs (the transformation coefficient of the circuit is half as much as the transformation coefficient of one CT). In series connections of the CT, the load is proportionally divided between all CTs, which is why such a connection allows the use of low-power transformers to supply the load, consuming much power. Parallel connections are used for loads requiring increased current, or to obtain nonstandard transformation coefficients.

The error of VTs is determined by two components: by no-load current and by load current. Both cases involve additional energy losses. In the former case, these are losses caused by the magnetization of iron of the magnetic core. In the latter case, losses in the winding copper are caused by load current passing. The error from no-load current is usually much less than the error from load current; nevertheless obligatory checks of VTs imply no-load current measurements because it characterizes the state of the iron and the winding of the transformer. Like in the CT, there are two error components in the VC: by

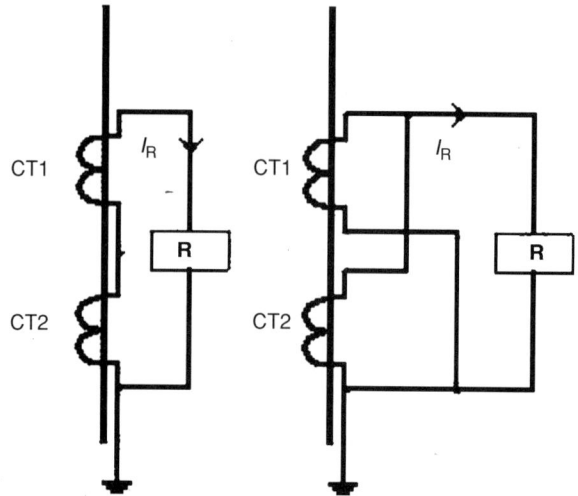

FIGURE 10.25
Series and parallel connections of CTs in the
same phase.

voltage and by angle (the phase displacement between primary and secondary voltage). The angle error much depends on the character of the load. If the load is active (cos $\varphi = 1$), the angle error is negative. When the load is inductive (cos $\varphi = 0.5$), the angle error turns to positive and increases linearly as the load increases. To lower, the VT voltage error for a nominal load, secondary voltage is artificially increased and one introduces a certain initial positive error maximum permissible during no-load operation. As the load increases, this initial positive correction is gradually compensated due to the negative error, and when it reaches the nominal load, the total resultant error turns out to be minimal. To unify and standardize the VT, it is usually designed for a secondary voltage of 100 or $100/\sqrt{3}$ V. As has already been mentioned, VTs are prone to so-called "ferro-resonance," which can disable the transformer.

Ferro-resonance is a nonlinear resonance phenomenon that can affect power networks. The abnormal rates of harmonics and transient or steady-state overvoltage and over-currents that it causes are often dangerous for electrical equipment. VT operation failures in distributing power circuits and in generator voltage circuits are quite common. On evidence derived from various sources, about 10% of installed VTs (for 6 to 36 kV) fail to operate annually in circuits with short circuit current to ground equal to 10 A. The parent cause of this phenomenon is the thermal destruction of the high-voltage winding of the VT by large currents, resulting in transformer core saturation and a drastic drop of its reactance (the main component of the impedance).

Usually the core saturation occurs during oscillation in the circuit, formed by circuit capacitance and transformer nonlinear inductivity. Such an oscillation process is initiated by unstable single-phase short-circuit to ground (Figure 10.26); partial phase operation of the VT itself results from the blow-out of fuses in high-voltage circuits; VT operation on "idle" buses; partial phase operation of the power transformer accompanied by over-voltages on the VT, etc.

The oscillations frequency may vary for different oscillation origins, parameters of the specific VT and specific circuits. In large VTs designed for 160 to 400 kV and more, the case in point is sub-harmonic frequencies 10, 12.5, 16.6, 25 Hz. In 6 to 10 kV distribution circuit VTs, the resonance frequency is due to attain 150 Hz.

It is worth mentioning that the processes occurring in the VT in these operation modes depends on a certain combination of the VT and the circuit parameters and their rating. Accounting for it is by no means easy (manufacturing variability of VT

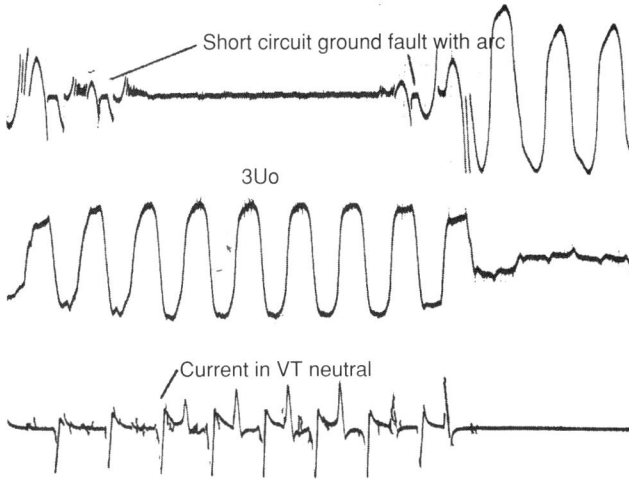

Short circuit ground fault with arc

3Uo

Current in VT neutral

FIGURE 10.26
Oscillograms of current and voltage in a
VT during a single line to ground short
circuit.

parameters; circuit parameters variation; undefined parameters in transient regimes that
caused the oscillation process, etc.). Therefore, different hardware is used to protect the
VT from those regimes that impede or disrupt resonance once it occurs. For example,
some manufacturers offer special ferro-resonance protecting devices for large VTs rated
for 160 to 400 kV. The protecting device has an analyzer of the current spectrum in the VT
circuit, which detects sub-harmonics emerging at low frequencies and subsequently
generates an activation command in parallel with the secondary VT winding special
throttle and active resistance of 0.3 to 0.7 Ω, which disrupts the oscillation process
(Figure 10.27). Unfortunately, this device may have designer's defects and may work
very unreliably.

For medium-size VTs rated for 6 to 24 kV, simpler means are used. For example, for
efficient protection against ferro-resonance of transformers with an ''open triangle'' type
winding, it is a good practice to connect a 5 to 150 Ω resistance in parallel to this winding.
Another common way of protection against ferro-resonance is to include a 3 to 5 kΩ
resistor to the VT neutral terminal.

The author has not encountered reports that analyze the influence of such protection
against ferro-resonance means on VT operation errors, but experience shows that designers ratings of VT loads (relay protection circuits, measurements, and registration of
energy consumption) take into account only the nominal power range of the VT for

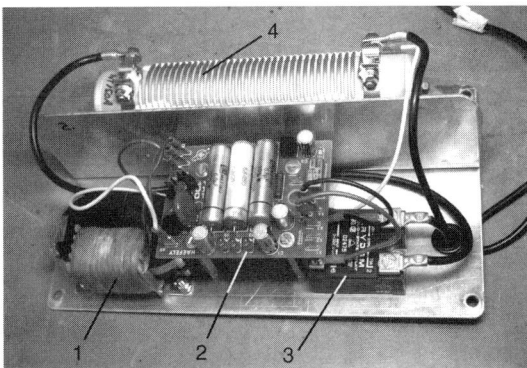

FIGURE 10.27
Device for VT protection against ferro-resonance
(Haefelly). 1 — Chock of low frequency filter
(16 Hz); 2 — electronic control board; 3 — solid-
state contactor; 4 — low resistance noninductance
resistor (0.5 Ω).

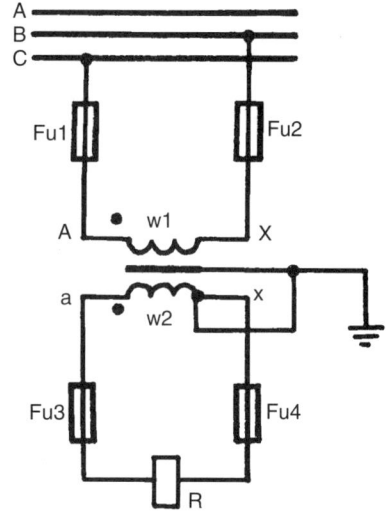

FIGURE 10.28
Connection scheme of a single VT of the 3 to 36 kV class for interphase voltage.

which its nominal accuracy level is preserved. However, some Western manufacturers supply VTs with antiferro-resonance resistors, which load the VTs to 60 to 80% of their nominal power. The result is overloaded voltage circuits; consequently VT operation error goes beyond the accuracy grade, and power consumption registration circuits powered by the VT result in heavy losses. There is no doubt that resolution of this problem will be provided by simple automatic devices, which connect resistors to the VT circuits only in response to ferro-resonance.

Some simple automatic devices for VT protection from ferro-resonance are suggested by the author in his books: *High-Voltage Automatic Devices with Reed Switch* (2001, Haifa) and *Protection Devices and Systems for High-Voltage Applications* (2003, Marcell Decker, New York).

VTs for voltage up to 36 kV can be connected to interphase voltage, or between the phase and the ground. As a rule they are protected by fuses (Figure 10.28). High-voltage fuses (Fu1 and Fu2) cannot protect the VT when the circuit of low-voltage load is over-loaded because of high internal resistance of the transformer. That is why additional fuses are installed in the low-voltage area. VTs of a higher class are connected phase-to-ground and do not contain fuses.

10.3   Instantaneous Current and Voltage Relays

The simplest and most widely used type of relay protection is the so-called "overcurrent cut-off" or "instantaneous current relay." As one can conclude from the name, a relay designed for such protection must pick-up immediately without a time delay (TD) when current overpasses the predetermined (by adjusting the relay) value.

10.3.1   Protective Relays of the Electromagnetic Type

Electromagnetic relays of such type energized when the predetermined current threshold is overpassed (so-called "current relays," which differ from standard electromagnetic

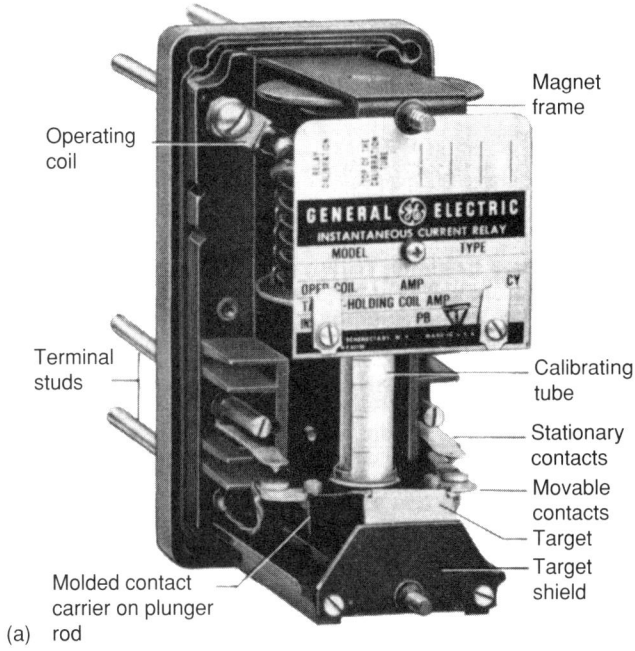

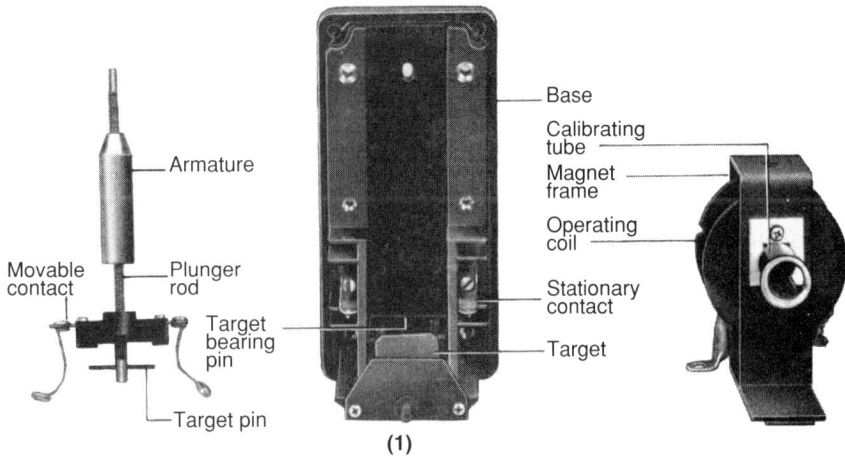

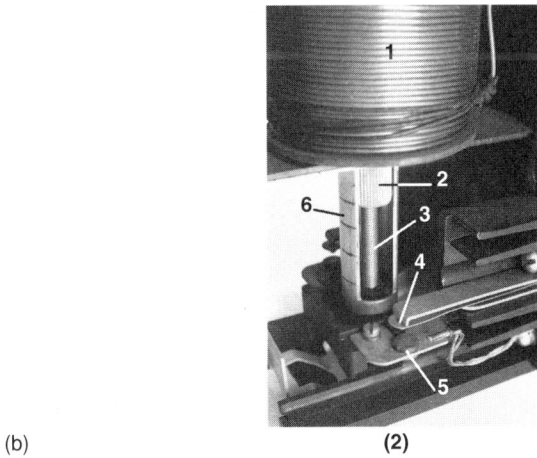

Operating coil

Magnet frame

GENERAL ⊕ ELECTRIC
INSTANTANEOUS CURRENT RELAY
MODEL   TYPE

Terminal studs

Calibrating tube

Stationary contacts

Movable contacts

Target

Molded contact carrier on plunger rod

Target shield

(a)

Armature

Movable contact

Plunger rod

Target bearing pin

Target pin

Base

Calibrating tube

Magnet frame

Operating coil

Stationary contact

Target

(1)

(b)

(2)

(*Continues*)

FIGURE 10.29 (*Continued*)
(a) Instantaneous current relay, PJC type (GE), with a retractable armature of the solenoid type. (b) Construction of elements and parts of the instantaneous current relay, PJC type. 1 — Operating current coil; 2 — knob-nut for adjustment position of armature in coil; 3 — plunger rod; 4 — fixed contact; 5 — movable contact; 6 — calibration scale. (c) Relay of PJC type with three independent units and three additional auxiliary relays with indicating flags. (General Electric Instruction GEH–1790A).

relays only by the winding containing a small number of turns wound around with thick wire, and by the unit of the adjusting pickup threshold).

Like in standard electromagnetic relays, in such current relays different types of magnetic systems described above are used. One of the simplest is a magnetic system of the solenoid type with a retractable armature (Figure 10.29). When current reaches a certain value, the armature is retracted to the coil and closes the contacts. When the current decreases, force of gravity makes the armature return to its initial position. The pick-up current is adjusted by revolving an adjusting nut (2) (Figure 10.29b). The initial position of the armature in the coil changes, and so does the threshold of the relay pick-up. The scale (6) is used for rough estimation of pick-up current. Relays of this type have been produced by the General Electric Company for tens of years, since the middle of the last century, and work successfully in thousands of electric power stations and substations around the world. The same principle is used to produce modifications of relays, with two and three independent relays mounted in the same case (Figure 10.29c), current relays with integrated auxiliary multi-contact relays.

Relays with three units, each of which is adjusted to a certain pick-up current, are used for selective three-stage protection of electric power lines. Every current protection device has a certain service area (work zone) on the protected line. This is caused by the fact that the wire of the power line has a certain impedance which considerably limits short-circuit current, if the point of fault is tens of kilometers of distance from the place of the current relay installation. In cases of such remote short circuits, the protective relay may not detect the fault (Figure 10.30). In the given example, when the short circuit occurs at the beginning of the line (point 1), current of about 1600 A flows through the relay (we mean both the relay and the CT). When the short circuit occurs at point 2, the current passing through the relay decreases to 900 A, and at the remote short circuit at the point 3, this current (less than 600 A) approximates the pick-up threshold of the relay (500 A). This relay will not respond at all to short circuits at points 4 and 5 because these faults are out of relay working zone. But how do we provide protection for a power line outside the working zone? The solution is, together with a relay providing instantaneous (INS)

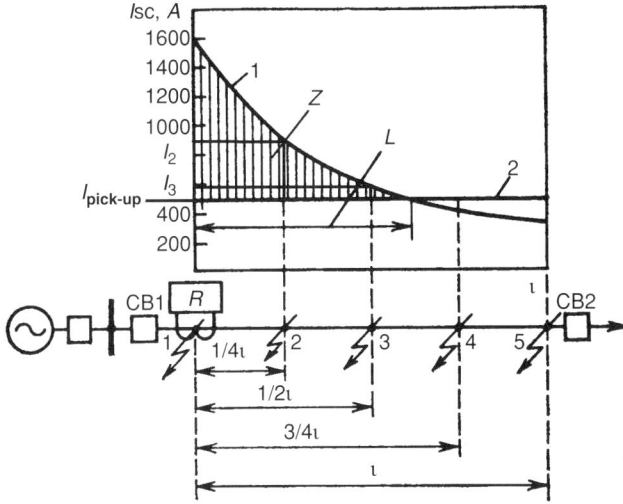

FIGURE 10.30
Diagram explaining the principle of operation of the overcurrent instantaneous (cut-off) relay on a portion of a long power line. I_{SC} — Short-circuit current at the point of the installation of the relay R; I_{PICKUP} — pick-up current of the relay; L — maximum area of the protected zone Z; CB — circuit breaker; I_2 — current passing through the relay when short circuit occurs at point 2; I_3 — current passing through the relay when short circuit occurs at point 3; 1 — curve of short-circuit variation along the line; 2 — zone of insensitivity of the relay.

current cut-off, an additional current relay with a TD is installed at the same place (Figure 10.31).

When short circuit occurs both protections installed in the substation (1) start to work in the cut-off, both the instantaneous cut-off and the TD. Obviously, the instantaneous cut-off will pick up first and disable the whole line with the help of switch CB1.

The TD relay will not pick up because of its TD, but if short-circuit current is not enough for the instantaneous cut-off pick-up (that is if the short circuit is out of its working zone), it is only the TD relay that will pick up because it is adjusted to a smaller current trip than the instantaneous relay current cut-off (that is, it has a larger zone of operation). First, it is necessary for protection of that part of line (L1), which remains out of the zone of operation of the cut-off relay INS (the TD Zone), and secondly it is needed for backup protection of the remote part of the line when the relay or the high-voltage circuit breaker malfunctions on other portions of the line. For example, if switch CB2 is not switched off for some reason, when a short circuit occurs on the part of the line L2, the

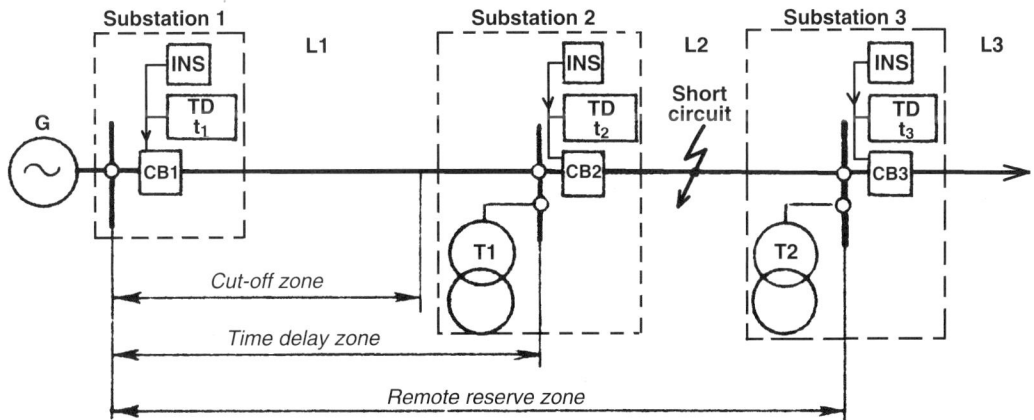

FIGURE 10.31
A composite single-line diagram of overcurrent protection for high-voltage transmission line. G — power source (power station); CB — circuit breakers; INS — instantaneous (cut-off) current relays; TD — time delay current relays; T — power transformers; L — parts of transmission line.

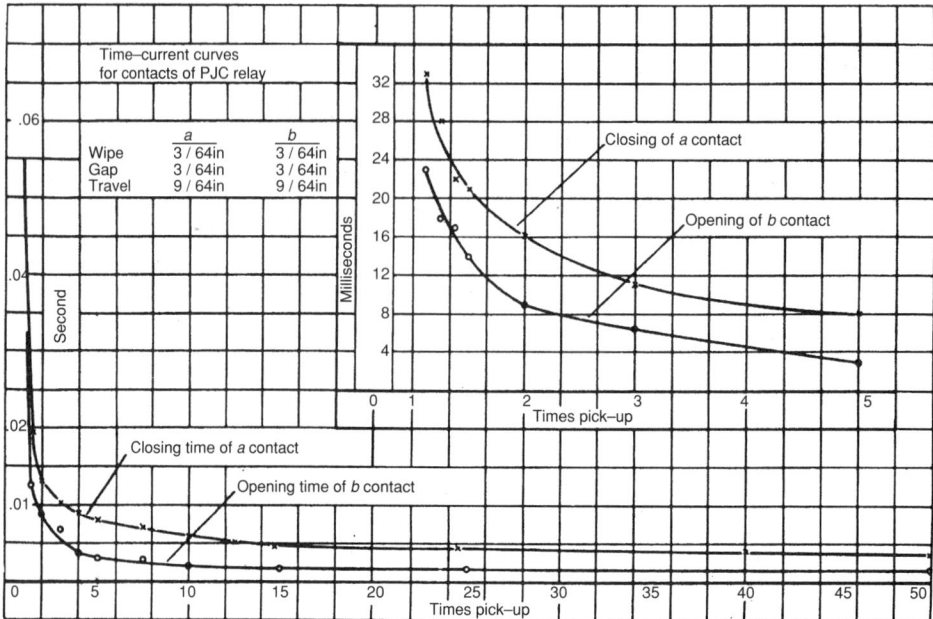

Time–current curves
for contacts of PJC relay

	a	b
Wipe	3 / 64in	3 / 64in
Gap	3 / 64in	3 / 64in
Travel	9 / 64in	9 / 64in

.06

.04

.02

.01

Second

Milliseconds

32
28
24
20
16
12
8
4
0

Closing of *a* contact

Opening of *b* contact

Times pick–up

0 1 2 3 4 5

Closing time of *a* contact

Opening time of *b* contact

0   5   10   15   20   25   30   35   40   45   50

Times pick–up

FIGURE 10.32
Time–current characteristic of the instantaneous current relay PJC. (General Electric Instruction GEH–1790A.)

second stage of the TD of the TD relay, installed in the first substation, will pick up and will disable the whole line with help of switch CB1. However, to prevent switching off of CB1 before CB2, and CB2 before CB3 in the normal mode, selectivity of operation protection must be provided.

The general principle here is as follows: the farther the protection from the power supply is, the smaller pick-up current it must be designed for, and the more TD it must have ($t_1 < t_2 < t_3$, Figure 10.31). The condition of selectivity of operation protection that must be provided is that the portion of the line closest to the fault from the power supply, is the portion that must be disabled. If the short-circuit current still does not disappear, the second portion, which is more distanced from the place of fault and closer to the supply, etc. Such an approach allows us to provide maximum survivability of the line, and maximum resistance of the line to faults.

Despite the term "instantaneous current" relay, it does have its own nonzero pick-up time, like any other relay. Moreover, the pick-up time of this relay depends on the current value in the winding. This should not surprise you if you recall the fact that the pick-up time of any electromagnetic relay decreases when the ratio of the voltage applied to the winding (or of the flowing current) with regard to its pick-up voltage (current) increases. Taking into account that a current relay works with high ratios of current in the winding, one can easily guess what kind of time–current characteristic it will have (Figure 10.32). It is worth mentioning that the construction of a relay with an armature of the plunger type can be used not only as a current relay, but also as a voltage relay — with winding, of course, which is wound around with a great number of turns of thin section. GE produced the so-called Ground Detector PJC type for the detection of grounds on ungrounded AC generator field circuits.

This relay consists of two plunger-type units (such as shown in Figure 10.29), a transformer, and a diode bridge, all in one case. The relay is operated from a grounded voltage source connected through one plunger-unit coil to the machine field circuit

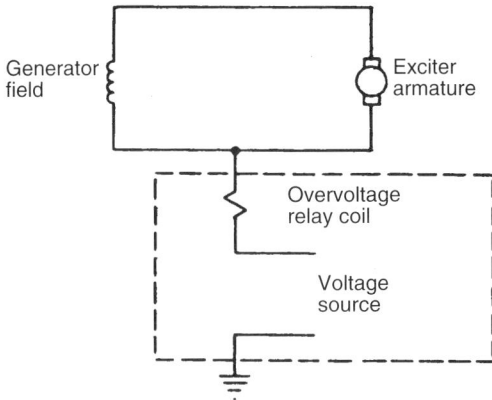

FIGURE 10.33
Typical external connections PJC relay, for detection of grounds on ungrounded AC generator field circuits. (General Electric Instruction GEH–1790A.)

(Figure 10.33). When a ground on the normally ungrounded field completes the circuit, the unit picks up.

The AC supply voltage is transformed, rectified, and filtered, producing a DC operating voltage for the relay with a ripple of one-half volt or less. This is applied between the ungrounded field and the ground. An indicating light on the front of the relay shows that the DC operating voltage is available.

One of the plunger-type units picks up if a ground develops on the field circuit. The other removes the operating unit from the grounded field and closes a contact for tripping or alarm. Reset is either by hand or electrically, through the test–reset switch. An AC machine field circuit is usually operated ungrounded, and a single ground does not damage the machine; however, a second ground can cause considerable damage, so protective equipment is therefore recommended to detect the first ground. The PJC type relay (Figure 10.33) serves this function. It can be used either to sound an alarm or to remove the load from the machine. The PJC type relay may be used with machine fields rated 375 V or less. It should not be applied where the exciter reverse voltage can rise above 500 V.

Voltage relays with a so-called "floating core" (a light core vertically "floating" in the magnetic field of quite a large coil) were investigated by Igor Gurevich (Kharkov, former U.S.S.R.) in the 1970s (Figure 10.34). He designed and constructed a lot of voltage relays

FIGURE 10.34
Igor Gurevich (1919–2003), designer of voltage relays with a floating core.

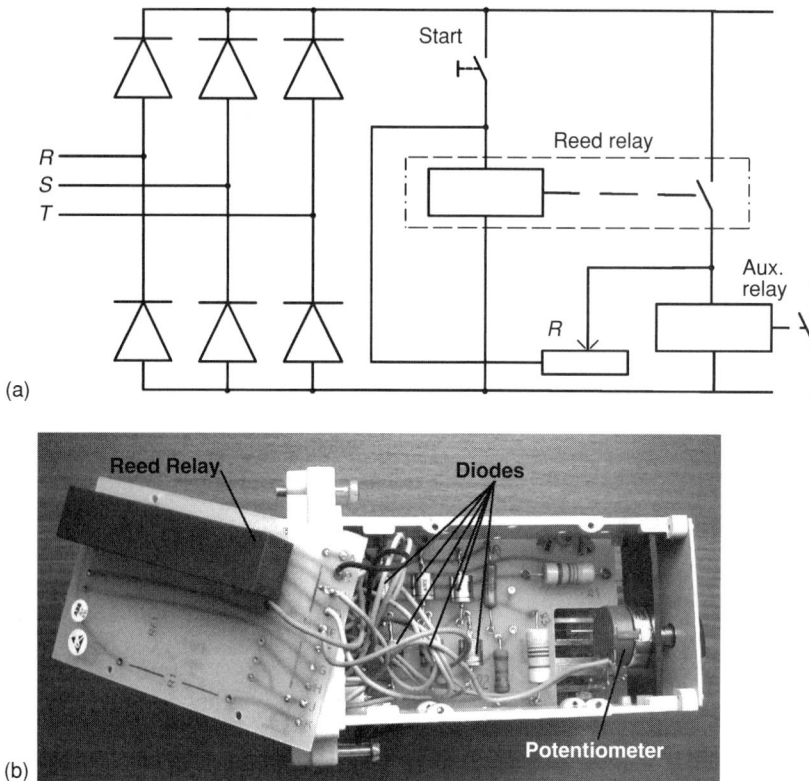

(a)

(b)

FIGURE 10.35
Simplified circuit diagram (a) and internal design (b) of the RXOTB-23 undervoltage relay.

based on the similar principle, with different types of the contact systems: from a standard microswitch to a mercury contact, a reed switch, and an optical photosensor. Such relays were used in different automatic control systems; automatic control with voltage level control, in which one such device was used both as a relay of overvoltage and a relay of undervoltage.

The RXOTB-23-type relay (ABB) has a very original principle and a simple design. This is a three-phase under-voltage relay that operates for symmetrical or asymmetrical voltage drops, or for phase failures (interruption in the AC supply). The RXOTB-23 is used, among other uses, as protection for control equipments and thyristor converters.

Three-phase input voltage is rectified in a six-pulse bridge having avalanche diodes, and is then supplied to a measuring dry-reed relay via a voltage divider having a variable resistor. The operating value is set with the aid of the resistor to between 50 and 100% of the rated voltage, and when any of the incoming voltages drop below the set value, the dry-reed relay disengages and interrupts the supply to the output auxiliary relay (Figure 10.35). After activation of the reed relay by the "start" push-button (and release "start" push-button, of course) its coil appears under the lowered voltage because of a voltage drop on the potentiometer. In this condition the slightest voltage reduction will be enough for the release of the reed relay. The level of release voltage is adjusted by the potentiometer.

Relays with a turning armature (Figure 10.36) were also widely used as instantaneous current and voltage relays. A Z-shaped turning armature was used in the ET-520-type relay, a predecessor of the RT-40 relay.

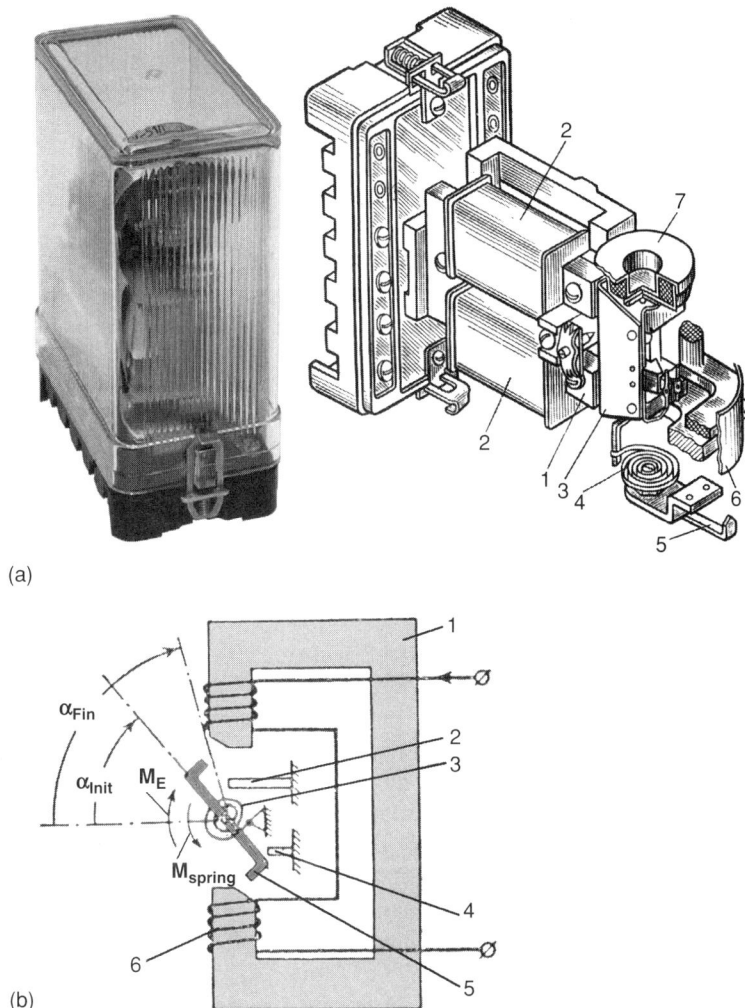

(a)

(b)

FIGURE 10.36

(a) A PT-40 type instantaneous current relay with a turning armature ($158 \times 130 \times 86\,$mm), produced in the former U.S.S.R., and then in Russia, for the past 50 years. The first modification of this relay was called ET-520. 1 – Poles of the II-shaped magnetic core; 2 — current coil; 3 — turning Γ-shaped armature; 4 — spiral spring; 5 — lever adjusting the spring tension (adjustment of pick-up current); 6 — scale; 7 — brake drum with sand. (b) Construction if the magnetic system of the ET-520 type current relay (Russia). 1 — Magnetic core; 2 and 4 — stops; 3 — spiral spring; 5 — Z-shaped turning armature; 6 — current coils. α_{Init} and α_{Fin} — initial and final angle of armature turning, respectively; M_E and M_{Spring} — electromagnetic moment of rotation and counteractive spring moment, respectively.

This old relay had lower pick-up power than the PT-40-type relay (and therefore created a smaller load for the CT). The turning armature of both relays is constructed in a particular way, which provides increased reset ratio.

Use of a magnetic system of this type in measuring current and voltage relays is not a Russian invention. A current relay with a magnetic system similar to the famous one produced by Siemens was already described by Manfred Schleieher in his book: *Die moderne Selektivschutztechnik und die Methoden sur Fehlerortung in Hoschspannungsanlagen*, Berlin, Verlag von Julius Springer, 1936. That magnetic system was most likely invented long before 1936 if one takes into account the fact that the first protective relays were developed by the industry already at the beginning of the 20th century.

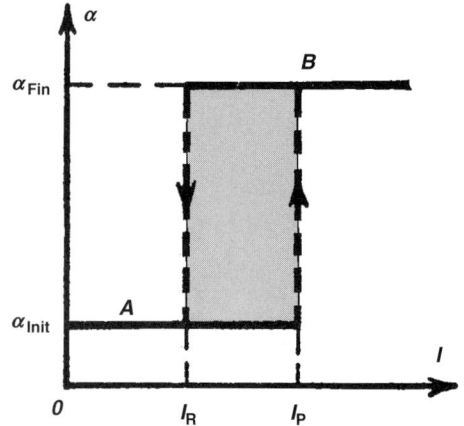

FIGURE 10.37
Typical characteristic of relay devices. A and B — extreme steady positions of the relay; α_{Init} and α_{Fin} — initial and final armature angle; I_P and I_R — pick-up and release parameter (current).

When we considered magnetic systems of standard (not measuring) electromagnetic relays, we mentioned that to increase the pick-up reliability (accurateness) of such a relay, an input value (of current or voltage) exceeding by 1.2 to 1.8 times the pick-up value of the relay must be applied to its winding. This excess is called the pick-up safety factor. Measuring relays adjusted for a certain level of input value must provide accurate and reliable pick-up at a gradual increase of the input value, and when it reaches the established level without any safety factor. Theoretically any relay by definition can take up only extremely steady positions, A and B (Figure 10.37). Here is how it happens: current in the relay winding creates a magnetic field in the core and in the working gap between the armature and the core. The magnetic field in the gap creates electromagnetic torque, affecting the movable armature in the direction in which the moving armature will reduce the air gap.

When the current in the winding reaches a value equalling pick-up current, the electromagnetic torque M_E of the relay will reach the nominal value M_{Enom} (Figure 10.38), and the armature will begin to move. As can be seen in Figure 10.38, in the process of moving of the armature the electromagnetic torque M_{ETrip} spontaneously

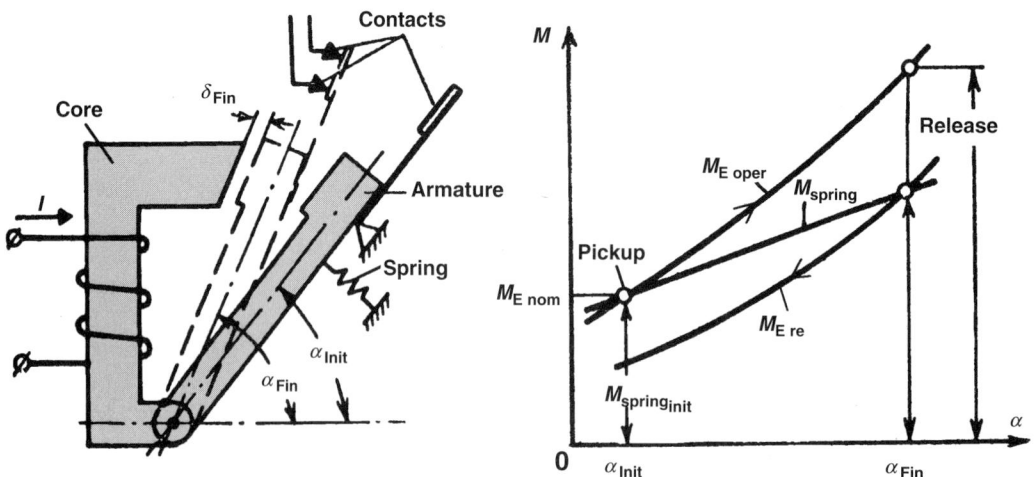

FIGURE 10.38
Pick-up process of an electromagnetic relay of the clapper type.

increases, while the current value in the coil remains the same. This happens because changing of the armature position (an increase of angle α) leads to a reduction of the air gap δ in the magnetic system (that is, reduction of the magnetic resistance of the circuit and an increase of the magnetic flux), so we have a circuit with positive feedback. As is well known such systems, when out of balance because of slight changes of the input value, cannot stop until they reach the new steady state (electronic circuits of this kind are called triggers).

Changing of the electromagnetic torque of the relay in the process of changing of the armature angle (or of the air gap in the magnetic system) is called the tractive effort characteristic of the relay. Changing of the counter torque of the spring in the process of changing of the armature angle (or of the air gap in the magnetic system) is called the mechanical (or load, or force, or displacement) characteristic of the relay.

It is clear that the relay will be in an ON state until its tractive curve will be higher than the load one. The intersection of these characteristics at a common point, means balance is achieved at that point, that is hovering of the relay in the intermediate position.

As can be seen in Figure 10.38, such balance is possible only at the initial point, when current has not yet started flowing in the winding and there is still no armature movement. As soon as $M_{\text{E oper}}$, is greater than $M_{\text{spring init}}$, the armature pickups and begins to move and these characteristics do not intersect anywhere. When current in the relay winding is reduced to the initial pick-up current value, that is when $M_{\text{E oper}}$ is reduced to $M_{\text{E re}}$, the tractive effort curve of the relay (now $M_{\text{E re}}$) is lower than the load curve M_{spring}, and the armature returns to its initial position.

The air gap in the magnetic system increases, the magnetic flux is automatically reduced (at constant value of current in the winding) and the relay quickly returns to its initial state.

It is obvious that the quality of the pick-up process of the relay and the release ratio (the width of the loop in Figure 10.39) is determined by the coordination of these characteristics. Throughout the long history of the development of relays there have been many attempts to create constructions with numerous springs and additional ferromagnetic elements, in an attempt to combine relay characteristics in the best possible way.

However, all of these previous attempts turned out to be too complicated and unreliable. Only traditional clapper-type relays, and relays of the solenoid type remain in practical use, although these too are still by no means the best solution, if one takes into account coordination of characteristics. In this sense more complex relays with a Z-shaped turning armature, possessing specific relay characteristics, appeared to be of a

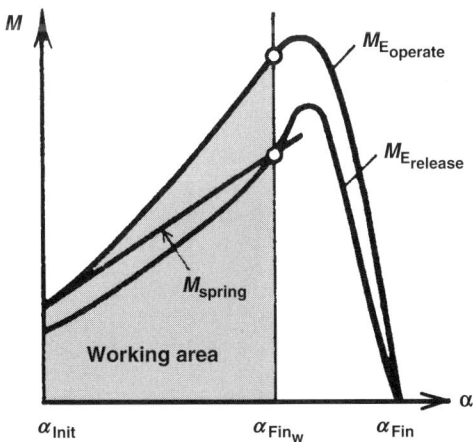

FIGURE 10.39
Characteristics of a relay with a magnetic system with a Z-shaped turning armature.

much higher quality (Figure 10.39). In relays of the clapper and solenoid types the air gap in the final position of the armature is minimal and the electromagnetic torque is maximal when the relay has picked up. This means that for return (switching OFF) of the relay, a considerable reduction of current in the coil is required. In contrast, in relays with a Z-shaped turning armature, there is no such sharp intensification of the electromagnetic moment when the air gap is reduced, because when the armature turns, the electromagnetic torque arm is reduced simultaneously. In the extreme armature position ($\alpha_{Fin} = 90°$, Figure 10.36b) when the electromagnetic force developed by the coil is maximal, the electromagnetic torque decreases to zero (since the entire electromagnetic force is applied along the armature and the arm of force equals zero) and there is no value that can make it turn. Of course, the working area of the relay does not include this extreme armature position ($\alpha_{Fin} = 90°$, Figure 10.36b). It is limited by the position $\alpha_{FinW} < 90°$ because when $\alpha_{Fin} = 90°$ the armature cannot create contact pressure. Nevertheless, the electromagnetic moment in the switched-ON position of this relay appears to be less than in relays of other types. This means that for reset of this magnetic system a slight reduction of current in the coil is required; in other words this magnetic system has a very high release ratio (a narrow loop on the characteristic, Figure 10.37).

The PT-40-type relay with a half-Z-shaped armature has similar positive properties, which is why it has been applied in industry and in the electric power industry for such a long time. In the PT-40-type relay, there is also an additional component on the common axle with a turning armature: a brake drum (7) (Figure 10.36c) with radial partitions filled with dry quartz sand. This component is a damper, reducing sharp acceleration of the movable system, vibration of contacts when they collide, and vibration of the armature caused by the alternating magnetic field. Vibration is extinguished by friction between the grains of sand.

Two relay windings can be connected to each other in series or parallel, thus allowing a change of the pick-up threshold of the relay by two times. The PT-40-type relay is produced for nominal currents from 0.2 to 200 A. The pick-up time of the relay does not exceed 0.1 sec at $1.2I_N$ current and 0.03 sec at $3I_N$ current. For application at currents protractedly exceeding pick-up current (up to $30I_N$), the relay must be supplied with an integrated saturating transformer, a diode bridge rectifier. This modification of the relay is called PT-40/1D.

As in the cases considered above PT-40 current relays also serve as the base for voltage relays (PH-51, PH-53), differing from PT-40 relays only by the winding. Separate coils in this relay allow creating on the basis of this construction a relay picking up from the difference of magnetic fluxes created by the currents in the relay coils. In the PH-55-type

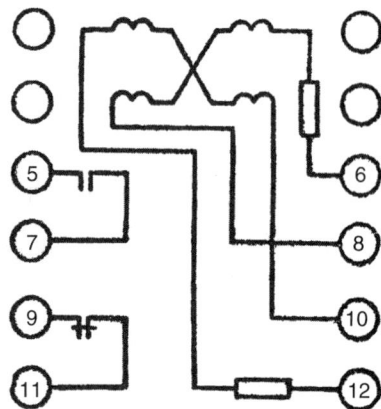

FIGURE 10.40
Circuit diagram of PH-55-type relay.

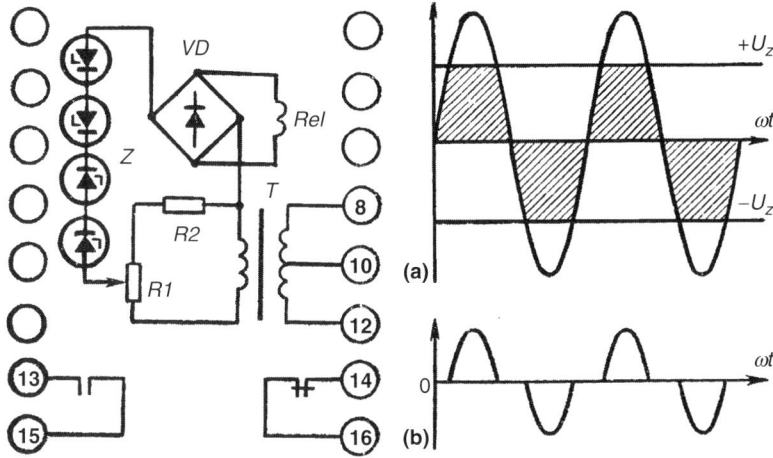

FIGURE 10.41

PH-58 type voltage relay with increased release ratio, produced on the basis of the magnetic system of the PT-40 relay. (a) — voltage form on Zener diodes Z; (b) — voltage on the input of the rectifier VD.

relay of synchronism control (Figure 10.40), each coil has two insulated half-windings with equal total section of wire copper.

The lower winding of one of the coils is linked with the upper half-winding of the other coil. Such linking allows us to obtain two insulated windings with strictly equal parameters, and a coupling coefficient between the windings, approximating 1. Each of the windings is connected to one of the synchronized voltages through the additional resistor. Resistances, the number of turns of the windings and the polarity of their connection are chosen in such a way that when nominal voltages coinciding by phase are applied to both windings, the total magnetic flux created by the windings will be full compensate and there will be no electromagnetic torque in the movable system of the relay. When vectors of synchronized voltages diverge, or one of them decreases, the total magnetic field will be unbalance and the relay picks up.

For the cases when a voltage relay with an increased release ratio (0.85 to 0.9) is needed, the winding of this relay is connected to Zener diodes (Figure 10.41). The pick-up

FIGURE 10.42

Instantaneous electromagnetic over-current (RXIC1) and overvoltage (RXEC1) relay, produced by ASEA (ABB) from the year 1970. (ABB Relay Units and Components. Buyer's Guide, 1990.)

FIGURE 10.43
Instantaneous current relay of the HFC type with a clapper
armature (General Electric).

threshold of the relay is adjusted with the help of a potentiometer (R1) supplied with a
scale.

Simple instantaneous current and voltage relays were also produced on the basis of
electromagnetic relays with a clapper armature (Figure 10.42). The pick-up threshold was
adjusted in such relays by changing the spring stretch. Current relays were produced for
nominal currents from 0.05 to 50 A, and voltage relays-for nominal values from 5 to 200 V.
The size of the relay is $67 \times 41 \times 135$ mm. General Electric is still producing relays of such
type (Figure 10.43). The HFC relay is a group of three independent relays equipped with a
latch and a pointer flag.

It should be noted that relays of clapper type used as current relays have simpler
constructions than relays with a turning armature (like PT-40), but they do not work as
well when ratios of overload current are not enough. When current gradually increases in
the relay winding and its value approximates the pick-up current value, the armature of
such relays begins vibrating too much.

Large power relays with pick-up currents up to 630 A have the same constructive
scheme (Figure 10.44). The REV-800 relay is designed for work in DC circuits. Nominal
pick-up currents of different types of relays vary from 1.6 to 630 A. The pick-up threshold
is adjusted by changing the tightness of the restorable spring. The size of the relay is $155 \times
190 \times 180$ mm, and the weight is 3.5 kg.

Relays with similar constructions are also produced for work in AC circuits with nominal
pick-up currents of up to 1500 A. Such relays belong to the so-called "primary" (see above)

FIGURE 10.44
Power open electromagnetic current relay, REV-800, with
a clapper armature (Russia).

FIGURE 10.45
Primary instantaneous DC relays (Elektroba AG, Switzerland).

class of relays, which are switched to the circuit of high current directly, without auxiliary CTs. It should be noted that the percentage of primary relays with regard to the total number of produced relays is quite small, but nevertheless some companies continue producing such relays, mostly for DC circuits where it is impossible to use CTs (Figure 10.45). Different variants of these relays have pick-up currents from 1 to 6000 A. Pick-up time does not exceed 10 msec and the error is ± 5% from setting. The magnetic system consists of a U-shaped iron with a hinged-armature. The latter actuates the contacts. The relays can be equipped with a mechanical latch and indicator flag, which can be reset by a button.

Relays of similar construction, designed for protection of DC and AC circuits, are produced by Siemens (Figure 10.46). The 3UG1 instantaneous electromagnetic over-current relay (Figure 10.46) consists of a high-armature magnetic system mounted on a moulded plastic base, together with a single-pole auxiliary switch.

With currents below the operating value, the armature is held in the rest position by a spring. Given an overcurrent equal to the preset value, the armature is instantaneously attracted and the auxiliary switch actuated. The armature returns to its rest position once the overcurrent has dropped to less than 50% of the lowest setting.

Automatic switches with an electromagnetic trip (without a thermal bimetal element) also belong to protective instantaneous current relays (Figure 10.47). Such relays cannot be used for protection from overloading because they do not contain bimetal elements with a

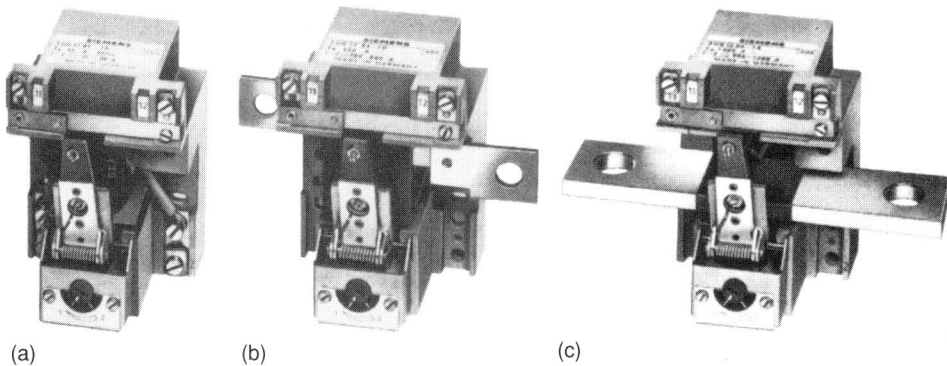

(a)   (b)   (c)

FIGURE 10.46
Single-pole primary instantaneous DC and AC relays of the 3UG1 type for currents 0.4 to 1300 A (Siemens catalogs).

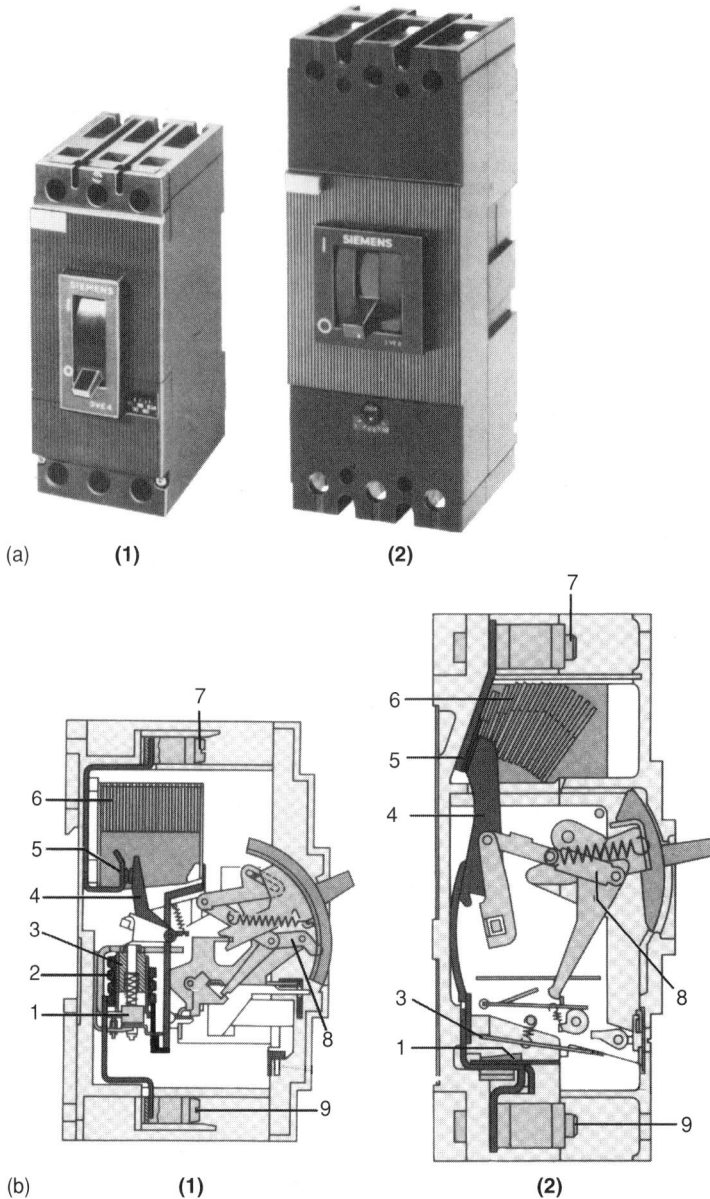

FIGURE 10.47

(a) Automatic circuit-breakers of the 3VN4 and 3VN6 types (Siemens). (b) Construction of automatic-circuit-breakers 3VN4 (1) and 3VN6 (2), with electromagnetic trips of the solenoid type (1) and the clapper type (2) armature. 1 — Instantaneous electromagnetic over current release; 2 — coil; 3 — armature; 4 — moving contact; 5 — fixed contact; 6 — arc quenching device; 7 and 9 — screws for fixing of external conductors; 8 — breaker mechanism. (Siemens catalogs).

TD dependent on current. The main function of these relays is instantaneous switching off of short-circuit currents with high ratios (5 to 20) with respect to the nominal current.

Like in the case of thermal relays, there are a lot of constructions of such instantaneous current relays, but their basic elements are the same: a current coil (or a portion of a bus bar for strong currents) connected in series with the protected load; a retractable (solenoid) or clapper armature, a spring, a trip mechanism and a contact system with an arc-suppressing unit.

10.3.2 Electronic Current and Voltage Relays

In the 1970s many key international companies developed electronic protective relays (Figure 10.48a and b) while continuing active production of electromagnetic relays.

The printed circuit board, with cheap electronic components installed on the board and soldered by fully automatic systems functioning without man's control, was considered to be more progressive and cheaper for production than precision electromechanical units of manual assemblage and adjustment. Other difficulties for the user, for example the need to locate malfunctions in electronic relays, the ways of repairing them, resistance of relays to overloads, overvoltages and interferences — all of these inevitable concomitants of electric networks, were not taken into account.

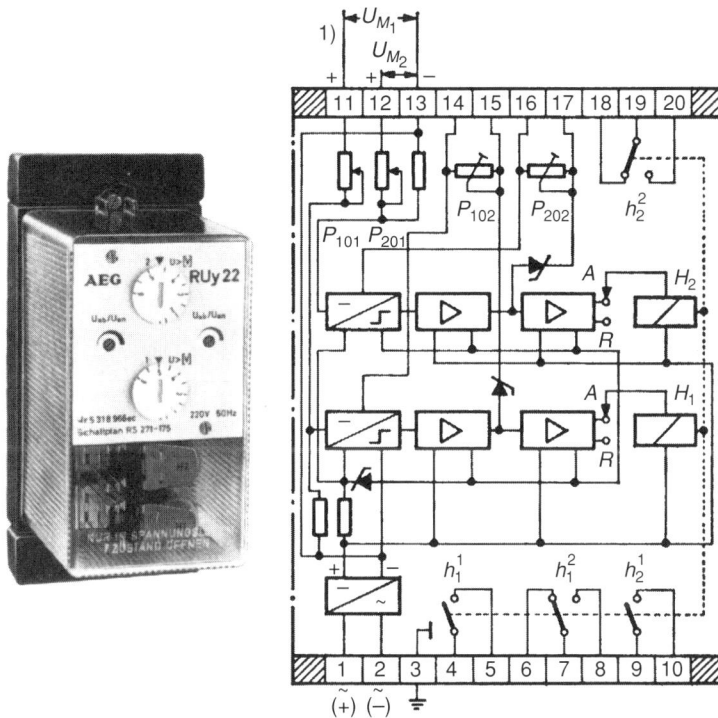

(a)

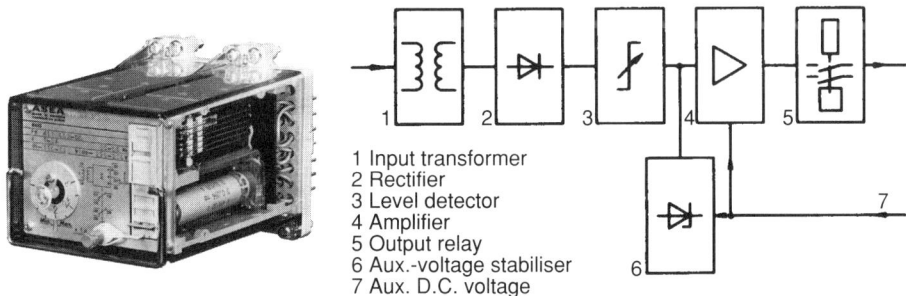

1 Input transformer
2 Rectifier
3 Level detector
4 Amplifier
5 Output relay
6 Aux.-voltage stabiliser
7 Aux. D.C. voltage

(b)

FIGURE 10.48
(a) Over-and undervoltage electronic relay of the RUy22 type (1974, AEG). (AEG Mebrelais Schutzrelais Reglar Elecktrisher Groben 1975). (b) Electronic instantaneous over and undercurrent relay RXIG-2 type (1982, ASEA). (*ABB Relay Units and Components. Buyer's Guide, 1990.*)

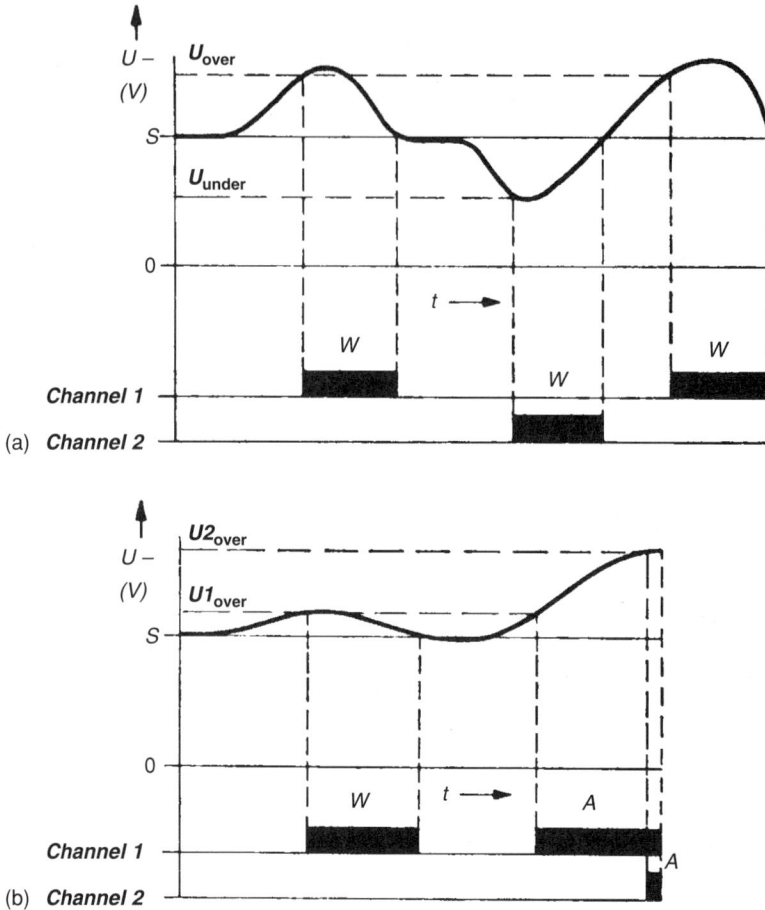

FIGURE 10.49

Diagram of function of a two-channel electronic current (voltage) relay. (a) Relay controlling the over-and undervalue. (b) relay controlling the two-stage overvalue.

As a rule, such relays contained an electronic measuring body in which the input values (current or voltage) were converted to low-level voltage, proportional to the input value and compared by level with the voltage on the electronic reference element. If the measured voltage is higher (or lower) than the reference voltage, a signal appears at the output of the compared element. It is amplified by an electronic amplifier and is applied to the winding of the electromagnetic output relay.

Simple types of such relays contained one channel with described elements, and more complex ones — two similar channels (like the RUy22-type relay produced by AEG — Figure 10.49). In earlier relay constructions, these channels had very simple circuits with two to three transistors, with a reference element on a Zener diode D1 (Figure 10.50). The ASEA Company produced RXIK-1-type relays based on similar principles of operation and made as a separate construction (Figure 10.51); however, they contained neither an individual output point electromagnetic relay nor the power source required for relay operation. This was a separate module which could work only with other modules containing the missing components, which well fit the concept of modular construction of relay protection systems, "COMBIFLEX," actively developed by the ASEA at that time.

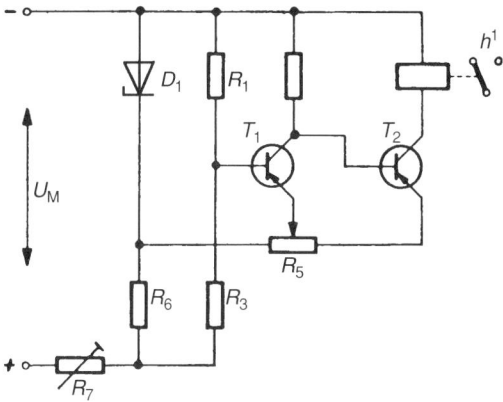

FIGURE 10.50
Circuit diagram of one channel of an RUy22-type relay (AEG).

(a)

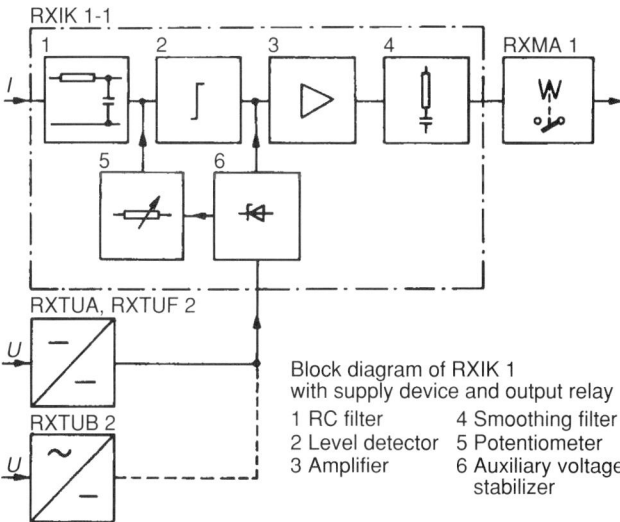

Block diagram of RXIK 1
with supply device and output relay

1 RC filter   4 Smoothing filter
2 Level detector   5 Potentiometer
3 Amplifier   6 Auxiliary voltage
                       stabilizer

(b)

FIGURE 10.51
RXIK-1-type modular relay containing only an electric transducer in its case (1974, ASEA). (*ABB Buyer's Guide, 1990.*)

According to this concept, the company did not aim at producing a protective relay with complex functions as a whole, but rather built such complex protective systems out of separate simple relay-"blocks." Perhaps that idea was right in 1970s, but it proved to be a failure when units of miniature specialized processors, capable of functioning as a cupboard full of such COMBIFLEX "units," began to appear.

The main drawbacks of the so-called "static" electronic relays based on this principle, were limited sensitivity to changing of the input signal, and a not very high release ratio (0.7 to 0.8), caused by hysteresis of the Zener diode.

The so-called dynamic relays are free from this drawback. They have appeared recently and have practically replaced semiconductor static relays in protective systems. The general principle of functioning of such relays is that during operation, a special threshold circuit (a comparator, a univibrator, or a trigger) is switched ON every half-period, when the amplitude of the input current (voltage) reaches the value equaling the reference voltage, and returns to its initial position (switched OFF) when the sinusoid of the input current (voltage) passes through the zero value, or when the input signal changes its polarity (Figure 10.52). Thus during the whole period when the affecting input value exceeds the prescribed level, the sensitive element of the relay is automatically switched from one mode to another, synchronously with the sinusoid of the controlled current. To rule out the possibility of influence of signal oscillations on the stability of the state of the output unit, a pulse stretcher (or an integrator) is used. It is based on a capacitor charging

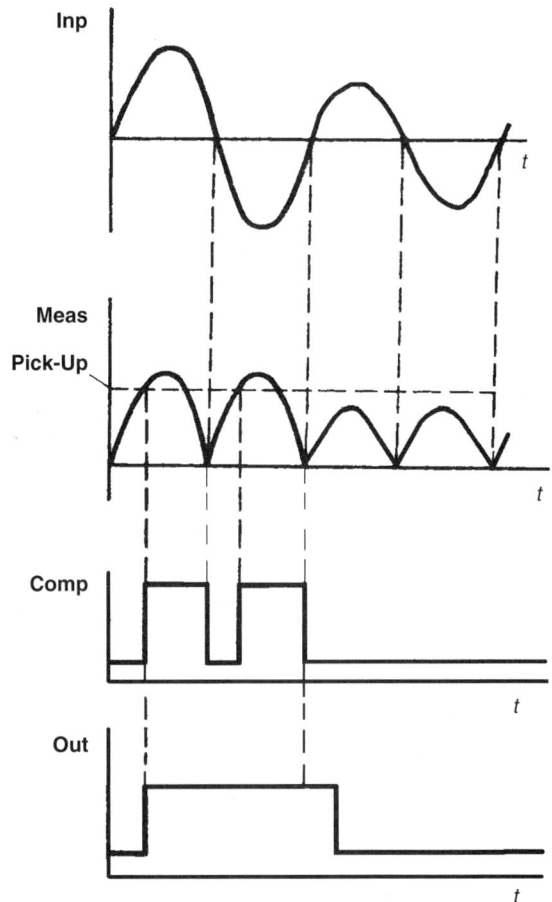

FIGURE 10.52
Simplified diagram of the function of an electronic relay of the dynamic type. **Inp** — input value (current, voltage); **Meas** — measured voltage obtained by rectification and proportional transformation of the input value; **Pick-Up** — comparator pick-up threshold; **Comp** — voltage at the output of the comparator; **Out** — output voltage of the electronic circuit.

(a)

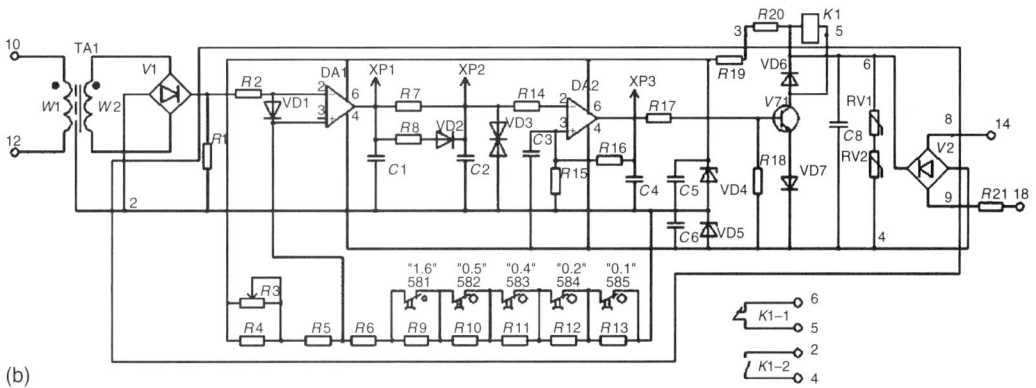

(b)

FIGURE 10.53

(a) Dynamic semiconductor current relay of the RCT type (1990, Russia). (b) Circuit diagram of universal relay RCT-type (for current) and RCH-type (for voltage).

during the ON-state of the threshold circuit and maintaining the output element in the ON-state during the pause between pulses. To increase noise-immunity of the relay, an RC-circuit is cutting between the comparator and the output amplifier. It delayed picking up of the output unit up to a few tens of milliseconds. If the high level input signal is of short duration, which is typical of noise, the output unit will not have time to pick up during the noise.

The release ratio of this circuit equals 1, since the relay returns to its initial position every time the amplitude of the sinusoid of the controlled current is not sufficient for repeated (during each half-period) starting of the threshold element. This has nothing to do with the hysteresis of elements of the circuit.

One can adjust the pick-up threshold of the relay by changing the reference voltage value with the help of a potentiometer. Practically all electronic current and voltage relays produced in 1980–90's on, are based on a similar principle (Figure 10.53).

10.3.3 Reed Switch Current Relays

For construction of instantaneous overcurrent and overvoltage relays reed switches were also used, although not so widely as relays of the types mentioned above.

According to the information we have, the simplest reed switch current relays (Figure 10.54) were produced only in the former U.S.S.R., and are still produced in Russia. These are simple constructions containing a large reed switch installed with a possibility

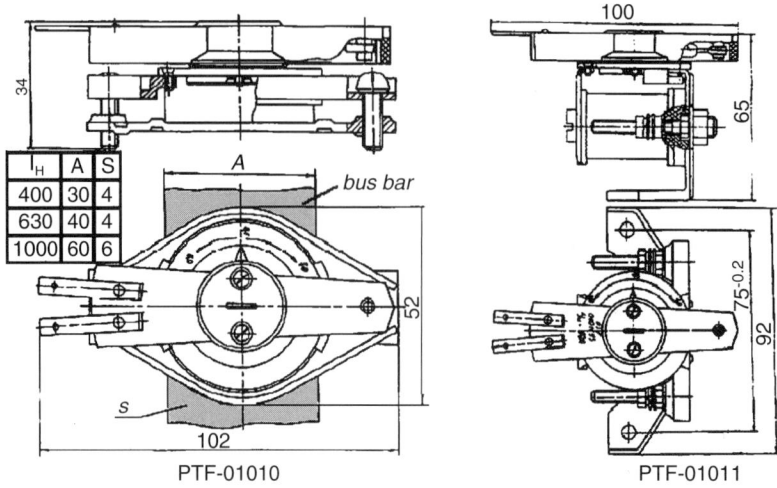

I_H	A	S
400	30	4
630	40	4
1000	60	6

PTF-01010   PTF-01011

FIGURE 10.54
Reed switch current relays with a bus bar section, and with a coil. Sizes of bus bar are shown in the table (to the left).

of turning on the bus bar section or on the coil. As the reed switch turned relatively of magnetic field supply (bus bar or coil), its sensitivity to the magnetic field of current passing through the bus bar or the coil changed (see Figure 5.76). Such relays are designed for work in DC circuits only.

Nominal values of currents for different modifications of these relays lie within 400 to 1000 A (with bus bar) and from 1.6 to 250 A with coil. A series of protective hybrid (reed-electronic) current relays was designed by the author of this book in the 1980–90's, and was put into production under his direction. These relays are still produced in the Ukraine under the brand "Quasitron" by a private company, "Inventor" (Kharkov).

The "Quasitron" is a multipurpose protection relay based on a hybrid (reed-electronic) technology, with very high noise immunity (Figure 10.55a). A current sensor may be mounted into the relay unit (as shown in Figure 10.55a) or mounted outside the relay unit on an additional plate (Figure 10.55d). One relay unit may be used simultaneously with several different types of current sensors, each of which has a different current pick-up value.

All sensor outputs are connected to the relay unit via low-voltage wires. The relay unit has three time–current characteristics (T1–T3) (Figure 10.56), one of which can be selected by a customer by means of jumpers on resistors R1, R2 (see Figure 10.55b). For this purpose one (or two) jumper(s) may be cutting.

As the circuit configuration (Figure 10.55b) becomes simpler, it does not contain ICs; its active solid-state components (transistors) do not constitute a threshold element and are merely used as a simple amplifier. An interface between the electronic circuit and the outside network bus is implemented via an insulated interface, based on a reed switch (K1), which also plays the role of threshold elements and starts vibrating with double network frequency when the relay picks-up. The contact erosion-free capacity of the reed switch (about 10^6 to 10^8 operations) along with the short period of the maximum current relay's ON-state ensures the required commutation resource of the relay.

The amplifying module of the base circuit is nothing more than a compatibility link between the integrating couple (L1–L2C1), and the output auxiliary relay (K2) provides

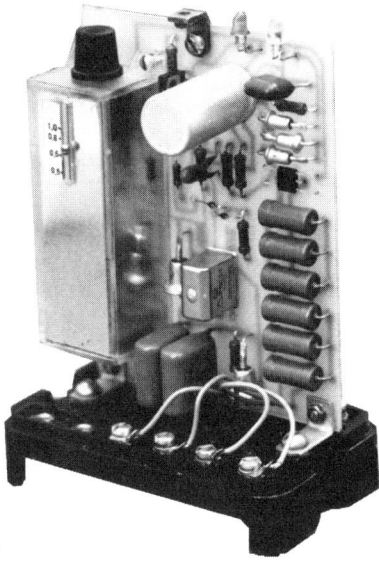

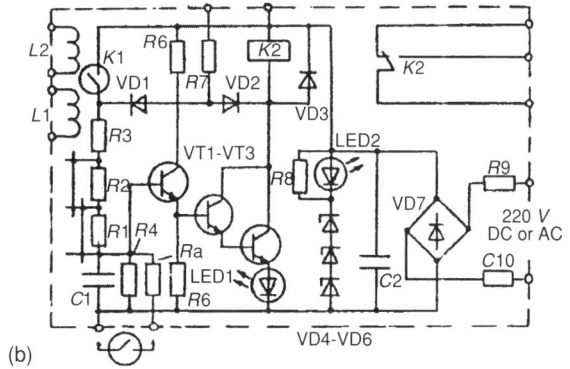

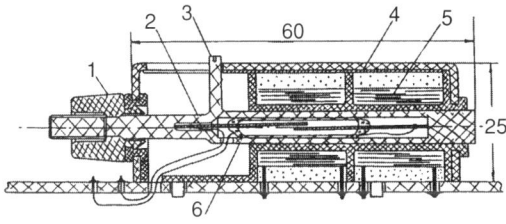

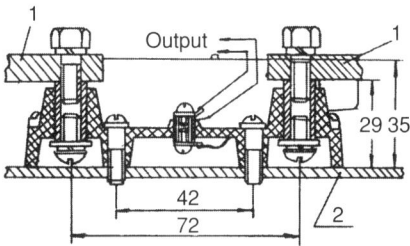

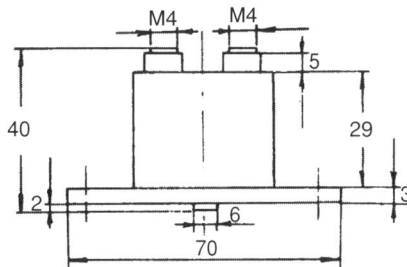

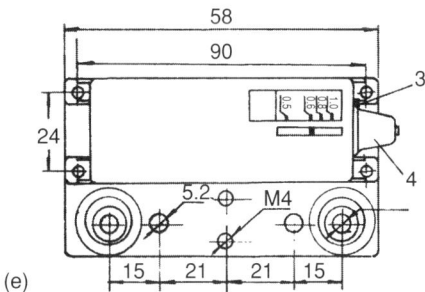

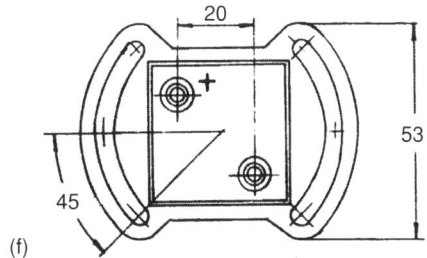

(a)

(b)

(c)

(d)

(e)

(f)

(*Continues*)

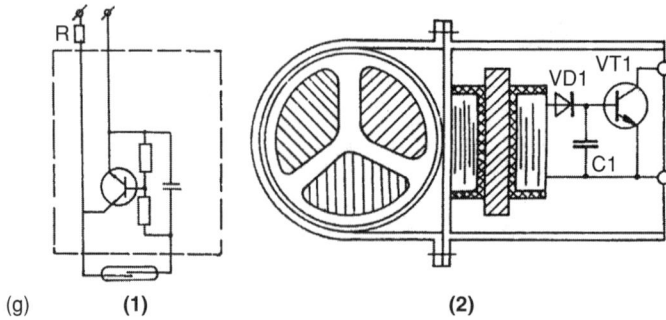

(g)   (1)   (2)

FIGURE 10.55 (*Continued*)
(a) Hybrid over-current protection relay of the "Quasitron" series (without protection lid). (b) Circuit diagram of "Quasitron" relay. K1 — Reed switch; L1, L2 — input current coils; K2 — output auxiliary relay. (c) A "Quasitron" imbedded current sensor, with adjustable current trip level. 1 — Limb; 2 — movable dielectric capsule; 3 - level indicator of current pickup; 4 — ferromagnetic screen; 5 — coil; 6 — reed switch. (d) External low voltage current sensors for "Quasitron" relays. 1 — Cutting-circuit sensor type 1 for current pick-up 0.01 to 100 A; 2 — sensor type 2 for bus bar and cable installation (30 to 5000 A). (e) Outside dimensions of external low voltage current sensor, Type 1. 1 — External wires of current circuit; 2 — plate; 3 — fixative element; 4 — limb; "output" is connected to relay unit. (f) Outside dimensions of sensor type 2 for bus bar and cable installation. (g) Circuit diagrams of type 2 sensors. 1 — for current level 100 A and more; 2 — for low current levels.

for the stability of the ON-state of the relay under the K1 (reed switch) vibration conditions.

The "Quasitron" is a relay of the dynamic type (see above), with a high release ratio (0.85 to 0.95). This is caused by the algorithm of its operation: return of the relay to the initial position is not connected with hysteresis of the reed switch, as it follows the current sinusoid and switched OFF forcedly every half-period. This is perhaps the only one type of electro-mechanic relay (the reed switch is an electro-mechanical element) capable of working in the dynamic mode peculiar to electronic relays. The high frequency and short pulse interference at the relay input cannot migrate to the electronic module since K1, being the interface link, does not react to the high frequency control signals due to the inherent inertia. Neither does it respond to the transient interference from the power circuit commutation, therefore the whole relay becomes very robust to power circuit pulse interference. The effect of the magnetic component of the dissipation fields can be

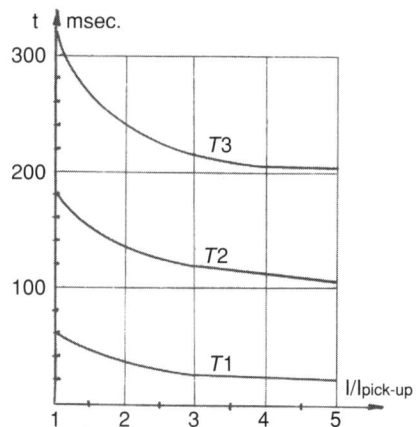

FIGURE 10.56
Time–current characteristics of "Quasitron" relay.

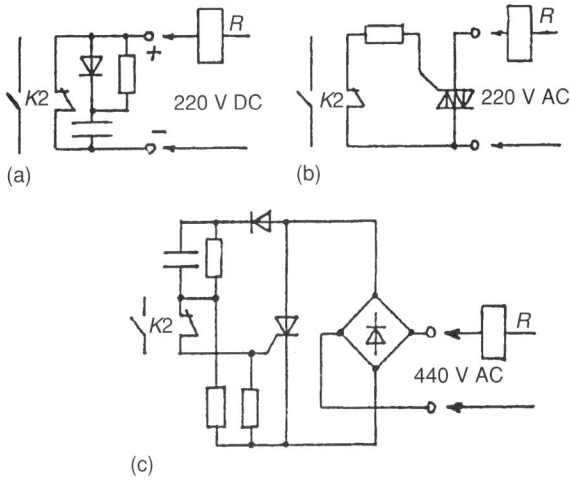

FIGURE 10.57
Output modules for "Quasitron" relay. (a) With spark protection, for DC load with large inductance; (b) with power amplifier, for power AC load (up to 500 VA); (c) for AC load, connected to power supply with voltage more than switching voltage of output auxiliary relay. K2 — contact of output auxiliary relay, mounted on PCB in relay unit; R — load of resistive or resistive-inductive type.

neutralized by introducing a ferromagnetic screen into the relay design (see Figure 10.55c). This 1.5 mm screen shields the reed switch in the fields with an intensity much higher than that of the dissipation fields under actual operating conditions.

For different applications of the "Quasitron" device several types of output modules are available (Figure 10.57). For "Quasitron" relays not only were low-voltage current sensors designed, but also high-voltage current sensors that can be installed directly onto the high-voltage line without CTs (Figure 10.58). Without a doubt, one advantage of this sensor is its compactness and the possibility of direct installation on the high-voltage wire. This is important from the point of view of cost reduction of high-voltage equipment, and from the point of view of minimization of compact distributing devices of the cabinet type. On the other hand, that compactness and the lack of necessity for high-

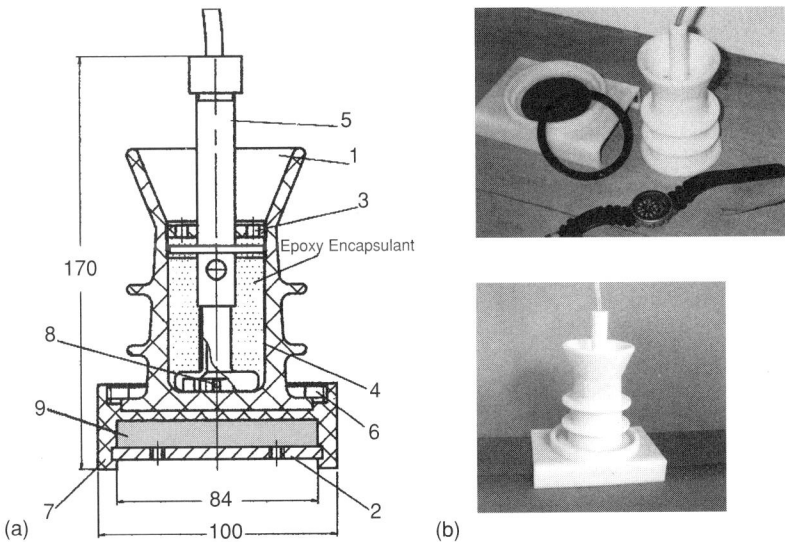

FIGURE 10.58
(a) Construction of the high-voltage current sensor of the "Quasitron" relay. 1 — Main insulator; 2 — fixative plate; 3 — inside nut; 4 — semiconductive cover; 5 — bushing; 6 — fixative nut; 7 — fastener; 8 — reed switch; 9 — bus bar. (b) External design of the high-voltage current sensor of the "Quasitron" relay.

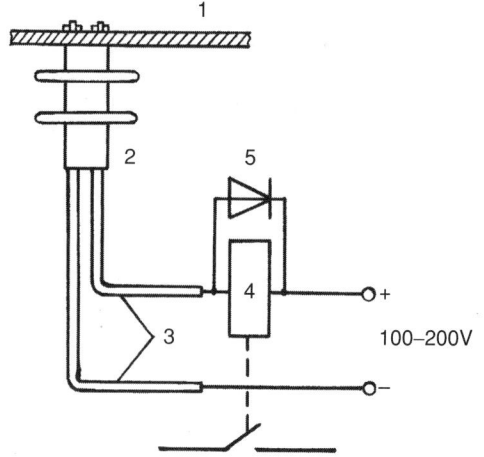

FIGURE 10.59
Principle of application of a single AC relay with high-voltage insulation. 1 — Conductor line; 2 — relay; 3 — high-voltage wire; 4 — low-voltage auxiliary relay; 5 — diode.

voltage CT applications do not allow modernization of existing compact distributing devices, since there no place for an additional CT was envisaged.

Entire independence on the one hand and full compatibility on the other hand allow the relay to provide complex protection simultaneously of all circuits (both low-and high-voltage ones, with different current settings) of the complex electrical installation with the help of only one "Quasitron" relay unit.

In the case of a single AC relay when picking up excess of a given setting in the high-voltage circuit and switching of the auxiliary relay of the low-voltage circuit (Figure 10.59) is required, it is possible to use a high-voltage sensor with an integrated electronic filter, transforming the reed switch vibration at its pick-up to a standard signal of the "ON–OFF" type (Figure 10.60).

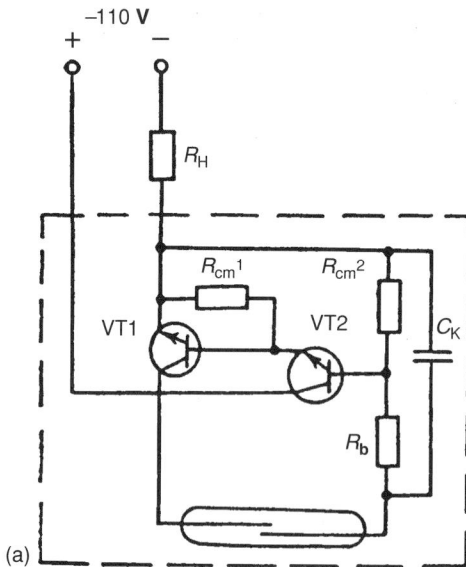

FIGURE 10.60
Electronic filter integrated to the high-voltage current sensor. After the sensor has been assembled, all inner space is filled with epoxy resin.

10.4 Current Relays with Independent "Time-Delays"

10.4.1 Relays with Integrated Clockwork

When protective relays began to be used in the power industry it became clear that just instantaneous relays are not enough for effective protection. As mentioned above, only a set of relays with different and at the same time matched pick-up currents, and different TDs, can provide selectivity of protection on long power lines.

Already at the beginning of the last century direct-action relays (that is relays directly affecting the driving gear of the switching apparatus), with the simplest integrated clockwork providing the required TD, were produced (Figure 10.61). The pick-up time of such a relay is determined by the balance of three constituents: power of attraction of the armature to the core, opposing force of the spring, and the position of the lower end of the spring depending on the anchor clockwork.

The Anchor mechanism (anchor) (Figure 10.62) consists of an anchor wheel, a fork, and a balance (double pendulum) — the part of the clockwork transforming energy from the main part of the driving wheel (in clocks — a spring) to pulses transmitted to the balance to maintain a strictly determined oscillation period necessary for uniform rotation of the pinion mechanism.

The principle ideas of such mechanisms were laid already in the 17th century, but its modern design derives from efforts in the 18th and 19th centuries of many watchmakers, the especially distinguished ones being Abraham-Lui Breget and Jorge Leshaut. Even the famous playwright Pierre Ogusten Bomarschet turned his attention to improvement of this mechanism — the main clock unit, determining its accuracy.

When the running torque affects this mechanism from the direction of the driving wheel, the anchor fork (3) moves the teeth of the anchor wheel (4) one by one at a certain speed that

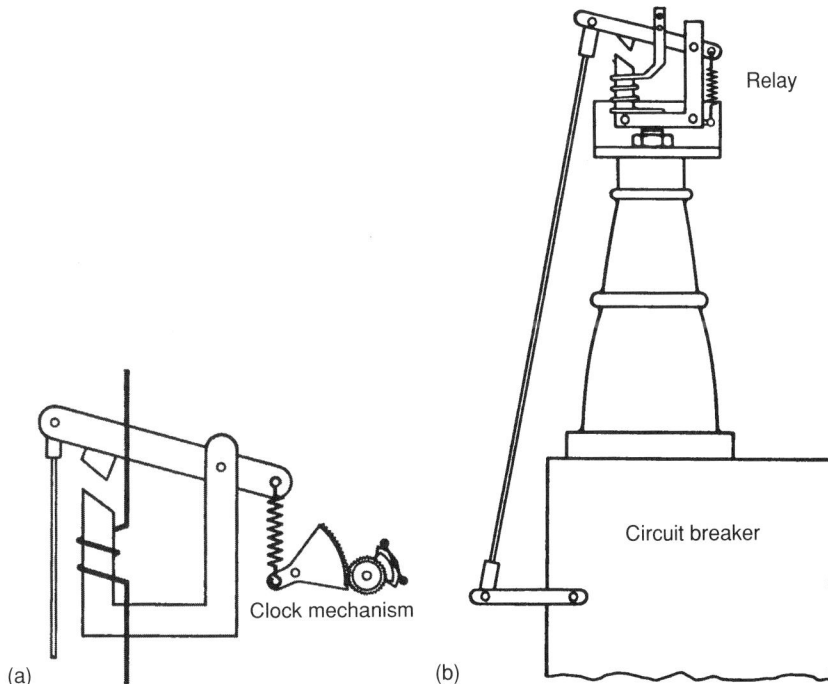

FIGURE 10.61
Protective direct-action current relay with an integrated anchor clockwork.

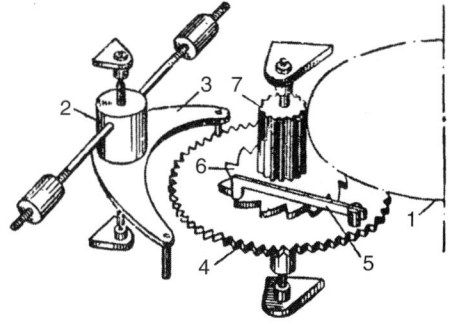

FIGURE 10.62
Anchor clockwork. 1 — Driving wheel; 2 — balance-wheel;
3 — anchor fork; 5 — ratchet spring; 6 — ratchet-wheel;
7 — pinion.

does not depend (much) on the running torque. In this relay, shown in Figure 10.61, the spring was selected in such a way that the electromagnetic torque developed by the coil was enough for full overcoming of its resistance at high current ratios only (more than 2 to 3). In this case, the armature overcame the resistance of the spring and was immediately attracted to the pole of the magnetic core before the clockwork started operating. At currents in the coil equaling 1.1 to 1.5 of the nominal value the armature began moving but the relay was not activated because of increasing resistance of the spring. The lower end of the spring activates the anchor mechanism which slowly releases the lower end of the spring by reducing its tension and allowing the armature to go on moving. The speed of the lower end of the spring is constant and does not depend much on the tautness of the spring. Obviously, when initial current is strong in the relay winding, the degree of the initial

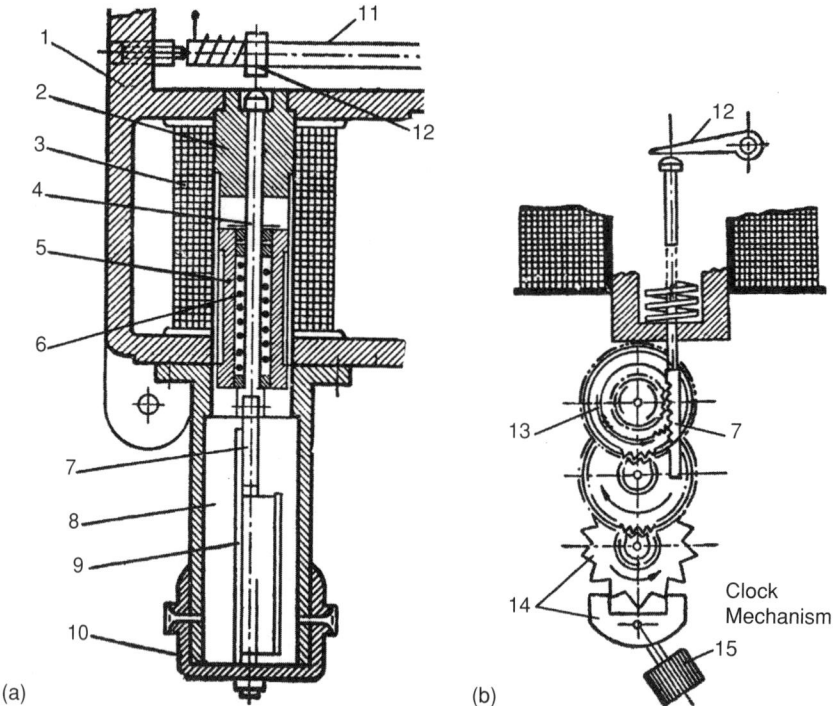

(a)   (b)

FIGURE 10.63
KAM-type direct-action current relay with an integrated clockwork (1937, "Elektroapparat" plant, Russia).
1 — Magnetic core; 2 — stopper; 3 — current coil; 4 — striker-pin; 5 — retractable armature (hollow cylinder);
6 — spring; 9 — rack; 8 — space for the clockwork; 9 — clockwork; 10 — cover; 11 — breaking shaft; 12 — rod
of the breaking shaft; 13 — ratchet-wheel; 14 — anchor mechanism; 15 — pendulum.

tension of the spring will be greater and the armature will move closer to the pole of the magnetic core. This means that at high current the armature has less travel to its final position which is why the time of operation of the anchor mechanism, releasing the end of the spring at a constant speed, will be less, meaning that the TD of the relay (until its complete pick-up) will be less.

Relays with similar principles of operation (only not of clapper type, but with a retractable armature) were produced in the 1930s in the former U.S.S.R. by the Leningrad plant "Electroapparat" (Figure 10.63). In this relay when current equaling or exceeding setting current passes through the coil (3), the armature (5) starts going up into the coil and also carries the striker-pin (4) through the spring (6). As this pin is linked with the rack of the clockwork, it slows down movements of the armature in the upper direction, as in the example considered above. When the armature (5) reaches its extreme upper position, it turns the rod (12) of the shaft (11) with the help of the striker (4) and the shaft turns and releases the latch of the trip, thus putting into action the breaking mechanism of the switch. At high current ratios, the armature immediately reaches its top position and activates the breaking mechanism of the switch with the help of the striker-pin (4). Thus the relay has two areas on its time characteristic (Figure 10.64): a TD dependent on the current and a constant delay caused by mechanical displacement of construction elements.

The relay coil has several taps and by switching of them, one can choose one of the characteristics. In addition, by displacing the whole clockwork up or down of the rod (7) one can change the TD value of the clockwork. Release ratio of the relay in the dependent part of the characteristic equals 0.6 to 0.8, and for work in the independent part of the characteristic — 0.8 to 0.9. Minimal pick-up time in such construction of the relay is 0.7 sec in the independent part of the characteristic and almost 1 sec — in the dependent one. These are quite high values (several times higher than modern protective systems provide) for effective protection. The KAM relay was produced until 1940, when it was slightly modified and was produced commercially in the U.S.S.R. by several plants under the names RTB, RTM, and RMB. The problem of low performance remained in all of these relays.

Strange as it might seem, nowadays that computers and microprocessors are widely used in relay equipment, the RTB relay with its almost 70-year history (basically a KAM relay with an enhanced clockwork) is still produced toady, by the Repair Enterprise Lenenergo (St Petersburg, Russia).

Some current relays produced contain such a set of elements in their construction that it may be difficult to define to what class of relays they belong to. For example, the primary direct-action relay of the MUT-1 type (Figure 10.65) contains an integrated CT, an

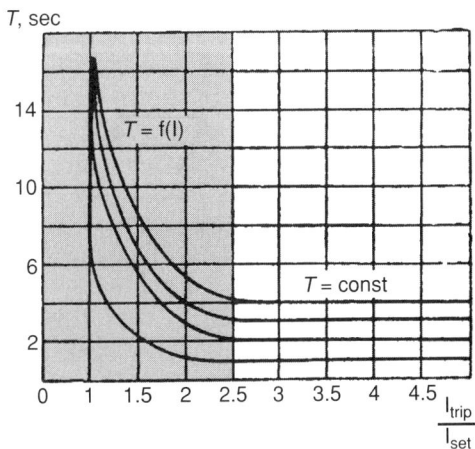

FIGURE 10.64
Time–current characteristics of a direct protection relay of the KAM type.

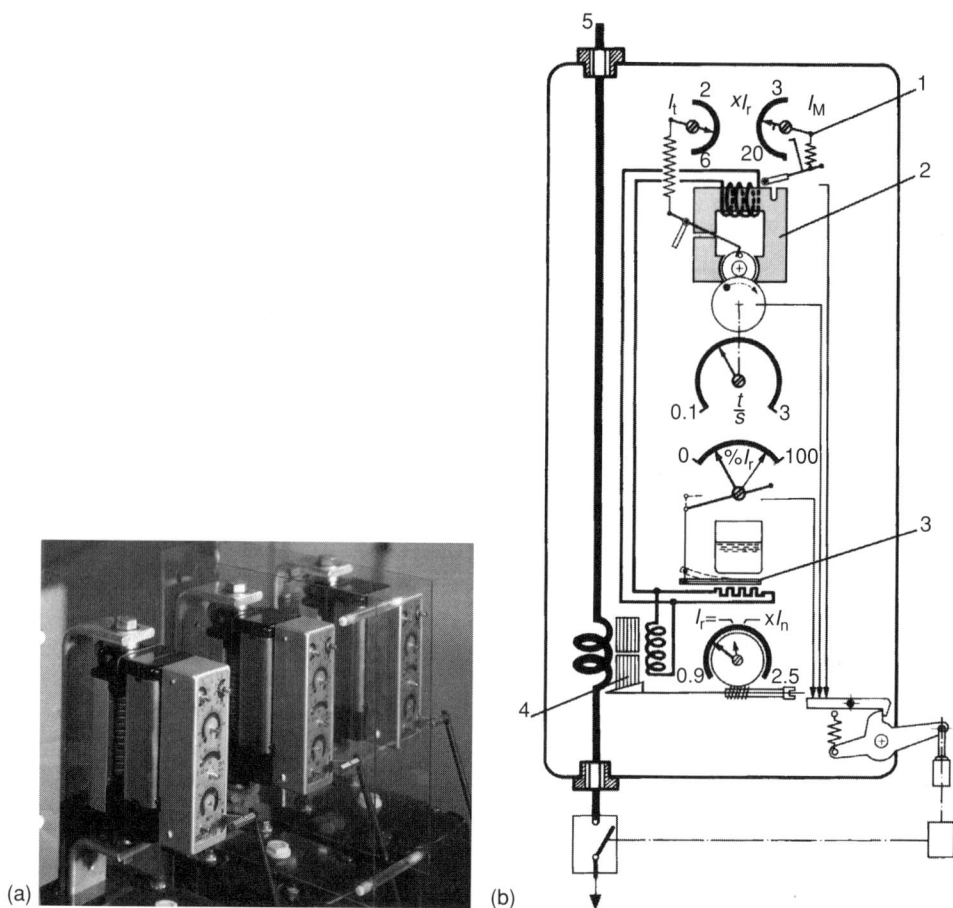

FIGURE 10.65

(a) Three primary direct-action current relays of MUT-1 type (Sprecher + Schuh). (b) Construction diagram of the combined primary direct-action current relay of the MUT-1 type. 1 — Electromagnetic instantaneous element with adjustable ratio of pick-up current; 2 — TD element (synchronous motor); 3 — bimetal element; 4 — integrated CT; 5 — conductor line.

instantaneous element, a TD element based on a synchronous motor, and a thermal (bimetal based) element. The nominal current of the MUT-1 relay lies within 1.5 to 300 A, the adjuster of current ratio of instantaneous cutoff — within 2 to 20, pick-up current of the thermal element — within 0.9 to 2.5 of the nominal current, and pick-up time of the thermal element — 15 to 120 min.

In the 1960–70's many leading companies produced indirect-action current relays with an integrated clockwork. A typical example of such a relay is the widespread RSZ 3gk type-relay, produced by AEG (Figure 10.66). The RSZ 3gk-type relay is quite a large and heavy device (its size is $342 \times 165 \times 152$ mm and its weight 6.8 kg), containing three current units with independent adjusting consisting of clockwork, three units with adjustable instantaneous pick-up thresholds and the clockwork itself (Figure 10.66a).

Current relays (or units of relays) with a TD are usually indicated by the symbol $I>$ and instantaneous relays (or units of relays) by the symbol $I\gg$. These symbols are quite logical and easily memorized. The former characterized the relay picks up at a current exceeding a certain value (as a rule this is a relay with a TD) and the latter characterized the relay picks up at a current much exceeding a preset value (this is a characteristic of instantaneous relays). Relays of maximum and minimum voltage are marked similarly ($U>$, $U<$).

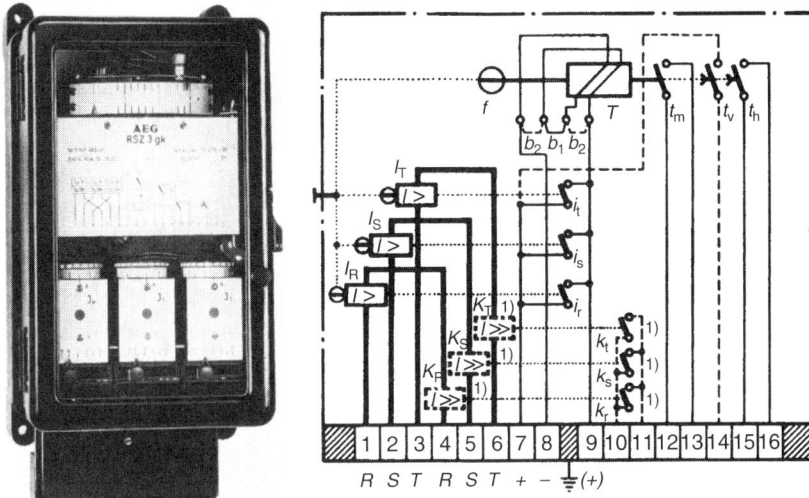

(a)

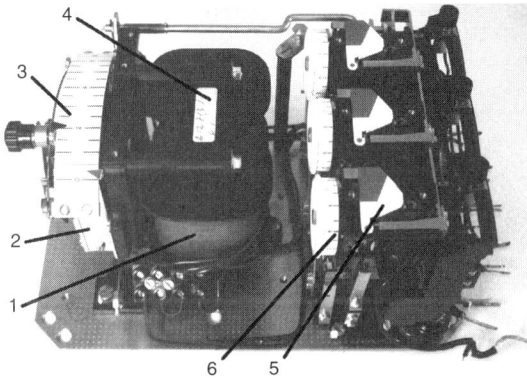

(b)

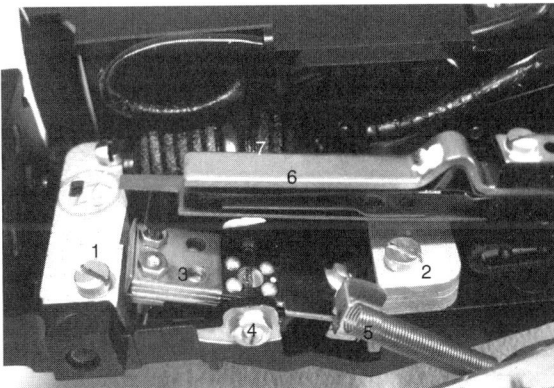

(c)

FIGURE 10.66

(a) Three-phase current relay of the RSZ 3gk type, with integrated clockwork (AEG, 1975). (b) RSZ 3gk relay construction. 1 — Coils of the electromagnet of the clockwork; 2 — clockwork; 3 — scale of the clockwork; 4 — magnetic core of the electromagnet; 5 — turning flag opening the red sector when the relay pick-ups; 6 — scale of adjustments of the current unit $I>$. (c) Magnetic system of a current unit $I>$, with a turning Z-shaped armature. 1 and 2 — Terminals of the Π-shaped magnetic core; 3 — Z-shaped turning armature; 4 — rotation axis of the armature; 5 — spring; 6 — contacts holder; 7 — coil.

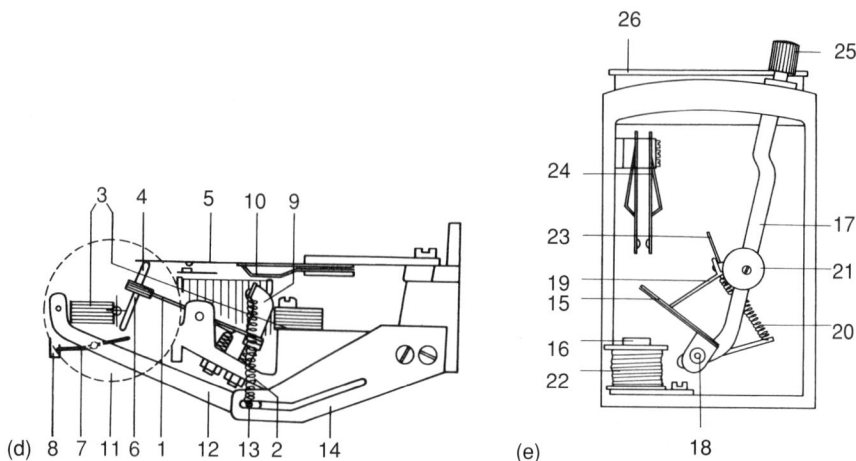

FIGURE 10.66 (*Continued*)

(d) Construction of the current unit *I>*. 1 — Turning armature; 2 — restorable spring; 3 — terminals of the ∏-shaped magnetic core; 4 — contact pusher; 5 — movable contact; 6 — pusher of the position flag ; 7 — intermediate flag pusher; 8 — prominent part of the flag; 9 — angle bar linking the armature with the spring; 10 — contact limiter; 11 — limb of pick-up current adjustment with a scale; 12 — tie-rod of adjusting element of the spring tension; 13 — moving element of the spring; 14 — post with a notch. (e) Construction of instantaneous unit (*I≫*) 15 — clapper type armature; 16 — pole of the core; 17 — pick-up threshold arm; 18 — rotation axis of the armature; 19 — lever; 20 — spring; 21 — eccentric disk for adjustment of the relay scale; 22 — coil; 23 — contacts pusher; 24 — contact system; 25 — lever handle for pick-up current adjustment; 26 — scale. (AEG 1975).

Such signs can be usually found on the relay itself, near the corresponding element of adjustment. They are also indicated on schemes, where the letters *R, S, T* (Figure 10.66a) are used to indicate the three phases of AC systems.

As can be seen from the scheme, when the current unit *I>* picks up, it switches ON the electromagnet of the clockwork by its contacts, and the contacts of the current unit *I≫* go out directly to the terminal block where they are used for switching of external circuits.

The magnetic system of the current unit *I>* (Figure 10.66b) is constructed with a Z-shaped turning armature, which has been known and in use for more than 70 years now and which continues to give a good account of itself in current relays (the peculiarities of relays with such armatures were considered above).

The placement of the both current units, *I>* and *I≫*, in constructions, are on both sides of the same plastic carcass, in the form of a common block. All these blocks, relating to all three phases, are installed on a common heelpiece. Unlike relays with a Z-shaped armature described above, which were supplied with a very soft spiral spring requiring protection from mechanical effects, in this construction quite a heavy coil spring of cylindrical shape is used and its power is transmitted to the armature with the help of a simple and reliable mechanical system, with the possibility of slide and regulation of its tension degree (Figure 10.66d).

To adjust the required pick-up current the limb (11) is turned and the proper section of its scale placed in front of the stationary pointer. When the limb turns the finite element of the spring (13) moves in the notch of the post (14) and the tension of the spring changes.

The instantaneous current unit *I≫* works only under high current ratios and has a considerably simpler construction, of the clapper type (Figure 10.66e). When the lever (17) changes its position, the armature (15), together with the spring (20), turns, so the pick-up current is adjusted not by changing of the spring tension as in the previous case, but by the reducing of the air gap between the armature and the terminal of the magnetic core.

10.4.2 Current Relays with Electronic Time-Delay

As it can be seen from Figure 10.66d, the clockwork, together with the electromagnet, occupies half of the inner space of the relay. This is the heaviest element of the construction: it weighs about 2 kg. It is probably also worth mentioning that this is the most expensive unit of the relay. That is why it would be only natural to look for the solution to replace this unit with a semiconductor time unit.

This solution was accepted and production of the RSZ3yk-type current relays was arranged. These relays were very much like the relay considered above, except that the clockwork and the powerful electromagnet were replaced with a simple (and light) printed circuit board with an electronic time relay based on the RC-circuit and electronic amplifier, already well known to us (Figure 10.67b) . Another difference of this relay from

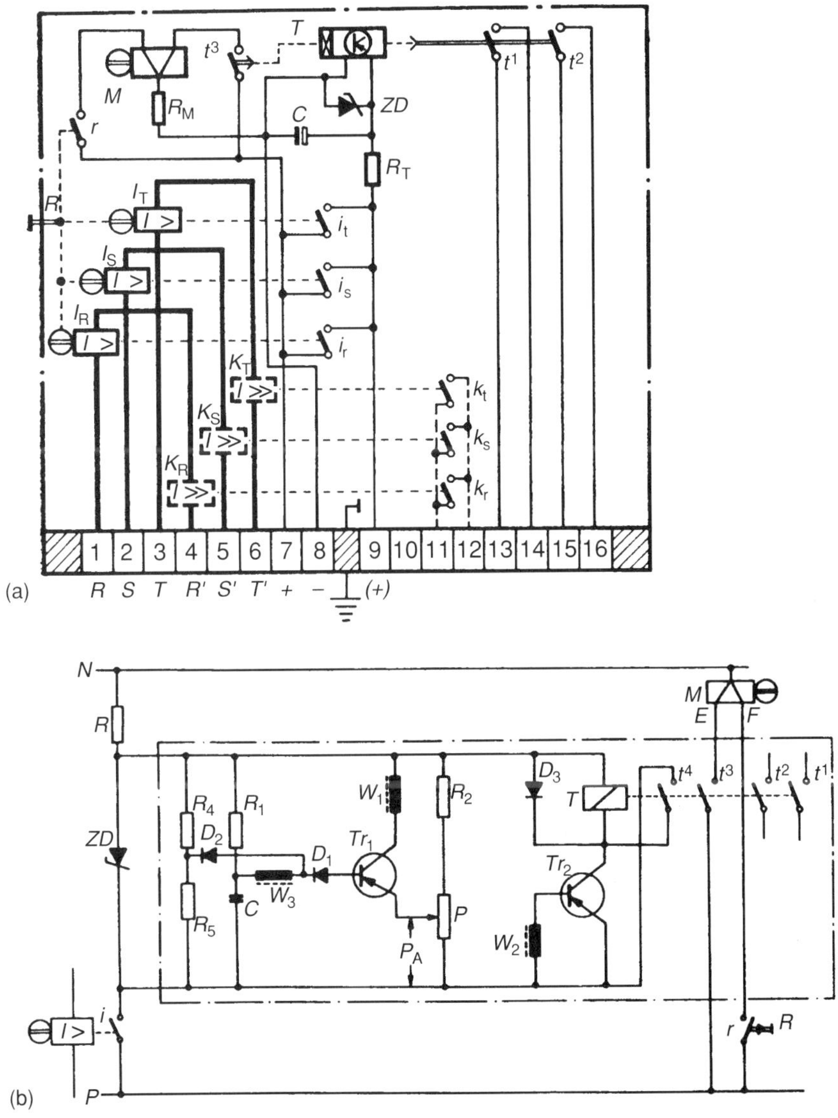

FIGURE 10.67
(a) RSZ3yk-type current relay with a semiconductor element of TD (T) instead of a mechanical one. Produced by AEG. (b) Circuit diagram of the TD element of the RSZ3yk-type current relay (AEG 1975).

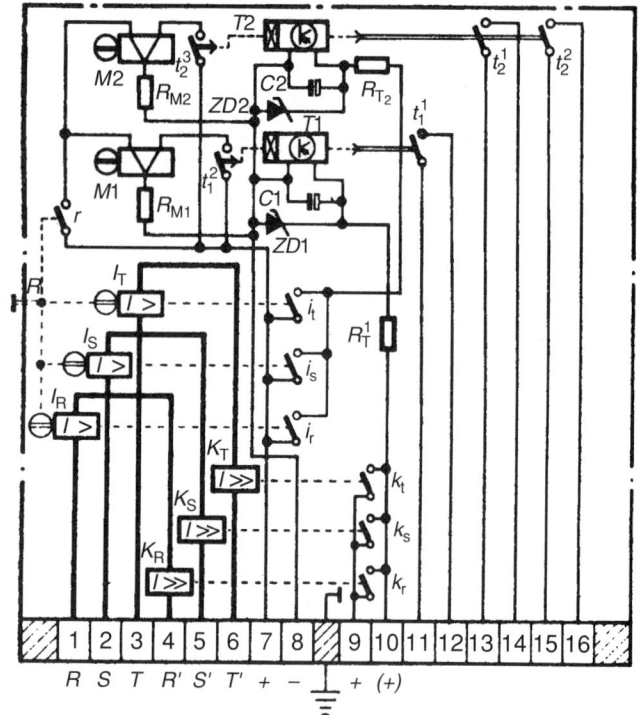

FIGURE 10.68

The RSZ3yk-type current relay, with two TD elements. (AEG 1975).

the previously considered one is an additional indicating relay (M) opening the colored flag at the pickup.

The use of a simple and light TD element enabled installation of two such elements with different TDs in the same case of the current relay (Figure 10.68). One of them (with a longer TD) runs from the current relay $I>$ as in all previously considered cases, and the other (with a shorter TD) runs from the relay $I\gg$, allowing for adjustment of the pick-up time of the relay $I\gg$ within some narrow limits at the stable value of this TD.

This development of current relays, that is the transition from a heavy and expensive mechanical time element to a one based on a semiconductor, seems quite logical. The only astonishing fact is that all of these modifications of relays have been produced by the AEG Company, parallel to each other for many years!

Analyzing the tendencies of development of protective relays from the appearance of transistors and then chips, one notices that not only the AEG Company but also many other companies produced in parallel both mechanical relays designed tens of years ago, and the most modern (for that time) transistor relays, than microelectronic ones, and finally microprocessor ones. How can this phenomenon be explained? The answer to this question is probably evident: constant demand for electromechanical relays. Why should such a demand still remain high, is another question requiring separate treatment and we will return to it later. Logically, if one follows tendencies of technical developments in the 1960–70's, right after the hybrid relays considered above (containing both mechanical and electronic elements), fully electronic relays without electromechanical sensitive current units should have appeared.

10.4.3   Electronic Current Relays with Independent Time-Delay

We have already considered principles of construction of electronic relay circuits turning from one stable state to another when the input value of the given pick-up threshold is

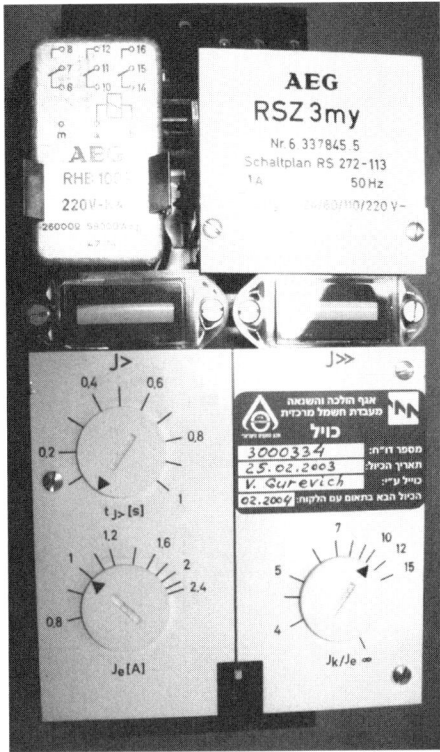

FIGURE 10.69
Fully electronic current relay, with an integrated TD element of the RSZ 3my type (AEG). Front panel view. To the left above — output electromagnetic relay. $J>$ and $J\gg$ equal $I>$ and $I\gg$, respectively; $t_{J>}$ (s) = scale of TD in seconds; J_e (A) = scale of pick-up currents in amperes (for the unit $I>$); J_k/J_e = ratio of the pick-up current (for the unit $I\gg$).

exceeded. Semiconductor current relays are based on the similar principles. The construction of such relays contains a metal or plastic case with input CTs (as a rule there are three of them) and several printed circuit boards with electronic components. Usually, the auxiliary neutral electromagnetic relay is used as an output element (Figure 10.69).

It should be noted that production of the relay is a high-quality one. The output electromagnetic relay (in Figure 10.69 at the upper left) is used in its sealed modification. Printed circuit boards with discrete elements are covered (by dipping) with a thick layer of high-quality varnish protecting the board from humid air (Figure 10.70). As it can be seen from the circuit shown in Figure 10.71, the three-phase current converted to voltage

FIGURE 10.70
Electronic current relay of the RSZ 3my-type made on discrete electronic components. View from printed circuit boards.

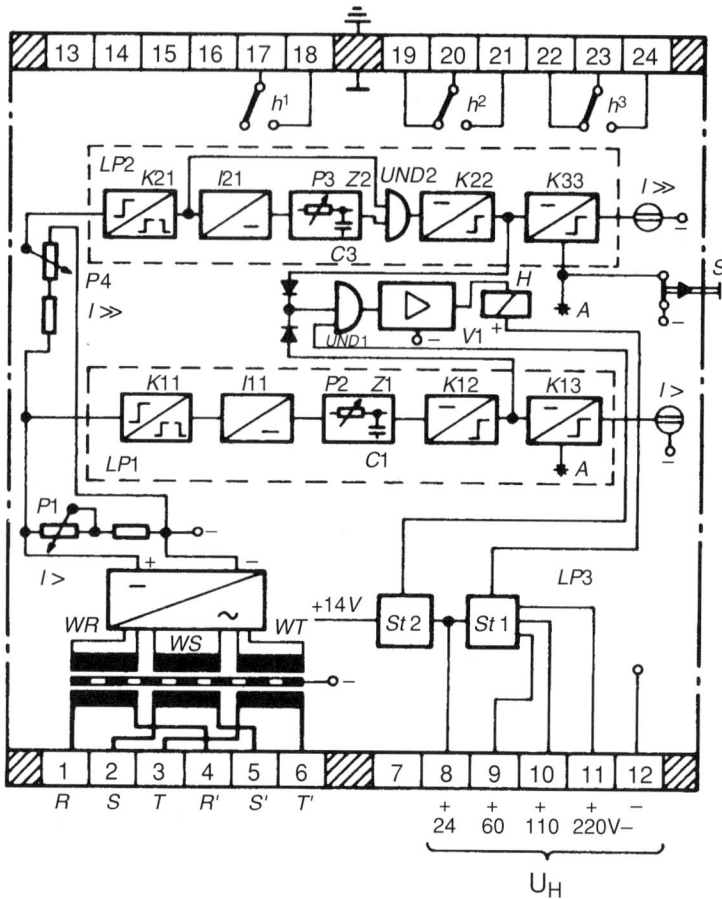

FIGURE 10.71
Circuit diagram of RSZ-3my type current relay (AEG 1975).

with the help of the input transformers WR, WS, and WT is rectified and applied to two current relays: $I>$ placed on the printed circuit board LP1, and $I\gg$ placed on the second printed circuit board LP2. Pick-up currents of these relays are adjusted by potentiometers P1 and P4, respectively. On the third printed circuit board LP3, there are stabilizers St1 and St2, allowing internal circuit supply from an external source with voltage of 24, 60, 110, and 220 V. The TD is provided by units Z1 (for $I>$) and Z2 (for $I\gg$), working on the principle of RC-circuit charging. Charging time (that is TD) is adjusted by potentiometers P2 and P3 (it is only potentiometer P2 that is brought out to the front panel).

The author had a chance to test such a relay after it had been used in an electric power station. The results were good for a relay which had served for 29 years, without repair, and show that a well-proven circuit, use of high-quality electronic components and properly chosen modes of operation for them allow even relatively simple electronic devices without modern chips or microprocessors, to remain quite competitive, at least from the point of view of stability of parameters and reliability of operation. Moreover, according to the personal experience of the author, modern microprocessor relays do not always provide as reliable protection of electric units as these old relays based on discrete elements.

Because of this and some other reasons quasi-electronic current relays on discrete electronic elements of the "Qusitron" series considered above, still remain promising, and a TD feature can also be added to this promising construction (Figure 10.72). The TD body in this construction is made not on the basis of an RC-circuit as in the

(a)

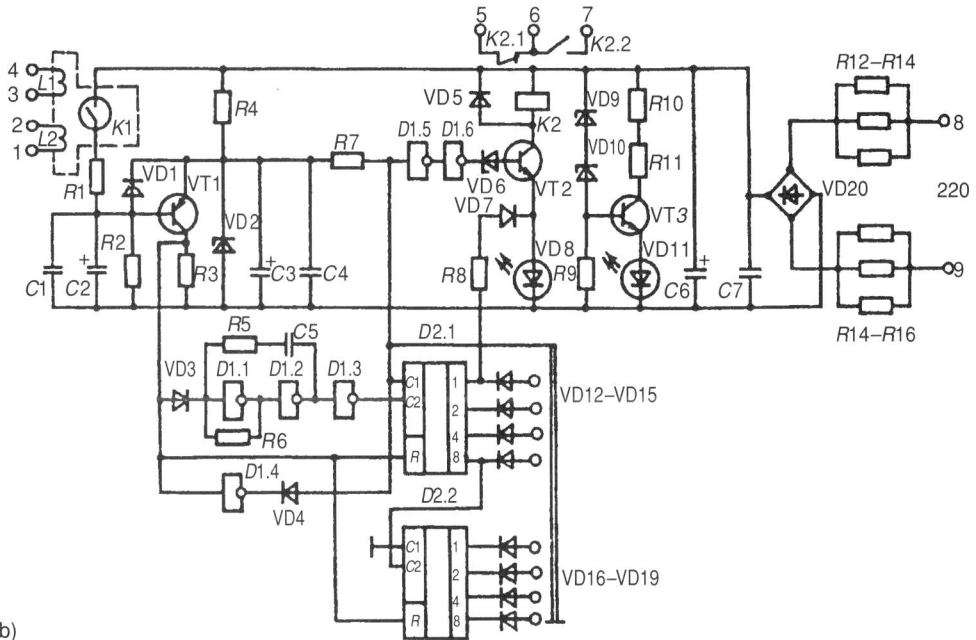

(b)

FIGURE 10.72

Universal hybrid current relay "Quasitron-T" with an integrated TD element, designed by the author. External design (a) and circuit diagram (b).

constructions considered above, but in the form of a pulse generator (multivibrator) on the chip D1 (elements D1.1 and D1.2) and a counter on the chip D2 counting these pulses. The time dial is adjustable, with a range of 0.1 to 25 sec ($\pm 2.0\%$), with a grade of 0.1 sec

10.5   Current Relays with Dependent Time-Delays

The types of secondary relays considered above have an independent on current (fixed) TD, which is not always enough for effective protection from overloading. Much better protection would be provided by a TD that depends inversely on the current value of overloading. At small ratios of overloading the time will be just a few or tens of seconds

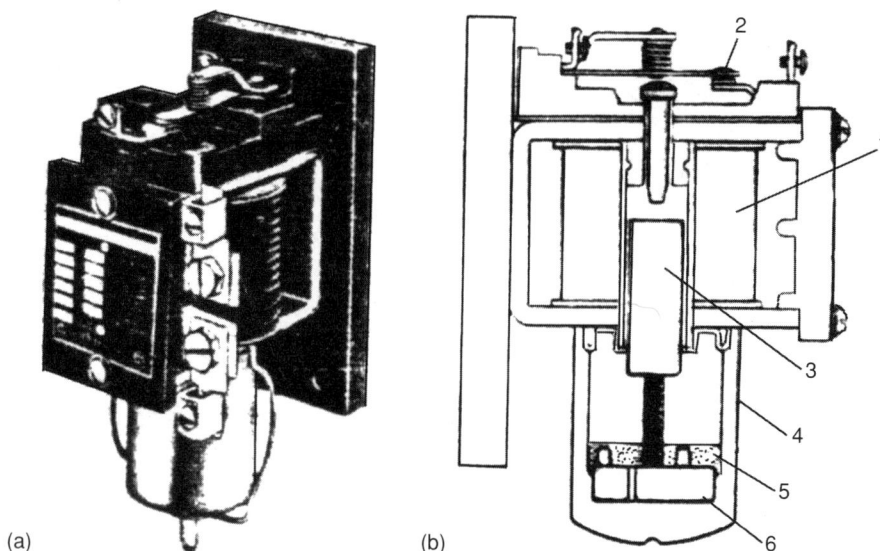

FIGURE 10.73
Inverse time–current relay "Bulletin 810" type (Allen Bradley, Inverse Time Current Relays Instruction Bulletin 810, 1976). 1 — Operating coil; 2 — contacts; 3 — core; 4 — dashpot; 5 — silicon fluid; 6 — piston.

and at higher ones — fractions of a second. The transition must be smooth: the greater the current, the quicker the relay pick-ups. This is how effective and widely used thermal relays work (see above).

Realization of this problem led to the appearance of relays with inverse dependent time–current characteristics. Nowadays such characteristics are implemented in two types of relays: induction ones (with a disk or a cup rotating in the magnetic field), and electronic ones. however, from the very first years of relay development up to the 1970s and 1980s, there were attempts to create relays with inverse time–current characteristics based on other principles (Figure 10.73).

10.5.1   Relays with a Liquid Time-Delay Element

The Bulletin 810 (produced by the Allen Bradley Company in 1976) has inverse time–current characteristics that are dependenton the viscosity of the fluid in the dashpot; however, unlike thermal relays minimum operating current is independent of ambient temperature change or cumulative heating. Current through the Bulletin 810 operating coil imparts an electromagnetic force on the movable core. The vertical position of the core in the coil is adjustable, thereby providing an adjustable trip point. When the coil current increases to the trip point, the core raises to operate the contact mechanism. TD is provided by a silicone fluid dashpot mounted below the core and coil assembly. An adjustable valve in the dashpot piston provides for TD adjustment. Upward motion of core and piston is dampened through the use of the silicon fluid dashpot. The core rises slowly until the piston reaches an increased diameter in the dashpot, where it is free to trip the contact with quick action. The time and current required to complete this cycle are inversely related as shown by the time–current curves (Figure 10.74).

Standard models of the Bulletin 810 are automatically released as soon as the current through the coil is decreased to approximately 20% of the pick-up current. The core is

Time - Minutes

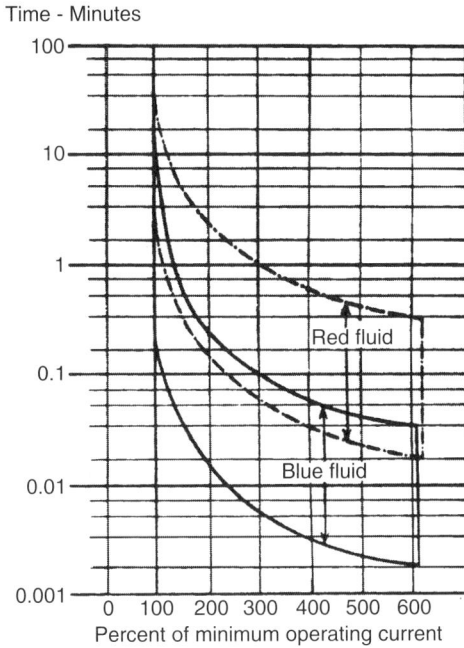

FIGURE 10.74

Inverse time–current curves of Bulletin 810 for silicon fluid with high (red) and low (blue) viscosity.

TABLE 10.1

Time–Temperature Relationship (+40°C Reference)

Ambient temperature (°C)	0	−10	−20	−30	−40
Operating time correction factor	2.25	1.80	1.45	1.20	1.0

designed to drop quickly, returning the contacts to their normal position. A check valve allows the position to bypass the fluid in its return to the bottom of the dashpot so there is no waiting period as with thermal relays.

The minimum operating current (100% on the time–current characteristics graph) is independent of ambient temperatures at the relay; however, the operating time at over-current varies directly according to the viscosity of the silicone fluid. Since the viscosity varies inversely with ambient temperature, the operating time is also inversely affected. The time–temperature table shows the proper correction factors to be applied to the operating times for various temperatures (Table 10.1).

10.5.2 Induction Relays: Design and Characteristics

Induction relays are based on a different principle than the other devices considered above. As it is known, if one places a conducting element in an alternating magnetic field an electric current will be created in this element. This current creates its own magnetic field, biased on 90° from the external field that has created it.

These two magnetic fields can interact with each other and as a result a torque appears that can be used for closing of contacts. In induction relays, a coil serves as a source of an alternating magnetic field and the conducting element is made in the form of a light aluminum disk, a hollow rotor in the form of a cup or a frame that is free to rotate around its axis. In order for the torque to appear, the movable elements must start moving, and for

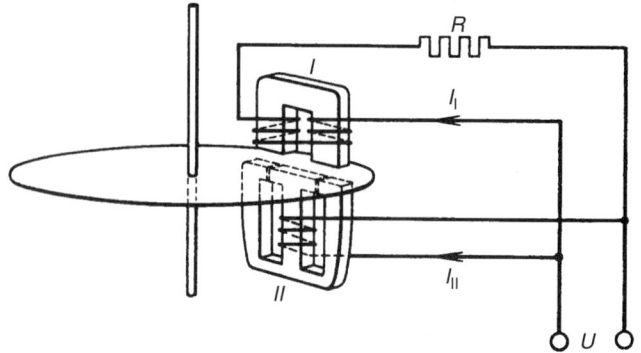

FIGURE 10.75
An induction magnetic system of the tangential type with a rotating disk.

that it must be affected by not less than two alternating magnetic fluxes displaced in space and in time from each other (that is out of phase). These requirements can be implemented in different ways. For example, if one places two coils, one on each side at the edge of the aluminum disk, and switches one of them with a greater number of turns (that is with greater inductance) directly to the source of the input voltage (U) and the other one with a less number of turns in series to the additional active resistance (R) (Figure 10.75), the magnetic fluxes created by the upper and lower coils will be biased in phase.

One can also obtain magnetic fluxes biased in phase by arranging the windings in space in some special way (Figure 10.76). It should be noted that Westinghouse already

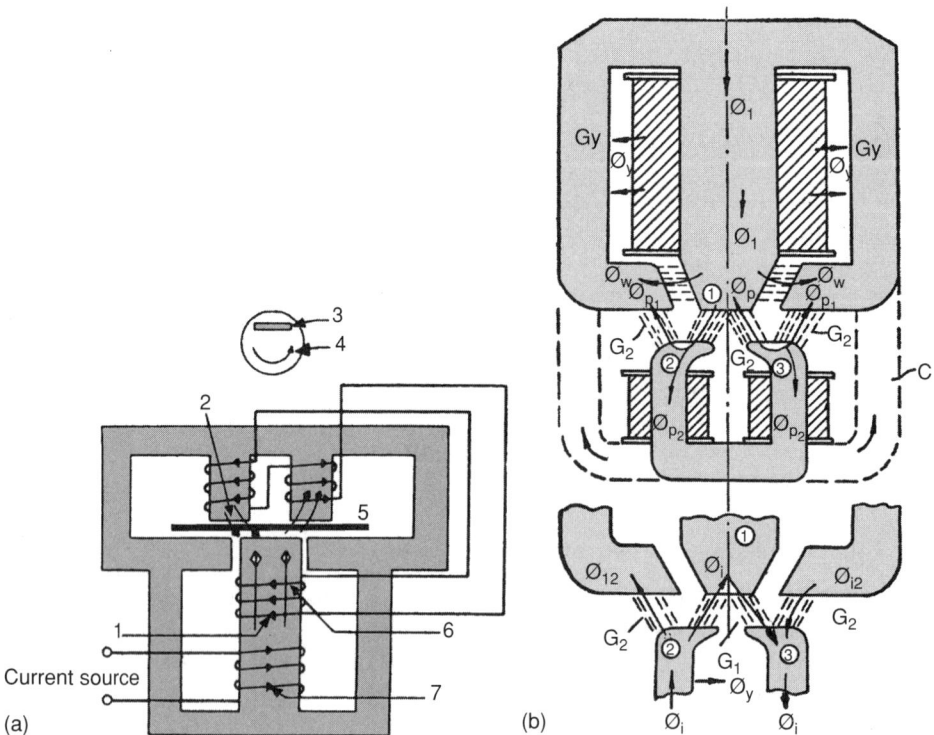

FIGURE 10.76
(a) Construction of the magnetic system of a CO-type induction relay (Westinghouse). 1 — Secondary current; 2 — upper pole flux; 3 — pole electromagnet; 4 — direction of disk rotation; 5 — disk; 6 — main pole flux; 7 — main pole current. (b) Magnetic fluxes in the induction magnetic system of the so-called "tangential" type with a rotating disk.

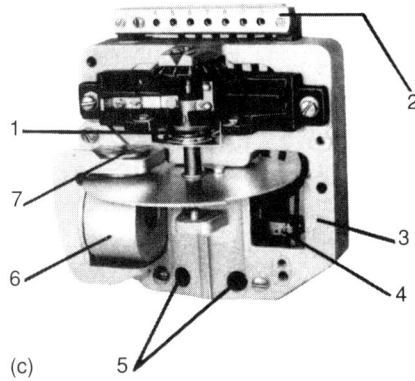

(c)

FIGURE 10.76 (*Continued*)
(c) External design of the magnetic system of a CO type relay (Westinghouse). 1 — Spring tension adjustment; 2 — tap block; 3 — rigid one-piece die-cast frame; 4 — tap value adjustment; 5 — magnet plugs (in electromagnet); 6 — damping magnet; 7 — magnet adjustment.

produced the first induction relays of this type in 1901, and CO-type relays have a history dating back to 1914. Relays of this type have been produced for decades and were a kind of standard on which new projects were based until the 1950s (Figure 10.76b).

This arrangement allows the magnetic fluxes of the coils (Figure 10.76c) to run through the disk edge, causing disk rotation as a result of the interaction with its magnetic field, occurring on the section near the disk edge.

Another example is the so-called shielded magnetic system (Figure 10.77). The shield here is a copper-shading ring put on the shaded pole of the magnetic core. We have already considered this construction in the section devoted to magnetic systems of the electro-

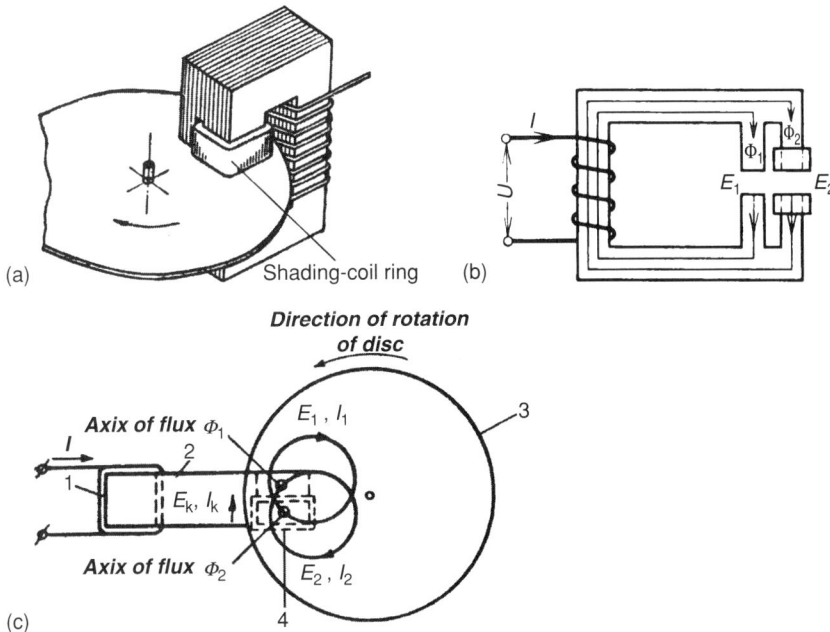

FIGURE 10.77
(a) Induction magnetic system of the shielded type: External view and design. (b) Induction magnetic system of the shielded type: Operating principle. 1 — Coil; 2 — magnetic core; 3 — disk; 4 — shield (shading-coil ring).

magnetic relays. In a standard clapper-type AC relay, this ring prevents the armature from vibrating because the magnetic flux created by this ring is biased by 90° from the main magnetic flux, which is why the total magnetic flux affecting the armature never reaches zero as the main flux changes sinusoidaly. The induction magnetic system uses the property of this ring to create additional flux biased by 90°, for torque on the aluminum disk. As the rotating disk crosses the magnetic fluxes (setting it in motion), there a so-called cutting current appears, preventing disk movement according to the Lenz law. It not only prevents movement, but also takes on a stabilizing affect at higher rotation rates of the disk, the greater the cutting current is. To intensify this effect an additional permanent magnet, the poles of which cover the disk edge, is used in some induction relays. The exposure degree of the magnet on the disk depends on its power and position on the disk. In relays with a permanent magnet, there is usually an adjusting mechanism providing radial displacement of this magnet within some limits.

Instead of a disk, a hollow rotor in the form of a cup can be used in induction relays. This rotor has a much smaller diameter and arm of force than the disk, which is why two sources of the magnetic field biased in space and in time are not enough for its rotation, unlike the case with a disk. A magnetic system of at least with four poles is used for rotor rotation (Figure 10.78). Such a magnetic system implies the use of two coils arranged on the core arms in such a way that the pole axes of these coils intersect at the right angle. This means that the magnetic fluxes of these coils are also biased at this angle in space. The phase shift between currents in these coils is determined either by parameters of current (voltage) sources, to which these windings are connected or created artificially with the help of a capacitor (if only one source is used).

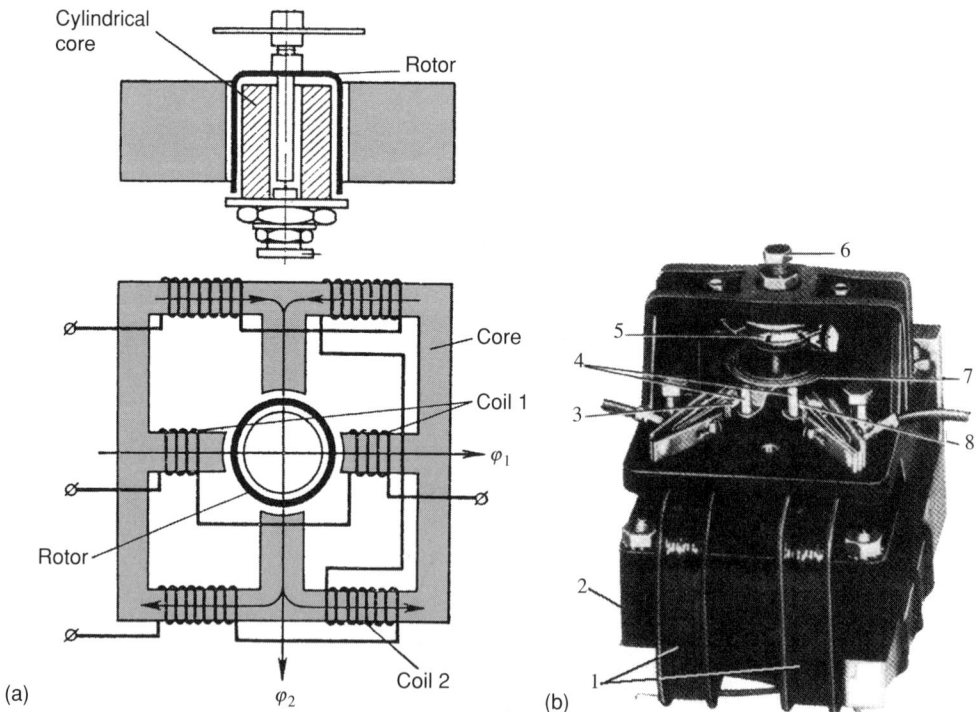

FIGURE 10.78

(a) Construction diagram of the induction magnetic system of a four-pole type with a rotating rotor. (b) External design of the unit of the induction magnetic system of a four-pole type with a rotating rotor. 1 — Coils; 2 — ferromagnetic core; 3 — normally open stationary contact; 4 — moving contacts; 5 — spring adjusting ring; 6 — upper pivot assembly; 7 — upper control spring; 8 — normally closed stationary contact.

As the torque in induction systems is the function of frequency, amplitude of magnetic fluxes, and phase angle between them, these systems can be applied in relays of different purposes:

- Current and voltage relays
- Frequency relays
- Active and reactive power relays
- Impedance relays, etc.

It should be noted that protective relays based on these principles have been in existence for 75 years already. Apparently current relays were the first application of induction magnetic systems. Earlier constructions (Figure 10.79) already contained practically all the elements of modern relays but were not good enough.

For example, early models (Figure 10.79) had quite a primitive reduction gear with a large gear ratio, allowing a great number of disk revolutions until the relay picked up. This construction is made in the form of a thread (2), which is wound around the axis while the disk is rotating and which pulls the movable contact to the stationary one.

Also in the 1940–50's, relays based on an induction system of tangential type were also produced (Figure 10.80). The principle of operation of them has been considered above. In this relay, the thread has already been replaced with a real worm-gearing.

Induction relay RIK type with very complex kinematics to have no parallel, designed by ASEA (Sweden) in 1930's was a model for modern Russian RT-80 series relays (Figure 10.81).

The disk axis in this construction is fixed with the help of bearings (4 and 5) in the frame (5), which can also rotate in bearings 7 and 8. That is, the disk in this construction can move. In the initial position the spring draws off the frame (6) in such a way that the worm (3) does not touch the toothed quadrant (17).

At a certain current value in the winding, due to the factors described above, the disk begins to rotate. When the disk reaches a certain rotation speed, depending on the

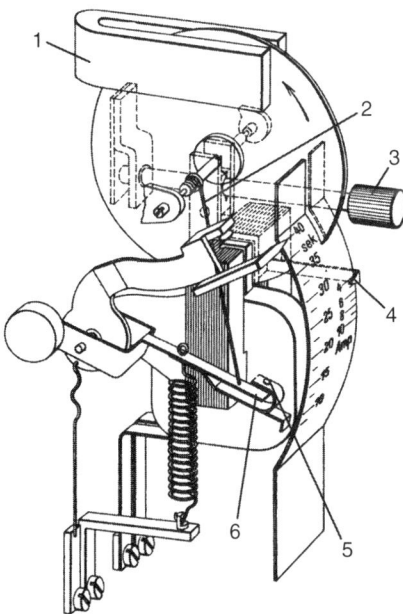

FIGURE 10.79

Induction current relay with a vertical rotating disk, produced by Siemens in the 1930s. 1 — Permanent magnet; 2 — thread; 3 and 4 — setting handles of pickup current and TD; 5 and 6 — contacts of the relay.

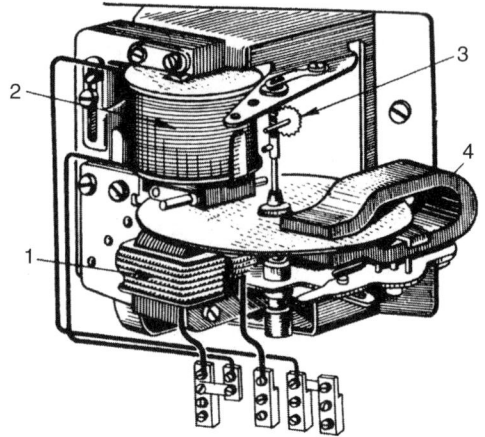

FIGURE 10.80
A protective relay on the basis of an induction system of the tangential type, produced in the 1940–50's. 1 — First coil; 2 — second coil; 3 — worm-gear; 4 — permanent magnet.

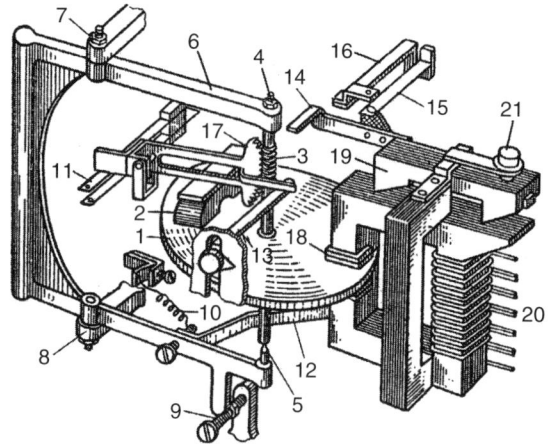

FIGURE 10.81
Construction of an induction current relay of the IT-80 (PT-80) series, produced in Russia. 1 — Aluminum disk; 2 — permanent magnet; 3 — worm; 4–8 — bearings; 9 — limiter; 10 — spring; 11 — low-power signal contacts; 12 — ferromagnetic plate; 13 — stop; 14 — lever; 15, 16 — main contacts; 17 — toothed quadrant; 18 — shading-coil ring; 19 — armature rotating around the axis; 20 — coil outlets; 21 — adjusting screw.

current value in the winding, all forces affecting the disk including the magnetic field (2) lead to some resultant mechanical force applied to the disk axis. This force is transmitted to the frame (6), which rotates quite quickly in bearings 7 and 8, together with the rotating disk until the worm (3) touches and hooks the toothed quadrant (17). At that moment the TD of the relay starts counting. When the frame (6) rotates and the ferromagnetic plate (12) approaches the magnetic core, it is picked up by the magnetic field of dispersion and pressed to the magnetic core, thus providing fixation of the position of the frame (6) and reliable operation of the worm-gearing. When the disk rotates further, the toothed quadrant (17) with the pusher raises and closes the main contacts (15 and 16) affecting the lever (14). As the lever (14) is linked not only with the contacts, but also with the left end of the rotating armature (19) this end also goes up, bringing the right end near the special ledge of the magnetic core. At a certain gap the right end of the armature is taken by the magnetic field and safely engages with the magnetic core, providing reliable fixation and good pressure on the main contacts. When the current in the winding increases, the working torque affecting the disk increases proportionally to the current square, and then considerably more slowly because of saturation of the magnetic core (when the magnetic core is saturated, the increase of current in the coil does not lead to an increase of the magnetic flux in the working gap). Accordingly, the pick-up time of the relay is sharply reduced at first when the current goes up (the dependent part of the characteristic: $I>$), and then

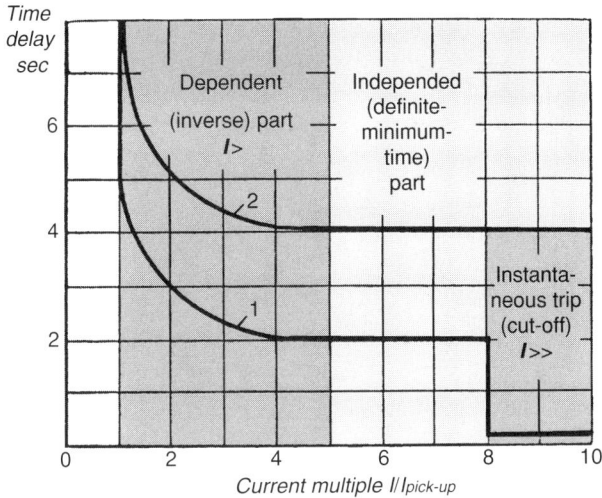

FIGURE 10.82
Time–current characteristics of the PT-80 relay. 1 — For setting $t = 2$ sec and multiple cut-off setting of 8; 2 — for setting $t = 4$ sec (without cut-off unit).

becomes practically constant (the independent part of the characteristic) (Figure 10.82). For relays of this type, the independent part of this characteristic starts approximately at an eight-to-tenfold pick-up current.

At such current ratios the right end of the armature (19) is immediately attracted to the magnetic core, providing closing of the main contacts long before the disk starts to rotate. The pick-up current of this part of the relay ($I\gg$) is adjusted by changing the air gap with the help of the screw (21).

There are 12 modifications of relays of the PT-80 series, for currents ranging from 2 to 10 A, and time settings from 0.5 to 16 sec.

Induction current relays with dependent characteristics, of the IAC and IFC series (a modernized IAC relay with its size reduced by 25%) produced by General Electric Co. for many years, have considerably simpler constructions (Figure 10.83). According to GE statistics, the total working time of these relays, used in various power units all around the world, is 15 million relay-hours.!

Unlike the previous construction, with complex kinematics in which the disk made many revolutions until the contacts closed, IAC relays are very simple (Figure 10.84) and do not contain any mechanical transmission. The required TDs at low currents are provided due to the very small rotation of the disk. Instantaneous cut-off is provided by a separate relay (8) of the clapper type. The contact, closed by the disk, switches the auxiliary relay (3) with its powerful output contacts, and a drop-out flag signals the pick-up of the relay.

The nominal pick-up current is fixed with the help of a simple switch of coil taps (tap setting). The current at which the disk begins to move (pick-up) is adjusted by changing the tension of the spiral spring. The time dial setting is chosen in accordance with the initial position change of the unit with the spring and the movable contact. The distance that the movable contact must travel before closing, and therefore the operate time of the relay, also changes. One can make an additional adjustment of the operate time by shifting the permanent magnet along the disk radius. A lot of credit goes to the designers of this relay who created a reliable construction, having stable characteristics, that has been used all over the world for decades, with a minimal set of elements.

The protective properties of induction current relays are determined by the form of their time–current characteristics, which may differ (the slope of the characteristic, its curvature) in various relays of different types. The characteristics of protective relays

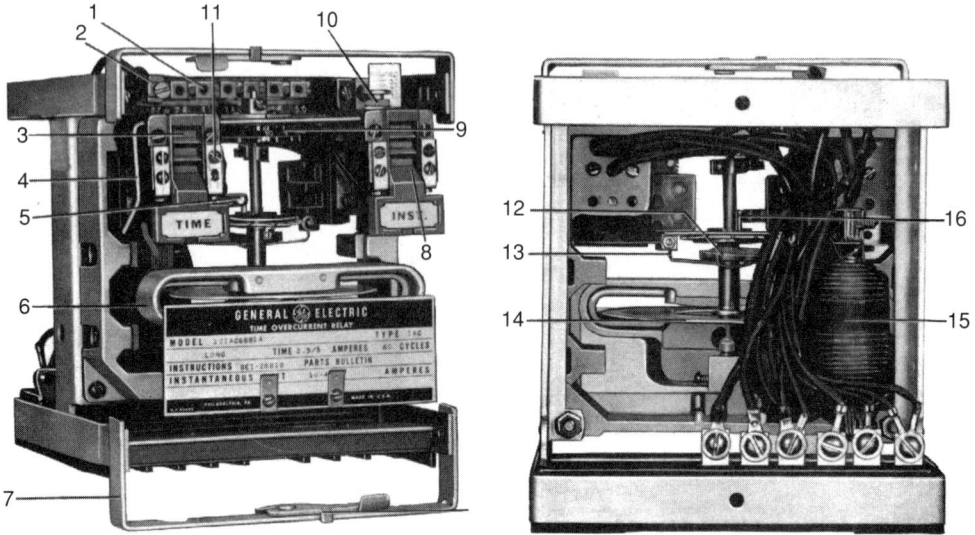

FIGURE 10.83

Induction current relay with dependent characteristic of the IAC series, produced by General Electric Co. 1 — Tap block; 2 — sliding lead; 3 — target seal-in unit; 4 — reset armature; 5 — stationary contact of induction unit; 6 — drag magnet; 7 — cradle; 8 — instantaneous unit; 9 — time dial; 10 — adjustable core; 11 — target seal-in unit coil taps. 12 — control spring; 13 — control spring support; 14 — disc; 15 — coil with U-share magnetic core; 16 — moving contact of induction unit. (General Electric Time Overcurrent Relays).

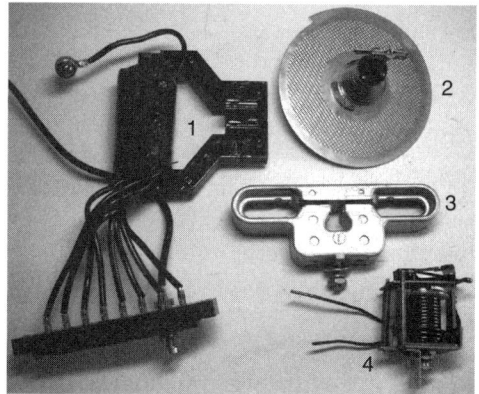

FIGURE 10.84

The basic set of elements of the IAC relay. 1 — Coil with U-share magnetic core; 2 — disc with spiral spring and movable contact; 3 — magnet; 4 — instantaneous unit.

must be well matched with the parameters of the protected object, which is why it is important to distinguish between these characteristics. Western producers apply a special classification that includes six types of characteristics: inverse (or normal inverse); very inverse; extremely inverse; short-time inverse; medium-time inverse; long-time inverse (British Standard 142). The most commonly used and popular are the first three (Figure 10.85 and Figure 10.86).

The *normal inverse* (or inverse) characteristic is most suitable for systems where there is a large variation of fault-current for different fault locations (i.e., the source impedance is much smaller than the line impedances). The inverse characteristic enables improved utilization of the protected object's overload capacity, and increased cold-load pick-up capability compared with the definite-time characteristic.

Very inverse characteristic: The operating time is more dependent on fault-current magnitude; therefore this characteristic is suited for systems where there is a fairly large

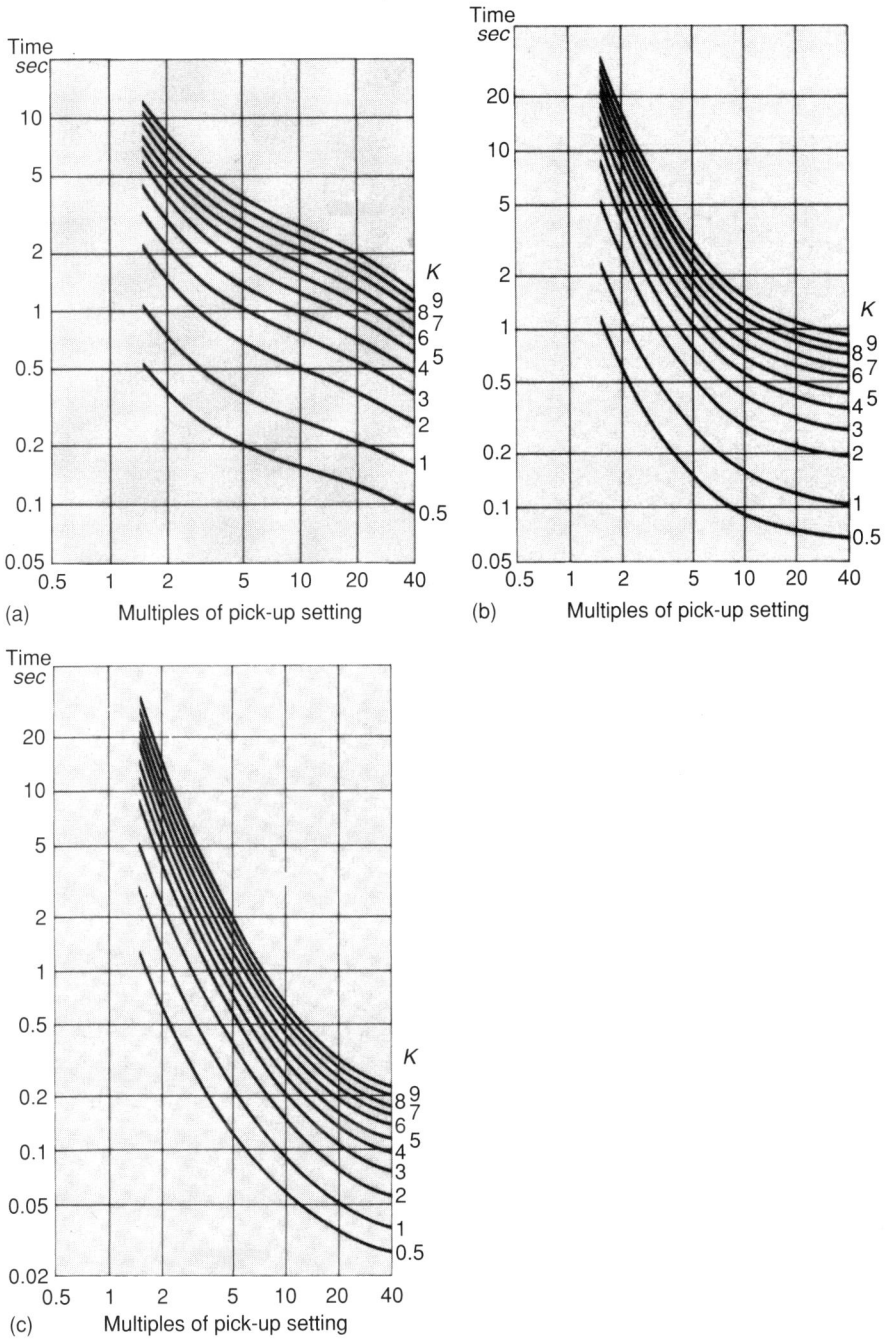

FIGURE 10.85
(a) Time–current characteristics of the ''inverse'' type (the example used here is an IFC-51 relay, produced by GE). K — time dial setting. (b) Time–current characteristics of the ''very inverse'' type (this example is an IFC-53 relay, also produced by GE). (c) Time–current characteristics of the ''extremely inverse'' type (another example of a IFC-53 relay, produced by GE).

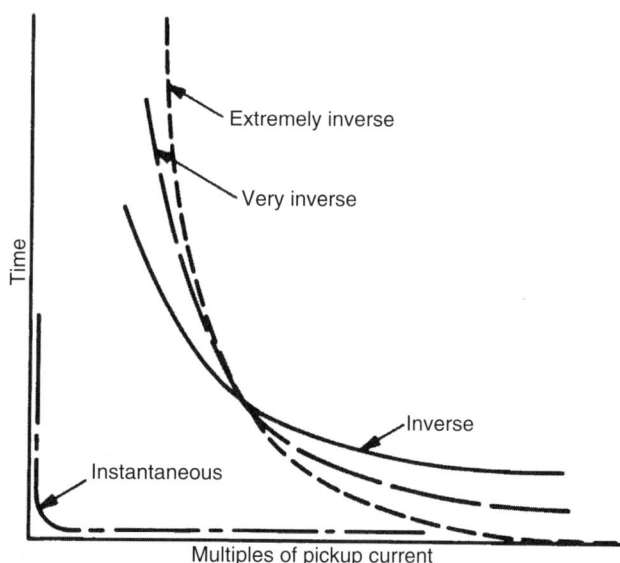

FIGURE 10.86
Characteristic set of most popular types, shown in the same scale for comparison with each other.

variation in fault-current for different fault locations (i.e., the source impedance is less than the line impedances). Where line impedance has a large influence on the fault-current level the very inverse characteristic often permits faster overall fault clearance than do normal inverse characteristics.

The use of inverse-time or very-inverse-time relays for the protection of long lines is generally more satisfactory than for short lines. This is because on a long line there are substantial differences in short-circuit currents, and hence the operating time (depending on whether the fault is near the remote end or the local end of the protected line). This makes it easier to obtain proper time coordination.

Extremely inverse characteristic: The operating time is very dependent on fault-current magnitude. These characteristics are intended for coordinating with fuses on distribution or industrial circuits, and where an extended service outage results in a heavy accumulation of loads of automatically controlled devices (such as water pumps, refrigerators, water heaters, oil burners, etc.). Such load accumulations often produce inrush currents considerably in excess of the full-load feeder current for a short time after the feeder is re-energized. The extremely inverse time relay characteristic permits successful pick-up of these loads and at the same time provides adequate fault protection which can be made selective with fuses and cut-outs in other parts of the circuit.

They are used in situations requiring a high degree of overload capacity utilization, and where cold-load pickup or energizing transient currents could be a problem. They are also suitable for providing coordination in networks where current changes for different fault locations are small but finite and distinguishable.

It is interesting to compare the efficiency of a protection relay having such characteristics, to that of a relay having definite-time (independent) characteristics (see above). Let us remind ourselves that a relay with a definite-time characteristic will have an operating time is independent of the fault-current magnitude. This characteristic is mainly suitable for use in systems where the fault-current magnitude is relatively constant for different fault locations (i.e., source impedance is much larger than line impedances). It also simplifies selectivity planning in conjunction with other relays having instantaneous or definite-time characteristics (Figure 10.87). TD selection should be begun at the farther-

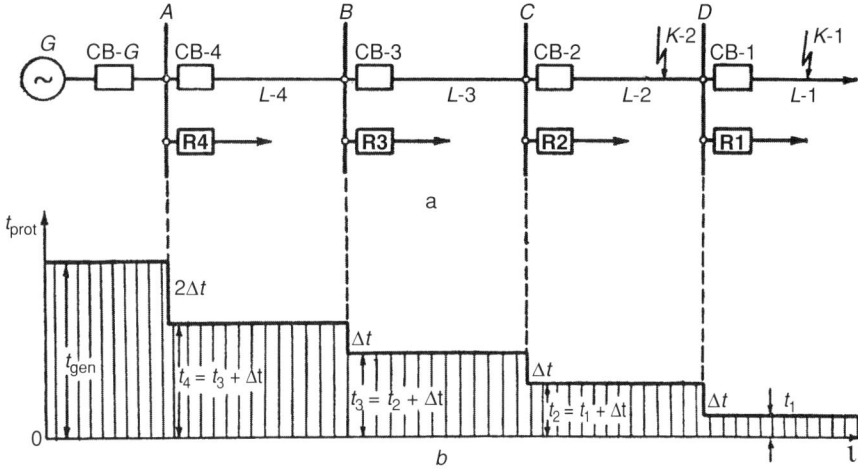

FIGURE 10.87
Radial power transmission line with usage protective relays with an independent TD (a "time-graded" or "stepped" TD).

most circuit section and finally completed at the power source (Figure 10.87). Let us select the minimum possible TD t_1 for the far-end circuit section *L*-1. By reason of selectivity considerations, in order that a fault at point *K*-1 will not lead to disconnection of circuit section *L*-2, the TD for circuit section *L*-2 must be somewhat greater:

$$t_2 = t_1 + \Delta t$$

where Δt = time-delay step or increment necessary as a reliable margin against operation of protection on the succeeding circuit section.

On circuit section L-3 the overcurrent protection should operate in a similar fashion with the TD:

$$t_3 = t_2 + \Delta t,$$

etc.

Figure 10.87 shows a diagram with the TDs selected for a system of power circuits and a generator. We can point out here that in order to increase the reliability, the generator protection has a TD setting which includes a double time-delay step, that is

$$t_G = t_4 + 2\Delta t$$

generally Δt = 0.4 to 0.5 sec.

The major advantage afforded by use of overcurrent relays having inverse-time characteristics is that their TDs are approximately inversely proportional to the short-circuit current or overload current. This property provides the possibility of obtaining simple and quick-acting protections for individual radial circuits against short circuits.

Figure 10.88 shows a system of protection time-delay characteristics in the form of a function $t = f(t)$. They clearly show with what TD the faults are cleared at the various circuit points, not only by the individual protection of a given circuit section (main protection) but also by the protection of the next power-supply-side circuit section (back-up protection). This is another advantage of these relays.

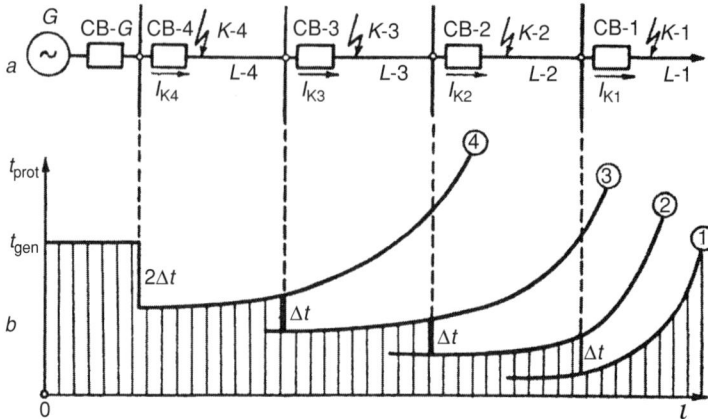

FIGURE 10.88
A radial power transmission line with usage protective relays with a dependent TD (an inverse-TD).

Any comparison of Figure 10.87 and Figure 10.88 makes it evident that the nature of the TDs is identical in both cases. In protections using relays with inverse-time characteristics, a fault at the far end of the circuit section is cleared with a greater TD than a fault at the near end (refer to Figure 10.88). To provide an allowance for the greater inertia error of induction-type relays, the time-delay step Δt is equal to 0.7 to 0.8 sec.

As a whole, the application of inverse-time relays results in higher TD levels. This is a disadvantage of these relays. As can be seen, the relay with an extremely inverse characteristic (1) has a smaller effective area than the relay with a normal inverse characteristic (4).

Induction systems with a rotating disk have a certain inertia (because of the large diameter of the disk) which plays no role when these systems are used for relays with a dependent TD; however, these systems cannot be applied in relays with a small TD because the minimum pick-up time of the system with a disk exceeds 0.1 sec.

A modern induction magnetic system (Figure 10.89) of the four-pole type with a rotating rotor of small diameter (as the so-called "Ferraris motor") provides a minimum pick-up time of 0.02 to 0.04 sec and is used for current relays with small TDs.

In May–June 1885 Galileo Ferraris (Figure 10.90a) conceived the idea that two out-of-phase, but synchronized, currents might be used to produce two magnetic fields that could be combined to produce a rotating field without any need for switching or for moving parts. This idea, which is common place to electrical engineers now, was a complete novelty in the 1880s. Ferraris published it in a paper to the Royal Academy of Sciences in Turin in 1888.

Ferraris devised motor (Figure 10.90b) using coils (as electromagnets) at right angles and powered by alternating currents that were $90°$ out of phase, thus producing a revolving magnetic field. The motor, the direction of which could be reversed by reversing its polarity, proved the solution to the last remaining problem in alternating-current motors. The principle made possible the development of the asynchronous, self-starting electric motor that is still used today.

At the time Ferraris seems not to have thought that his principle would lead to a motor for industrial purposes, but he suggested that it could be used as the basis of a meter for alternating current measurements. Ferraris did not want to take out a patent on his inventions and refused a large sum from an American company, because he thought that the discovery can be put in the service by everyone: "I am a professor, not

FIGURE 10.89
Rotor and a ready-assembled four-pole induction system. 1 — Magnetic core; 2 and 3 — coils; 4 — pusher of the movable contact; 5 — movable contact; 6 — stationary contact; 7 — rotor; 8 — spring.

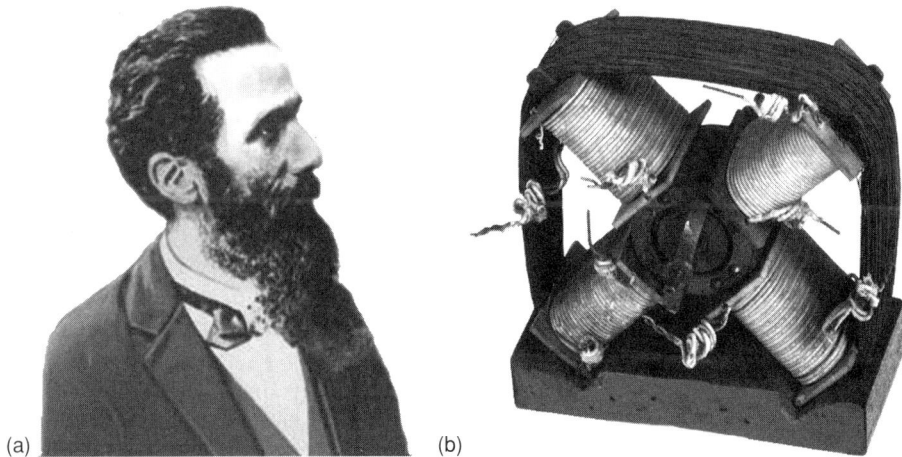

FIGURE 10.90
(a) Galileo Ferraris (1847 to 1897), Italian physicist who studied optics, acoustics, and several fields of electrical engineering. (b) Ferraris motor. Revolving rotor can be seen in central part of construction.

an industrialist," he said with regard to the offer (now, there are several International Patent Classification indices for devices with Ferraris motor principles: H01H53/12, G01R11/36, G01R5/20).

A typical case of application of such a magnetic system with a rotor is a relay of the CHC11A type, produced by General Electric. When using the CHC11A relay in circuit-breaker-failure back-up schemes, the relay may be called on to carry maximum fault current for some fraction of a second before the fault is cleared. For this reason, the short-time current capability of the relay should be noted. This is particularly true of the hinged-armature unit (without an inverse TD). CHC11A relays have inverse time characteristics for very short TDs (Figure 10.91a).

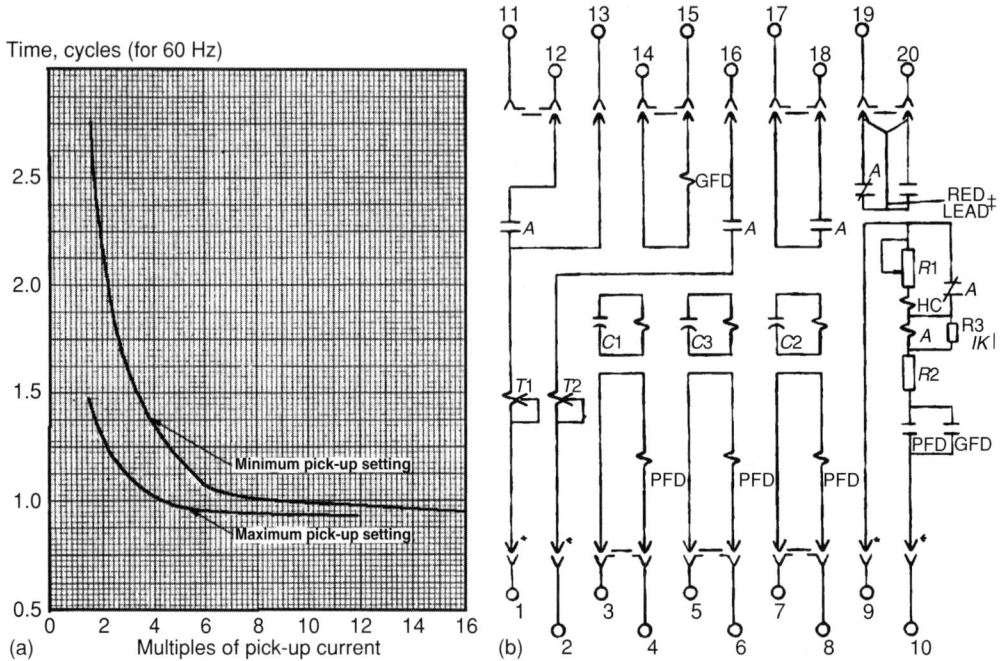

FIGURE 10.91

(a) Time–current characteristics of the CHC11A-type relay. The time is expressed in cycles of AC with a frequency of 60 Hz. Duration of one cycle is $t = 1/60$ sec. (b) Circuit diagram of the CHC11A-type relay in a three-phase modification (General Electric). PFD — Phase-fault detector (induction cup unit); GFD — ground-phase detector (hinged-armature).

The short-time rating of the cup unit is so high that it will probably never be a limiting factor. While the 2 to 8 A cup unit is continuously rated for 5 A, it is capable of carrying 8 A continuously. This is important in multi-breaker bus arrangements, where bus current that the relay may be connected to receive can exceed 5 A during maximum load conditions.

The phase bias required for torque in the four-pole induction magnetic system of the CHC11A-type relay is created with the help of capacitors (C_1, C_2, and C_3 in Figure 10.91b).

The induction cup unit is intended for multiphase faults and the small hinged-armature for ground faults.

10.5.3 Electronic Current Relays with Dependent Characteristics

TDs dependent on current are implemented in such relays with the help of the RC-circuit considered above. The charging rate of the capacitor *with constant power supply voltage* is known to be determined by the so-called time constant $\tau = RC$. If one switches a threshold element picking up at a certain voltage level parallel to the capacitor, one can change the TD value by changing the resistance of the resistor R and the value of the capacitor C until this threshold element picks up. This principle is basic for the operation of the time elements with independent characteristics in current relays. Mind the words above in italics "with constant power supply voltage." This is the requirement for a fixed TD determined only by the parameters R and C. The voltage on the capacitor U_C (that is on the threshed element) increases according to the exponential law (Figure 10.92):

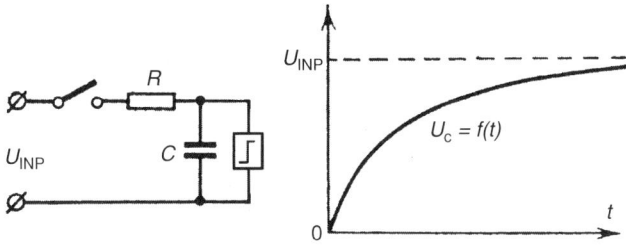

FIGURE 10.92
Basic principle of construction of the TD unit on the basis of an RC-circuit.

$$U_C = U_{INP}\left(1 - e^{-\frac{t}{\tau}}\right),$$

where is the U_{INP} is the input voltage applied to the RC-circuit; t is the charging time of the capacitor to the voltage U_C.

The threshold element picks up when the voltage on the capacitor U_C reaches the pick-up voltage of the threshold element U_P, and the TD (t_P) to the pick-up equals:

$$t_P = RC \ln \frac{U_{INP}}{U_{INP} - U_P}$$

As can be seen from the latter formula, the TD depends not only on the time constant value (that is on parameters R and C), but also on the input voltage value U_{INP}, which is why special means of stabilization of the input voltage are applied in relays with an independent TD.

On the contrary relays with a dependent TD make use of this property of the RC-circuit. In the latter case, current is converted to voltage (for example, with the help of the input transformer) and is applied to the RC-circuit with the threshold element. The greater the input current (that is the voltage applied to the capacitor) is, the quicker it is charged and the smaller the TD will be until the pickup of the threshold element. This is what is needed for protective current relays with a dependent *TD* (Figure 10.93). In the initial mode, when current is lesser than a certain threshold value, the contact (*S.R.*) of the start relay short-circuits the winding (*w2*) of the transformer (*T3*) through the current-limiting resistor (*R4*). As a result, there is no voltage, neither on the charging RC-circuit nor on the winding (*w3*) of this transformer. The start relay picks up at the certain value of the input current and de-shunts (by its contact *S.R.*) the rectifier (VD2), through which the capacitor (*C2*) is charging. As the capacitor is charging, the current used by it from the winding (*w2*) decreases and the voltage on the winding (*w3*), which is rectified, filtered and applied to

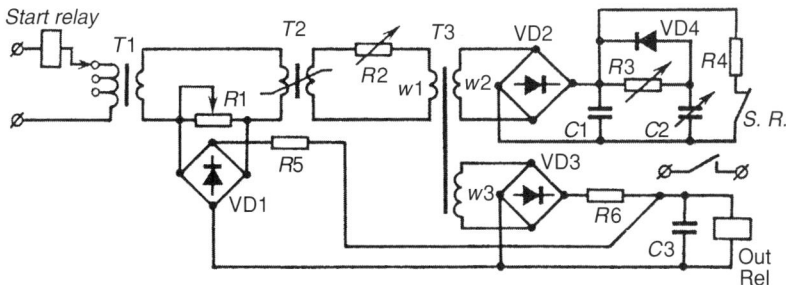

FIGURE 10.93
Circuit diagram of a simple electronic current relay with a dependent TD, based on an RC-circuit.

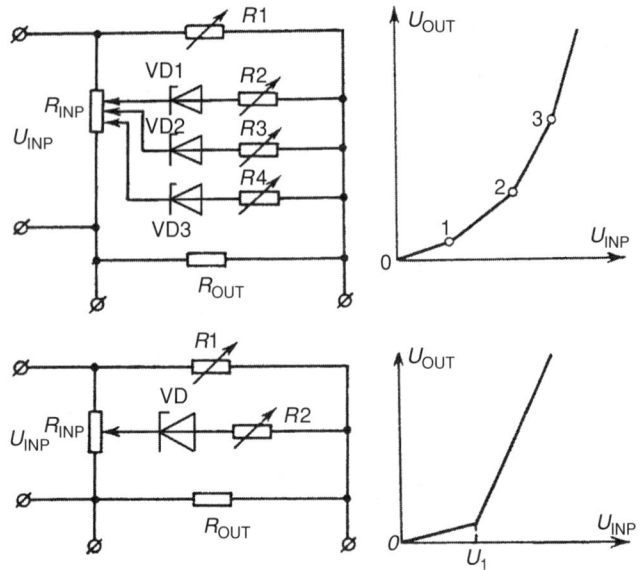

FIGURE 10.94
Principle of forming of a specific characteristic with the help of Zener diodes (*VD*) and resistors: above — with several stages; below — with one stage.

the winding of the output relay, increases. At a certain voltage level this relay picks up. If the input current is great, the voltage drop on the resistor (*R1*) becomes great enough for pick up of the output relay before the capacitor (*C2*) is charged. This forms the independent part of the characteristic. The dependent part of it is formed when the voltage applied to the capacitor (*C2*) is proportional to the input current. At very high current values the transformer (*T2*) is saturated and the voltage in the RC-circuit does not increase further, while the current goes up. This allows stabilization of the dependent part of the characteristic.

If one cut in an additional converter of the input voltage on Zener diodes, for instance, with different voltages at the input of those devices (Figure 10.94), it is possible to make the voltage applied to the capacitor charge change, depending on the input current (voltage) according to a specific low. If the input voltage is less than the drop-out voltage of the Zener diodes it is applied to the charging RC-circuit through the resistor (*R1*), which corresponds to the first section on the curve. If the input voltage (current) increases and reaches the drop-out voltage of the Zener diodes (VD1), resistor *R2* appears to be switched parallel to resistor *R1*. At that point the resultant resistance decreases and the curve passes to the second section, and so on. As a result of this process it is possible to obtain different forms of time–current characteristics of the relay. This principle has been in use for quite a long time already, since the 1960s and 1970s (Figure 10.95).

The relay's input has a built-in CT with several secondary windings with taps. Over the switch for the setting of the operating current, the voltage is taken out across a resistor. When this voltage, rectified, smoothed, and compared with above-mentioned voltage, exceeds the reference voltage, the starting relay picks up. At the same time, the RC circuit starts charging up. For the inverse time lag relay the required time characteristic is obtained through a combination of Zener diodes and resistors used in the mentioned RC circuit. When the capacitor in the RC circuit is charged up to a certain voltage level, the tripping relay also picks up. In the three-phase design, the measuring circuit acquires a voltage that is proportional to the largest of the three currents.

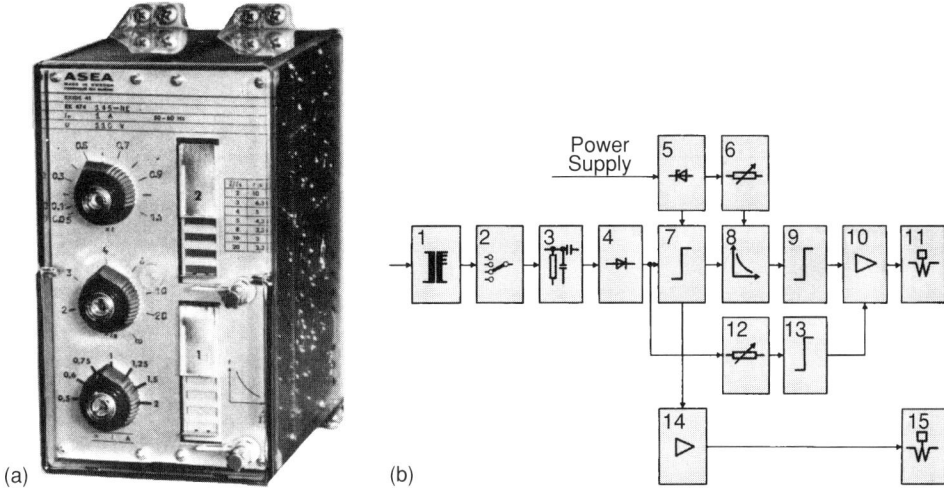

FIGURE 10.95

An electronic relay of the RXIDE-4 type, with a dependent time–current characteristic. External view and block-diagram. 1 — Input transformer; 2 — current setting; 3 — RC-filter; 4 — rectifier; 5 — stabilizing circuit; 6 — time setting; 7 — level detector; 8 — time circuit; 9 — level detector; 10 — amplifier; 11 — tripping relay; 12 — instantaneous setting; 13 — level detector; 14 — amplifier; 15 — starting relay. (*ABB Buyer's Guide 1990.*)

Instantaneous operation is obtained by means of the rectified part of the voltage from the transformer being compared to the reference voltage — when the latter is exceeded an operating impulse is given to the tripping relay.

In later constructions of relays, the required time–current characteristics are formed with the help of a specialized microprocessor synthesizer. The time–current characteristics of such relays match exactly the characteristics of induction relays. This relay of the IC91 type, produced by ABB, can serve as a good example of such a relay (Figure 10.96).

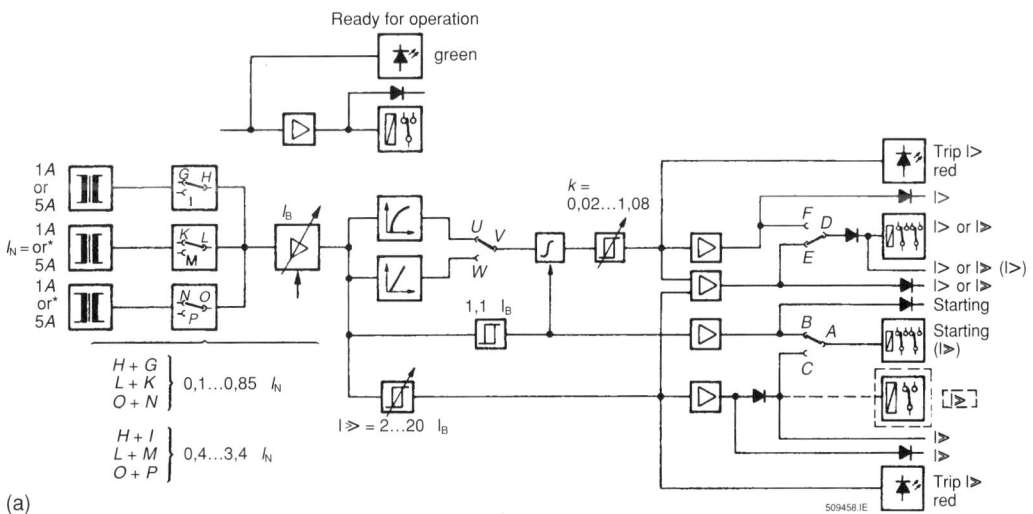

FIGURE 10.96

(a) Block-diagram of a microprocessor current relay of the IC91 type with a dependent characteristic.

(Continues)

(b)

FIGURE 10.96 (*Continued*)
(b) External design of a microprocessor current relay of the IC91 type, with a dependent characteristic (ABB Buyer's Guide 1990).).

10.6 Harmonic and Voltage Restraint Relays

When a power transformer is energized, current is supplied to the primary, which establishes the required flux in the core. This current is called a "magnetizing inrush" and it flows only through CTs in the primary winding. This causes an unbalanced current to flow in the coil of the current relay, which would cause faulty operation if means were not provided to prevent it.

Power system fault currents are of a nearly pure sine waveform, plus a transient DC-component. The sine waveform results from sinusoidal voltage generation and nearly constant circuit impedance. The DC-component depends on the time in the voltage cycle at which the fault occurs, and upon the circuit impedance magnitude and angle.

Transformer magnetizing inrush currents vary according to the extremely variable exciting impedance resulting from core saturation. In modern transformers, larger than approximately 10 MVA and with orientated sheet-metal, the amplitude of the inrush current can be five to ten times of the rated current when it is connected to the high voltage side, and 10 to 20 times the rated current when it is connected to the low voltage side. The amplitude and duration of the inrush current (up to some few seconds) is dependent on the design of the transformer, its connection, and its neutral earth point, as well as the short-circuiting effects of the network.

These currents have a very distorted waveform made up of sharply peaked half-cycle loops of current on one side of the zero axis, and practically no current during the opposite half cycles. These two current waves are illustrated in Figure 10.97. Any current of distorted, nonsinusoidal waveform may be considered as being composed of a direct-current component plus a number of sine-wave components of different frequencies: one belonging to the fundamental system frequency, and the others, called "harmonics," having frequencies which are 2, 3, 4, 5, etc., times of the fundamental frequency. The relative magnitudes and phase positions of the harmonics, with reference to the fundamental, determine the waveform. When analyzed in this manner the typical fault

Typical offset fault current

Typical transformer magnetizing inrush current

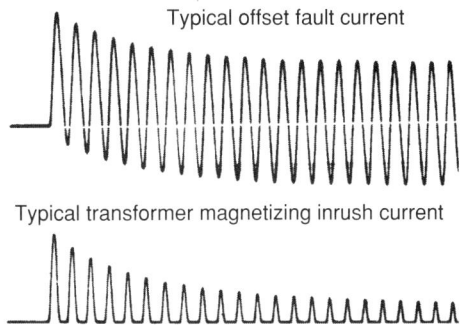

FIGURE 10.97
Typical fault and transformer magnetizing inrush currents.

current wave is found to contain only a very small percentage of harmonics, while the typical magnetizing inrush current wave contains a considerable amount (Table 10.2).

The high percentage of harmonic currents in the magnetizing inrush current wave affords an excellent means of distinguishing it electrically from the fault current wave. In special current relays the harmonic components are separated from the fundamental component by suitable electric filters.

The simplest type of relay with a harmonic restraint characteristic is the PT-40/F-type relay, differing from the PT-40-type relay considered above by a filter formed by inductance of the transformer (T) and the value of the capacitor (C) (Figure 10.98).

Currents of the highest harmonics are closed through the capacitor (C) and not applied to the winding of the relay. The parameters of the elements are chosen in such a way that the pick-up current of the relay at a frequency of 150 Hz is no less greater than eight times that of the pick-up current of the relay at a frequency of 50 Hz. In this way different sensitivities of relays to short-circuit currents (50 Hz) and current rushes of magnetization of the transformer (over 100 Hz) are provided.

Another widespread principle of construction of relays with harmonic restraint is to use a quickly saturated transformer (QST) cut in between the standard current relay and the current source (Figure 10.99). Such a QST is usually made on a triple core (Figure 10.99b), and contains two short-circuited windings ($w3$ and $w4$) in addition to the working winding ($w1$) and the secondary winding ($w2$).

In the normal mode of operation, when the current I_1 is sinusoidal, the magnetic flux in the left rod of the transformer equals the sum of magnetic fluxes $F1$ and $F2$ created by the primary winding w1 and the short-circuited winding $w4$. At the nominal current value I_1 these fluxes create current I_2 in the secondary winding ($w2$), enough for pick-up of the

TABLE 10.2

Harmonic Analysis of a Typical Transformer Magnetizing Inrush Current Wave

Wave Component	Ratio of Amplitude of Harmonic Component to Amplitude of Fundamental (%)
Fundamental	100
Direct current	57.7
Second harmonic	63.0
Third harmonic	26.8
Fourth harmonic	5.1
Fifth harmonic	4.1
Sixth harmonic	3.7
Seventh harmonic	2.4

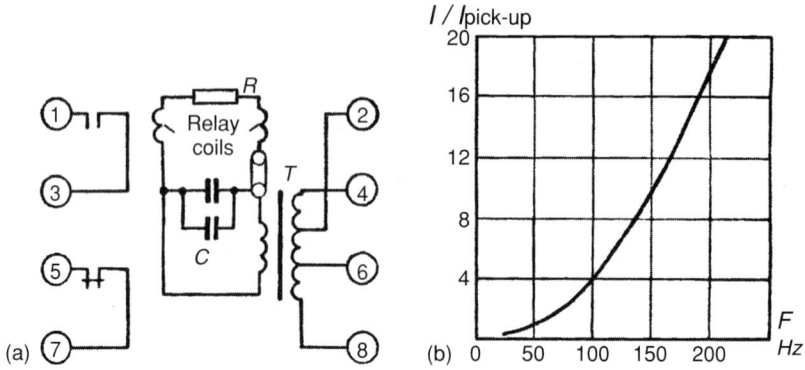

FIGURE 10.98

Circuit and frequency characteristic of the simplest PT-40/F-type relay with a harmonic restraint characteristic.

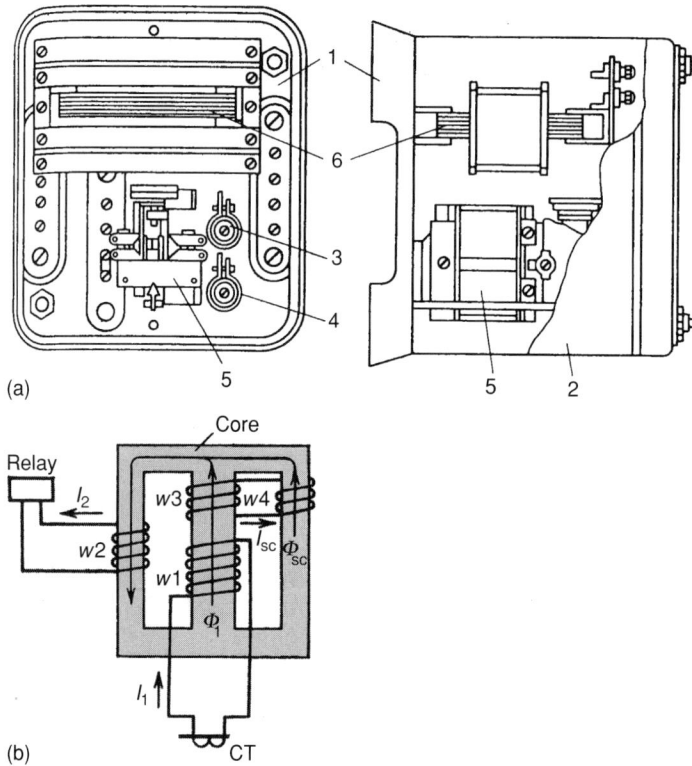

FIGURE 10.99

(a) Current relay of the PHT-565 type (Russia) with harmonic restraint characteristic produced on the basis of a QST. 1 — Heel piece; 2 — case; 3 and 4 — adjusting resistances; 5 — integrated. PT-40 relay (see above); 6 — QST. (b) A simplified circuit of an RNT-565 current relay with a QST.

relay. When the nonsinusoidal magnetizing inrush current starts flowing in the primary winding of the QST, its core is quickly saturated, since this current is unidirectional. As a result, the current transformation to the secondary winding $w2$ is especially reduced due to a decrease of the constituent of this current, obtained from the short-circuited winding. The current I_2 remains very small (not enough for a pick-up of the relay) even at a high amplitude of the input nonsinusoidal current. The principle of offset from rushes of

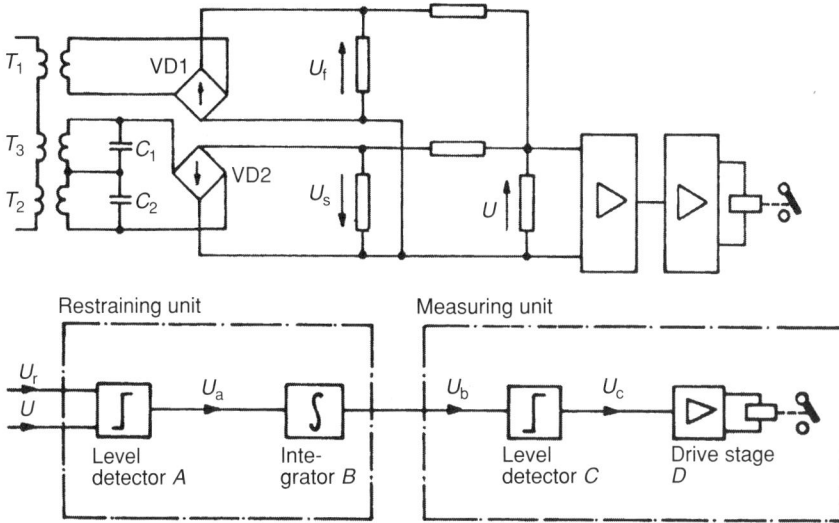

FIGURE 10.100
Block-diagram of harmonic restraint electronic overcurrent relay RAISA type (ABB).

magnetization current with the help of a filter is also used in electronic current relays (Figure 10.100).

The restraining method used in the RAISA relay is based upon the fact that the second harmonic (100 Hz) is relatively much larger in the switching surge than in the short-circuit current. The restraining voltage U_S is obtained via transformers T_3 and T_2. The inductance of the transformers and the capacitance of the capacitors C_1 and C_2 are tuned for resonance at the second (100 Hz) and the fifth (250 Hz) harmonic components. The second harmonic component is used to restrain the relay from the magnetizing inrush current. The fifth harmonic component is incorporated to attain the desired insensitivity for higher harmonics. The transformers supply the rectifier bridge VD2, which provides the counteracting voltage U_S for restraining. Voltages U_f and U_S are summated in accordance with their signs to a resultant voltage U, which is supplied to the measuring circuit.

Voltage restraint overcurrent relays are relays, the current pick-up of which changes because of voltage level, not harmonics as the in cases considered above. It is constructed as a standard current relay of the induction type with a rotating disk, but it is also equipped with an additional voltage coil which creates voltage restraint (Figure 10.101 and Figure 10.102). Why are such relays needed?

The voltage restraint overcurrent relay was designed for the purpose of providing external-fault back-up protection for the generator. System fault back-up protection should be provided at the source of faulty currents, the generator. Such protection should protect against the generator continuing to supply short-circuit current to faults in the adjacent system element, because the fault may not have been removed by other protective equipment. The current source for the voltage restraint relay should be a CT at the neutral end of the generator windings, or at the line-side CT. Phase-to-phase voltage should be obtained from the generator potential transformers. Loss of potential to the voltage restraint relay will cause the relay to trip if the generator load current, expressed in relay secondary amperes, is greater than the zero voltage pick-up current of the relay.

FIGURE 10.101
Time–overcurrent relay with a voltage restraint of the IJCV51 type (GE). 1 — Potential restraint coil; 2 — potential restraint resistor; 3 — current coil. (General Electic Type IJCV Relays with Voltage Restraint).

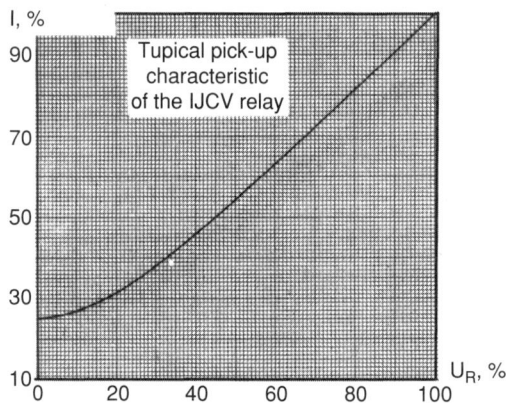

FIGURE 10.102
Typical pick-up characteristic of the type IJCV relay. U_R — percent of rated voltage across restraint circuit; I — current in percent of tap value.

10.7 Pulse Current Relays

The so-called pulse current relay is energized by very short current pulses with duration of just a few microseconds of input. Such relays must have a very small TD, sufficient to pick-ups when partial breakdown of isolation in high-voltage vacuum tube is occurred, or to operations with arresters, when current carry in lightning rods (when lightning strikes), etc., and also must have a long output signal, compatible with ordinary electro-mechanical devices

In micro-electronic devices, especially high-frequency ones, electric signals of small duration are frequently used; therefore electronic circuits for pulse stretch are very widely used in these devices. An example of such a circuit is the RC-integrator (Figure 10.103).

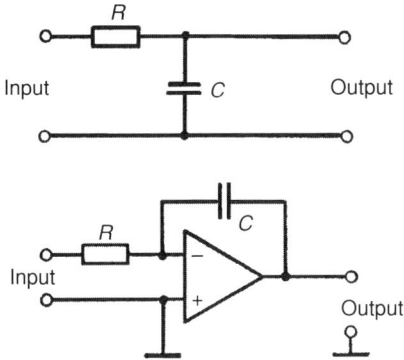

FIGURE 10.103
Examples of simplest RC-integrators, widely used in electronic devices.

For the same purpose the so-called "monostable multivibrator" (or "waiting multivibrator," "one-shot multivibrator" or "univibrator") is frequently used (Figure 10.104). Many companies produce special microchips that carry out functions of a monostable multivibrator. In powerful impulse devices in the electric power industry such high-sensitivity devices do not always work well, as they tend to pick-up under the external electromagnetic influences, and not from the input signals. In addition, such elements are not yet relays and additional circuits required: power supply, amplifying circuits, input and output devices. As far as is known to the author, there are still no universal pulse relays for high currents suitable for every such application available today in the market, therefore in each concrete case designers must invent, design, and develop such a relay in order to solve each specific problem.

The author solved one such problem when he developed a variant of the pulse relay for high currents (Figure 10.105). As a result, this very reliable relay, with a very low input resistance (that is, not sensitive to electromagnetic influences) was developed. This relay,

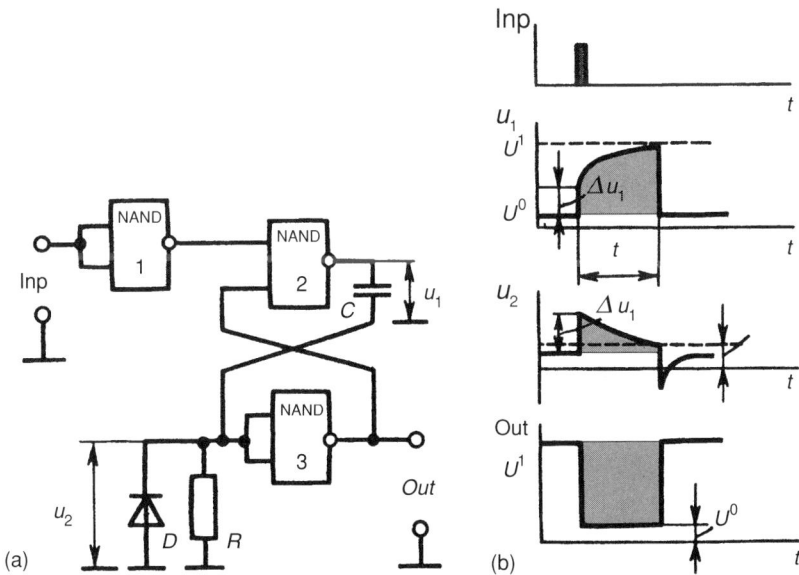

FIGURE 10.104
Univibrator: circuit diagram with NAND logical elements and oscillograms.

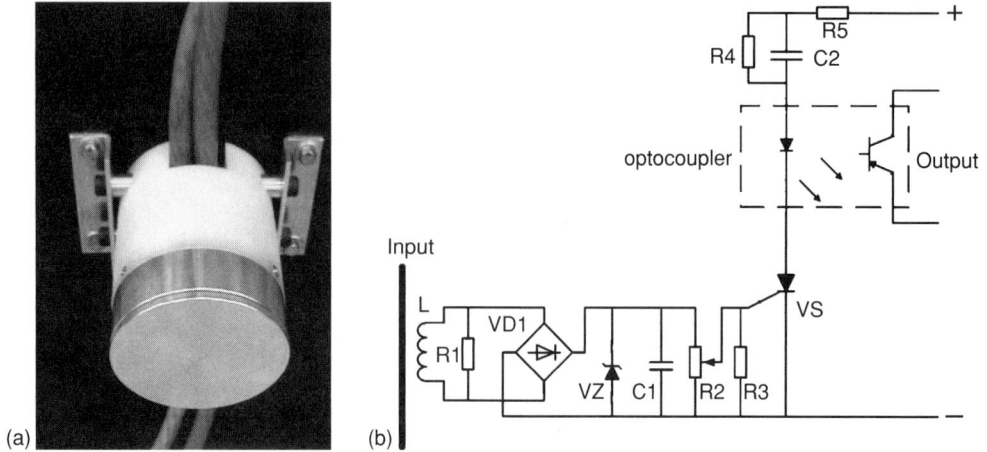

FIGURE 10.105
External view (a) and circuit diagram (b) of the pulse relay for high currents.

fixed on a thick copper conductor is capable of pick-ups from short pulses of current of only 5 to 10 μsec duration, with amplitudes from several tens of amperes up to several tens of thousands of amperes.

In tests, the high-voltage pulse generator of this relay made single current pulses with amplitudes ranging from 200 up to 20,000 A and standard pulse shape with $T1/T2 = 8/20$ (Figure 10.106). The exact threshold of the pick-up of this relay is adjusted by an internal trimmer. The duration of the output pulse depends on the capacity of capacitor C2 and can change from tens of milliseconds within a few seconds.

Three independent identical units were used in the specific design of the relay developed by the author, adjusted on three different thresholds of pick-ups: 300, 1000, and 10,000 A (Figure 10.107). Each unit contains a thyristor VS1, which will be actuated by input pulse with level, exceeded of preset (by trimmer R2) value and switched ON during few microseconds. Capacitor C2 start charging across the open thyristor. Thyristor remains in open state during capacitor charging time and turned OFF after capacitor C2 will be fully charged. In other words, weight of output pulse (up to 150 V, 0.1 A) is equivalent

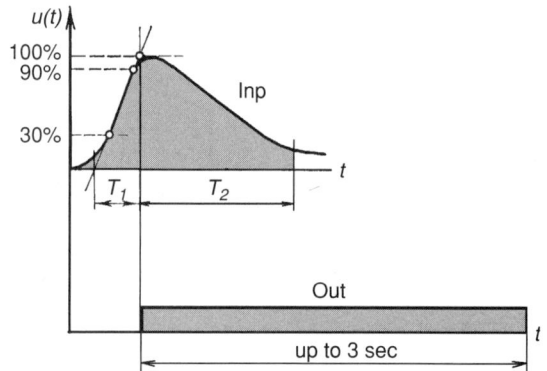

FIGURE 10.106
Input pulse shape applied to relay during test and signal on relay output ($T_1/T_2 = 8/20\,\mu$sec).

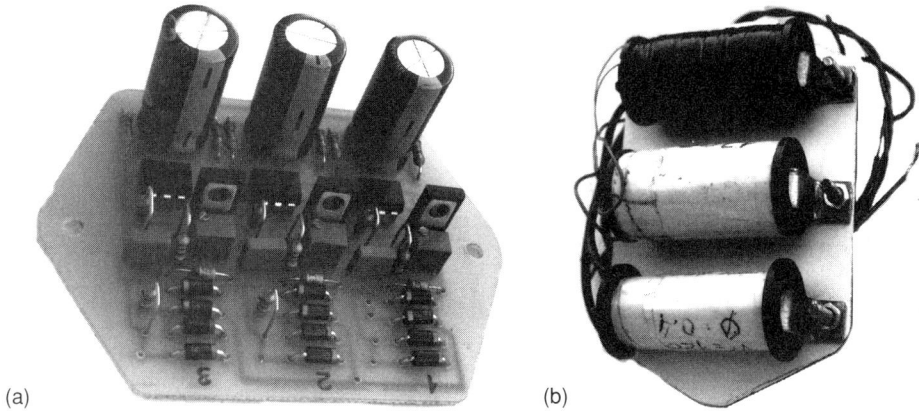

(a)   (b)

FIGURE 10.107
Printed circuit board (a) and input coils unit (b) of the multistage pulse current relay, developed by author.

to the capacitor time charging. Relay returns to its initial state after capacitor $C2$ will discharge across resistor $R4$.

If the printed circuit board and input coil are placed in a high-voltage insulating case (like that are used in high-voltage reed relays of the RG-series, described above) and filled with epoxy resin, the result will be a fine high-speed protective relay with high-voltage isolation.

11

Power and Power Directional Relays

Very often even in technical literature "power relays" and "power directional relays" are mixed up. Some books do not distinguish between these two notions, claiming that power relays respond only to the direction (sign) of the power applied to their clips, and not to the power value, which can vary within wide ranges. In other books, one can find claims that power relays respond both to the value and to the direction of power.

In fact, neither of these claims is exactly so. There are power relays controlling both the value and the direction of the power applied to them, and there are also power relays responding only to the direction of power flux. These are two different types of relays with different constructions and different characteristics.

11.1 Induction-Type Relays

The operating element of the power relay is of the watt-metric type, similar to that used in the standard watt-hour meter. It is provided with one current and one voltage coil on a common magnetic core. Interaction of the two fluxes developed by the coils produces a torque in the aluminum disk, causing the disk to rotate. An Alnico magnet provides effective damping of the disk so that the characteristics are accurate throughout the timing range. When the power flow equals or exceeds the power setting of the relay and is in the proper direction for tripping, the disk begins to turn. The contacts are geared to the operating shaft and close at the end of disk travel, at the zero time-level position. The time required to close the contacts is dependent on the magnitude of power and the time-level setting. The relay is, in effect, a contact-making wattmeter.

Power relays are commonly used (as over-power relays) for protection against excess power flow in a predetermined direction. Such a need arises, for example, in the case of a small generating station which has its own local load and a normally closed emergency tie to a large power source. The small station has enough capacity to supply its own load but cannot supply an appreciable amount of power into the large system. In such a case, a power relay can be used to trip the emergency tiebreaker if power in excess of a predetermined amount is fed into the large system for longer than the given length of time.

The ICW-type relay, produced by General Electric (Figure 11.1), is a typical example of such a power relay. Power direction relays are used in protective systems as units that determine the direction of power passing through a protected line, where damage has occurred to the protected line or to other outgoing lines adjoined to the same substation (Figure 11.2).

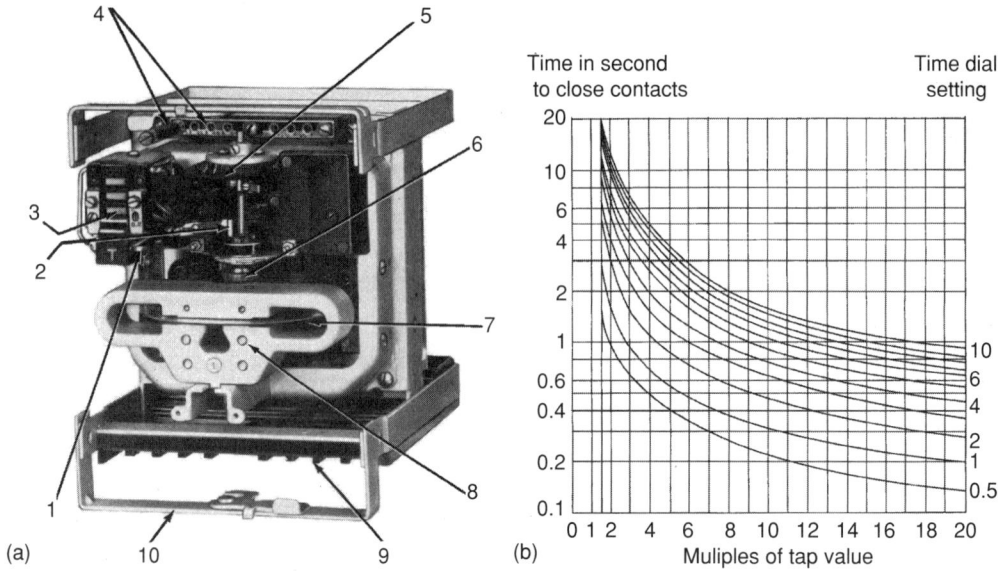

FIGURE 11.1

(a) ICW51 type power relay (General Electric), based on an induction system with a rotating disk. 1 — Contact (silver); 2 — seal-in contacts; 3 — target; 4 — tap block for power setting; 5 — time dial; 6 — control spring; 7 — aluminum disk; 8 — Alnico magnet; 9 — relay drawout contacts. (b) Time–watt curves of ICW51 relay. (General Electric Type ICW Owerpower Relays GEA-3417D.)

If a short circuit occurs, for example, at point *A* (Figure 11.2), the power (S_A) will pass from the source (*G*1) to the closing point (*A*), through the installation place of the relay. If the short circuit occurs at point *B*, the power (S_B), in the opposite direction, will pass through the installation place of the relay — that is from the source (*G*2) to the point of the short circuit (*B*).

To determine the direction of power, the exact values of power are not important. They can vary widely and are needed only for the estimation of the minimal power required for relay pick-up. This power is called as the pick-up power of the relay. *Power direction is determined by the angle between current and voltage.*

The power direction relay is based on an induction magnetic system with a rotating rotor (considered above) in the form of an aluminum cup (Figure 11.3), as such relays must be fast-acting and as has already been mentioned the rotating disk does not provide the required speed of operation. In addition to relays having a single-phase magnetic system with four poles, there are also three-phase power direction relays with eight poles (Figure 11.4).

FIGURE 11.2

Connection of the power direction relay and its application for determination of damages in the line.

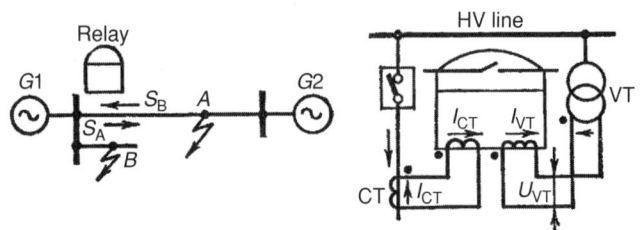

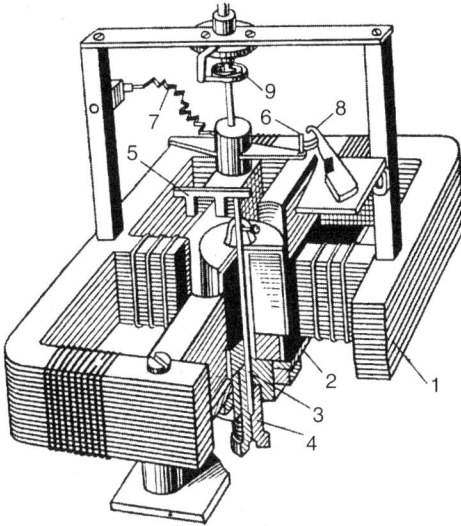

FIGURE 11.3
Construction of a power direction relay of the RBM type (Russia), based on an induction system with a rotating rotor. 1 — Magnetic circuit; 2 — drum; 3 — shaft; 4 — bearing; 5 — stopper for limiting drum rotation; 6 — moving contact; 7 — current lead; 8 — stationary contact; 9 — spiral spring.

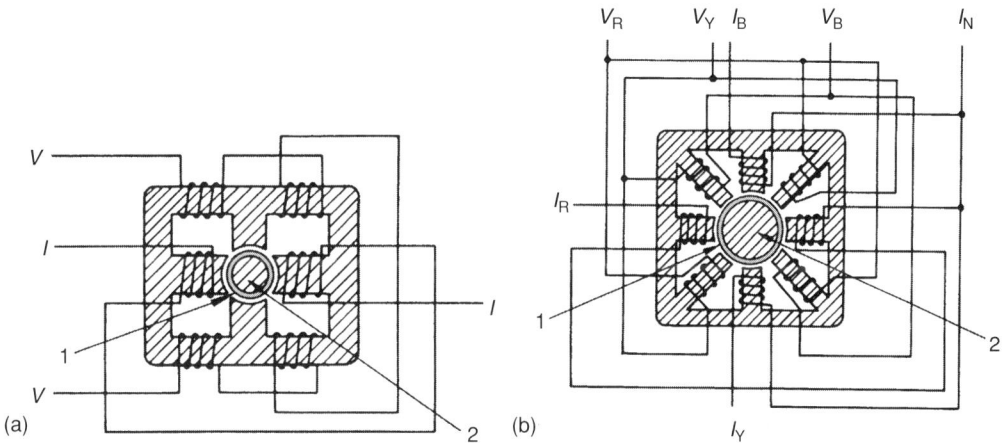

FIGURE 11.4
Single-phase (a) and three-phase (b) magnetic systems of the power direction relay with a rotating rotor. 1 — Aluminum rotor; 2 — stationary section of the magnetic core of cylindrical form.

11.2 Characteristics of Power Direction Relays

As it has been considered above, perpendicular sinusoidal magnetic fluxes, F_I and F_U (created by current and voltage coils), induce eddy currents, i_i and i_u, respectively, in the rotor body. These currents interact with the magnetic fluxes F_I and F_U, inducing them to create a constant rotating torque (M) on the rotor:

$$M = kF_I F_U \sin \phi$$

where ϕ is the angle between the fluxes F_I and F_U and therefore between the currents in the current (I) and voltage (I_U) windings.

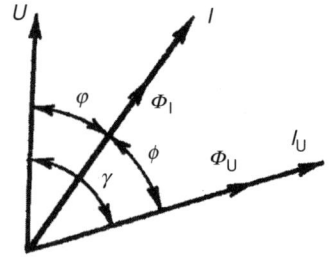

FIGURE 11.5
Vector diagram of the induction power direction relay. U, I — voltage and current applied to the relay inputs; I_U — current in the voltage winding of the relay; γ — interior angle of the relay (the angle between the applied voltage and current in the voltage winding).

As the maximum value, $\sin \phi$ takes place when $\phi = 90°$, one can say that the maximal torque (corresponding to the maximal sensitivity) will be developed in the relay when $\phi = 90°$. This is the angle between the current on the current coil (I) and the current in the voltage coil (I_U), and not between the input current and input voltage (Figure 11.5).

In the formula shown above, if the magnetic fluxes F_I and F_U are replaced with the current I and voltage U (proportional to them), and the angle ϕ with angle $\gamma - \varphi$ equaling it, one obtains a general formula for the torque on the rotor, expressed through the input current and input voltage:

$$M = kIU \sin (\gamma - \varphi) = kS$$

where S is the full power on the relay input.

The interior angle of the relay γ is determined by the constructive parameters of the relay and may vary synthetically. If the voltage winding is made in such a way that its pure resistance is less than its reactance ($R \ll X$), the current in the voltage winding (I_U) will be behind the applied voltage (U) by an angle approximating $90°$ (that is $\gamma = 90°$), and one will obtain the following formula for the torque:

$$M = kIU \sin (90° - \varphi)$$

or taking into account that $\sin(90°-\varphi) = \cos \varphi$:

$$M = kIU \cos \varphi = kP$$

where P is the active power.

Relays reacting to active power are called *active power relays or cosine relays.* And vice versa, if pure resistance of the voltage winding is much higher than the reactance ($R \gg X$), the current in this winding (I_U) will practically coincide in phase with the voltage (U), the angle between them will equal $0°$ (that is $\gamma = 0°$), and the torque (M) will be:

$$M = kIU \sin (0 - \varphi)$$

or taking into account that $\sin (0 - \varphi) = -\sin \varphi$:

$$M = -kIU \sin \varphi$$

To obtain positive torque, outlets of the voltage circuit of this type of relay are made with opposite polarity. In this case:

$$M = kIU \sin \varphi = kQ$$

where Q is reactive power.

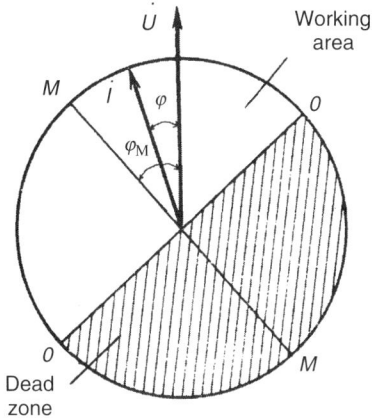

FIGURE 11.6
Characteristic of a power direction relay.

Relays responding to reactive power are called *reactive power relays* or *sine relays*. At intermediate values of the interior angle (γ), the relay responds to both power constituents and is called as a *mixed type relay*. These relays are most widely used in the power industry. Characteristic of the power direction relay looks rather unusual in comparison with the characteristics of the other relays considered above (Figure 11.6). How such a characteristic is obtained?

First, one lays off the vectors of the actual current (I) and voltage (U) applied to the relay inputs, taking into account the actual angle (φ) between them. The vector of the voltage is the basis for counting. Then, through changing of the current phase (by turning vector I), the maximal sensitivity of the relay is achieved.

The angle between current and voltage at which the sensitivity of the relay is maximal is called the *angle of maximal relay sensitivity* (φ_M), and the line drawn at this angle through the beginning of the vector of the voltage (M–M) is called the *line of maximal torques*. In contrast to that, an additional line drawn at a right angle to the line of maximal torques is called the *line of zero torques*. As it can be seen from the characteristic, this is a very important line separating the *working area* of the relay from the *dead zone* (which is sometimes called the *zone of dead band of the relay*) and in which the relay will not work. What happens to the relay if the current vector crosses the line 0–0? Nothing terrible. The torque affecting the rotor (and the contacts through it) will change its polarity to the opposite one, that is, it will become negative. If the torque is directed to closing of contacts in the working area in the dead zone, it will be directed to repulsion of contacts. There are so-called *double-acting relays* in which the movable contact can move to both sides from an initial neutral position, closing the left-hand or the right-hand contact corresponding to direct or reverse direction of the power flux. Such relays are applied in double-side supply electric power systems.

When the power flowing through the installation point of the relay changes its direction, the angle between the current and voltage also changes. This allows use of the relay sensitive to this angle as a device responding to the direction of the power flux.

Another important characteristic of power direction relays is their volt–ampere characteristic (Figure 11.7). We have already mentioned that the power direction relay reacts to the direction of the power flux, and not to the power value; however, for normal relay operation a certain minimal power needed for rotor rotation, not equaling zero, must obviously be applied to the relay inputs. When there is a short circuit, voltage may considerably decrease in the circuit, which is why when one chooses a relay for a particular electric circuit, one must be sure that the proportion of voltage and current

FIGURE 11.7
Volt–ampere characteristic of the power direction relay.
1 — Actual; 2 — theoretical; $U_{0\,min}$ — minimal voltage
level permissible in emergency modes for the given
relay (~2.6 V for RBM-type relay, for example); I_{SAT} —
core saturation current.

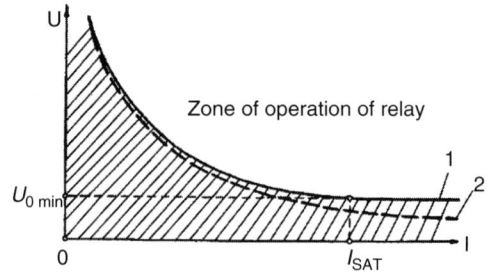

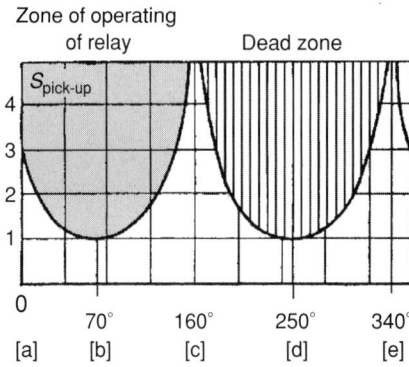

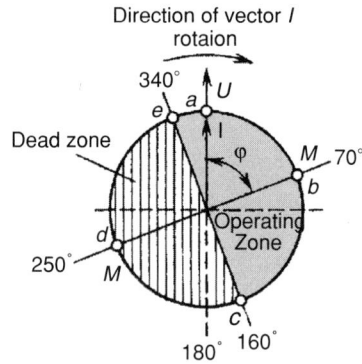

FIGURE 11.8
Dependence of pick-up power of a power direction relay of the RBM-171/1 type on the angle between current
and voltage. The torque–angle curve is shown for explanation.

will provide the minimum required power for relay operation in every emergency
mode. This proportion is described by the volt–ampere characteristic (Figure 11.7).
For reliable relay operation, its working area must be higher than its volt–ampere
characteristic.

As the pick-up power of the relay (S_{trip}) is measured within some limits depending on
the angle between the current and voltage, the corresponding dependence for every
particular relay can be plotted (Figure 11.8a), though it is not of that great importance
and is needed just for understanding the principle of operation of the relay.

11.3 Electro-Dynamic-Type Relays

A whole series of relays including RP-1-, RP-2-, RPA-, and RPF type devices, produced by
ASEA in the 1950–60's, had a peculiar (at least rear) construction. These were power
electro-dynamic relays (Figure 11.9 and Figure 11.10).

These relays, which work on the electro-dynamic principle, are provided with an iron
core (1), see Figure 11.9a, which is excited by the current in the coils A_1–A_3 and A_2–A_4. In
the air gap flux of the iron core, the voltage coil (2) moves, this coil is being fixed to one
end of the balance arm (5). The latter also supports the moving member (4) of the contact
and a soft iron pin (6), which under the action of a small permanent magnet (7) provides
the arm with the torque that determines the operating value of the relay, and retains the

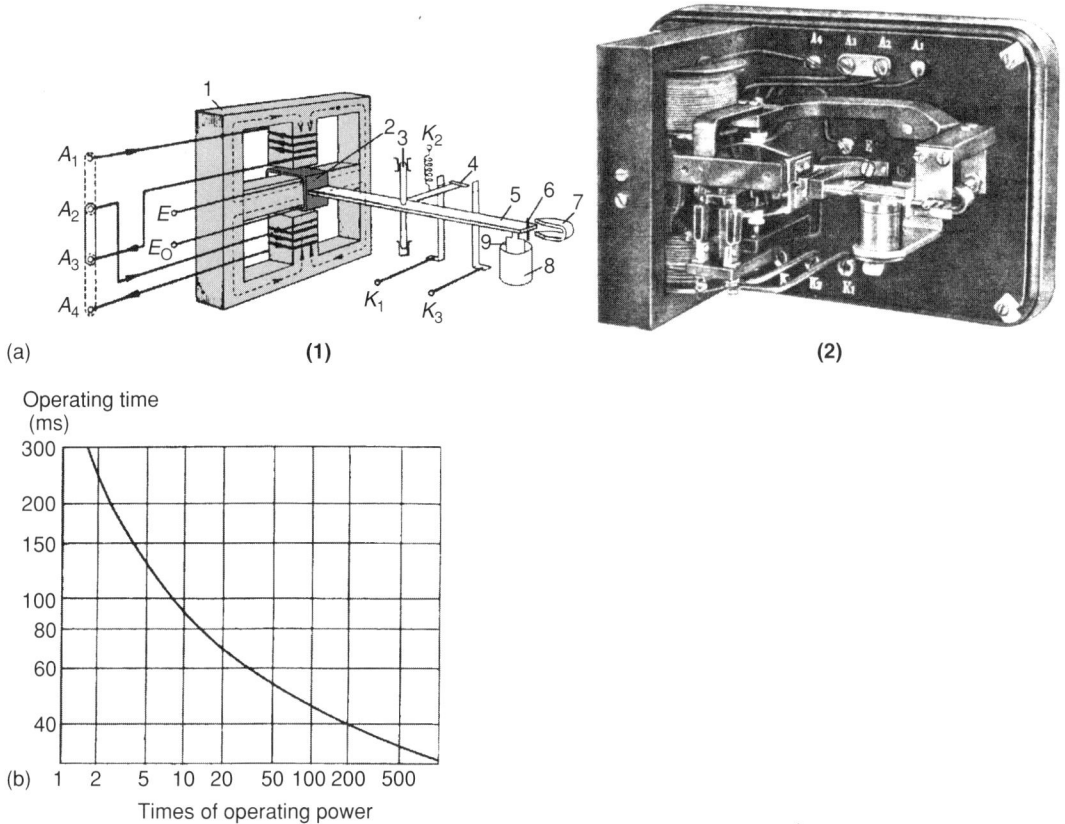

(a)   (1)   (2)

(b)

FIGURE 11.9

(a) Construction diagram and external design (with the cover off) of the single-phase power direction electro-dynamic relay of the RP-1 type. (ASEA). 1 — Iron core; 2 — moving coil; 3 — pivot for balance arm; 4 — moving contact member; 5 — balance arm; 6 — soft-iron pin; 7 — permanent magnet; 8 — dashpot with oil; 9 — metal vane. (b) Operating time for electro-dynamic power direction relay of the RP-1 type. (ABB 1968 Power Relays Types Catalog RK 51-1E).

arm in a defined initial position before operation. A damping device consisting of a metal vane (9), which moves in a dashpot (8) containing oil, prevents vibrations and hunting. When the two systems are supplied with current and voltage that have such a magnitude and phase angle that sufficient torque is obtained, the contacts (K1–K2 or K2–K3) close, depending on the direction of the power. The torque is largest when currents in the current coil and the voltage coil are in phase. If a phase shifting element is introduced into any of the circuits, it is possible to obtain the desired value of the angle between the voltage circuit voltage and the current of the current circuit at which maximum torque is obtained.

Because the RP-1 is not provided with any phase shifting element, the torque is proportional to the cosine of the angle between the currents in the two circuits. Due to this, the characteristic angle is $0°$ with a series resistance in the voltage circuit.

The mode of operation described above is most suitable for any directional-power relay in which the moving system is not influenced by any other mechanical forces than those the permanent magnet produces. On the other hand, when the type RPF relay is utilized as an over-, under- or regulating-power relay, it is provided with a spring and a graduated scale with a pointer by which the electro-dynamic forces can be balanced at a set power value.

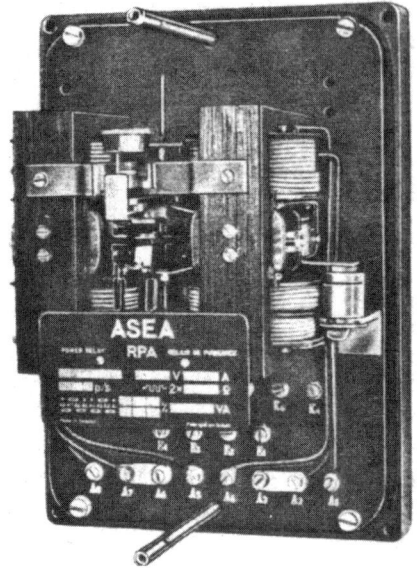

FIGURE 11.10
A two-phase electro-dynamic power direction relay of the RPA type, with cover removed. (ASEA (ABB) 1968 Power Relays Types Catalog RK 51-1E.)

The series resistors for the RP-1 and RPF relays are separately mounted. When the current coils are connected in parallel, these values are doubled. When the electro-dynamic relay RP-1 is used and the protection is based on the active component of the fault current, for example, in compensated power networks, the voltage coil of the relay is connected in series with a resistor. In the case of networks having a free neutral, where the protection is based on the capacitive grounding current, the voltage coil of the relay is connected, on the other hand, in series with an inductance.

Also available is a two-phase electro-dynamic power direction relay (RPA type, Figure 11.10), which consists of two single-phase systems with a common shaft and contact. It is connected in accordance with the two-wattmeter method, and can consequently be used in asymmetrically loaded networks without a neutral.

11.4   Electronic Analogs of Power Direction Relays

As for all other types of relays, there are also electronic analogs of power direction relays. In any such a relay, the input current and input voltage are first converted to two low-level voltages, which are then applied to the circuit, determining the coincidence (or noncoincidence) of these voltages in phase. Such circuit is called a *phase comparator* (Figure 11.11).

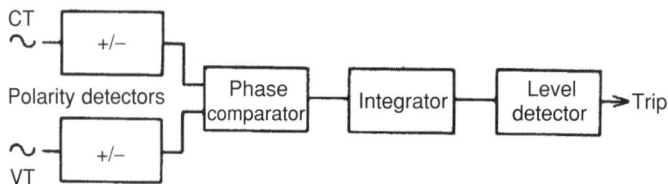

FIGURE 11.11
Block diagram of an electronic analog of a power direction relay.

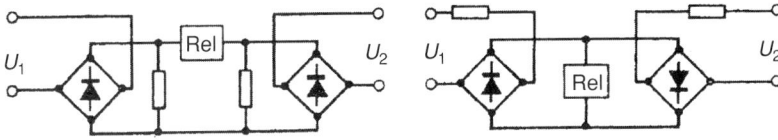

FIGURE 11.12
Circuit of phase comparators on rectifying bridges.

In the simplest case, the circuit containing the two rectifying bridges and a threshold element cut in on the difference in voltage of these bridges can serve as this *phase comparator* (Figure 11.12). Let us remind ourselves that the voltage at the input and output of the rectifying bridge looks like in Figure 11.13.

The difference of voltages applied to the threshold element can be zero (other conditions being equal) only when the rectified voltages are in phase or in opposite phase, and maximal difference in these voltages occurs when the phases are displaced by 90° between them (Figure 11.14). As a threshold element, one uses special highly sensitive electromagnetic relays with filters, providing leveling of the current in the winding or electronic switches.

In a ring phase-sensitive circuit (Figure 11.15), diodes play the role of the switching elements, which are either in the open or closed position, that is, they either let current pass or not, depending on the polarity of the applied voltage.

Power direction relays with diode phase comparators were produced in the 1970s in many countries and were very popular due to their simplicity. This simplicity can be seen in the RXPE-4-type relay, produced by ASEA (Figure 11.16), which serves as a good example.

Phase comparators may be based not only on diodes, but also on transistors (Figure 11.17). The current through transistors VT1 and VT2 (Figure 11.17) is initially balanced by the potentiometer (R_B) bridging their emitters, so that zero potential exists across VT3. An input voltage which raises the potential of A relative to B (A positive relative to B) causes VT2 to conduct more than VT1, which draws current through the base-emitter junction of VT3, which is then turned ON. In consequence, VT4 turns ON and produces an output voltage. When A is negative relative to B, the output is zero.

Lately, electronic devices of phase comparison have become very popular. They are based on the measurement of coincidence time by the signs of the two voltages, one of which is proportional to the current, and the other to the voltage (Figure 11.18). This time is fully dependent on the angle of phase bias between the current and voltage, which is why it is a parameter applicable in power direction relays.

In the circuit in Figure 11.18a, both transistors (VT1 and VT2) are fully enabled since there are no input signals, and are shunting the capacitor C. When input voltages E_1 or E_2

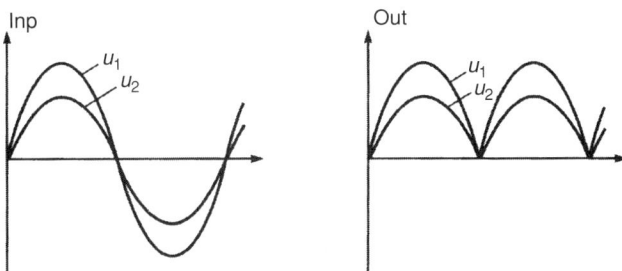

FIGURE 11.13
Curves explaining the process of rectification of alternating current with the help of diode bridges.

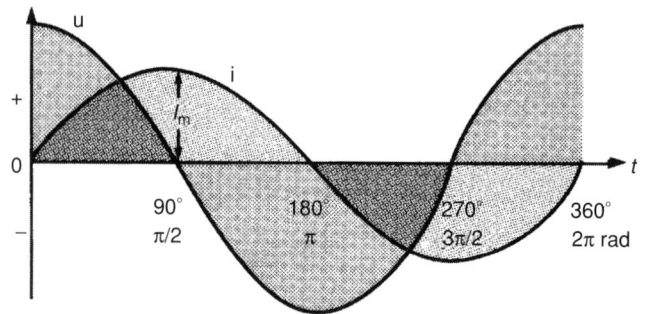

FIGURE 11.14

Phases of current and voltage at the input of the circuit providing a maximal output signal.

are applied, the corresponding transistor will be switched OFF. The voltage on the capacitor will always equal zero, if one of the transistors is constantly enabled, but if at any moment both transistors happen to be in state OFF, the capacitor C will begin charging during the period of time that they are in this state. The average voltage to which it can charge will be proportional to the time of coincidence of voltages E_1 and E_2, or to the angle of phase bias between the current and voltage. In fact this circuit is a transistor analog of the logical element "OR" with a capacitor at the output.

In Figure 11.18b, there are circuit transistors VT1 and VT2, connected in series, which will be open (that is, the part of the circuit between points 1 and 2 is shunted) only when both transistors are opened, and vice versa, voltage between points 1 and 2 occurs only when both transistors are disabled (closed). This is nothing else but the logical scheme "AND." As is apparent, the breadth (duration) of the output pulse will be inversely proportional to the time of coincidence of the open state of transistors. The circuit shown in Figure 11.18c works similarly.

The same principle is applied to power direction relays based on a thyristor (Figure 11.19), patented by the author of this book in 1977. The principle of operation of such a relay is based on the fact that in order to open (switch ON) the thyristor on DC, two voltages must be applied to it: one between the anode and the cathode, and the other one between the cathode and the gate. The polarity of these voltages must be strictly defined; otherwise the thyristor will not be opened. In AC circuit, when these two voltages fully coincide in phases, the thyristor will be energized with a minimal angle of lag and will provide maximal average voltage in the load (for example, in the output relay winding). If there is a phase bias between these voltages, the thyristor will be switched ON with a certain delay with regard to the phase of the voltage applied to it, and as a result, the average voltage in the load will decrease proportionally to the phase bias. At angles approximating 170°, the thyristor will not be opened at all. Use of two thyristors connected inverse-parallel and of two output relays (their windings are marked as R1 and R2) allows us to obtain a double-acting relay for operation in double-side-supply power

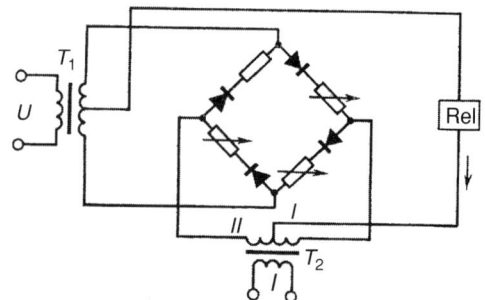

FIGURE 11.15

Ring phase-sensitive circuit on diodes.

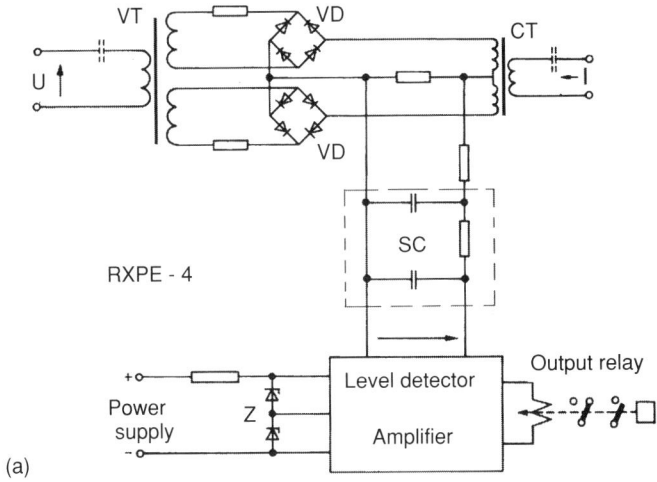

(a)

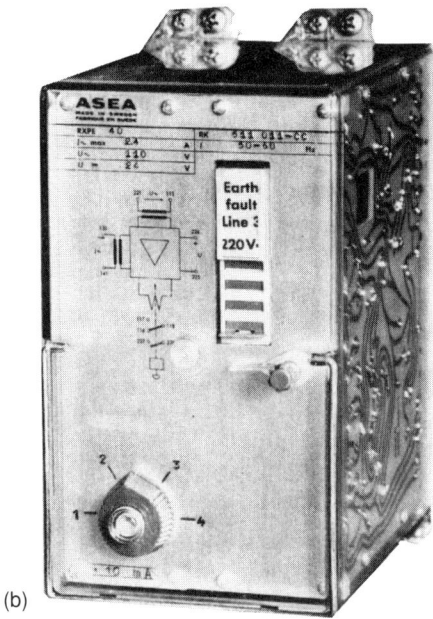

(b)

FIGURE 11.16
RXPE-4-type power direction relay, based on a diode phase comparator. VD — diode bridges; SC — smoothing circuit.

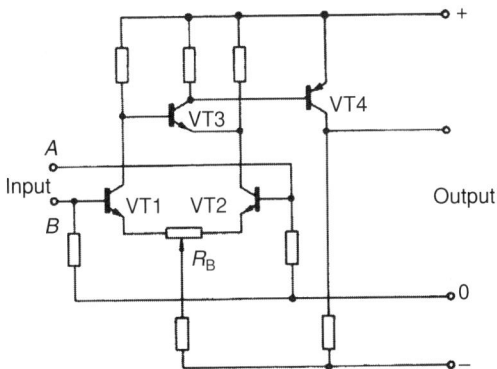

FIGURE 11.17
Phase comparator based on transistors.

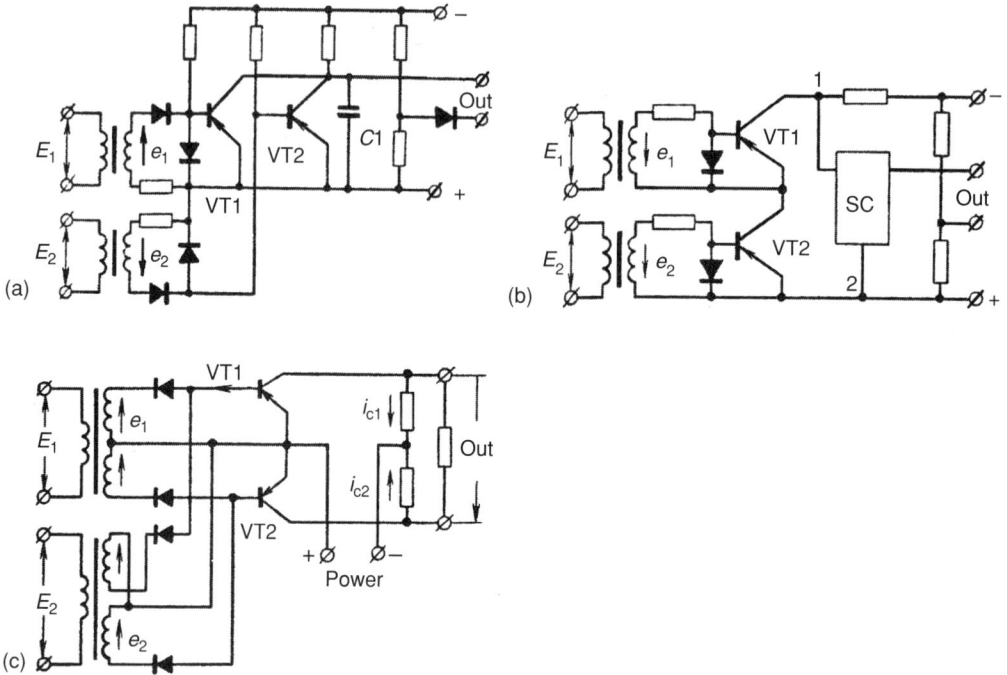

FIGURE 11.18
Electronic circuits determining the time of coincidence of (polarity) signs of input signals proportional to current (E_1) and voltage (E_2). SC — smoothing circuit.

networks. Capacitors in this circuit smooth voltage pulses in the load (the relay windings).

Generally, one can suggest a lot of variants of electronic circuits reacting to the phase difference between current and voltage. For instance, it is quite easy to implement the circuit measuring the time interval, when the sinusoids of voltages E_1 and E_2 (Figure 11.20a) pass through zero. Practically this can be implemented with the help of an electronic timer on a chip, which is energized when the sinusoid E_1 passes through zero, and stops when the sinusoid E_2 passes through zero. The phase bias between current and voltage will be determined by this very period of time.

Some industrial models of power direction relays are also based on the principle of the measurement of time of sign (polarity) coincidence of the input voltages, as in

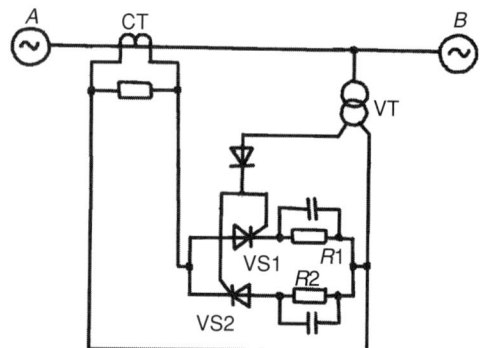

FIGURE 11.19
Circuit of a power direction relay based on thyristors, suggested by the author of this book (U.S.S.R. patent 641536, 1977).

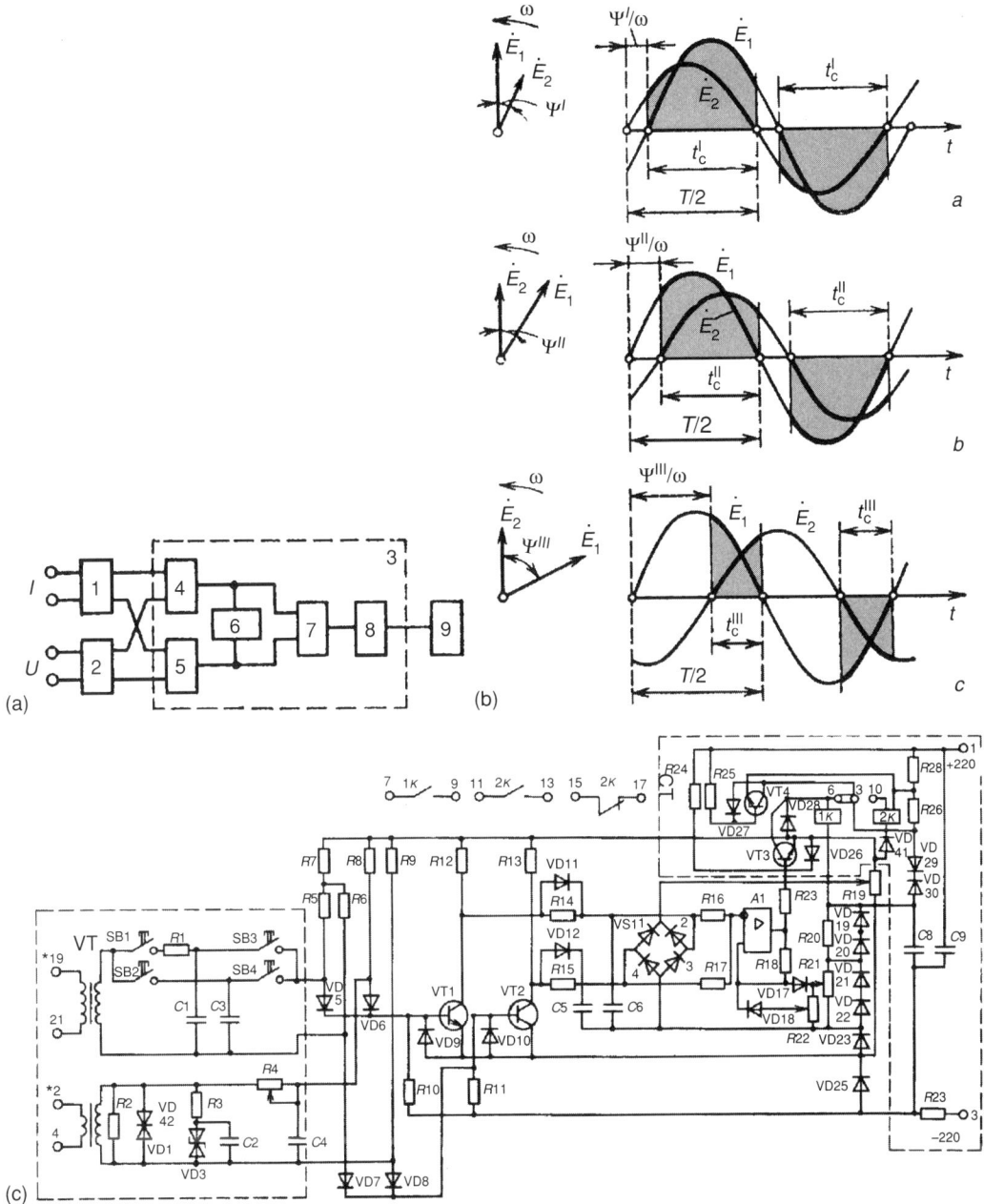

FIGURE 11.20

(a) Simplified block diagram of a electronic power direction relay of the RM-11 type (Russia). (b) Time diagrams of a power direction relay based on the principle of time measurement of coincidence of single-polar current and voltage values. t_c = Coincidence time. (c) Circuit diagram of an electronic power direction relay of the RM-11 type (Russia).

Figure 11.20, for example, an RM-11-type relay produced in Russia. In this device input, units 1 and 2 convert input current and input voltage to low-level voltages (E_1 and E_2) proportional to them, which are applied to the phase comparator (3). Values with positive signs go to measuring unit 4 and those with negative signs to measuring unit 5. In these units, the coincidence times of the positive and negative values of the voltages E_1 and E_2

are separately determined. Output voltages of units 4 and 5, limited by level with the help of a limiter (6), are summed up in a summation unit (7) and are applied to the threshold element (8) with an output electromagnetic relay (9) at the output. The full circuit scheme of the relay is shown in Figure 1.20c.

Coincidence time of positive values E_1E_2 and negative values E_1E_2 is measured by a sophisticated method based on alternating enabling and disabling of transistors VT1 and VT2 if instantaneous values (when signs) of both input voltages coincide. When the corresponding transistor is turned ON, there is a current pulse the breadth (duration) of which will be proportional to the duration of coincidence of the voltage polarities. These pulses are used to charge the capacitors C5 and C6, the voltage on which will be proportional to the time of coincidence of the signs of the input voltages. At that point, these voltages are summed up and the value is compared with the prescribed level; if it is exceeded, then the relay picks up.

12

Differential Relays

12.1 Principles of Differential Protection

The term "differential" itself signals that the chapter will be concerned with relays responding to differences of actuating quantities; and that is true. Differential protection compares two (or more) currents to locate a fault; which actually makes current protection. In comparison with other types of protection, differential current protection possesses an absolute selectivity in the sense that it operates smartly only in those cases where the fault is within the protected zone, and does not operate at all if the fault is out of its zone. The zone of the differential relay is limited by a part of the electric circuit between the current transformers (CTs), to which the relay is connected. Due to such high selectivity of protection, there is no need to activate a delay for the relay pick-up, which is why all differential relays are high speed. That being so, extraordinarily high selectivity and high speed of operation are the distinguishing features of differential protection.

Differential current protection is applied for sections of power lines and some important elements of the power-supply system such as generators, transformers, reactors, and power electric motors. In addition to the protection from current overloading, differential protection is also used for localization of insulation damages in high-voltage equipment (generators, for instance).

Differential protection of power lines is divided into longitudinal protection and transverse protection. The former refers to protection of longitudinal sections of single lines (which is of course why it is called "longitudinal"), and the latter to the protection of parallel lines (comparing currents in these parallel lines).

Longitudinal differential protection operates on the principle of comparison of magnitude and phase of the currents entering and leaving the protected circuit section or element. To accomplish differential protection, two CTs (CT1 and CT2), having identical ratios of transformation, are interposed in the circuit at both ends of the protected circuit element. These CTs have their secondaries interconnected by the connecting leads as shown for one phase in Figure 12.1. The differential relay Rel is arranged in parallel with the CT interconnection leads.

This scheme, based on the circulation of currents, was first established at the end of 19th centuries by Merz and Price and called the "Merz–Price differential scheme." This fundamental principle has formed the basis of many highly developed protective arrangements.

If the secondary currents of CTs CTl and CT2 are denoted, respectively, by I_1 and I_2, and the positive direction of the current in the relay is taken to be that of current I_1, during normal operating conditions on line AB (here and hereinafter vector values):

$$I_{REL} = I_1 - I_2$$

This relation is valid by virtue of the fact that the impedance of the winding of relay Rel (usually a current relay) is considerably less than that of the current-transformer secondary windings, and it can hence be considered that the secondary currents flow through interconnection leads and complete their circuits through the relay winding.

In the ideal case, during normal operating conditions and in the event of external short circuits (outside the zone of protection), the secondary currents of CTs CTl and CT2 will be equal in magnitude and opposite in direction. Because of this, no current will flow in the differential relay circuit or:

$$I_{REL} = I_1 - I_2 = 0$$

In the actual circuit, a current of unbalance will flow in the relay circuit because of the unequal magnitude of the CTs' currents at one and the same primary current. This problem will be studied in greater detail below.

When a short circuit occurs within the zone of protection and the power supply comes from one end (Figure 12.1b), the relay circuit carries a current:

$$I_{REL} = I_1 - 0 = I_1$$

The relay then operates to transmit an impulse to trip the circuit breaker at the supply end.

When a short circuit occurs within the zone of protection and power is supplied from both ends (Figure 12.1c), the sum of the secondary currents will flow through the relay, or:

$$I_{REL} = I_1 + I_2$$

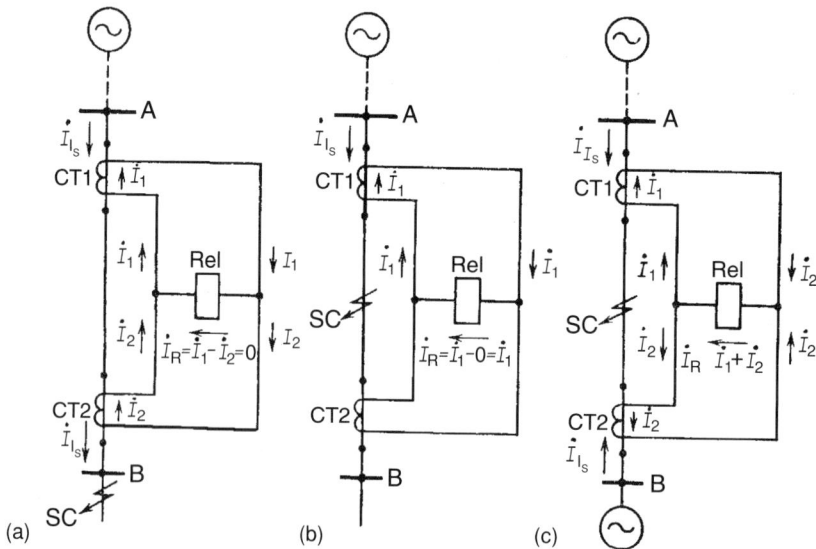

FIGURE 12.1
Circulating-current longitudinal protection performance during fault outside (a) and within zone of protection (b, c) (b) case of single-end supply; (c) case of two-end supply.

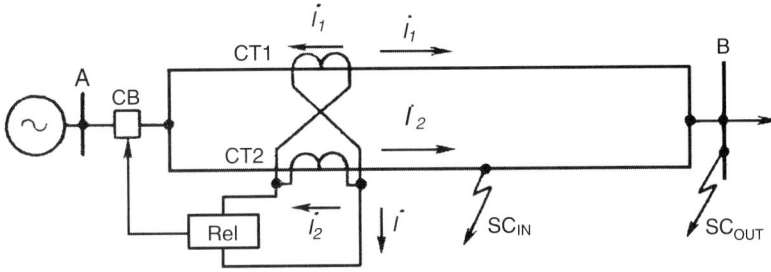

FIGURE 12.2
Single-phase circuit diagram of transverse differential protection for two parallel lines with single-end supply.

(these currents, in general, being different in magnitude). The relay in this case operates to transmit impulses to trip the circuit breakers at both ends of the faulted line.

The differential scheme studied here is called a "circulating-current circuit" due to the fact that the current continuously circulates in the interconnecting leads of this circuit. If two identical power lines are run in parallel from one substation to another and are connected to the buses by a common circuit breaker, a common protection is installed for both lines and will trip out the circuit breaker when a short circuit occurs on any one of the lines. For protecting two parallel high-voltage lines of wide applications, transverse differential current protection is implemented, based on a comparison of the magnitude

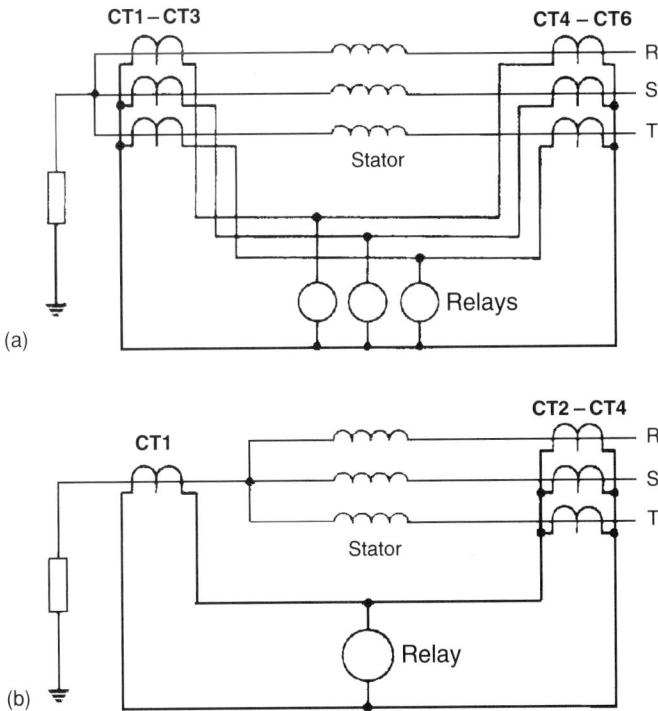

FIGURE 12.3
Three-phase version of the scheme of differential protection of the generator's stator with high-impedance differential relays. (a) Phase and earth fault protection; (b) restricted earth fault protection.

and phase of the currents in the parallel lines (Figure 12.2). As already mentioned above, the differential protection is applied not only for protection of segments of lines, but also for protection of such important system components as transformers, reactors, generators, switchgears, power motors, etc. (Figure 12.3).

12.2 High-Impedance Differential Relays

The most simple design of this type is the so-called "high-impedance differential relays," used for protection of concentrated objects such as switchgear, bus bars, generators, reactors, and power motors. The following limitations on CTs are required for applying these relays:

1. All CTs in the differential circuit must have the same ratio.
2. All CTs should be operated on their full winding (i.e., tap connections must be given special consideration).
3. Inlet and outlet current levels of protection object must be equivalent.
4. No other equipment, including no other types of protection devices, can be connected to the CT used for high-impedance differential relays.

As already mentioned above, under external fault conditions if the CT has no error, the current in the secondary of the CT in the faulted line is equal and opposite to the vectorial sum of the currents in the secondaries of the remaining CT in the same phase. No current flows in the relay and the voltage that appears across the paralleling points is zero. Unfortunately during fault conditions, CTs do not always perform ideally, since core saturation may cause a breakdown of ratio. Such core saturation usually is the result of a DC transient in the primary fault current, and may be aggravated by residual flux left in the core by a previous fault.

The worst condition of unbalanced secondary currents is realized when the CT in the faulted circuit is completely saturated and none of the other CTs suffer a reduction in ratio. Under this condition the saturated transformer secondary winding presents an impedance which is practically equal to its DC resistance, since the leakage reactance of the full winding of the toroidally wound CT can be neglected and a secondary current will be forced through the saturated CT equal to the sum of the secondary currents in the remaining paralleled CTs, less the current through the high-impedance relay which is negligible. The maximum voltage across the relay under external fault conditions therefore, will be the resistance drop of the theoretical secondary current which flows through the leads and secondary winding of the saturated CT in the faulted line.

It is obvious that under external fault conditions, no higher voltage than this can exist, since either a reduction in ratio of any of the other CTs or a diversion of current through the relay will reduce the current through the saturated transformer secondary. Whatever secondary voltage is generated by the core flux of the saturated CT will also reduce the voltage across the paralleling points.

Therefore, in order to prevent incorrect operation under extreme values of external faults, it is only necessary to set the pick-up of the relay above this maximum through-fault voltage, which can readily be calculated from the resistance of the CT leads and

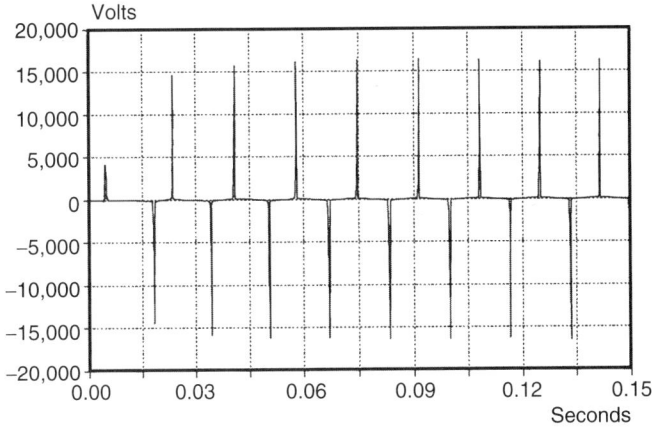

FIGURE 12.4
Secondary voltage on open circuit CT during an internal fault in the protected zone, without over-voltage limiting.

secondary winding, and the maximum short-circuit current. A factor of safety of 2:1 can be used in making this setting while still retaining good sensitivity for internal faults.

When an internal fault occurs, the secondary current from the source CT is limited by the impedance of the idle CTs and the voltage relay, which are all connected in parallel and are all high-impedance elements. Under this condition, the voltage that appears across the paralleling points approaches the open-circuit voltage of the CT secondaries. This voltage is reduced by the shunting effect of the exciting current for the idle CTs and whatever current flows through the relay.

When these shunting effects are taken into account in typical applications, it is found that the net voltage applied to the relay due to a comparatively low internal fault is greatly in excess of the pick-up value that is established for the maximum external fault condition. Since the source CTs are practically open circuited during an internal fault, it is necessary to limit the secondary voltage under this condition to relieve insulation stresses (Figure 12.4). For this purpose a special precision nonlinear resistor is connected in parallel with the relay.

General Electric Co. used nonlinear resistor, called Thyrite® (developed by GE in 1918) (Figure 12.5). Other companies used nonlinear resistor, called Metrosil®, or Ceramsil® (Figure 12.6 and Figure 12.7). These resistors are proportioned to shunt negligible current

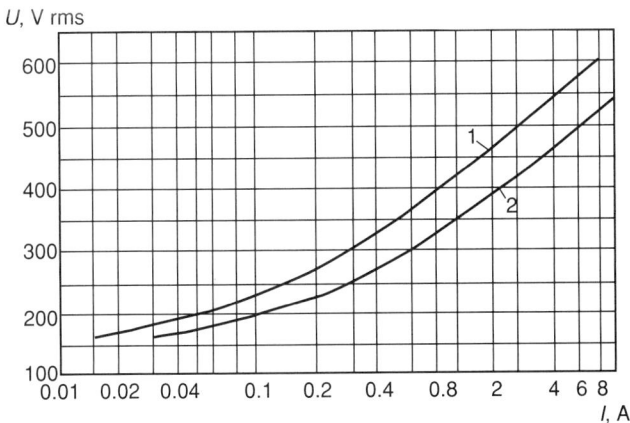

FIGURE 12.5
Volt–ampere characteristic of nonlinear resistors Thyrite. 1 — For one four-disk Thyrite stack; 2 — for two four-disk Thyrite stack.

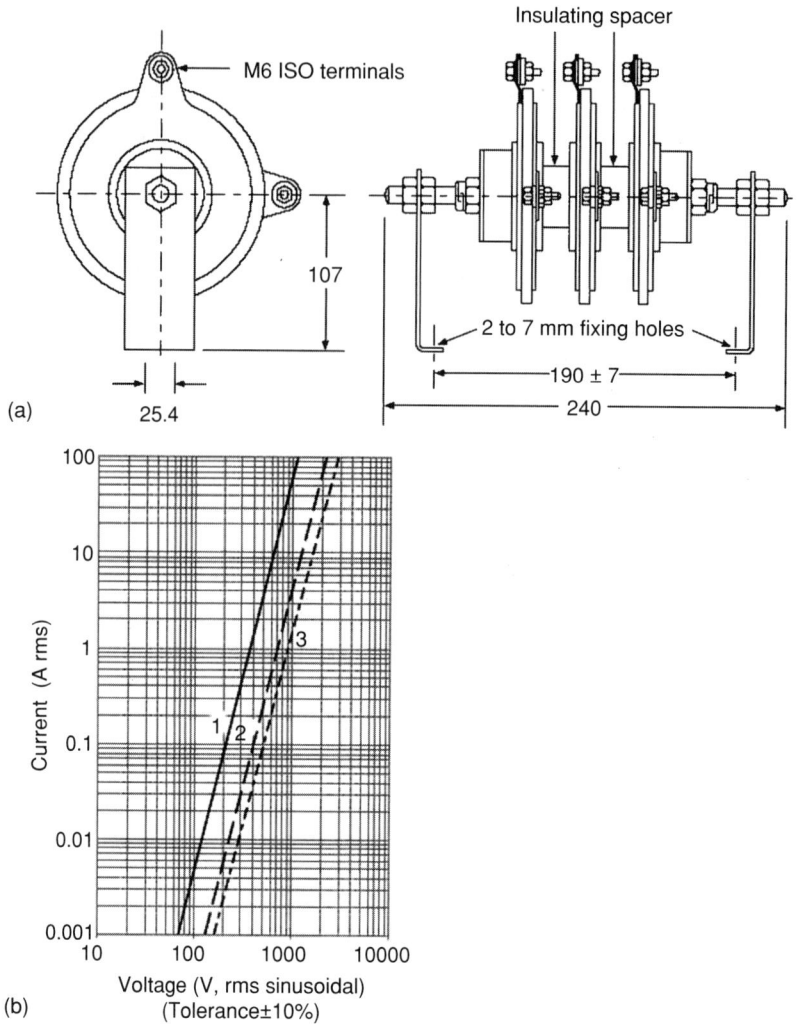

FIGURE 12.6

(a) A nonlinear resistor Metrosil: one three-disk stack (ALSTOM). (b) Volt–ampere characteristics of nonlinear resistor Metrosil (ALSTOM). 1 — For relay setting range 25 to 175 V; 2 — setting range 25 to 325 V; 3 — for setting range 100 to 400 V.

around the relay at voltages near the pick-up value, and to prevent a rise in voltage above a predetermined level by permitting a high current to flow on severe internal faults.

These are not just simple nonlinear resistances of the varistor type, but are precision elements with precise volt–ampere characteristics, used when the pick-up parameters of the relay are calculated. Unlike standard varistors, these resistors have considerably more quiet characteristics, and can operate on the nonlinear sections, thus causing greater power releases on them. Usually these resistors withstand not more than 70 to 100 ms in this mode, which is enough for the relay pick-up. On one hand, these resistances effectively limit voltage in secondary circuits of the CT when a high-impedance differential relay (Figure 12.8) is used, and on the other hand they provide pick-up of the relay shunted by this resistance at a certain voltage level.

The working element of such a high-impedance differential relay is a voltage relay. Many companies produce such relays both of electromagnetic and electronic types. For

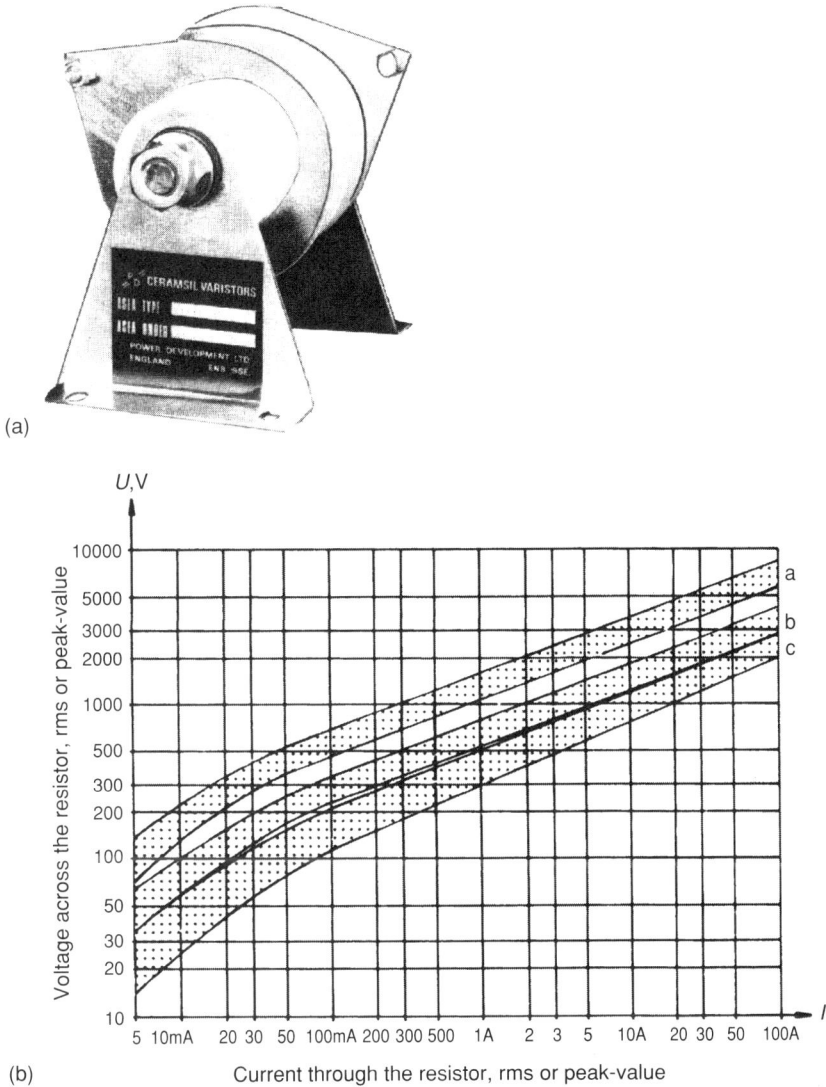

FIGURE 12.7
(a) Precision voltage dependent resistor, called a Ceramsil (ABB). (b) Volt–ampere characteristics of a Ceramsil for different resistor types (a, b,c).

example, earlier relays produced by General Electric were of a plunger type with a retractable core (Figure 12.9). As shown in Figure 12.9, the PVD11 relay consists of two plunger-type operating units: a "low-set" voltage unit (called as "device 87L") and a "high-set" current unit ("device 87H").

Device 87L is an instantaneous voltage unit with a high-impedance operating coil. To cancel the effect in the secondary of the DC component of an offset wave, which may be exaggerated by cumulative residual magnetism from previous faults, a circuit that is tuned for resonance at system frequency is used in series with the coil. In order to permit adjustment of the relay without affecting the tuning, a rectifier is interposed between the tuned circuit and the relay coil (Figure 12.10).

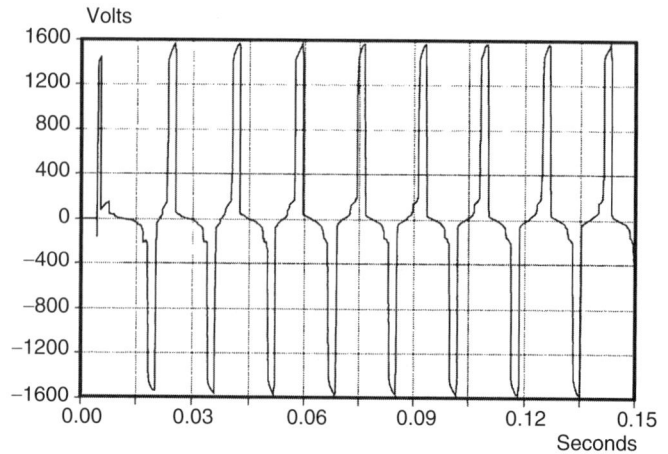

FIGURE 12.8
Secondary voltage on open-circuit CT during an internal fault with over-voltage limiting by a Metrosil (ALSTOM).

Device 87H is an instantaneous over-current unit with a low-impedance coil, which is connected in series with Thyrite resistor disks. The Type PVD II C relay has a stack of four Thyrite disks in its voltage-limiting circuit, and is intended for use with CTs having a 5-A secondary. The 87H unit, when set with the proper pick-up, may be used to supplement the voltage unit, 87L, and/or implement breaker failure protection when a suitable timing relay and other auxiliary devices are provided by the user. The required setting of the 87H unit is related to the actual setting of the 87L unit.

During an internal fault, current will flow in the Thyrite stack, causing energy to be dissipated. To protect the Thyrite from thermal damage, a contact of the lockout relay

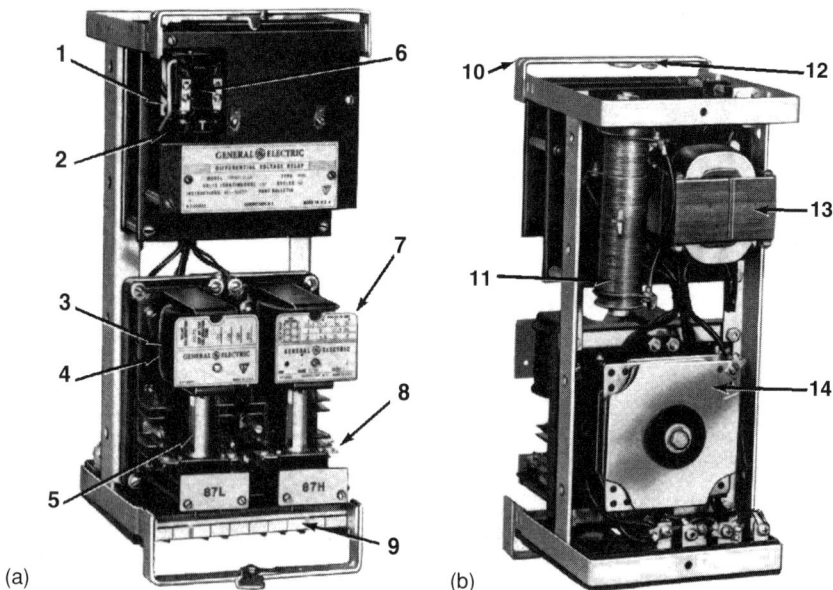

FIGURE 12.9
(a) and (b) High-impedance differential relay PVD-11 with plunger-type operating units. 1 — Seal-in coil with tap; 2 — seal-in contacts; 3 — low-set unit; 4 — operating coil; 5 — calibrated plunger rod; 6 — target; 7 — high-set unit; 8 — self-aligning contacts; 9 — marking strips; 10 — steel cradle; 11 — rectifier bridge (assembly); 12 — latches; 13 — reactor; 14 — Thyrite assembly. (General Electric Type PDV Instruction GEA-5449A.)

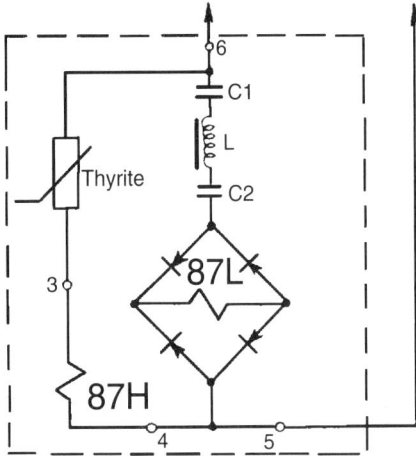

FIGURE 12.10
Circuit diagram of PVD11 relay. (General Electric Type PDV Instruction GEA-5449A.)

(a)

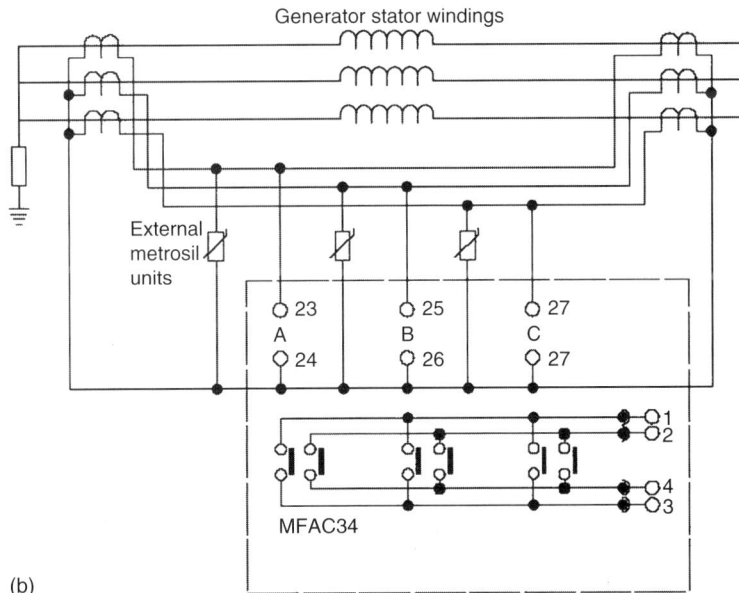

(b)

FIGURE 12.11
(a) Modern high-impedance MFAC34 type (ALSTOM) differential relays have a different construction of the case, but in all other respects they are similar to earlier PVD11 relays produced by GE. (b) Connection diagram for high-impedance differential relay MFAC34 type (ALSTOM).

FIGURE 12.12
The 2V73 type high-impedance differential relay, produced by Relay
Monitoring Systems Pty Ltd (Australia).

(outside auxiliary relay) must be connected to short out the Thyrite during an internal
fault; however, the 87H unit is not shorted so that the relay can continue to operate as an
over-current function, remaining is picked up until the fault is cleared. The 87H unit may
be used to implement breaker failure protection. The thermal limits of the Thyrite will not
be exceeded if relay time plus lockout relay time is less than four cycles.

Later high-impedance differential relays produced by General Electric (PVD21)
were based on attracted armature units, simple and robust. Similar relays were and are
still produced by many companies at present. Thus the ALSTOM company (France)
which has already been mentioned, produces high-impedance differential relays of
the MFAC type (Figure 12.11), CAG34, based on the same principles as the relays
described above.

The Australian company "Relay Monitoring Systems Pty Ltd" (RMS) produced high-
impedance differential relays of the 2V73 type for one phase (Figure 12.12), and the
2V47K6 type for three phases. These relays are intended for various items of power
system plants including generators, bus bars, and motors. The 2V73 is also suitable for
restricted earth fault applications. The relay-measuring element is basically an attracted
armature unit of simple and rugged construction powered from a bridge rectifier. Each
phase of the relay can be set from 25 to 325 V AC in 50 V steps by using the front panel
mounted selector switches. A capacitor is connected in series with the operating coil to
make the relay insensitive to the DC component of fault current. The setting can thus be
calculated in terms of RMS AC quantities without regard for the degree of offsets
produced by the point on wave at which the fault occurs. An inductor connected in series
with the capacitor forms a resonant circuit tuned to the relays-rated frequency. As an
alternative to the simple high-impedance relays, biased systems can be used. Usage of
such systems is particularly relevant for protection of power transformers.

12.3   Biased Differential Relays

In modern power systems, transformers with ratings of more than about 1000 kV A are
protected against internal short-circuits by differential protection. The heart of this form
of protection is the differential relay in which the currents on the primary and secondary

Differential (unbalance) current

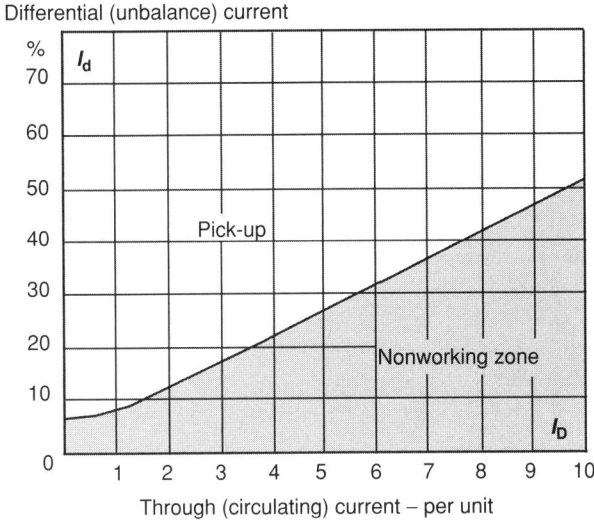

FIGURE 12.13
Typical bias characteristic of percentage differential relay.

sides of the transformer to be protected are compared with respect to magnitude and phase relationship.

In normal operation, the ratio between the primary and secondary currents is constant at any instant, apart from the magnetizing current which appears on one side but which only amounts to a few percent of the rated current of the transformer, depending on the transformation ratio of the transformer. The phase relationship between the two currents is fixed by the vector group of the transformer. On occurrence of a fault inside the protected zone, restricted by the location of the CTs necessary for connecting the differential protection on the high- and low-voltage sides of the transformer, this ratio of the currents, and in some circumstances their phase relationship, is disturbed. The unbalanced current can be evaluated directly as a criterion of the fault. The differential relay responds to this current by closing its tripping contacts, causing the circuit breakers on the high- and low-voltage sides of the transformer to open.

Even under normal working conditions unbalanced currents (spill currents) appear, their magnitude depending on the individual ratio and phase-angle errors of the CTs employed. They generally increase as the load on the transformer increases. They attain particularly high values when through faults outside the protected zone the CTs tend to saturate; and also in the case of tap-changing transformers, when (as is usual) the CTs are not adjusted when the ratio of power transformation is changed.

To compensate for these influences, the differential relay is stabilized, that is, the relay is given a characteristic basically similar to that in Figure 12.13. This shows the unbalance (differential) current I_d necessary to operate the differential relay in relation to the circulating current I_D. Differently, the increase of a circulating current results in a desensitization of the relay to a differential current.

At low values of the circulating current the curve does not rise more steeply than is necessary with a view to the spill currents, and changes its slope only when the circulating currents are such that the CTs are approaching saturation.

Figure 12.14 is a schematic diagram of a differential protective system stabilized in this way. The measuring circuit of the differential relay contains a bias system (H) and an operating system (A), which work in opposition. The CTs in the same phase on the high- and low-voltage sides of the transformer are connected in series through the bias system

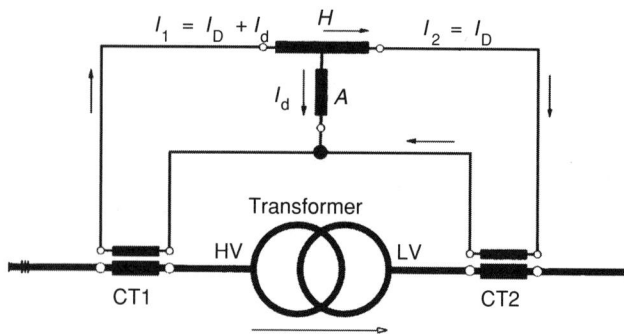

FIGURE 12.14
Schematic diagram of transformer biased differential protection. I_d — Unbalance (differential) current; A — operating winding; I_D — circulating current; H — bias winding. (AEG Transformer Diffential Relay RQ4 1959.)

(H) of the relay, and the operating system (A) is connected across the bridge. As will be seen in Figure 12.14, the vector sum of the two current-transformer currents:

$$|I_1 + I_2| = |2I_D + I_d|$$

acts on the bias system in a restraining sense, and the vector difference on the operating system in the tripping sense:

$$|I_1 - I_2| = I_d$$

The secondary currents of the CTs:

$$I_1 = |I_D + I_d| \quad \text{and} \quad I_2 = I_D$$

are so matched, if necessary, by the inclusion of intermediate CTs to compensate for differing current-transformer secondary currents, and in some cases, dissimilar phase angles if the protected transformer is of a vector group causing a phase difference. That is so because in normal working, a spill current flowing through the operating system is not sufficient to operate the differential relay. The desired relationship between the unbalanced current and the circulating current is obtained by suitable design of the bias and operating systems.

In early electromagnetic differential relays (e.g., the QS4, produced by AEG) the comparison of the vector sum:

$$|I_1 + I_2|$$

and the vector difference:

$$|I_1 - I_2|$$

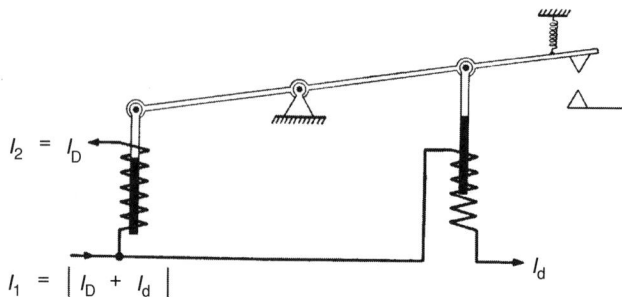

FIGURE 12.15
Schematic diagram of an early electromagnetic percentage differential relay of the QS4 type.

was effected by means of a mechanical balanced beam system as represented in Figure 12.15, in which the bias and operating system, in the form of electromagnets, act in opposite senses on a pivoted beam.

One of the disadvantages of the arrangements of this kind is that a separate beam system is necessary for each of the three conductors and the power consumption is relatively high.

12.4 Electromagnetic Percentage Differential Relay

In later designs (for example, the RQ4 relay, produced by AEG in the 1950–60's, Figure 12.16) of transformer differential relays, the currents from the bias transformer T_h and the operating transformer T_d are rectified, the DC outputs being compared in a bridge circuit across which is a sensitive moving-coil relay Rel (Figure 12.17).

The advantages of this arrangement are firstly that the characteristic (unbalance current in relation to circulating current) can be largely matched to the requirements of the transformer differential protection by means of simple elements, and secondly that the power consumption is considerably less, with much shorter operating times.

When the power transformer is energized, current is supplied to the primary, establishing the required flux in the core. This current is called the "magnetizing inrush" (which may reach a value several times the rated current of the transformer) and it flows only through the CTs in the primary winding. This causes an unbalance current to flow in the coil of the differential relay, which would cause false operation (trip) if means were not provided to prevent it.

The inrush current differs from the other currents due to internal faults in the transformer in that it has a considerably higher harmonic content. This property can be utilized

FIGURE 12.16
The RQ4 differential relay, produced by AEG in the 1950–60's. (AEG Transformer Differential Relay RQ4 1959.)

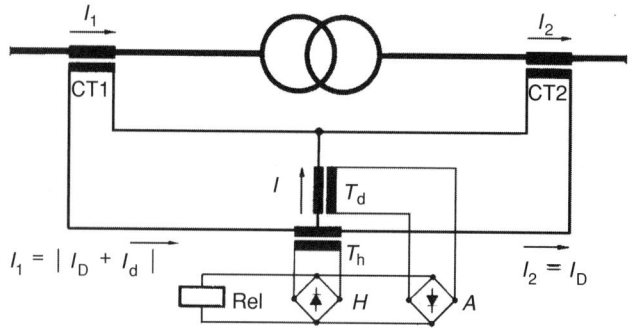

FIGURE 12.17
Schematic diagram of a later designed electromagnetic percentage differential relay of the RQ4 type. (AEG Transformer Differential Relay RQ4 1959.)

to stabilize the differential protection against the undesired effect of the inrush current by means of a suitable blocking element, so that it is possible to dispense with the time-lag arrangements formerly employed.

In the RQ4 transformer differential relay a measuring element for determining the magnitude of the unbalance current I_d and a blocking element are combined in a single case. The measuring and blocking elements are in the form of moving-coil relays. The contacts of each relays (NO-contact of the measurement relay and NC-contact of the blocking relay) are connected in series and only permitting tripping when the unbalance current I_d is of a certain value and has not too high a harmonic content. Each of the two relays is provided with an instantaneously-acting contact-pressure reinforcing device, which ensures complete absence of bounce and acts as a safe and robust auxiliary relay.

When the transformer is switched ON, the inrush current causes the moving-coil relay of the blocking element to operate so rapidly as to open the tripping circuit (NC-contact) before tripping is effected by closing of the NO-contact of the measuring element. The contact of the blocking element does not reclose until the relay of the measuring element has returned to the off position after the inrush current has died away. On the occurrence of genuine faults within the protected zone of the differential protection, however, the contact of the blocking element remains closed because of the preponderance of the fundamental over the harmonics in the short-circuit current, so that instantaneous tripping is initiated.

Start relay operating times are thus obtained even with small unbalanced currents. With three-pole faults and I_D/I_d greater than 80%, they are less than 100 ms and with two-pole faults only slightly more.

The blocking element incorporated in the relay is the supplementary blocking relay, which has already proved highly satisfactory in service; this has been used as a separate supplementary relay both in conjunction with the well-known quotient differential relay QS4 and also with other modern relays. For these purposes, it is still used as a separate relay.

Figure 12.18 is a schematic diagram of the blocking element, the function of which is to prevent tripping by inrush current and to permit tripping only on the occurrence of genuine faults. The two three-winding CTs T1 and T2 in the unbalance current circuit feed an Ohmic shunt R_1 and a reactive shunt L_2, respectively, which are connected through a nonlinear resistor R_{12} and a high-pass filter F to a DC bridge, across which is the moving-coil relay Rel. The current which flows in the left arm of the bridge through

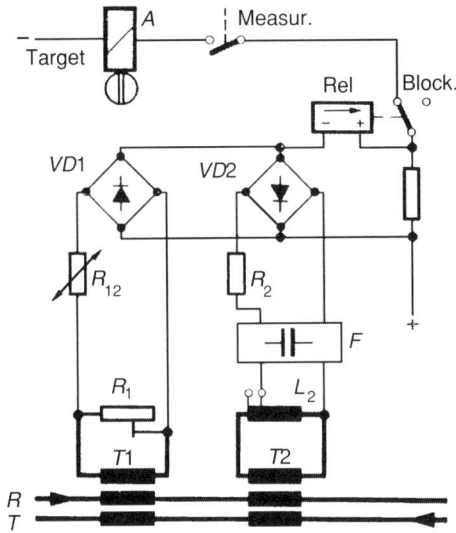

FIGURE12.18
Schematic diagram of blocking element. (AEG Transformer Differential Relay RQ4 1959.)

the rectifier VD1 tends to keep the contact of the moving-coil relay closed, while the current which is passed through the rectifier VD2 tries to open the contact. The high-pass filter is so designed that in the right arm of the bridge, all harmonics are effective and by the provision of suitable damping, the effect of the second and third harmonics is increased. Thus the second harmonics, which is particularly pronounced in the inrush current of transformer (about 30 to 70% of the fundamental), is amplified for the purpose of blocking the differential relay against the effects of inrush currents. As in the left arm of the bridge, unlike the right, no accentuation of the harmonics is effected. They add little to the fundamental harmonic fed into the bridge by the rectifier VD1, so that the moving-coil relay Rel connected across the bridge responds essentially to the ratio of all the harmonics to the fundamental.

On the occurrence of a short-circuit across the terminals on the primary side of the transformer, when the fault current is limited only by the impedance of the supply network, the main CTs may become saturated so that they also produce harmonics. To ensure that the blocking element shall also be effective in case of such very high fault currents, the nonlinear element R_{12} is included in the left arm of the bridge of the blocking element. It consists of a suitably dimensioned resistor shunted by rectifier cells in anti-parallel.

Specialized differential relays for protection of power transformers with a sensitive element based on standard electromagnetic relays (or moving-coil relay, as in the case described above) are quite popular and were produced not only by AEG in 1960s to 1990s but also by many others, such as ASEA (ABB), Siemens, etc. These relays had similar circuits and principles of operation (Figure 12.19). As it can be seen from the circuit of connecting the relay to the power transformer, there is an additional transformer between one of the windings ("star") and a 1-4-7 outlet of the relay. This transformer has the same transformation ratio as the power transformer and is necessary for leveling of currents applied to the inputs of the relay from the same CTs installed from both sides of the power transformer. This is a standard method widely used in all types of differential relays for the protection of transformers.

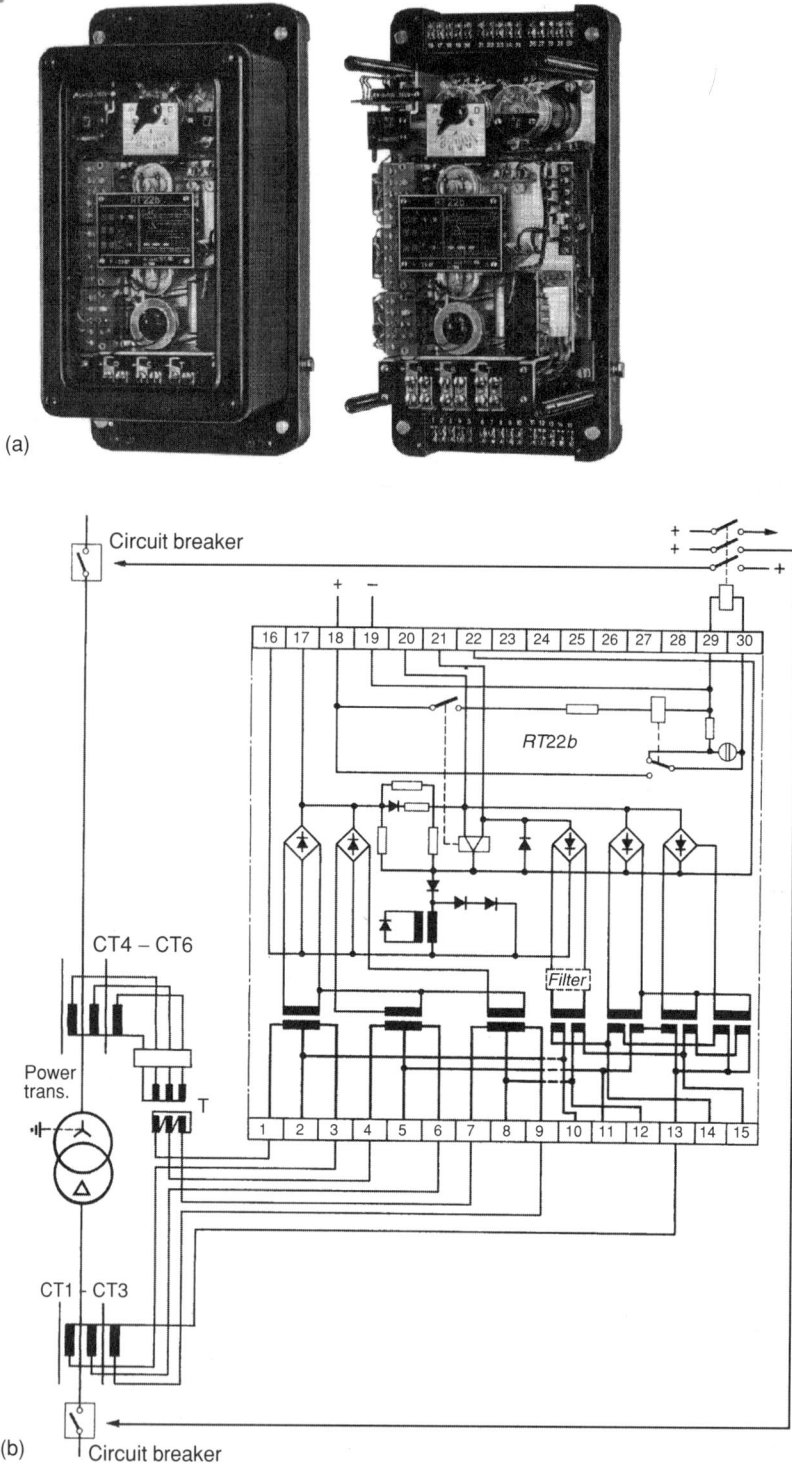

FIGURE 12.19

(a) Relay of the RT22b type for differential protection of power transformers (Siemens). External design. (b) Circuit diagram. (Siemens 1989 Protective Devices catalog NS 1-89.)

12.5   Induction-Type Differential Relays

Along with the principle of construction of differential relays for the protection of power transformers based on electromagnetic relays, described above, another approach was developed, based on application of an induction magnetic system with a rotating disk. It should be noted that induction-type biased relays are not more modern than the relays described above, they were developed parallel to and have been produced since the 1920–30's by the Westinghouse Company (Figure 12.20). Induction-type biased relays contain two electromagnets similar to those described above, operating on a single disc assembly and arranged to develop opposed torques. The operation magnet will be supplied with the differential current while the other will be energized by the biasing quantity. The windings will be proportioned to provide the desired bias ratio.

According to such principle, many types of differential relays were designed and produced during the past 50 to 70 years, for example, this percentage-differential relay of the IJD53 type (General Electric Co.) (Figure 12.21). The IJD53 relays contain two shaded-pole U-magnet driving elements acting on opposite sides of a single disk. One of these (the operating element) drives the disk in the contact-closing direction, and the other (the restraining element) drives the disk in the contact-opening direction. Since it is not always possible to provide CTs that supply equal secondary currents, the relays have tapped coils to permit balancing these currents.

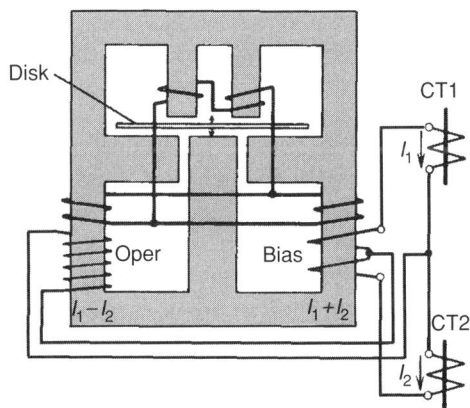

FIGURE 12.20
Induction-type biased relay, produced by Westinghouse since the 1920s to 1930s.

FIGURE 12.21
Single-phase percentage-differential relay IJD53C type (General Electric Co.). 1 — Seal-in coil; 2 — target; 3 — seal-in contacts; 4 — induction disk; 5 — taps in operating and restraining elements; 6 — contacts; 7 — slotted collar; 8 — spiral control spring. (General Electric Instruction GEA 3236B.)

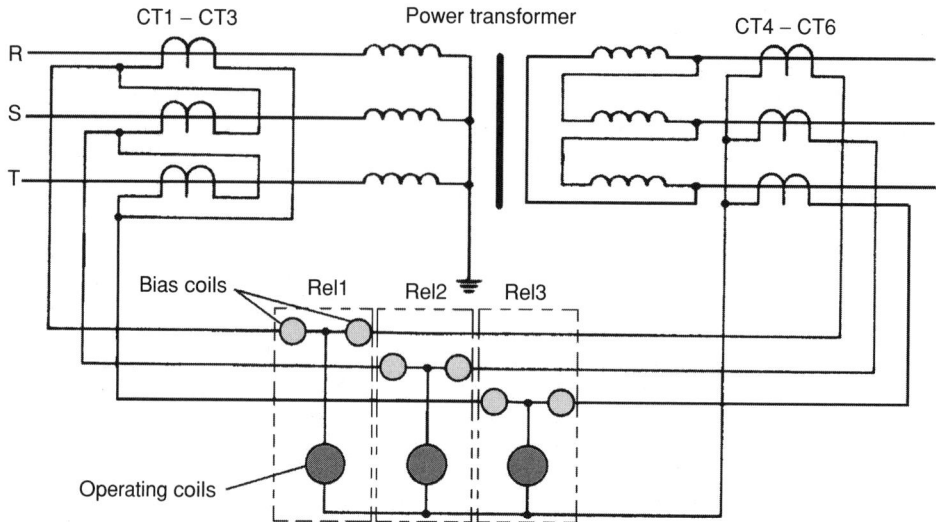

FIGURE 12.22

A typical elementary diagram: three single-phase relays used for differential protection of a two-winding power transformer.

Percentage-differential protection of a two-winding power transformer is shown in Figure 12.22. While this scheme balances the incoming current against the outgoing current, as is done in the protection of AC rotating machines, the CTs are in leads from different windings. Because of this, protection is provided for turn-to-turn faults as well as for faults to ground and for faults between phases or windings. Protection is also provided to the leads between the CT location and the power transformer.

The IJD53C relays will close contacts on a minimum operating current, with no restraint, of 0.4 times tap rating. This value, in conjunction with the short time delay provided by the relay, is usually not sufficient to prevent incorrect operation on magnetizing inrush currents that occur when the transformer is connected to the line or bus bar. It is probable that the current setting will not be high enough to take care of the magnetizing inrush, and therefore, it is recommended that auxiliary desensitizing equipment be added.

In high-speed differential relays, the induction magnet system with a rotated cup (rotor) is widely used also, for example, in CFD type relays, produced by the General Electric Co. (Figure 12.23). The CFD relays are of the induction cylinder construction. The unit consists of a multi-pole stator, a stationary central core, and a cup-like induction rotor. The cup rotates about a vertical axis in the air gap between the stator and core. The lightweight aluminum cylinder offers a high ratio of torque to inertia and results in a fast operate time. The axis of the cylinder is supported at the lower end by a steel pivot, which rotates against a selected sapphire jewel. The jewel is spring mounted to protect it from shocks. The upper end of the shaft is held in place by a polished steel pivot, which projects down through a bronze guide bearing mounted in the end of the shaft.

The stator of the induction unit is of the eight-pole construction, but uses only six of the poles in two sets of three. One set carries the current from the CTs in one phase on each side of the generator winding. The other set carries the difference current between the two CTs.

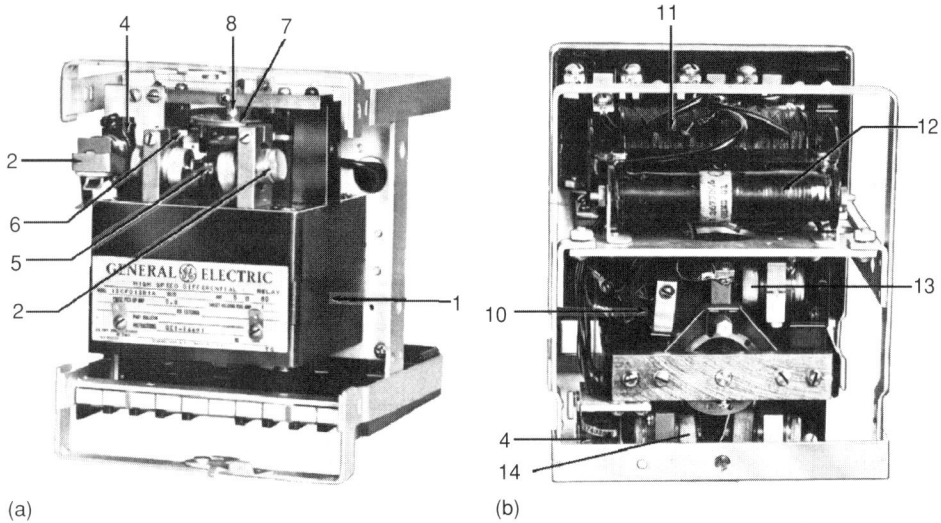

(a)   (b)

FIGURE 12.23

(a) High-speed differential relay of the CFD type removed from case, front view. Produced by the General Electric Co. 1 — Induction cup unit; 2 — shock backstop; 3 — target assembly; 4 — target coil; 5 — moving contact; 6 — stationary contact; 7 — control spring; 8 — top bearing stud. (b) High-speed differential relay of the CFD type, removed from case, top view. Produced by General Electric Co. 10 — holding coil; 11 — autotransformer; 12 — resistor; 13 — rear stationary contact; 14 — front stationary contact. (General Electric Instruction manual GEH-2057A.)

Type CFD differential protective relays function on a product restraint principle. The restraining torque is proportional to the product of the current entering one side of the protected equipment and the current leaving the other side. The operating torque is proportional to the square of the difference between the two currents. The operating and restraining torques balance when the differential current is 10% lesser than the other two, up to approximately normal current. This 10% "slope," as it is called, allows small differences to exist due primarily to CT errors. Above normal current, the differential current circuit will saturate before enough operating torque is produced to close the contacts on a 10% slope basis.

The CFD type relay is a cup-type induction unit. This type of construction results in a fast-operating protective device, even at currents only slightly in excess of the pick-up value.

12.6 Harmonic Restraint Differential Relays

The high percentages of harmonic currents in the magnetizing inrush current wave afford an excellent means of distinguishing it electrically from the fault current wave.

In the BDD type relays (Figure 12.24), the harmonic components are separated from the fundamental components by suitable electric filters. The harmonic current components are passed through the restraining coil of the relay, while the fundamental components pass through the operating coil. The direct current component present in both the magnetizing inrush and offset fault current waves is largely blocked by the auxiliary differential CT inside the relay, and produces only a slight momentary restraining effect.

Relay operation occurs on differential current waves in which the ratio of harmonics to fundamental is lower than a given predetermined value for which the relay is set (e.g., an internal fault current wave) and is restrained on differential current waves in which the ratio exceeds this value (e.g., a magnetizing inrush current wave).

In the BDD15B type relay, the through CT has two primary windings, one for each line CT circuit. Winding No. 1 terminates at stud 6 and winding No. 2 terminates at stud 4 (Figure 12.25). A full wave bridge rectifier receives the output of the secondary of the through current restraint transformer. The output is fed to a tapped resistor through the percent slope tap plate at the front of the relay. By means of the three taps a 15, 25, or 40% slope adjustment may be selected. Resistor taps are adjustable and preset for the given slopes. The right tap corresponds to the 40% slope setting. The output is rectified and applied to the restraint coil of the polarized unit. The differential CT secondary output supplies the instantaneous unit directly, the operating coils of the polarized unit through

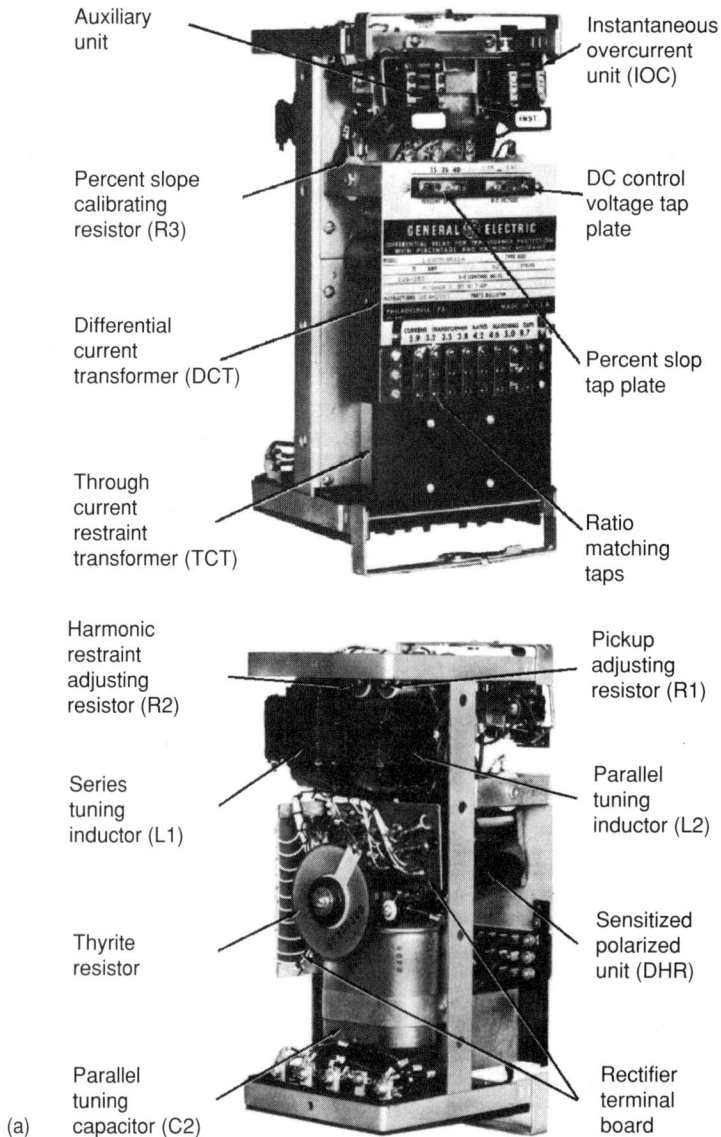

Auxiliary unit

Instantaneous overcurrent unit (IOC)

Percent slope calibrating resistor (R3)

DC control voltage tap plate

Differential current transformer (DCT)

Percent slop tap plate

Through current restraint transformer (TCT)

Ratio matching taps

Harmonic restraint adjusting resistor (R2)

Pickup adjusting resistor (R1)

Series tuning inductor (L1)

Parallel tuning inductor (L2)

Thyrite resistor

Sensitized polarized unit (DHR)

Parallel tuning capacitor (C2)

(a)

Rectifier terminal board

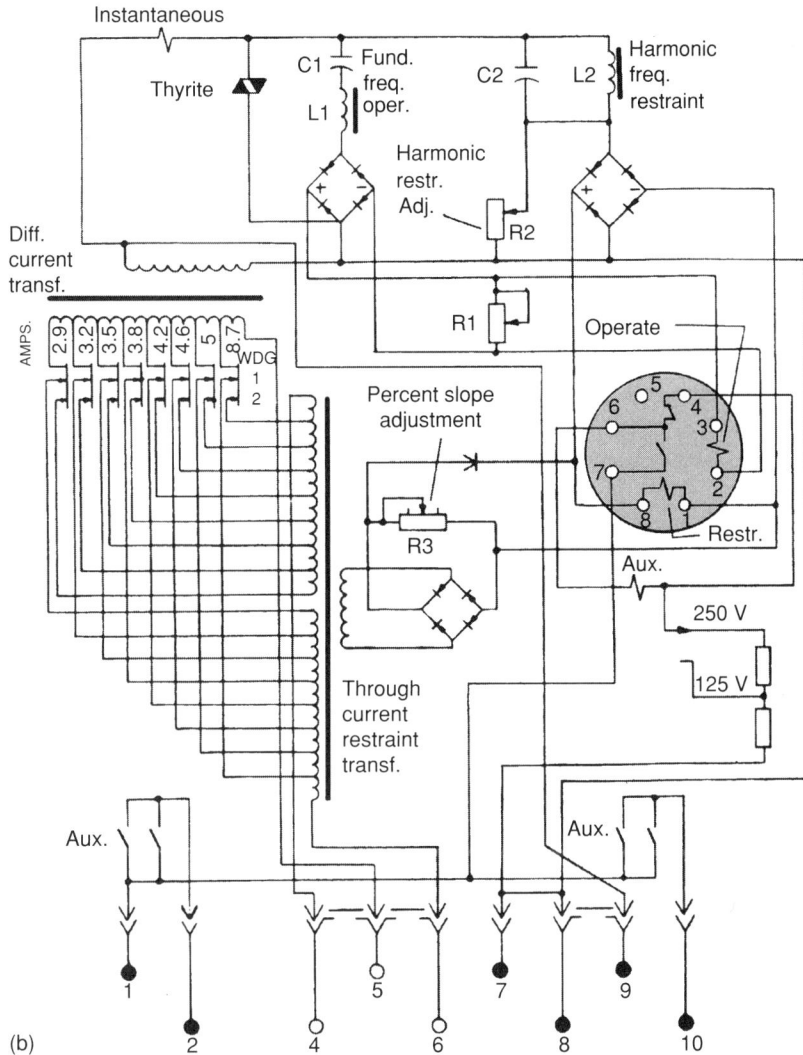

FIGURE 12.24 (*Continued*)
(a) Transformer differential relay with percentage and harmonic restraint of the BDD type (General Electric Co.,).
(b) Internal connection. In circle — high sensitive two-winding polarized electromagnetic relay. (General Electric Instruction manual GEH-2057A.)

a series tuned circuit, and the harmonic restraint circuit through a parallel resonant trap. The operating and restraint currents are each passed through a full wave bridge rectifier before passing through the polarized unit coils. The series resonant circuit is made up of a $5\,\mu F$ capacitor (C1) and a reactor (L1) which are tuned to pass currents of the fundamental system frequency and to offer high impedance to currents of other frequencies. Resistor R1 is connected in parallel on the DC side of the operate rectifier and can be adjusted to give the desired amount of operate current. The output of the rectifier is applied to the operating coil (2 to 3) of the polarized electromagnetic relay. The parallel resonant trap is made up of a $15\,\mu F$ capacitor (C2) and a reactor (L2) which are tuned to block fundamental frequency currents while allowing currents of harmonic frequencies to pass with relatively little impedance. Resistor R2 is connected in parallel on the AC side of the

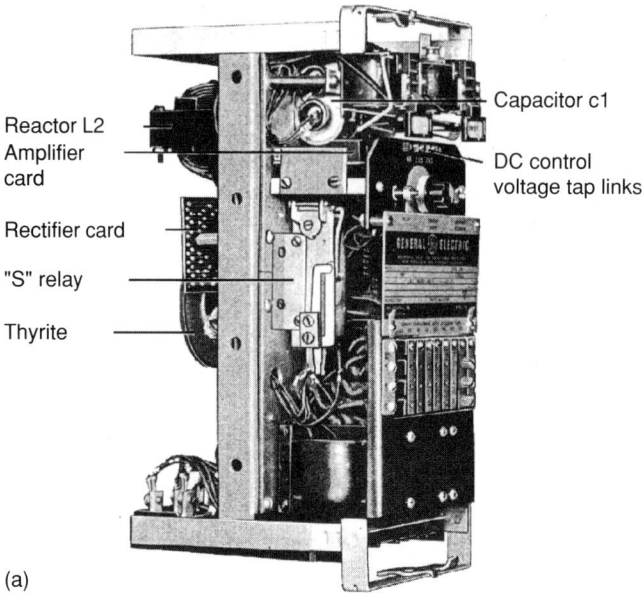

(a)

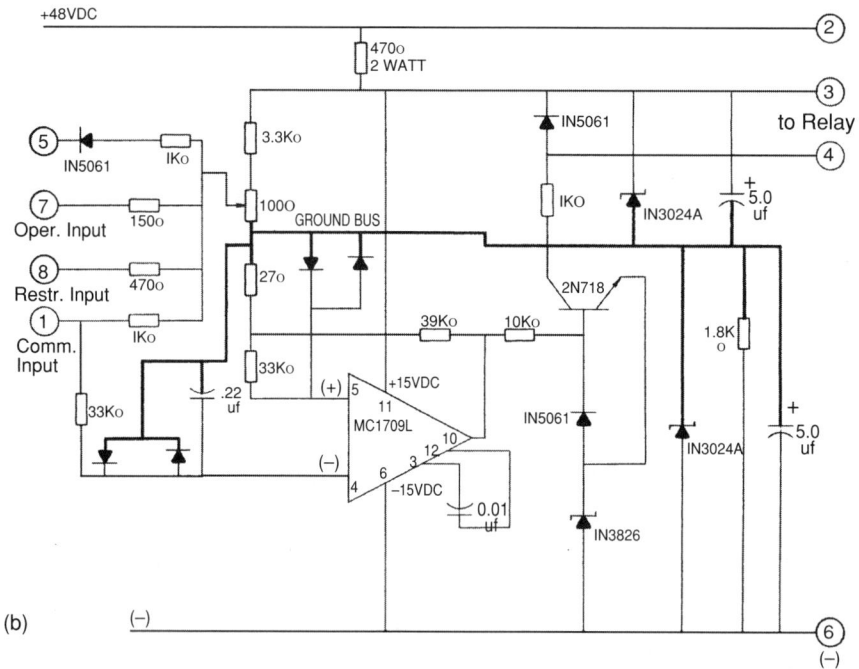

(b)

FIGURE 12.25

(a) Transformer differential relays of the STD15 type with percentage and harmonic restraint and electronic amplifier, instead of polarized relay. (b) Circuit diagram of electronic amplifier. (General Electric Instruction manual GEH-2057A.)

harmonic restraint rectifier and can be adjusted to give the desired amount of harmonic restraint. The output of the rectifier is paralleled with the through current restraint currents and applied to the restraint coil (1–8) of the polarized relay.

It is evident that if the differential current applied to the BDD type relay is of a sinusoidal wave form and system frequency, it will flow mostly in the operating coil circuit and will cause the relay to operate. If on the other hand the differential current contains more than a certain percentage of harmonics, the relay will be restrained from operating by the harmonic currents flowing in the restraint coil.

A Thyrite resistor connected across the secondary of the differential CT, limits any momentary high-voltage peaks which may occur, thus protecting the rectifiers and capacitors from damage without materially affecting the characteristics of the relay.

The instantaneous unit is a hinged armature relay with a self-contained target indicator. On extremely heavy internal fault currents, this unit will pick up and complete the trip circuit. The instantaneous unit target will be exposed to indicate that tripping was through the instantaneous unit.

Because of the saturation of the CTs and relay transformers at high fault currents, it is possible that less operating current will be provided from the differential-CT than the percentage slope tap would imply, and more harmonic restraint will be provided than the actual harmonic content of the fault current would supply. As a result, under conditions of a high internal fault current, the main unit may be falsely restrained.

Tripping is assured, however, by the over-current unit operation. Pickup is set above the level of differential current produced by the maximum magnetizing inrush current. The main operating unit of the BDD type relay is a sensitive polarized relay with components as shown within the large circle on internal connection diagrams (Figure 12.24c). The unit has one operating and one restraining coil. The relay is a high-speed low energy device, and its contacts are provided with an auxiliary unit whose contacts are brought out to studs for connection in an external circuit.

Precisely the same as the BDD relay, the electrical circuit and characteristics has a differential relay of the STD type (Figure 12.25). The difference consists only in replacement of the polarized relay assembly by an electronic amplifier, thus the STD type relay becomes an intermediate version, between electromechanical and electronic differential relays.

Transformer differential relays of the RADSB type (ASEA) with an electronic measuring element are based on a similar principle of operation, which differs from the relay considered above by its original construction, typical of the ASEA firm with its COMBIFLEX modular principle (Figure 12.26). As we have seen from the above

FIGURE 12.26
Transformer differential relays of the RADSB type with percentage and harmonic restraint and an electronic measuring unit. (ABB 1990 Relay Unites and Components buyers guide.)

examples, relays for differential protection of power transformers have specific characteristics and peculiarities that provide effective protection of transformers from emergency modes.

12.7 Pilot-Wire Relays

Relays with specific characteristics are required not only for the protection of power transformers. For the protection of power lines, applications of simplified circuits such as that shown in Figure 12.1 are not practical or reasonable when one takes into account that the length of the protected section of line may constitute tens of kilometers. In such cases instead of one relay as it is shown on Figure 12.1, two relays (Figure 12.27) are used, each affecting a power circuit breaker which is the nearest to it.

Pilot wires between the relays have high resistance exceeding by tenfold acceptable bounds for the load of even the most powerful CTs. For example, for a length of 10 km resistance of a pilot copper wire with a section of 1.5 mm^2 is 130 Ω, while the permissible load for CTs is 1 to 2 Ω. This difficulty can be overcome with the help of auxiliary CTs CT1–1 and CT2–1 (Figure 12.28). Use of the second relay connected parallel (according to circuits 12.27 to 12.28) causes considerable changes in the conditions of operation of the

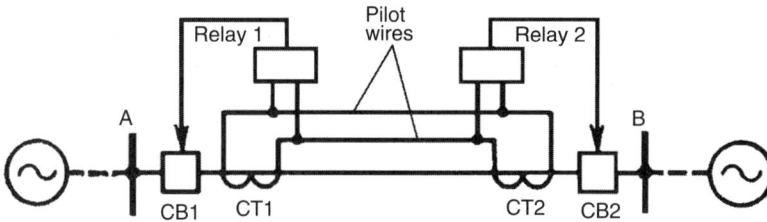

FIGURE 12.27
A Scheme of longitudinal differential protection with two relays, installed on both ends of the line.

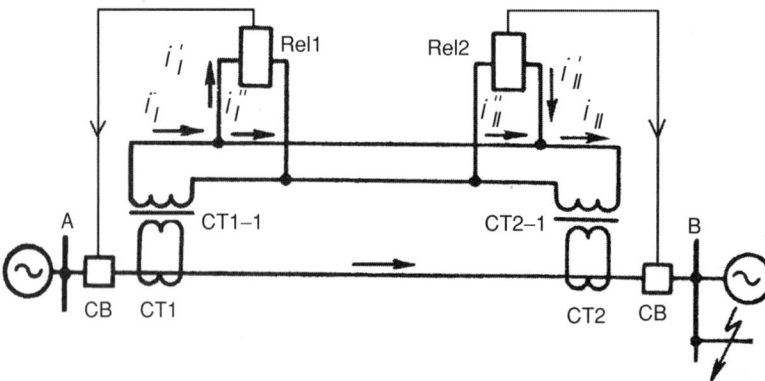

FIGURE 12.28
Application of auxiliary CTs (CT1–1 and CT2–1) for reduction of the load of the main CTs (CT1 and CT2).

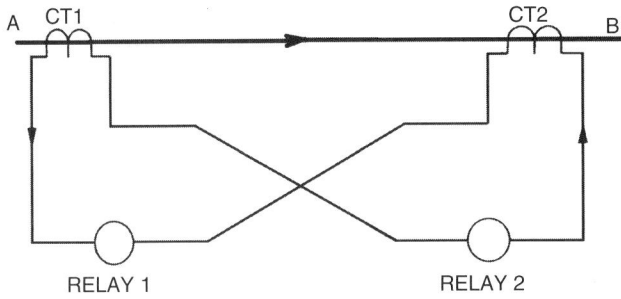

FIGURE 12.29
A basic scheme of differential protection based on a balanced-voltage system.

protection. Current applied from each of the CTs to each of the relays is distributed in inverse proportion to the resistance of their circuits. The circuit of the "farther" relay (for CT1, the "farther" relay is Rel2 and for CT2 — Rel1) includes pilot wires with high resistance, which is why current received to the "farther" relay is less than current received to the "nearer" relay. As a result, the currents applied to the relay will be unbalanced even if perfect (without errors) CTs are used.

Every protection of this kind, depending on its sensitivity, has a maximal allowable impedance value of pilot wires. If that impedance is exceeded, the protection will not operate properly, however, even if the impedance of the pilot wires equals zero (an absolutely hypothetical case), each of the parallel connected relays will receive only half of the current, which is why the sensitivity of such a protection decreases. Since there were many difficulties in implementing circuits based on the Merz–Price principle (see above) for longitudinal protection of lengthy power lines, another principle, based on a balanced-voltage system instead of a balanced-current system suggested in the Merz–Price circuit was used (Figure 12.29).

In the circuit with the balanced-current system, the secondary windings of the CT are connected to each other in phase and the main windings of the relay connected in parallel to these windings (Figure 12.28), while in the circuit with the balanced-voltage system (Figure 12.29), the secondary windings of the CT are linked with each other in a crisscross manner and the main winding of the relay connected to the windings of the CT in series.

In normal conditions in the circuit with the balanced-voltage system total impedance of the series circuit turns out to be very high, and voltages initiated in the secondary windings of the CT are mutually compensated, which is why there is no current in this series circuit (or in the windings of the relay, of course). This absence of current in the circuit and also in the pilot wires in the normal mode is a great advantage of the balanced-voltage system in comparison with the balanced-current system, as it reduces requirements for pilot wires and eliminates limitations on the size of the protected area. However, owing to the fact that a demagnetizing flux does not exist in the CT, and the entire current-transformer primary current is a magnetizing current, standard CTs are not suitable for such operating conditions, as they will then be operated close to an open-circuited secondary condition. For these cases incorporated in the scheme are auxiliary CTs that have their primaries connected in the main current-transformer secondary circuit (providing normal operating conditions for the main CTs). These auxiliary CTs are designed for continuous open circuited secondary operation.

In the circuit shown on Figure 12.29, and all following circuits in single-line modification, two pilot wires are shown. How can we deal with three-phase circuits? Can one use six such wires? Here is what the 41–658 "Type HCB Pilot-Wire Relaying and Pilot-Wire Supervision" directory, published in 1942 by Westinghouse says:

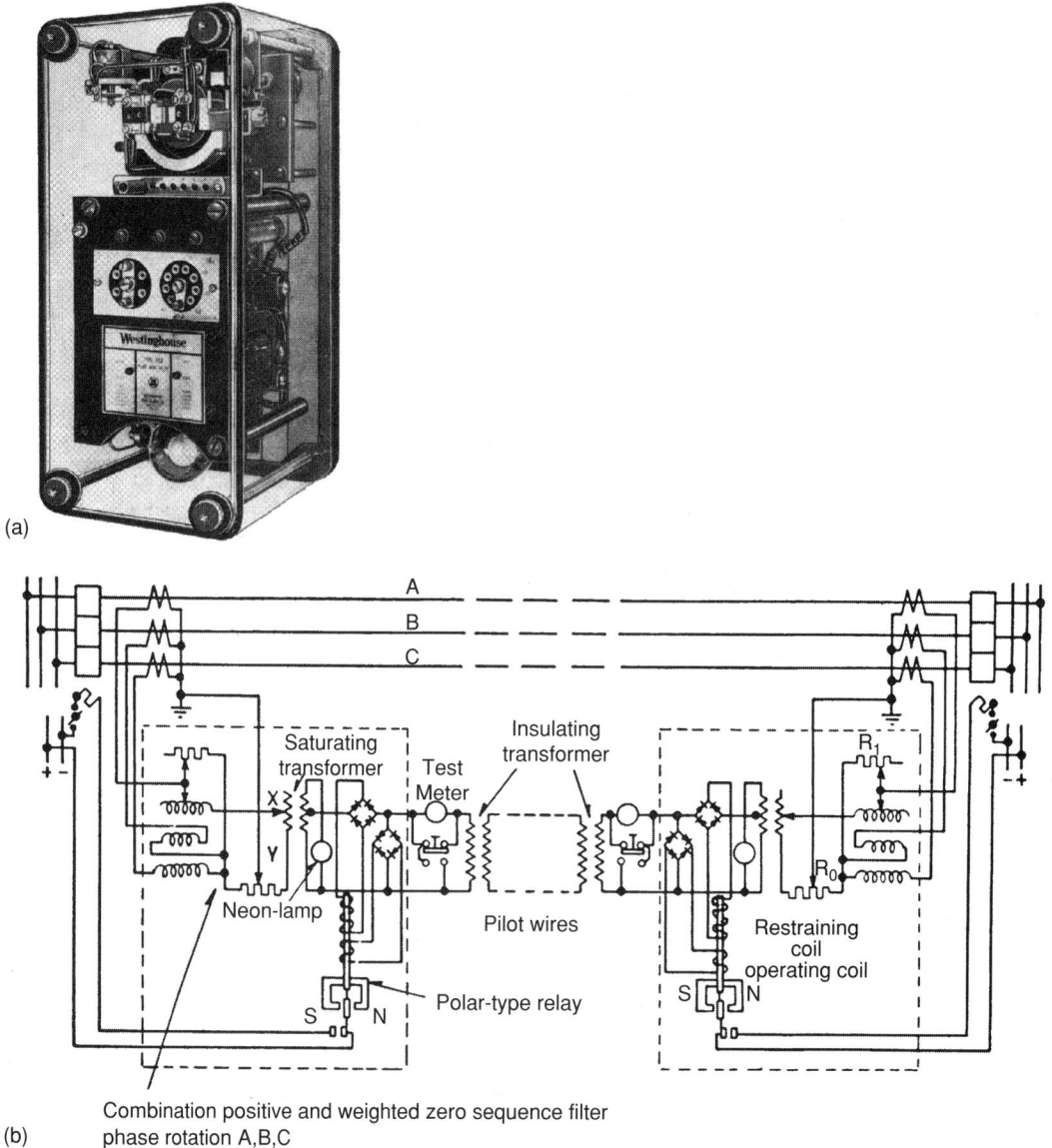

(a)

(b)   Combination positive and weighted zero sequence filter
             phase rotation A,B,C

FIGURE 12.30
Type HCB pilot-wire relay for longitudinal differential protection. External view (a) and circuit diagram (b) (Westinghouse, 1942).

The advantages of pilot wire relay protective schemes for transmission lines have been recognized for many years. Before the advent of the HCB relay, pilot wire schemes were complicated by the objectionable necessity of using multiple wire circuits, batteries, and several relays per line terminal. The type HCB relay equipment was designed to overcome these objections (Figure 12.30). This equipment gives complete high speed one cycle protection for phase and ground faults, yet uses only two pilot wires, does not apply battery voltage to the pilot wires, and has only a single moving element at each end of the line section.

This scheme can be applied to either two or three terminal lines, and to lines containing a power transformer. The same pair of pilot wires can also be used to transmit a remote

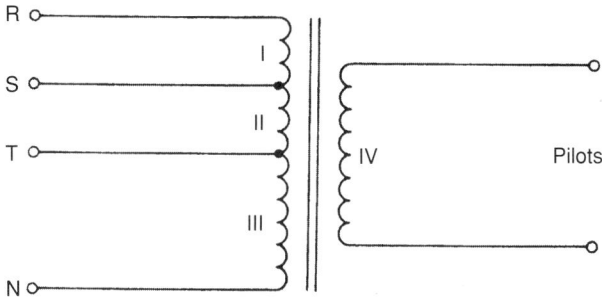

FIGURE12.31
Summation transformer for longitudinal differential protection.

tripping signal from one end of the line section to the other for special applications where this may be desired. Such special applications usually involve a power transformer in the protected section. The HCB scheme also leaves wide latitude with respect to the quality of CT with which it is used, since it does not require close matching of CT characteristics at opposite ends of the line section. The scheme can also be applied to systems having a ratio of 10 to 1 between minimum phase fault current and minimum ground fault current; that is, the zero sequence current can be one-thirtieth of the positive sequence current. The HCB scheme can also be used on privately owned pilot wires or on leased telephone wires. The HCB relay is normally supplied with a milliamperemeter and a test button to periodically check the condition of the pilot wires. Supervisory relays are available for continuous supervision of the pilot wires and for the remote tripping of circuit breakers over the pilot wire circuit.

So as the Westinghouse Company affirms, it was the first which managed to reduce the number of pilot wires to two ones in a three-phase circuit of longitudinal differential protection.

In modern circuits, a so-called "summation transformer" is used (Figure 12.31). This transformer has low-resistance primary windings (I–III), a high-resistance secondary winding (IV), and also a core with an air gap providing a linear characteristic. Such a transformer with an air gap is also called a "transactor" (a combination of the words "transformer" and "reactor"). The number of turns of the primary windings is taken in the following proportion:

$$I : II : III = 1 : 1 : 3$$

If the number of turns between the outlets of phases R–S and S–T is the same, the number of turns between the outlets of phases R–T will be doubled. This causes unequal sensitivity of protection to different combinations of damaged phases when a short-circuit occurs.

In some cases, this type of summation transformer is made as a part of the magnetic system of the relay (Figure 12.32). "Translay" is the trade name initially given to a biased, electromechanical balanced-voltage system introduced nearly 100 years ago, which is still providing useful service on distribution systems even though it remains fundamentally unchanged.

This electromechanical design derives its balancing voltages from the transactors incorporated in the measuring relay at each line end. The latter are based on an induction-type meter electromagnet as shown in Figure 12.32. The upper magnet carries a summation winding to receive the output of the CTs, and also a secondary winding which delivers the reference electromagnetic force. The secondary windings of the conjugate relays are interconnected, as a balanced-voltage system over the pilot channel, the lower electromagnets of both relays being included in this circuit.

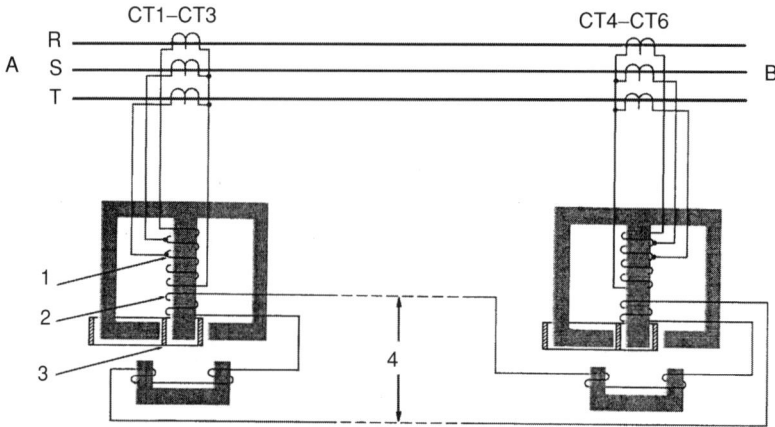

FIGURE 12.32
Circuit of the three-phase induction relay of HO4 type for the balanced-voltage longitudinal differential protection system (GEC Measurements, now ALSTOM). 1 — Summation winding; 2 — secondary winding; 3 — bias loop; 4 — pilot wires.

Through current in the power circuit produces a state of balance in the pilot circuit and zero current in the lower electromagnet coils. In this condition, no operating torque is produced. An in-zone fault causing an inflow of current from each end of the line produces a circulation of current in the pilot circuit and energization of the lower electromagnets, which co-operate with the flux of the upper electromagnets to produce an operating torque in the discs of both relays. An in-feed from one end only will cause relay operation at the feeding end but there is no operation at the other end because of the absence of upper magnet flux.

Bias torque is produced by a copper-shading loop fitted to the pole of the upper magnet, thereby establishing a Ferraris motor (see above) action that gives a reverse or restraining torque proportional to the square of the upper magnet flux value. A permanent magnet is fitted for damping, providing further improvement of both mechanical and transient stability.

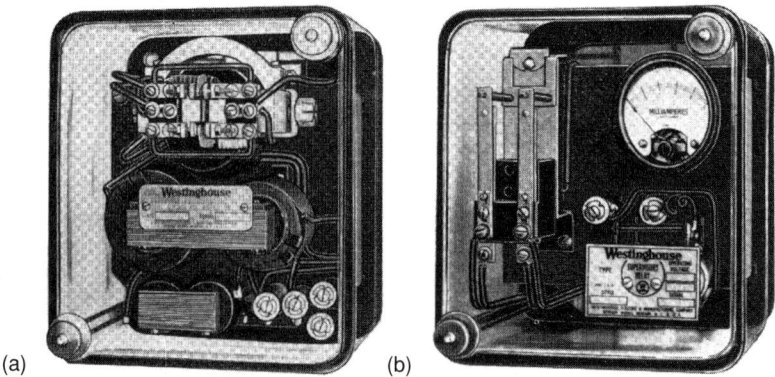

(a)   (b)

FIGURE 12.33
Pilot-Wire Supervision Relays (Westinghouse, 1942).

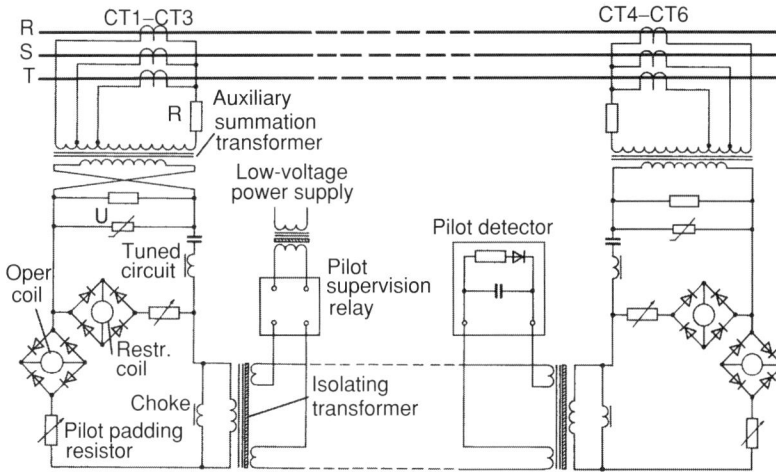

FIGURE 12.34
Modern differential protection relay type DS7 with pilot supervision relay SJA (ALSTOM).

Damage of pilot wires (breakage, short-circuit) may cause incorrect operation of protection, which is why in order to increase its reliability it is equipped with special devices controlling the state of these wires. As the producer assured, already in the first Westinghouse differential relay with two pilot wires-special measures for control of these wires were taken. These measures consisted of the use of special relays, so called "Pilot-Wire Supervision Relays," providing constant control of these wires (Figure 12.33). The principle of operation of such devices was that the source of direct current was switched to these wires. This direct current did not affect operation of the differential relay, but it allowed control of the state of the wires. This principle is still used today (Figure 12.34).

This scheme, shown in Figure 12.34, is in the higher speed class. It is also of the balanced-voltage type, but differs in its derivation of the reference voltage. An auxiliary summation transformer is loaded with a resistive shunt to provide voltage that is balanced over the pilot circuit, with a corresponding quantity at the other end of the zone. The measuring relay is a double-wound moving coil element of the axial motion type, the coils being energized through bridge rectifiers. One coil, connected in series with the pilots, observes any unbalance component. The other coil is connected in series with an adjustable resistance across the reference voltage, to provide restraint.

This scheme is suitable for use with pilots of up to 1000 Ω. The pilot loop, not including the relays, is made up to this value by padding rheostats, and the bias rheostat is also adjusted to give the correct degree of restraint, according to the length and capacitance of the pilot.

The pilots are compensated where necessary by shunt reactors at each end. Pilot isolating transformers used in high-induced voltage are expected and pilot supervision can be applied. This scheme will trip both ends of the circuit even if fault current is fed from one end only.

In cases where there are no power transformers (and therefore no problems with inrush currents) on the protected section of the line, it is more reasonable to apply simplified (and cheaper) differential relays containing no restraint coils. The RYDHL type relay, produced by the ASEA Company (ABB), known for decades, may serve as an example (Figure 12.35). This is really the simplest type of differential relay, containing very few

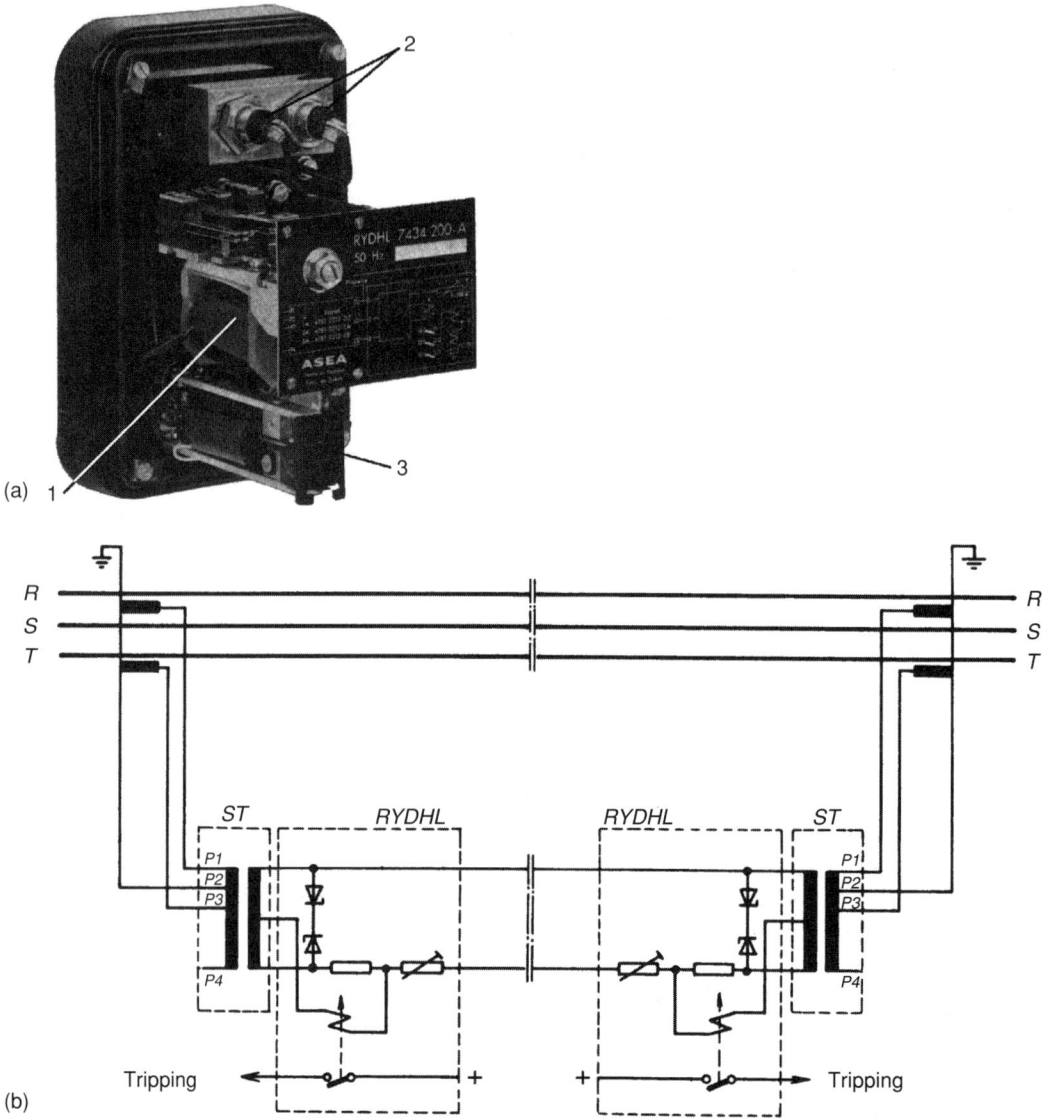

FIGURE 12.35

(a) Simplest pilot-wire differential relay of the RYDHL type without restraint circuits: external view with cover removed. 1 — Simple electro-magnetic relay; 2 — two Zener diodes; 3 — manually reset trip indicator (flag-relay). (b) circuit diagram (ASEA, 1969). ST — Summation transformer.

elements: an electromagnetic clapper type relay, two Zener back-to-back diodes for protection from overvoltages, a trimming resistor, and a pick-up indicator with hand reset. The summation transformer is a separate unit without any cover, which can be placed in a suitable position in the relay cubicle. Similar relays were also produced by many other companies (Figure 12.36).

The pilot links between relays have been treated as an auxiliary wire circuit that interconnects relays at the boundaries of the protected zone. In many circumstances, such as the protection of long transmission lines or where the route involves installation

FIGURE 12.36
Simplest pilot-wire differential relay of the RN22 type. (Siemens 1972
Protective Devices catalog NS 1-72.)

difficulties, it is too expensive to provide an auxiliary cable circuit for this purpose, and other means are sought.

It was offered to use the main line conductors as the interconnecting conductors of a longitudinal differential protection. The need for special interconnecting conductors (cables) then disappears and it hence becomes possible to set up a longitudinal differential protection on lines of any length. This is the basis of what are called "carrier-current protections." To make the transmission of commercial-frequency load current possible, and at the same time use the main line wires as the interconnecting conductors of the differential protection, it is necessary to use a current of higher frequency in order to be able to transmit current impulses from one end of the line to the other. For this purpose, it is usual to employ auxiliary current having a frequency 50 to 150 kHz, generated by a special high-frequency transmitter and received at the other line end by a high-frequency receiver. The protected power line must then be accordingly equipped to handle the high-frequency current within its confines, this equipment comprising high-frequency traps (HFT) interposed in the line conductors at both ends of the protected line and coupling filter (capacitors) (Figure 12.37a).

The HFT (filter) is an LC-circuit tuned to resonance at high frequency. It hence presents high reactance to the high-frequency carrier current, but relatively low reactance to the power-frequency current. The high-voltage coupling capacitor connects the high-frequency receiver–transmitter to one of the line conductors and simultaneously serves to isolate the receiver–transmitter from the high power-line voltage. It presents a relatively low reactance to the high frequency and a high reactance to the power-frequency.

The carrier channel is used in this type of protection to convey both the phase and magnitude of the current at one relaying point to another for comparison with the phase and magnitude of the current at that point using FM modulation or analog-digital converters and digital transmission.

One another type of protection (phase-differential) uses carrier techniques for the communication between relays is phase comparison protection, when the carrier channel is used to convey only the phase angle of the current at one relaying point to another for

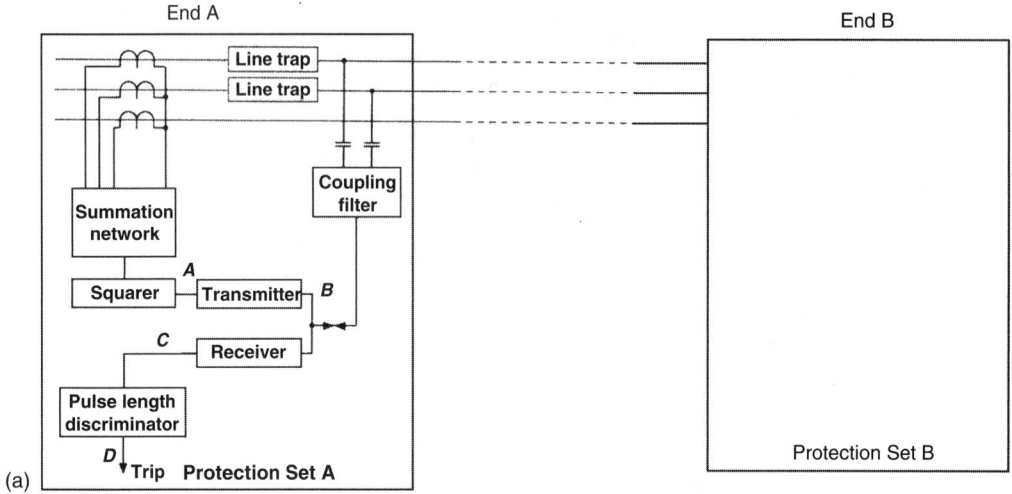

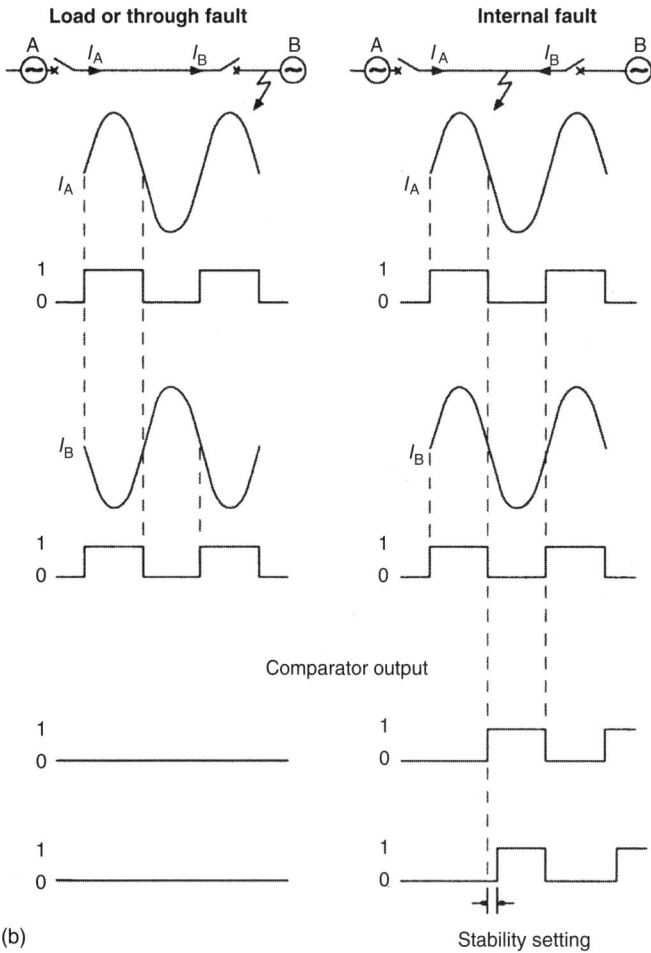

FIGURE 12.37
(a) Schematic diagram of carrier-current differential line protection; (b) Principles of phase comparison differential protection.

comparison with the phase angle of the current at that point. In this type of protection the carrier channel transfers a logic or on/off signal that switches at the zero crossing points of the power frequency waveform. Comparison of a local logic signal with the corresponding signal from the remote end provides the basis for the measurement of phase shift between power system currents at the two ends and hence discrimination between internal and through faults. If the phase relationship of through-fault currents is taken as a reference condition, internal faults cause a phase shift of approximately 180° with respect to the reference condition. Phase comparison schemes respond to any phase shift from the reference conditions, but tripping is usually permitted only when the phase shift exceeds an angle of typically 30 to 90°, determined by the time delay setting of the measurement circuit, and this angle is usually referred to as the stability angle.

13

Distance Relays

13.1 Principles and Basic Characteristics of Distance Protection

A number of cases in electric circuits of complex configuration with several power sources, neither maximal current protections nor directional protections (power direction relays), can provide selectivity of tripping a short circuit (SC). Why? Let us consider a simple example of a ring circuit with a two-way supply (Figure 13.1), provided with directional current protection (with each set of relays Rel containing current relays and power direction relays). The direction of the relay action and the directions of the power flows in the circuit when an SC occurs, are indicated by the arrows. As it can be seen in Figure 13.1a, an SC (SC1) creates conditions for picking up of sets of relays (Rel1, Rel2 and Rel3). Rel4 does not pick up because the direction of the power flow at that point of its installation does not agree with the standard installation of the power direction relay (from the bus bar to the line).

For selective deenergization of only Line 1, on which the SC occurs, relays Rel2 and Rel1 must pick up before relay Rel3 does. That is: $t_2 < t_3$. But if an SC (SC2) occurs on the other line (Line 2), for selective deenergization, the relays Rel3 and Rel4 must pick up before relay Rel2, that is, $t_2 > t_3$. Therefore, we can see that it is impossible to meet such opposing requirement with the help of current and directional protections. In such cases, so-called "distance protection" is used. Distance protection is that in which the time delay varies according to the distance to the point at which the SC has occurred (Figure 13.2).

If each of the relays installed along the line have time delays depending on impedance (distance), the relay which picks up first will always be the one that is nearest to the point of short circuiting. This is the main purpose of distance protection. In circuits with two-way supply, distance protection is directional (that is it responds only to one direction of the power flow).

An example of a distance protection scheme having coordinated characteristics of circuit breaker protection on a system with two-end power supply is seen in Figure 13.3. The time-delay characteristics of the odd-numbered relays 1, 3, and 5, operating on the fault-power flow in the left-to-right direction (denoted by the arrows at the circuit breakers) are represented on the top side of the axis, with those of the even-numbered protections below the axis. In the diagram, the shaded portions represent the tripping times of the distance protections on the corresponding lines. Thus, for example, when a fault appears on line BC (at point SC.), relays 3 and 4 operate with the minimum time delay t_1. If for one or another reason, these protections fail to operate, circuit breakers 1 and 6 of the preceding circuit sections from the power supply are tripped, with the time delay t_2 and t_3. This is how back-up of the protection on the adjoining circuits is achieved.

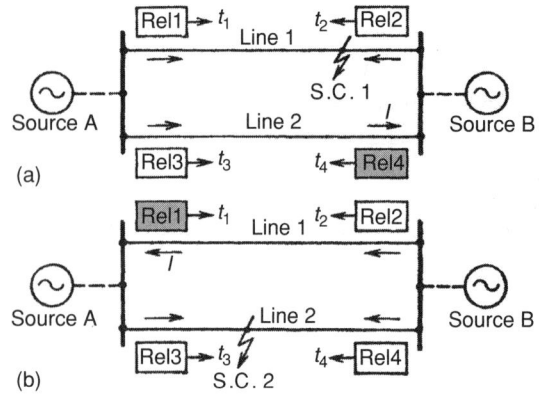

FIGURE 13.1
Looped power system with two sources of power supply. SC — short circuit points, t — time delays of relays.

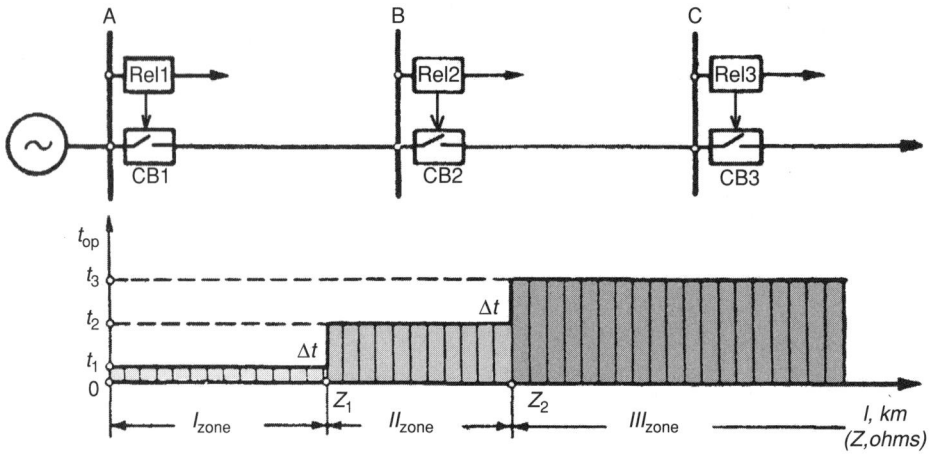

FIGURE 13.2
Principle of distance protection. Protective zones are shown for first (Rel1) relay. Each of the relays (Rel1–Rel3) has a same three-stage characteristic $\Delta t = f(Z)$.

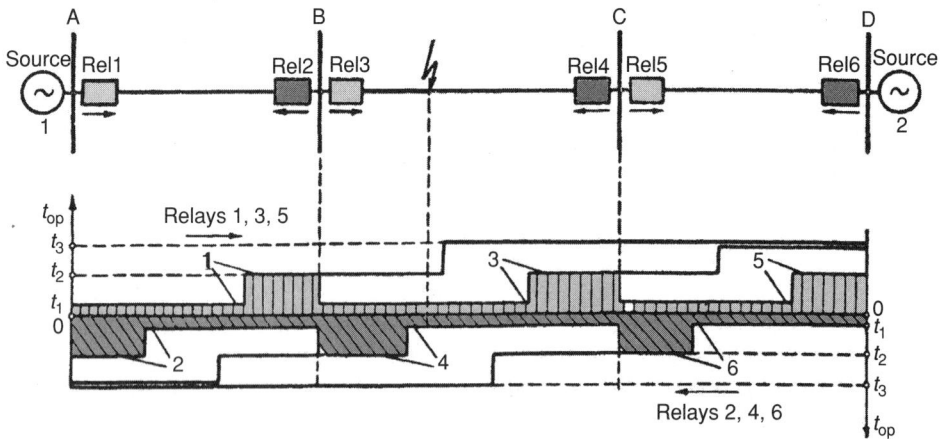

FIGURE 13.3
Principle of construction of directional distance protection with a time grading in power system with two-way supply.

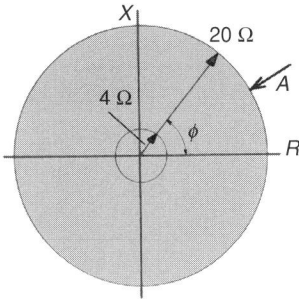

FIGURE 13.4
Nondirectional impedance relay characteristic.

Distance is determined in such protection by measuring impedance of the line up to the point of short circuiting. As impedance is a value determined by the proportion of voltage and current ($Z = U/I$), distance relays have measuring elements of voltage and current, and also a mechanism (or an electronic circuit) responding to the ratio of the voltage to the current, that is to some fictitious impedance seen at their terminals. This impedance may take the form of a full impedance (Z), a reactance (X), or a resistance (R). Distance relays may, hence, be subdivided into three types.

The impedance relay, which responds to the fictitious impedance:

$$Z = U/I$$

The operating characteristics of distance relays, as generally accepted, are plotted on the complex-quantity plane on which the resistance R is laid off on the real axis and the reactance X is laid off on the imaginary axis. The operating characteristic of an impedance relay on the complex-quantity plane is represented by a circle of radius $Z = Z_{oper}$ (Figure 13.4).

In 1937, M.J.H. Neher presented the first paper on distance-relay characteristics plotted on an impedance diagram (Neher J.H.A., Comprehensive method of determining the performance of distance relays, *AIEE Transactions*, vol. 56, 1937, pp. 833–844). Physically such characteristic means that the relay picks up as impedance decreases to a certain value (the radius of a circle) and does not depend on the angle ϕ between X and R (that is between the current and the voltage), in other words the sensitivity of the relay is the same for any angle ϕ. The circle is the working area of the relay. The relay picks up at the entering (of vector A) at any angle in this circle from the outside. If this type of relay is used for distance protection of the line, it must be equipped with a separate power direction relay. Then the total characteristic of this relay set will look like as in Figure 13.5. The relay picks up at the entering (of vector A) at angle interval $180°$ (in a half circle) only in the first quadrant and partly (depending on angle ϕ) in the second and fourth.

The characteristic of the directional impedance relay in Figure 13.6 has a form of a circle passing through the origin of the coordinates. As can be seen from the Figure 13.6, the length of the vector from the starting point (0) to the circle bound, is different for different angles. This means that the maximum sensitivity of the relay is achieved at a certain angle ϕ. This angle is called the angle of maximum sensitivity (or the angle of maximum torque) (Figure 13.7). The relay will not operate in the third quadrant. This means that it cannot operate if the power flow is directed from the line to the substation.

As the impedance relay picks up when impedance decreases to a certain threshold level, relays capable of picking up earlier, that is at greater impedance, are more sensitive, which is why the maximum radius shown in Figure 13.6 corresponds to the maximum

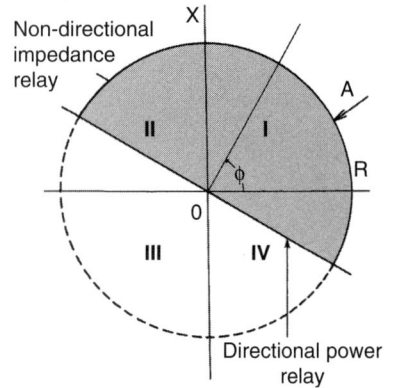

FIGURE 13.5
Relay characteristic of a device consisting of a nondirectional impedance relay and a directional power relay.

sensitivity of the relay. This means that at the initial point, the sensitivity (pick-up) of the relay to the impedance will be minimal.

Usually for all other types of relays, the following dependence is considered to be correct: the less the inlet actuating quantity (current, voltage, power, etc.) at which the relay picks up is, the higher will be the sensitivity of this relay. The directional impedance relay, as we see, operates like "vice versa": picks up when input value — impedance — decreases. To emphasize this peculiarity of the relay, it was called "MHO," that is OHM (the unit of measurement of resistance) in the reverse order. Of course, this expression belongs to slang but it is widely used in technical literature and specifications.

The MHO relay may have a characteristic displaced by 0-S with regard to zero (Figure 13.8), which is why this relay operates not only on the protected line, but also covers the buses supplying the line and some parts of other outgoing lines. Theoretically the characteristic of the MHO relay can be biased in any direction, but in practice only a two-directional offset is used (Figure 13.9).

In addition to the impedance relay, there are also reactance relays and resistance relays (Figure 13.10). The reactance distance relay, which responds to the reactive component of the fictitious impedance, or:

$$X' = U/I \sin \phi$$

The operating characteristic of a reactance relay, as represented on the complex-quantity plane, has the form of a straight line, parallel to the R-axis (Figure 13.10a). In this figure, the area that has been shaded is the zone of operation of a reactance relay of the under-type. It is clear that the reactance relay responds to X', irrespective of the value of angle ϕ.

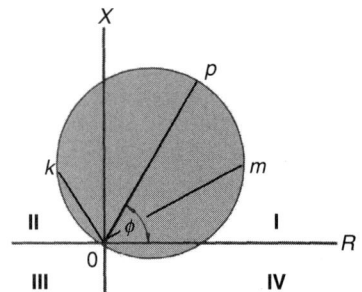

FIGURE 13.6
Characteristic of a directional impedance relay. $0-p > 0-m > 0-k$.
ϕ — angle of maximum torque.

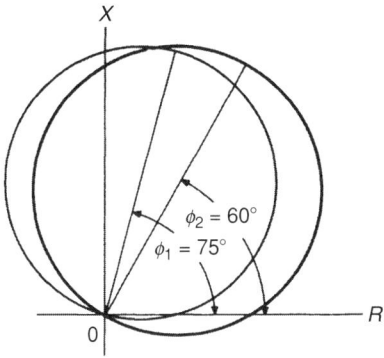

FIGURE 13.7
Variation of the angle of maximum torque in the MHO relay.

The resistance distance relay which responds to the active component of the fictitious impedance, or:

$$R' = U/I \cos \phi$$

The operating characteristic of a resistance relay as represented on the complex-quantity plane is given in Figure 13.10b.

Resistance distance relays have not found application, due to the fact that they are the most complicated in construction and may be replaced with full success by simpler impedance distance relays. It is important to note that the reactance (OHM) relays will operate for faults that plot anywhere below its characteristic line on the *R–X* diagram. Thus, the relay is not directional in itself, and therefore is always used in conjunction with a (directional) MHO relay (Figure 13.11). Its horizontal characteristic makes this relay insensitive to resistance, and therefore it measures only the reactive portion of the impedance from the relay location to the fault. It operates to trip if this reactance is less than the relay setting. The measurement of this unit is unaffected by arc resistance in the fault and it is eminently suited for application on lines where arc resistance can be appreciable compared to the protected line impedance. This is generally true for short lines. The first zones of these relays are set to reach about 80 to 90% of the distance to the far end of the line. The second zone is set to reach beyond the end of the line and the third zone may be set beyond that.

The other type of this class of relay is a three-zone single-phase directional distance relay. It is composed of three cup-type MHO units, one per zone, which are combined as shown in the *R–X* diagram of Figure 13.12 to provide three-zone protection for all multi-phase faults.

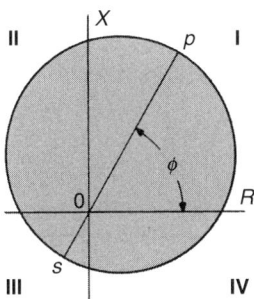

FIGURE 13.8
Characteristic of MHO relay with offset (0–s).

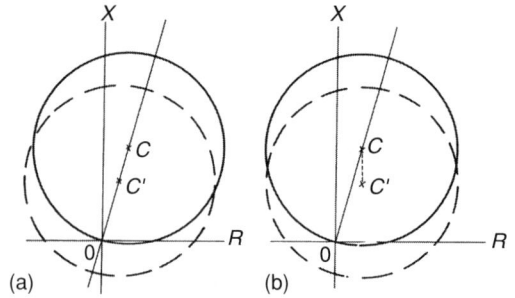

FIGURE 13.9
Building of MHO characteristics with offset. (a) Offset along maximum torque angle; (b) offset along X-axis; C — original center; C′ — offset center.

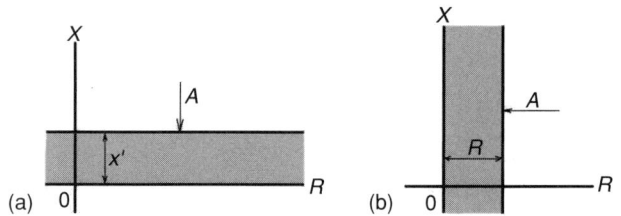

FIGURE 13.10
Reactance (a) and resistance (b) relay characteristics.

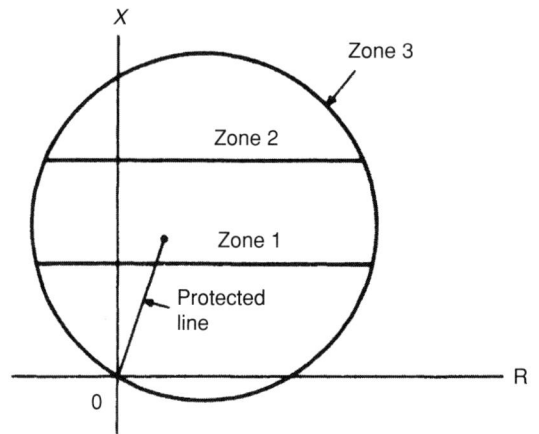

FIGURE 13.11
Typical characteristics of three-zone single-phase directional distance relays, which compose three reactance (OHM) units (straight lines) and one impedance (MHO) unit (circle).

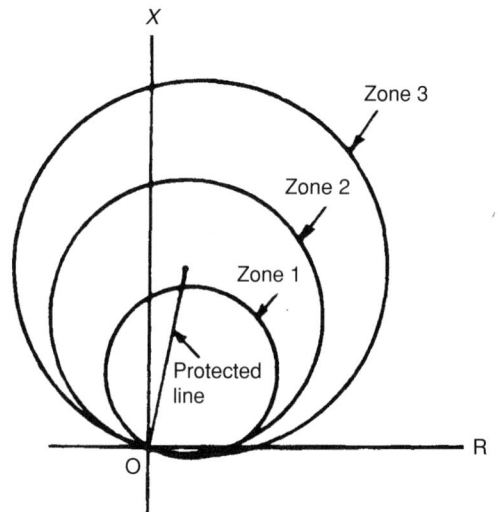

FIGURE 13.12
Typical characteristics of three-zone single-phase directional distance relays, which compose three MHO units (one per zone).

Because the characteristics of the MHO units pass through the origin of the $R–X$ diagram, they are inherently directional and provide their maximum reach in the general direction along the impedance angle of the protected line. Because of this, the relays have minimum exposure to system swings and are particularly well suited for applications on longer lines. The first zone unit may be set to reach as far as 80 to 90% of the distance to the remote terminal. The second zone is set to reach beyond the remote terminal and the third zone may be set beyond that.

13.2 System Swing

What is a "system swing" (or "power swing")?

In 1937 C.R. Mason presented for the first time an AIEE paper in which he analyzed relay performance during swing conditions (Mason C.R., Relay operation during system oscillations, *AIEE Transactions*, vol. 56, 1937, pp. 823–832). The results of this analysis were summarized in plots of relay torque as a function of the separation angle between the two generators of his equivalent system.

What is actually meant is a mode dealing with operating irregularity of synchronous operation of generators working for the common network (the so-called "asynchronous running"). This mode is accompanied by periodical flows of great current through the line and considerable voltage drops (that is "swings" or "system swings") — signs typical of SCs causing distance protections to respond. The impedance relay in itself is incapable of distinguishing between system swings and SCs without special blocking devices. To detect system swings the blocking unit is activated in different cases. One such case is the occurrence of dissymmetry in the circuit. An SC never occurs simultaneously between all three phases (three-phase SC). First, a single-phase ground SC or a short between two phases occurs, and then it may become a three-phase one. Even the contacts of a three-phase high-voltage circuit breaker, as it picks up, are not separated absolutely symmetrically. Unlike in cases of the SC mode, the network mode changes entirely symmetrically when a system swing occurs. This distinction is used in many types of distance relays produced in Russia as method of swing detection.

Detection of swing mode as rate of raise of power flow is quite popular in the West now: at an SC the power flow increases stepwise by the closure point, whereas at a system swing it varies slowly (low slip frequency) (Figure 13.13). As soon as a slow increase of the power flow is detected, an element blocking the relay operation is energized.

There is another method of detection of this mode, which is also quite widespread in practice. During a power swing, the change of impedance is slower than during a fault on the power system. The RXZD-4 relay is based upon this principle. The impedance measuring elements of the relay have an operating characteristic in the form of two concentric ovals in the $R–X$ plane (Figure 13.14). When a power swing occurs, the RXZD-4 measures the time difference between the impedance operating characteristic ZP_2 and ZP_1. If the time is longer than 50 ms, the power-swing-blocking relay operates and the output signal is maintained for approximately 2 sec. The power-swing-blocking relay has an 80° characteristic. The ratio between the major axis and the minor axis is 2/1. The set operate value corresponds to the outer oval ZP_2. The inner oval ZP_1 is fixed at 0.8 times of the set value of ZP_2.

How can one obtain relay characteristics in the form of circles, or is that just an abstraction which has nothing to do with reality?

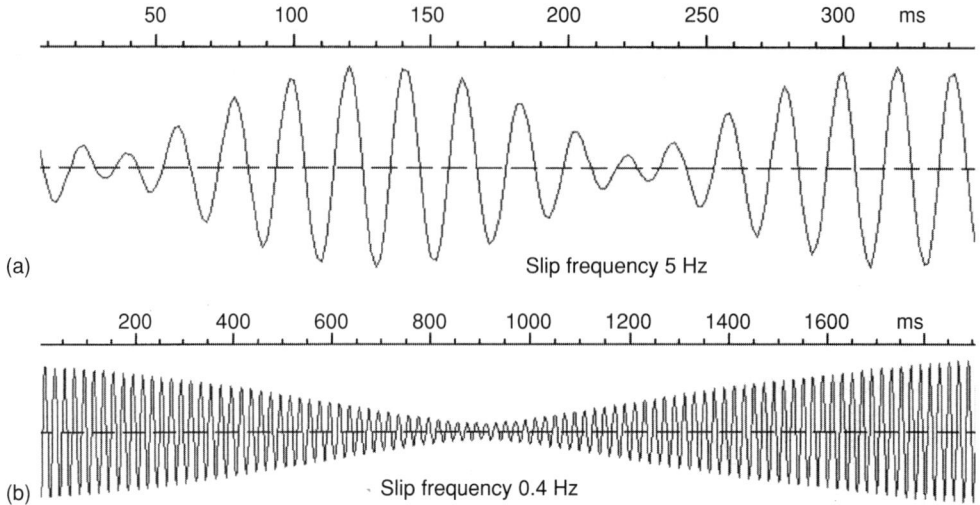

FIGURE 13.13
Power swing in a network with low slip frequency.

Let us consider this.

Let us take an impedance relay with MHO and OHM elements — the GCX-17 type relay for instance (see a description of the construction below) and apply to its inlets (terminals 5, 7, 8, 10, 17, and 18) currents and voltages according to the scheme of testing of this relay given in the Instruction Manual (Figure 13.15).

Let us then choose some points (enough to build a relay characteristic), say five points. As can be seen from typical characteristics of the MHO relay, the points are placed in the quadrants I and II, that is within 0 to 180°. Dividing 180° into six equal parts we will have 30°. Let us draw five rays: a, b, c, d, and e, through each 30°, starting from 0 on the vector plane *X–R* (Figure 13.16).

Let us then apply currents and voltages according to the scheme considered above to the relay. We set a certain constant current value, 10 A, for instance, and we will continuously vary voltage with the help of a variac and record pick-up voltage for selected values of the angle ϕ.

As a result, we will obtain five values of pick-up voltage for the following important angles: 0, 30, 60, 90, and 120°. Then by Ohm's law we will calculate impedance for the current and voltages for each angle value: $Z = U/2I$ (doubled current value is used

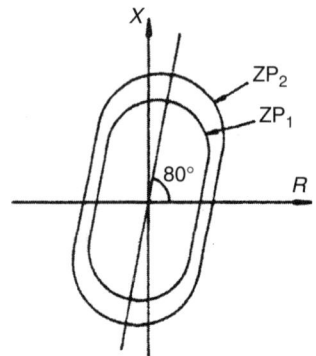

FIGURE 13.14
Operating characteristics of a power-swing-blocking relay of the RXZD-4 type.

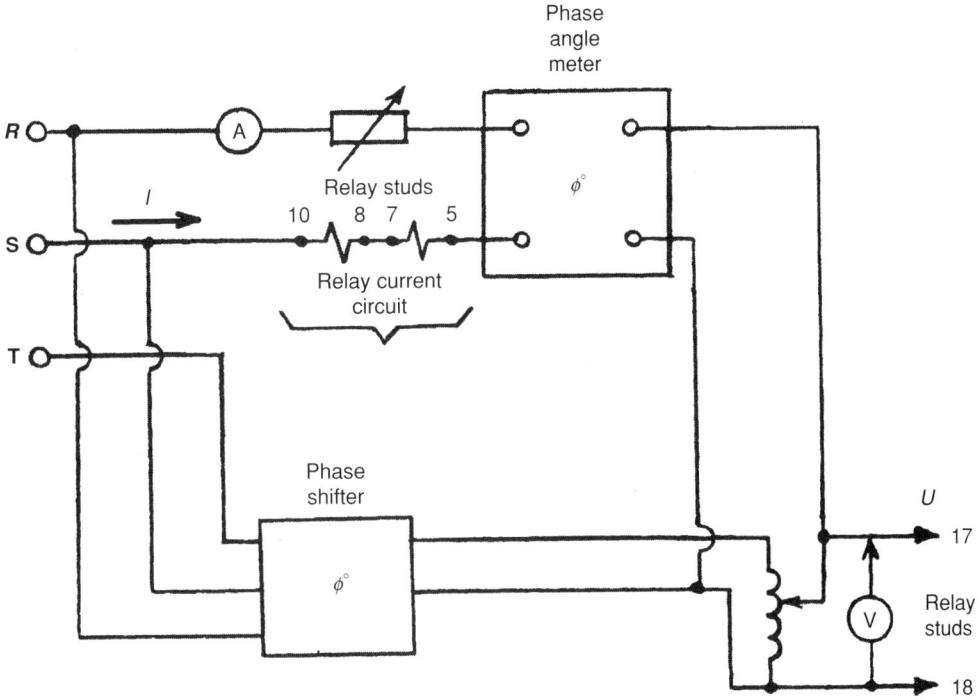

FIGURE 13.15
Scheme of reading of characteristics for the MHO OHM GCX-17 type relay. (General Electric Operating Principles...GET 6658.)

because of series connecting of current windings in this experiment). We lay off the values as segments a, b, c, d, e on the rays corresponding to the angles. If we draw a smooth curve through these dots, we will have a circle.

In the GCX-17 type relay only one element of reactance measurement (OHM unit) for all three zones is used. For work in one of the zones only the settings of this element need to be switched with the help of an additional electromagnetic relay. The OHM unit is tested at constant voltage, changing only the angle. As a result one obtains similar values of pick-up current regardless of the angle (this proves that the characteristic of the OHM unit is a straight line). Having calculated the reactance value by Ohm's law and laid off

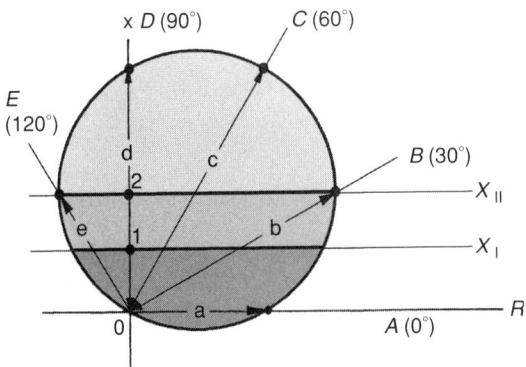

FIGURE 13.16
Construction of the characteristic of the impedance relay with elements of MHO and OHM.

the obtained value on the X-axis, we will have point 1. The line X_I passing through this dot will be the first stage of the OHM unit. In order to obtain the second-stage characteristic, we close and fix the contacts of an additional electromagnetic relay and repeat the experiment. As a result, one obtains a second value independent of the angle, and a second line X_{II} corresponding to the second zone.

13.3 Principles of Distance Relay Construction

How are these relays, capable of measuring resistance, constructed?

The principle of operation of the distance relay is very much like that of the current restraint relay in which the restraining is carried out by voltage (Figure 13.17). In such a relay, the greater the current and the lower the voltage, the greater torque the final control element of the relay will have, but current increase and voltage decrease in the circuit, according to Ohm's law, lead to a decrease of resistance of the circuit. It turns out that such a relay picks up as the resistance of the controlled circuit decreases to a certain threshold. This is what is actually called a resistance relay. In fact the resistance relay is the one that responds to the ratio of one input quantity (voltage) to the other one (current); that is it compares voltage and current.

These so-called "balanced-beam relays" of the simplest type (shown in Figure 13.17) provided comparison of voltages and currents only by value and did not take into account the angle between them, which was an essential disadvantage of this type of relay. A more complex construction (Figure 13.18) allowed a comparison of current and voltage not only by value, but also by phase.

Nevertheless, impedance relays of the induction type turned out to be the most widespread (Figure 13.19) and relays of this type were produced by many companies 1920–30's on (Figure 13.20). The first of these were relays with a rotating disk, and then more modern cup-induction relays were introduced (Figure 13.21).

Let us return to the characteristics of relays considered above. What is the construction difference between a reactance unit (OHM unit) and a directional impedance unit (MHO unit)?

On closer examination (Figure 13.21), it turns out that there are not so many differences in fact. What strikes the eye is that in the reactance unit the current winding is arranged

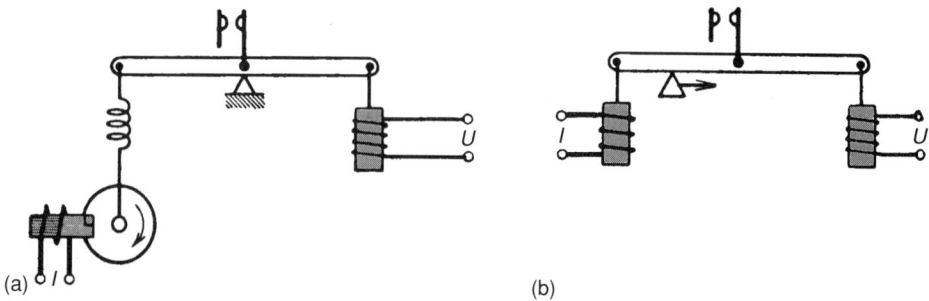

FIGURE 13.17
Principle of impedance relay construction with a balanced beam (balanced beam type relay) (a) Combined relay (induction disk and an electromagnet); (b) electromagnetic relay.

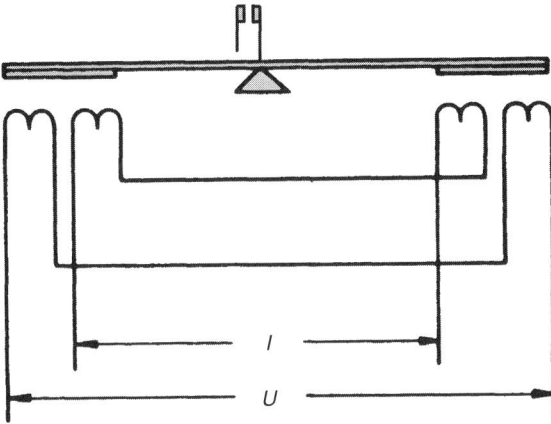

FIGURE 13.18
Balanced-beam relay operated as an amplitude and phase comparator.

on three poles and the voltage winding is fully placed on one pole, while in the MHO unit, vice versa.

One relay may contain both an OHM unit and a MHO unit at the same time, as in the GCX type relay for instance (Figure 13.22), or three MHO units, as in the GCY type relay.

The MHO unit of the relay is the four-pole induction cylinder (cup) construction. The two side poles, energized with phase-to-phase voltage, produce the polarizing flux. The flux in the front pole, energized with a percentage of the same phase-to-phase voltage, interacts with the polarizing flux to produce restraint torque. The flux in the rear pole, energized with the two line currents, associated with the same phase-to-phase voltage, interacts with the polarizing flux to produce the operating torque.

The primary purpose of the MHO unit in the relay is to provide the directional discrimination that is necessary since the OHM unit is inherently nondirectional. The MHO unit directional characteristic is such that it will operate correctly for either forward or reverse faults at voltages down to 1% of the rated voltage over a current range of 5 to 60 A. A secondary purpose of the MHO unit is to measure fault impedance for the third zone of protection.

The OHM unit is also the four-pole induction cylinder (cup) construction. The front and back poles, energized with delta current, produce the polarizing flux. The side poles are energized with a voltage equal to the difference between the operating quantity (IZ_T) and the restraint voltage (E), where I is the delta current and Z_T is the transfer impedance of

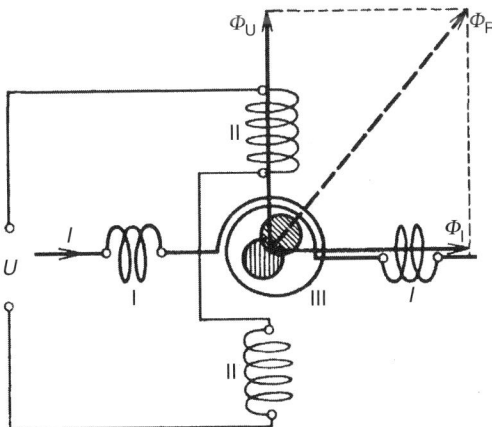

FIGURE 13.19
Principle of construction of an impedance relay of the induction type (principle of Ferraris motor, see above).

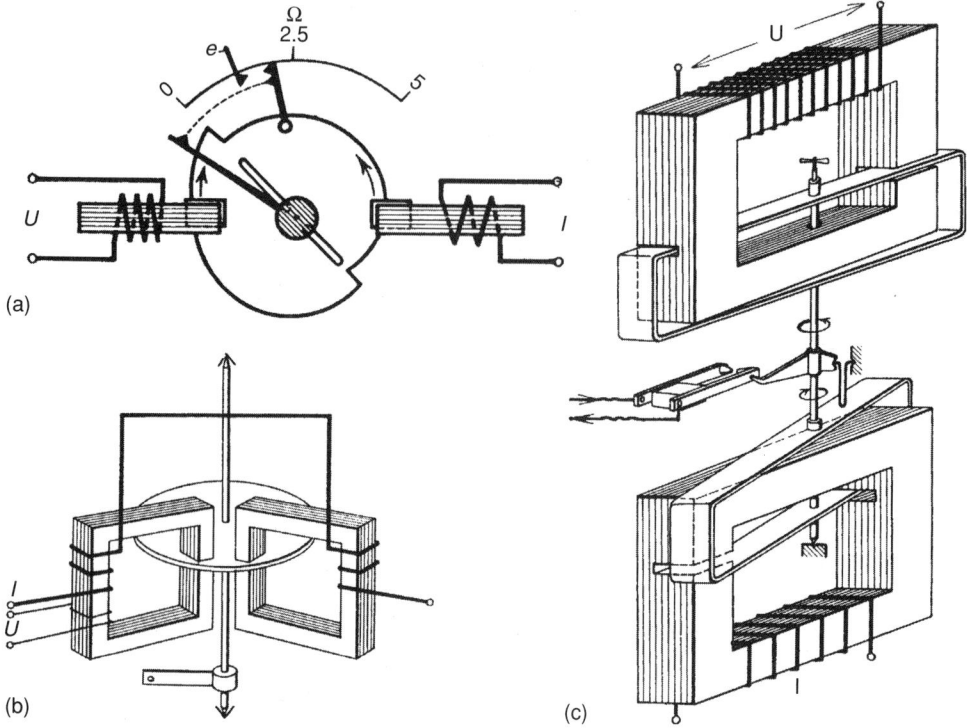

FIGURE 13.20

Construction diagrams of impedance relays of the induction type with a rotating disk, produced in the 1920–30's of the last century. (a) Relay produced by Siemens; (b) relay produced by Reyrolle; (c) relay produced by Oerlikon.

the transactor. The OHM unit reactance characteristic, when represented on the *R–X* diagram, is a straight line parallel to the *R*-axis. The unit will operate for fault impedances lying below its characteristic and hence is nondirectional.

During normal conditions when load is being transmitted over the protected line, the voltage and current supplied to the tile unit present impedance that lies close to the *R*-axis, since the load will be very near unity power factor, in contrast with the reactive power that flows during fault conditions. Impedance near the *R* axis will be below the OHM unit characteristic and hence the OHM unit contact will be closed. This will cause no trouble, however, since the directional MHO unit contact will not be closed for this condition.

The Ohmic reach can be extended by setting the restraint tap leads on the lower percentage position on the tap block. The setting of the two tap leads marked No. 1 determines the reach of the instantaneous or first zone, and the setting of the two tap leads marked No. 2 determines the reach of the intermediate or second zone.

The OHM unit transfer auxiliary is a simple type relay. The unit that is mounted at the top of the relay end is used to change the setting of the OHM unit to provide a second step of transmission-line protection. Its operation is controlled by a SAM type timing relay (see above). The normally-closed contacts of the transfer auxiliary provide the circuit with instantaneous tripping used for faults in the first step of line protection. If the fault is beyond the first zone of protection the transfer auxiliary changes the setting of the OHM unit by switching to the No. 2 taps on the autotransformer, from which a smaller potential is supplied to the unit potential restraint windings. This extends the ohmic reach of the

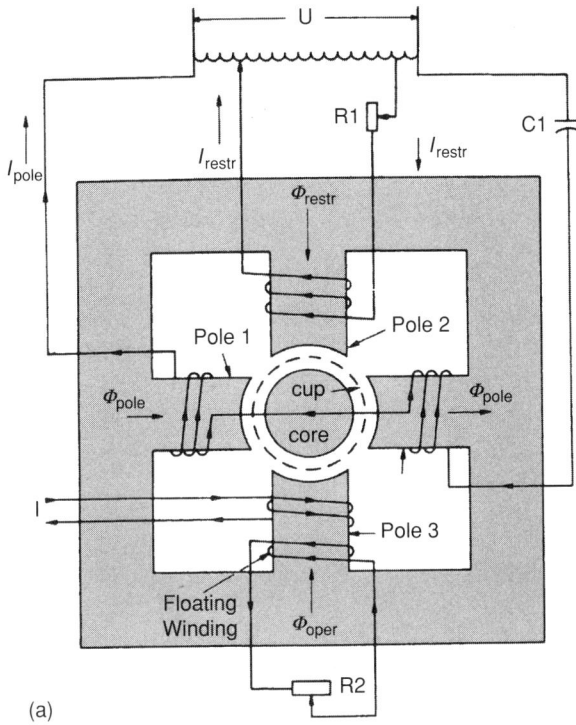

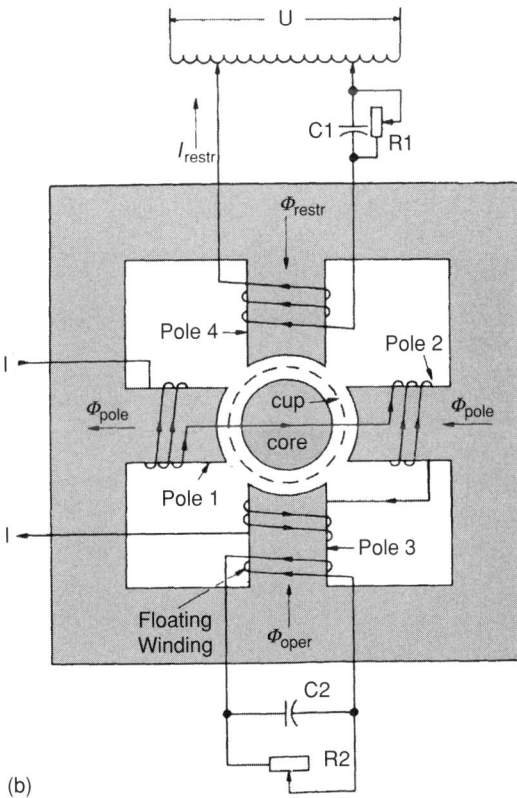

FIGURE 13.21
Simplified construction diagrams of modern cup-induction relay (a) for a basic directional impedance unit (MHO unit) and (b) for basic reactance unit (OHM unit).

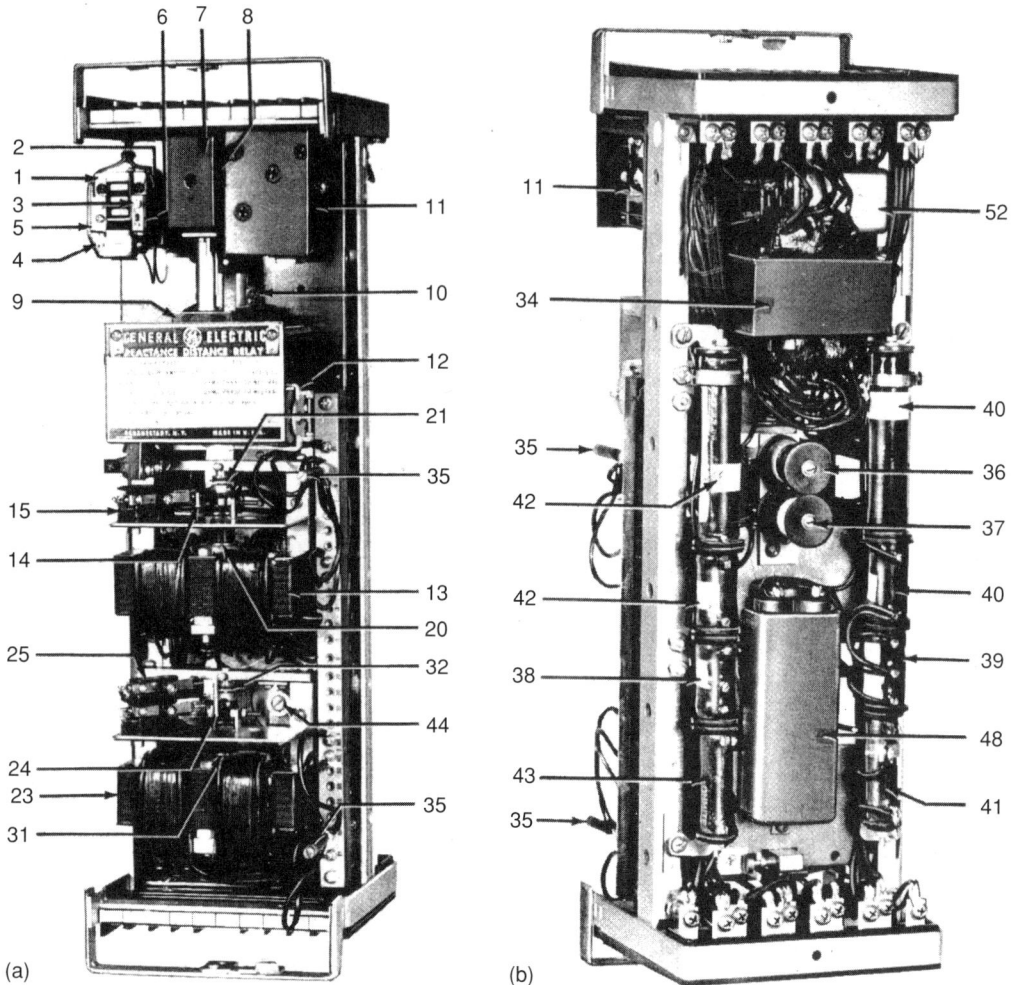

FIGURE 13.22
GCX-17 type directional-distance relay with OHM and MHO units and instantaneous overcurrent unit. (a) Front view; (b) back view. Seal-in unit with target (1): 2 — operating coil; 3 — tap screw; 4 — moving contact assembly; 5 — stationary contact — left; 6 — stationary contact — right. Over current unit complete (7): 8 — operating coil; 9 — moving contact; 10 — stationary contact. Auxiliary unit: 11 — co-ordination unit; 12 — transfer relay. OHM unit: 13 — operating coil assembly; 14 — moving contact and spring assembly; 15 — stationary contact block assembly; 20 — cylinder (cup) and shaft assembly; 21 — adjusting arm and bearing stud. MHO unit: 23 — operating coil; 24 — moving contact and spring assembly; 25 — stationary contact block assembly; 31 — cylinder (cup) and shaft assembly; 32 — adjusting arm and bearing stud. Miscellaneous: 34 — transformer and tap block assembly; 35 — tap plug; 36 to 44 — resistors; 48 and 52 — capacitors (General Electric Co.).

OHM unit and enables it to operate for faults in the second zone of transmission-line protection.

The KPC (KPC-111, KPC-112, KPC-121, KPC-131, KPC-132, KPC-142, KPC-143) type impedance relays produced for many years in the former U.S.S.R. were also based on the induction cup-rotor magnetic system (Figure 13.23). In this relay, a magnetic flux Φ_1 passes through the pole tips I–I, created by winding I which connected across the voltage:

$$U_1 = U + E_i$$

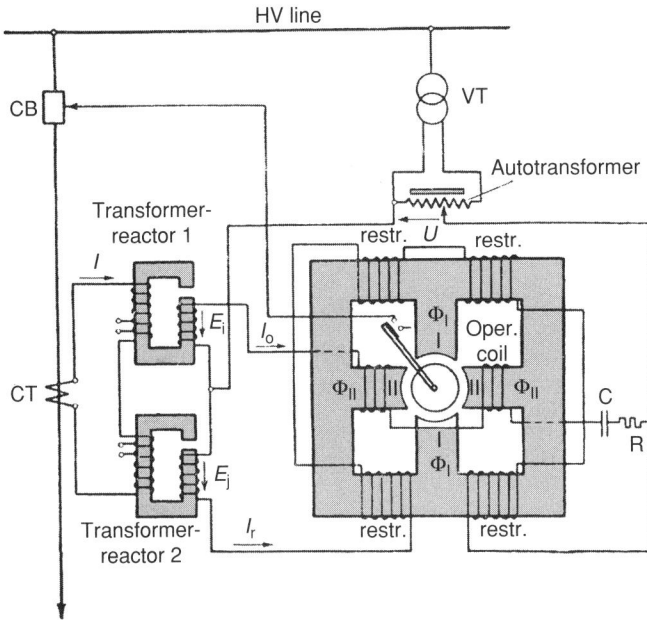

FIGURE 13.23

Schematic diagram of a KPC-112 type MHO relay with induction cup-rotor magnetic system, based on Ferraris motor (Russia).

where U is the voltage applied to the relay from the bus voltage transformer through the auxiliary autotransformer and E_i is the the e.m.f. proportional to current in the protected line ($E_i = kI$).

The magnetic flux Φ_{II} which passes through pole tips II–II is created by an operating coil to which the voltage is applied:

$$U_2 = U - E_i$$

Connected in series with the operating winding are capacitor C and active resistance R. Due to the existence of core air gaps in the auxiliary transformer-reactors (transactors), the e.m.f. E_i, fed into the circuit of operating and restraint windings, is directly proportional to the current I.

Under the action of voltages U_1 and U_2, the currents flow through the operating and restraint windings:

$$I_r = (U + E_i)/Z_r$$
$$I_o = (U - E_i)/Z_o,$$

where Z_r and Z_o are the impedances, respectively, of restraint and operating windings.

The fluxes Φ_I and Φ_{II}, due to currents I_r and I_o, are shifted in space by 90° and also differ in phase by some angle. This angle between fluxes Φ_I and Φ_{II} and, consequently, the sense or sign of the operating torque, is a function of the angle between the voltages U_1 and U_2. The value of this latter angle depends upon the preselected parameters of the relay (for example, the characteristics of windings, capacitor C), and likewise, on the relation of U to E_i.

KPC-131 and KPC-132 type relays are cut into current difference of two phases and the linear voltage between them, and respond to a decrease below the fixed threshold value of impedance at the inputs of the relays as two or three-phase SCs occur.

13.4   Why do Distance Relays Need "Memory?"

If there are three-phase SCs close to the place of installation of the relay, all voltages turn to zero and all currents change the phase step-wise. There is no polarizing voltage vector with regard to which one could fix this change of the current phase. This is the so-called "dead zone" of the relay, within which it cannot operate. To reduce the dead zone at three-phase SCs one uses a slight displacement of the MHO characteristic of the relay with regard to the origin of coordinates along the maximum sensitivity axis towards the quadrant III (Figure 13.9a) (unfortunately, this causes deterioration of other characteristics of the relay).

A more cardinal solution is to use a "memory" device, which memorizes the voltage phase at the point of installation of the relay up to the moment of short circuiting, and applies this voltage to the polarizing winding of the relay during of a SC. The simplest "memory" device is made in the form of an RCL-circuit (Figure 13.24), capable of reserving energy and then returning it back to the circuit in the form of damped oscillations. Parameters of this circuit are chosen in such a way that the discharge current of capacitor C is of oscillatory character with a frequency of 50 Hz. At close SCs, when $U \approx 0$, the energy stored in the "memory" circuit is enough for the relay pick-up. It is worth mentioning that the level of voltage received from the "memory" element" is much lower than the normal one and that is why this voltage cannot be used for a precise determination of the resistance value up to the fault location. The "memory" element is used to obtain information only about the voltage phase preceding the moment of short circuiting. In such emergency mode the impedance relay actually turns to the power direction relay, allowing it to coordinate the protection operation correctly.

Such "memory" elements are used in the KPC-131 and KPC-132 type relays and also in some other types of distance protection relays. Aharon Bresler (1889–1951), a Russian engineer and inventor, suggested a relay (known as a "Bresler's relay") without a dead zone (Figure 13.25). This is a three-phase directional impedance relay responding to two-phase SCs between any phases (with and without "ground") without switchings in the circuit of measuring currents, and voltage transformers that provide a two-stage distance protection.

The relay has windings $w1$ and $w2$ which are supplied by voltages U_1 and U_2 correspondingly. Each of these voltages is a difference of the linear voltage at the point of installation of the relay and the compensation voltage, which equals the voltage drop caused by the SC current in the given resistance of the protected line, that is:

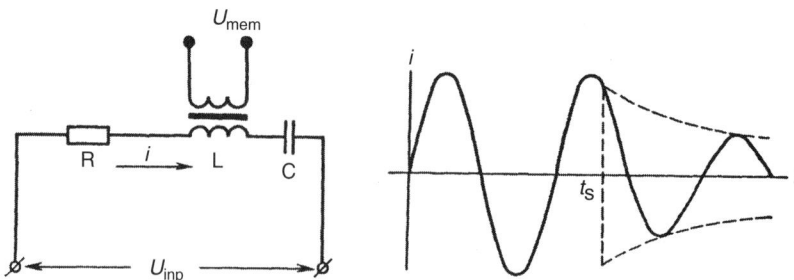

FIGURE 13.24

Construction of the "memory" of the simplest type and the form of its output voltage. U_{inp} — input voltage applied to the circuit in prefault conditions; U_{mem} — output voltage of the "memory" reproduced by the circuit after U_{inp} disappears at a SC; t_s — SC moment.

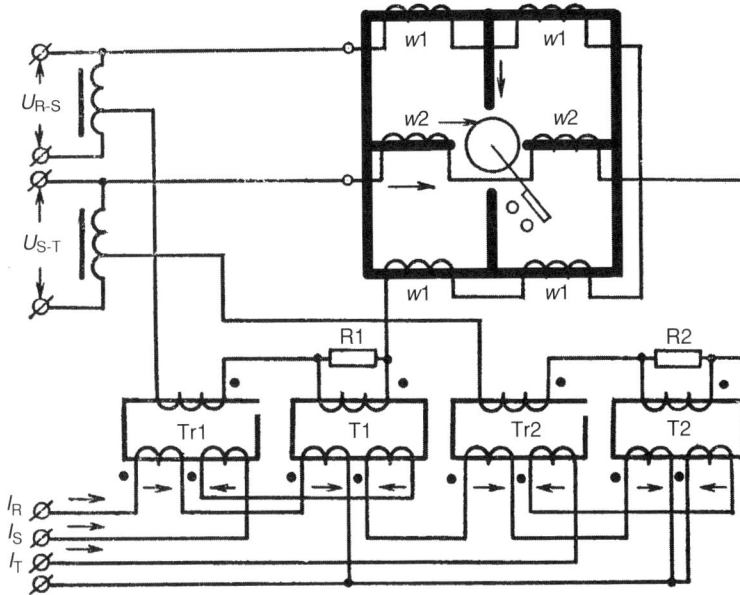

FIGURE 13.25
Induction directional impedance relay of A. Bresler's system. T1, T2 — transformers; Tr1, Tr2 — transactors.

$$U_1 = kU_{R-S} - Z\,(I_R - I_S); \quad U_2 = kU_{S-T} - Z\,(I_S - I_T)$$

Due to the fact that the linear voltage is applied to the relay windings between the damaged and undamaged phases, in a two-phase SC the relay has no "dead zone." The relay also does not respond to symmetrical changes of currents in phases of power network and that is why no special blocking of the system swing is required. Another important peculiarity of this relay is that it does not respond to the load, which is why such a relay can operate in distance protection independently, without a starting element.

The relay designed by A. Bresler in the 1940s was produced by the Cheboksar Electrical Equipment Plant in the former U.S.S.R. for many decades (including in the 1970s and 1980s) under the brand KPC-121. In the book *"Protection Relays,"* published in 1976 by the leading experts of this plant, Bresler's name is unfortunately not mentioned when his relay is described.

In many distance protection relays, only one element of impedance measurement for all zones is used in order to make them cheaper. In such relays the impedance element is constantly switched ON with the setting for the first zone. If a SC occurs outside the first zone, the starting relay automatically switches (with a time delay) the setting of the impedance element for the operation in the second, third, and even fourth zone. Current relays and impedance relays are mostly used as starting relays of distance protection. The main disadvantage of current starting relays is that they respond equally both to SCs and to swings and great load currents. Their advantages: simplicity and low cost. Such starting elements are applied in short lines with voltage not higher than 30 to 40 kV. In all other cases, the impedance relay is used as a starting element.

Nondirectional impedance relays with a circular characteristic are simpler than directional ones but respond to system swing, to currents and loads practically like current relays do. In contrast to the latter, though, they are more sensitive to SCs because they respond not only to the increase of current but also to voltage decreases during the SC.

The area of application of such starting relays is restricted to circuits with a voltage of 30 to 40 kV and to short lines 110 to 160 kV with small loads. Directional impedance relays (MHO relays) are more sensitive to SCs than to the load. This is caused by dependence of the pick-up threshold of the relay from the angle between current and voltage. During SCs in the line this angle of the input terminals equals the so called "angle of line resistance" and usually is 65 to 80°, which is quite close to the angle of maximum sensitivity of the MHO relay.

At great loads of line caused by flowing of high active power through it, the angle of resistance is less than during an SC, and is usually 10 to 40½. At such angles, the sensitivity of the relay decreases by 20 to 50%, thus allowing the relay to distinguish SCs from a great load. The MHO relay is less prone to malfunctions caused by the system swing then the nondirectional relay, because it can pick up only when the impedance vector is in the first quadrant. At any other positions of the impedance vector, the relay simply cannot pick-up.

13.5   Distance Relays with Higher Performance

Relays with an elliptic characteristic instead of a circular one have an even better offset from great currents. The disadvantage of relays with characteristics in the form of ellipses or a lenses is that they are greatly affected by increased transient resistances at the site of short circuiting. This can lead to incorrect measurement of the distance to the fault point (and therefore to the wrong choice of the relay operation zone) which is why producers of relays try to make the performance of relays higher with the help of different sophisticated constructions (Figure 13.26). One can also offset the starting impedance relay from great load currents in distance protection by using an additional blocking relay with a so-called "blinder characteristic" (Figure 13.27).

The term "blinders" as it applies to phase distance relays has the same significance as when it is applied to a horse. In the case of the horse, blinders limit his vision to a narrow beam in the direction in which he is facing. In the case of a distance relay, blinders limit the operation of the distance relays to a narrow beam that parallels and encompasses the protected line. In general, relay blinders are required with MHO units only where long lines are involved and the resulting MHO unit settings are large enough to pick up on maximum full load currents or minor system swings.

The blinder and MHO unit contacts are interlocked in the trip circuit in such a way that tripping can only occur in the fault impedance plots inside the MHO characteristics and between blinders A and B. Actually the blinders are nothing more than reactance units similar to those of Figure 13.27 that have been rotated by modifying the power factor angle of the restraint circuit of the units. The A blinder operates for faults that plot to its right. The B blinder operates for faults that plot to its left. The overall effect of the blinders is to restrict the operating zone to an area on the *R–X* diagram that parallels the protected line and thus makes the combination relatively insensitive to system swings and immune to operation on full load.

One pair of blinders is required per phase, thus three pairs are needed per terminal on a three-phase system. Due to such a characteristic, the probability of malfunctioning of the relay because of system swings and great load currents is minimal, as all these exposures are out of the working zone of the relay.

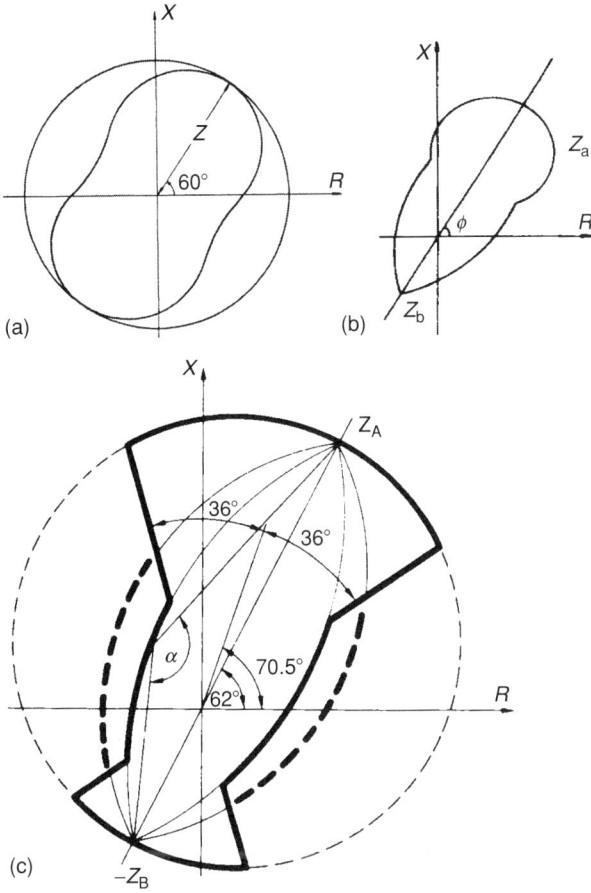

FIGURE 13.26

Improved variants of the elliptic characteristic of a starting distance protection relay. (a) Characteristic of a nondirectional starting relay of the RXZF type (ASEA); (b) modified lens characteristic of a directional relay of the impedance type RXZK (ASEA); (c) sector-shaped characteristic of a starting directional relay of the microprocessor distance protection LZ95 type (ABB).

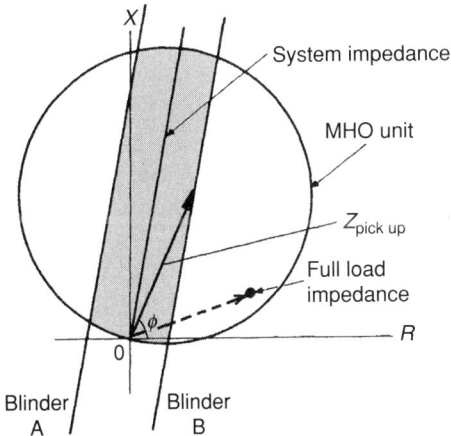

FIGURE 13.27

Characteristic of the combination relay switching the directional impedance relay (MHO unit), and a relay with a "blinder" OHM characteristic.

As the blocking relays ("blinders") power direction relates with angles of internal displacement, 60 and 30° can be used. There are special relays with such a characteristic, such as the CEX-57 type relay for instance (Figure 13.28). CEX57 type relays are high-speed induction cup-type devices with OHM unit characteristics that can be set parallel to

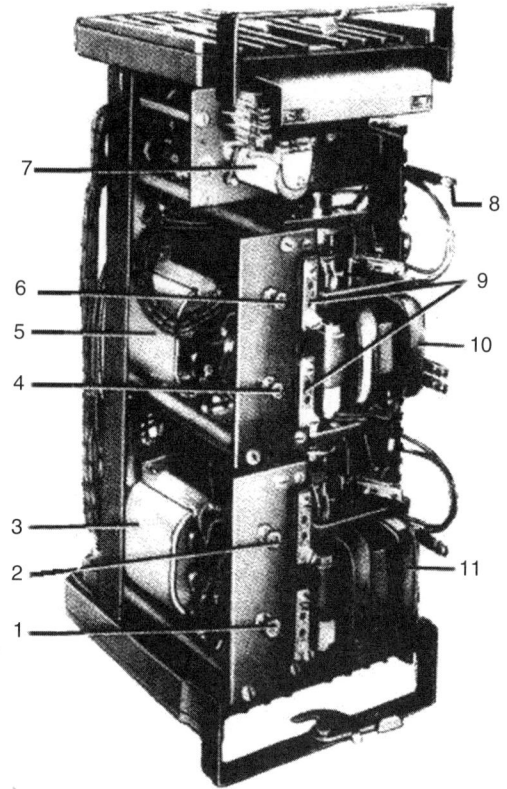

FIGURE 13.28
Angle-impedance relay with "blinders characteristic" of the CEX-57 type (General Electric). 1 — Potentiometer P4; 2 — potentiometer P2; 3 — capacitor C2; 4 — potentiometer P3; 5 — capacitor C1; 6 — potentiometer P1; 7 — auxiliary relay; 8 — restraint tap plug; 9 — current tap block; 10 — upper induction-cup unit; 11 — lower induction-cup unit. (General Electric Angle-Impedance Relays. Instruction GEK-497788B.)

the impedance of a transmission line. These relays are designed for use with other protective devices in "blinder" applications to restrict the tripping area of the tripping units used in a protective relay scheme.

CEX-57 type relays each contain two cup-type units (upper and lower), similar to the cup-type unit shown in Figure 13.28. The CEX-57F relay also contains an auxiliary telephone-type unit.

Either three CEX-57D or three CEX-57F relays are required. Basically, tripping will be permitted only when the fault impedance plots within the reach of the MHO tripping function AND both of the OHM units. Since the right hand OHM unit will operate only for faults to the left of it, both units can operate simultaneously only for faults that plot between them. The tripping function (MHO) will provide correct directional action, and limit the reach in the forward direction.

The units in the CEX-57 Type relay are four-pole induction cylinder (cup) units with schematic connections, as shown in Figure 13.29. These units measure impedance at an angle. The two front coils and the two back coils are energized with delta currents to produce polarizing flux. The same delta current flows through the operating coil to produce an operating flux. The phase-to-phase voltage is applied to the restraint coil to produce a restraint flux.

The angle-impedance unit characteristic is a straight line when plotted on R–X diagram (Figure 13.30). The shorted distance from the characteristic to the origin (Z_M) is the minimum relay reach, which is determined by a set of relay taps. The angle of maximum torque (ϕ) is the angle that the reach Z_M leads the R-axis. This angle is adjustable in CEX-57 from 5 to 35° lead. The reach of the angle-impedance unit at any angle is given by the following equation:

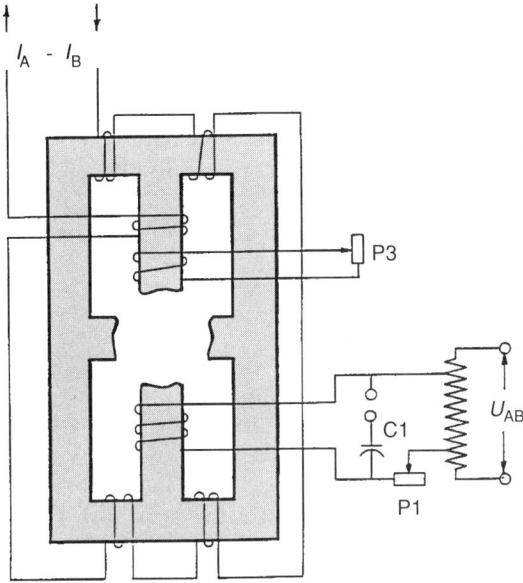

FIGURE 13.29
Schematic connections of the induction unit in a CEX-57 relay.

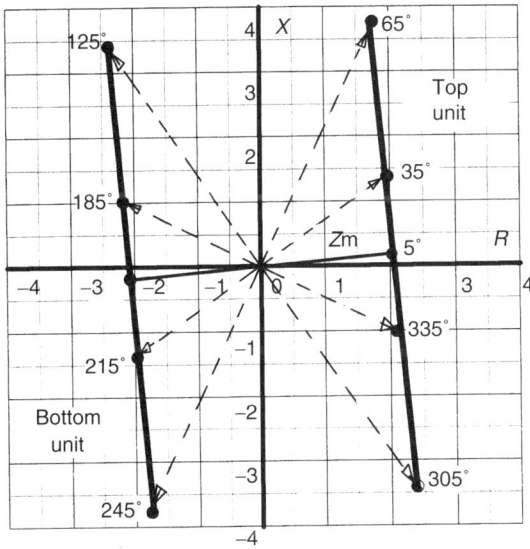

FIGURE 13.30
The CEX-57 relay characteristic on an R–X diagram. $Z_M = 2.08\,\Omega$ — minimal relay reach; $\phi = 5°$-angle of maximum torque.

$$Z = \frac{Z_M}{\cos(\theta - \phi)} - \text{for upper unit}$$

and

$$Z = \frac{Z_M}{\cos(\theta - \phi + 180°)} - \text{for lower unit}$$

where θ is the angle I_{AB} leads U_{AB} (see Figure 13.29).

The lower unit is identical to the upper unit, except that it is polarized to have maximum torque 180° from the upper unit. In fact, the characteristic of the relay can be constructed by the fixed values Z_M and ϕ. One draws a ray advancing the axis R by the angle ϕ through the zero point, then lays off sections with the value Z_M on this ray to both sides of the zero point, and then draws straight lines perpendicularly to these sections and through their ends so that they form "blinders" on the characteristic of the relay. Theoretically this is the characteristic which proper relay functioning must have. One can obtain this characteristic experimentally if the relay is connected according to the scheme suggested by the producer (Figure 13.31).

A constant voltage of 120 V is applied to it and one measures the current while trying to make the relay pick up at certain fixed angles between the current and voltage. Impedance is calculated by the formula:

$$Z = U/2I$$

The values we have obtained are laid off on the rays corresponding to the given angle, forming sections. Straight lines drawn through the ends of these sections are called "blinders."

A relay with a polygonal (quadrilateral) characteristic (see Figure 13.32) is even more effective. Usually microprocessor relays have characteristics of this type. Such characteristic is provided with forward reach and impedance reach settings that are independently adjustable. It therefore provides better resistive coverage than any MHO-type characteristic for short lines. This is especially true for earth fault impedance measurement, where the arc and fault resistance to earth contribute to the highest values of fault resistance. To avoid excessive errors in the zone reach accuracy, it is common to impose a maximum resistive reach in terms of the zone impedance reach. Recommendations in this respect can usually be found in the appropriate relay manuals.

Quadrilateral elements with plain reactance reach lines can introduce reach error problems for resistive earth faults where the angle of total fault current differs from the angle of the current measured by the relay. This will be the case where the local and

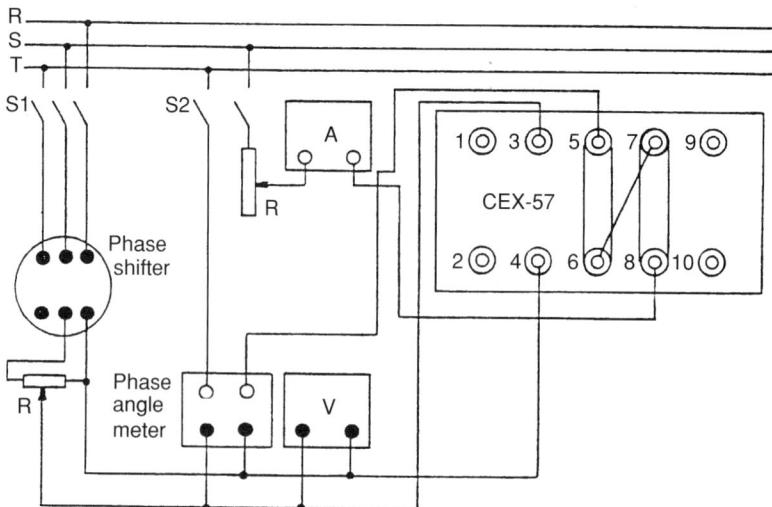

FIGURE 13.31
Test connection for type CEX-57 relays. (General Electric Angle-Impedance Relays. Instruction GEK-497788B.)

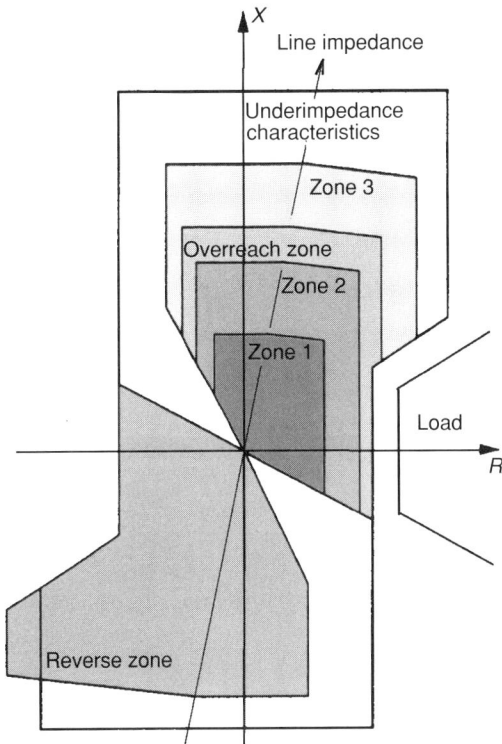

FIGURE 13.32
Quadrilateral characteristic of distance relays.

remote source voltage vectors are phase shifted with respect to each other due to prefault power flow. This can be overcome by selecting an alternative to use of a phase current for polarization of the reactance reach line. Polygonal impedance characteristics are highly flexible in terms of fault impedance coverage for both phase and earth faults. For this reason, most digital and numerical distance relays now offer this form of characteristic.

Such characteristic provides the required sensitivity of the relay, and at the same time has the best offset from exposure of the load resistance and the system swing.

From the above explanations it is clear that the function of three-stage distance protection is carried out not by one, but by a group of relays, among which there should be starting relays for each phase, relays for measuring of impedance, several timers (each responsible for a certain zone), and auxiliary relays. Because of the complexity, distance protection devices based on electromechanical relays are usually occupies a whole separate cabinet.

In the 1970's the Brown Bowery Co. came up with a unique design to cope with the complexity of electromechanical-relay-based distance protection devices, the LZ31 type relay, which unites in one design all of the necessary elements for three-stage distance protection, (Figure 13.13). In this device a pick-up of one of the starting elements (2) is achieved when the impedance of a line drops below a certain threshold value. The starting relay (Figure 13.13b) energizes a cam-timer (4) and the first contact (through the auxiliary relay) connects the first tap of the transformer (5) to the measurement-relay of impedance (1). If the voltage on the winding is not enough for operation of the relay (1), a cam-timer (4) continues a time-count and connects the second tap of the transformer (5) to the measuring relay (1). If this is still not enough, the cam-timer (4) continues to operate until it reaches its final position and energizes the output trip relay. Thus the time delay

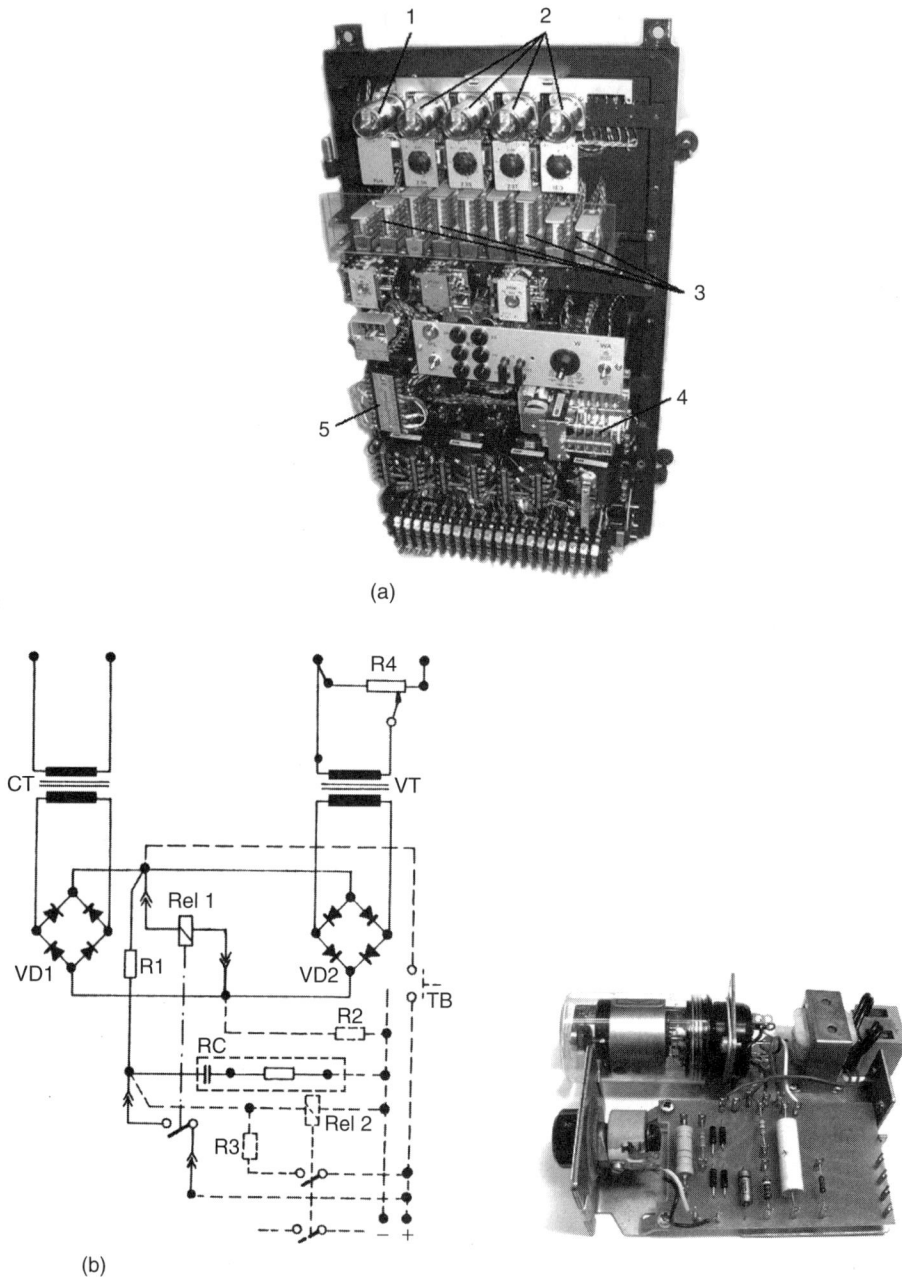

(a)

(b)

FIGURE 13.33

(a) Distance relay LZ31 (BBC), dimensions: 778x484x206 mm, weigh 44 kg. 1 – measuring relay; 2 – phases and ground impedance starting relays; 3 – auxiliary relays; 4 – cam-timer with independent adjustable time for each step; 5 – multi coil transformer. (b) External view and connection diagram for impedance relay unit of LZ31 protection device (starting and measuring relays have same construction). CT and VT-auxiliary current and voltage transformers; VD1 and VD2 – rectifier bridges; Rel1-moving-coil relay; RC-spark suppression RC-circuit; Rel2 – output relay; TB-test button; R4 – potentiometer for setting pick-up value.

that corresponds to the most distant third zone will be the greatest. If at the very first connection of the transformer (5) to the relay (1) the voltage level appears to be sufficient for a pick-up of the relay (1), the output trip relay will be enabled and the process will stop. This corresponds to the first zone (the closest distance to the short circuit point) and the smallest time delay. The output voltage of the transformer (5) depends on the tap number and the voltage in the high-voltage line at the moment of short circuit (that is, proportionally to the impedance of that section of line, up to the short circuit point).

13.6 Electronic Analogs of Impedance Relays

As in all the other cases considered above, there are electronic analogs of electromechanical impedance relays. The simplest device of this type is a so-called relay with a detector circuit for comparison of current and voltage by absolute values. In fact, this is a direct electronic analog of the "balanced-beam relays" considered above (Figure 13.34). In the circuit of current balance, the output relay Rel is connected in parallel with the rectifier to the difference of rectified currents.

The current in the output relay $I_{REL} = |I_2| - |I_1|$

In circuits based on voltage balance, the relay will operate only when $|U_2| > |U_1|$ Such impedance relays are nondirectional and have a characteristic in the form of a circle with its center at the origin of the coordinates. These relays are used as a starting element, switching ON the measuring unit of the distance protection.

Russian starting relays of distance protection of the DZ-1 type, produced in the 1960–80's, were based on the balance of currents (Figure 13.35). The output relay is constructed on the basis of a highly sensitive polarized relay with moving coil (see below) or in the form of a standard electromagnetic relay with an electronic amplifier, working at only one polarity of the applied voltage (current). That is, the outlet relay will work if $|I_2| > |I_1|$ and will not pick up if $|I_2| < |I_1|$.

The RAZOG type distance protection (ASEA), widely produced in the years 1970–90's, can serve as an example of relays based on the balance of voltages (Figure 13.36). This

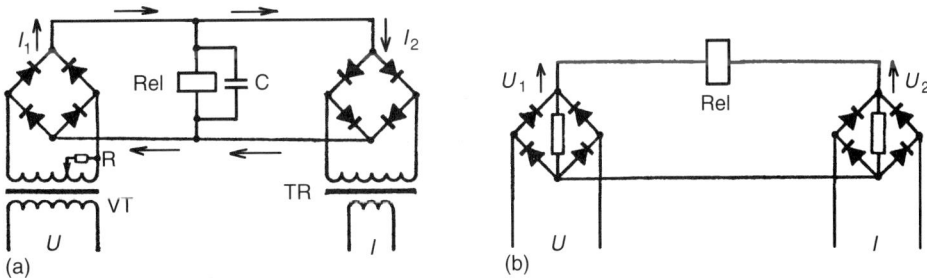

FIGURE 13.34
Principle of construction of a nondirectional impedance relay with a detector circuit for comparison current and voltage by absolute values. (a) By balance of currents; (b) balance of voltages; TR — transactor; C — capacitor leveling voltage pulses (ripple).

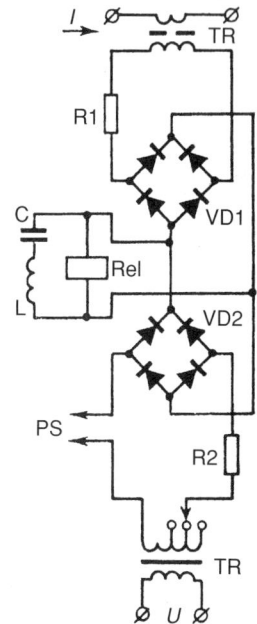

FIGURE 13.35
Circuit diagram of a starting relay of distance protection of the DZ-1 type
(Russia), based on the balance of currents (voltages). TR — transactor; Rel —
highly sensitive relay with moving coil (M237); PS — power supply required for
the relay reset. Resistors R1 and R2 reduce the nonlinearity of resistance of the
diodes of the rectifying bridges.

distance protection contained a great number of separate relays connected with each
other, so as to form a block diagram of the protection (Figure 13.36b). A nondirectional
electronic impedance relay of the RXZE type (Figure 13.37), based on the balance of
voltages mentioned above, is used as a starting element in this device.

The current quantity (*I*) (Figure 13.37) is transformed via transactor Tr1, through the
diode bridge (VD1), after which an unsmoothed DC voltage is obtained across R1, directly
proportional to the supplied current. The voltage quantity (*U*) is transformed via transformer Tr2 to two circuits where one of the circuits produces a smoothed and the other
circuit an unsmoothed direct current.

The smoothed DC is produced by a six-pulse connection, which consists of an RC
circuit plus six diodes. These are denoted as R5, C1, and the diode circuit VD3
(Figure 13.37c). The capacitor (C2) is used for smoothing the six-pulse voltage, which in
turn gives a smoothed current through R2 and R3.

The unsmoothed direct current is produced by the diode bridge (VD2). The capacitor
C3 is used to phase-shift the current about 30°, which is obtained by comparison with
resistor R4. This current produces an unsmoothed voltage across R2. This phase-shift of
the current quantity will cause the major axis of the characteristic to lie at about 60° in the
first quadrant of the *R–X* plane. Other angles for the characteristic are, as such, obtained
by altering the phase-shift of the unsmoothed current.

The above-mentioned currents are summed at point 6 (Figure 13.37c) and produce a
summation voltage across R2 proportional to the supplied voltage. The characteristic,
with this circuit, becomes an oval, which lies symmetrically around the origin in the
impedance plane.

If the unsmoothed direct current (unit III, Figure 13.37b) is eliminated, that is, not
allowed to be summed at point 6, the voltage at that point will become a smoothed DC
voltage and the characteristic will then become a circle, having its center at the origin in
the impedance plane. By means of simple reconnection, circular, or oval characteristics
are obtained, as shown in Figure 13.37d.

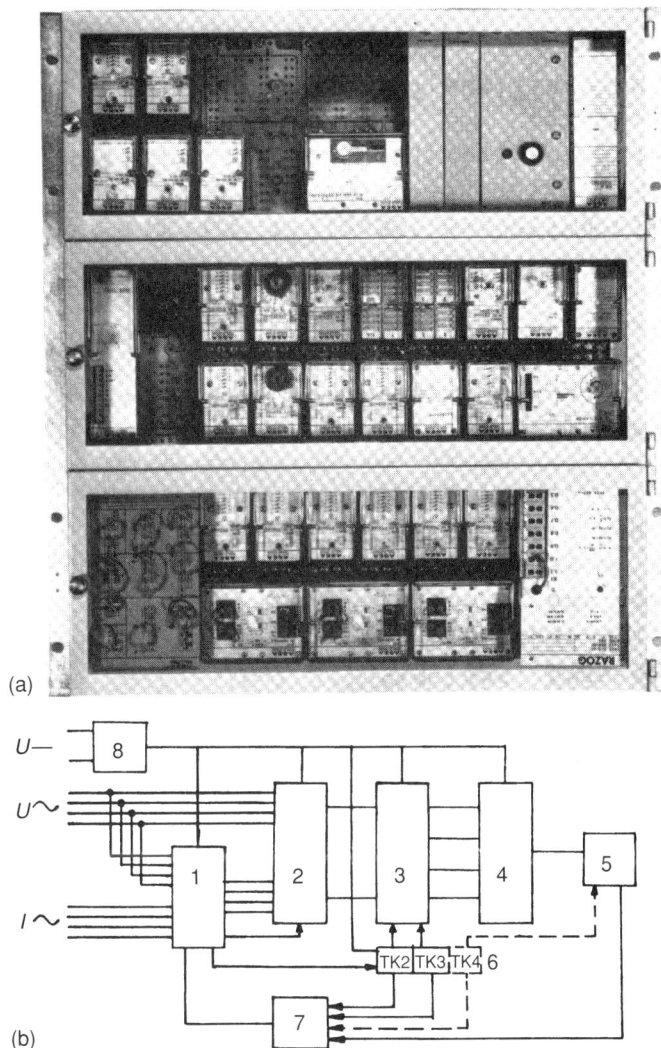

FIGURE 13.36

(a) The board of the RAZOG type distance protection. (b) Block diagram for a RAZOG type distance relay. 1 — Starting elements; 2 — relays for phase-switching; 3 — compensating circuits and setting unit; 4 — measuring element; 5 — tripping relays; 6 — time-lag relays; 7 — indicators; 8 — auxiliary power supply. (ASEA (ABB) 1975 Distance Relay Type RAZOG.)

The Rel, the null-detector that is connected to points 2 and 6, senses the voltage difference between these two points. The detector gives an output signal when the instantaneous voltage at point 2 exceeds the instantaneous voltage at point 6. An auxiliary relay can be used as an output relay. The detector requires a separate auxiliary DC voltage.

As a rule this detector is constructed on the basis of a very sensitive relay — of a magneto-electric type (with moving coil), or a standard electromagnetic relay with an electronic amplifier. The RXZF is a single-phase relay. In three-phase circuits one applies a set including three such relays, three intermediate relays and a time relay (Figure 13.38).

Directional impedance relays were also constructed on the basis of the balance of voltages (Figure 13.39). In this relay the required angle of maximum sensitivity is obtained due to displacement of the input current phase with the help of a TR1 transreactor by the angle equaling the angle of resistance.

This angle in a real construction can be adjusted within some limits due to variable resistors switched to the secondary windings of the transreactor TR1 (not shown on the scheme). The additional transreactor (TR2), tuned to resonance to the circuit frequency (50 or 60 Hz) is switched to an additional source of voltage biased by 90° with regard to the voltage U, and works as a "memory" element (mentioned above) during three-phase SCs.

As it has already been mentioned above, in detector circuits of impedance relays either a special highly sensitive magneto-electric relay or an electronic relay (usually an electronic amplifier based on transistors or operational amplifiers). It is quite interesting to learn that such famous and prospective (from the point of view of new technologies) companies such as Siemens, produced impedance relays with an amplifier based on vacuum tubes in the middle of the 1970s (Figure 13.40). Such a relay was dependent on a time characteristic

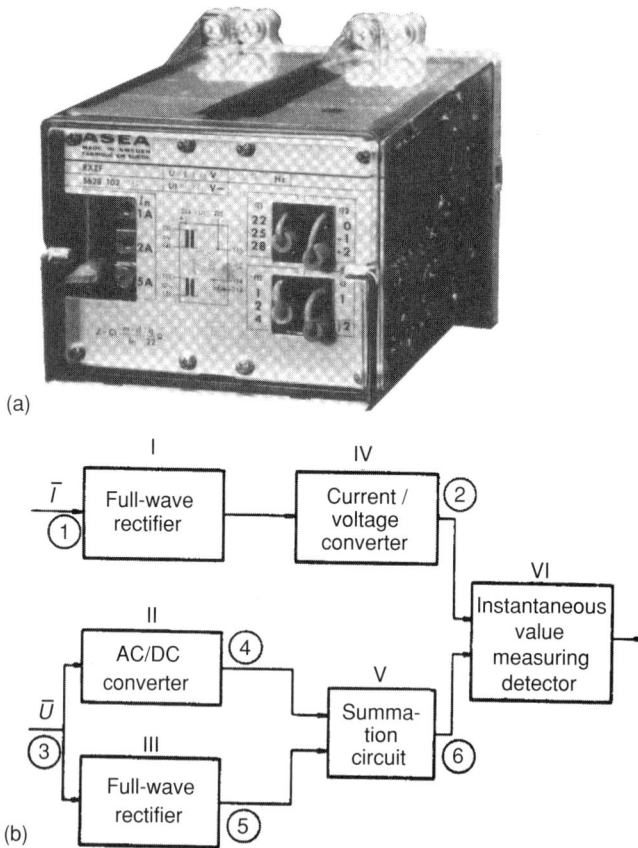

FIGURE 13.37
(a) Nondirectional single-phase electronic impedance relay of the RXZF type, based on the principle of the balance of voltages. (b) Block diagram of the RXZF relay.

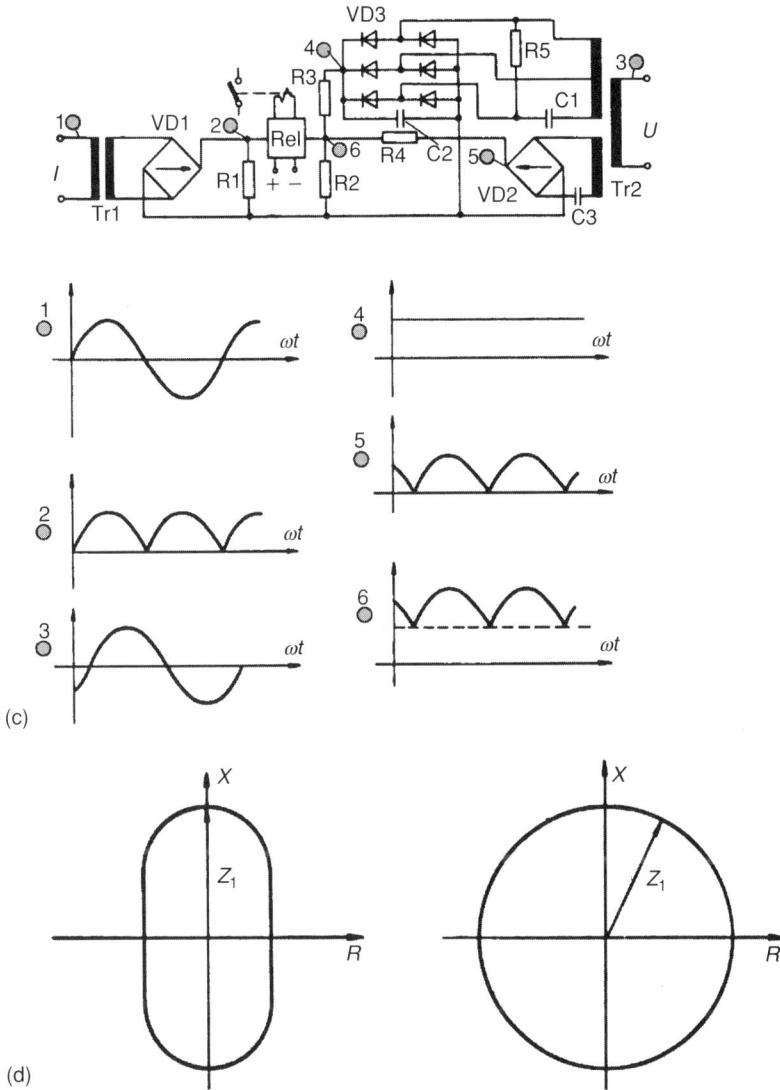

FIGURE 13.37 (*Continued*)
(c) Principal diagram of an RXZF relay. (d) Operating characteristics of the RXZF relay. (ASEA (ABB) 1970.)

(based on charging–discharging of the RC-circuit) and did not require the use of a separate time relay as in distance protections of other types. The high and stable input resistance of an amplifier based on a vacuum tube allowed a dependence on time characteristics similar to those of induction relays.

As has already been mentioned above, the distance protection of power networks is carried out by a whole set of different relays in which the impedance relay is only one of many other components, although the main one. In the 1950–60's, a set of electromechanical relays providing distance protection of power lines would occupy a whole cabinet. In the 1970–80's, these became more compact, the constructions being based both on electromechanical (Figure 13.41) and electronic relays (Figure 13.42).

FIGURE 13.38
This three-phase impedance-measuring protective RAKZA type relay (ASEA), included three RXZF relays, three auxiliary RXMA1 relays, and a time-delay relay of the RXKB type. (ASEA (ABB) 1977 Catalog RK 65-52E.)

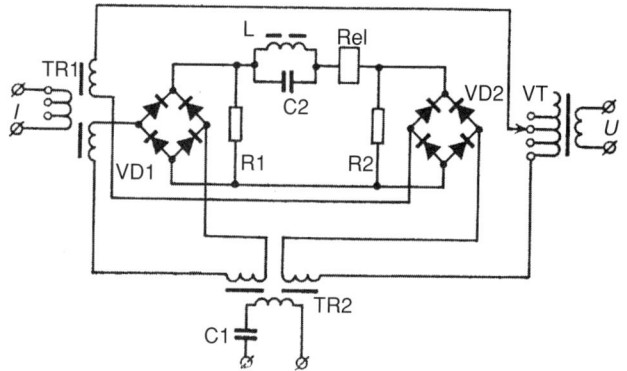

FIGURE 13.39
Circuit diagram of a directional impedance relay based on the balance of voltages (DZ-2, Russia).

(a)

FIGURE 13.40
R1Z80 type impedance relay on vacuum tubes (Siemens, 1972). (a) External design.

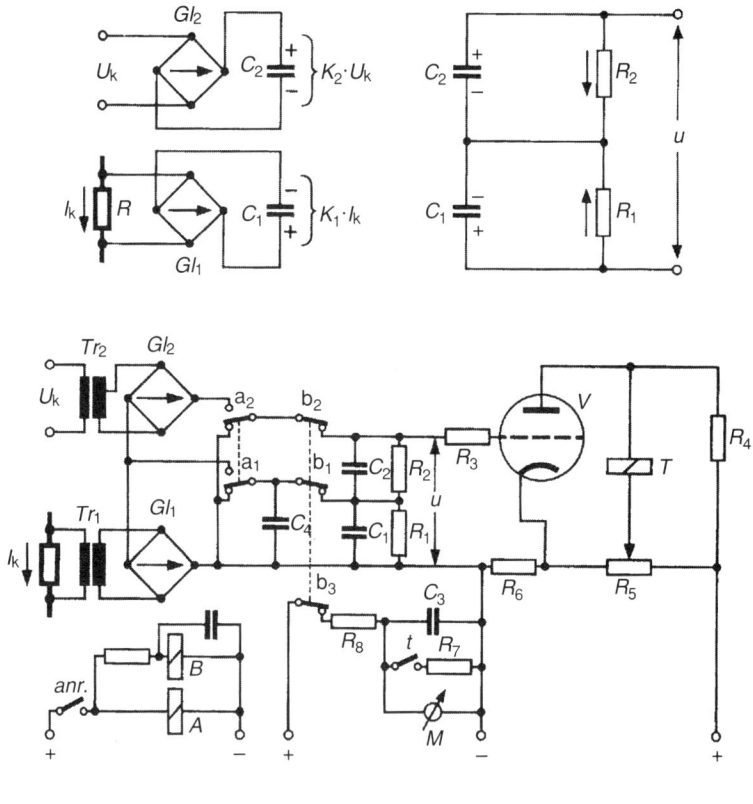

FIGURE 13.40 (*Continued*)
(b) Fragments of the electronic circuits. (Siemens, 1972 protective devices catalog.)

FIGURE 13.41
Electromechanical relay of distance line protection of the LZ-31 type (Brown Bowery Co.).

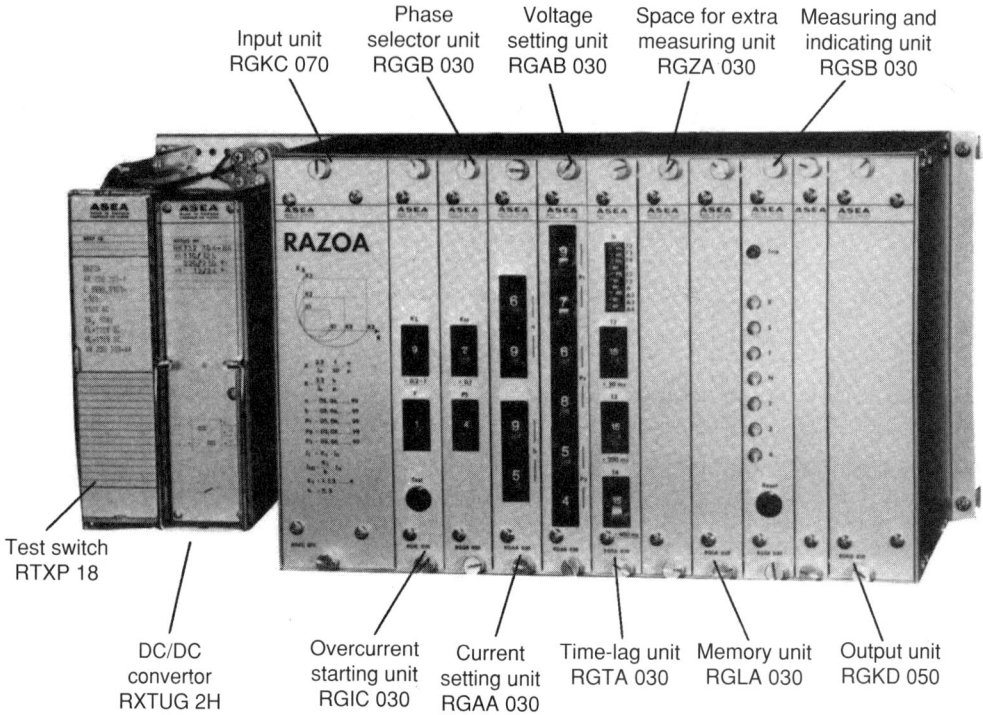

Input unit RGKC 070 — Phase selector unit RGGB 030 — Voltage setting unit RGAB 030 — Space for extra measuring unit RGZA 030 — Measuring and indicating unit RGSB 030

Test switch RTXP 18

DC/DC convertor RXTUG 2H — Overcurrent starting unit RGIC 030 — Current setting unit RGAA 030 — Time-lag unit RGTA 030 — Memory unit RGLA 030 — Output unit RGKD 050

FIGURE 13.42
RAZOA type electronic relay of power line distance protection (ASEA). (ASEA (ABB) 1981 Distance Relay Type RAZOA.)

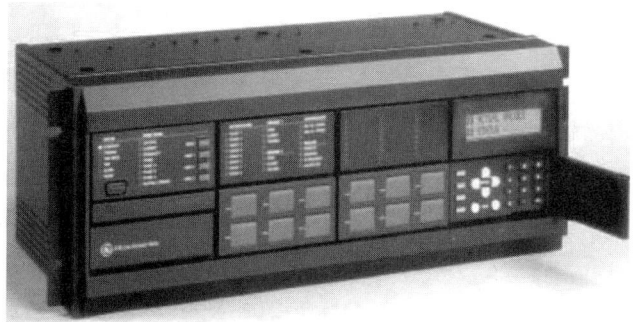

FIGURE 13.43
Microprocessor relay of the D-30 type for distance protection of power lines. (General Electric 2004 Online internet catalog.)

Testing and adjustment of such relays are rather difficult. For testing of stability of the relay to transients, real transitive process in a real network is writing into fault recorder as a special file. This file by means of a computer is loading into special power simulator which converts the file into currents and voltages, applied to relay, which are fully complying with real transient (Figure 13.44).

FIGURE 13.44
Testing of distance relay LZ-31 type by means of transient simulator 1 — Relay LZ-31; 2 — three-phase power simulator DOBLE-2500; 3 — PC with transient file, previously loaded from fault recorder.

Lately practically all new types of distance protections are made on the basis of micro-processors (Figure 13.43), however, microprocessor relays and microprocessor distance relays in particular can hardly be called "relays," but will be discussed in Chapter 15.

14

Frequency Relay

14.1 Why is it Necessary to Control Frequency in Electric Networks?

Voltage frequency is the most important figure in a power network. First of all, the speed of rotation of electric motors, and therefore the performance of the machines and mechanisms, depends directly on frequency.

Second, generators in power stations are designed for work at fixed frequencies. Deviations from these fixed frequencies, in either direction, by 5 to 10% can lead to a sharp intensification of vibration in a large-tonnage rotor (Figure 14.1), to premature failure of the generator, to sharp productivity slowdowns in the following: powerful pumps that deliver water to the boiler, fans of the air injection systems, cooling systems pumps and many other important systems of power plants.

Third, if there are several generators in the power network, their work must be synchronized by frequency with a high degree of accuracy.

Fourth, any decrease of voltage frequency caused by an overload of generators is inadmissible in itself. Even a slight excess of power consumption over generator power can lead to a significant voltage frequency drop in the power system (Figure 14.2). When there is such a frequency drop below a certain critical level, usually some of the customers, or even a line and whole subsystem, is automatically disabled in order to maintain serviceability of the generators and the network.

Frequency decreases because of power system overload, while a frequency increase is evidence of a power excess. Power excess occurs in the system when one or several hard loaded lines are suddenly disabled. Surplus power is directed to other lines, causing dangerous power flows that can lead to a power system breakdown. Such an accident, followed by a frequency excess of up to 63 Hz, took place on August 14, 2003 during the biggest power system breakdown in the U.S.A. That is why it is so important to control voltage frequency.

Like all other parameters of electric circuits, frequency too is controlled by special relays.

14.2 Charles Steinmetz — Inventor of the Frequency Relay

The basic principle of a frequency relay circuit was patented by Charles Steinmetz in 1900 (Figure 14.3).

FIGURE 14.1
The rotor of the generator of a steam turbine (647 MW, 22 kV) during repairs.

Charles Proteus Steinmetz was a giant of a pioneer in the field of electrical engineering. Charles Steinmetz (originally named Karl August Rudolf Steinmetz) was born in Breslau, Prussia (now the city Wroclaw, Poland) on April 9, 1865. He studied in Breslau, Zurich, and Berlin. Shortly after receiving his Ph.D. in 1888, Steinmetz was forced to flee Germany after writing a paper criticizing the German government. Charles Steinmetz was an active socialist and held strong antiracist beliefs.

Thomas Edison founded the General Electric Company in 1886 and wanted to hire Steinmetz. In 1893 the newly formed General Electric Company purchased Eickemeyer's company, primarily for his patents, but Steinmetz was considered one of its major assets. In 1894 Steinmetz was transferred to the main General Electric plant at Schenectady, New York. After studying alternating current for a number of years, Charles Steinmetz patented "A system of distribution by alternating current (AC power)," on January 29, 1895. Steinmetz retired as an engineer from General Electric to teach electrical engineering at that city's Union College in 1902. General Electric later called him back as a consultant.

Charles Steinmetz died on October 26, 1923 and at the time of his death, held over 200 patents.

Frequency relays of the slow-speed (induction disk) type were commercially available in 1921, and the high-speed (induction-cup) type was put into use in 1948.

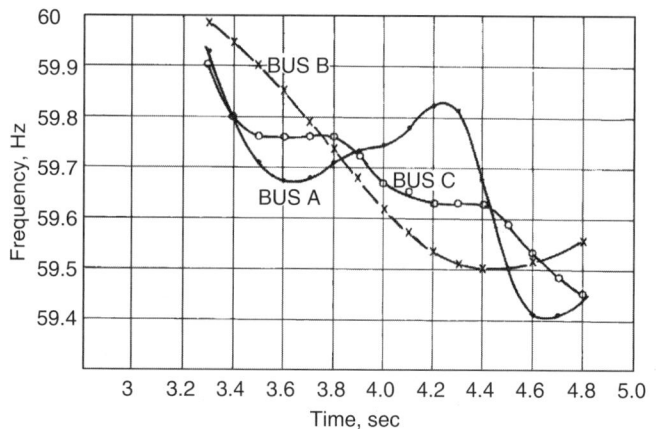

FIGURE 14.2
Time–frequency characteristics of a power system after a 5% loss in generation (nominal frequency 60 Hz).

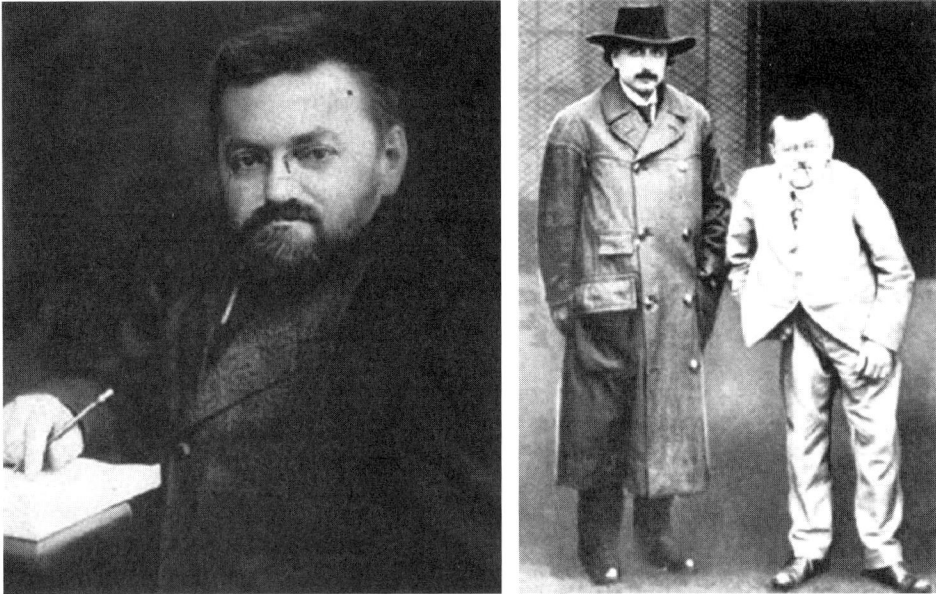

FIGURE 14.3
Charles Proteus Steinmetz (left). In the picture (right) he is next to Albert Einstein in 1921.

14.3 Induction Frequency Relays

Discrimination between normal frequency and abnormal frequency in the induction disk type relay (Figure 14.4) is accomplished by the opposite variation in impedance with the frequencies of two circuits, one circuit containing the coil of one U-magnet connected directly to the voltage supply and designated as the inductive circuit, the other circuit containing the coil of the remaining U-magnet in series with an external capacitor connected to the same supply voltage and designated as the capacitive circuit (since the capacitive reactance predominates at normal frequency).

In the *under-frequency* relay, the coil of the operating U-magnet composes the inductive circuit, and the coil of the restraining U-magnet in series with the capacitor composes the capacitive circuit. At normal frequency the torque produced by the current through the capacitive circuit (restraining U-magnet) is greater than the torque produced by the current (operating U-magnet). A decrease in the frequency of the supply voltage is accompanied by a decrease in the impedance of the inductive circuit permitting an increase of the operating current, while the impedance of the capacitive circuit increases, thereby reducing the restraining current. Thus as the supply frequency is decreased the operating U-magnet overcomes the restraining U-magnet and the relay operates.

The *over-frequency* relay differs from the under-frequency relay in that the operating U-magnet coil is in the capacitive circuit and the restraining U-magnet coil forms the inductive circuit. Consequently, the torque of the inductive element is adjusted to preponderate at normal frequency.

The electromagnet has potential windings on both the upper and lower poles. The under-frequency relay is so designed that at normal frequency (60 or 50) cycles the upper pole current leads the lower pole current and two out of phase fluxes thus produced act to give contact opening torque on the disk. When the frequency drops, the phase angle of the

(a)

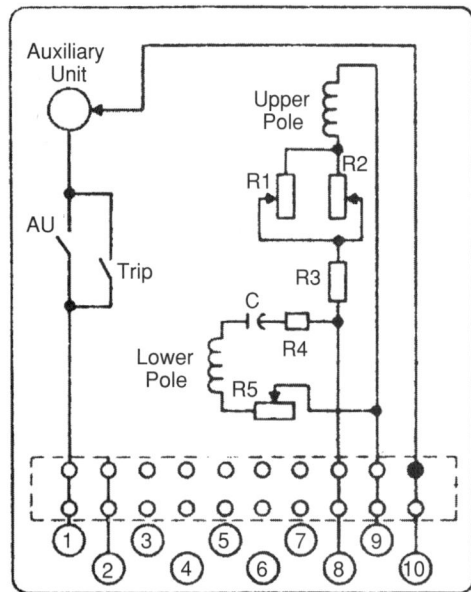

FIGURE 14.4

(a) Type CF-1 induction frequency relay for under- or over- frequency protection without case (Westinghouse, 1963). 1 — Frequency setting rheostat; 2 — time dial; 3 — moving contact; 4 — stationary contact; 5 — indicating contactor switch. (b) Circuit diagram of frequency relay CF-1 type.

(b)

lower pole circuit becomes more leading, until at the frequency setting of the relay the lower pole current begins to lead the upper pole current, and the relay torque is reversed to the tripping direction. The lower the frequency, the greater the phase angle displacement and hence the faster the relay trips. The relay has inverse time characteristics (Figure 14.5). An adjustable resistor in the upper pole circuit is provided to set the frequency at which the relay trips.

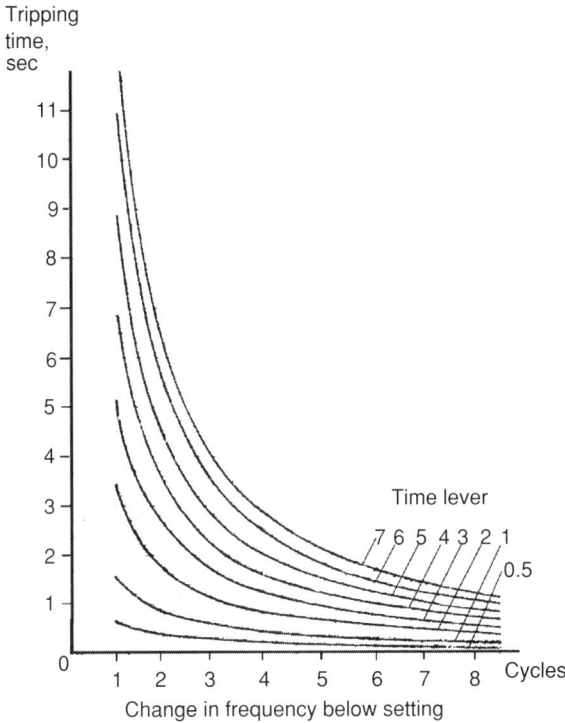

Tripping time, sec

Time lever

Change in frequency below setting

Cycles

FIGURE 14.5

Typical time–frequency curves of the CF-1 relay.

The CF-1 type relay is available in two forms — either as an under-frequency relay or an over-frequency relay. The disk rotation of the under-frequency and the over-frequency relays is in the same direction. Where operation on both under-frequency and over-frequency is desired, two relays are required, one of each form.

Frequency relays with a rotating disk produced by some other firms in the 1950-70's, have a similar construction and principle of operation (Figure 14.6).

Unlike the CF-1 type relay, the IJF relay does not contain an integrated capacitor. For over-frequency relay (IJF51A) the lower coil is the operating coil and upper coil is the restraint coil. For under-frequency relay (IJF51B) the lower coil is the restraint coil and upper coil is the operating coil. The IJF52A type relay is an over-frequency and under-frequency relay having double throw contacts. The left contacts close on under-frequency and the right contacts close on over-frequency

High-speed frequency relays of induction-cup type were put into use in 1948. One of the first relays of this type was a CFF type relay (Figure 14.7). The CFF type under-frequency relay is a high-speed, induction-cup type. Its basic principle of operation is the use of two separate coil circuits (Figure 14.8) which provide increasing phase displacement of fluxes as the frequency decreases, thereby causing torque to be developed in the cup unit to close the tripping contacts. The quantity of torque produced is proportional to the sine of the angle between these two fluxes.

As the frequency decays the angular displacement increases, thereby increasing the torque produced. If the frequency decays rapidly the torque will increase rapidly and cause the relay to close its contacts in less time.

Due to application of a high-speed induction-cup rotor instead of a disk the speed of response of this type of the relay has been considerably increased (Figure 14.9). The relay operating time is an important factor, since the under-frequency condition will develop as a *rate- of- change of frequency* (*ROCOF*). While a constant ROCOF on a power system is

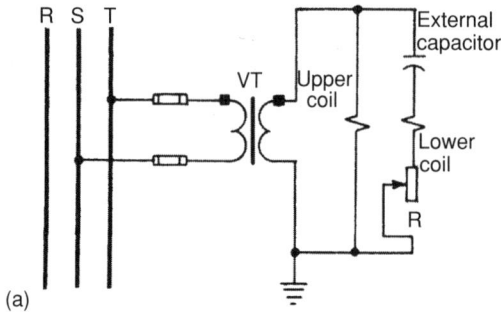

(a)

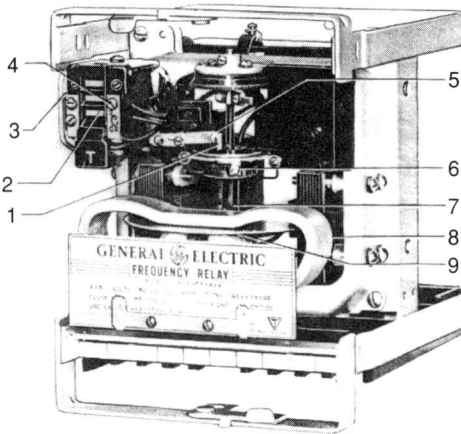

(b)   **(1)**

(2)

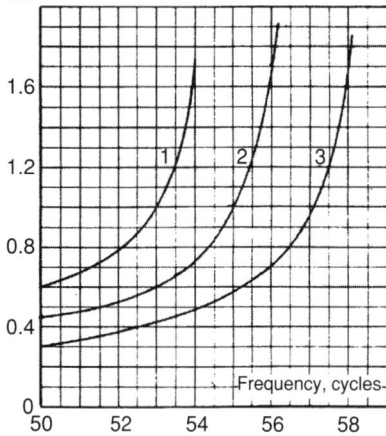

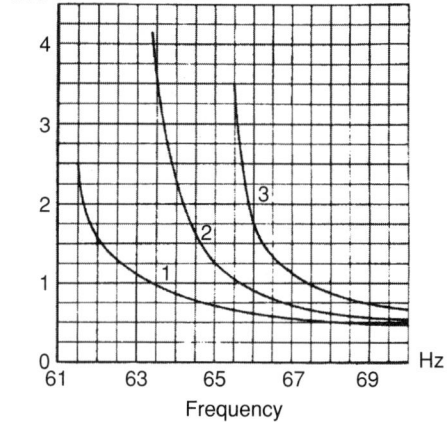

(c)

FIGURE 14.6

The IJF type induction disk type frequency relay (General Electric Co.). (a) Circuit diagram and external connection. (b) Front (1) and rear (2) views. 1 — Moving contact; 2 — target; 3 — seal-in; 4 — seal-in unit tap selector; 5 — stationary brush and contact assembly; 6 — control spring and adjusting ring; 7 — shaft; 8 — drag magnet; 9 — disk; 10 — lower coil; 11 — adjustable resistor; 12 — upper coil. (c) Type IJF51B (left) and IJF51A (right) time–frequency characteristics.

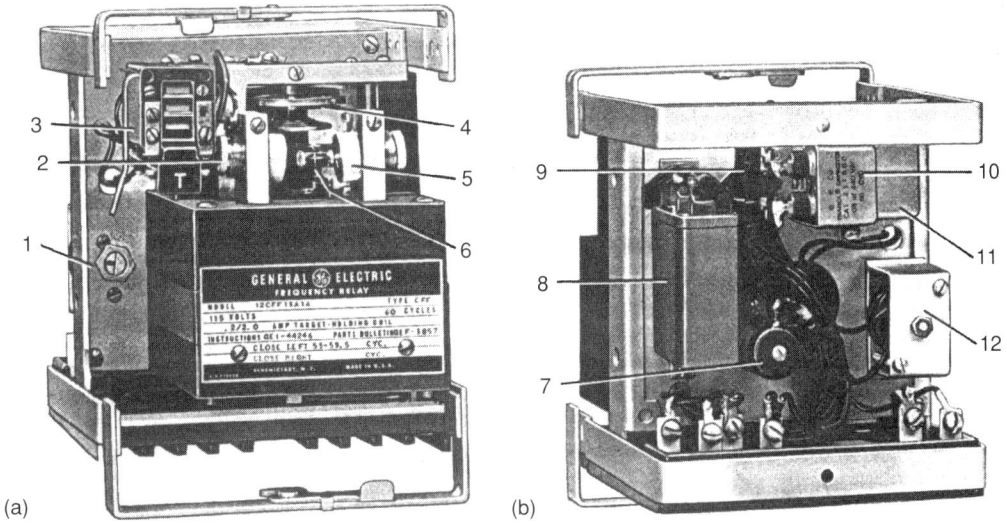

FIGURE 14.7
High-speed induction-cup under-frequency relay of the CFF13A type, removed from case. Front (a) and rear (b) views (General Electric Co.). 1 — Variable reactor for pick-up adjustment; 2 — left stationary contact barrel; 3 — target and seal-in unit; 4 — control spring adjusting ring; 5 — right stationary contact; 6 — moving contact assembly; 7 — resistor R1; 8–11 — capacitors; 12 — adjustable reactor. (General Electric. Under-Frequency Relay Type SFF13A.)

seldom experienced, it is believed that these time curves offer a more realistic way of analyzing the problem.

A serious under-frequency condition on a system is likely to be accompanied by low voltage. The voltage response of a frequency relay is therefore important. The relays considered above had quite considerable dependence of pick-up frequency on the applied voltage, which was an essential disadvantage of this type of relays. In contrast to them the CFF relays are remarkable for their increased stability of parameters.

The CFF relay setting is continuously adjustable over a range of 56 to 59.5 Hz. Relay models are provided with compensation for voltage variation and self-heating; repeatability of set points is held within 0.25 Hz over the normal temperature range from −20 to +55°C, and AC input voltage variations from 50 to 110% of rating (Figure 14.10).

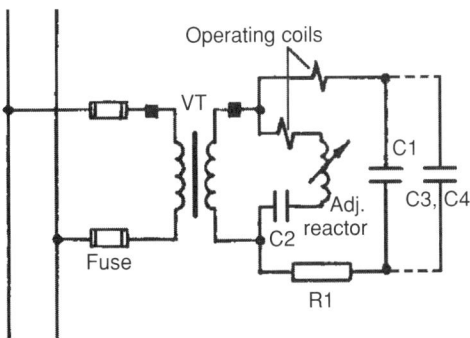

FIGURE 14.8
Internal connection diagram of a CFF13A type relay. (General Electric Under-Frequency Relay Type SFF13A.)

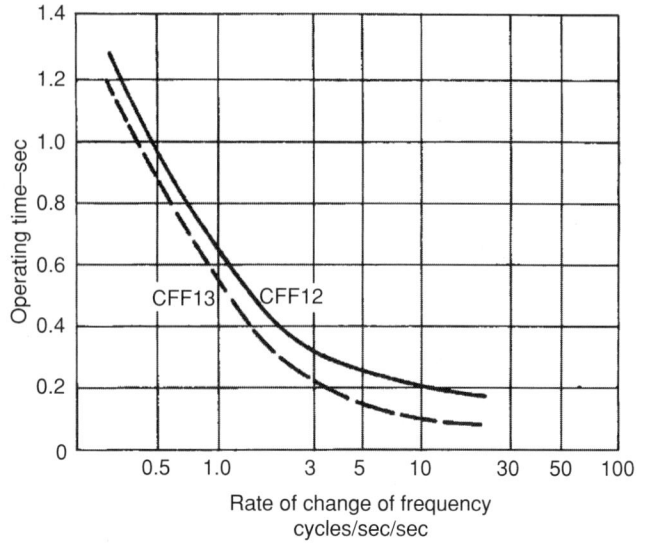

FIGURE 14.9
Time–frequency characteristics of high-speed induction-cup relays of the CFF type. (General Electric. Under-Frequency Relay Type SFF13A.)

Induction frequency relays produced by many firms have similar constructions and principles of operation, for example Russian relays of the {ИВЧ-011 and ИВЧ-3 type, and RFA type relays produced by ASEA, etc.

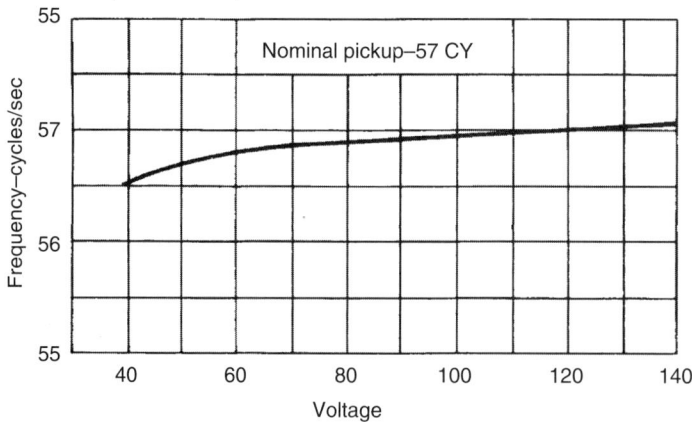

FIGURE 14.10
Variation in frequency relay pickup with applied voltage for CFF12 type relay. (General Electric. Under-Frequency Relay Type SFF13A.)

14.4 Resonance Relays

So-called resonance frequency relays have a much simpler construction (Figure 14.11). Such relays contain three elements: a simple electromagnetic relay (Rel), a capacitor (C), a resistor (R), and a reactor. A transformer (Tr) is used to choose the level of supply voltage.

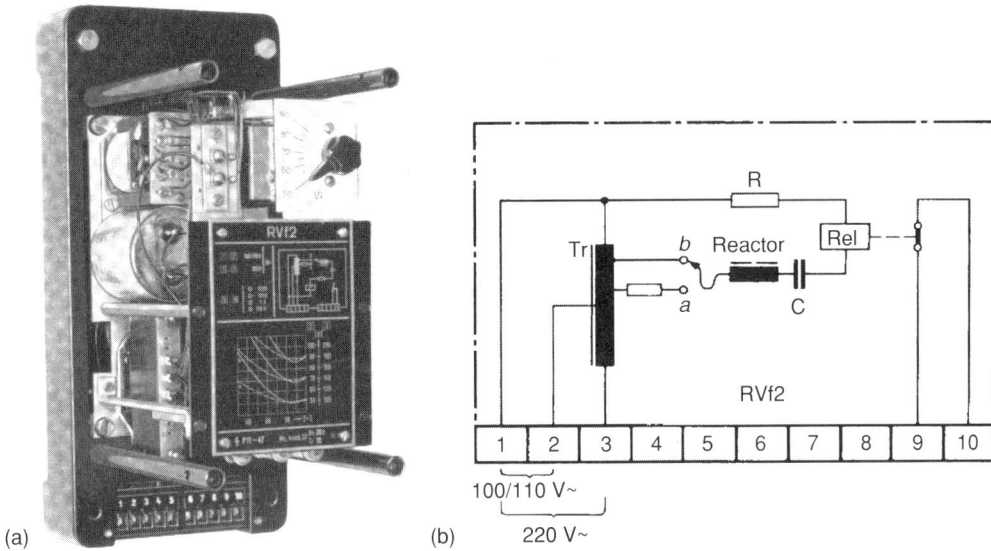

FIGURE 14.11
RVf2 type resonance frequency relay with case off (a), and its circuit diagram (b). (Siemens, 1972.)

The basic principle of this very simple relay is an application of a RLC-circuit tuned to resonance to the required frequency. If the input voltage deviates from resonance frequency, current jumps in the circuit and the relay picks up. The transformer inductance must be also taken into consideration when the circuit is adjusted. Moreover, one can adjust the form of the frequency characteristic by switching the outlets of the transformer ("a" and "b"). Apparently such a relay is simpler, cheaper and perhaps more reliable than an inductance one.

Parameters such as sensitivity and precision of operation of this relay depend much on the properties of the reactor and the individual characteristics of the electromagnetic relay Rel. If the quality of construction is high enough this relay may be quite competitive in its parameters with many more complex and expensive induction relays.

14.5   Electronic Frequency Relays

It is only natural that as in all other types of protective relays, frequency relays also have electronic analogues produced by all of the major (and not only) producers of protective relays. Semiconductor frequency relays are more precise than inductive ones, have less temperature dependence, and are less sensitive to sharp voltage variations at input. In the former U.S.S.R. electronic under-frequency (РЧ-1) and over-frequency (РЧ-2) relays were already produced in 1971. A bit earlier many Western companies started producing such relays, sometimes simultaneously with production of nonelectronic frequency relays.

In such a relay (Figure 14.12), the voltage of the circuit U is applied through the isolating transformer (1) and the filter suppressing high harmonics, to a phase shifter containing two frequency-dependent measuring elements (3), (4) with similar construction but with different adjustment parameters and a resistance divider (5).

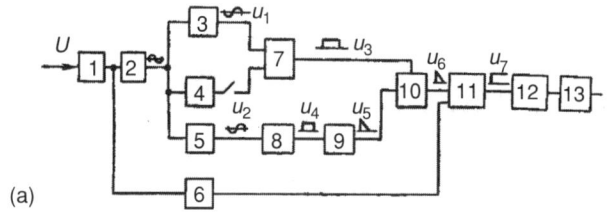

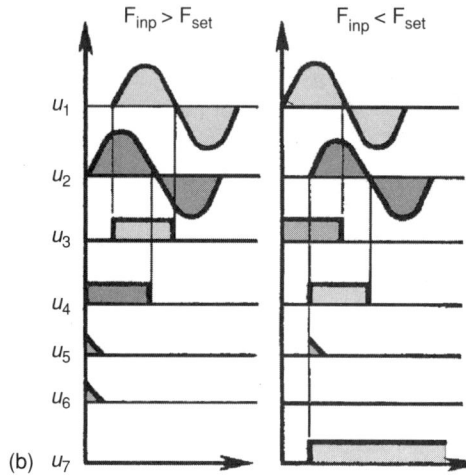

(a)

FIGURE 14.12
Structural scheme (a) and diagram of signals (b) of electronic frequency relay of the РЧ-1 type. 1 — Input transformer; 2 — filter of high harmonics; 3 and 4 — measuring elements; 5 — resistance divider; 6 — starting element; 7 and 8 — pulse shapers; 9 — differentiator; 10 — logical element "PROHIBITION"; 11 — pulse stretcher; 12 — amplifier; 13 — final control element.

(b)

Measuring elements transform change of frequency into change of the phase angle. They are made in the form of a series of resonance circuits with a resistance divider (Figure 14.13).

The voltage U_1 on the resistor R3 is proportional to the current flowing through the reactor L and capacitors C1 and C2. It is in phase with this current. The voltage U_2 on resistor R2 is proportional to the voltage applied to the LC-circuit and is in phase with it. Voltage on the inductance L leads the current in the circuit by an angle of 90° and the voltage on the value C lags from this current by the same angle. Parameters L and C can be chosen in such a way ($2\pi fL = 1/2\pi fC$, where f is frequency of supply voltage) that at a certain frequency of the supply voltage (U_{inp} = 50 or 60 Hz) the leads angle equals the angle of lag. In this case the LC-circuit works as a standard resistor with very low

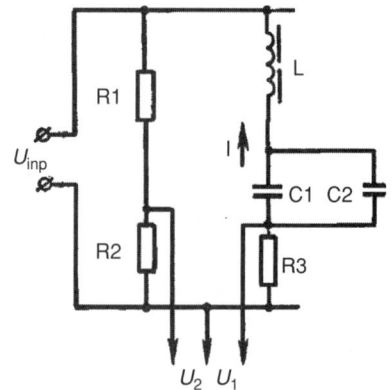

FIGURE 14.13
Simplified construction diagram of the measuring unit of a frequency relay.

resistance. This operation mode is called *resonance,* but as soon as the frequency of supply voltage deviates from the resonance frequency to which the relay is adjusted, the LC-circuit begins to reveal its properties. When frequency decreases, capacitive reactance increases and dominates in the circuit (that is, phase displacement between voltages U_1 and U_2 occurs) and when frequency increases, the inductive one does as well (the phase displacement has the opposite angle).

The resistance divider (5) is used for creation of reference voltage (u_2) with regard to which angles of phase displacement of voltages U_1 and U_2 are measured (in the block diagram these voltages are marked by u_1). The outlet voltage of the frequency-dependent measuring element (3) and the outlet voltage of the reference element (5) are applied to the inputs of the pulse shapers (7 and 8) in which alternating sinusoidal voltage is transformed to rectangular pulses (u_3 and u_4) respective to duration close to the duration of a sinusoid half-period of alternating voltage. The positions of the pulses formed by the reference voltage and the pulses formed by the output voltage of the frequency-dependent element in time are determined by the ration of frequency of this reference voltage to the frequency of the voltage of the circuit. One must determine the moment when these pulses do not concur. This will mean that the frequency of the voltage in the circuit differs from the natural frequency to which the relay is adjusted. Concurrency or nonconcurrency of pulses in time (synchronization) is controlled by the logical element (10). At that point the output pulse is stretched, amplified, and is applied to the output electromagnetic relay. With the help of an additional external contact the second frequency-dependent measuring element (4), used for the return of the relay to the initial position at a certain circuit frequency, can be started.

RXFE-4 type relays have a similar construction and principle of operation (Figure 14.14). The RXFE-4 is a static, instantaneous frequency relay module available in both under-frequency (99.8%) and over-frequency (100.2%) models. Power to the static circuits is supplied from the measured voltage, thus a separate auxiliary supply is not required. A blocking circuit prevents unwanted operation when the measured voltage is switched ON or OFF.

The RXFE-4 operates on the principle of comparing the phase angle of the current in a tuned LC-circuit with the current in a purely resistive circuit, both circuits being supplied by the measured voltage after it has undergone transformation and filtering.

The resonant frequency of the LC-circuit and thus the operating value of the relay are continuously adjustable within a range of approximately 12% of rated frequency. The setting knob, located at the lower left of the relay module front plate, can be reached by inserting a small screwdriver through a normally plugged hole in the clear plastic cover.

A feedback circuit, which modifies the resetting level after operation of the output relay, provides a stable operating band of 40 to 70 Hz, depending on the measured voltage magnitude. A red target becomes visible upon operation of the output relay and is hand-reset by a knob on the front.

Digitalization of information has allowed considerable improvement of the parameters of the frequency relay, although their construction has become, of course, much more complex. The FCX103 type relay can be given as a good example of this (Figure 14.15). This relay is made of big discrete electronic components and has a modular construction typical of the 1970s. Its main features: one to four independent tripping levels; highly accurate frequency measurement (± 0.03 Hz); a wide setting range (39.1 to 65 Hz in 0.1 Hz increments); dimensions: $270 \times 210 \times 269$ mm.

The FCX103 type frequency relay can be delivered with one to four output stages and its tripping values and delay times can be individually selected for under- and over-frequency protection. The auxiliary supply for the electronic circuits is derived from the measured quantity. A matrix plug-board (see Figure 14.15b) is provided for setting

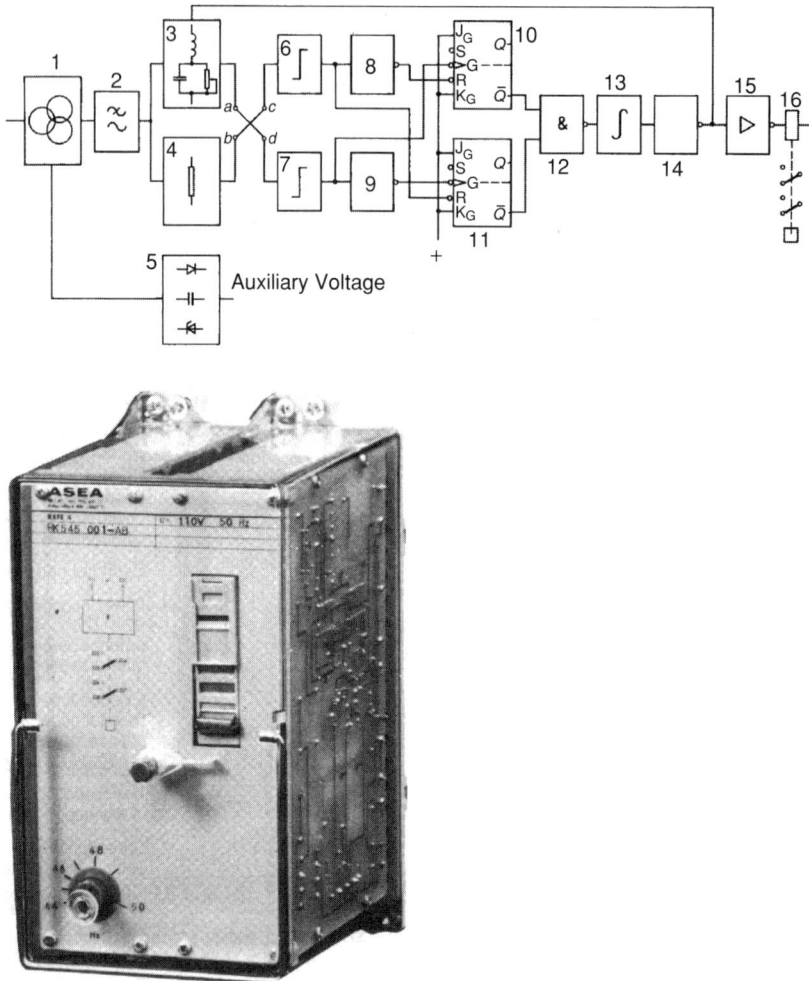

FIGURE 14.14

RXFE-4 type static frequency relay (ABB, 1990). 1 — Transformer for measured voltage and auxiliary voltage; 2 — low pass filter; 3 — LC-circuit with facilities for setting the operating value; 4 — R-circuit; 5 — auxiliary circuit; 6 and 7 — level detectors; 8, 9, 13, and 14 — inverting circuits; 10 and 11 — JK-triggers; 12 — "NO-AND" logical circuit; 15 — amplifier; 16 — output relay. (ASEA (ABB), 1982.)

the tripping levels of the various output stages, and the corresponding codes can be obtained from a table.

A high degree of accuracy is attained by using a crystal oscillator as a reference and employing digital techniques to compare the unknown period with the reference period. The number of oscillations of the quartz reference during each period of the system frequency is counted. At the end of the period, the relay decides whether the frequency of the system is greater or lesser than the relay setting. This decision is stored for at least 150 msec and tripping if during this time the measurements of all subsequent periods indicate the same result. The stable 100 kHz sinusoidal signal from the quartz oscillator (1) (see Figure 14.15c) is transformed by the shaper (2) into a square-wave signal, which can be counted by the binary counter (3). A square-wave signal is also derived from the system voltage via the shaper (4). Each positive flank of this signal causes the monostable multi-vibrator (5) to produce an impulse of 10 μsec which upon being applied to the counter resets it to zero.

The count reached by the counter immediately prior to being reset is proportional to the period of the system frequency. A decoder (7) is set by means of a plug-board (6) (Figure 14.15b) to detect the desired under- or over-frequency.

The binary setting (corresponding to the period of the pick-up frequency) depends on which combination of direct and inverted outputs from the counter are applied to the decoder (7), for example, for a period of 20 msec (50 Hz) the direct and inverse

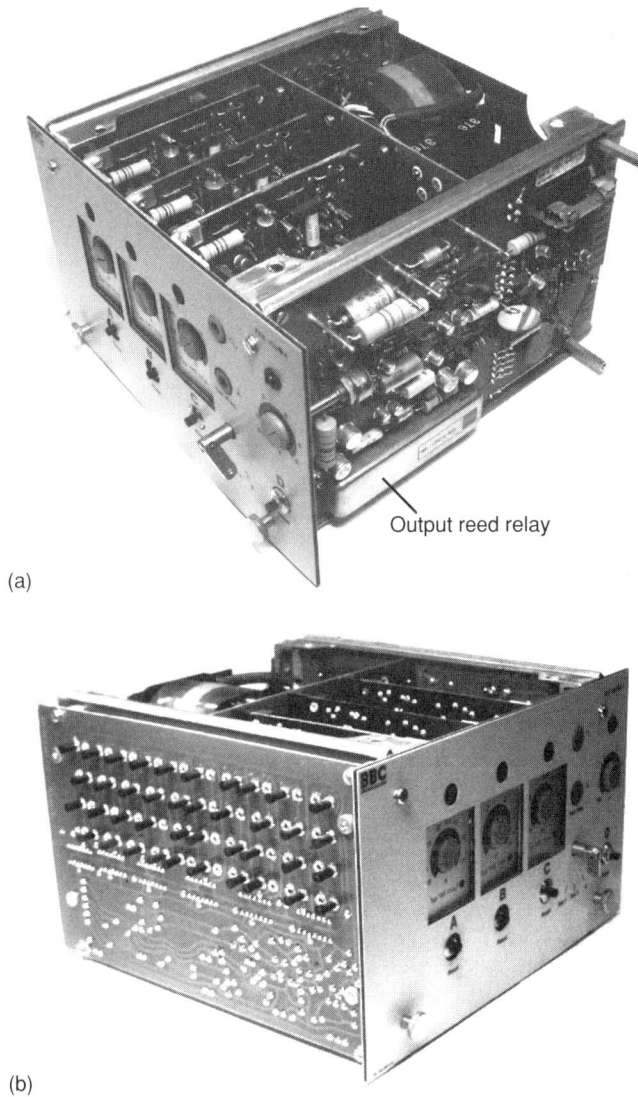

Output reed relay

(a)

(b)

FIGURE 14.15
(a) Relay of the FCX103 type without cover (Brown Bowery Co., 1973). (b) view of frequency selector plug-board.

(Continues)

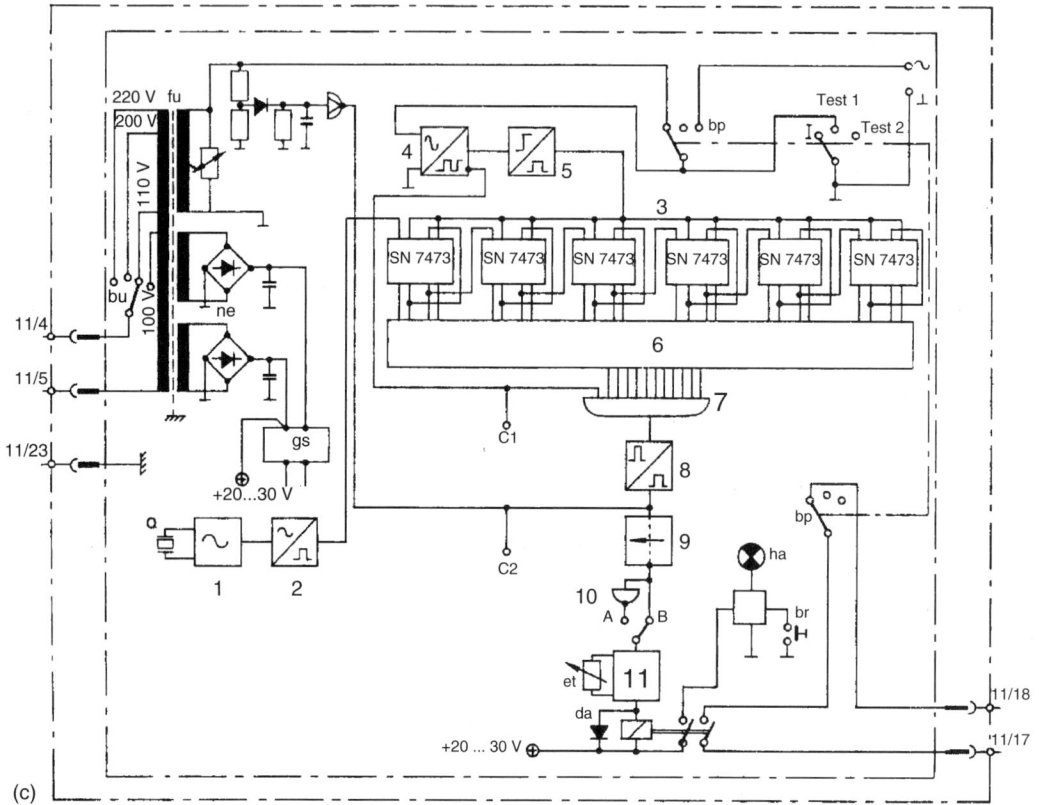

FIGURE 14.15 (Continued)
(c) Block diagram (one output designed) of an FCX103 relay (BBC).

counter outputs must be decoded such that at a count of 2000 (2000 periods of the 10 μsec long 100 kHz quartz signal are equal to 20 msec) a digital "1" is produced at the output of the "AND" gate. At frequencies above the setting (shorter period), the counter will be reset before it reaches the decoder count and no output will be generated by the "AND" gate. At lower frequencies, however, a 10 μsec impulse is produced by the "AND" gate once every period (upon reaching the decoder count, an impulse of this duration is applied to all the inputs of the "AND" gate). The short-duration impulses, produced when the frequency being supervised falls below the setting, are lengthened by the monostable flip-flop (8) and transformed into a continuous signal by the pulse stretcher (9). This signal controls the timer (11) either directly or via the inverter stage (10). With the inverter connected in series, the relay registers an over-frequency condition. Underfrequency is registered when the pulse stretcher is connected directly to the timer.

Each FCX103 can be fitted with from one to four tripping or output stages. The basic measuring unit includes the tripping stage "D" as standard, the plug-in additional stages "A," "B," and "C" being inserted as required.

In 1977 a modification of this relay (FCX103B) appeared. Instead of one of the blocks (YAT 111) it contained a special plug-in part YAT 115 based on integrated circuits which allowed this relay to measure the rate of change of frequency (df/dt).

Fundamentally the manufacturer is against the use of so-called *df/dt ancillary units* which measure frequency deviation because such relays are very sensitive to switching operations and sudden changes in system voltage caused by such operations can also cause it to pick up, however, the df/dt measurement can be used as an additional feature when the frequency decreases considerably, that is, when there is a large energy deficit not only can the load corresponding to the first shedding stage (10 to 20%) be shed, but also that of the second or third stages. This combination accelerates the action of the relay in the event of sudden overloading of the network.

Setting range pick-up value for frequency deviation of new YAT 115 integral circuit based unit, is 0.1 to 9.9 Hz/sec, accuracy ± 0.05 Hz/s. Relays of this type have been in use for more than 15 to 20 years and still meet all the requirements.

Modern technical integrated circuits of high integration levels allow production of relatively simple and compact electronic frequency relays (Figure 14.16), not only by the leading relay producers but also by small companies. The most modern static under-frequency relay (Figure 14.17) employs digital counting techniques to measure system frequency. Basically this relay consists of a highly stable, crystal-controlled oscillator, which continuously supplies 2 MHz (5 MHz in some relay types) pulses to a binary counter. The counter, in conjunction with other logic circuitry, determines system frequency by counting the number of 2 MHz pulses that occur during a full cycle (one period) of power system voltage. For any preset frequency a specific number of pulses should occur during a one-cycle period. If the number of pulses is less than this specific number, this indicates that the system frequency is above the setting. Conversely if the number of pulses is greater than this specific number, it indicates that the system frequency is less than the setting.

For reasons of safety, an under-frequency indication must occur for a minimum of three consecutive cycles before the relay produces an output. This minimum time can be extended to 80 cycles by means of an adjustable auxiliary timer. If the system frequency recovers for even one cycle during the timing period, the timing circuits will be reset and the relay will immediately start monitoring system frequency again. The relay operating time is independent of the rate of change of the system frequency. The static under-frequency relay is an extremely accurate and stable device. It can be adjusted to a frequency range of 40 to 70.9 Hz in increments of 0.01 Hz, and its setting is accurate within ± 0.005 Hz of the desired set point. This accuracy is maintained over an ambient temperature range of -20 to $+60°$C and is independent of voltage over the range of 30 to 120% of rating. Some models are provided with an under-voltage detector, which blocks operation of the relay when the applied voltage falls below the set level of the detector.

The static relay has a minimum operating time of three cycles, as described previously, when the output is a thyristor. Most models provide electromechanical contact outputs and in these models the minimum operating time is increased to four cycles simply because of the operating time of the output telephone relay. A compromise solution is to use reed switch output relays (as in the SFF type relays).

Microprocessor frequency relays produced by the General Electric Company, Basler, have a similar construction and external design. Heavy steel cases of similar size and with similar attached elements to their electromechanical analogues, simplified applications of the new equipment and allowed replacement of some electromechanical relays with microprocessor ones in functioning power stations and substations. Such ideology was typical of the initial stage of development of microprocessor relay protection. More lately, relay producers have diverged from this ideology and now are producing microprocessor relays in cases of any shape and size (Figure 14.18).

(a)   (b)

(c)   (d)   (e)

FIGURE 14.16
This is how modern small-size frequency relays based on integrated circuits (produced by many firms) look.

FIGURE 14.17
Static digital frequency relay of the SFF204 type (GE) in
a standard steel case (without front cover) with pull-out
printed circuit boards.

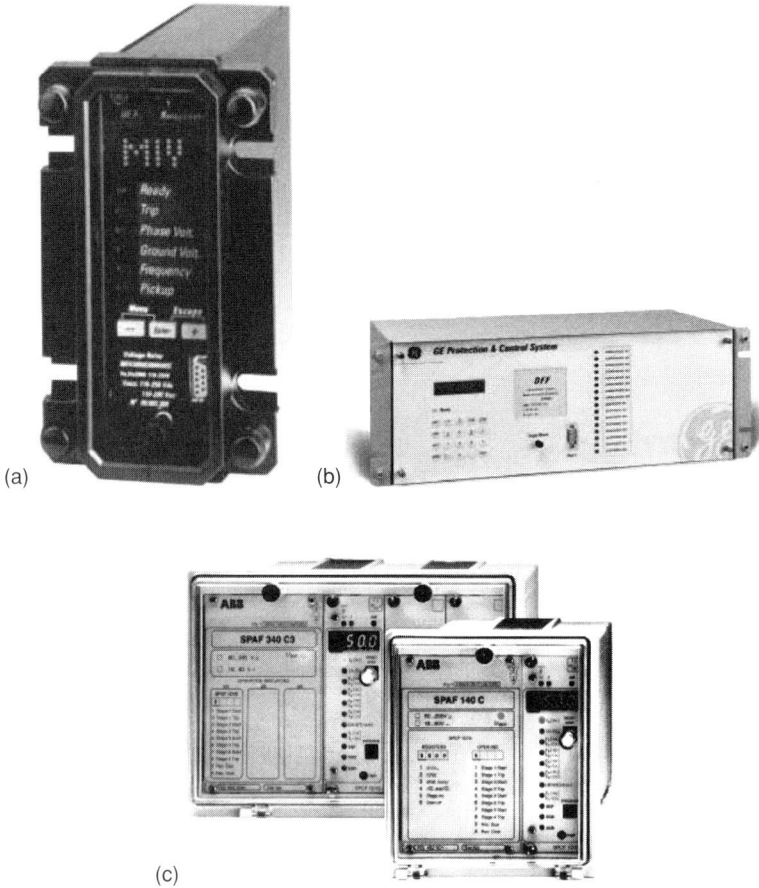

FIGURE 14.18
Microprocessor based frequency relays of the MIV (a) and DFF (b) type, produced lately by the General Electric Co. (Gneral Electric 2004 online catalog) and the SPAF type (c), produced by ABB. (ABB Relay Units and Components Buyers Guide 1990.)

Why it was necessary to replace cheap and reliable nonmicroprocessor semi-conductor frequency relays providing accuracies of ± 0.03 to 0.05 Hz with large and expensive microprocessor relays with an accuracy of ± 0.005 Hz, and if such accuracy is really required in practice when do we have to apply frequency relays in electric circuits, are quite different questions. The author still fails to find a satisfactory answer to those two questions.

15

Microprocessor-Based Relays: Prospects and Challenges

15.1. Is It a Relay at All?

Microprocessor systems are similar to simple digital computer systems (Figure 15.1), in which the microprocessor performs the timing and control of the system and carries out all arithmetic and logical operations. The system memory may be Read Only Memory (ROM) for dedicated applications or Random Access Memory (RAM) for the storage of data and programs, or a combination of both. System memory stores the program to be executed and the data relevant to the specific task.

The microprocessor communicates with the system memory by means of a bus system. The same bus system permits communication of the microprocessor with the interface adaptor, or input and output (I/O) unit, which makes possible the transfer of data and control signals to and from the system.

As it can be seen from Figure 15.1, the microprocessor is quite a complex device with specific terminology and principles of operation that have nothing in common with the protective relays considered above. The question arises if the "microprocessor-based relay" is a "relay" in the full sense of the word. On closer examination, it turns out that the "microprocessor-based relay" is a small computer in which the output circuits (usually built-in CT or VT — Figure 15.2) have matched parameters with external current and voltage transformers, with a program stored in memory, allowing processing of input signals in such a way that operation of this or that type of protective relays can be modeled. With the help of a basic universal microprocessor one can create any relay by just making certain changes in the program, at least that is how it used to be at the initial stage of development of microprocessor-based equipment. For example, when the first universal programmable microcomputer (Elektronika-60) appeared at the end of the 1970s in the former U.S.S.R., a whole series of different protective relays was designed for the power industry, but input circuits of other types could also be installed and other programs set up on the same device (for example, a program directing a telescope to the sky in such a "relay"), making it not a "relay" in the full sense of the word.

So it turns out that the microprocessor becomes a "relay" only when it is based on a program of a "relay." This sounds quite strange. Our computer does not turn to a canvas or a palette, only because we run PhotoShop® or CorelDraw® and start drawing a picture, although the computer does allocate us a zone for drawing (a virtual canvas) and a tool for color selection (a virtual palette) and a whole set of different virtual brushes, so just as in this example, a "microprocessor-based" relay is really only a "virtual" relay.

Opinions are sometimes expressed that protective devices now available on the market are in fact only single-purpose devices designed for execution of a limited set of functions,

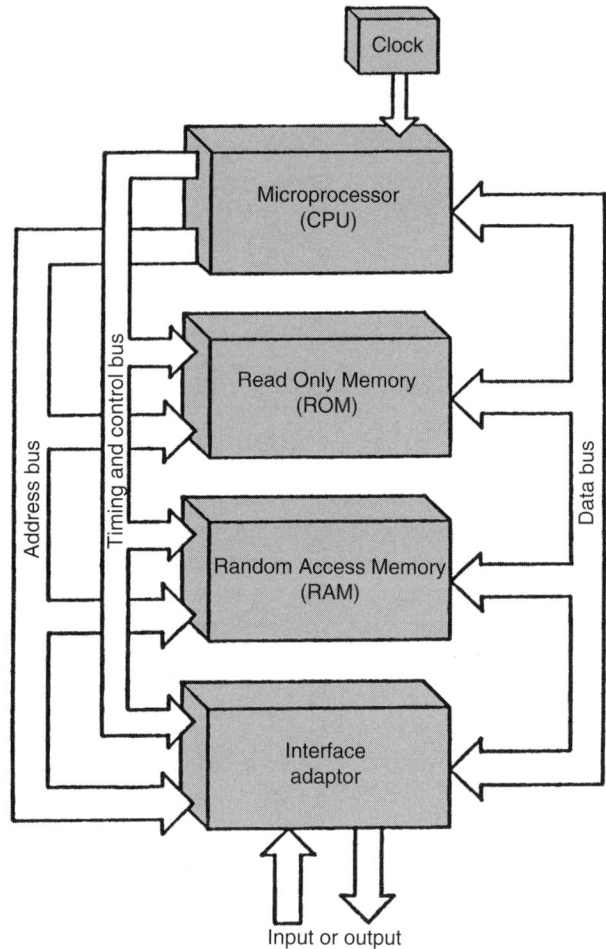

FIGURE 15.1
A structure of a typical microprocessor system.

typical of relays of some particular type. Such devices have names corresponding to the name of a relay of a particular type, like Frequency relays, for instance, and one can communicate with such a device only with the help of a special program which specially created for this particular device, taking into account all of its peculiarities. Actually the relay is programmed by inputting certain pick-up thresholds, time intervals, and algorithms of choosing of the proper type, among all possible types of working characteristics, but in this case, limitations are set not for the microprocessor (for which it is all the same, whatever signals to process), but for ROM containing the program of this microprocessor and the number of I/O channels. If in devices performing the function of protective relays one uses not ROM, but Erasable Programmable Read-Only Memory (EPROM) or Electrically Erasable Programmable Read-Only Memory (EEPROM) and a pocket programmer that allows recording to ROM of any algorithm of a microprocessor operation, one will obtain a universal protective relay instead of a frequency relay. It will not differ practically from modern universal Programmable Logic Controllers with digital and analog inputs such as Modicon® family (Gould Modicon) or SIMATIC® family (Siemens) and many others. Each such device may contain tens of input modules for transformation of signals to Boolean or hexadecimal code, tens of virtual timers of different types, comparators, counters of different configurations, different types of triggers, univibrators, a great number of memory registers used for recording of

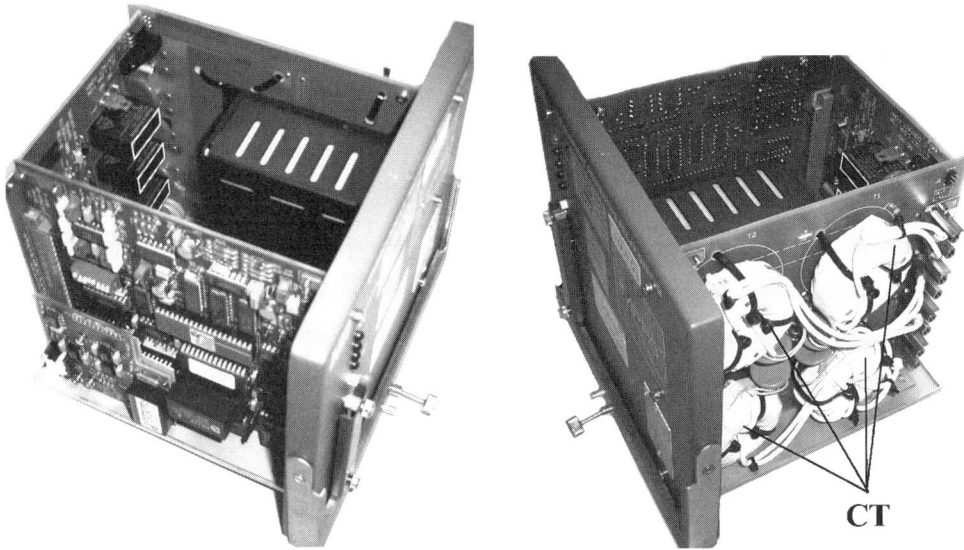

FIGURE 15.2
Microprocessor-based current relay IMPRES (ABB).

intermediate results, powerful output modules, etc. Using this set of virtual elements in a computer program running in Windows®, one can draw very complex automation systems (much like in graphics editors) which are then loaded to the controller. Having chosen the option "simulation," one can see on the display how this automation system works in real-time operation modes, or in emergency modes modeled purposely. Today a special computer program is used for work with each type of such controllers; however, scientists are trying to create a universal program allowing work with controllers of different types.

Of course there are devices with much lower performance capabilities, based on quite "simple" controllers and miniature EEPROM (Figure 15.3). But actually "low" performance capabilities are in fact not as low as it might seem. The EEPROM 24LC256 is capable of both random and sequential reads up to a 256 kb boundary. Functional address lines allow erase/write cycles = 1,000,000; data retention > 200 years; a standard 8 up to 8 devices on the same bus for up to 2 Mb address space; pin DIP package. The PIC16C73B microcontrollers have high performance RISC CPU; 35 single instructions to learn; up to 20 MHz operating speed; 56 kb words of program memory; 192×8 bytes of data memory (RAM); 3 timers, 5 analog or digital channels; 11 interrupt sources; a 28-pin DIP package.

The "simplicity" of these devices is as relative as their "low" performance capabilities. For example, the Data Sheet for microchip PIC16C73B type, containing only brief descriptions and specifications of the device, has 189 pages!

Apparently the internal architecture and principles of operation of microprocessor-based devices have little in common with devices called "electric relays." To illustrate this fact one can mention a well-known complex universal microprocessor-based relay of the REL-316 type (ABB), designed for distance protection of power lines and for differential protection. This relay appears to be used quite often as a substation controller and not as a protective relay, since it is based on a powerful universal microprocessor 486 series supplied with a great number of logical inputs and relay outputs.

Microcontroller with A/D converter
PIC16C73B type

EEPROM 24LC256 Series

FIGURE 15.3
Universal microcontroller with analog or digital converter of the PIC16C73B and EEPROM 24LC256 type (Microchip Technology, Inc.).

As follows from the facts considered above, in the author's opinion, construction and principles of operation of microprocessor-based devices, including protective relays, should be considered not in a book devoted to electrical relays, but in technical computer literature, however, since these virtual microprocessor-based relays are widely used as protective relays, it is still worthwhile to consider some important aspects of practical use of these devices.

First, we will consider those numerous advantages of microprocessor-based "relays" which are usually indicated in advertisements.

15.2   Advantages of Microprocessor-Based "Relays"

1. Many microprocessor-based relays allow us to record and then replay modes preceding or functioning during breakdowns, for the analysis of emergency situations.

Well, were power-engineering specialists really deprived of this possibility before? Are not there a great number of various loggers of emergency modes and of relay pick-ups? The ABB, Siemens, NxtPhase, Areva, RiS, Dewetron GmbH Company alone offers tens of variants of loggers and analyzers of various different emergency modes (Figure 15.4).

2. Microprocessor-based relays allow us to change pick-up settings with the help of a computer and to turn from one characteristic to the other using only software tools.

This is really more convenient than to adjust the relay with the help of potentiometers and a screwdriver, but how often does one have to adjust setting modes of the relay during 20 to 25 years? Two times? Three times?

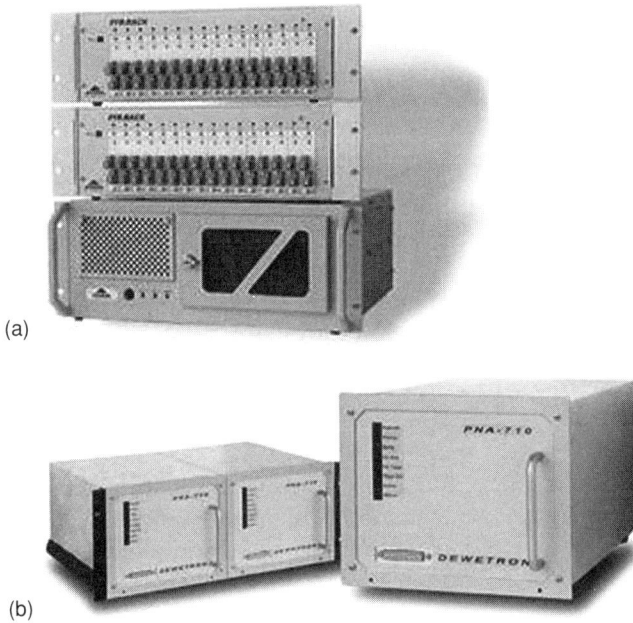

(a)

(b)

FIGURE 15.4
(a) PFR-700 type power fault recorder (Dewetron GmbH, 2004); (b) PNA-710 type power network analyzers (Dewetron GmbH, 2004).

3. Microprocessor-based relays allow us to provide all the information regarding their state to remote dispatching centers through special communication channels.

Had not remote multi-channel systems of data transmission (SCADA, for instance), transmitting information about the pick-up of every electromechanical relay to the dispatching desk, been used before microprocessor-based relays appeared?

4. Microprocessor-based relays allow us to change configuration of the relay protection set: to switch some functions ON or OFF (that is to switch ON or switch OFF some relays) by software means with the help of an external computer.

This is really much more convenient than to install separate relays and remake the assemblage in relay protection boards, but again the same question arises: How often does one actually need to resort to such operations? Once (or twice under the most adverse conditions) for the whole service term of the relay (20 to 25 years)?

5. Microprocessor relays are less prone to dust, increased humidity, aggressive gas and vapors than electromechanical relays.

The author wonders if the author of this thesis has ever been to modern halls (or rooms) of relay protection in power stations or substations. It seems that he has not, otherwise he would have been aware that, first, electromechanical protective relays have been produced for decades in heavy hermetic cases of metal and glass that are well protected from dust and other negative environmental factors. Second, modern halls of relay protection are separate clean enclosed spaces equipped with air-conditioners maintaining stable conditions regardless of conditions outside. Microprocessor-based relays are installed in similar halls.

6. A small microprocessor-based relay can replace a whole set of standard electromechanical relays. In the first place, this applies to complex distance protections. Thus you can save expensive space occupied by cabinets with relay protection.

It is true that complex microprocessor relays occupy smaller areas of mounting by five to ten times less than a set of standard relays with similar functions. It is also true that boards with microprocessor-based protections occupy less space by several times than conventional ones, but the tricky question is: What part of space of the power station or substation can one actually save if one replaces electromechanical relays with microprocessor-based ones? One hundred thousandth? Or one millionth?

7. Microprocessor-based relays are more sensitive to emergency modes than electromechanical ones.

This is also absolutely true, as all the arguments considered above given by advocates of microprocessor-based relays. The question is whether such high sensitivity and accuracy are really required in relay protection of power units. For example, let us take microprocessor-based frequency relays picking up when frequency diverts by 0.005 Hz, and standard analog electronic relays with a pick-up accuracy of 0.01 to 0.05 Hz (for different models). The author wonders if anywhere in the world there is a power station or substation with frequency relays performing some operations in the power system at a frequency error of 0.005 Hz from the nominal value? In many cases, even sensitivity of standard electromechanical or analog electronic relays is excessive and one has to coarsen it artificially. Can relay protection of power units face the problem of low sensitivities of the relay?

8. Higher reliability of static microprocessor-based relays in comparison with electromagnetic relays containing elements moving mechanically.

At first sight it may really seem uncontestable that a static device without movable elements is much more reliable than a complex mechanism with numerous interacting elements, but only on the face of it. On closer examination it appears that things are not so simple.

First, the number of pick-ups (that is the movements of movable elements) of electromechanical protective relays is paltry in comparison with their service life. Referring to his personal experience, the author can say that he has come across such cases when relays with original (factory) defects have been exploited for more than 10 years. The fact that these defects have not been discovered for 10 years proves that during all this time the relay never picked up (and also that it is inadmissible to check relays so rarely!). Is it really worth speaking about mechanical wear in such cases?

Second, the number of elements from which a microprocessor-based relay is constructed is by hundreds and thousands times more than the number of elements from which an electromechanical relay is made. The reliability theory says that there is an inversely proportional dependence between the number of elements and the reliability of complex systems. As far as reliability of the elements is concerned, everything is also not as simple as that. In the electromechanical relay affected by external factors capable of causing damage, there are only coils of electromagnets and insulation of internal installation wires. These are very reliable and stable elements, but if it was a question of improving their reliability, the coils could be impregnated with epoxide resin in vacuum and internal wiring in Teflon insulation could have been used. In microprocessor-based relays, practically all electronic elements are affected by the supply voltage, and a part of them by input current or voltage. Some elements are constantly in the mode of generating signals. Some components (electrolytic capacitors, for example) wear considerably under constant exposure to working voltage. As far as integral circuits (IC — basic active elements of microprocessor-based relays) are concerned, they are the main cause of relay malfunctions (Figure 15.5 – Matsuda T., Kovayashi J., Itoh H., Tanigushi T., Seo K., Hatata M., Andow F. Experience with maintenance and improvement in reliability of microprocessor-based digital protection equipment for power transmission systems. Report 34–104. SIGRE Session, 30 August–5 September 1992, Paris). One of the major problems of complex electronic devices is aging of their components, bringing on changes in their parameters, during their lifetime. As a rule, the lifetime of such devices usually does not exceed 10 to 15 years. At about that time we begin to encounter various failures, malfunctions, and disturbances that are sometimes very difficult to locate in such complex devices (such as in the microprocessor relays, for example), and even if we do successfully diagnose a malfunction, it is not always possible to repair it (continuing with the above example, printed circuit boards on the surface mounting microelements for instance — standard technology for microprocessor relays). In such situation, it is possible to replace

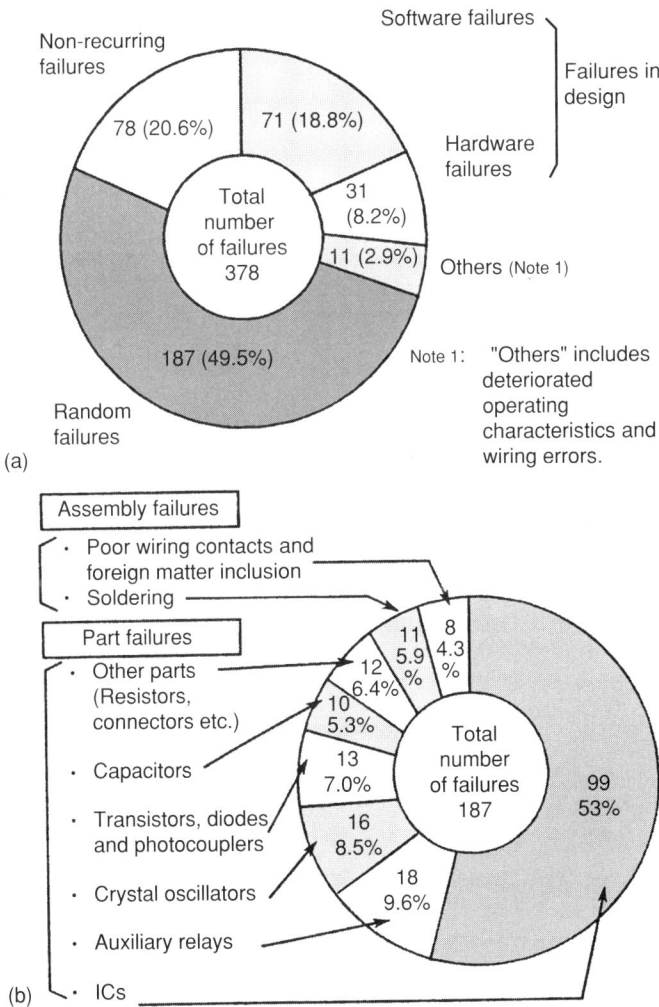

FIGURE 15.5

Statistics of malfunctions of 378 microprocessor-based relays produced by leading Japan companies. (Matsude T., Repart 34–104, SIGRE, Session 30).

damaged PCB only entirely, which cost sometimes makes a significant part of relay cost. With all due respect to our new and modern technologies, it should be noted here that previously far more "simple" electromechanical protection relays, some produced as many as 40 and 50 years ago from materials and according to technologies of that time, continue to work reliably today in many power systems (in Russia, for example).

Mass maintenance of microprocessor relays presents many problems, with their built-in switching power supply. Such power supplies recomplex devices (see Figure 15.6, for example) and with the addition heavy duty, continuous work, and exposure to spikes, harmonics, etc., they often fail.

Power supplies of microprocessor devices frequently create problems that designers did not foresee at all when developing these devices. The author experienced such a problem when a breakdown of one of the minor elements of a microprocessor device produced a short circuit of power supply. The microprocessor instantly gave a set of uncoordinated commands, which led to simultaneous disconnecting of all the power transformers of a large class 161 kV substation. Analysis of the reasons for this failure established that the short circuit of the power supply had come in the current limiting mode, as is necessary for high-grade power supplies. Current limitation is provided by fast decrease of output voltage level so that the output current does not exceed the maximal value allowed for the power supply. In the presence of large-capacity capacitors in the power supply, this voltage reduction occurred relatively slowly: during 0.5 to 1 sec. During this time the microprocessor, whose voltage supply had essentially been reduced, started to "go around the bend" and have sufficient lengthy time to give out complete commands, causing and leading to the serious failures.

FIGURE 15.6
Built-in switching power supply of a REL-316 type microprocessor relay (ABB), dimensions: 270 × 230 mm.

In order to prevent this, in our opinion, power supplies for microprocessor devices must be completed with so-called "crowbar protection" — a simple circuit (a thyristor, for example) which provides instantaneous short circuiting of the output of the power supply, whenever the emergency mode is enacted.

Protection functions of the important object (high-voltage line, power transformer, bus bar system, and generator on power plant) have been divided between five and six separate relays till an era of microprocessor devices. Failure of one relay yet did not lead to malfunction of all protection system completely. In one microprocessor device, functions of many relays are concentrated. For example, only the microprocessor device such as REG-216 carries out functions: differential protection, inverse-time overcurrent, negative phase-sequence, overvoltage, distance protection, underimpedance, overload, overtemperature, frequency, rate-of-change frequency, overexitation, etc. In such device failure of any common element, for example, the power supply, the microprocessor or its auxiliary elements leads to malfunction of protection system completely.

One of the serious problems which have been found out by the author that discrepancy between switching capabilities of subminiature electromagnetic relays (using as output elements in microprocessor protection devices [Figure 15.7]) and real conditions. Researches executed by the author have shown, that as output elements of the microprocessor protection devices produced by all leading companies are used subminiature electromagnetic relays which are not intended for switching inductive loading with currents about 2 to 5 A (coils trip of high-voltage circuit breakers or auxiliary lockout relay) at 125 and the more so 250 V DC. These subminiature relays work with a huge overload and can be damaged at any moment.

Internal constant monitoring of the condition of main units, even separate important elements of microprocessor protection device promoted by manufacturers as great progress in protection technique allowed the maximum protection reliability. Actually, the statement that internal self-diagnostics of microprocessor relays allows increasing reliability of relay protection is not correct. There is no connection between failure intensity of elements of microprocessor relays and the information on happened failures. The actual fact appears no more than an advertizing gimmick.

For example, in protection device MiCOM P437 constant monitoring serviceability of each of output electromagnetic relays (so anyway, the manufacturer asserts) is available. But who can explain, how (even only theoretically) it is possible to supervise serviceability

FIGURE 15.7
Input/output board of microprocessor protection device with subminiature electromagnetic relays as output elements.

(that is ability to close normally open contact when control current pass across the coil) or absence of the welding of normally closed contact of the relay without its pick-ups? As a more detailed consideration of this question, it appears that continuity of the coil is constantly only supervised (by passing through them weak pulses of a current which are not causing operations of the relay). What can happen with no energized coil? For what is such monitoring necessary?

That is why it is only natural that research performed by the Research and Development Division of the Israel Electric Corp. (Aspects of Digital Protective Relaying. RE-626, 1991) led to the following conclusion:

> Microprocessor relay reliability is lower than that of electromechanical and static relays: Microprocessor relay components tend to fail more often than those of conventional relays. This disadvantage is not compensated by a self-monitoring function, especially in unmanned substations. Maloperation or a failure to operate as a result of an internal relay failure may occur before the arrival of staff, after receiving an alarm signal.

The experience in this area of such a huge country as the former Soviet Union and present Russia is very interesting. More than 1.6 million relay protection and automation devices, mostly Russian made electromechanical relays (Electrical Devices Plant in Cheboksary), are installed in the Russian power supply system which is one of the oldest and biggest in the world. Of those, microprocessor protection systems constitute less than 1%. Even in the Moscow power system — one of the most advanced and well provided power systems — microprocessor devices amount to only 3,000 out of a total of 170,000 relay protection devices used there, which is at the most 2%.

Moreover, according to specialists' evaluations, about 80% of the relay protection devices in Russia have been operating for 20 years, and some relays even for over 50 years, while their service life should not exceed 15 years. It is worth mentioning that with all of this going on, the true response factor of this protection remains stable, and is over 99%! This fact suggests that Russian electromechanical protection relays are highly reliable and that even in the 21st century such protection can be successfully used in the world's largest power supply system. Such situation is characteristic not only for Russia, but also for other old power systems. In spite of the fact that microprocessor relays exist in the market any more for the first 10 years, rates of replacement of electromechanical or old static relays remain, on the average, very low. According to the publication of Basler Electric Co. (Johnson G., Thomson M. *Reliability Considerations of Multifunction Protection*), it will take about 70 years to replace all the predecessor relays with modern microprocessor.

Yet it is clear that the life-time of the major part of the protection relays has expired and they are called for replacement. Because of aging of bronze and brass subjected to permanent mechanical stress, and pealing of the insulation material of relay wires, they are only microns far from short circuit. At the same time, the opinion of many Russian specialists in the field of power engineering says that the total replacement of traditional Russian electromechanical relays by compatible digital parts made in Europe and U.S.A. will result in a drastic increase of emergency cases in power supply systems. Moreover, unlike their Western colleagues, the Russian manufacturers of low voltage equipment advance a sound belief in rational implementation of technical innovations in such specific fields as relay protection. They believe that in the next 3 to 5 years, all of the obsolete equipment should be replaced with traditional time-proved electric equipment, and only then should carefully planned installation of digital devices begin in power installations. This should be done selectively rather than totally, and moreover that the microprocessor relays be backed up with new generation electromechanical protection systems.

Leading manufacturers of microprocessor devices, such as ABB, Siemens, and Alstom have successively penetrated the Russian market, however, at the moment even they show certain caution, taking into account the scale of the Russian power system and the possible loss in case of a major crash. ABB, for example, purchased in Russia sets of microprocessor protection relays for its project at one of the biggest power stations in Kyrgyz, and then connected them in parallel with its own (Russian) electromechanical relays. A similar solution was used in some other objects, for example, at the Zubkovsky substation at the center of Moscow.

In spite of such very careful treatment of microprocessor protection devices in Russia and their very limited use, the rate of these device failures in the Russian power supply system still turned out to be twice as high as that of the traditional protection devices. (Based on material of the 15th Scientific and Technical Conference, "Relay protection and automation of power supply systems"). The following data were brought up: Within 3 years, from 1999 to 2001, 100 out of 23,264 operations of relay protection devices at the Novosibirsk power system were false responses. Part of these false responses occurred with modern equipment that was only put into operation late. In another case, false responses of Siemens microprocessor devices at one of the thermal power stations of the Moscow power system "Mosenergo" resulted in disconnection of all of these protection systems, which remained unconnected for more than 2 years. At another substation of the same power system, microprocessor relays were damaged by a stroke of lightning. At a power station in Kostroma-microprocessor device failures were triggered by static voltage from synthetic carpets.

Taking into account that safe operation of the entire power supply system depends on the relay protection system, the above-mentioned cases are more than just unfortunate incidents. This is a problem that needs to be immediately addressed. Moreover, it is clear that as a result of such incidents which occurred within the last few years, the excitement about implementation of intellectual digital technology in Russia has declined. At present, 70% of the Russian low voltage industrial equipment market (which also includes relay protection devices) belongs to foreign companies. Because of serious accidents that occurred in the Western power supply systems, this fact does not inspire the Russian specialists: They "do not want to repeat somebody else's mistakes," as mentioned in one of the publications.

These were some of the so-called "advantages" of microprocessor-based protective relays. Let us take a look at their disadvantages.

15.3 Disadvantages of Microprocessor-Based "Relays"

1. Impact of electromagnetic disturbances from the power supply network on the operation of the relay.
Blackout
A blackout results in the total loss of utility power.

- Cause: Blackouts are caused by excessive demand on the power network, lightning storms, ice on power lines, car accidents, construction equipment, earthquakes, and other catastrophes.
- Effect: Current work in RAM or cache is lost. Total loss of data stored on ROM.

Noise
More technically referred to as electromagnetic interference (EMI) and radio frequency interference (RFI), electrical noise disrupts the smooth sine wave one expects from utility power.

- Cause: Electrical noise is caused by many factors and phenomena, including lightning, load switching, generators, radio transmitters, and industrial equipment. It may be intermittent or chronic.
- Effect: Noise introduces malfunctions and errors into executable programs and data files.

Sags
Also known as brownouts, sags are short-term decreases in voltage levels. This is the most common power problem, accounting for 87% of all power disturbances according to a study by Bell Labs.

- Cause: Sags are usually caused by the startup power demands of many electrical devices (including motors, compressors, elevators, and shop tools). Electric companies use sags to cope with extraordinary power demands. In a procedure known as rolling brownouts, the utility will systematically lower voltage levels in certain areas for hours or days at a time. Hot summer days, when air-conditioning requirements are at their peak, will often prompt rolling brownouts.
- Effect: A sag can starve a microprocessor of the power it needs to function, and can cause frozen keyboards and unexpected system crashes, which result both in lost or corrupted data. Sags also reduce the efficiency and life span of electrical equipment.

Spikes
Also referred to as an impulse, a spike is an instantaneous, dramatic increase in voltage. A spike can enter electronic equipment through AC, network, serial, or communication lines and damage or destroy components.

- Cause: Spikes are typically caused by a nearby lightning strike. Spikes can also occur when utility power comes back online after having been knocked out in a storm or as the result of a car accident.
- Effect: Catastrophic damage to hardware occurs. Data will be lost.

Surge
A surge is a short-term increase in voltage, typically lasting at least 1/120 of a second.

- Cause: Surges result from presence of high-powered electrical motors, such as air conditioners. When this equipment is switched off, the extra voltage is dissipated through the power line.
- Effect: Microprocessors and similar sensitive electronic devices are designed to receive power within a certain voltage range. Anything outside of expected peak and RMS (considered the average voltage) levels will stress delicate components and cause premature failure.

Many cases of malfunctions and even damages of microprocessors caused by spikes and surges are described in literature. For example, mass malfunctions of microprocessor-based time relays occurred in nuclear power plants in the U.S.A. (Information Notice No. 94–20: *Common-Cause Failures Due to Inadequate Design Control and Dedication*, Nuclear Regulatory Commission, March 17, 1994). A review of these events indicated that the microprocessor-based timer or relay failed as a result of voltage spikes that were generated by the auxiliary relay coil controlled by the timer or relay. The voltage spikes, also referred to as "inductive kicks," were generated when the timer or relay time-delay contacts interrupted the current to the auxiliary relay coil. These spikes then arced across the timer or relay contacts. This arcing, in conjunction with the inductance and wiring capacitance generated fast electrical noise transients called "arc showering" EMI. The peak voltage noise transient changed as a function of the breakdown voltage of the contact gap, which changed as the contacts moved apart and/or bounced. These noise transients caused the microprocessor in the timer or relay to fail.

The organization of the supply system of relay protection is also very important. Power units are supplied by powerful accumulator batteries with a constantly connected charger, or by an uninterrupted power supply (UPS) cushioning the negative impact of the factors listed above, however, the same system supplies driving gears of power switches and many other devices, causing spikes. Besides, investigations of UPS systems (*The Power Protection Handbook* — APC, 1994) have shown that at certain conditions noise spikes and high harmonics can get into microprocessors through grounded circuits and neither UPS nor filters can prevent this. In addition, UPS devices have their own change-over times. Usually specifications for UPS indicate a switch delay time of 3 to 5 msec, but in fact under certain conditions this time may increase by a factor of more than 10. In a normal mode, the load in some types of UPS devices ("OFF-line" types) is usually feed through a thyristor switch (bypass) which must become enabled when voltage diverts below 190 or above 240 V. After that the load is switched to the output of the inverter supplied from the accumulator battery. The time of disabling of the thyristor switch is summed up from the time of decrease of current flowing through the thyristor to the zero value t_0 and turn-OFF time t_q, after which the thyristor is capable of withstanding the voltage in the closed (OFF) position. For different types of thyristors $t_q = 30$ to 500 μsec, and t_0 depends on the proportion of induction and pure resistance of the circuit of the current flowing through the disabled thyristor (if the circuits break). As a result of switching OFF of the input circuit breaker or pulling out of the UPS supply cable from the terminal block $t_0 = 0$ to 5 msec, which will provide very high performance, however, at actual interruption of supply a break usually occurs in circuits of a higher voltage level, and not in the circuit for 120/220/380/400 V. The thyristor switch is then shorted to the second winding of the network transformer and the load connected to it. If the beginning of the supply interruption coincides with the conductivity interval of the thyristor, the duration of the process of current decaying may exceed 400 msec, and changeover to reserve power supply (inverter) may be delayed for inadmissibly long times. If the beginning of supply interruption coincides with the no-current interval, switching requirements are similar to those at input cable break. It follows that at interruptions of supply and voltage falls-through coinciding with the interval of current flowing, the working source will changeover to the reserve supply for inadmissibly long times (Dshochov B.D. Features of electrical power supply of means of computer networks. *Industrial Power Engineering*, 1996, N2, p. 17–24). The user who sets up and checks the UPS usually reaches conclusions about its serviceability by switching the input circuit breaker OFF. As shown above, this does not always correspond to conditions of actual transient processes (short circuiting of the input to the induction shunt), which creates the possibility that after installation and successful check of the UPS by switching OFF of input circuit breaker, some voltage falls-through (coinciding with intervals of current flowing) may lead to disabling ("suspension") of the microprocessors. That is why in some types of UPS of the "OFF-LINE" category in order to provide high performance during change-over in the short-circuit mode, high-speed electromechanical relays instead of thyristor ones are used.

To determine the actual time of changeover of the UPS one should make an experiment imitating short circuits of the working source, to inductively active a shunt while a special measuring system records the transient processes. But who runs such experiments and where?

Thereupon, another aspect of the problem gains our attention: suspensions and malfunctions of the operation of the microprocessor of the UPS in emergency modes on high-voltage circuits. When the control microprocessor malfunctions, alternation of switching-ON and switching-OFF of power semiconductor elements of the inventor may be disturbed and short-circuit loop making, followed by automatic switching-OFF of the input circuit breaker of the UPS. This same phenomenon can happen to automatic chargers whose microprocessors are supplied from an external auxiliary UPS. Such incidents quite often occur in practice, but nobody yet has concerned himself with a serious analysis of the reasons. It is quite possible that the reasons for such emergency switching of UPS, and of the chargers, are similar to those for the case considered above.

2. Microprocessor-based relay protections, especially complex ones such as distance protections, do not always operate adequately in complex breakdowns or on boundaries of protection zones and cannot always trace transient processes correctly and in proper time. In practice, one often comes across breakdowns and malfunctioning of complex microprocessor-based protections in exploitation conditions. If the relay is tested on a standard laboratory test bench with standard signals at its inputs, it will operate precisely and reliably. The problem is that it is impossible to simulate all possible combinations and signal distortions that may take place in real situations on a test bench. It is also impossible to foresee all such situations when the relay is designed. This situation is similar to when a properly functioning powerful PC equipped with an undamaged powerful software shell (such as Windows®) suddenly buzzes at a certain instruction set, or if several programs run simultaneously. In most cases, it is impossible to foresee and prevent such situations. The working group of the U.S.A. and Canada has published the report on the reasons of well-known accident (August 14, 2003) in which it is ascertained, that one of the reasons of occurrence of computer "suspension" of control system and occurrence of emergencies in a power supply system of the company "First Energy" in U.S.A. In electromechanical relays, such situations are impossible. Therefore, many researchers insist that at the further wide introduction of digital techniques in protective relaying, it is necessary to provide additional independent (reserved) not digital protection relays for emergency modes.

3. A strange phenomenon exists whereby high-speed microprocessor-based protections respond to the emergency mode much more slowly than electromechanical ones. In one of the power systems, for reliability improvement microprocessor-based and electromechanical distance relays are switched in parallel. When emergency situations were analyzed, more than once it turned out that the electromechanical relay had picked up and tripped the circuit breaker before the microprocessor-based relay responded. This may be explained by the fact that unlike an electromechanical or analog electronic relay, the microprocessor-based relay operates with input values discretely. It "picks" current

values of input quantities and copies them into the buffer, then picks another set of input values in a certain time interval and compares them with those stored in the buffer. If the second set is identical to the first, the input values are directed to the microprocessor for processing. In general, for pick-up of an electromechanical or instantaneous electronic relay 10 to 15 msec are enough, while for a microprocessor-based relay 30 to 40 msec are required. Actually full operating time of microprocessor relay frequently reaches up to 50 to 80 msec for complex failures. So it often turns out that the superior performance of the microprocessor-based relay indicated in the advertisement of the producer is not provided in practice. In transient emergency modes, the microprocessor has to process great sets of information in a real-time mode, accompanied by quick and considerable changing of input signals. For this, it requires certain time (sometimes hundreds of milliseconds). Moreover, if after the starting of the microprocessor, the situation changes (for example, a single phase short circuit to the ground turned to the two-phase and then to the three-phase one), the starting process of calculation is interrupted and all calculations must be performed from the very beginning.

4. There are essential differences in operation of electromechanical and microprocessor-based relays caused by their different susceptibility to harmonics, saturation, and other wave distortions.

It is well known that at great ratios of short-circuit currents, current transformers considerably distort the curve of the output current applied to the relay. The problem of deterioration of accuracy is relevant for all types of relays, including electromechanical ones. Electromechanical relays produce torque that is proportional to the square of the flux produced by the current. These relays respond to the current squared or to the product of the currents produced by the input quantities. Since root-mean-square (rms) is defined as the average of the integral of the square of the current, these relays are said to be rms responsive. For most microprocessor relays, all quantities other than the fundamental component are noise. These relays used digital filters to extract only the fundamental, and either attenuate or eliminate harmonics (Zocholl S.E. and Benmouyal G. how microprocessor relays respond to harmonics, saturation, and other wave distortions. Meta world Schweitzer Engineering Laboratories, Inc., Summer 2003). The fast fourier transformation (FFT) is a very useful tool for analyzing the frequency content of stationary processes in microprocessor relays. Protection algorithms based on FFT have serious disadvantages including the neglecting of high-frequency harmonics, when dealing with nonstationary processes (magnetizing inrush and fault currents) for determining the frequency content. Furthermore, different windowing techniques should be applied to calculate the current and voltage phasors and this causes significant time delay for the protection relay. In this case, accuracy is not assured completely. For example, in cases of influence of inrush current on transformer differential relay with harmonic restraint, the relaying information is contained in the system fundamental and the harmonics only interfered. It is somewhat surprising that the digital filter will faithfully extract the fundamental from any waveform that is periodic at system frequency. The distance elements, in another example, did not operate because no voltage depression accompanied the high-current signal. However, sensitive settings caused the negative-sequence directional to identify a forward fault.

5. Considerable complication of exploitation of the protective relay: apparently, testing, and adjustment of microprocessor-based protections with the help of a computer (or even without it) require some new level of training of specialists and more time (what we mean here is that a technician or an engineer does not have to adjust the same relay every day, but they have to learn everything about it from the very beginning and to

gain an understanding of testing methods). It is enough to look through Instruction Manuals of these devices, which are almost as thick as this book, to realize this, and as far as trouble tracing and repair of such devices go, this is practically impossible during the exploitation. An article by John Horak (Basler Electric) "Pitfalls and Benefits of Commissioning Numerical Relays," *Neta World, Summer 2003* tackles the problems arising during testing of microprocessor-based relays. The acceptance test is a step-by-step procedure published in the relay's instruction manual that checks that the relay's measuring elements, timing elements, status inputs, contact outputs, and logic processing system are functional, and that relay performance is within the manufacturer's intended specifications, using settings and logic defined by the manufacturer's test procedure. The test will include calibration checks involving secondary current and voltage injection. The relay is not field calibrated since, generally, only factory processes can calibrate numerical relays. In the process of working through these tests, one will learn a bit about the relay and will perform the value of showing that the relay is functioning correctly. The acceptance test does not make one completely knowledgeable of the relay, so some time should still be set aside for further investigation of the relay as the commissioning program proceeds.

Modern microprocessor-based systems (as line current differential protection, for example) are complex devices that include sophisticated protection algorithms and intense communications. As a result, performance testing of such complex systems may create a problem particularly because expensive and specialized equipment is required. Basic validation testing may be performed using phasors and test sets as far as the protection functions are considered; and a local loop-back procedure as far as the communications are considered. True performance testing requires either a real-time digital simulator or a playback system capable of driving several sets of three-phase currents and voltages (two- and three-terminal testing).

Testing the communication channels for high noise, bursts, channel asymmetry, channel delay, etc., is a field that does not belong to traditional relay testing. This requires new expertise and specialized test equipment. Due to the complexity of modern current differential relays, it is highly beneficial, if not crucial, to conduct performance tests involving both protection and communication functions particularly if difficult system conditions or poor communication channels are anticipated.

The increasingly large weight of the "human factor" in the operation of microprocessor relays created many more opportunities for additional mistakes, particularly during the programming and testing stages of the relay. Many interrelated functions and parameters controlled by one microprocessor-based relay lead to the necessity of artificial coarsening and even to entire disabling of some functions to test the other ones. After testing, one should not forget to input the previous settings of the relay. Such problems do not exist in electromechanical relays. In instruction manuals for many such relays, it is indicated that the settings of the relay may be changed during testing of the relay, which is why after that one should carefully check them.

In addition, the interfaces of many modern programs are often not too friendly, and the internal logic that works with them can sometimes warrant anguish! Many new programs (including from some very well-known companies!!) are simply "raw" and contain a lot of bugs. Who can know what will occur if even one bug starts to control relay protection?

6. Information redundancy. Many digital relays have too many variants of parameters for setting such which are not unequivocally necessary for relay functioning. Especially it concerns the devices with complex functions, such as distance protection with their one hundred set parameters. A function for 15 to 20 light-emitting diodes located on the

forward panel of the relay; a degree of brightness of the screen; color of a luminescence of the screen; color of the reports of information display; time of preservation of the data on the screen; and many other parameters with numerous variants which can be chosen from library of parameters. Frequently, these variants are superfluous. For example, in microprocessor protection device MiCOM P437, only the fuse supervision algorithm for voltage transformer can be chosen on four different variants! Such obvious redundancy leads to great number of settings, variants passes for the protection device with complex functions. It increases a error probability because of the "the human factor." The problems pertinent to the human factor grow repeatedly if the same group of people should serve the relay of the different manufacturers having various programs with different interfaces, different principles of a parameter's choice, at times adjustments, even different names and designations of the same main parameters.

7. Possibility of intentional remote actions to break the normal operation of the microprocessor-based relay protection (Electromagnetic Weapons, Electromagnetic Terrorism). The theory behind the E-bomb was proposed in 1925 by physicist Arthur H. Compton not to build weapons, but to study atoms. Compton demonstrated that firing a stream of highly energetic photons into atoms that have a low atomic number causes them to eject a stream of electrons. Physics students know this phenomenon as the Compton Effect. It became a key tool in unlocking the secrets of the atom. Ironically, this nuclear research led to an unexpected demonstration of the power of the Compton Effect, and spawned a new type of weapon. In 1958, nuclear weapons designers ignited hydrogen bombs high over the Pacific ocean. The detonations created bursts of gamma rays that, upon striking the oxygen and nitrogen in the atmosphere, released a tsunami of electrons that spread for hundreds of miles. Street lights were blown out in Hawaii and radio navigation was disrupted for 18 h as far away as Australia. The United States set out to learn how to "harden" electronics against this electromagnetic pulse (EMP) and develop EMP weapons.

Now, intensive investigations in electromagnetic weapons field are being carried out in Russia, the U.S.A., England, Germany, and China. In the U.S.A. such research is carried out by the biggest companies of the military–industrial establishment, such as TWR, Raytheon, Lockheed Martin, Los Alamos National Laboratories, the Air Force Research Laboratory at Kirtland Air Force Base, New Mexico, and many civil organizations and universities.

In the 1990s, the U.S. Air Force Office of Scientific Research set up a 5-year Multidisciplinary University Research Initiative (MURI) program to explore microwave sources. One of those funded was the University of New Mexico's Schamiloglu, whose lab is located just a few kilometers down the road from where the Shiva Star sits behind tightly locked doors.

The German company "Rheinmetall Weapons and Munitions" has also been researching E-weapons for years and has test versions. The EMP shell was designed following revelations that Russia was well ahead of the West in the development of so-called radio-frequency (RF) weapons. A paper given at a conference in Bordeaux in 1994 made it clear that the Russians believed it possible to use such weapons to disable all of an enemy's electronic equipment. Written by Dr. A.B. Prishchipenko, Deputy Director of Scientific Center "Sirius," member-correspondent of the Russian Academy of Military Sciences (Figure 15.8) and entitled "Radio Frequency Weapons on the Future Battlefield," it described Soviet research dating back to the late forties, provoking near panic among western military planners (A.B. Prishchepenko, V.V. Kiseljov, and I.S. Kudimov, Radio frequency weapon at the future battlefield, Electromagnetic environment and conse-

FIGURE 15.8
Dr A. Prishchipenko, deputy director of Scientific Center "Sirius," member-correspondent of the Russian Academy of Military Sciences.

quences, *Proceedings of the EUROEM94, Bordeaux, France, May 30–June 3,* 1994, part 1, pp. 266–271). It gave credence to the nightmare scenario of a high-technology war in which all the radio, radar, and computer systems on which their weapons depended would be disabled, leaving them completely defenseless. Then 2 years ago, it emerged that the Russians had developed an electromagnetic device, a so-called E-Bomb, capable of disabling electrical and electronic systems, which could be carried in a briefcase. Amid intelligence reports showing that the Irish IRA had discussed the possibility of paralyzing the city of London with an E-Bomb, British research in that technology was stepped up.

Today in Russia electromagnetic weapons are being developed by huge research and production institutions like the Scientific Association for High Temperatures (OIVT), consisting of the following Moscow organizations: the Institute of High Temperatures of Academy of Sciences, the Institute of Thermal Physics of Extremal States, the Institute of Theoretical and Applied Electrodynamics, the Research-and-Development Center of Thermal Physics of Impulse Excitations, and the proving ground in Bishkek, in addition the All-Russian Scientific Research Institute of Experimental Physics in Sarov (Arzamas-16) in the Nizhni Novgorod region, the All-Russian Scientific Research Institute of Technical Physics in Snezhinsk (Chelyabinsk-70). In spite of the economic crisis in Russia and a lack of money for many military programs, the government allocates money to these institutions. For example, recently in Moscow for the Scientific Association OIVT, a new building with an area of 1500 m^2 has been built.

Lately, many projects of past age have been declassified and are freely sold today. For example, the Institute of High-Current Electronics of the Russian Academy of Sciences in Tomsk (HCEI SB RAS) offers at free sale ultra-wideband high-power sources of directional electromagnetic radiation (Figure 15.9). As the technology of military RF weapons matures, such weaponry also becomes affordable and usable by criminals and terrorists. Both cheap low-tech and expensive high-tech weapons exist. High-power sources and other components to build EM weapons are available on the open market and proliferate around the globe (Figure 15.10).

One potential ingredient made available by the military is old radars, sold when facilities close down. Anything that operates between 200 MHz and 4 or 5 GHz seems to be a real problem. The reason they are for sale is that they are not very effective. Radar technology has improved drastically, but the radar does not need to be the newest technology to cause problems to electronic equipment and systems that are not prepared for an intentional EM threat. Intentional EMI includes both pulses and continuous-wave signals, in two basic forms. One is high-power microwave (HPM), a continuous-wave signal at a Gigahertz, like radar. The other is ultra-wideband, which is essentially a fast pulse produced by a radar using pulse techniques rather than a continuous wave. These threats can be packaged in a mobile van or even a suitcase. The effective ranges decrease with size, but even a suitcase-sized threat is widely available. According to Peter Cotterill, managing director of MPE Ltd (Liverpool, U.K.), an electromagnetic bomb in a

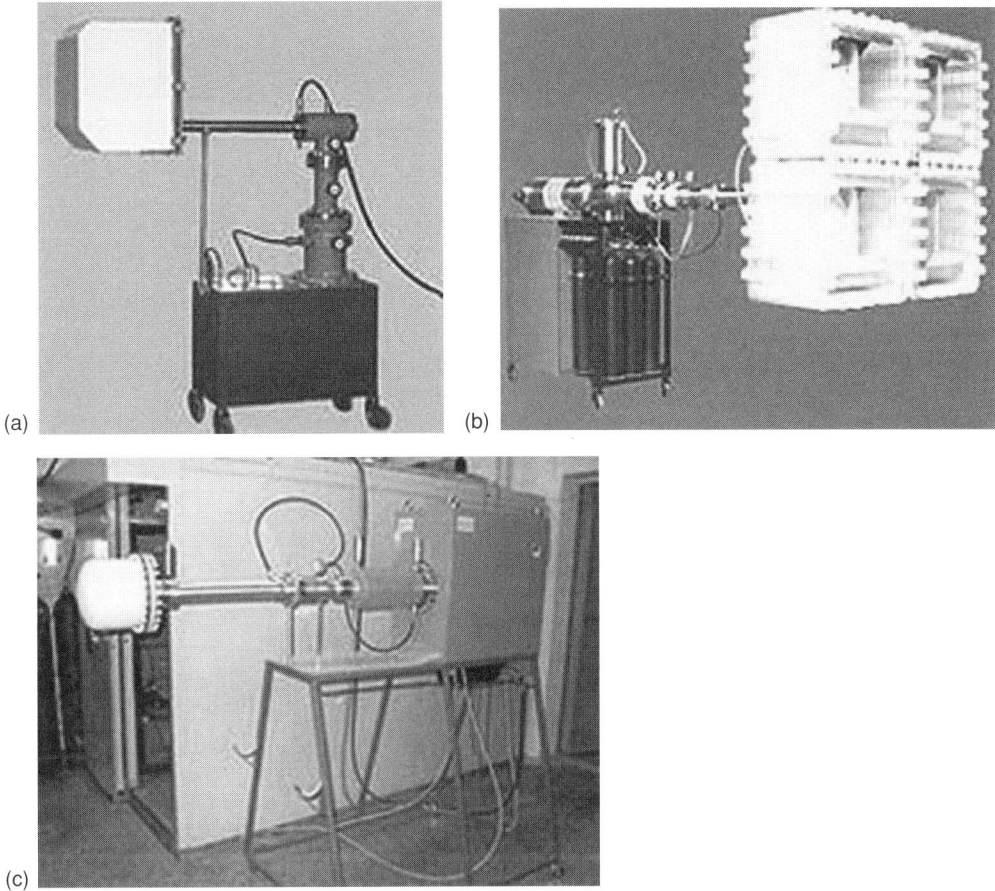

FIGURE 15.9
Compact ultra-wideband generators of directional pulse electromagnetic radiation with power output of 100 to 1000 MW (Institute of High-Current Electronics, Russia).

suitcase with a range possibly as high as 500 m can be purchased on the Internet at the cost of only $100,000. Terrorists could use a less expensive, low-tech approach to create the same destructive power. "Any nation with even a 1940s technology base could make them," says Carlo Kopp, an Australian-based expert on high-tech warfare. The threat of E-bomb proliferation is very real. *Popular Mechanics* estimates a basic weapon could be built for $400.

Nowadays there are no measures preventing the distribution of electronic weapons. Even if agreements on limitation of distribution of electromagnetic weapons are reached, they will not be able to solve the problem of accessibility of required materials and equipment. One cannot rule out the possibility of leakage of electromagnetic weapons technology from countries of the former U.S.S.R. to third world countries, or to terrorist organizations, as the former really face great economic difficulties. The danger of distribution of electromagnetic weapons is quite real.

Today it is possible to find finished drawings and descriptions of generators of directional high-frequency radiation based on household microwave ovens on the Internet (see: www.powerlabs.org, www.voltsamps.com, etc).

Problems of "electromagnetic terrorism" capable of causing man-caused accidents on a national scale similar to that which happened in New York in August 2003, were formu-

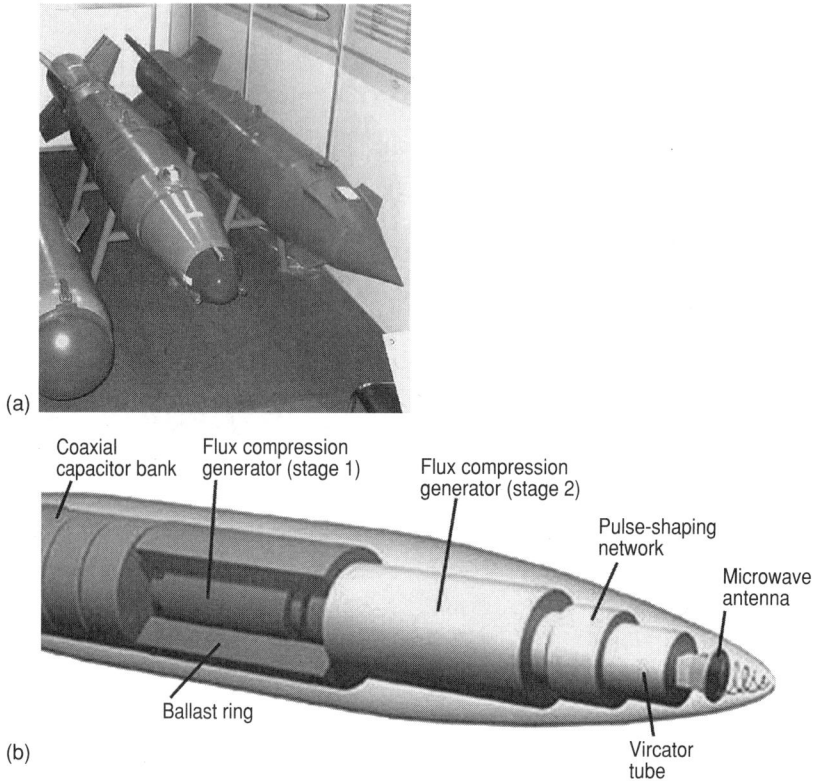

(a)

(b)

FIGURE 15.10
(a) Russian GPS-guided KAB-500S type electromagnetic bomb (right); (b) Typical construction of an E-bomb.

lated in an article by Manuel W. Wik (now chief engineer and strategic specialist on future defense science and technology programs at the Defence Materiel Administration, Stockholm) "Electromagnetic terrorism — what are the risks? What can be done?," Published in 1997 in the *International Product Compliance Magazine*. Here is what that article says on the subject:

> Although electromagnetic terrorism is not often discussed in public, as it is potentially an extremely sensitive issue, there needs to be wider public awareness of the threats posed and a better understanding of the consequent risk-management strategies required. Nevertheless, with the gradual development of smaller equipment that can be used to produce short, intense EMPs capable of damaging the controls of much electronic equipment, electromagnetic terrorism is increasingly something that needs to be considered during the compliance-planning route. Thus, although it is important that neither the details of electromagnetic (EM) interaction with particular systems nor specific vulnerabilities should be made public, public awareness of the potential threats and, indeed, a better understanding of the relevant risk-management strategies need to be more widely disseminated. Electromagnetic terrorism (EM terrorism) is the international, malicious generation of electromagnetic energy, introducing noise or signals into electric and electronic systems, thus disrupting, confusing or damaging these systems for terrorist or criminal purposes. EM terrorism can be regarded as one type of offensive information warfare. EM terrorism needs to be considered more carefully in

the future because information and information technology are increasingly important in everyday life.

Electronic components and circuits, such as microprocessors, are working at increasingly higher frequencies and lower voltages and thus are increasingly more susceptible to EMI. At the same time, there have been rapid advances in RF sources and antennae and there is an increasing variety of equipment capable of generating very short RF pulses that can disrupt sophisticated electronics. Intentional EMI poses a significant threat worldwide. Until recently, industry has been resistant to addressing the issue, but the International Electrotechnical Commission (IEC) is beginning to develop methods to fight criminal EMI.

The possibility of intentional EMI has come under the scrutiny of the United States Congress. Representative Jim Saxton of New Jersey and Representative Roscoe Bartlett of Maryland have held several investigations concerning this threat and have lobbied Congress for funds for appropriate research. As early as February 1998, Saxton began holding hearings on the proliferation and threat of RF weapons.

The issue of intentional EMI has also begun to be addressed at international conferences. The 1999 International Zurich Symposium on EMC held the first workshop on intentional EMI, with nearly 200 people in attendance. The 2001 Zurich Symposium was the culmination of several years of work in the field of intentional EMI. This symposium included the first refereed session on intentional EMI. The threat of intentional EMI is not limited to RF energy. Most of the emphasis in this area has been on RF fields but the issue of injecting directly into power and telecom systems has been overlooked. Yuri Parfenov and Vladimir Fortov, of the Russian Academy of Sciences Institute for High-Energy Densities, recently experimented with injection of disturbances into power lines outside a building and found that the signals penetrate very easily and at a high-enough voltage to cause damage to computers inside the building. Additionally, radiated fields often become a conducted threat due to coupling of RF energy to exposed wires.

It is astonishing that numerous research projects devoted to EM terrorism are concerned with the EMI impact on such objects as communication systems, telecommunications, air planes, computers, but there are practically no projects devoted to investigation of resistance of microprocessor-based relays to EMI, malfunctioning of which can lead to high consequences. However, it is obvious without any investigations that microprocessor-based relays are more prone to EMI impact than electromechanical and even analog electronic ones.

In addition, it turns out that "electromagnetic terrorism" is not the only form of modern remote terrorism to which microprocessor-based relays are prone. There are also electronic intrusions called cyber-attacks. A cyber intrusion is a form of electronic intrusion where the attacker uses a computer to invade electronic assets to which he or she does not have authorized access. The IEEE defines electronic intrusions as:

> Entry into the substation via telephone lines or other electronic-based media for the manipulation or disturbance of electronic devices. These devices include digital relays, fault recorders, equipment diagnostic packages, automation equipment, computers, PLCs, and communication interfaces.

A cyber-attack can be an intrusion as described above, or a denial of service attack (DOS) where the attacker floods the victim with nuisance requests and/or messages to the extent that normal services and functions cannot be maintained. A DOS attack is also called a flood attack. A distributed DOS attack (D-DOS) is a flood attack launched simultaneously from multiple sites.

Tools for attacking computer-based control equipment by telephone and network connection are free and widely available over the Internet. There are literally dozens of Websites devoted to hacking, usually providing downloadable programs or scripts to help the novice hacker get started.

Nowadays hackers' attacks are becoming terrorist weapons. Real cases of terrorist attacks of this kind are usually kept secret, but some are already known. For example, an attempt to damage the Israeli power system with the help of a hacker's attack was prepared by the "Special Services" of Iran for several months in 2003. Fortunately, the security service of the Israel Electric Corp. managed to block these attacks. As attacks of this kind to the main national computer systems of Israel have become more frequent, within Israeli Counter-Intelligence and Internal Security Service (SHABAK) there is a special subdivision for counteraction to such attacks.

But this problem is not only actual for Israel. The North American electric power network is vulnerable to electronic intrusions (a.k.a. cyber-attacks) launched from anywhere in the world, according to studies by the White House, FBI, IEEE, North American Electric Reliability Council (NERC), National Security Telecommunications Advisory Committee (NSTAC) KEMA, Sandia National Laboratories. At the heart of this vulnerability is the capability for remote access to control and protection equipment used by generation facilities and Transmission and Distribution (T&D) utilities. Remote access to protective equipment historically has been limited to proprietary systems and dedicated network connections. Now, however, there is an increased use of public telephone services, protocols, and network facilities, concurrent with a growing, more sophisticated, worldwide population of computer users and computer hackers which is why special services of many countries had to create special subdivisions to fight this dangerous phenomenon. In Russia, in particular, it is the Federal Agency of Governmental Communication and Information (FACI) and "Atlas" Scientific-Technical Center of Federal Security Service (FSB) that tackles these problems.

Is there a solution for this situation?

Probably yes, if:

- We completely replace all electric wires connected to microprocessor relays, including current and voltage circuits, with nonconductive fiber-optical wires.

- Use opto-electronic CT and VT, instead of traditional instrument transformers.

- Provide full galvanic separation from the power electric network by using a power supply of microprocessor relays to carry through the unit "motor generator."

- The relay should be placed in a completely closed metal case made with a special technology, used for ultrahigh frequencies in which there are no other kinds of the electric equipment.

This is the price necessary to pay for progress in the field of relay protection.

15.4 Summing Up

Some conclusions in brief:

1. Did microprocessor-based relays introduce any new functions for relay protection that were unknown before or impossible to implement with the help of traditional relays? On closer examination, it appears that the answer is NO.

Microprocessor-based relays only combined features of some relays adding some functions that used to be carried out by registration devices.

2. Do microprocessor-based relays provide a higher level of reliability of power supply? NO!

3. Did microprocessor-based relays make the work of the maintenance staff simpler? Obviously NO!

4. Do microprocessor-based relays have any uncontestable advantages? Again the answer would appear to be NO! Microprocessor relays have appeared as a result of developments in microcontrollers and not in order to improve conventional (static or electromechanical) relays. The behavior of conventional relays in operation continues to be excellent. Why do we need to make our life more complicated by using microprocessor-based relays, which on the one hand have no essential advantages in comparison with traditional ones, and on the other hand have many of their own unsolved problems?

It turns out that there is an important reason to use microprocessor-based relays, however it does not lie in the power industry field, but in the field of relay production (Schleithoff F.S. *Statischer Schutz im Mittelspannungsnetz,* "Elektrizitatswirtschaft," 1986, 85, No. 4, pp. 121–124). It appears that it is much more profitable to produce microprocessor-based relays than electromechanical or even analog electronic ones. This is explained by the possibility of complete automation of all technological processes and production and control of parameters of microprocessor-based relays. The following question is to the point here: Where do problems of producers concern development of correct technical politics in the power industry field? In fact, the largest international concerns, such as ABB, General Electric, Siemens, Alstom have become "trendsetters" in the power industry and now determine main tracks of development not only of relay protection, but also of the whole power industry. If in some years these companies stop producing all other types of relays except for microprocessor-based ones (and this is the main tendency today), this fact will not justify uncontestable advantages of such relays from the point of view of interests of power suppliers and of the whole society.

5. The transition to microprocessor relays (if inevitable!) should be complete, that is excluding teamwork with electromechanical relays, such transition should be carried out together with the replacement of traditional instrument transformers to optical, and full replacement of all electric wires connected to the relay to isolated optical wires. Microprocessor relays should be mounted in closed metal cases made with use of high-frequency technology. Relay power supplies should be carried out through the unit's "motor generator."

To neglect these requirements could lead to serious problems in the electric power industry in the near future. So we can see that indeed there are many new problems still not known in world of electromechanical relays.

16

Special Relays

16.1 Polarized Relays

A polarized relay is a sort of direct current (DC) electromagnetic relay with an additional source of a permanent magnetic field affecting the relay armature. This additional source of the magnetic field (called "polarizing") is usually made in the form of a permanent magnet.

Polarized relays have been known since the time of the first sounders (Figure 16.1). Peculiarities of polarized relays are, first, polarity of the winding switching and, second, very high sensitivity. The latter can be explained by the fact that the permanent magnet creates a considerable part of the magnetic flux required for the relay pick up, which is why the relay winding causes only a small additional magnetic flux, which is much less than in standard relays. Some types of polarized relays *can pick up from a signal of a few millionths Watts.*

The magnetic circuit of the polarized relay is, of course, more complex than the magnetic circuit of a standard electromagnetic relay (Figure 16.2), and such a relay is also more expensive than a standard one.

The magnetic flux (ϕ_m) of the permanent magnet passes through the armature of the relay branches into two parts: the flux (ϕ_1) passes through the left working gap and the flux (ϕ_2) passes through the right working gap. If these are the only magnetic fluxes (and there is no current in the coil), the armature of the relay will be to the left or to the right of the neutral position, since a neutral position is not stable in such a magnetic system.

When current appears in the windings w_1 and w_2, an additional magneto-motive force and the working magnetic flux (ϕ) both pass through the working gap. The force affecting the armature depends on:

- The current value in the winding
- The power of the magnet
- The initial position of the armature
- The polarity of the current in the winding
- The value of the working gap

At certain combinations of these parameters the armature of the relay turns to a new stable state, closes the right contact and the relay picks up.

There are several types of magnetic systems of polarized relays. The two most popular today are the *differential* and *bridge* types (Figure 16.3). In a relay with a differential magnetic system, the magnetic flux of the permanent magnet passing through the

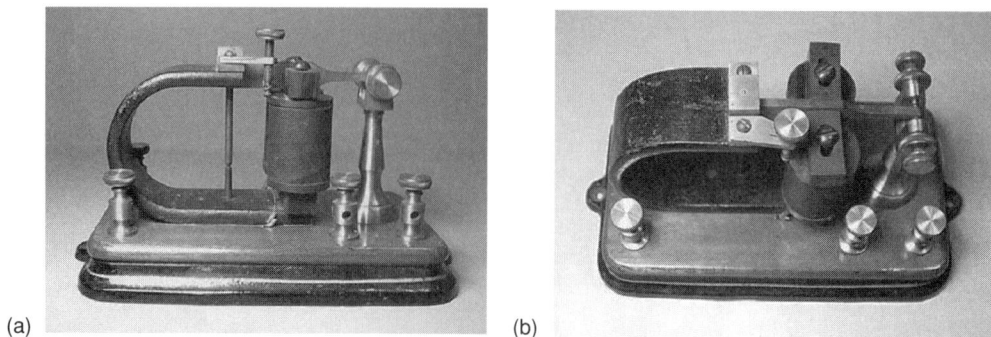

(a)　　　　　　　　　　　　　　　　　(b)

FIGURE 16.1
An ancient polarized relay of high sensitivity with a horseshoe-shaped permanent magnet (year approximately 1900).

armature of the relay is divided into two fluxes in such a way that in the left and in the right part of the working gap these magnetic fluxes are directed to opposite sides, that is the armature is affected by the difference of these two fluxes.

In the bridge magnetic system, the magnetic flux of the permanent magnet is not divided into two fluxes in the area of the working gap and the armature, but has only one direction. The field created by the coil is divided into two fluxes, which have opposite signs in the working gap area.

The first type of this magnetic system was widely used in polarized relays of normal size (Figure 16.4).

A later and modernized construction of this relay, produced in the 1970–80's, is shown in Figure 16.5.

There were also a lot of relays with bridge magnetic systems (Figure 16.6). One can affect the relay by adjusting the initial position of the armature of the polarized relay. The polarized relay is usually adjusted by screws with which one changes the position of the stationary contacts (Figure 16.7), and therefore the armature.

At neutral adjustment of the relay, when there is no current in the winding, the armature (together with the movable contact) remains in the position it has taken at pick up of the relay, that is in the right or left position. To switch to initial position one

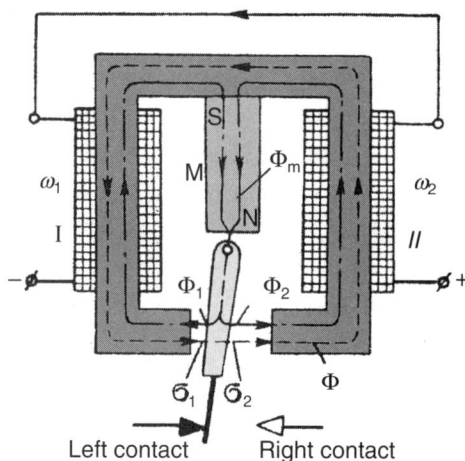

FIGURE 16.2
Simplified magnetic circuit of a polarized relay. M — Permanent magnet; ϕ_m — polarizing magnetic flux of the permanent magnet; ϕ_1 — magnetic flux of the left gap σ_1 between the armature and the magnetic core; ϕ_2 — magnetic flux of the right gap σ_2.

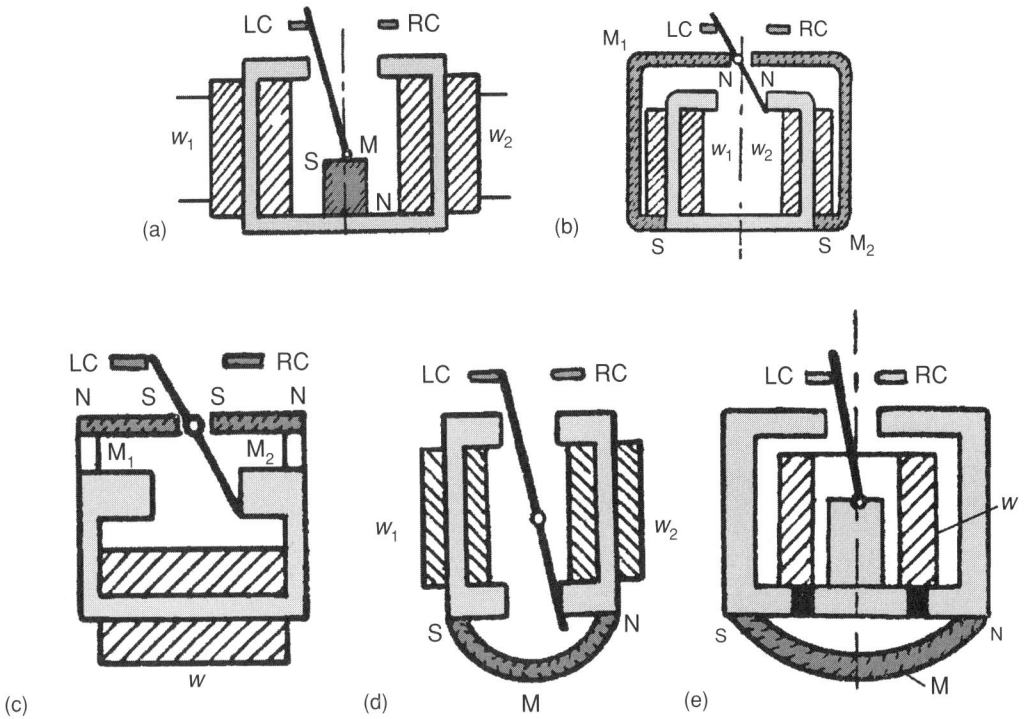

FIGURE 16.3
Schemes of the magnetic circuit of a so-called "differential" (a, b, c) and "bridge" (d, e) type. w — Coil; M — permanent magnet; LC — left stationary contact; RC — right stationary contact.

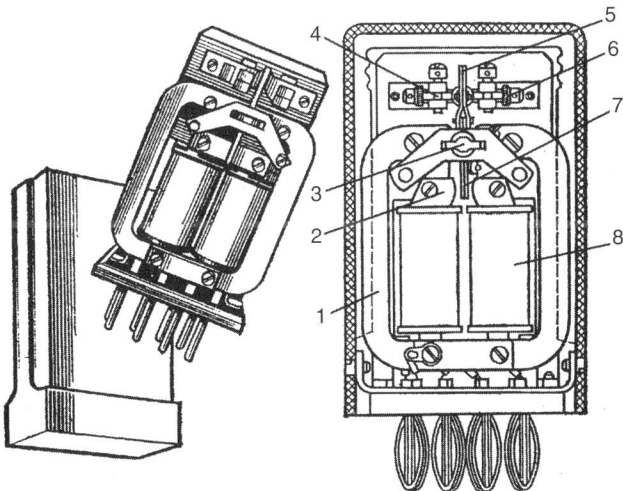

FIGURE 16.4
A polarized relay of the RP series with a differential magnetic system (corresponding to Figure 16.3b), produced in Russia in the 1950s. 1 — One of the two permanent magnets; 2 — pole of the core; 3 — rotation axis of the armature; 4 and 6 — adjusting screws of stationary contacts; 5 — movable contact fixed on the armature; 7 — armature; 8 — coils.

FIGURE 16.5
Construction of a polarized relay of the RP series, produced in 1970–80's (Russia). 1 — Heel piece with outlets of the relay; 2 — case from silumin (aluminum alloy); 4 — magnetic core; 5 — coil; 6 — pole lugs; 7 — steel insert; 8 — permanent magnet; 9 — fastening screws; 10 — armature; 11 — cramp; 12 — plat spring; 13 and 15 — fastening screws; 14 — ceramic board; 16 and 21 — screws adjusting the position of stationary contacts; 17 and 20 — lock screws; 18 — springs of the movable contact; 19 — movable contact.

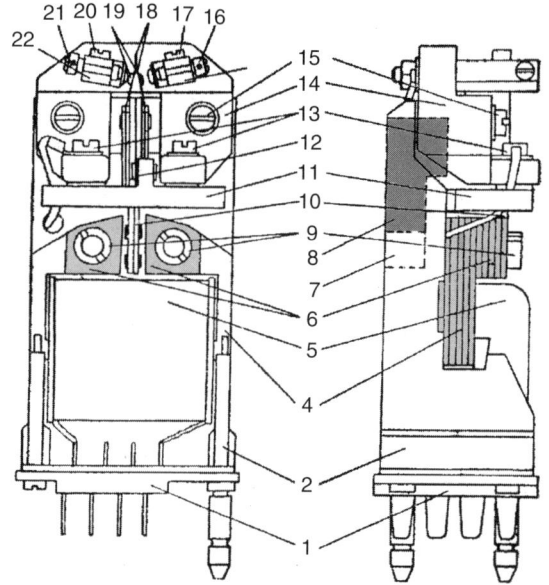

FIGURE 16.6
A relay with an aluminum case (off) produced by Sigma (1960–70's). The magnetic system is of the bridge type (Figure 13.3d). Size: 51 mm in diameter, the full length is 82 mm. 1 — Coil; 2 — armature.

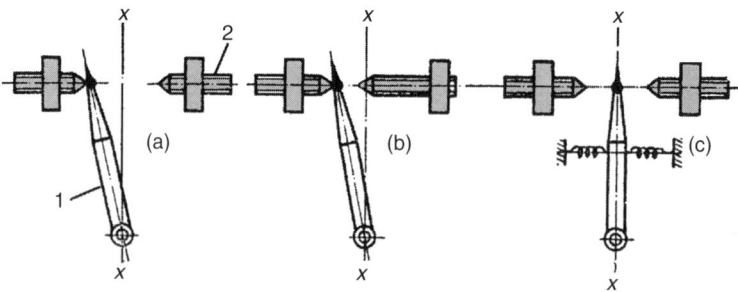

FIGURE 16.7
Types of adjustment of initial position of the armature of the polarized relay: (a) Neutral; (b) with predominance; (c) three-position. 1 — Armature with a movable contact at the end; 2 — adjusting screw with a stationary contact at the end.

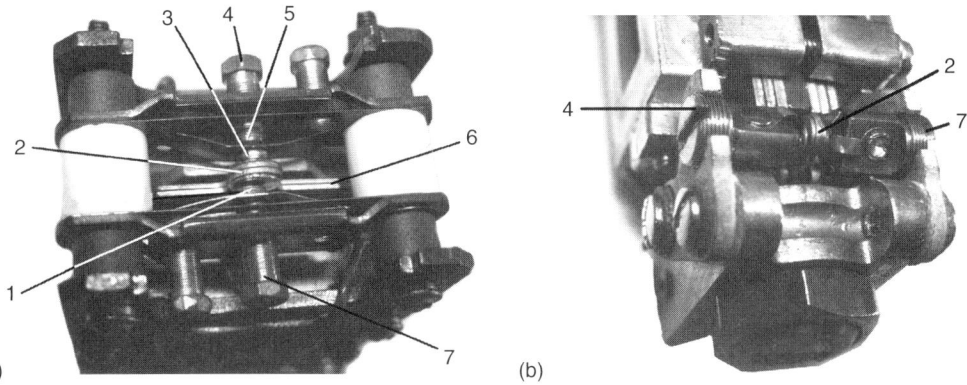

FIGURE 16.8
Construction of adjusting unit (enlarged) of different types of polarized relays, produced by Sigma (1960–70's, U.S.A.). 1 and 3 — Stationary contacts; 2 — movable contact placed on the end of the armature; 4 and 7 — adjusting screws for adjustment of initial position of stationary contacts and the armature; 6 — armature.

must apply voltage of the opposite polarity to the winding (if there is only one winding), or to the second winding (if there is one).

In this modification, there is no need for a restorable spring which also increases sensitivity of the relay. If self-reset of the relay to the initial position is required after the winding supply has been switched OFF, one uses adjustment with predominance to one of the poles by twisting in one of the adjusting screws, shifting the proper stationary contact outside the neutral line (Figure 16.7b).

In relays with three-position adjustment, the armature in the currentless state is retained in the intermediate position between the contacts (the first position) with the help of a spring (Figure 16.7c). When current flows in the winding, the relay switches to the left (the second position) or to the right (the third position), depending on the polarity of the voltage in the winding.

Such principle of adjustment of polarized relays is used by practically all producers of such relays (Figure 16.8). As it can be seen from Figure 16.8, in relays produced by Sigma contacts of the hard type have been used. This is the simplest, but not the only variant of contacts of polarized relays. In such relays they often use contacts of special types (Figure 16.9).

The main purpose for construction of these contacts is suppressing shock of movable contacts to stationary ones, due to the fact that kinetic energy is used for friction of flexible springs. This may be friction of spring ends (a) or of the additional spring covered by

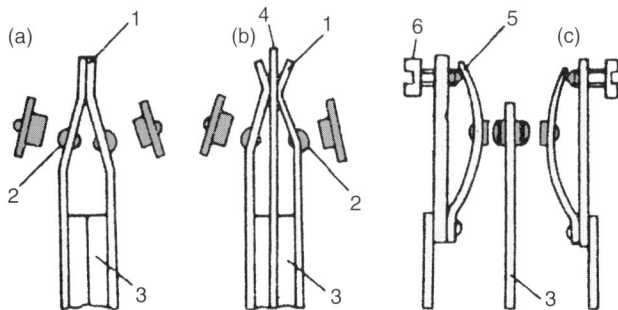

FIGURE 16.9
Contacts of special type, which are frequently used in polarized relays. (a) Flexible movable contact consisting of two springs and rigid stationary contacts; (b) flexible movable contact consisting of three springs and rigid stationary contacts; (c) rigid movable contact and flexible stationary contacts. 1 — Movable contact springs; 2 — contact straps; 3 — the end of the armature; 4 — additional auxiliary spring; 5 — stationary contact springs; 6 — supporting screws.

special material (b) or friction of the spring ends by stationary supporting screws (c) with the help of which the contact system can be adjusted to the moment until the bounce disappears fully.

16.2   Latching Relays

A latching relay is one that picks up under the effect of a single current pulse in the winding, and remains in this state when the pulse stops affecting it, that is when it is locked. Therefore this relay plays the role of a memory circuit. Moreover, a latching relay helps to reduce power dissipation in the application circuit because the coil does not need to be energized all the time (Figure 16.10).

As illustrated in Figure 16.10, the contacts of a latching relay remain in the operating state even after an input to the coil (set coil) has been removed. Therefore this relay plays the role of a memory circuit. As shown in Figure 16.10, the double-coil latch type relay has two separate coils each of which operates (sets) and releases (resets) the contacts.

In latch relays, two types of latching elements are usually used: magnetic and mechanical. A relay with magnetic latching elements is a polarized relay with neutral adjustment (see above). Unlike polarized relays, latching relays are not designed to be used as highly sensitive ones. Sometimes it is impossible to distinguish between a polarized relay and a latching relay, so all relays with a permanent magnet are just called "polarized relays."

Perhaps this is not correct, since the main quality of the polarized relays considered above is their high sensitivity. This determines the field of application of such relays. As far as latching relays are concerned, they do not possess extraordinarily high sensitivity (we cannot really speak about high sensitivity meaning the latching contactor capable of

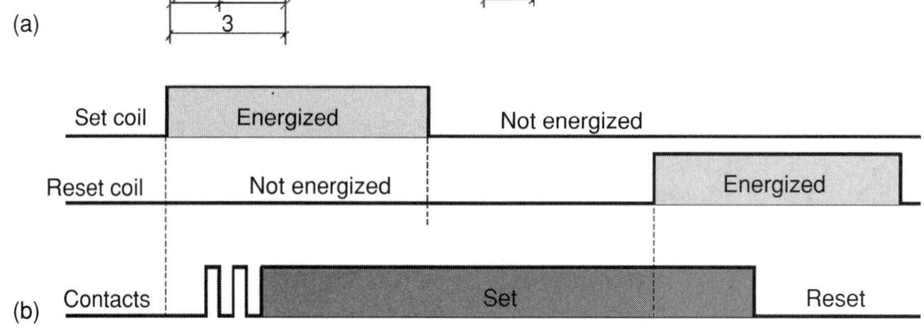

FIGURE 16.10
(a) Time chart of nonlatch relay. 1 — Current-rise time; 2 — bounce time; 3 — full operate time; 4 — release time.
(b) Time chart of double coil latch relay.

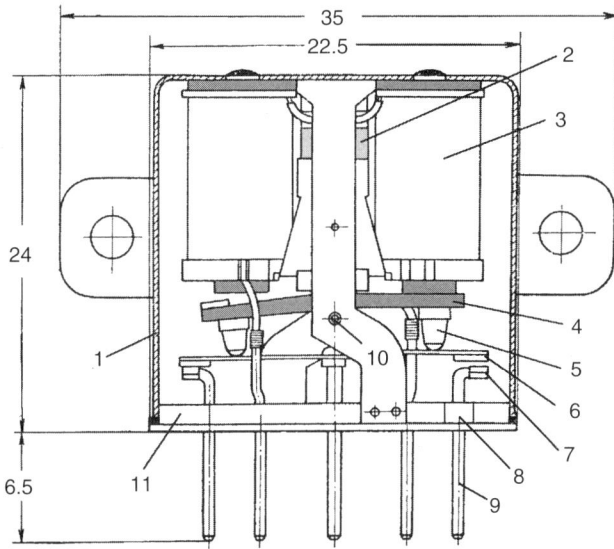

FIGURE 16.11
Construction of a miniature latch relay with magnetic latching of the RPS20 type (Russia). 1 — Hermetic brass case; 2 – permanent magnet; 3 — coil; 4 — flat symmetrical armature; 5 — pushers; 6 — movable contact; 7 — stationary contact; 8 — glass bushing; 9 — output prongs; 10 — rotation axis of the armature; 11 — heel piece.

switching currents of hundreds of Amperes). They are designed to be used when switching circuits affected by single pulse control signals and increased resistance to shocks and vibrations, preventing permanent consumption of energy from the power source, as elements of memory, etc.

In addition, the principle of their operation does not imply only *magnetic latching* of the position, and rules out a considerable part of such relays from the class of polarized relays. Like standard electromagnetic relays, latching relays are produced for all voltage and switched power classes: from miniature relays for electronics, with contact systems and cases typical of standard relays of the same class according to switched power (Figure 16.11 and Figure 16.12), up to high-voltage relays and high-current contactors.

Magnetic systems of latching relays, as mentioned above, are not distinguished by high sensitivity and are constructed in such a way in order to simplify and to minimize the relay (Figure 16.12). Practically, all western companies producing relays also design and produce latching relays (Figure 16.13).

Lately the famous Russian company "Severnaya Zaria" has also taken up designing miniature latching relays. Two types of such relays constructed by this company are shown in Figure 16.14. The RPK61 type relay is a double-coil low-profile relay whose dimensions are $19 \times 19 \times 12$ mm. The RPK65 is a single-coil (with changeable polarity) latching relay whose dimensions are $9.53 \times 9.53 \times 6.99$ mm.

The smallest latching relays in the world in standard metal cases of low-power transistors, are produced by the American company Teledyne Relays (Figure 16.15). Latching relays produced by Omron, Deltrol, and some other companies, have similar construction and external design. In these relays another scheme of magnetic circuit with a rocking armature of the clapper type (Figure 16.16) is used. It is frequently applied not only in micro-miniature relays, but also in large-sized relays designed for industrial purposes and for the power industry (Figure 16.17 and Figure 16.18).

Manufacturers have noted that these relays typically have high vibration and shock resistance. The large-sized latching relay produced by ASEA designed for application in power industry is based on a similar principle (Figure 16.18). Recently high power latching relays have become very popular and are produced in great numbers by many companies (Figure 16.19).

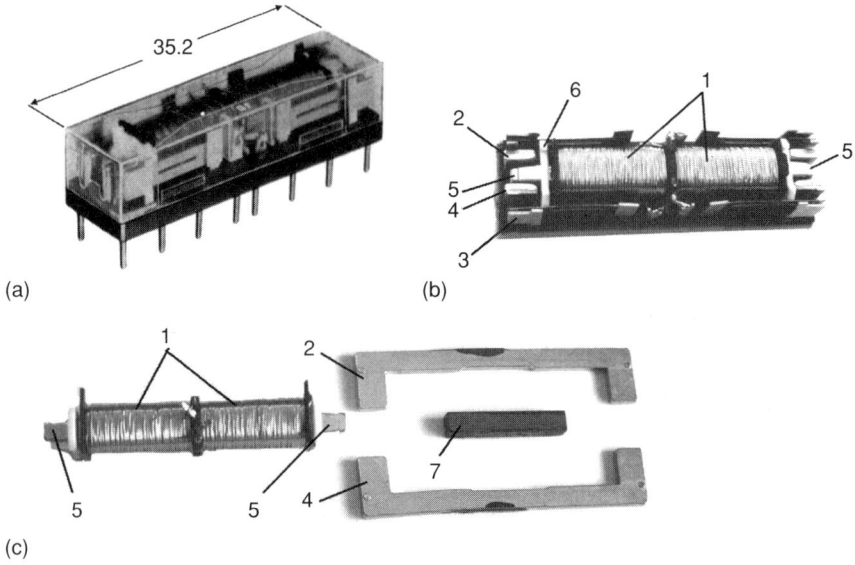

FIGURE 16.12
Construction of a miniature latch relay with a magnetic latching of the DS4 type, produced by Euro-Matsushita.
1 — Set and reset coils; 2 and 4 — plates of the magnetic core; 3 — contacts; 5 — ferromagnetic pole lugs;
6 — plastic pushers put on the pole lugs; 7 — permanent magnet placed in the centre of the coil.

So-called *throw-over relays* are considered to be a variant of latching relays (Figure 16.20). These relays have a scheme unusual for latching relays, of a magnetic circuit with a permanent magnet, but unlike the constructions considered above, they do not require pulse control signals from the control circuits. Current pulses for switching ON and OFF of the relay are formed by the relay itself due to the fact that its coil is supplied through

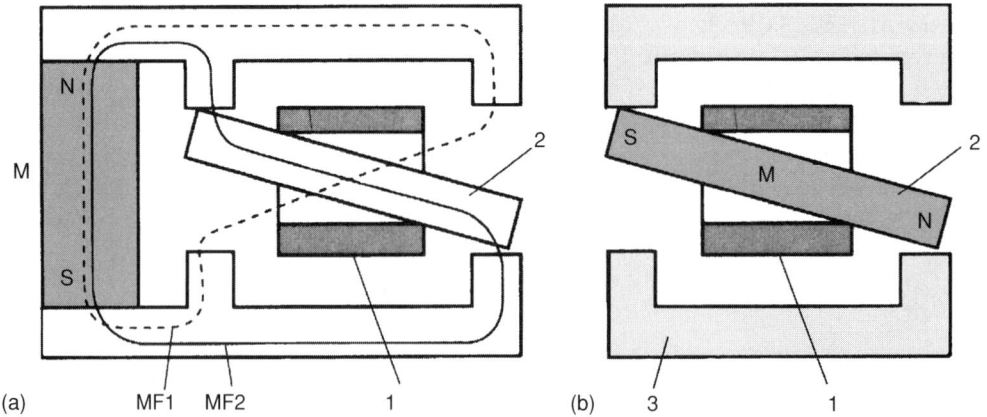

FIGURE 16.13
Popular types of the magnetic system of cheap miniature latching relays, in plastic cases produced by many companies. M — Permanent magnet; MF1 — magnetic flux in first position; MF2 — magnetic flux in second position; 1 — coil; 2 — rotating armature; 3 — yoke.

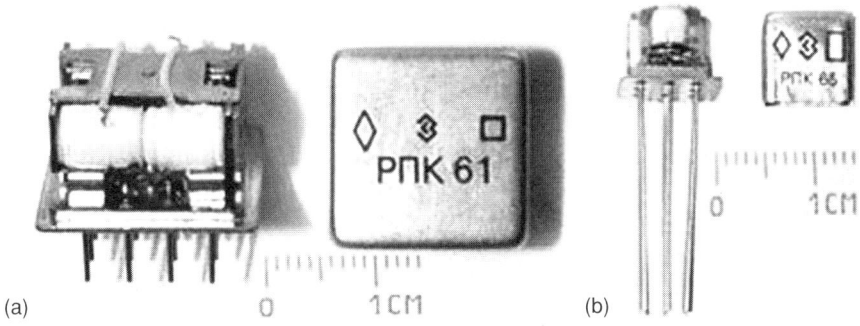

FIGURE 16.14
Miniature RPK61 type (a) and micro-miniature RPK65 type (b) latching relays, produced by the Russian company "Severnaya Zaria."

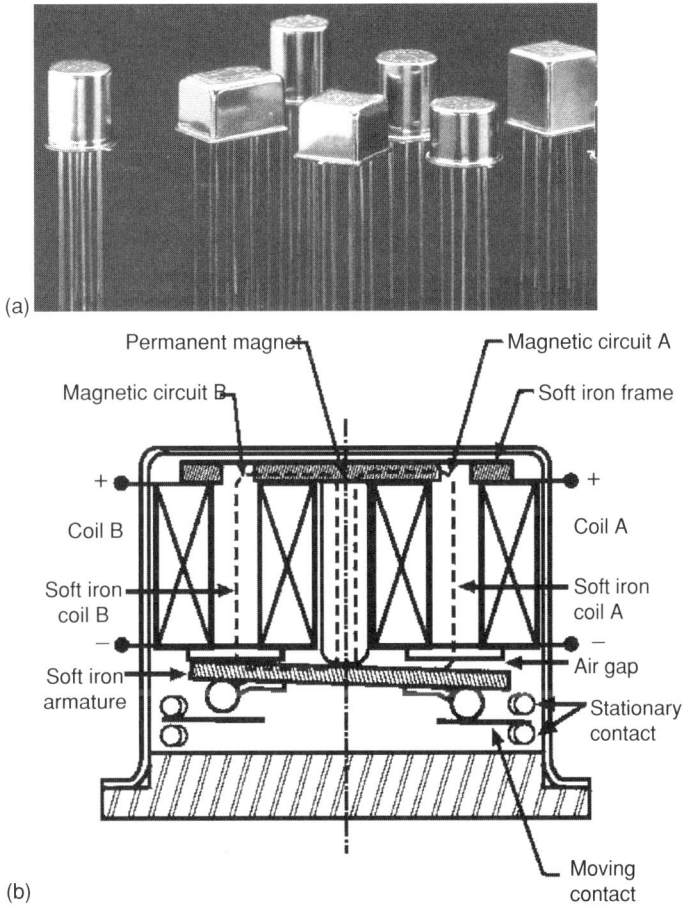

FIGURE 16.15
(a) The smallest latching relays in the world, produced by the American company Teledyne Relays. External design of relays in Centigrid[R] and TO-5 cases. (b) Construction of miniature latching relays produced by the Teledyne Relays (U.S.A.). (Teledyne 1999.)

FIGURE 16.16
Variant of a scheme of a magnetic circuit, widely used in production of micro-miniature and also a big-sized relays. 1 and 2 — Coils; 3 — permanent magnet; 4 — armature; 5 — left contacts; 6 — right contacts.

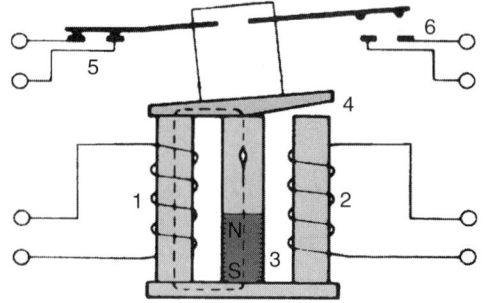

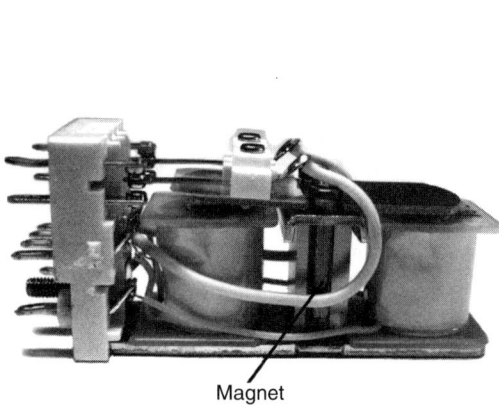

Magnet

FIGURE 16.17
Industrial latching relays of the RR2KP-U type (Idec Izumi Corp., Japan).

FIGURE 16.18
Large-sized multicontact latching relay of the RXMVB type (with the cover off) designed for application in power industry (ASEA). 1 and 5 — Coils; 2 — left contact system; 3 — armature clapper type; 4 — right contact system; 6 — permanent magnet.

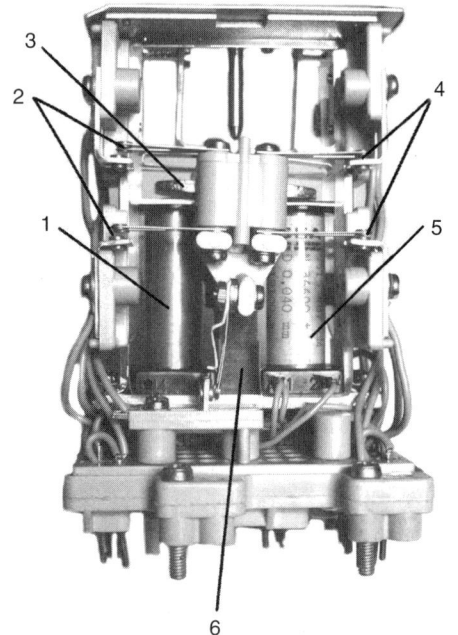

FIGURE 16.19
(a) A power latching relay with the possibility of manual reset for industrial applications, produced by Gruner AG (Germany) for switched currents from 10 to 200 A. (b) A power latching relay produced by KG Technologies (U.S.A.), for switched currents of 100 and 200 A. (c) A power latching relay produced by Babcock (U.S.A.), for nominal switched currents of 100 and 200 A.

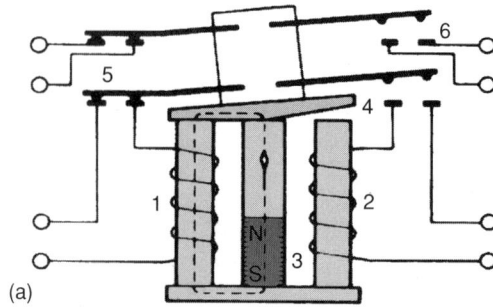

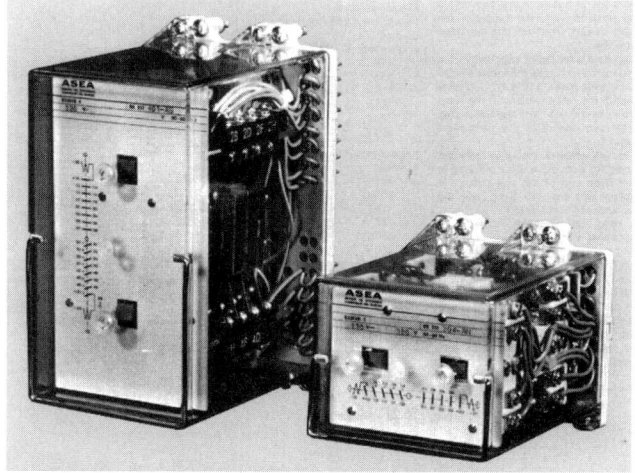

FIGURE 16.20
Throw-over relays, type RXMVB-2 and RXMVB-4 (ASEA). (ABB 1987 Technical Description B03-1510.)   (b)

normally closed (NC) contacts of the relay. As soon as voltage is applied to the input terminals of the relay, as in a standard relay, a current pulse, required for the pick up of the relay, is formed by relay contacts 4 and 5 (Figure 16.20).

After that, the supply circuit of the relay is automatically broken and the relay remains "ON," due to the restraining force of the permanent magnet. Relays type RXMVB-2 and RXMVB-4 are available for both DC and alternating current (AC) supply. The AC relay changes position during the first half cycle when the AC flows in such a direction that the force of the permanent magnet is overcome. The relays have indicating knobs which can also be used to operate the relay position manually.

Another popular type of latching relay with magnetic latching is the so-called *remanence* type. This relay consists of a coil and an armature made of a special ferromagnetic material on a nickel base, with admixtures of aluminum, titanium, and niobium, capable of becoming magnetized quickly under the effect of a single current pulse in the coil, and of remaining in its magnetized state when the pulse stops affecting it.

This type of relay contains a coil with one or two windings wound around it. In the first case, magnetization and demagnetization of the material of the core are carried out by current pulses of opposite polarities, and in the second case — by two different windings on the same bobbin, one of which magnetizes the core while the other one, which demagnetizes it, is a disabling one. The advantage of type of relay is that it does not require any special construction. One has only to make a core in the already existing construction of a standard relay of any type from remanent material, and the latching relay is complete!

This is why these relays typically have very similar designs to standard electromagnetic relays, and actually can exist in the design of any relay — large and small. The only problem that arises in application of such relays is the danger of remagnetization of the

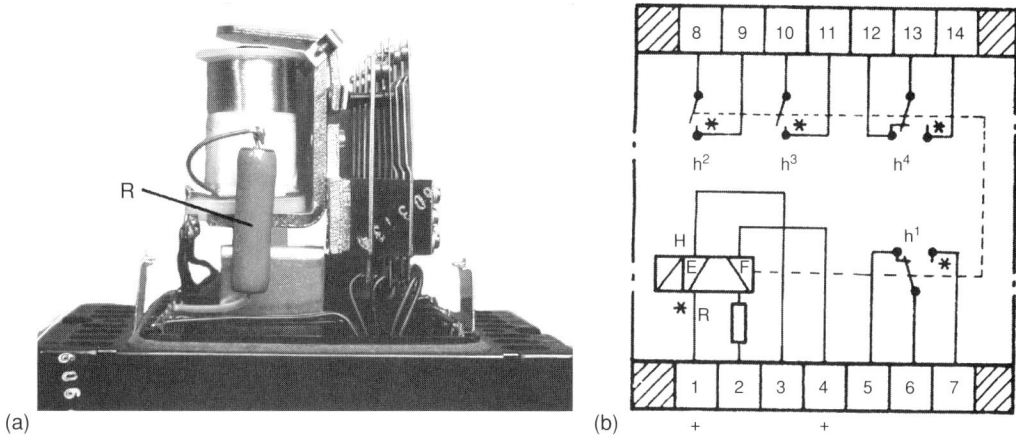

FIGURE 16.21
Remanence relay of the RHH-114 type (cover removed) with additional resistor (R), connected in series with a demagnetized (reset) coil. (AEG Haftrelais RHH-114.)

core by the reset winding if the magnetic field created by it coincides with the field of the set winding by value. In this case, the relay will be switched OFF for a short period of time and then it will be switched ON again.

For normal work it is required that the magnetic field of the reset winding be just strong enough for demagnetization of the core and no more, in order to prevent remagnetization. In order to make application of the relay easier for the user, these windings are made differently, or the reset winding is connected to an additional resistor built-in to the case of the relay (Figure 16.21). In this case at the same voltage value applied to set and reset, windings of different magnetizing forces will affect the core.

Miniature single-coil latching relays for operating voltage of 3 to 5 V are found in many applications, including signal routing, audio, and automotive systems. In these relays coil current must flow in both directions — through a single coil (Figure 16.24a). Current flowing from pin 8 to pin I causes the relay to latch in its reset position, and current flowing from pin I to pin 8 latches the relay in its set position.

A simple integral circuit of the MAX4820/4821 type (+3.3/+5 V, eight-channel, cascadable relay drivers with serial or parallel interface), produced by Maxim Integrated Products Ltd, driving up to four such single-coil (and also four ordinary dual-coil) latching relays, includes a parallel-interface relay driver (UI) with open-drain outputs (Figure 16.24b), and inductive-kickback protection. Latch any of the four relays to their set or reset positions by turning on the corresponding output (OUTX). That output is selected by asserting its digital address on pins A2 to A0 while CS is high. Activate the output by toggling CS. Both devices feature separate set and reset functions that allow the user to turn ON or OFF all outputs simultaneously with a single control line. Built-in hysteresis (Schmidt trigger) on all digital inputs allows this device to be used with slow rising and falling signals such as those from opto-couplers or RC power-up initialization circuits.

The MAX4820 features a digital output (DOUT) that provides a simple way to daisy chain multiple devices. This feature allows the user to drive large banks of relays using only a single serial interface (Figure 16.24c).

The principle of magnetic latching is widely used in many constructions of high- and low-power high-voltage relays (Figure 16.22 and Figure 16.23).

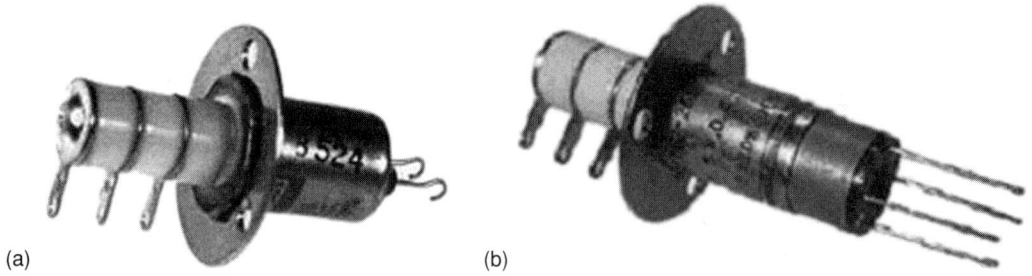

(a)   (b)

FIGURE 16.22
RFID-26S, 2 kV (a) and RF1J-26N, 4 kV (b) high-voltage relays, by Jennings Co. (U.S.A.).

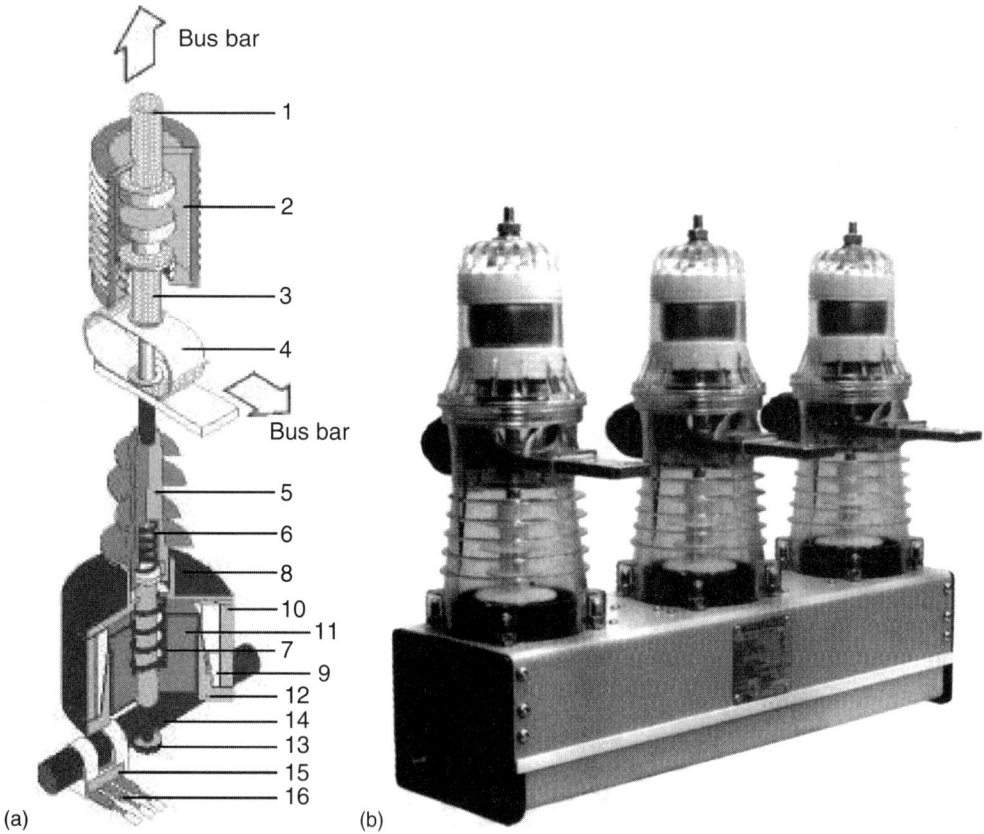

(a)   (b)

FIGURE 16.23
Magnet-latching power circuit breaker of the BB/TEL type (Tavride Electric, Russia). 12 kV; 1000 A; operating time 25 msec; 1 — stationary contact of the vacuum arc-suppressing chamber; 2 — vacuum arc-suppressing chamber; 3 — movable contact of the vacuum arc-suppressing chamber; 4 — flexible current-carrying bus bar; 5 — traction insulator; 6 — spring of compression; 7 — disabling spring; 8 — upper cover; 9 — coil; 10 — annular magnet; 11 — armature; 12 — lower cover; 13 — plate; 14 — shaft; 15 — permanent magnet; 16 — reed switches.

The suppressing chamber of the circuit breaker (Figure 16.23) is provided due to the fact that the reset spring (7) affects the movable contact (3) through the traction insulator (5). When the "ON" signal is given, the control unit of the circuit breaker forms a voltage pulse of positive polarity, which is applied to the coils (9) of the electromagnets. In the gap of the magnetic system, an electromagnetic attractive force appears and as it increases, it overcomes the power of the disabling spring (7) and the compression spring (6).

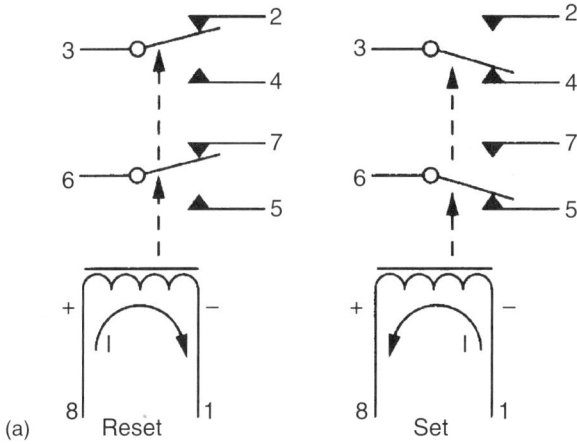

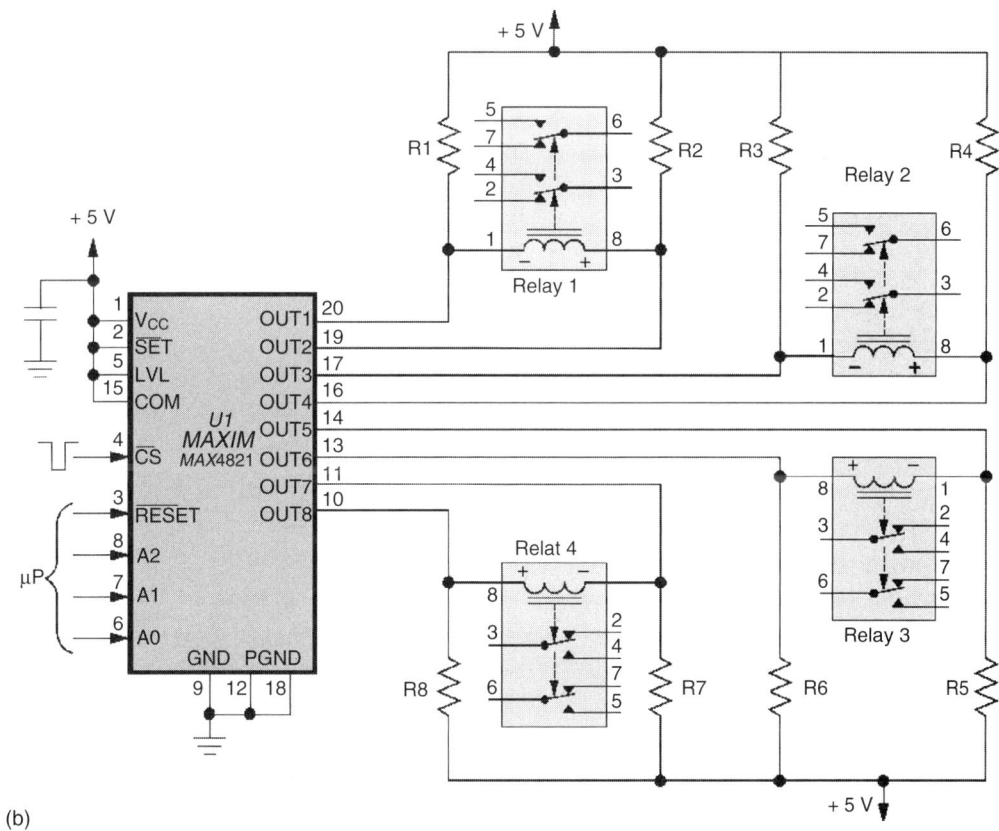

(a) Reset   Set

(b)

(Continues)

FIGURE 16.24 (*Continued*)
(a) Principle of operation of a miniature single-coil latching relay. The AROMAT AGN210A4HZ relay, for example, has such a principle. (b) The MAX4821 integral circuit easily drives four, single-coil latching relays. (c) Daisy-chain configuration for MAX4820. (MAXIM 2004.)

As a result of the impact of difference of those forces, the armature of the electromagnet (11) together with the traction insulator (5), and the movable contact (3) of the vacuum chamber (2), begins moving in the direction of the stationary contact (1), contracting the disabling spring (7). After the two main contacts close, the armature of the electromagnet continues moving up, contracting the compression spring (6). The movement of the armature continues until the working gap in the magnetic system of the electromagnet equals zero, thus the circuit breaker turns to the magnetic lock. Thus, the control energy for holding contacts 1 and 3 in a closed position is not consumed.

In the process of resetting of the circuit breaker plate 13, fitting the notch of shaft 14, turns this shaft by shifting the permanent magnet (15) installed on it and providing pick up of the reed switches (16), switching external auxiliary circuits.

When the "OFF" signal is applied, the control unit forms a current pulse, which has an opposite direction with respect to the making current, and smaller amplitude. The magnet (10) is demagnetized and the drive is unlocked. Affected by the energy accumulated in the disabling spring (7) and the spring of compression (6) the armature (11) moves down, striking the traction insulator (5) connected with the movable contact (3). The contacts 1 and 3 open and the circuit breaker disables the load.

Another variant of latch relays is a relay with mechanical latching. These are not as widespread as relays with magnetic latching, and are considered to be less reliable because of a mechanical unit that can wear out with time and get out of order. Nevertheless such relays are produced by a number of companies. The principle of construction of the mechanical latch is quite simple and sometimes quite original (Figure 16.25).

In Figure 16.25a, a latching relay with mechanical elements of blocking (metal pins) in its two positions is shown. When the relay is switched, the ends of the metal pins connected to the armature of the relay change places. Quite an original solution! In Figure 16.25b, one can see a latching relay with a plastic latch (placed in the center of the relay), made in the form of a tooth into which the curved metal plate jumps as the relay is switched.

The high-voltage latch relay produced by the Ross Engineering Corp. (Figure 16.26) is also made with a mechanical latching. So-called *lock-out relays* (Figure 16.27) are a variant of latching relays with mechanical blocking. The HEA type relays are applicable where it is desired that a number of operations be performed simultaneously. Some of the functions that can be performed by these relays are: tripping the main

FIGURE 16.25

(a) Principles of construction of latch relays with mechanical latching (Liberty Controls). (b) A latching relay with a plastic latch element (in center of relay) and two coils (for set and reset).

circuit breaker of a system, operating an auxiliary breaker and other relays, which, in turn, perform various functions. Another important use of the HEA type relays is in conjunction with differential relays, which protect transformers, rotating apparatus, buses, etc.

The HEA type relay is a high-speed, multicontact, hand reset auxiliary relay, provided with a mechanical target which indicates whether it is in the tripped or reset position. The HEA63 type hand or electrically reset auxiliary relay has the addition of a rotary solenoid that is used to electrically reset.

The current closing rating of the contacts is 50 A for voltages not exceeding 600 V. The contacts have a current-carrying capacity of 20 A continuously, or 50 A for 1 min. The interrupting rating of the contacts varies with the inductance of the circuit.

There are also latching relays of electronic types. The simplest type is a thyristor switched to a DC circuit. As has been noted above, this thyristor, opened by a pulse control signal, also remains in the open position after the control signal has stopped to affect it. One can use a solid-state switch — a triac, for work in an AC circuit but in that case it is the control circuit that performs the function of latching. Electronic relays based on this principle are produced by a number of firms (Figure 16.28).

FIGURE 16.26
High-voltage latching relay of the B-1001-E type. (Ross Engineering 2004.)

The NLF Series relays provide a *Flip-Flop* latching function with optical isolation between the solid-state output and the control voltage. This is a solid-state encapsulated relay for switching 1 to 20 A, with up to 200 A inrush current. If voltage to the output is maintained, each time control voltage is applied the output changes state and latches. It is designed for industrial applications requiring rugged reliable operation and long silent operation. The zero voltage switching NLF2 can extend the life of an incandescent lamp up to ten times. The random switching NLF1 is ideal for inductive loads. When fully insulated female terminals are used on the connection wires, the system meets the requirements for touch-proof connections.

The solid-state output is located between terminals 1 and 2 (Figure 16.28a), and is normally open (NO) (or NC) without control voltage applied to terminals 4 and 5. When momentary or maintained control voltage is applied to terminals 4 and 5, the output

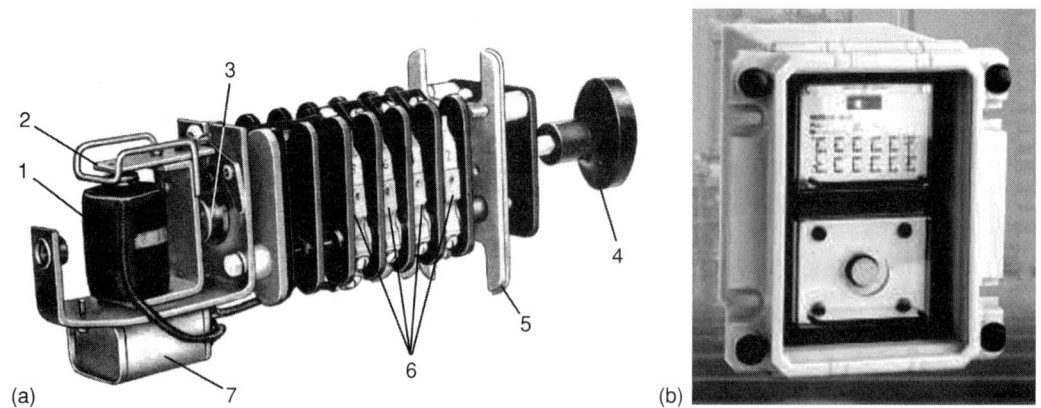

FIGURE 16.27
(a) A HEA series lock-out relay without cover (General Electric Co.). 1 — Coil; 2 — armature; 3 — latch mechanism; 4 — operating handle; 5 — front support; 6 — contacts; 7 — rectifier and resistor assembly.
(b) Lock-out relay of the RDB86 type with a new design. (General Electric Type HEA Auxillary Relays.)

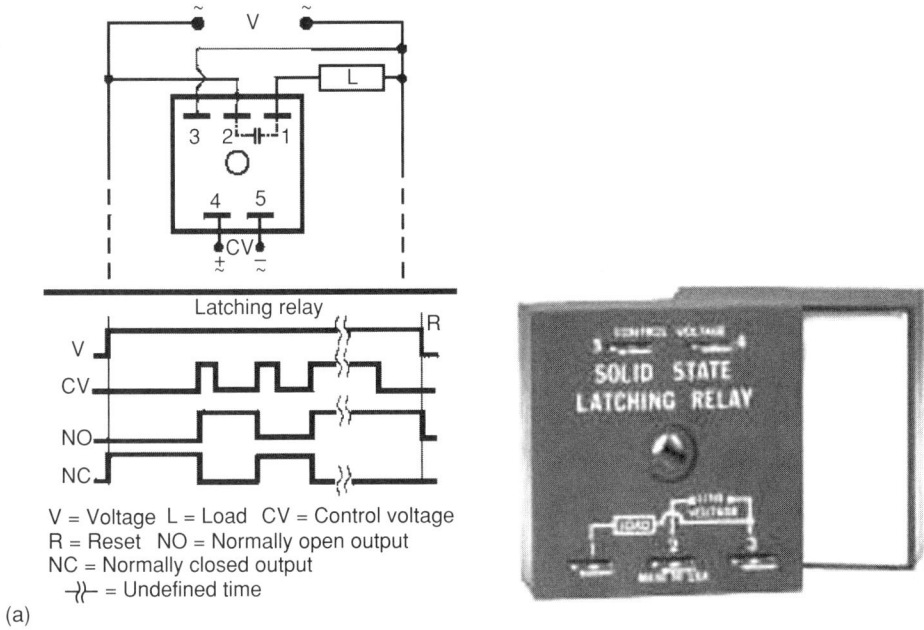

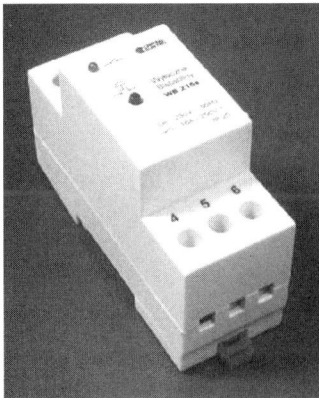
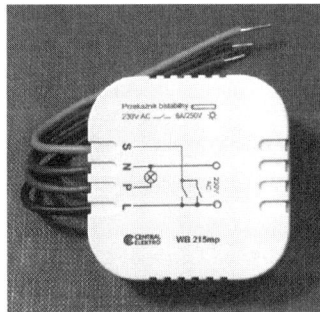

FIGURE 16.28
(a) Solid-state latching relay of the NLF type (ABB). (b) Solid-state latching relays of different types.

closes (or opens) and latches. If control voltage is removed and then reapplied, the output opens (or closes) and latches. The output transfers each time the control voltage is applied. For reset: remove and reapply control voltage. Reset is also accomplished by removing output voltage. Many other companies produce relays of various types (Figure 16.28b).

16.3 Sequence Relays

A sequence relay is sometimes called an *alternator, stepper, step-by-step, flip-flop,* or *impulse* relay. The relay has the ability to open and close its contacts in a preset sequence.

All sequence relays use a ratchet or catch mechanism to cause their contacts to change state by repeated impulses to a single coil. Usually, but not always, one pulse will close a set of contacts, the next will open them, and so on, back and forth. This alternating of open and closed states has many possible uses. A sequence relay requires a pulsed voltage to the coil of approximately 50 msec for each sequence to take place. When the coil is pulsed, the relay armature moves a lever that in turn rotates the ratchet and cams to the first position in the sequence. This position will remain as long as another pulse is not introduced to the coil. The relay is normally comprised of at least two sets of contacts to allow the contacts to alternate in combinations of open and closed states, with each pulse of voltage to the coil. The example of possible two-pole combinations would be where one pole remains open and the other pole is closed with the first pulse applied to the coil.

The second pulse could then reverse the above sequence. The third pulse could have both poles closed and the fourth pulse could open both poles. The above example could also have other sequences, depending upon the amount of teeth in the ratchet and the amount of lobes on the cams. Figure 16.29 shows an example of how cam placement on the contact blades can change the position of the contacts as the cams are rotated by the ratchet gear.

Typical applications of sequence relays (Figure 16.30) are remotely starting and stopping a conveyer from a single momentary push button. Several momentary push buttons might be wired in parallel to control the conveyer from a number of locations. Another

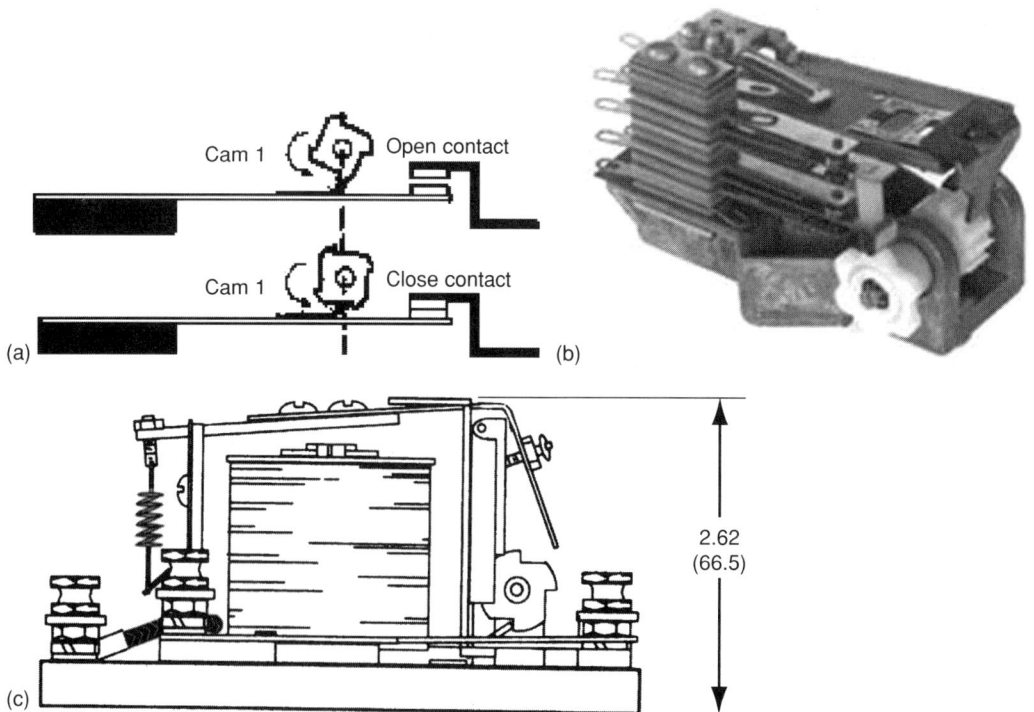

FIGURE 16.29
Operation principle of sequence relay.

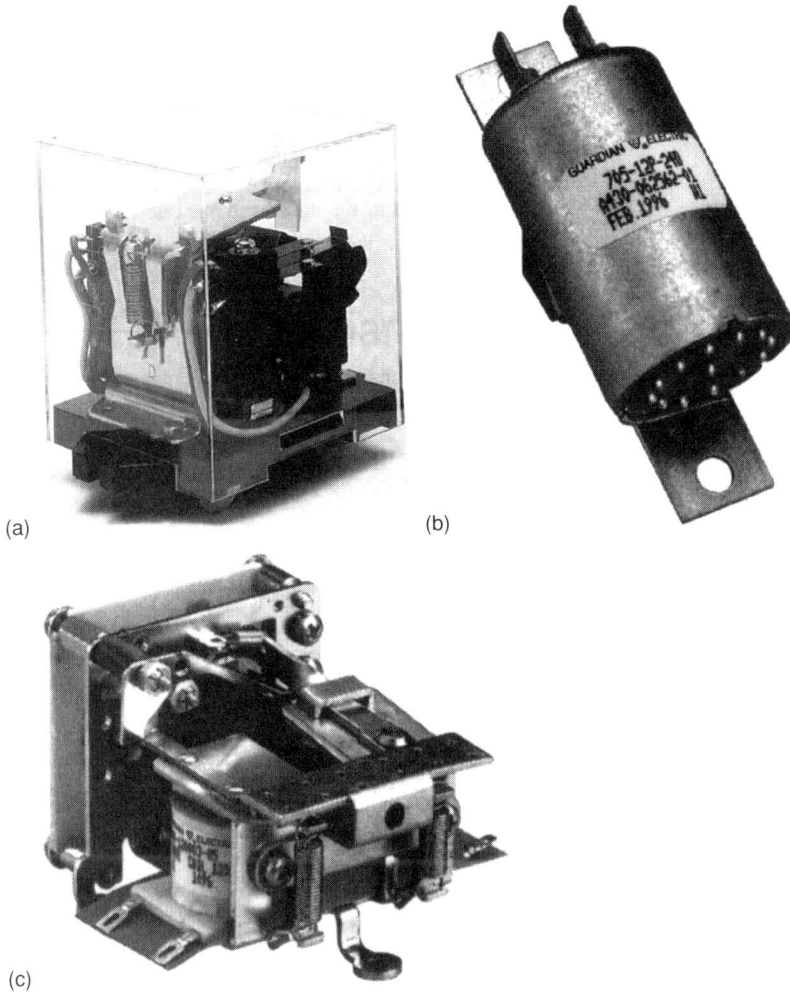

FIGURE 16.30
External designs of sequence relays produced by a number of companies. (a) Ratchet relay G40 type (Omron); (b) stepper relay 705 series (Guardian Electric); (c) stepper relay C85 (Magnecraft).

common use for sequence relays is a cascade starting a multiple HVAC, or other high start-up load systems to limit the high starting current.

One of the variants of sequence relays is a so-called *"step-by-step selector"* used in telephone communication for putting through the subscriber of the dialed telephone number. In this device (Figure 16.31) when each pulse is applied to the coil, the contact wipers rotate to one position, closing the corresponding contacts. When one connects the dialer of the telephone, the source of DC voltage and the step-by-step selector, and then if one dials, say "5," five current pulses will be applied to the coil of the step-by-step selector, the contact wipers will make five steps and stop in the position corresponding to the number "5."

Dial telephone systems derive their name from the use of a dial, or equivalent device, operated by a subscriber or operator to produce the interruptions of current that direct or control the switching process at the central office. The use of a dial for such purposes, however, is much older than the telephone. It was suggested by

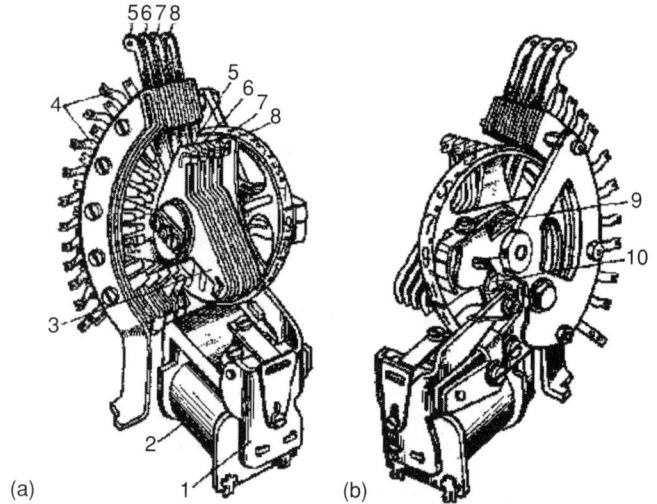

FIGURE 16.31

(a) and (b) Step-by-step selector used in telephone communication. 1 — Armature; 2 — electromagnet; 3 — three-rayed contact wipers; 4 — contact lamellas; 5–8 — inputs of the wipers; 9 — latch; 10 — ratchet gear.

William F. Cooke in 1836 in connection with telegraphy, and first used in Professor Wheatstone's dial telegraph of 1839. In the following years it was subject to many improvements, and was employed not only in dial telegraph systems but in fire alarm and district messenger systems as well. Figure 16.32 shows Froment's telegraph of 1851 transmitting and receiving dials.

The first dial telephone exchange patent, No. 222,458, was applied for on September 10, 1879, and issued on December 9, 1879, jointly, to M.D. Connolly of Philadelphia, T.A. Connolly of Washington D.C., and T.J. McTighe of Pittsburgh (Figure 16.33). Although this first system was crude in design and limited to a small number of subscribers, it nevertheless embodied the generic principle of later dial systems. At each station, in addition to the telephone, batteries, and call bell, were a reversing key, a compound switch, and a dial similar to that employed in dial telegraph systems, and on its face the numbers corresponded to the different stations of the exchange. At the central office were ratchet wheels: one wheel for each station, mounted one above the other on a common vertical shaft and carrying wiper arms which moved with the ratchets. Actuated by the circuit interruptions made by the calling subscriber dial, an electromagnet

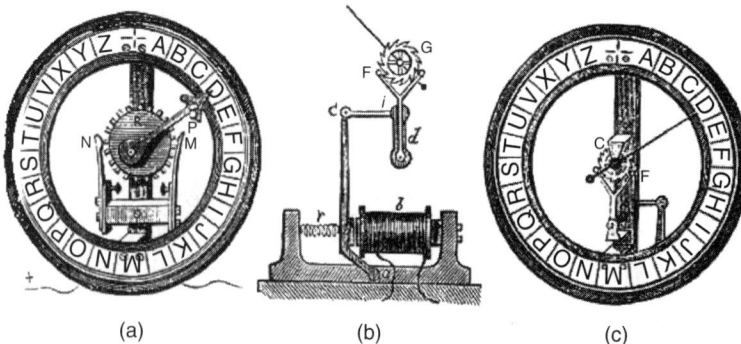

FIGURE 16.32

Transmitting dial (a) and receiving dial (c) used with Froment's alphabetical telegraph system of 1851 together with the electromagnet, ratchet, and pawl arrangement used with the receiving dial (b).

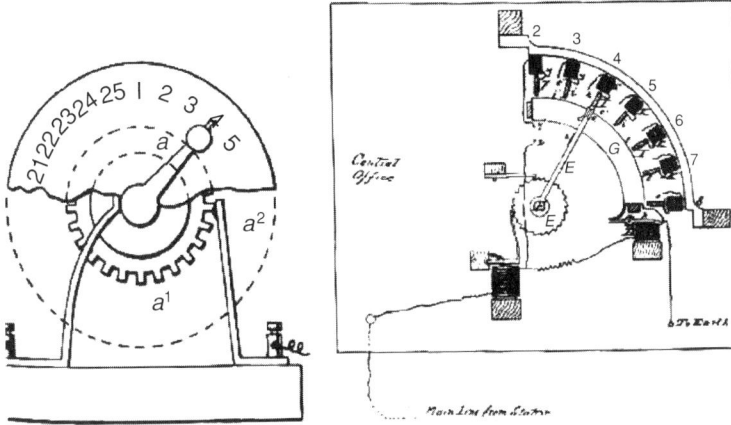

FIGURE 16.33
Fragment of Connolly-McTighe patent No. 222,458 (1879).

stepped the wiper arm around to engage the contact of the called subscriber line. Although the switching mechanism was relatively simple, various manipulations of the reversing key and compound switch were required.

Meanwhile, Almon B. Strowger, (Figure 16.34) Kansas City, U.S. is regarded as the father of automatic switching. Strowger developed a system of automatic switching using an electromechanical switch based around electromagnets and pawls. With the help of his nephew (Walter S. Strowger) he produced a working model in 1888 (U.S. Patent No. 447918, 1891). In this selector, a moving wiper (with contacts on the end) moved up to and around a bank of many other contacts, making a connection with any one of them. Since then, Strowger's name has been associated with the step-by-step selector (controlled directly from the dial of the telephone set), which was part of his idea. But Strowger

FIGURE 16.34
Almon B. Strowger.

was not the first to come up with the idea of automatic switching: it was first proposed in 1879 by Connolly and McTigthe, but Strowger was the first to put it to effective use. The 26 patents on the list that were issued between the Connolly and McTighe patent of 1879 and Strowger's patent No. 447918 of 1891 all related to the operation of small exchanges and for the most part employed complicated electromagnetic step-by-step arrangements, constantly running synchronized clockwork mechanisms, reversals of current direction, changes in current strength, and the like.

Together with Joseph B. Harris and Moses A. Meyer, Strowger formed his company, the "Strowger Automatic Telephone Exchange," in October 1891. In the late 1890s, Almon B. Strowger retired and eventually died in 1902. In 1901, Joseph Harris licensed the Strowger selectors to the Automatic Electric Co. (AE); the two companies merged in 1908. The company still exists today as AG Communications Systems, having undergone various corporate changes and buyouts.

Later electromagnetic step-by-step arrangements were widely used in many fields of engineering such as measuring engineering (in systems of information acquisition from measured objects), etc. Such devices were also produced in special modification for military purposes. In telephone communication such devices lasted until the 1970–80's, being replaced in recent years by quasi-electronic and then by purely electronic devices.

16.4 Rotary Relays

Rotary or *motor-driven* relays are relays in which forward movement of the armature and contacts is replaced by rotary movement. In fact this is a standard multicontact rotor switch with an electromagnetic drive instead of a manual one (Figure 16.35, Figure 16.36).

What is its purpose? The point is that in standard relays movable internal elements (the armature, contacts) may spontaneously shift (and contacts may close) when the relay is affected by considerable accelerations, caused by quick moving in space or by shocks or vibrations of considerable amplitude. Such effects usually take place in airborne military equipment installed in aircraft or in missiles. In addition, there are a number of important surface facilities; the normal functioning of which must be provided with the great ground shaking caused by close explosions or earthquakes. Nuclear power plants also belong to this category of facilities, for example. All equipment for such facilities is built

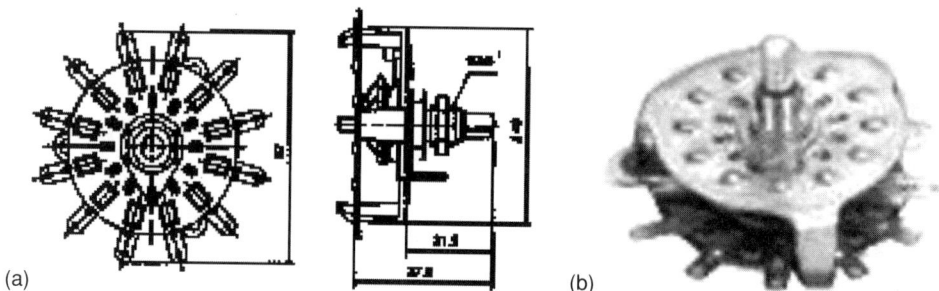

(a)   (b)

FIGURE 16.35
Rotor switch with a manual drive.

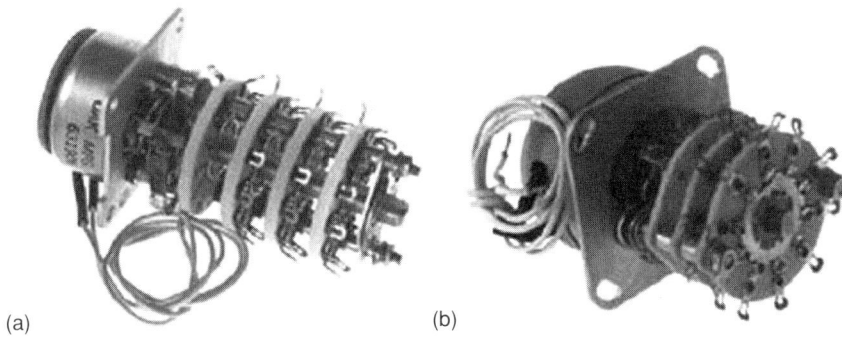

FIGURE 16.36
Rotor relay — is a motor-driven rotor switch.

according to specific requirements. Rotor relays (Figure 16.37, Figure 16.39) are quite frequently used in facilities of this kind.

Rotor relays can be both nonlatching and latching. The nonlatching relay has two coils connected in series inside the relay, which, when energized, rotate the relay rotor shaft, which operates the contacts through a shaft extension. The stator faces and stop ring limit the rotor movement to a 30° arc. Two springs return the rotor to the stop ring and the contacts to their normal positions when the coils are deenergized. The nonlatching MDR series relays have two positions: "energized" and "deenergized" (Figure 16.38). Each relay in the MDR latching series has two sets of series coils that provide a latching

FIGURE 16.37
Rotor relays of the MDR series (Potter & Brumfield Co.).

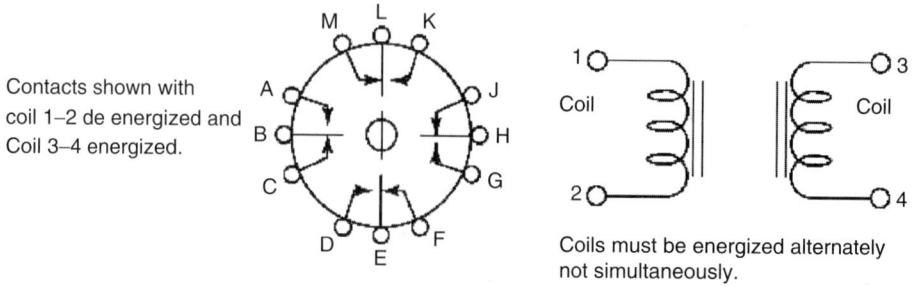

Contacts shown with
coil 1–2 de energized and
Coil 3–4 energized.

Coils must be energized alternately
not simultaneously.

FIGURE 16.38
Circuit diagram of rotary relay.

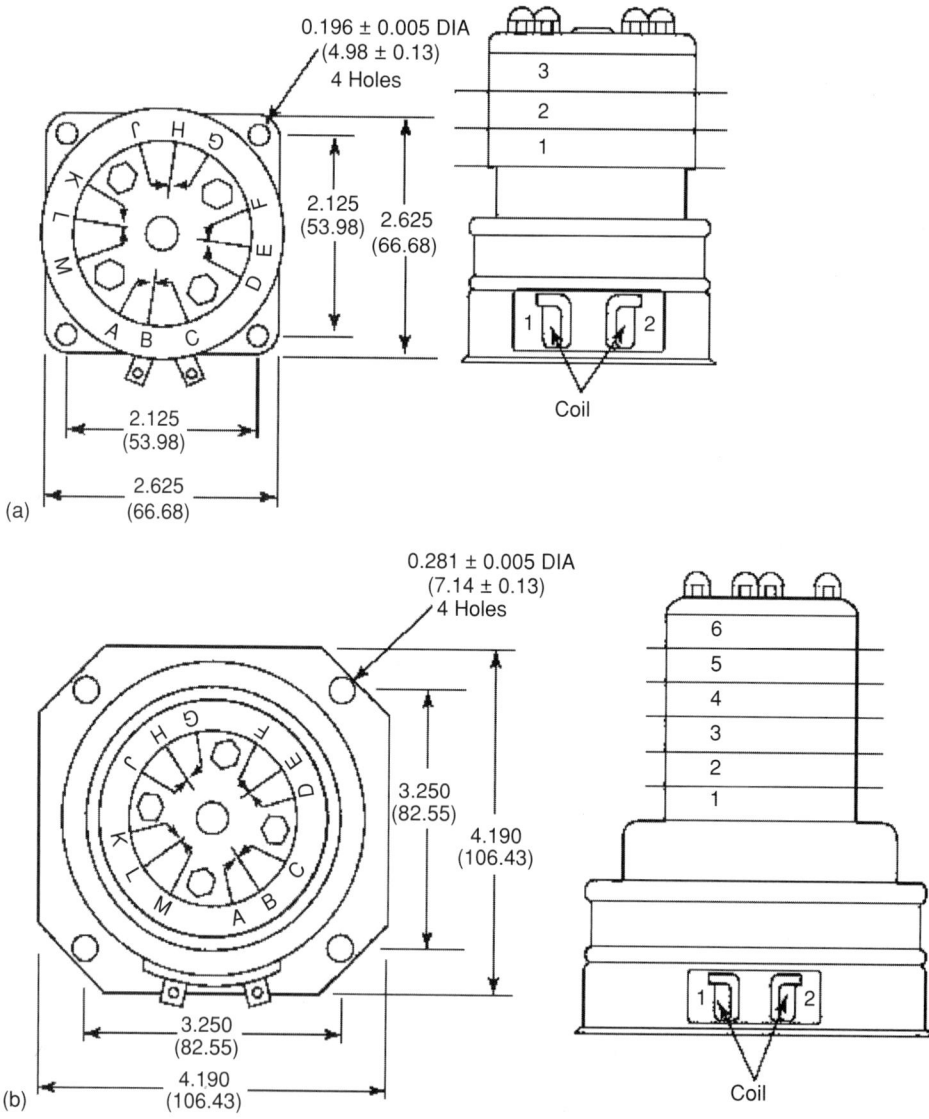

FIGURE 16.39
Overall size of relay of the MDR series with 4 to 12 changeover contacts (a) and 16 to 24 changeover contacts (b).

two-position operation. When one set of coils is energized, the rotor shaft rotates through a 30° arc, changing the state of the contacts. The other set of coils must be energized to return the relay to its original position. Relays of this type are capable of switching currents up to 10 A with voltage up to 115 or 3 A with voltage of 440 V AC (or 28 V DC).

16.5 Moving-Coil Relays

Relays of this type have a quite unusual external design, sometimes resembling a vacuum tube (Figure 16.40) or a measuring device (Figure 16.41). It is only natural that such a relay resembles a measuring device because in fact it is a highly sensitive measuring mechanism, with very sensitive contacts (Figure 16.42). The functioning of this device is based on the interaction of the magnetic field of the permanent magnet with the current in the winding. The winding is wound around a light aluminum bobbin of rectangular shape (a frame) placed in the gap between the permanent magnet and the core ring, Figure 16.43.

When current is applied to the winding, it creates its own magnetic field, interacting with the field of the permanent magnet and tending to swing the frame around its axis. If one links this frame with a pointer, one will obtain a measuring device (galvanometer). If one fixes a contact instead of a pointer, one will obtain a moving-coil relay.

The phenomenon of interaction of two conductors with current, and later of a current conductor with a permanent magnet, was investigated by many scientists in the 19th century. Instruments to measure the passage of an electric current depend on a serendipitous observation. In the course of a lecture demonstration in April 1820, the Danish natural philosopher, Hans Christian Orsted (1777–1851), passed a heavy current through a metallic wire, and noticed that a nearby compass needle was affected. In the early summer, the experiment was repeated under controlled conditions, and on July 20th he published (in Latin) the first paper on electromagnetism. His discovery of the interaction of the magnetic field produced by the current with the compass needle also provided the

FIGURE 16.40
Moving-coil relay of the E51-1 type with glass cover (BBC).

FIGURE 16.41
Moving-coil relay of the Z_T type (BBC).

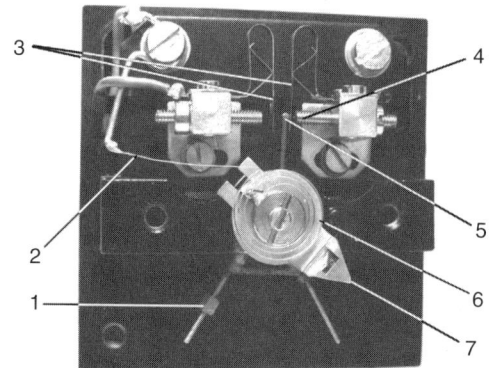

FIGURE 16.42
Contact system of the moving-coil relay. 1 — Balance beam; 2 — outlet of the movable contact; 3 — stationary contact springs; 4 — first adjustable stop; 5 — movable contact; 6 — concentric spring; 7 — indicator of pick-up current.

FIGURE 16.43
Magnetic system of the moving-coil relay. 1 — Core ring; 2 — aluminum frame with the winding; 3 — permanent magnet.

FIGURE 16.44
Sturgeon's galvanometer.

mechanism for the measurement of the electric current. Shortly after, Andre Marie Ampere (1775–1836) suggested that this effect could serve as the basis for measuring the electric current.

The basic galvanometer, devised by the British physicist William Sturgeon (1783–1850) in 1825, allowed all of the various combinations of current and magnetic needle directions to be tried out. By making suitable connections to the screw terminals, current can flow to the right or to the left, both above and below the needle. Current can be made to travel in a loop to double the effect, and, with the aid of two identical external galvanic circuits, the currents in the two wires can be made parallel and in the same direction (Figure 16.44). Note that the wires are insulated from each other where they cross.

In 1882, due to efforts of the French engineers Jaques-Arsène d'Arsonval and Marcel Deprez (Figure 16.45), the galvanometer looked like a measuring device used not as a physical curious thing or a visual aid for lectures in physics, but as a measuring device for practical needs.

(a)   (b)

FIGURE 16.45
(a) Jaques-Arsène d'Arsonval and (b) Marcel Deprez.

Electric Relays: Principles and applications

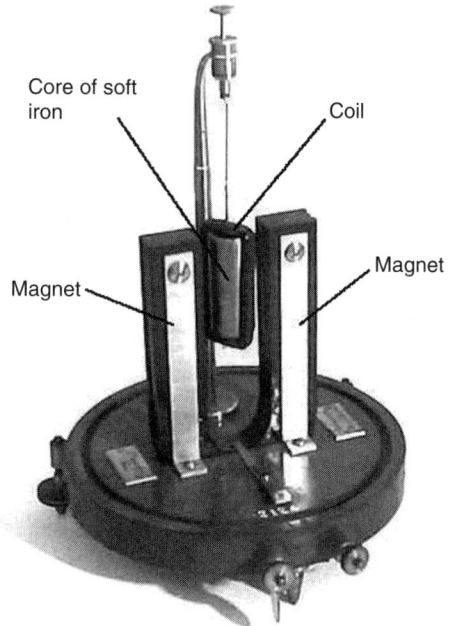

FIGURE 16.46
The D'Arsonval–Deprez galvanometer.

Jacques D'Arsonval (1851–1940) was a director of a laboratory of biological physics and a professor of experimental medicine, and one of the founders of diathermy treatments (he studied the medical application of high-frequency currents). Marcel Deprez (1843–1918) was an engineer and an early promoter of high-voltage electrical power transmission.

The galvanometer proposed by d'Arsonval in collaboration with Deprez is also defined as a mobile-bobbin (or moving-coil) galvanometer and differs from those with a mobile magnet in that it is based on the interaction between a fixed magnet and a mobile circuit, followed by a measurement of the current. Among the advantages of this type of galvanometer is higher sensitivity based on the strong magnetic field inside the bobbin. All mobile-bobbin instruments, both portable and nonportable, are derived from this galvanometer. In the D'Arsonval–Deprez design (Figure 16.46) the coil has many turns of fine wire and is suspended by a flat ribbon of wire, which serves as one lead-in wire. The connection to the lower end of the coil is provided by a light, helical spring that provides the restoring torque. The electromagnetic torque is greatest when the magnetic field lines are perpendicular to the plane of the coil. This condition is met for a wide range of coil positions by placing the cylindrical core of soft iron in the middle of the magnetic gap, and giving the magnet pole faces a concave contour. Since the electromagnetic torque is proportional to the current in the coil and the restoring toque is proportional to the angle of twist of the suspension fiber, at equilibrium the current through the coil is linearly proportional to its angular deflection. This means that the galvanometer scales can always be linear, a great boon to the user.

There are alternative constructions of the magnetic system in which the core and the permanent magnet are replaced by each other, such as putting the frame with the winding on an iron core, with the magnet placed outside (Figure 16.47), and also constructions with axial moving of the frame with the winding.

There are constructions in which instead of a magnet, the second winding is used as a source of a permanent magnetic field. In this case, the relay picks up at a certain

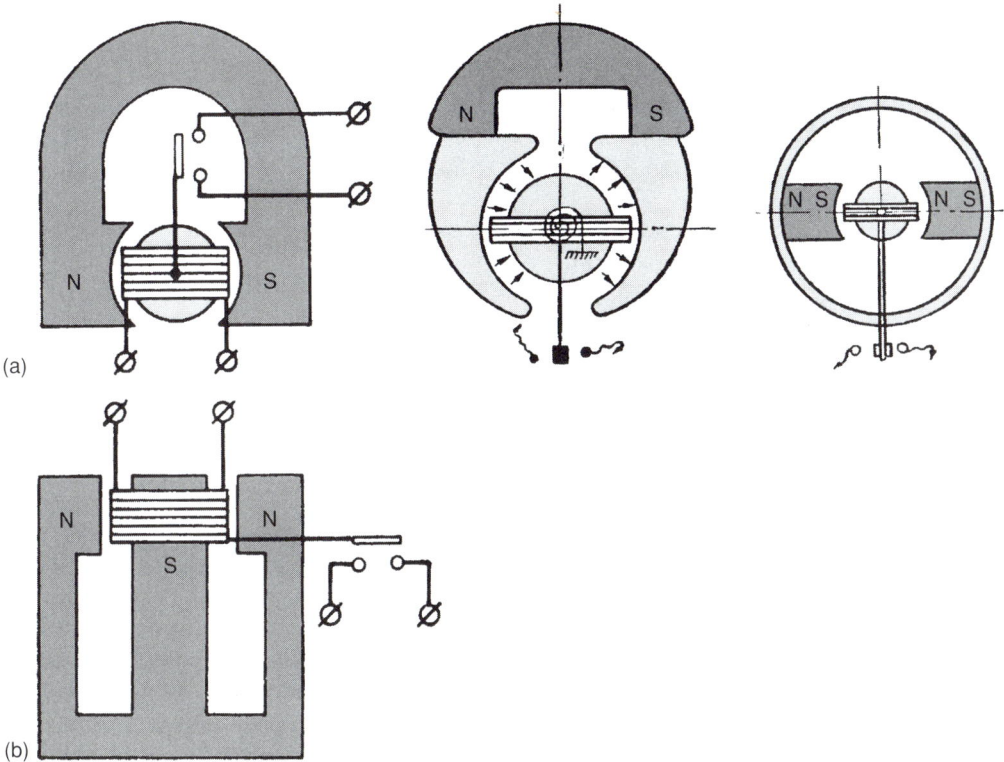

FIGURE 16.47
Alternative constructions of the D'Arsonval mechanism. (a) With external magnet; (b) with axial moving of the frame with a winding.

interaction of two currents (that is, currents flowing in two windings). Such a relay is called *electrodynamic*.

As both pointer-type devices and moving-coil relays are based on the similar D'Arsonval mechanism, it is only natural that there was an attempt to combine two types of these devices in one device. Such hybrid devices are called *control-meter relays* (Figure 16.48). These devices operate on an optical principle associated with the meter mechanism. A light source (infrared LED) and a phototransistor combination are positioned by the set point mechanism. An opaque vane is attached to the meter mechanism so that when the indicating pointer reaches the set point the vane intercepts the light from the source. This interruption changes the state of a phototransistor attached to the set pointer and switches an electronic circuit that either energizes or deenergizes the output relay. As

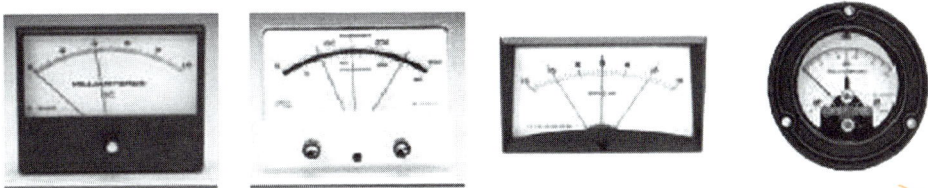

FIGURE 16.48
Control-meter relays (Beede Co.).

long as the indicating pointer remains above the set pointer, the electronic circuit remains switched. As the indicating pointer falls below the set pointer, the electronic circuit automatically returns to its former state.

Relays with a D'Arsonval mechanism are noted for the highest sensitivity among all types of electromechanical relays. Pick-up power of some types of such relays is only 10^{-7} to 10^{-8} W. Pick-up time is 0.05 to 0.1 sec. Contact pressure and therefore switching capacity are very low.

Lately, because of rapid development of electronics, there has been a tendency to use high-sensitive electronic amplifiers with standard electromagnetic relays at the output, instead of a moving-coil relay, which is why moving-coil relays have not been so widely produced and applied in recent years.

16.6 Amplifier-Driven Relays

Due to the development of semiconductor electronics and miniature transistors, in particular with working voltages of hundreds of volts, capable of amplifying signals by tens of thousands of times, moving-coil relays are not as popular as high-sensitive relays. Cheap miniature electronic elements for amplification of the control signal in combination with standard electromagnetic relays have superseded complex high-precision mechanics.

Most often high-sensitive relays are used as part of other complex devices. In such cases, there are no problems with several additional elements constituting a simplest amplifier working in the key mode. Very often single bipolar (Figure 16.49), or field-controlled (Figure 16.50) transistors are used as amplifiers.

Diodes switched parallel to the winding of the relay are necessary for preventing damage to transistors by over-voltage pulses occurring on the winding of the relay at the moment of blocking of the transistors (transient suppression).

Two back-to-back transistors (Figure 16.49b and Figure 16.50), are used to control a double-coiled latching relay of high sensitivity. Especially for electromagnetic relay control, a set of amplifiers on the basis of Darlington's transistor is produced in a standard case of an integrated circuit (Figure 16.51). The eight n–p–n Darlington connected transistors in

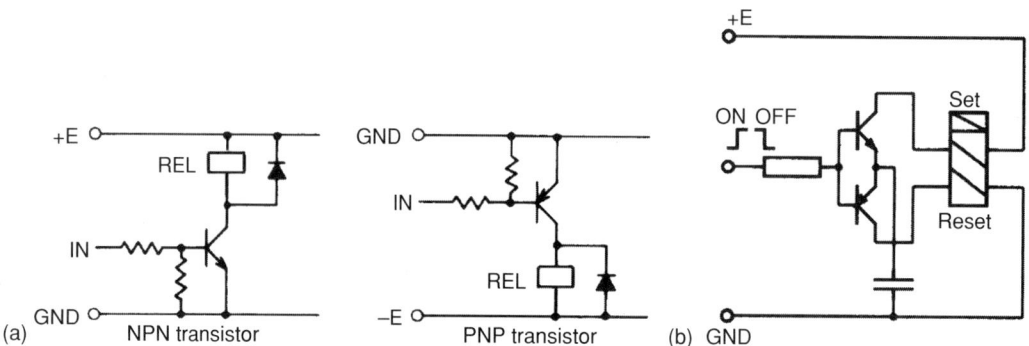

FIGURE 16.49
Amplifier-driven relays on bipolar transistors.

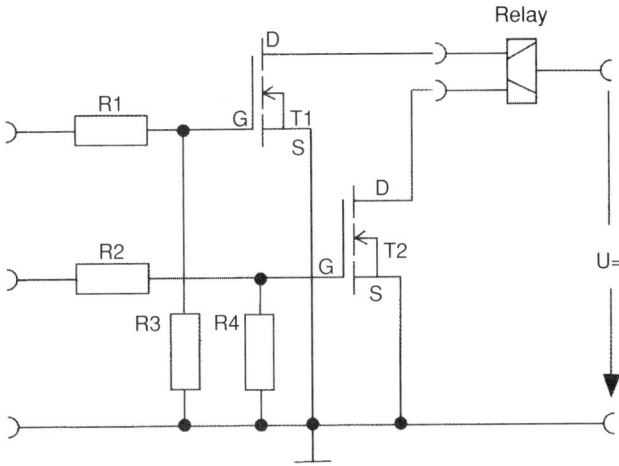

FIGURE 16.50
Amplifier-driven latching relay on FET transistors.

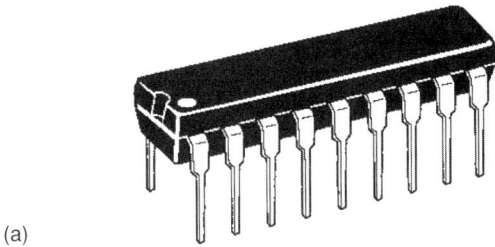

(a)

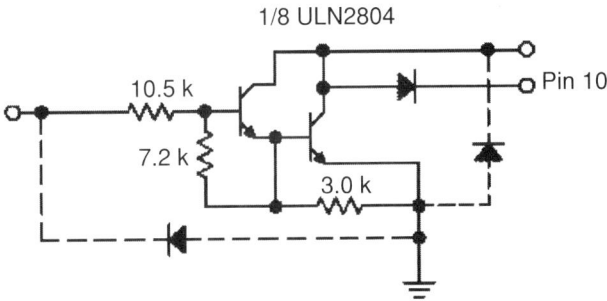

1/8 ULN2804

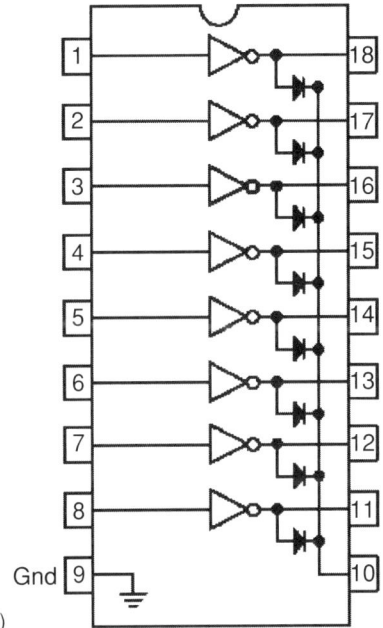

(b)

(c)

FIGURE 16.51
UNL2804 type chip (Motorola).

this family of arrays are ideally suited for interfacing between low logic level digital circuitry (such as TTL, CMOS, or PMOS/NMOS) and the higher current or voltage relays for a broad range of computer, industrial, and consumer applications. All these devices feature open-collector outputs and free wheeling clamp diodes for transient suppression. The ULN2803 is designed to be compatible with standard TTL families, while the ULN2804 is optimized for 6 to 15 V high-level CMOS or PMOS.

Amplifier-driven relays are produced by some firms in the form of independent fully-discrete devices with the electronic amplifier built-in directly to the case of the electro-magnetic relay (Figure 16.52). The Type 611 relay is an amplifier-driven relay that allows an input signal as low as 12 mW to control a double pole-double throw 10 A output. This

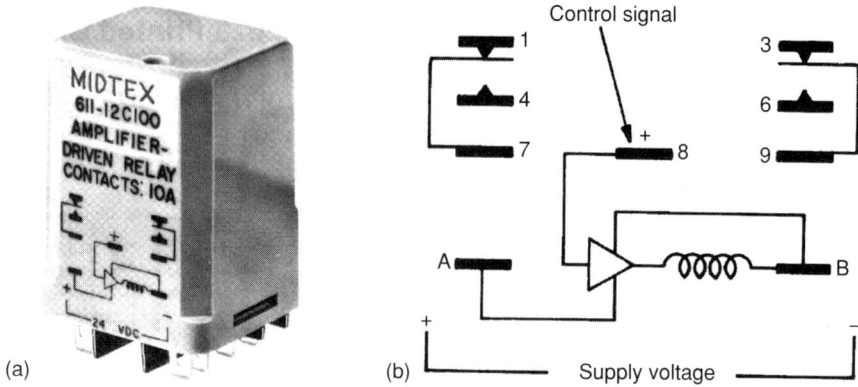

FIGURE 16.52
Amplifier-driven relay 611 type (Midtex).

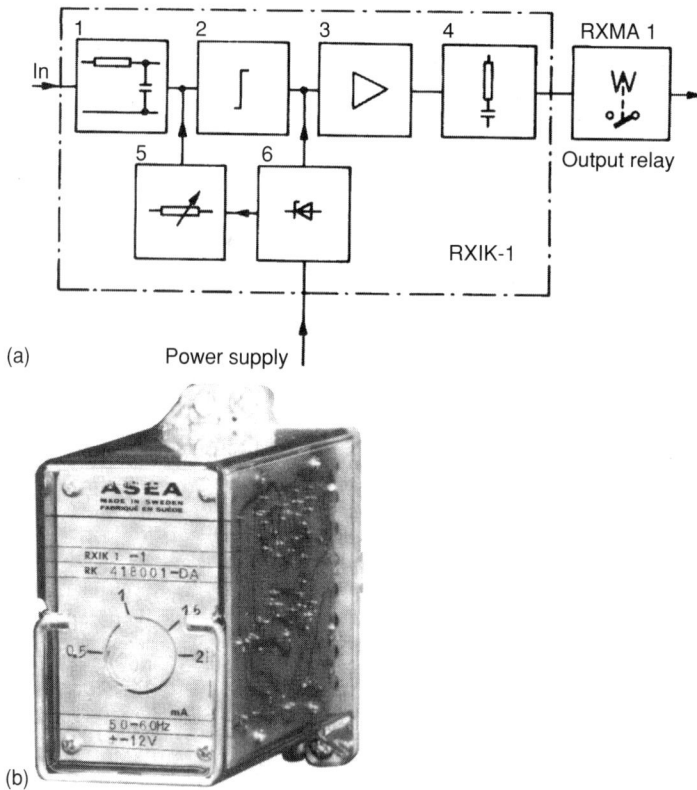

FIGURE 16.53
High-sensitive instantaneous AC and DC relay of the RXIK1 type (ASEA, ABB). 1 — RC filter; 2 — level detector; 3 — amplifier; 4 — smoothing filter; 5 — potentiometer; 6 — auxiliary voltage stabilizer (Zener diode).

low control sensitivity permits direct interfacing with most types of logic. The 611 is packaged in the popular Midtex Type 157 relay enclosure for panel mount, socket plug-in, or direct wiring.

Another example is an RXIK-1 type relay (Figure 16.53). The RXIK1 includes a level detector with an amplifier, a potentiometer, RC-circuit and Zener diode.

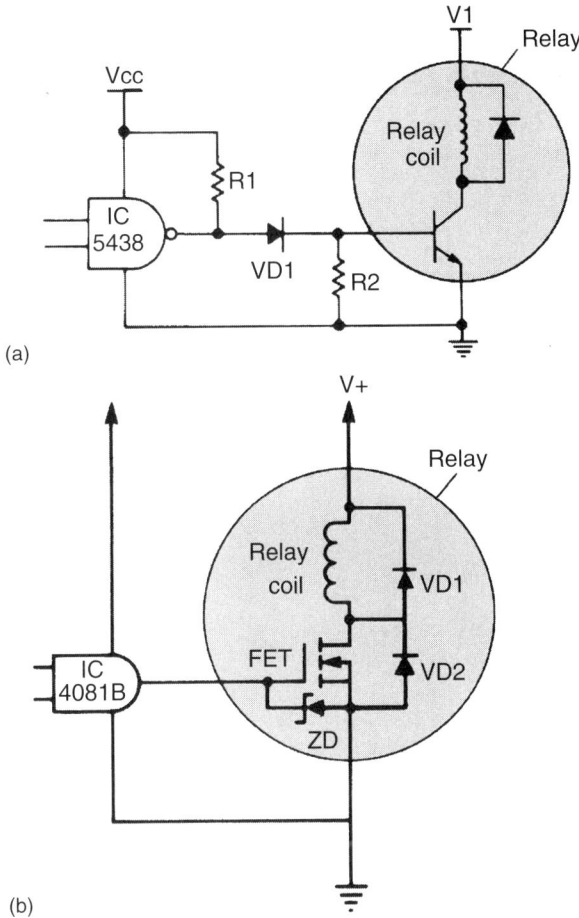

FIGURE 16.54
Amplifier-driven super miniature relays on bipolar (a) and FET (b) transistors, produced by Teledyne Relays Co. (a) Circuit diagram for relay series: 411, 412, 431, 432. (b) Circuit diagram for relay series: 116C and 136C. (Teledyne 1999 catalog.)

The operating value is set (0.5 to 2 mA) with a knob at the front. To reduce the risk of undesired operation upon high frequency signals, the input has an RC-circuit which causes the operating value of the relay to increase with the frequency. The output circuit has a smoothing filter, which ensures the function of the external output relay (RXMA-1). The measuring input is connected with the terminal for the auxiliary voltage; therefore the auxiliary voltage ought to be supplied via galvanically isolated converters. RXIK1 can be used for measuring in both voltage and current circuits.

The RXIK1 is used where high sensitivity is required (20 µW power consumption), where the RXIK1 constitutes the measuring unit. When the shunt is connected, the relay can be used to measure large DC. The shunt wires may be long and have small area, since the relay power consumption is very low. In conjunction with additional units, special versions of the RXIK1 may be used in, for example, over-current relays, neutral-point voltage relays, and differential relays, and will continue to operate even at very low frequencies.

Even the smallest electromagnetic relays in the world, for military and industrial applications, produced by Teledyne, are placed in transistor cases (TO-5 type packages, see above) and are produced with built-in amplifiers (Figure 16.54). Similar relays are produced also by some other companies (Figure 16.55).

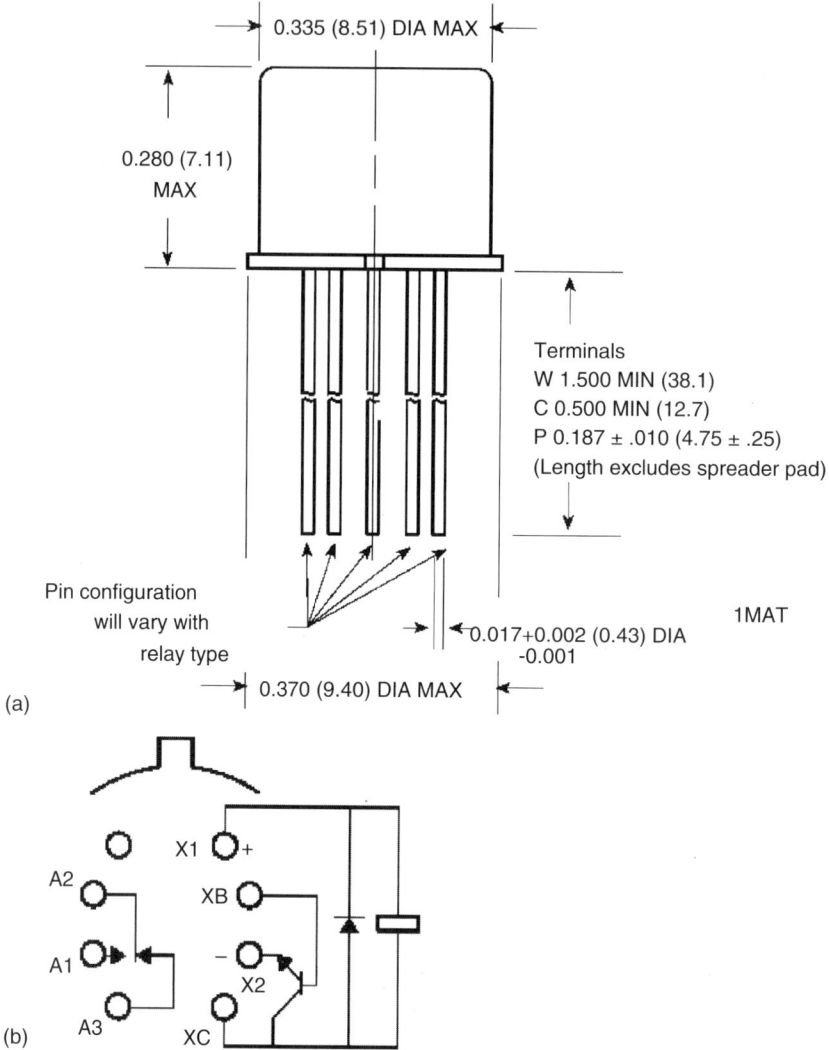

FIGURE 16.55
Amplifier-driven super miniature relay 1 MAT type in a standard transistor case, produced by Communications Instruments, Inc. (U.S.A.). Some parameters: Commutation: 1 A 28 VDC or 250 mA 115 VDC. Mechanical life: one million operations. Consumption power: 400 mW.

16.7 Magneto-Hydro-Dynamic Relays

The basic element of the magneto-hydro-dynamic (MHD) relay is an MHD pump (Figure16.56a). Usually mercury is used as conducting liquid. When electric current is applied between the electrodes (that is through the mercury) in the direction perpendicular to the direction of the magnetic field of the magnet, a magnetic field appears and interacts with the magnetic field of the permanent magnet. As a result of the interaction of the two magnetic fields an electromagnetic force (EMF) (*F*) and the corresponding pressure (*P*) tend to push out the mercury from the overlapping area of the magnetic fields. As a result, the mercury starts moving rapidly along the channel.

On DC:

$$P = BIk$$

where

B — magnetic induction in the channel
I — current through electrodes
k — the coefficient of proportionality.

On AC:

$$P = BIk \cos \gamma$$

where

γ — the angle of phase displacement between current I and the magnetic flux.

On AC, the traveling magnetic field is created by a three-phase current (the current in each phase is displaced with regard to current in the next phase by an angle of 120°). It is enough to create a traveling magnetic field near the channel with mercury, and there will be AC in the mercury, which is why there is no need to use electrodes linked by working liquid (mercury) to enable the current flow through the mercury.

Current induced in mercury creates its own magnetic field, interacting with the field causing it. As a result, the mercury is affected by some force, carrying it along the channel. The pump using induced currents is called an *induction pump* (Figure 16.56).

This device is capable of transferring electro-conductive liquid when there is an input signal (current, voltage) and can be easily used as a relay. One only has to place the contacts in the way of the mercury (Figure 16.56c). This idea was implemented by the Ukrainian inventor Dr. Barinberg from Donetsk Polytechnic University in the 1980s.

When there is no current in the control windings of these relays, the mercury level is the same in both parts of the hermetic container 4. Contacts 3 are placed below contacts 2 and plunged into the mercury. Contacts 2 do not touch the mercury. When current is applied to the control windings of the relay, the mercury moves from one part of the container 4 to the other. Contacts 3 open and contacts 2 close, and the relay picks up. When there is no current in the winding, the mercury returns to the initial state, affected by the force of gravity.

In the AC relay (Figure 16.56c) one can attach an additional transformer, matching voltages and currents of the power source of the relay and providing galvanic insulation of control circuits from contact circuits. A lot of interesting relay constructions can be implemented on the basis of the MHD pump, for example, a high-voltage relay (Figure 16.56d). When current is applied to the windings, the traveling magnetic field causes them to make the mercury move from container 5 to container 1. As a result, the gas pressure in the right tube decreases, and increases in the left tube. The mercury in 3 is squeezed from the left bend to the right one and closes the contacts.

The displacement degree of the mercury (and therefore the number of closed contacts) depends on the current value in the winding. One can obtain a relay with a time delay dependent on current by increasing the hydraulic resistance of the channels.

Unfortunately, Dr. Barinberg failed to develop his inventions to the industrial level, and as far as the author knows lives in Germany now. It seems that on the basis of

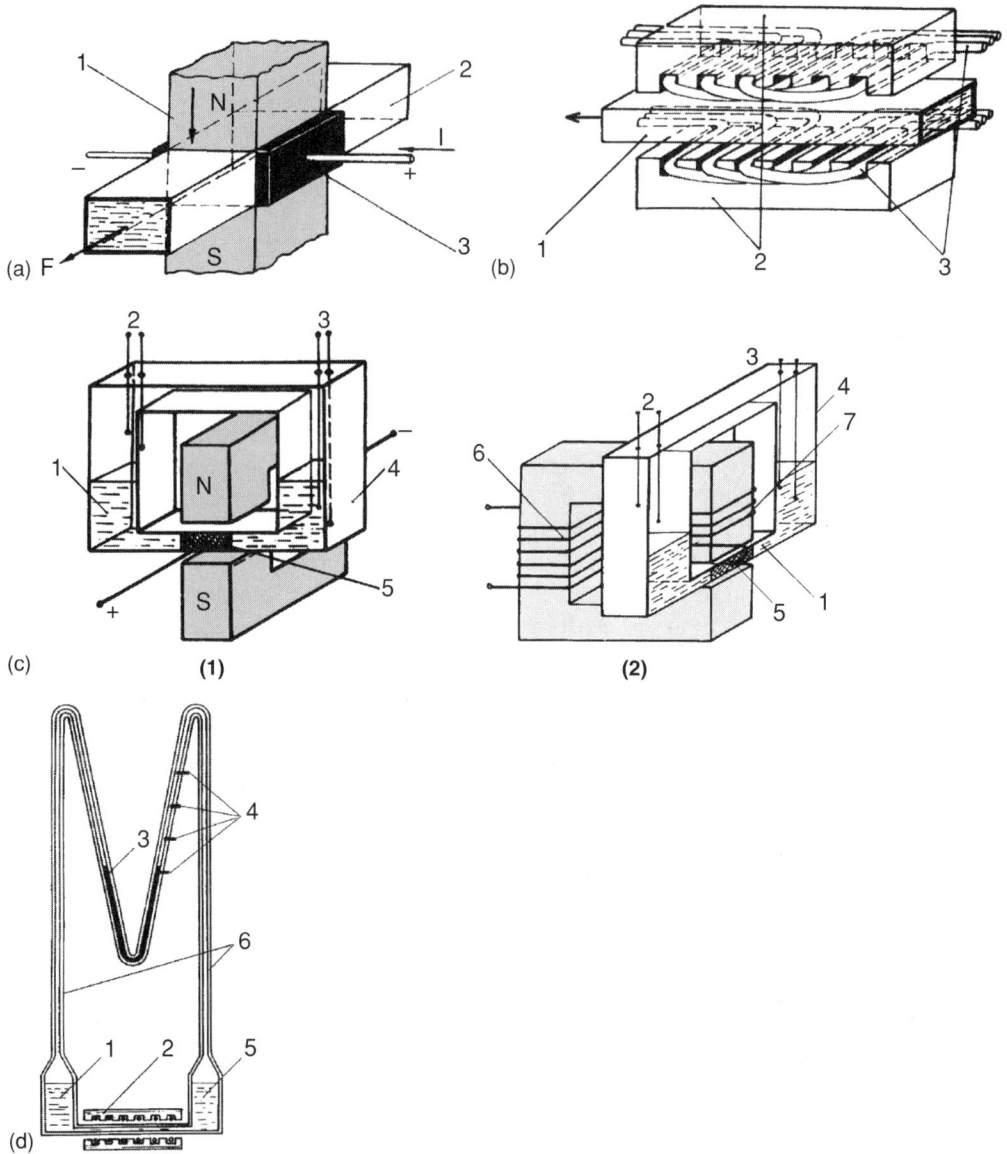

FIGURE 16.56

(a) MHD pump of conduction type. 1 — Permanent magnet; 2 — channel filled with conducting liquid; 3 — electrodes contacting with conducting liquid. (b) MHD pump of induction type. 1 — Channel with mercury; 2 — inductors; 3 — conductors forming a three-phase winding. (c) DC (left) and AC (right) MHD relay. 1 — Mercury; 2 and 3 — contact outlets; 4 — hermetic container for mercury; 5 — current-carrying electrode; 6 — primary winding of the additional matching transformer; 7 — secondary winding of the transformer switched to the electrodes 5. (d) High-voltage relay based on the MHD pump. 1 and 5 — Containers with mercury; 2 — source of the traveling magnetic field; 3 — mercury; 4 — contacts; 6 — insulation tubes filled with gas.

Dr. Barinberg's ideas one can construct quite compact and simple relays in solid metal–ceramic cases with various properties and practically unlimited service life.

16.8   Annunciator Target Relays

An *annunciator relay* (*target relay, signal relay, flag relay*) is a nonautomatically reset device that gives a number of separate visual indications of the functions of protective devices and which may also be arranged to perform a lock-out function.

In other words, target relays are used in relay protection and automation systems as an indicator of pick up of other relays. Since target relays do not have an automatic reset to the initial position, they are elements of memory storing the fact of pick up of some protective relay even if this pick up was momentary and the protective relay has returned to its initial position.

Target relays can be switched in series or parallel. In the former case, the winding of such a relay is made with low resistance (as a current one) and is switched in series with the current coil of the trip of the protective relay. In the latter case, it is a high-resistance coil switched parallel to the voltage coils of the protective relays.

The target relay has quite a simple construction containing a movable mechanical element (a shutter, flag, or disk), which is retained in the initial position with the help of a latch (Figure 16.57 and Figure 16.58).

At a short-term pick up, the armature of the target relay is attracted to the core and releases the latch. A white or colored flag drops or turns into a cut on the front board, thus becoming invisible. In passing it can close or open contacts. The flag is returned to the initial position with the help of a manual-reset mechanism. Sometimes target relays are based on standard multicontact electromagnetic relays already containing the unit, with a drop-out shutter and a reset mechanism (Figure 16.59).

The RXSF-1, for example, consists of two electromechanical relays with indicating flags. Standardly, each relay is provided with a red indicating flag (in certain cases yellow) but can be fitted with yellow or white flags if so required. The indicating flags are reset either manually with an external knob or automatically, that is, when the flag follows the movement of the armature. No-voltage relays have only automatic resetting. In cases

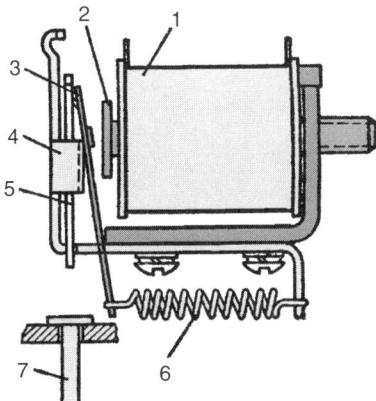

FIGURE 16.57
Construction of the target relay with a "dropping" flag. 1 — Coil; 2 — core; 3 — armature; 4 — inspection hole; 5 — drop flag; 6 — spring; 7 — pusher of the manual flag reset.

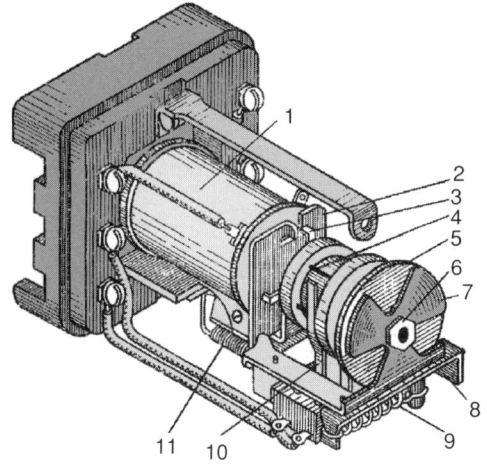

FIGURE 16.58
Construction of a target relay of the RY-21 type with a turning flag (Russia). 1 — Coil; 2 — armature; 3 — pin of the latch; 4 — contact bridge (movable contact); 5 — indicating disk with an eccentric load painted in sectors in white and black colors; 6 — axis of rotation the indicating disk; 7 — black stationary shutter with three cutout sectors; 8 — reset lever of the indicating disk; 9 — restorable spring; 10 — stationary contact spring; 11 — counter spring of the armature.

where an indication is not required a flag-locking strip can be supplied as an accessory. This may only be used on manually reset flags.

The RXSF-1 is also available as a no-voltage relay, that is, indicating when the voltage supply is interrupted. This indication can be obtained for both relays of the unit. Each relay can be supplied with 2 to 4 twin-type contacts.

Certain RXSF-1 relays can be supplied with a 4 to 5 sec time-lag on pick up (bimetallic contact) by making an external connection on the rear of the terminal base. There are constructions in which the traditional shutter is replaced with an indicating element of another type — for instance, with a colored pin (Figure 16.60), jumping put of the case of the target relay as it picks up. The relay is returned to the initial position by burying this pin in the relay case.

Target seal-in units are provided in many protective relays (Figure 16.61). These units provide a visible target to indicate that trip current has flowed. They also contain seal-in contacts, which shunt the flow of trip current away from the restraint spring in a

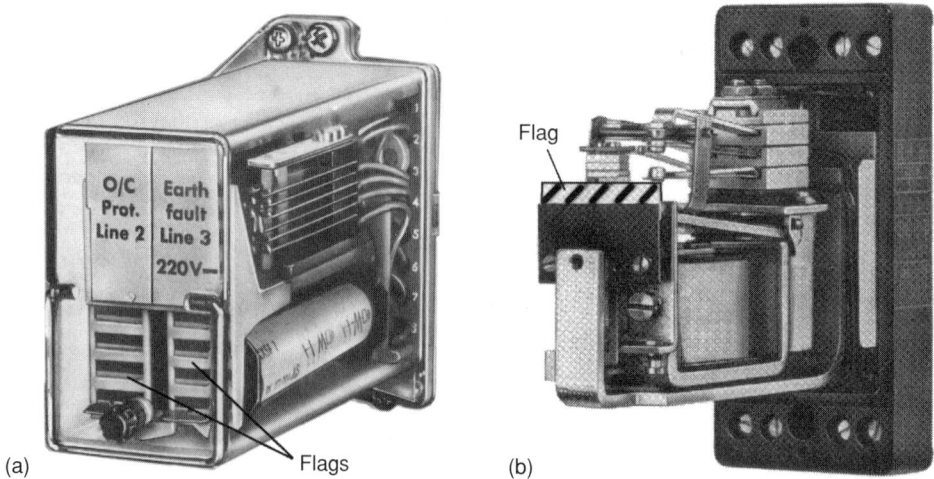

FIGURE 16.59
Target relays based on standard electromagnetic relays. (a) Dual-target relay of the RXSF-1 type (ASEA); (b) target relay RH-35 type (Siemens).

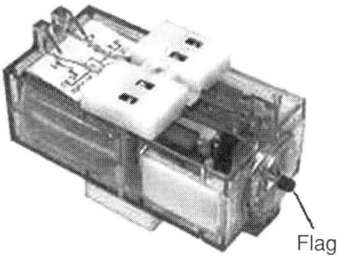

FIGURE 16.60
REY-11 target relay (Russia) with an indicating flag in the form of a colored pin.

Flag

protective relay. When the protective relay operates, its contacts close to initiate the flow of current in the trip circuit. In induction disc relays, that contact is mounted on the disc shaft and is connected to the trip circuit through the restraint coil spring. That spring has only a short time rating so the trip current must be shunted away from it to prevent overheating. When the disc contact closes and trip current begins to flow, it flows through the coil of the target seal-in unit. This causes the hinged armature unit to operate and close its seal-in contact, which then shorts out the disc contact and restraint spring. The seal-in contact remains closed until the circuit breaker trips.

The operation of the armature of this unit also sets a target flag to indicate that tripping has occurred. The flag remains set (red color showing) until manually reset from the front cover of the relay. On some units, a second electrically separate contact is also supplied. There are also target relays of the electronic type (Figure 16.62). The annunciator target relay (ATR) from Electroswitch is a compact, reliable solid state replacement or alternative for electromechanical devices currently used in many utility applications. Accepting an input signal for a variety of devices, the ATR will perform two basic functions. It will illuminate a bright LED to indicate a trip event, and also sense signals to activate up to two other devices within the system.

Once a trip signal has been detected, the ATR will latch on, keeping the LED lit until it has been manually reset. The target LED is highly visible even when viewed from extreme angles. It is designed for long life (100,000 h), available in a variety of colors, and

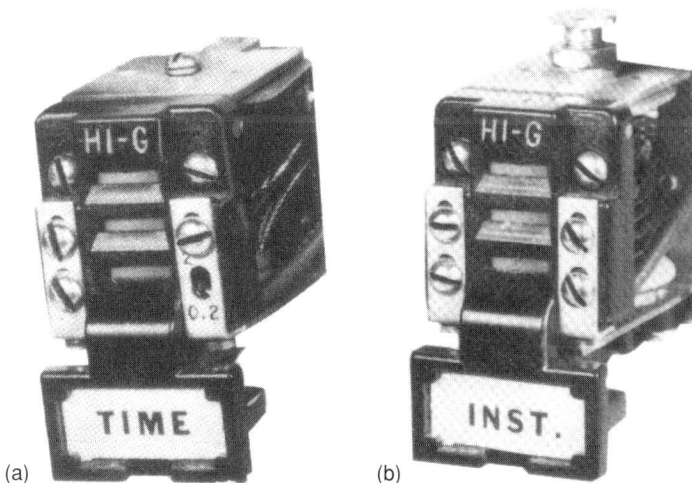

(a)   (b)

FIGURE 16.61
Target seal-in units.

FIGURE 16.62
Electronic target relay (Electroswitch, U.S.A.).

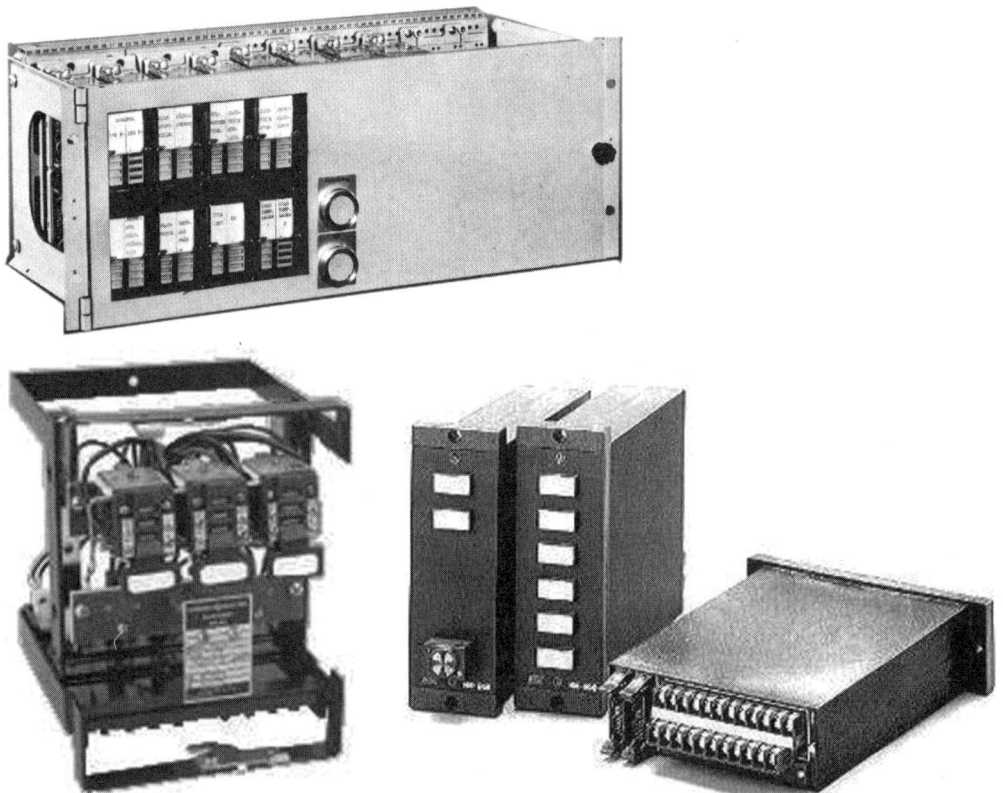

FIGURE 16.63
Units from a set of target relays.

replaceable from the front of the unit. The initial offering senses voltage and has an operating range of 37 to 140 V DC. Trip time can be specified from 0.001 to 0.100 sec by the user. Unless specified, a 0.500 sec response time is preset.

When a trip signal is received, a digital algorithm is used to validate the trip with high reliability. In its tripped state, the LED remains lit. A nonvolatile memory assures that the

ATR will retain in its state even through power outages. It will return to normal only when manually reset.

On power units (power stations, substations) a great number of protective relays are applied. In order to block and indicate the state of most of them, target relays are used. That is why such target relays are combined in units and even turned to indicator boards with legends indicating the functional peculiarity of each of them (Figure 16.63).

16.9   Flashing-Light Relays

Flashing-light relays (or *Flashers*) are used to produce the flashing light of signal lamps that, due to that flashing, attract more attention than permanently switched-on lamps. Such relays are widely used to control single signal lamps, and as a part of multivalve signal boards (Figure 16.64).

In Figure 16.65a there is a simplest relay circuit for a flashing light. Normally, when there is voltage in the circuit of the relay, current flows around KL1 and its contacts are open. When the signal is applied through signal lamps, fore example HLT, to the tie ($\sim$) *EP*, relay KL2 picks up, breaking the supply circuit of relay KL1, which links by its contacts the tie ($\sim$) *EP* with tie 0. As a result, phase voltage is applied to the lamp (HLT) and it becomes light up. Relay KL2 appears to be shorted by the contacts of KL1 and is no longer relevant. Relay KL1 picks up again, the lamp (HLT) is switched in series to KL2 and therefore it goes out. Then the process starts all over again.

The required intervals between lightings of the lamp are provided by the capacitors C1 and C2, switched parallel to the windings of the relay.

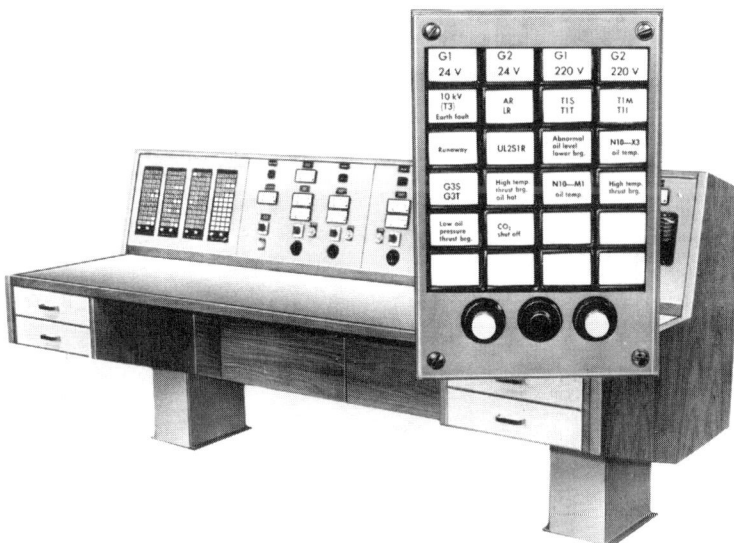

FIGURE 16.64
Signal board with lamps and flashing-light relays (ABB).

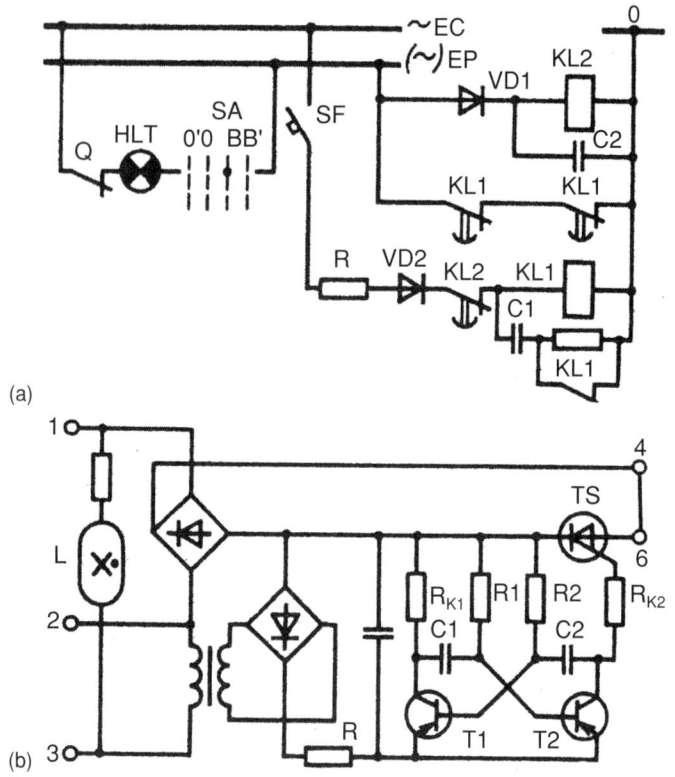

FIGURE 16.65
Simplest circuits of flashing-light relays: (a) Based on electromagnetic relays; (b) on semiconductor elements.

The thyristor (TS) in the circuit (Figure 16.65b) is enabled (ON) when a current pulse produced by an multivibrator formed by transistors T1 and T2 flows through its gate. The frequency of interruption and relative pulse duration are determined by the frequency of the multivibrator and depend on its timing circuits. The thyristor is disabled (OFF) when there is no current in the gate circuit, and when current flowing through the main junction anode–cathode equals zero.

In electronic relays of this type, different chips are applied (Figure 16.66). The TDE1767, for example, is a monolithic IC designed for high-current and high-voltage applications, specifically to drive lamps, relays, and stepping motors. These devices are essentially blow-out proof. The output is prompt and protected from short-circuits with a positive supply or drive. In addition, thermal shut down is provided to keep the IC from over-heating. If internal dissipation becomes too high, the driver will shut down to prevent excessive heating. The output stays null after the overheating is off, if the reset input is low. If high, the output will alternatively switch on and off until the overload is removed. These devices operate over a wide range of voltages, from standard 15 V operational amplifier supplies to the single +6 or +48 V used for industrial electric systems. Input voltages can be higher than in the V_{CC}. An alarm output suitable for driving an LED is provided. This LED, normally on (if referred to ground), will die out or flash during an overload, depending on the state of the reset input. Output current is up to 0.5 A without an external amplifier, and up to 10 A with the addition of an output transistor (Figure 16.66c).

Flashing-light relays based on the principles described above are produced by many companies (Figure 16.66–Figure 16.68). The RXSU device (Figure 16.67) contains

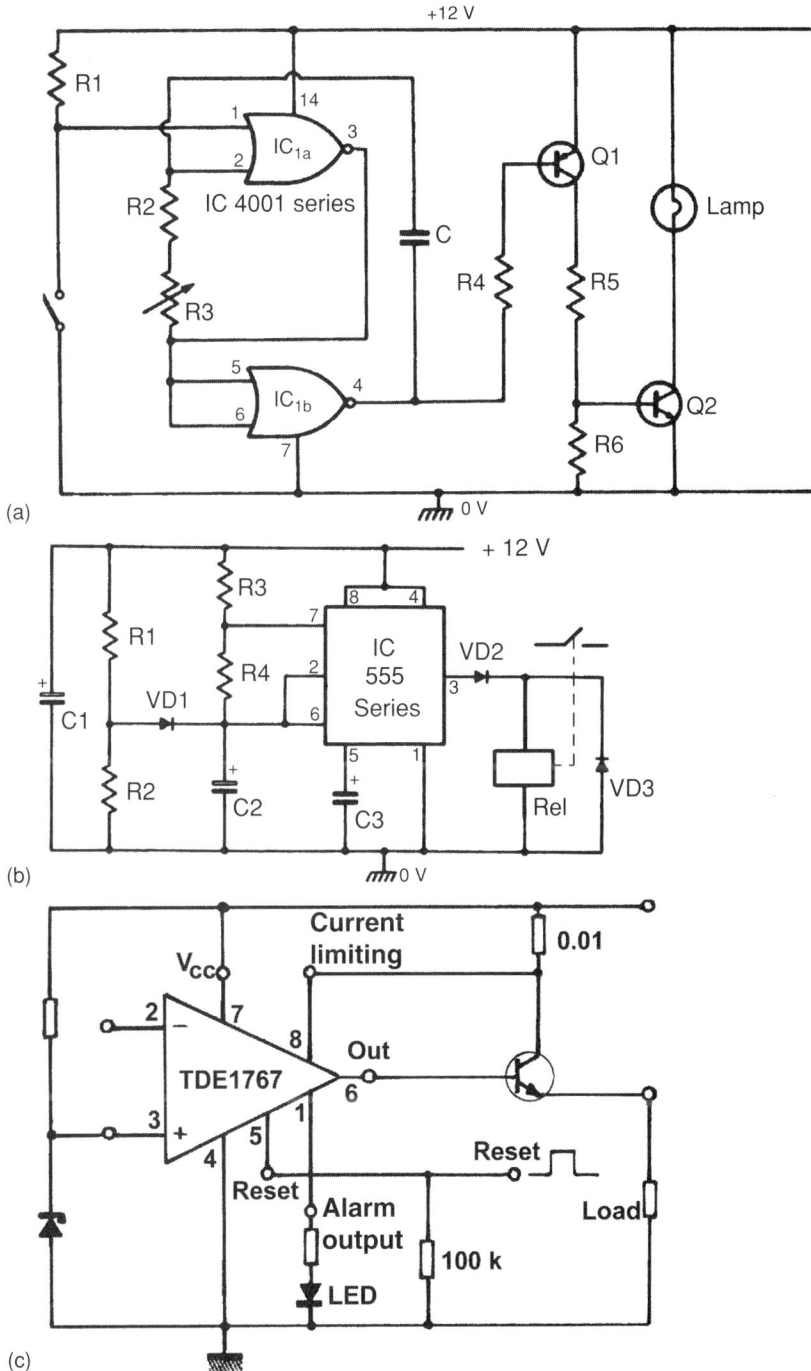

FIGURE 16.66

(a) Contactless flashing-light relay with a multivibrator, based on chip 4001, and a transistor amplifier at the output.
(b) Flashing-light relay on the basis of the IC 555 series, with an electromagnetic relay at the output. (c) Electronic
flashing relay with protected power output, based on the specializing TDE1767 type IC (SGS-Thomson).

(*Continues*)

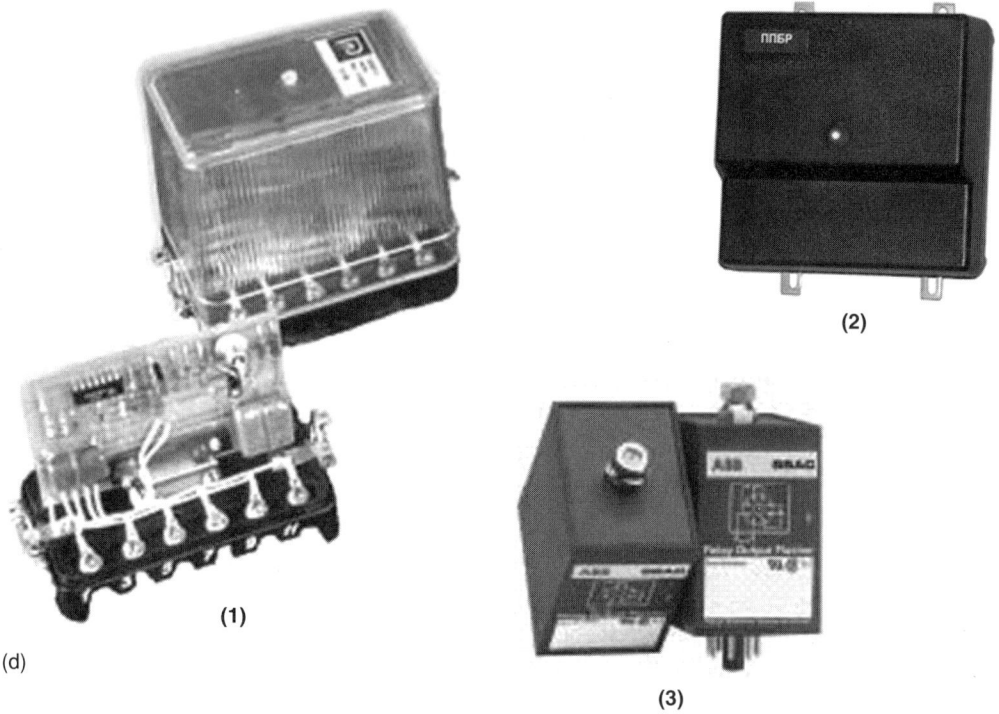

(2)

(1)

(d)

(3)

FIGURE 16.66 (*Continued*)
(d) Flashing-light relays based on IC. (1) and (2) — PPBR types (Russia, 2002); (3) — FS500 series (ABB).

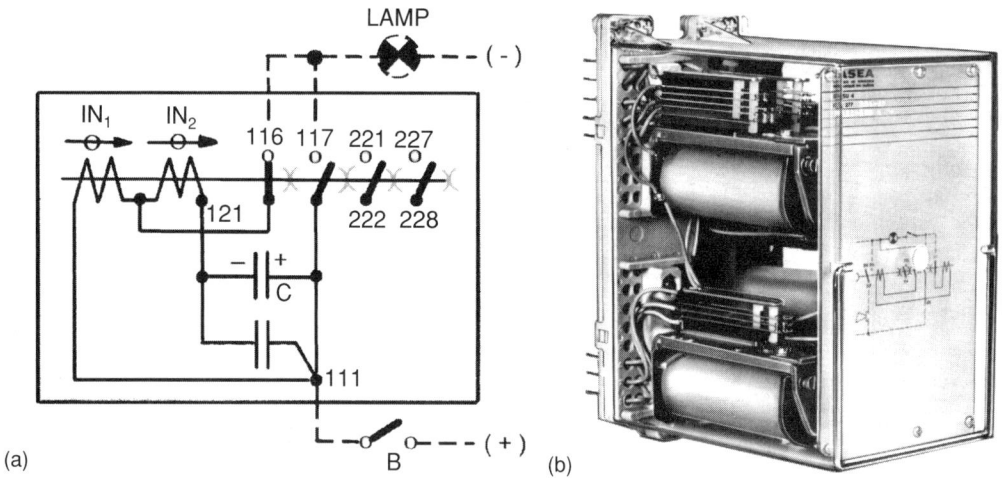

(a)

(b)

FIGURE 16.67
Flashing-light relay of the RXSU type, on electromechanical relays and capacitors. (ASEA (ABB) 1975.)

one relay with two approximately identical windings. Capacitor C is connected in series with one of the windings (see figure alongside). In addition the RXSU 4 has an auxiliary relay with a time-lag on dropout, which is picked up while the flashing-light relay is working.

When contact B closes, the relay coil is connected in series with the signal lamp. The voltage drop across the coil is so large that the signal lamp does not light up. To begin

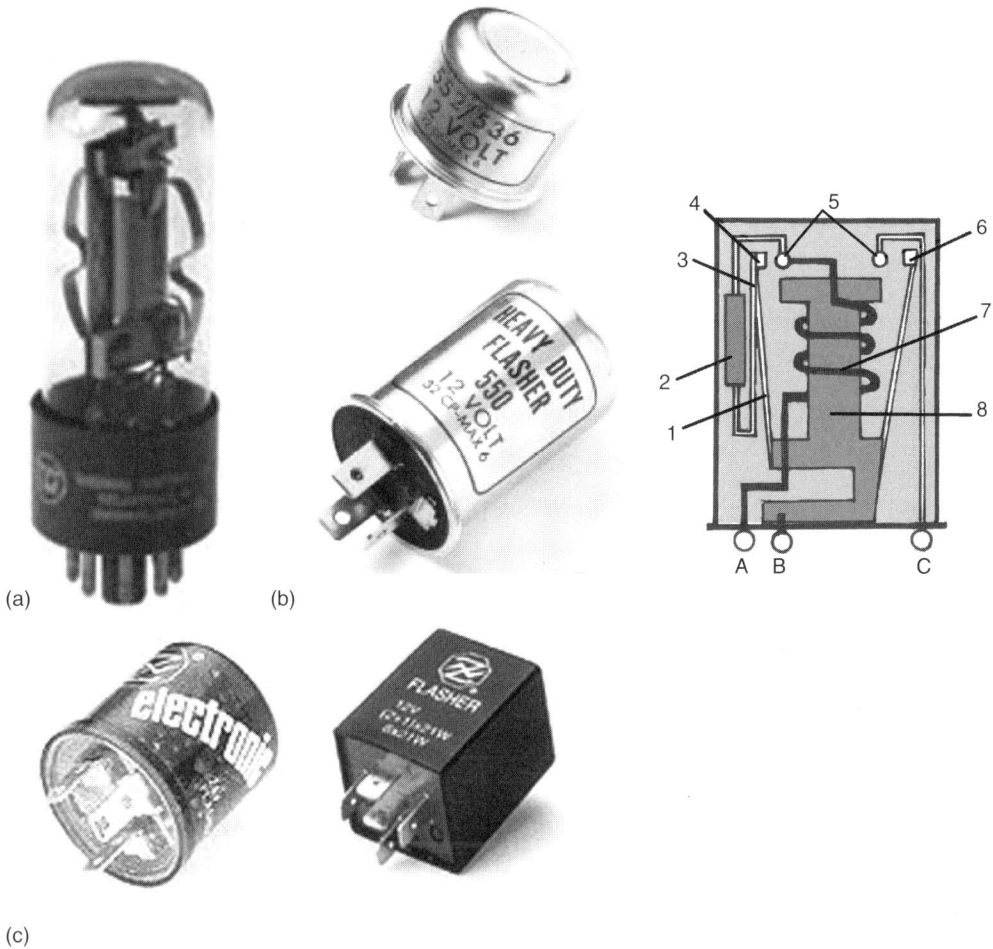

FIGURE 16.68

(a) Early thermal flashes relay of the 26F60T type, on basis of vacuum electron tube. (b) Thermo-magnetic automotive flashers in aluminum covers and it's design. 1 — Nihrom thread (20% Cr, 80% Ni); 2 — resistor (5 to 10 Ω); 3 — attracted armature; 4 — main moving contact; 5 — stationary contacts; 6 — auxiliary moving contact; 7 — coil; 8 — ferromagnetic core. (c) Electronic automotive flashers (Zung Sung Enterprise, Taiwan).

with, the ampere turns IN_1 and IN_2 are approximately equal but oppose each other, and the relay does not pick up. IN_1 is constant whereas IN_2 decreases as the current through the capacitor decreases. When the difference between IN_1 and IN_2 has become sufficiently large, the relay picks up and the lamp lights via contact 111–117. Contact 116 breaks at the same time and the capacitor discharges through both windings in series, causing IN_2 to change direction and ensuring that excitation suddenly becomes intense, so that the relay picks up distinctly. When the capacitor is discharged, the relay drops out and contact 116 makes again. Since IN_2 changes direction once more, the relay also drops out rapidly and cleanly. The sequence is repeated as long as contact B is closed. The signal lamp lights up and goes out via contact 111–117, and continues to flash as long as the relay is working. The relay is designed to give even flashing, that is, the light and dark intervals are equal in length. The flashing frequency is approximately 600 or 100, or approximately 40 flashes per min (rapid flashing and slow flashing, respectively). A flashing-light relay for AC supply has a built-in rectifier.

Automotive flashers take a separate place in this group of relays. Along with electronic devices (the principle of operation of which was described above), in automotive flashers the thermal principle is widely used. Very many companies are engaged in production of these relays.

Early thermal flashes relays were designed in bodies of vacuum electron tubes and outwardly were very similar to them (Figure 16.68a). Later on, many companies began making such relays in simple aluminum (and even plastic) covers (Figure 16.68b). Thermal (Bi-Metal) flashers are mechanical relays that operate through the heating and cooling of a bi-metal strip, opening and closing the contacts, causing the lamps to "flash."

The thermo-magnetic flasher works as follows: At actuation of the flasher, the current is applied on terminal B, from it to core 8, across nihrom thread 1 into armature 3, resistor 2 into the coil 7 to terminal A, then into the switch and further, in lamps of the parking lights and rear canopy.

In connection with that, resistor 2 is also for the purpose of burning of lamp filaments by partial light. At a passing of current through nihrom thread 1, it is heated and lengthened. The current, passing through coil 7, creates in core 8 magnetic fields. The aim of these fields is to attract armature 3 to core 8. As soon as thread 1 is lengthened to a definite value, contact 4 of an armature connecting with stationary contact 5 and resistor 2 is shunted (switched OFF) from the series circuit. The filaments of lamps thus ignite with full heat. The lamps will shine brightly as long as the thread 1 does not cool down and armature 3 is not removed from the core. The contacts thus will be disconnected, and resistance will also join in a circuit of lamps. Further all will repeat. One cycle lasts 0.6 to 0.8 sec (70 to 100 flashings in minute).

Simultaneously with armature 3, the addition armature with contact 6 is attracted to another side of core 8. This contact 6 actuates an indicating light in automotive control panel. At burnout of one of main lamps, the indicating light does not ignite. This is a very old principle that for tens of years has been widely applied in automotive flashers. During the last 10 years, such thermo-magnetic devices are being displaced with electronic ones.

Not only are totally electronic devices on IC relevant to electronic automotive flashers, but also simple devices, founded on charge–discharge of capacitors (similar, figured in Figure 16.67). Such a flasher is comprised of an electromechanical relay with an opposing A and B coil. A capacitor prevents one coil from pulling in relay contacts until it is fully charged. The charging and discharging of the capacitor, along with a resistor, determine the flash rate and set the duty cycle (ON/OFF time).

Specializing companies produce a wide range of automotive flashers of all types, for example Amperite, Littelfuse, Zung Sung Enterprise, and others. As can be seen from the examples considered above, flashing-light relays are quite simple and compact devices that are not noted for great originality. But it turns out that among these relays, there are also some very imaginative and mind-boggling constructions (Figure 16.69).

The principle of operation of this relay lies in periodic oscillations of the mercury filling the lower part of a U-shaped glass tube, when mercury is periodically squeezed from the right bend to the left, closing contacts A and C and returning, opening these contacts. The mercury is displaced by gas filling the free space in the section II, above the right mercury pile. The heater W, heating the gas, is also placed there. When part of the mercury is squeezed out from the right part of the U-shaped tube, contact B, through which the contact is supplied, opens, the gas cools down, and the mercury gradually returns to its initial state.

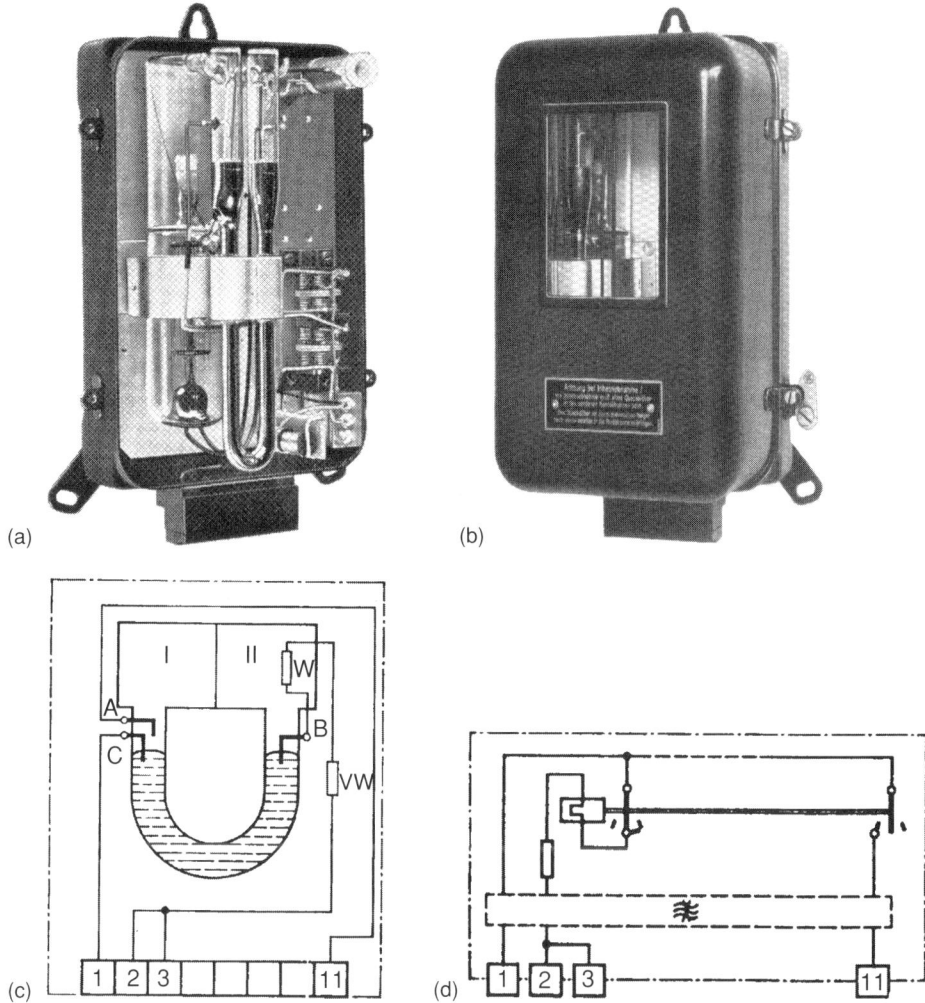

FIGURE 16.69
Mercury flashing-light relay produced by Siemens. External design and circuit diagram. (Siemens 1989 catalog.)

In what century do you think such relays were produced? You are most likely to be mistaken! Siemens produced such relays in 1972 (!), at the very same time when there were so many of the simplest constructions, of small size and weight, on the market. Perhaps this relay was a breakthrough in engineering in the 19th century. It is a very strange to see such device in middle 1970s of the 20th century.

16.10 Buchholz Relays

Protection relays of the Buchholz principle have been known for over 60 years. This *Gas Detector Buchholz Relay (Gas Relay, Buchholz Relay)* is used to protect equipment immersed in liquids by monitoring the abnormal flow or its absence, or an abnormal formation of gas by the equipment (most faults in an oil-filled power transformer are accompanied by a generation of gas). These relays are normally used in transformers with expansion tanks. They collect gas that is gradually released due to small internal problems such as bad connections, small arcs, etc. until the volume of gas operates a switch, which then gives an alarm signal. The gas can then be collected and analyzed to determine the nature of the problem.

The Gas Detector Relay also responds to larger internal faults, where larger amounts of gas are released. It will detect a larger flow of gas and operate a switch to give a trip signal, which can be used to trip the primary device and deenergize the transformer until the severity of the fault can be determined. The Gas Detector Buchholz Relay is normally installed between the main tank of the transformer and the oil expansion tank. (Figure 16.70) The Gas Relay has one (Figure 16.72) or, generally two independent contacts (Figure 16.71) (a twin-float relay) coupled to the float and the deflector or float, respectively. One of the contacts operates according to the gas accumulation and the other with sudden variations of the insulating liquid flow. It has two opposing viewing glasses with graduated scales, indicating the accumulated gas volume.

When the item of equipment being monitored is operating normally, the Buchholz relay is completely filled with oil and the up thrust on the floats keeps them at their top limit or "rest" position. If a fault causes gas to be generated slowly, the gas bubbles eventually accumulate in the Buchholz relay. The resulting fall in the oil level brings the float down and the permanent magnet incorporated in the float operates a contact until the response position is reached, which usually causes an alarm signal to be initiated.

The amount of gas that accumulates is approximately $200 \, cm^3$. The bottom float is unaffected by the gas. Any excess gas generated escapes to the conservator and this prevents tripping of the transformer by the bottom contact system, which is intended only to detect serious internal faults. When an alarm is given, the gas in the equipment must be examined without delay in order to prevent the fault from possibly becoming more serious. In the event of a leak in the transformer, the fall in oil level causes the top float to move downwards and the top contact system responds in the same way.

If the loss of oil continues, the conservator, pipe work, and the Buchholz relay itself, drain through the Buchholz relay, which causes the bottom float to fall and operate the contact to trip the transformer. Any sudden pressure surge in the transformer produces a surge flow of oil in the pipe to the Buchholz relay. The baffle plate, which is suspended in the oil flow, responds to a velocity of 100 cm/sec and works through two actuating levers to move the bottom float to a position where it triggers the contact system. The bottom float latches in the response position and this holds the contact system in the trip position. The float can be unlatched and returned to its original position by quickly rotating the test button to the stop.

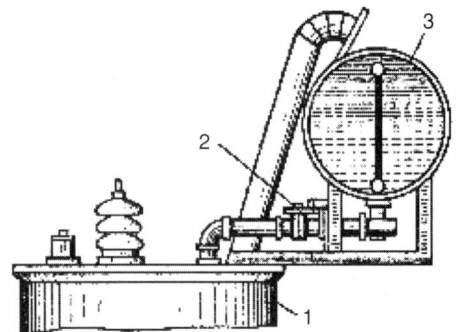

FIGURE 16.70
Mounting of a Buchholz relay on a power oil-filled transformer. 1 — Transformer tank; 2 — Buchholz relay; 3 — oil expansion tank (conservator).

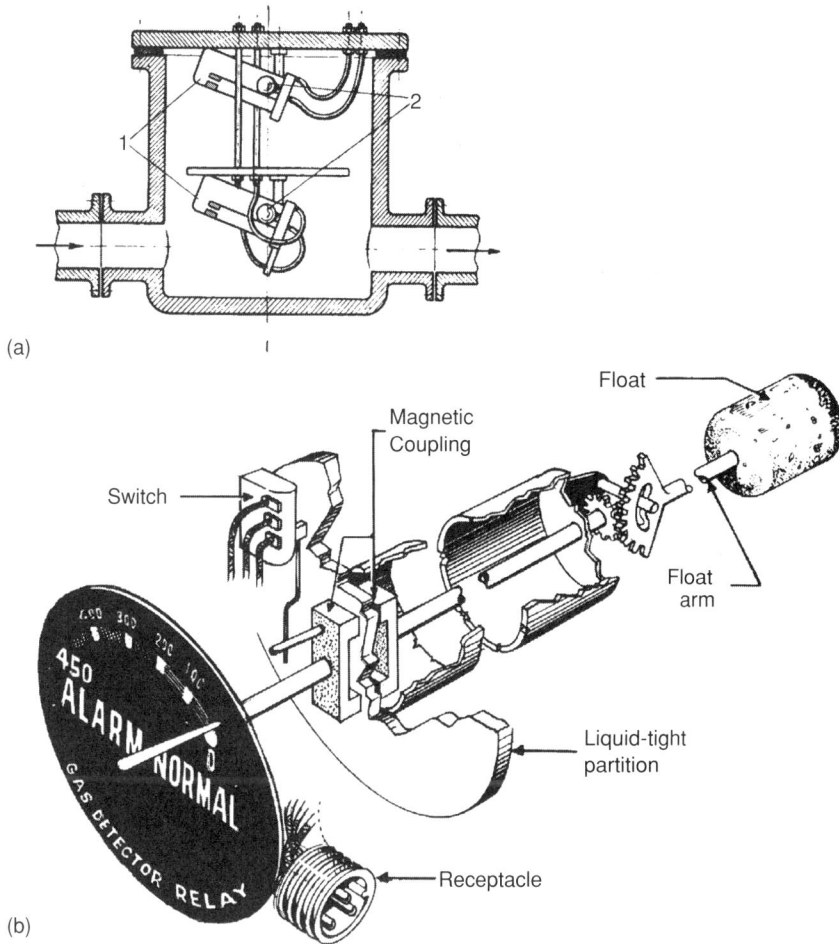

FIGURE 16.71
Earlier constructions of gas detector relays: (a) With overlapping iron balls (2), placed inside a hollow float with contacts (1); (b) with a float and a train linking the float with the contacts and also with a pointer indicator. (General Electric gas detector relay.)

Relays with hollow floats turned out to be not very reliable, because of a great number of cases of depressurization and filling of the float with oil, causing the "float" to stop being a float and the relay to stop functioning.

The construction of the Buchholz relay with cup-shaped elements, eliminating problems of this kind, was considered quite reliable. Relays of this type were produced by the Zaporozhye Transformer Plant (in the former U.S.S.R.) in 1970–80's. In these relays in a normal mode (when the entire relay volume is filled with oil), cups 2 and 3, also filled with oil, are retained in the initial position (shown in Figure 16.73), that is, they set against the stops with the help of the springs 8.

The oil pressure affecting the cups from all sides is balanced, and only the dead weight of the light aluminum cups is compensated by springs. When the upper part of the case is filled with gas, the level of oil decreases and the cup filled with oil, affected by the force of gravity of the oil, turns, closing the contacts. Serious damages in the transformer entail

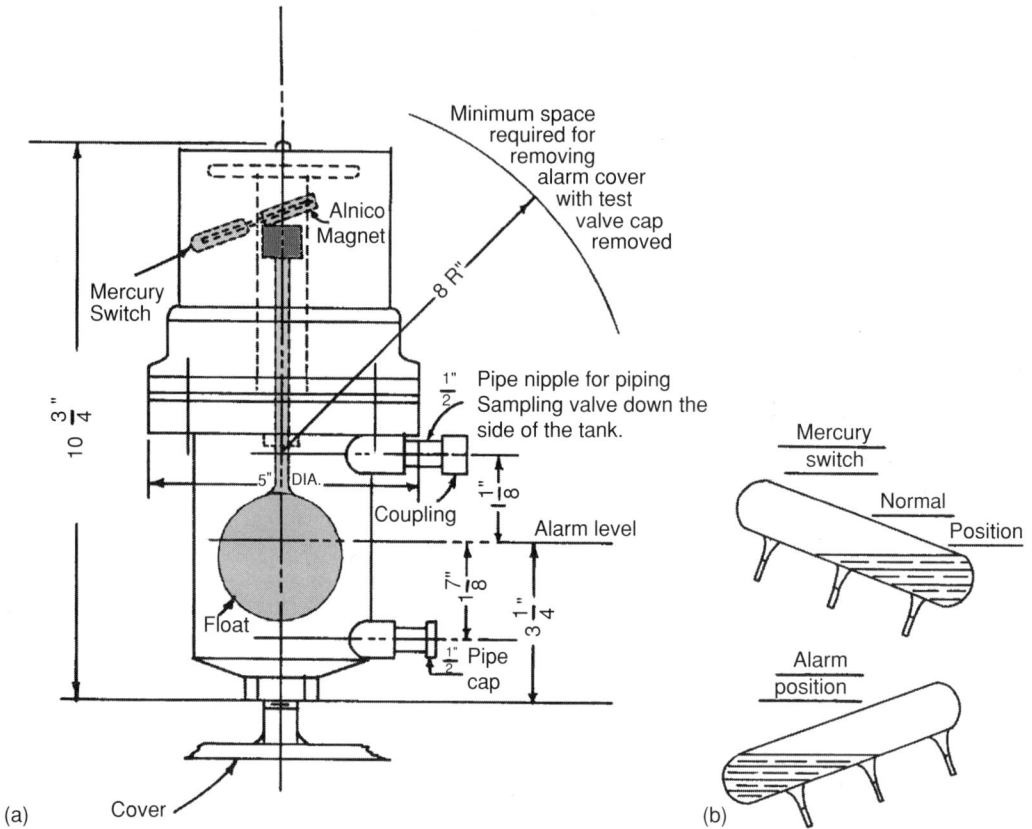

FIGURE 16.72
Later construction of gas detector relays based on mercury switches. (General Electric gas detector relay.)

violent gassing. In such cases the oil stream, directed into the conservator through the Buchholz relay installed on its way, affects blade 4 (Figure 16.73), causing rotation of the cup and the closing of the contacts.

When dry-reed switches controlled by a permanent magnet appeared, practically all producers of Buchholz relays started to apply these elements (Figure 16.74), and molded floats from foam plastic.

Various modern constructions have negligible construction differences (Figure 16.75), for example, translation coaxial displacement of floats (Figure 16.76) instead of an angular one.

Two essential parameters of the Buchholz relay are sensitivity and noise-immunity. These two parameters are equally important (malfunctioning of the relay or its false operation, and disabling of the power transformer are equally adverse) but their functions are not. It is impossible to considerably increase the sensitivity of the Buchholz relay without deterioration of its noise-immunity, which is why adjustment of such a relay is always a compromise between its sensitivity and noise-immunity.

False pick ups of the relays cause some specific problems during short circuits in the high-voltage line (that is outside of the transformer), which is why no additional gassing occurs inside the transformer and the Buchholz relay must not pick up. If such relay does pick up, the maintenance staff is faced with a dilemma: either to ignore such pick up of the Buchholz relay and switch the transformer under voltage, or to disable the trans-

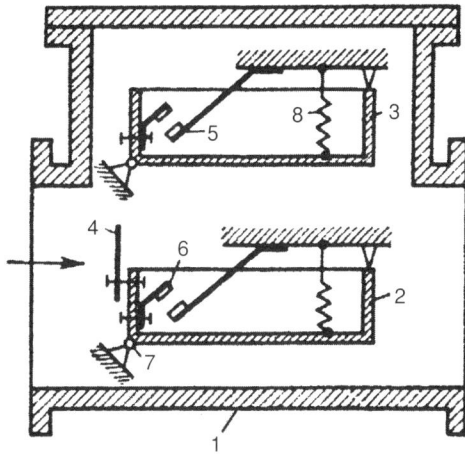

FIGURE 16.73

Construction of a Buchholz relay with cup-shaped elements. 1 — Case; 2 and 3 — open flat-bottomed aluminum cups; 4 — paddle; 5 — stationary contact; 6 — movable contact; 7 — axis of rotation of the cup; 8 — spring.

former and examine its insulation. In either case, this is quite a crucial decision because one runs a risk of very great potential damage.

Examination of a situation like that which the author happened to witness has shown that the prime certification center KEMA, conducting tests and certification of power transformers for the power industry, does not check and fix the state of the Buchholz relay when transformers are tested for resistance to short-circuit currents. Thus it remains unclear how this or that type of the Buchholz relay will operate during exploitation in

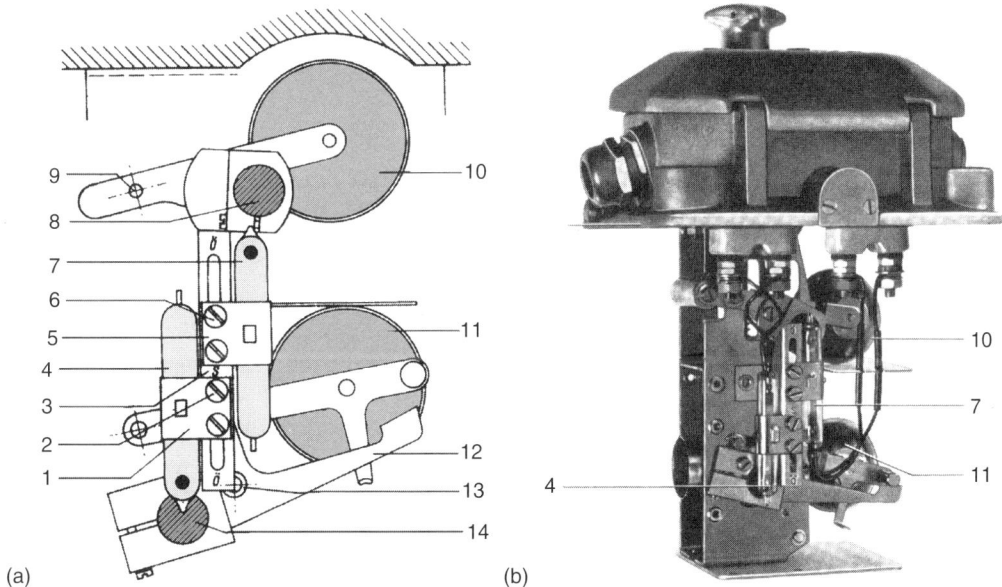

(a)   (b)

FIGURE 16.74

Buchholz relay of the RT 823 type, with dry-reed switches, operated by a moving permanent magnet (Siemens). 1 and 5 — Brackets for rearrangement of contact (reed switch) tubes; 2 and 6 — screws; 3 — slide rail for contact tube rearrangement; 4 and 7 — dry-reed switch tubes; 8 and 14 — permanent magnets; 9 — fulcrum of float; 10 — alarm float; 11 — trip float; 12 — tripping lever; 13 — bearing.

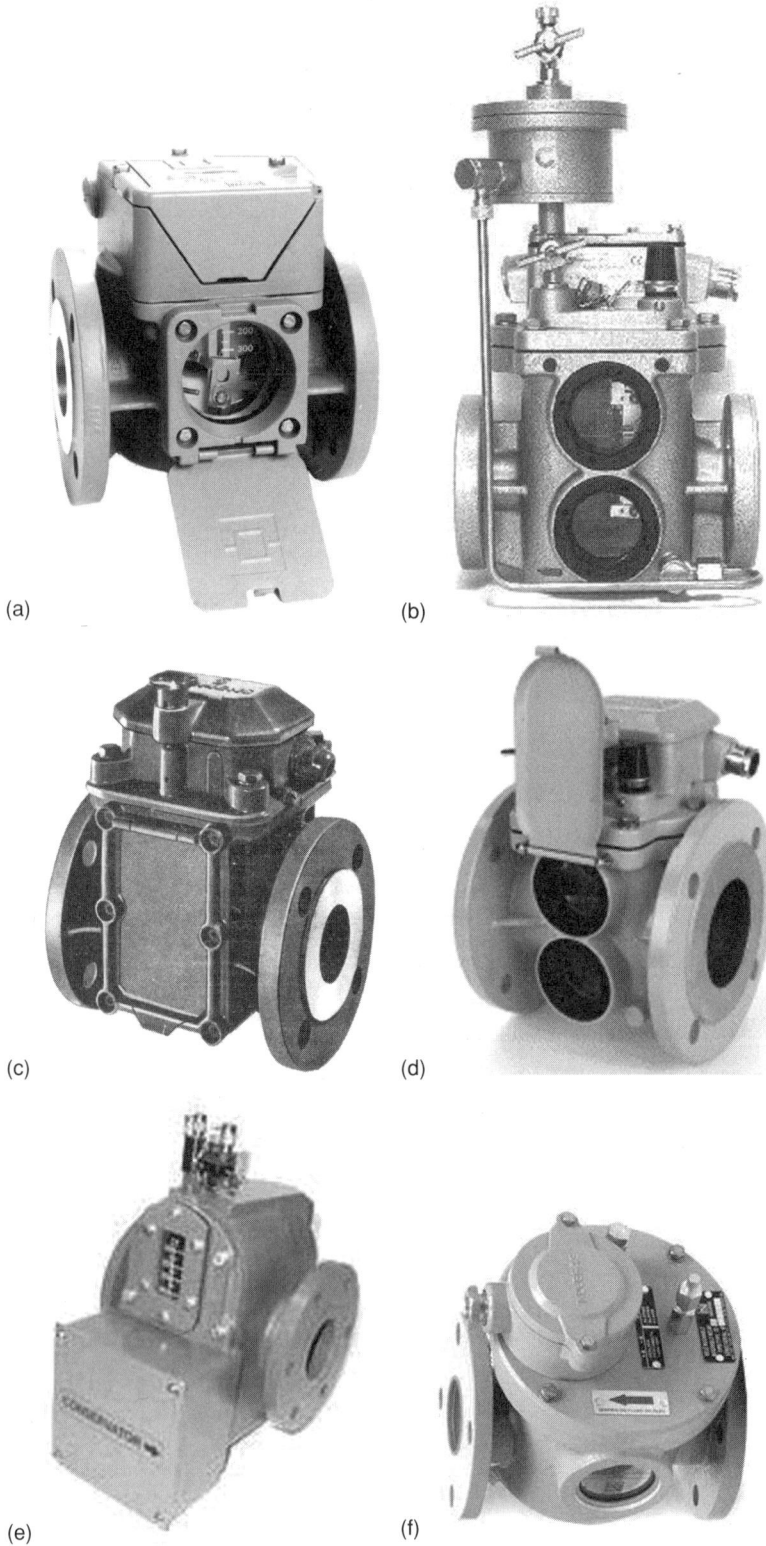

(a)

(b)

(c)

(d)

(e)

(f)

FIGURE 16.75
Modern types of Buchholz relays, produced by different companies.

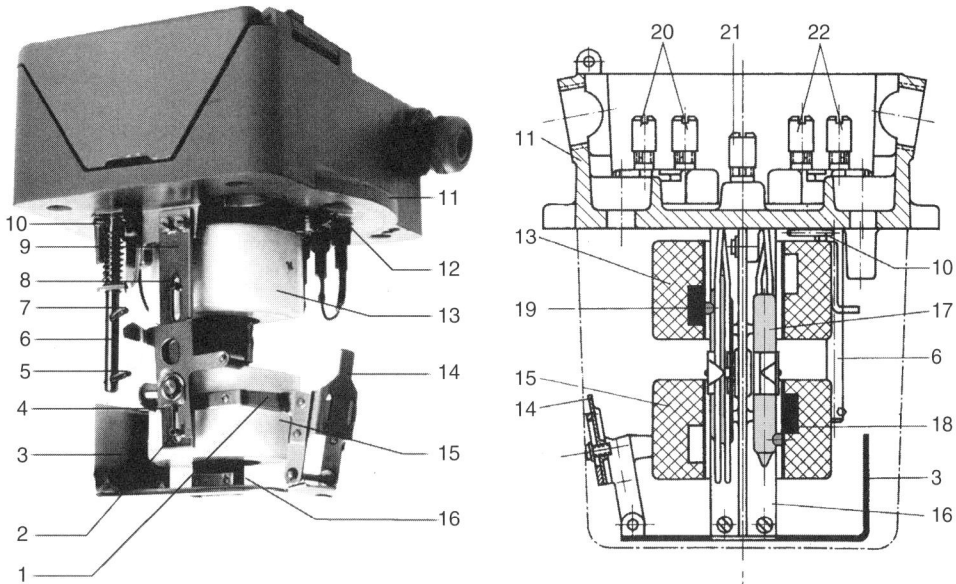

FIGURE 16.76

Modern Buchholz relay of the DR-50 type, with translating floats and permanent magnets molded into solid floats (Siemens). 1 — Actuating lever; 2 — location pin; 3 — flow guide plate; 4 — latch for bottom contact system; 5 — bottom test pin; 6 — test rod; 7 — release pin; 8 — location pin; 9 — guide rail; 10 — top test pin; 11 — housing cover; 12 — triple bushing; 13 — top float (alarm); 14 — baffle plate; 15 — bottom float (trip); 16 — strut (T-section); 17 — dry-reed switch for bottom float (trip); 18 — permanent magnet in bottom float; 19 — permanent magnet in top float; 20 — alarm terminals; 21 — earthing terminal; 22 — trip terminal. (Siemens. Twin Float Buchholtz relays.)

real circumstances, when short-circuit currents cause a sudden displacement and change of size of the windings of the transformer, leading to hydraulic shock in the oil and a shock wave in the Buchholz relay. Unfortunately, even modern constructions are prone to such false pick ups and the maintenance staff can only attempt to surmise the real reasons for picks up of the relay. One more problem is not adequate reaction of the relay at earthquake when displacement of internal parts of the transformer leads to occurrence of a wave in oil and to moving floats of the relay (and actuating contacts, of course).

16.11 Safety Relays

The technical requirements applicable to the design of control systems are stipulated in European Norm (EN) Standard EN 954-1, "Safety-related components of control systems." This standard is applied following the assessment of the overall risk to EN 1050, "Risk Assessment." European standard EN 954-1 stipulates that machines can be classified into five categories, whereby the safety circuits must be designed in accordance with the requirements of the relevant category. Moreover, standards VDE 0113, part 1, EN 60204, Part 1, and ICE 204 1.10 apply to control systems required to perform safety-related tasks. All safety relays can be used on the basis of their classification into the risk categories in EN 954-1, are approved by the employers' liability insurance associations and/or the German Technical Inspection Authority (TÜV), and comply with the requirements of EN 60204,

Part 1. EN 954-1, in order to achieve as high a level of personal safety and machine protection as possible. This classification is preceded by a risk assessment to EN 1050, which allows for various criteria such as the ambient conditions in which the machine is operated.

Control reliability information can also be found in documents published by the American National Standards Institute (ANSI) and Occupational Safety and Health Administration (OSHA). ANSI is an institute that provides industry guidance through their published machinery standards. OSHA is a U.S. government agency responsible for labor regulations. These organizations have provided the following definitions for control reliability. "Control Reliability" means that, "the device, system or interface shall be designed, constructed and installed such that a single component failure within the device, interface, or system shall not prevent normal stopping action from taking place but shall prevent a successive machine cycle." (ANSI B11.19-2003 "Performance Criteria for Safeguarding"; ANSI B11.20-1991 "Machine Tools — Manufacturing Systems/Cells — Safety Requirements for Construction, Care, and Use.") In addition, OSHA 29 CFR 1910.217 states that, "the control system shall be constructed so that a failure within the system does not prevent the normal stopping action from being applied to the press when required, but does prevent initiation of a successive stroke until the failure is corrected. The failure shall be detectable by a simple test, or indicated by the control system."

Unfortunately, there is more than a little confusion in the world of safety terms and codes, but the discussion of these problems is beyond the framework of our book The risks and dangers, and the possible technical measures to reduce these risks and dangers, are stipulated in the subsequent assessment of the overall risk.

Electric relays contain many parts, which are subject to dynamic, electrical, or thermal wear. There are many applications where safety is very critical and it is important to use electrical equipment, ensuring that dangerous machine movement cannot occur when a fault is detected with the moving relay contacts during the cycle in which the fault is indicated.

In order to assure safe function, especially in the event of a failure, appropriate controls are built into the circuits of safety devices. Relays with forced guidance contacts play a decisive role in preventing accidents in machines and systems. Safety control circuits enable switching into a failsafe state. Forcibly guided contacts monitor the function of the safety control circuits. For this safety function, all assumed faults that can occur must already have been taken into consideration and their effects examined. Standard EN 50205, "Relays with forcibly guided contacts," contains current internationally defined design requirements. Relays with forcibly guided contacts that comply with EN 50205 are also referred to as "safety relays."

Safety relays (Figure 16.77) are used in all control circuits for safeguarding devices such as interlocks, emergency stops, light screens, safety mats, and two-hand controls, to comply with the control reliability requirement. Safety relays have *Positive-Guided* (*Force-Guided* or *Captive,* IEC 60947-1-1) contacts, which are very different from conventional relays. The actuator (mechanical linkage) for the positive-guided relay is placed much closer to the contacts than on conventional relays. This placement of the actuator, and the lack of gap tolerance on the positive-guided relay, insures a consistent relationship between the NO contacts and the NC contacts: contacts in a contact set must be mechanically linked together so that it is impossible for the NO Make contacts and the NC Break contacts to be closed at the same time. The relays can provide positive safety for the NO and NC contacts, which assure that the *NO contacts will not close before any NC contact opens.* Therefore, if one of the contacts welds due to abnormal conditions in the control circuit, the other contacts will also remain in the same position as when the welding occurred. The positive-guided relay is guaranteed to maintain a minimum 0.5 mm distance between its

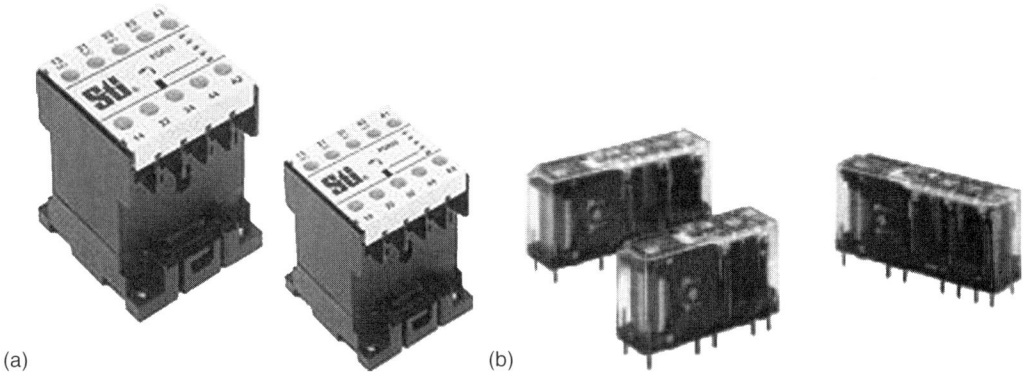

FIGURE 16.77
Positive-guided relays with mechanically-linked contacts conforming to IEC 60947-1-1 as required for safety-related control systems: (a) FGR type (STI Scientific Technologies GmbH) (b) G7SA type (Omron).

NC contacts when the NO contact is held closed. This characteristic makes the positive-guided relay a preferred relay when designing safety circuits for generating a safe output.

Safety relays are specifically designed to monitor safety-related control systems, such as emergency-stop circuits, safety mats and bumpers, security doors, standstill, over-travel monitors; two-hand controls, etc. Special expansion units are available if a large number of safety circuits are required.

In accordance with EN954-1, "Safety-related components of emergency stop and monitoring of guard doors control systems," the classification is made on the basis of one of five possible risk categories. These categories: B, 1, 2, 3, and 4 (the highest), then indicate the requirements applicable to the design of the safety equipment, whereby B as the basic category describes the lowest risk and stipulates the minimum requirements. Thus, for instance, category 2 requires compliance with the requirements of B and the use of time proven safety principles. Moreover, the safety function must be checked at appropriate intervals by machine control.

The choice of emergency stop of the machine is determined by its risk assessment. In EN 60204-1 the Stop Function is divided into three categories. Emergency stops must conform to category 0, category 1, or category 2.

Category 0 — STOP:
Shut-down by immediate switching off of the power supply to the machine drives (uncontrolled shut-down)

Category 1 — STOP:
A controlled shut-down, whereby the power supply to the machine drives is maintained in order to achieve the shut-down, and the power supply is only interrupted when the shut-down has been achieved.

Category 2 — STOP:
A controlled shut-down, whereby the power supply to the machines is maintained.

Stop functions of category 1 or 2 or both must be provided when this is necessary for the safety and functional requirements of the machine. Category 0 and category 1 stops must be able to function independently of the mode, and a category 0 stop must have priority. Stop functions must take place by deenergizing the corresponding circuit and have priority over the associated start functions.

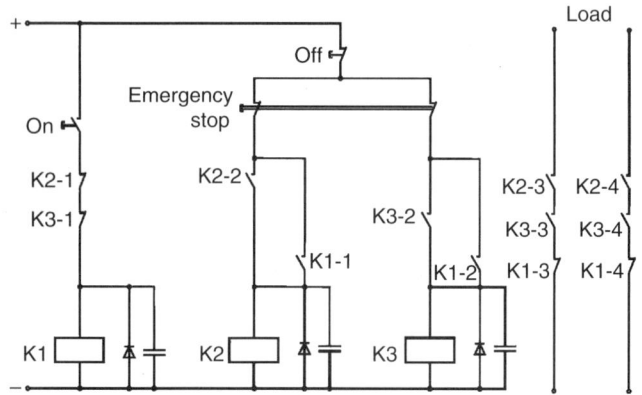

FIGURE 16.78
Basic connection diagram for safety emergency-stop relays.

In a redundant circuit (e.g., EMERGENCY STOP), it is possible for cross-circuiting to go unnoticed. If an additional fault then occurs, the safety device ceases to be effective. This is exactly what must not happen in a category 4 circuit. In other words, a cross-circuit does not cause the EMERGENCY STOP switch to be bridged. Other possible faults are what cause this. For this event, the safety relays are equipped with cross-circuiting detection.

Due to such switching (Figure 16.78), no damages of any of the relays can lead to lack of emergency switching of the load. Many companies around the world produce safety relays of different types. The Allen-Bradley Company produces the greatest number of variants of relays of this kind. Some variants of safety relays produced by Allen-Bradley under the trade brand "Minotaur" are briefly considered below.

The MSR5T relay (Figure 16.79.) has one NC single channel input for use with gate interlocks, and emergency-stop buttons, in lower risk applications. The MSR5T has output monitoring that can accommodate an automatic or manual-reset function. Automatic or manual-reset can use a jumper or can be used to check operation of the contacts. The MSR5T has three NO safety outputs and one NC auxiliary output. The safety outputs have independent and redundant internal contacts to help ensure the safety function. The

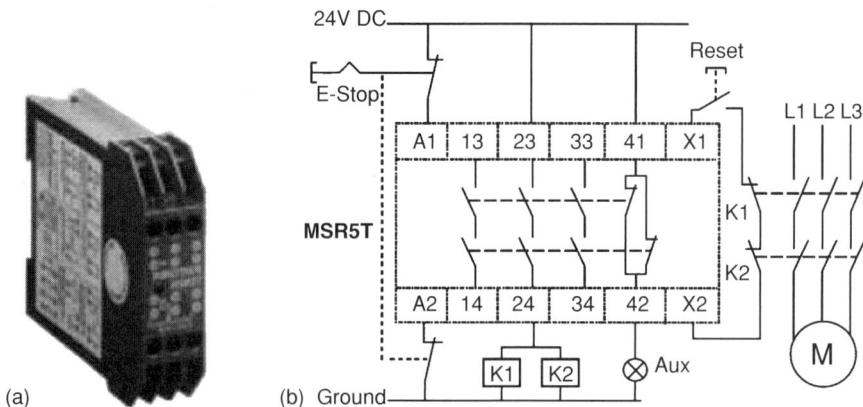

FIGURE 16.79
A dual-channel E-stop relay, MSR5T, with manual reset and monitored output. (Allen Bradley [Rockwell Automation 2004].)

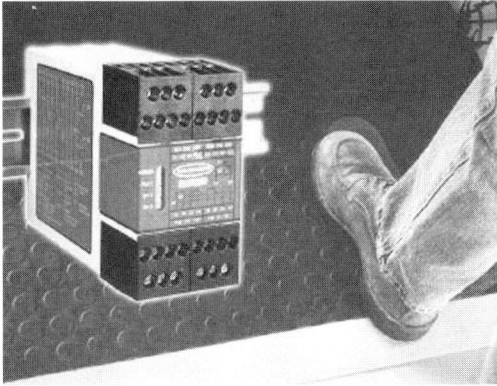

FIGURE 16.80
Safety-mat relay of the SM-GA-5A type (Banner Engineering Co.)

auxiliary contact is a nonsafety output, intended to provide an external signal regarding the status of the safety outputs.

The *Safety-mat relay (Monitor* — Figure 16.80) feature provides a uniform activation threshold (on/off signal) throughout the entire mat surface area. Modern design of uniform activation also provides a guarding system that contains no dead zones. This provides the user with a much safer guarding system, as well as compliance with domestic international standards.

The relay's sensor circuit monitors the contact plates of the safety mat and consists of bi-polar (diverse) redundant channels that issue the stop command (i.e., open the safety outputs) when the two channels are shorted together, as individual steps onto the safety mat. The relay provides the redundant safety outputs required to create a control-reliable safety circuit. Contacts include four redundant, forced-guided (positive-guided) outputs rated at 6 A. One NC output monitors status, and two auxiliary solid-state outputs indicate the state of the internal relays and power supply. The safety mat relay offers two primary functions. It monitors the contacts and wiring of one or more safety mats, preventing machine restarts in the event of a mat or module failure. It also provides a reset routine after an operator steps off a mat (per ANSI B11 and National Fire Protection Association (NFPA) 79 machine safety standards, via selectable Auto-Reset or Monitored Manual Reset modes).

The MSR23M (Figure 16.81) control relay is designed to monitor four wire safety mats that are connected together to form a safeguarded zone. The size of the safeguarded zone is limited by the total input impedance (100 Ω maximum), created by the wiring and connections.

The controller is designed to interface with the control circuit of the machine and includes two safety relays to ensure control redundancy. The controller detects a presence on the mat, a short circuit, or an open circuit. Under each of these conditions, the safety output relays turn off. When interfaced properly, the machine or hazardous motion receives a stop signal, and an auxiliary output turns ON.

Special safety relays are available for *two-hand controls* (Figure 16.82). Two-hand circuits (Switch-1 and Switch-2) require simultaneous operation by both hands to initiate and maintain the operating status of a machine. As a result the operator is protected, because he cannot reach the danger zone during hazardous procedures. The electronic safety relay monitors whether or not both buttons are operated within 0.5 sec of one another.

The MSR22LM safety-monitoring relay (Figure 16.83) is designed to monitor light curtains with the added features of muting and *presence sensing device initiation* (PSDI). It provides an output to a machine control system when the light curtain is clear. When

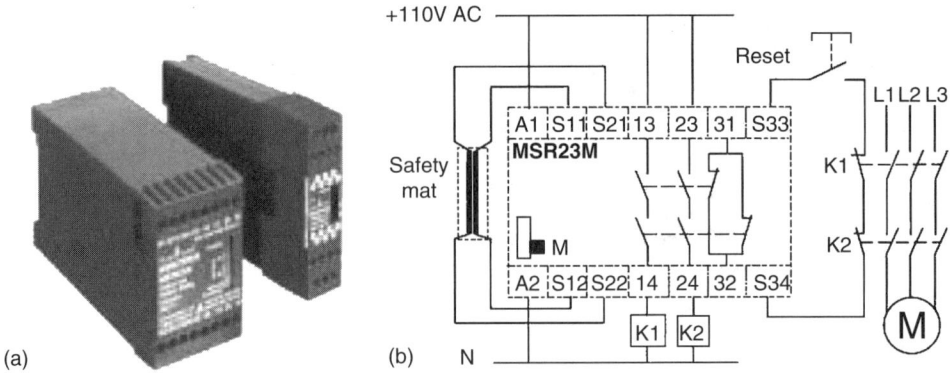

FIGURE 16.81
Safety-mat relay of the MSR23M type, with dual-channel monitored output and manual reset. (Allen Bradley [Rockwell Automation 2004].)

the inputs to the MSR22LM are closed (conducting), the output relays are closed if the monitoring circuit is satisfied. The MSR22LM has three sets of dual channel inputs. This allows it to operate in four different configurations:

1. Monitors up to three light curtains in guard only mode.
2. Monitors up to two light curtains with two muting sensors (only one curtain muted).
3. Monitor one light curtain with four muting sensors.
4. Monitors up to three light curtains with PSDI (only one curtain initiated).

The MSR21LM uses microprocessor-based technology to offer a wide variety of advanced safety solutions in a small 45 mm DIN rail mounted housing. Internal selector switches provide for easy selection of up to ten different applications. Four LEDs give

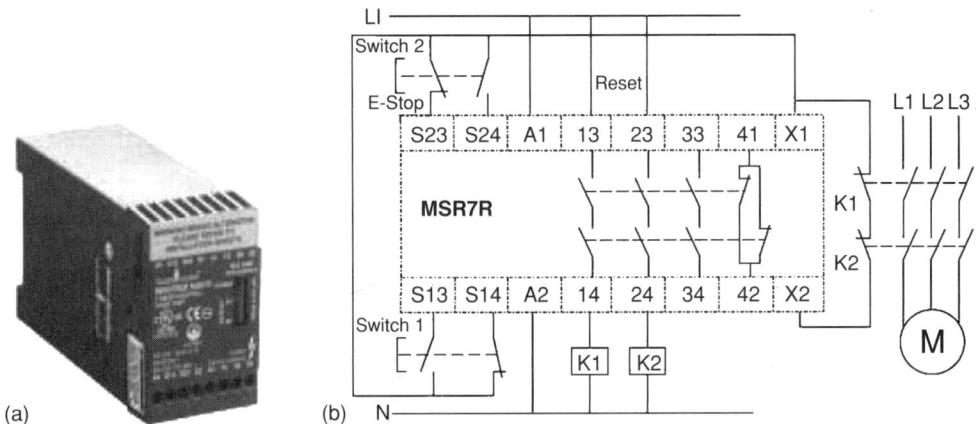

FIGURE 16.82
Two-hand safety relay of the MSR7R type, with dual-channel monitored output and automatic reset. (Allen Bradley [Rockwell Automation 2004].)

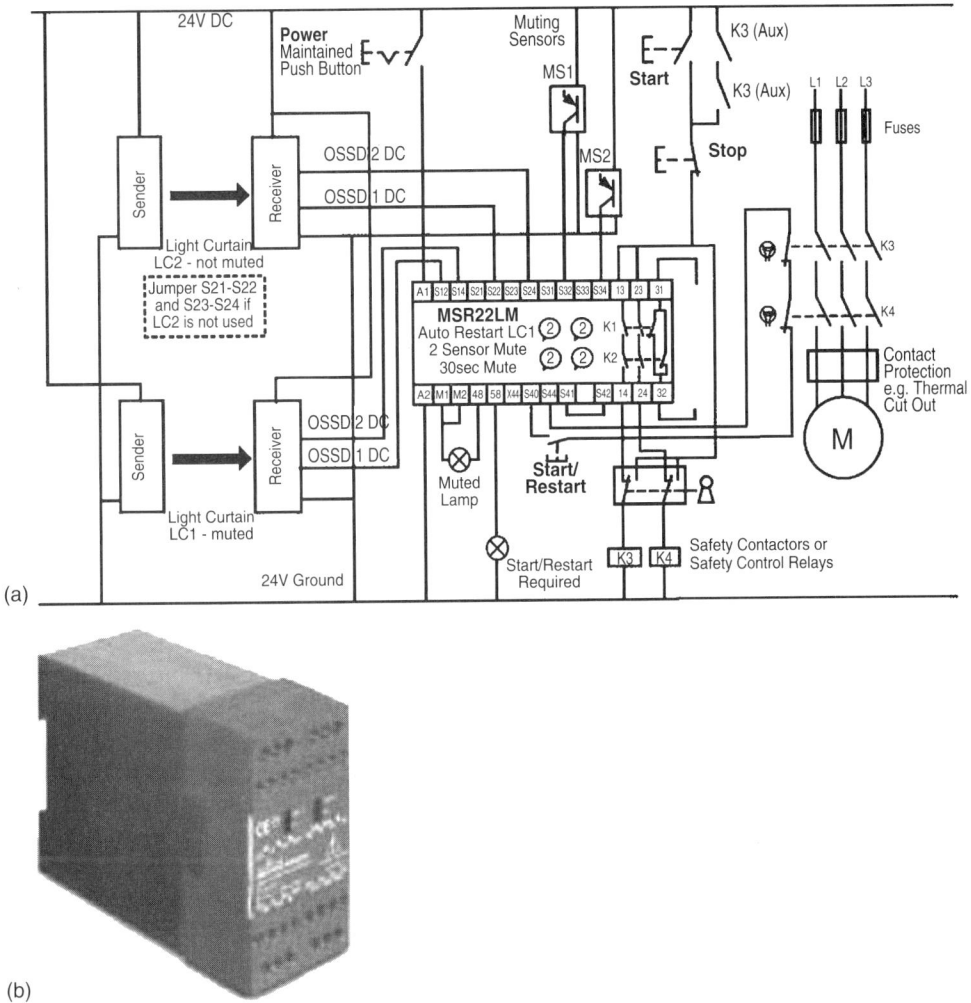

(a)

(b)

FIGURE 16.83
The MSR22LM safety monitoring relay. (Allen Bradley [Rockwell Automation 2004].)

operational status as well as diagnostic information. Removable terminals reduce wiring and installation costs when replacement is necessary.

If the stopping time of the machinery is unpredictable, use a *Standstill Relay* (Figure 16.84). This relay measures the back EMF of the connected motor from the terminals of one stator winding. When the EMF has decreased to almost zero, this device detects that the motor has stopped and energizes its output relays.

In addition, the FF-SR05936 monitors the connections to the motor for broken wires on terminals Z1 and Z2. If an open (line break) is detected, the output relay contacts the latch in the deenergized position, as if the motor was running. After the break has been repaired, the relay is reset by momentarily removing power to the module.

We have considered only a few main types of safety relays produced by several companies. There are several other types of such relays and the production of them has turned into quite a large industry, in which tens of major and many smaller companies work.

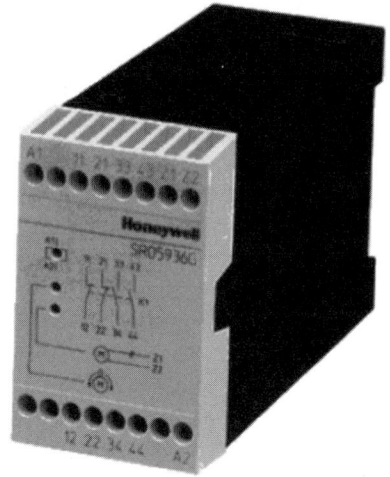

FIGURE 16.84
Standstill relay of the FF-SR05936 type (Honeywell).

16.12 Ground Fault Relays

A *Ground Fault Relay* is a device that is intended to trip out an electricity supply in the event of a current flow to earth. As such, it can provide protection from harmful electric shocks in situations where a person comes into contact with a live electrical circuit and provides a path to earth. These devices KILL THE CURRENT before the current KILLS YOU! Typical examples of this occurring are with the use of faulty electrical leads and faulty appliances.

Fuses or overcurrent circuit breakers do not offer the same level of personal protection against faults involving current flow to earth. Circuit breakers and fuses provide equipment and installation protection and operate only in response to an electrical overload or short circuit. Short-circuit current flow to earth via an installation's earthing system causes the circuit breaker to trip, or a fuse to blow, disconnecting the electricity from the faulty circuit, however, if the electrical resistance in the earth fault current path is too high to allow the circuit breaker to trip (or a fuse to blow), electricity can continue to flow to earth for an extended time. Ground fault relays detect a very much lower level of electricity flowing to earth and immediately switch the electricity OFF.

In various countries they have different names for ground fault relays: In *Germany* and in *Austria*: "Fehlerstrom-Schutzschalter" or "Fehlerstrom-Schutzeinrichtung" (Schutzschalter — protective switch, Schutzeinrichtung — protective device), and they also use the abbreviation FI (F — Fehler — fault, error, escape, I — indication of current in electrical engineering); in *France*: "Disjoncteur Differentiel" (differentia switch) or — DD in the abbreviated form; in *Great Britain*: "Earth Leakage Circuit Breaker," ELCB or ELB in abbreviated form; in the *U.S.A. and Canada*: "Ground Fault Circuit Interrupter," GFCI or GFI in abbreviated form; in *Israel*: "Mimsar Phat" ("Mimsar — relay, "Phat" — remainder, remaining).

At present, the International Electrotechnical Commission applies a joint name for all types of devices of this kind: *Residual Current Devices (RCD)*. There are also some derivatives of this mane:

- Residual current circuit breaker (RCCB) — mechanical switch with an RCD function added to it. Its sole function is to provide protection against earth fault currents.

- Residual current Breaker with overcurrent protection (RCBO) — an overcurrent circuit breaker (such as an MCB) with an RCD function added to it. It has two functions: to provide protection against earth fault currents and to provide protection against overload currents.

An RCD should be fitted to socket outlets installed in wet environments and outside, and must be provided on sockets in premises of public entertainment such as clubs, village halls, and pubs:

- SRCD — socket outlet incorporating an RCD
- PRCD — portable RCD, usually an RCD incorporated into a plug
- SRCBO — a socket outlet incorporating an RCBO

The basis for protective switching OFF as an electro-protective means is the principle of limitation (due to rapid switching OFF) of duration of current flowing through the human body when one unintentionally touches elements of charged electrical installation. Of all known electro-protective means, the RCD is the only one that protect a human being from electric current upon direct touch of a current-carrying unit.

Another important property of the RCD is its capability to provide protection from ignition and fires taking place in units because of various insulation damages, electric wiring and electrical equipment failure. Short circuits occur, as a rule, due to insulation defects, earth connections, and earth current leakages. In addition, the energy released at the point of insulation fault at leakage current flowing may be enough for ignition. Released power of just 50 to 100 W can ignite a fire, depending on the material and service life of the insulation. This means that timely pick up of a fire-preventive RCD with a pick-up threshold of 300 to 500 mA will prevent a power release, and therefore the ignition itself (RCDs for such pick-up currents cannot protect people from electrical shock).

The first construction of an RCD was patented by the German firm RWE (Rheinisch — Westfalisches Elektrizitatswerk AG) in 1928 (DR Patent No. 552678, 08.04.28). It suggested using the well-known principle of current differential protection of generators, power lines, and transformers for protection of people from electrical shocks. In 1937 Schutzapparategesellschaft Paris & Co. produced the first functioning device of this type, based on a differential transformer and a polarized relay with a sensitivity of 0.01 A and speed of operation of 0.1 sec. In the same year a volunteer conducted a testing of the RCD. The experiment was successful, the device picked up in time and the volunteer experienced only a slight electrical shock (though he refused to take part in further tests). During the next few years, with the exception of the war years and the postwar period, much research was done concerning the impact of electric current on human beings and development of electro-protective means; including first and foremost, the development and implementation of the RCD. In the 1950s it was stated that the human heart is most prone to electric current impact. Fibrillation (irregular twitching of the muscular wall of the heart) may occur even at small current values. Assumptions that asphyxia, muscle paralysis, and cerebral affection were primary reasons for the lethal outcome of electric current impact, ceased to have significance. It was also determined that the impact of electric current on the human organism depends not only on the current value but also on its duration, its route through the human body, and to a lesser degree on current frequency, form of the curve, pulsation factor, and some other factors. Results of research of the impact of electric current on humans are given in numerous publications, and serve as a basis for today's existing standards. One should pay special attention to fundamental

TABLE 16.1

Affection of Electric Current on the Human Organism.

Current (mA)	Effect	Result
0.5	Is not felt. Slight sensation by the tongue, finger-tips or through the wound	Safe
3	Sensation similar to that of ant bite	Safe
15	No chance to drop the conductor if one happens to touch it	Unpleasant, but not dangerous
40	Convulsions of the body and the diaphragm	Risk of asthma for a few minutes
80	Vibration of the ventricle of heart	Very dangerous immediate lethal outcome

research carried out in the 1940–50's at the University of California (Berkeley) by the American scientist Charles F. Dalziel. He conducted a set of experiments with a large group of volunteers to determine electrical parameters of the human body and the physiological impact of electric current on human beings. Results of his investigations are considered to be classical and are still significant at present. The electric current is considered to effect the human organism in the following way (see Table 16.1 and Table 16.2).

In the 1960–70's the RCD began to be actively implemented in practice all over the world, first place in countries of Western Europe, Japan and the U.S. At present, according to official statistics, hundreds of millions of RCDs successfully protect life and property of citizens of the U.S., France, Germany, Austria, Australia, and some others from electrical shocks and ignitions. The RCD has become a usual and compulsory element of any industrial or social electrical installation, of any switchboard panel, in all mobile dwellings (caravans in camping areas, commercial vans, junk food vans, small temporal electrical installations of external units fixed during festivals), hangars, and garages. RCDs are also integrated to switch sockets and units through which electrical tools or household devices are exploited in dangerous humid, dusty, etc., places with conductive floors. Insurance companies take into account insurance of RCDs installed on the unit and their state. Statistically, at present every resident of the countries mentioned above has at least two RCDs.

TABLE 16.2

Another Classification of Threshold Levels for 60 Hz Contact Currents.

Current (mA)	Threshold Reaction or Sensation
	Perception
0.24	Touch perception for 50% of women
0.36	Touch perception for 50% of men
	Cannot let-go
4.5	Estimated let-go for 0.5% of children
6.0	Let-go for 0.5% of women
9.0	Let-go for 0.5% of men
	Heart fibrillation
35	Estimated for 0.5% of 45 lb children
100	Estimated for 0.5% of 150 lb adults

Nevertheless, tens of major firms such as Siemens, ABB, GE Power, ABL Sursum, Baco, Legrand, Moeller, Merlin-Gerin, Cutler-Hammer, Circutor (In Russia: Gomel Plant "Electrical Equipment," Kursk Public Corporation "Electrical Device," Moscow Electrical-Type Instruments Plant, Cheboksar Electrical Equipment Plant) continue to produce such devices in different modifications, constantly improving their engineering factors.

Functionally the RCD can be defined as a high-speed protective relay, responding to differential current in conductors conducting electric power to a protected electrical installation. A differential measuring current transformer is its essential component.

Regardless of the function of the RCD it is switched ON in such a way that all working load currents, both phase current(s) and midpoint wire current, must pass through its most sensitive element, the differential current transformer. When it is switched ON accordingly, and when there are no faults, the magnetic fields created by all of these currents are mutually compensated (the algebraic sum of all currents passing through the transformer equals zero), and there is no voltage on the output (secondary) winding of the transformer. When the insulation is damaged (Figure 16.85) or one touches one of the phase conductors (Figure 16.86), additional leakage current through this damaged insulation or through the human body occurs. This current upsets the general balance of currents flowing through the transformer. Since it flows in only one direction, between the phase and the ground, it is not compensated by the reverse current of the neutral wire. This additional current is the "*residual current*," inducing EMF in the second winding of the transformer. At that point the induced voltage is applied to the executive relay, which picks up at a certain level of input voltage (proportional to the residual current) and defuses the circuit. Technically, construction of executive relays can be divided into electromechanical and electronic ones.

Electromechanical relays do not depend functionally on the voltage of the power supply (so-called: "*Voltage Independent*" or *VI* — type). The source of energy necessary for protective functions (that is disabling operations) is the differential current to which it responds. Such relays are based on sensitive direct-action polarized latching relays (Figure 16.87 and figure 16.88).

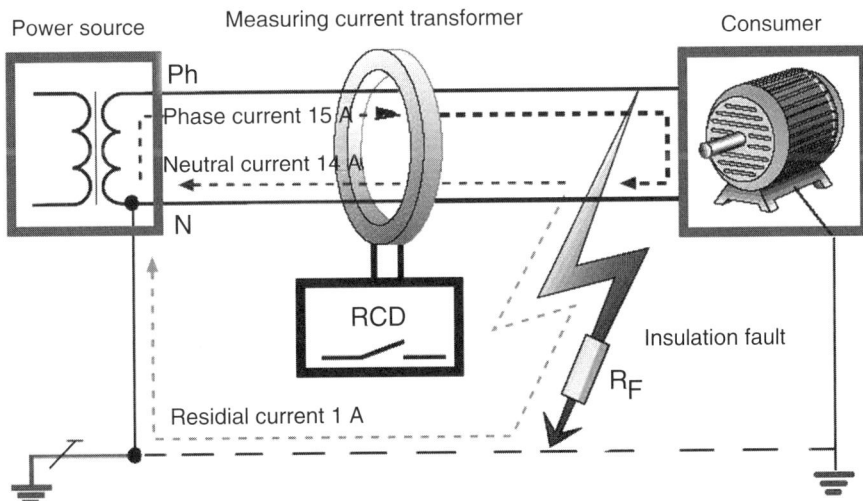

FIGURE 16.85
Principle of operation of a fire-preventive single-phase RCD device for pick-up current of 300, 500, and 1000 mA.

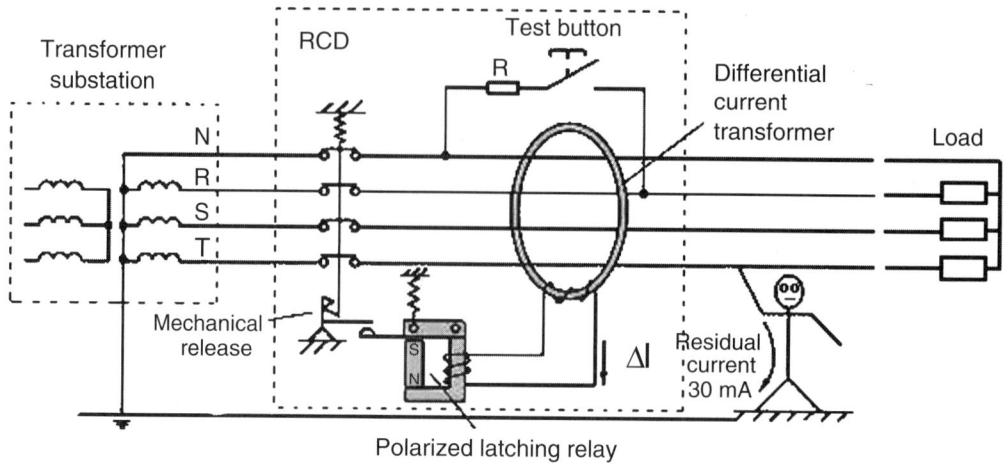

FIGURE 16.86
Principle of operation of a three-phase RCD device protecting from electrical shocks, with pick-up settings of 10, 30, and 100 mA.

Electronic executive relays depend functionally on the voltage supply (being *"Voltage Dependent"* or of the *VD* — type) and their switching-OFF mechanism requires electrical energy, which they obtain from the controlled circuit. The executive relay affects the disabling mechanism, which contains a contact group and a drive.

The principle of operation of the RCD can lead to the following conclusion: An RCD will significantly reduce the risk of electric shock, however, an RCD will not protect against all

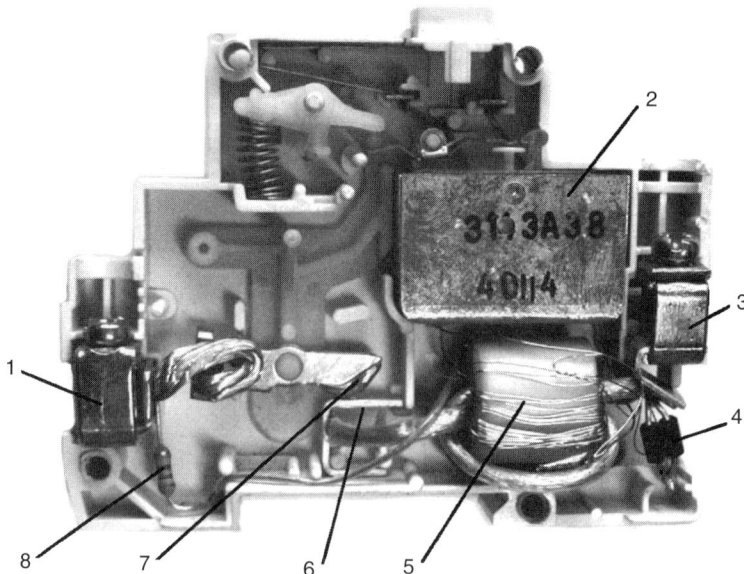

FIGURE 16.87
Construction of a single-phase electromechanical RCD made in the form of an additional section of a protective relay (automatic circuit breaker) of combined action. 1 and 3 — Terminals for connecting to the external circuit; 2 — executive relay; 4 — rectifying diodes for supply of the executive relay; 5 — differential current transformer; 6 and 7 — contact system; 8 — resistor of the tested circuit.

FIGURE 16.88
Construction of an electromechanical executive relay (enlarged). 2.1 — Coil; 2.2. — magnetized core keeping the armature of the relay on attracted position; 2.3. — armature-pusher; 2.4. — spring; 2.5 — spring tension (pick-up threshold) regulator; 2.6. — outlets of the coil.

instances of electric shock. If a person comes into contact with both the Active and Neutral conductors while handling faulty plugs or appliances causing electric current to flow through the person's body, this contact will not be detected by the RCD unless there is also a current flow to earth. On a circuit protected by an RCD, if a fault causes electricity to flow from the Active conductor to earth through a person's body, the RCD will automatically disconnect the electricity supply, avoiding the risk of a potentially fatal shock.

To check the serviceability of the RCD it is supplied with a test button creating a nonbalanced current (flowing through the differential transformer only in one direction) limited by the resistor R to the level of nominal pick-up current. This current affects the device in the same way as leakage current between phase and ground.

Electronic RCDs are constructed from standard elements (Figure 16.89) and special integrated circuits (Figure 16.90). Electromechanical RCDs are considered to be more reliable than electronic ones. In European countries — Germany, Austria, France — electrical specifications permit only applications of the first type of RCD — voltage-independent ones. RCDs of the second type can be applied in circuits protected by electromechanical RCDs only as additional protection for ultimate consumers, for example for electric tools, nonstationary electric transmitters, etc. The most essential drawbacks of electronic RCDs are considered to be malfunctioning during frequent and most dangerous, in terms of possibility of electric shocks, fault of the electrical installation — during breaks of the zero (neutral) conductors in the circuit up to the RCD in the direction to the power supply. In this case the "electronic" RCD will not function without a supply and potential dangerous for a human is applied to the electrical installation through the phase conductor.

That is why in constructions of many "electronic" RCDs, there is a function of switching OFF from the circuit of the protected electrical installation when there is no voltage of the supply. Constructively, such function is implemented with the help of an electromagnetic relay operating in the mode of self-holding. The power contacts of the relay are in the "ON" position only when current flows through its winding. When there is no voltage on the input outlets of the device, the armature of the relay falls OFF, the power contacts open and the protected electrical installation is defused. Similar construction of the RCD provides secure protection of human-beings from electrical shocks in electrical installations if a break of the zero conductor takes place. It is also notable that electronic RCDs are prone to impact of moisture and dust, which can delay the pick-up time of the device. Unstable voltage in circuits and voltage drops have a negative impact on electronic RCDs and can lead to their malfunctioning.

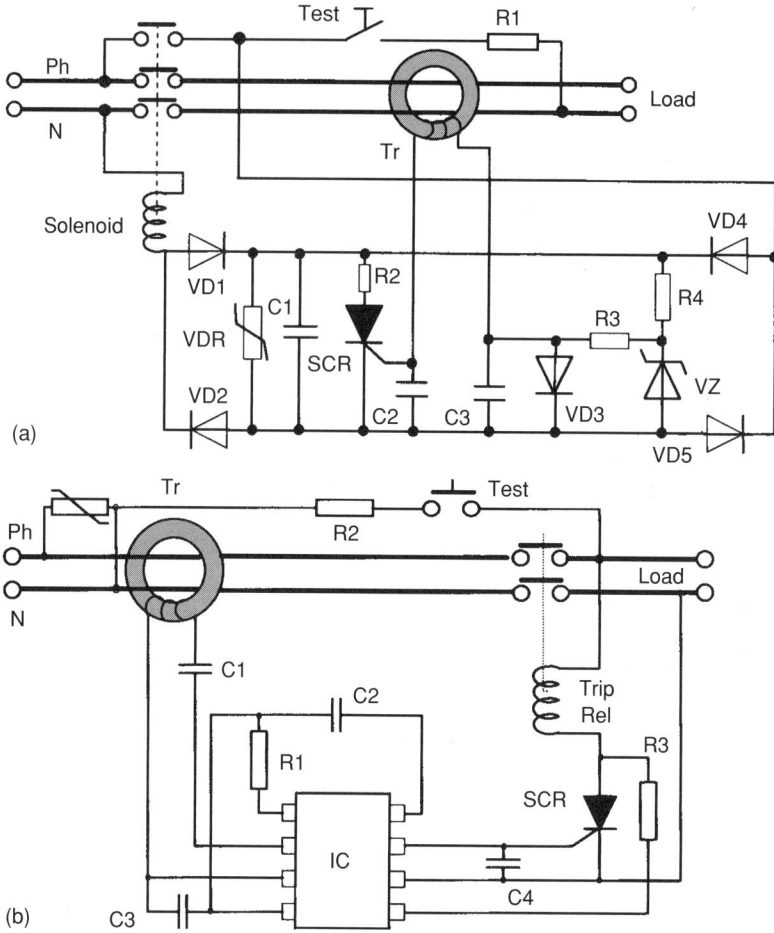

FIGURE 16.89
Electronic RCD based on standard elements.

As already mentioned above, RCDs differ from each other by sensitivity (pick-up current value). In accordance with the IEC Standard 1008/1009 of 30 mA, sensitivity for domestic and personal protection is a tolerance of 30 mA plus zero and minus 50%, that is a range from 15 to 30 mA. RCD units are manufactured to operate in a tolerance band of 19 to 26 mA. For personal protection 30 mA offers a high degree of protection and will operate by cutting off the earth fault current well within the time specified in IEC Publication 1008/1009.

Lower sensitivities (above 30 mA current trip) are sometimes used for individual circuits where there is less chance of direct contact such as in hot water tanks on a roof or under-floor heating. These earth leakage units from 100 to 375 mA provide reasonable protection from the risk of electrical fires, but it should be noted that under certain circumstances a current of less than 500 mA flowing in a high resistance path is sufficient to bring metallic parts to incandescence and can start a fire.

In addition to the pick-up current value, RCDs also differ by pick-up time. Standard pick-up time of an RCD must be 30 to 40 msec, but there are devices with pick-up time delays (the "G" and "S" type) that are designed for selective work in a group of several protective devices. RCDs also differ according to the type of current to which they

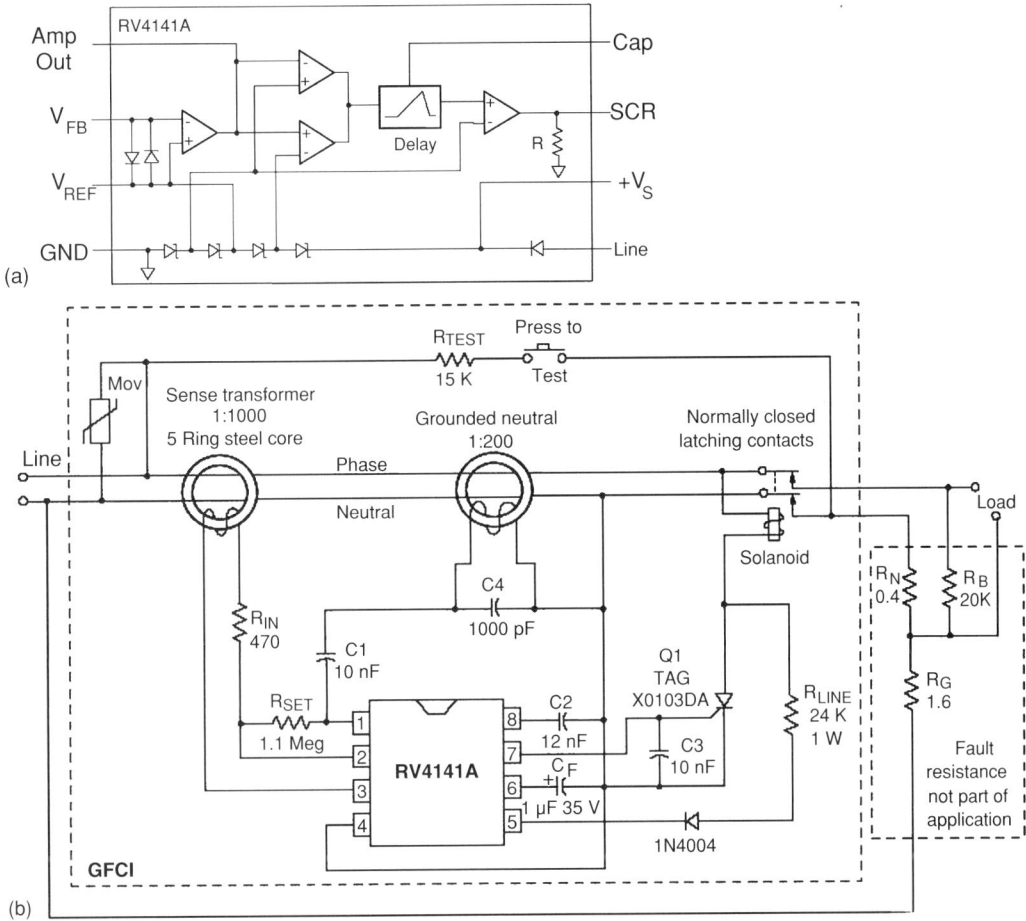

FIGURE 16.90
Electronic RCD based on a special integrated circuit (the U.S.A. patent 3,878,435).

respond. Class "AC" devices are used where the residual current is sinusoidal. This is the normal type, which is most widely in use. Class "A" types are used where the residual current is sinusoidal and/or includes pulsating DCs. This type is applied in special situations where electronic equipment is used. Class "B" is for specialist operations on pure DC, or on impulse direct or AC.

RCDs are constructed in cases very similar to those of the thermal and electromagnetic protective relays (automatic switches) considered above (Figure 16.91). RCDs are frequently combined with these switches in the same case, in such a way that the mechanism opening the power contacts runs when affected by any of three elements — a coil with a core of current cutoff responding to the short-circuit current, a bi-metal plate responding to the overload currents, and a polarized electromagnetic trip responding to differential current.

The RCD type numbers indicate over-current trip, residual current trip, and the number of poles. RCDs produced by the Cutler–Hammer company have quite an unusual external design (Figure 16.92). In such devices, the equipment wires connected to the terminals must pass through the window of the external CT.

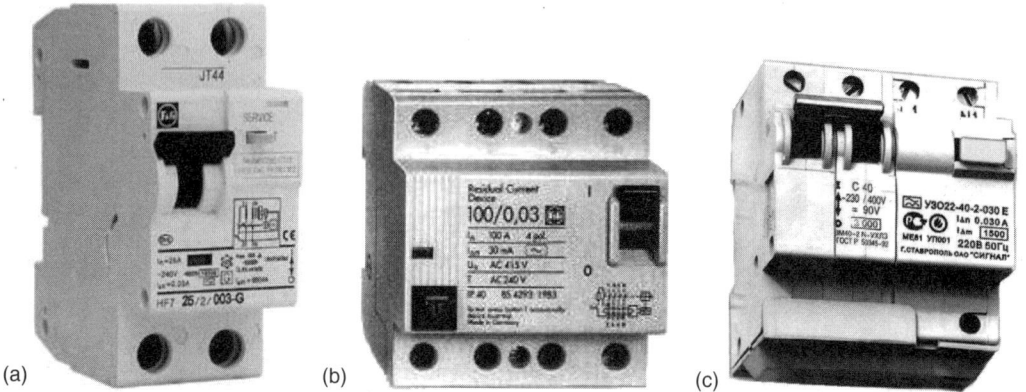

FIGURE 16.91
RCD devices of different types: 1 — HF7-25/2/003 type (Moeller); 2 — NFIN-100/0.03/4 type (Commeng Enterprise Corp.); 3 — Y3O22-40-2-030 type ("Signal" plant, Russia).

During exploitation of RCDs, there are sometimes problems connected with false pick ups, which can give a lot of trouble (see Ward P., Demystifying RCD's, *Irish Electrical Review*, December 1997). Usually, false pick ups are caused by transient processes, over-voltages, dissymmetry, spikes, inrush current, etc. These are typical problems of electromagnetic compliance (EMC) common for many types of electrical equipment. Sometimes problems occur because of the wrong choice of an RCD. IEC recommends choosing a RCD in such a way that its nominal residual current trip is three times as high than the actual leakage current through the insulation at the place of installation of the RCD. But even the correct choice of an RCD does not guarantee normal operation. As the RCD is incapable of distinguishing between constantly flowing leakage current through the insulation to the ground and from an emergency current of closing through the human body, and responds to the sum of both currents, a situation may occur when, after a lapse of time (from the time the RCD was installed) the state of the insulation deteriorates (increased temperature, moisture, aging) and gradually the current passing through it increases. As an RCD with a nominal current of 30 mA can pick up within a

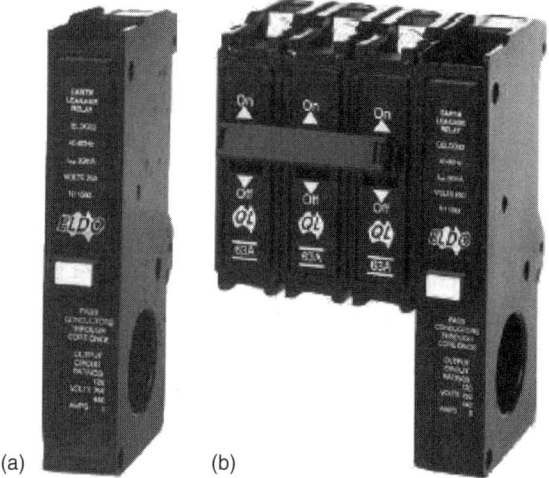

FIGURE 16.92
ELDO type (left) and combined QELDO (right) RCD devices with external differential current transformers. (Cutler-Hammer 2004 online catalog.)

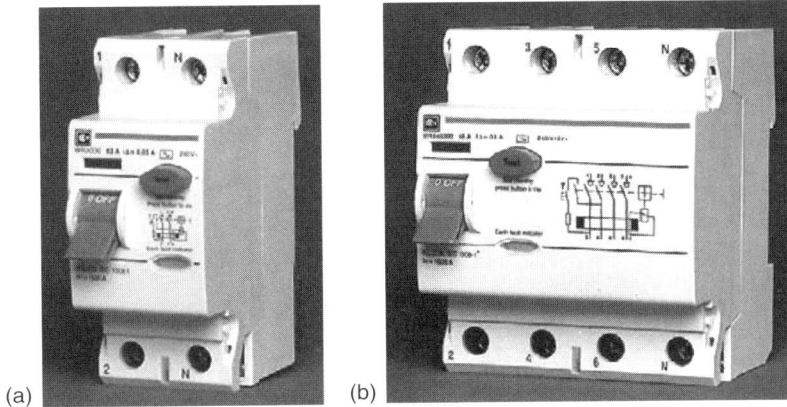

FIGURE 16.93

Two-and four-pole RCD devices of the WR type, independent of line voltage, 23 to 63, 30 mA. (Cutler–Hammer 2004 online catalog.)

range of 15 to 30 mA (see above), even a slight increase of leakage current — for instance, by 5 mA (from 10 to 15 mA), may cause false pick ups of even a well-functioning RCD chosen in accordance with all requirements.

The problem is aggravated because of the wide use of filters in electrical equipment, which are designed to eliminate radio interference. Such filters create increased leakage currents between the phase wire and the ground.

Short-term increases of potential of the grounded circuits when short-circuit currents or stray currents pass through them, can cause false pick ups of the RCD. In other words there are a lot of reasons for false pick ups of well-functioning RCDs. Some of them can be compensated by using RCDs with increased noise-immunity, for example, of the WR type produced by the Cutler-Hammer Company (Figure 16.93). In these RCDs, a filter device is incorporated as standard to protect against nuisance tripping due to transient voltages (lightning, etc.) and transient currents (from high capacitive currents).

Of course such devices are more expensive than those produced in China, but in some cases this can be the only solution allowing us to avoid much trouble.

16.13 Supervision Relays

What is the relay-supervisor? For whom or what reason does it supervise? Main purpose of such relays is a continuous monitoring of serviceability of important units (or important electric parameters of power applied to such units). The trip coils and the power supply of high-voltage circuit breakers in electrical networks; power supply circuits for sensors of fire-alarm systems; phase sequence and phase losses in power supply for electric motors; insulation level of electric equipment, etc., concern for such units and parameters. Supervision relays also detect interruptions, too high resistances caused by galvanically bad connections, increased transfer resistance in the contacts, welding of the control contact, disappearing control voltage, and voltage failures in the relay itself.

(a)

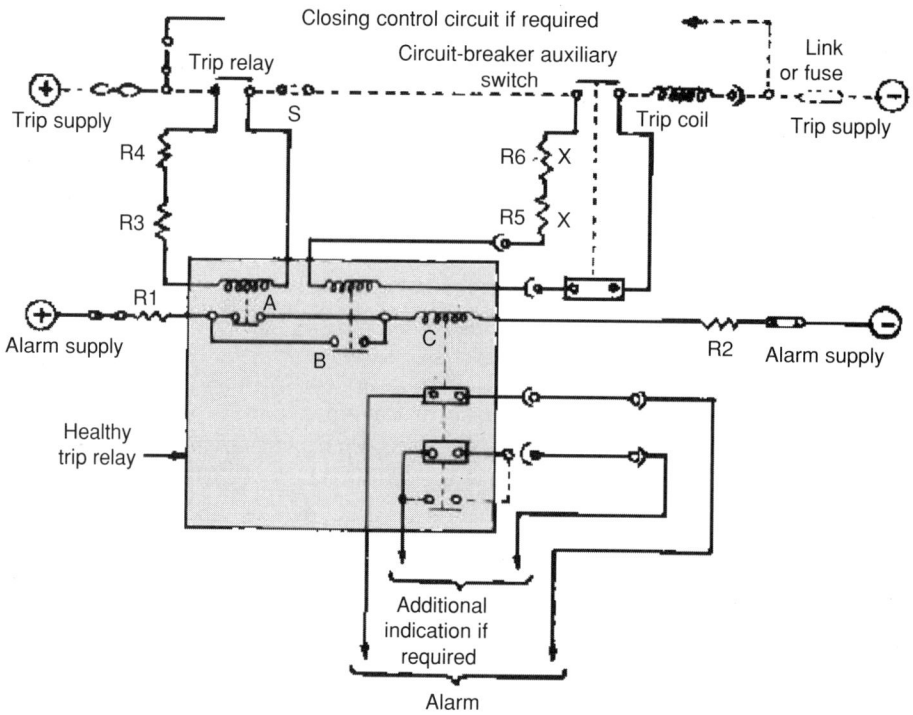

(b)

FIGURE 16.94
(a) Simplest circuit breaker supervision relay, based on electromechanical relays (Easun Reyrolle, India).
(b) Circuit diagram and external connection of a B51 supervision relay (called: "Healthy trip relay" in manufacturer documentation).

For example, a relay with a high coil resistance and suitable pick-up voltage, connected in series with the trip coil of a high-voltage circuit breaker, should be considered for red light indication and auxiliary contact to allow remote supervision of the trip coil (Figure 16.94). The B51 (Figure 16.94b) is the simplest circuit breaker supervision relay, comprising three attracted armature relays, continuously excited during normal operation by the external DC power supply voltage and slugged at the coil base to provide a delay on drop-off. The alarm contacts will not close in less than 400 msec after failure of the trip circuit.

The contacts are self-reset. The flag, fitted only to the output relays, may be hand reset or self-reset and indicates that the relay is deenergized. To prevent the supervision relay from providing spurious alarm signals, for instance, at circuit-breaker operation, the measuring current of the control circuit is supplied with external resistors (R3–R6), which are current limiters. The resistance of the components in the circuit across the trip–relay contacts is such, that accidental short-circuiting of any one will not result in trip-coil operation. With both trip–circuit supervision relay coils and limiting resistors connected in series, the current through the trip-coil will not exceed 10 mA. With only one trip-circuit supervision relay coil and a limiting resistor connected in series, the trip-coil current will not exceed 20 mA.

If the breaker-tripping coil does not tolerate this, the electromechanical auxiliary relay can be replaced by an optocoupler (Figure 16.95) in order to further reduce the current in the tripping circuit. The modern SPER series supervision relays, produced by ABB (Figure 16.96), are used for monitoring important control circuits such as circuit breaker and disconnector control circuits, signaling circuits, etc., in power installations.

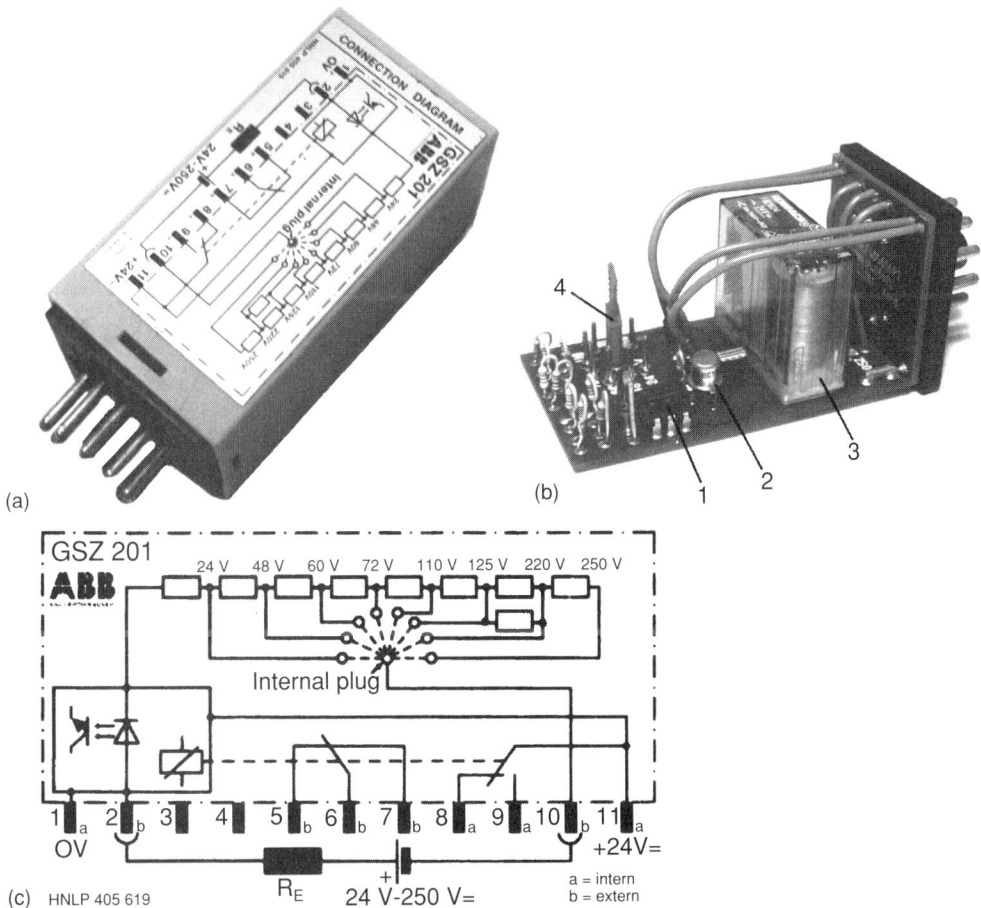

FIGURE 16.95
Circuit-breaker supervision relay GSZ201 type with optocoupler for reduce input current (ABB). 1 — Optocoupler, 2 — amplifying transistor, 3 — output relay, 4 — internal plug for manual switching of voltage in supervision circuit (R_E — supervision circuit). (ABB ASEA Relays buyers guide 1985–1986.)

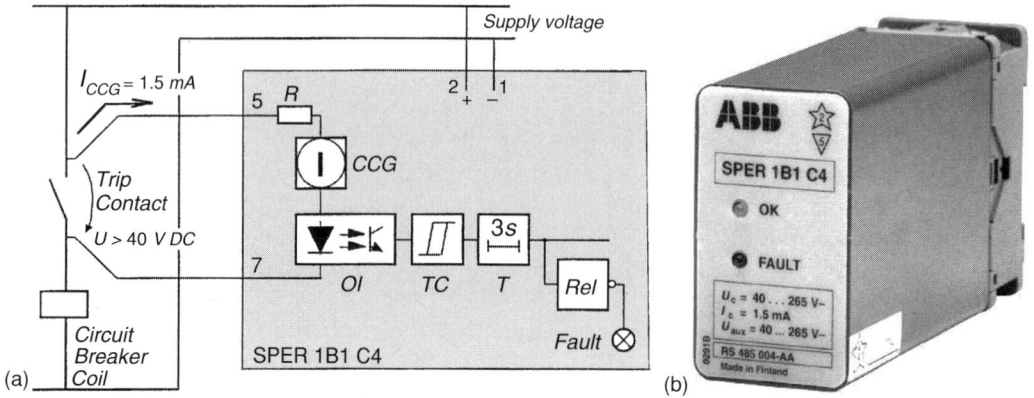

FIGURE 16.96
Modern supervision relay of the SPER type, produced by ABB. CCG — Constant current generator; OI — opto-isolator (optocoupler); TC — triggering circuit; T — time circuit; Rel — output relay; Fault — LED indicator.

One contact circuit is monitored by one relay. If several branches of a circuit are to be monitored, the required number of relays can be connected to the same control circuit.

The constant current generator (CCG) of the driver circuit feeds a small I_{CCG} current of some 1.5 mA, depending of the relay type used, through the circuit to be monitored. The contact inputs 5 to 7 are connected over the NO trip contact, so the measuring current flows between the poles of the control voltage.

The driver circuit of the relay operates independently of the measuring circuit and the output circuit, so different voltage levels are permitted. Should the auxiliary voltage supply be interrupted, the indicator LEDs of the supervision relay go out and the changeover contacts of the operated output relay operate without a time delay in the measuring circuit. The contact operation of the relay is the same as for a fault in the circuit monitored.

To avoid spurious CB tripping, for instance in the event of a short circuit in the control circuit, the CCG circuit of the SPER relay contains an internal current limiting series resistor (R).

Similar supervision relays are producing by many companies; see Figure 16.97, for example. For continuous monitoring of a trip circuit (that is, while the breaker is open as well as closed), a bypass resistor must be added in parallel to the 52a contact as shown below. This technique maintains current flow through the circuit when the breaker is open.

The EN 60204 standard "Safety of Machinery" (and also some medical applications with similar requirements), stipulates that auxiliary circuits must be protected with an earth-leakage supervisor in order to increase operating safety. The *insulation supervising relay* of type RXNA4 (Figure 16.98), is used for insulation supervision and as an earth-fault relay in single and three-phase networks with an isolated neutral, such as on board ships.

In comparison with an earth-fault relay consisting of a voltage-measuring unit connected between the neutral and earth of the system, the RXNA4 has the following advantages:

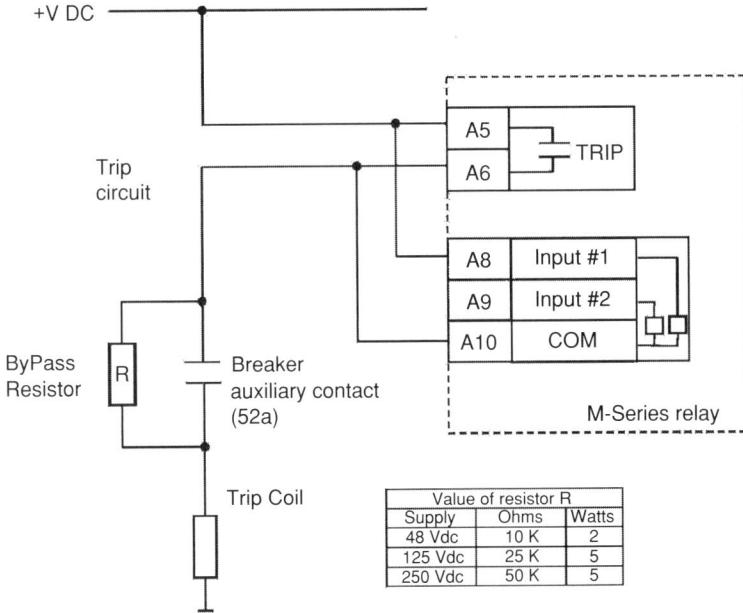

FIGURE 16.97
Trip circuit supervision for M-Family Relays (General Electric Co.).

- The operating value is not influenced by the capacitance to earth
- Operates even during a symmetrical drop of the insulation resistance
- Higher internal impedance

This implies that the RXNA4 can be used to supervise the insulation to the ground without any dangerous earth currents arising on the occurrence of a single-phase earth fault. A milliampermeter, which continuously indicates the value of the insulation resistance, can be connected to the relay.

For voltage supply purposes the RXNA4 has a fully isolated input transformer that is connected either to the supervised network or to a separate source of voltage. After transformation, the connected supply is rectified and stabilized into two voltages of 24 V each.

One voltage is used partly as an auxiliary supply to the static circuits, and partly for supplying the output relay. The other voltage is connected between the supervised AC network and the earth. On the occurrence of an earth fault, a DC current flows from the relay through the point of the fault to earth, and back again to the relay. This current is measured by a level detector the operating value of which is sleeplessly adjustable with a knob located on the front of the relay. When the insulation resistance drops below the set value, the output relay picks up after approximately 2 sec. Simultaneously, a light emitting diode lights up on the front of the relay.

A low-pass filter is incorporated in the input of the relay to prevent the operation of the relay from being influenced by the AC voltage of the network. A pushbutton is incorporated in the front of the relay, to facilitate the testing of the relay's operation. When the pushbutton is depressed the measuring input is short-circuited and the relay picks up.

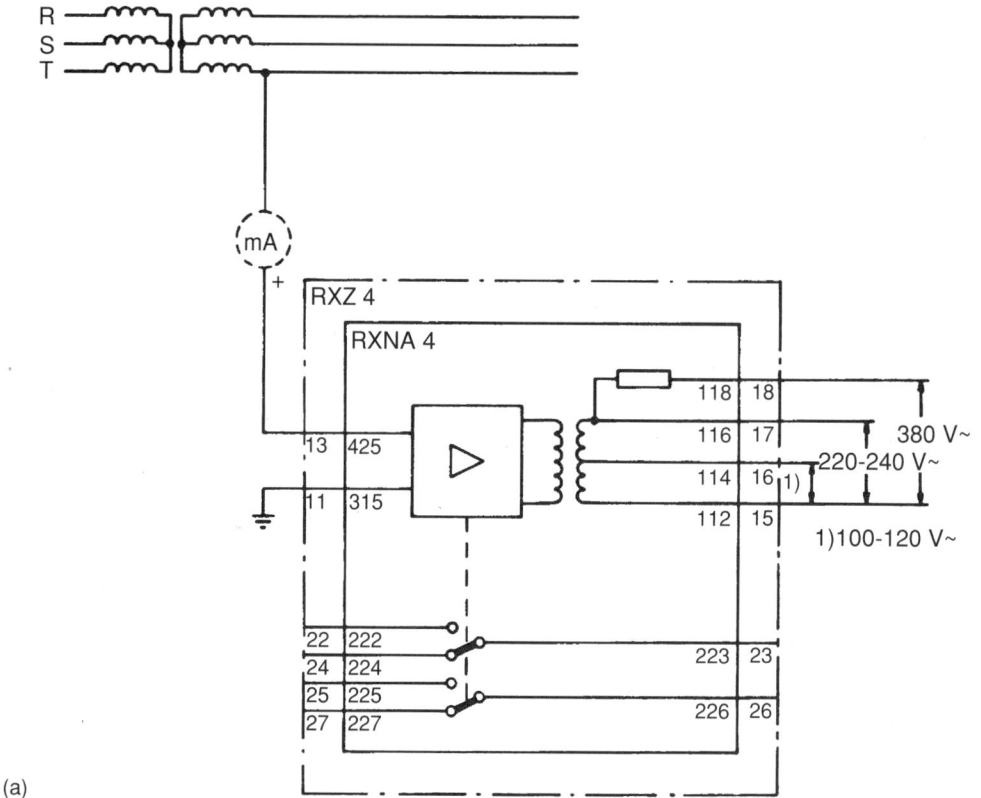

(a)

(b)

FIGURE 16.98
Insulation supervising relay, type RXNA4. (ABB ASEA Relays buyers guide 1985–1986.)

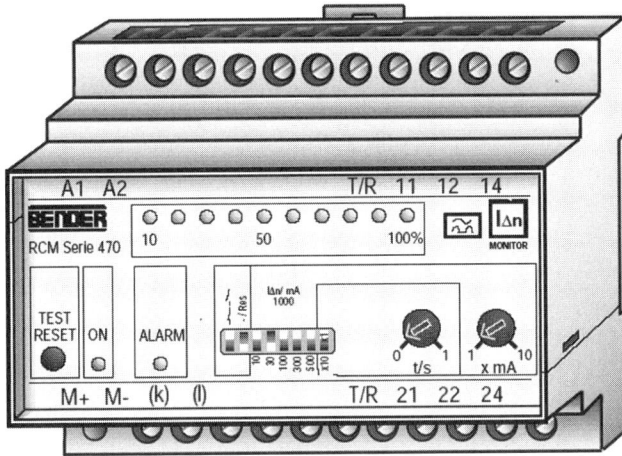

FIGURE 16.99
RCM series ground fault supervision relay, manufactured by the BENDER company.

Many companies specialize in manufacturing insulation supervision relays, for example BENDER. The BENDER RCM series (Figure 16.99) is specially designed to provide advanced warning of developing faults without the problems usually associated with high-sensitivity nuisance tripping. The RCM470LY and RCM475LY are IEC755 Type A Ground Fault Relays that can detect sinusoidal AC ground fault currents and pulsating DC ground fault currents. The response value ΔI is sleeplessly adjustable between 6 mA to 600 A or 10 mA to 10 A, and the delay time can be adjusted between 0 and 10 sec. The relay is equipped with an LED bar graph indicator. An external analog meter can be connected, and by using an optional external transducer a 4 and 20 mA signal is available. Meter indication is from 10 to 100%, where 100% is equal to the alarm set-point value.

The RXNAE 4 supervision relay (Figure 16.100) is primarily used for *monitoring arcs* on the commutators of DC machines. Substantial arcs can result in flashovers, which can damage the commutator.

Light-sensitive detectors placed in the machine activate the RXNAE 4, which instantaneously trips. The light detector contains a phototransistor which reacts to the infra-red radiation from an arc, and provides current to the supervision relay. The current is measured in the relay by a tripping detector and a signal detector. The tripping detector

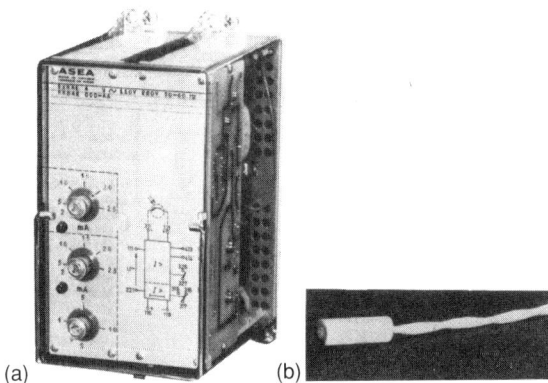

FIGURE 16.100
The RXNAE 4 type supervision relay with an infrared sensor for monitoring arcs. (ABB ASEA Relays buyers guide 1985–1986.)

is set for instantaneous tripping for severe arcs. It has two outputs, one a static high-load output and the other a dry-reed relay output. The static output triac is used for tripping of the main circuit breaker (AC or DC), while the relay output is used for signaling.

If the power supply to a four-wire (separately powered) smoke detector fails, or if the wires are cut, the detector will not work. To prevent a tragedy, we need to know immediately when there is a failure in the power system.

The EOL-1224RLY (Figure 16.101) is an end of line power supervision Relay used for the supervision of four-wire smoke detector voltage. If the power is interrupted, the EOL-1224RLY will cause your alarm panel to indicate a "trouble" condition. This is a requirement of the NFPA.

This supervision relay is called "End of Line" because it is installed at the end of the detector power circuit. A break in the detector power circuit or a loss of power deenergizes the power supervision relay, opening the contacts and causing a trouble annunciation at the fire alarm control unit.

A *Phase sequence supervision relay* (Figure 16.102), is used for monitoring the clockwise rotation of movable motors for which the phase sequence is important, such as with pumps, saws, and drilling machines. The phase sequence supervision relay detects the timed sequence of individual phases in a three-phase supply. In a clockwise phase sequence, contacts 11 to 12 and 21 to 22 are open, and contacts 11 to 14 and 21 to 24 are closed. In an anticlockwise phase sequence, contacts 11 to 12 and 21 to 24 are open, and contacts 11 to 14 and 21 to 22 are closed.

The EMR4-A *unbalance supervision relay,* (Figure 16.103) with its 22.5 mm module width, is the ideal protective device for supervision of phase loss. The detection of phase loss on the basis of phase shift means that reliable phase loss detection is ensured and overloads even when large amounts of energy are regenerated to the motor. The EMR4-A relay can be used for protecting motors with a rated voltage of $U_N = 380$ to 415 V at 50 Hz, and provides:

- Phase loss detection even with 95% phase regeneration
- Phase sequence detection
- ON-delay of 0.5 sec
- LED status indication.

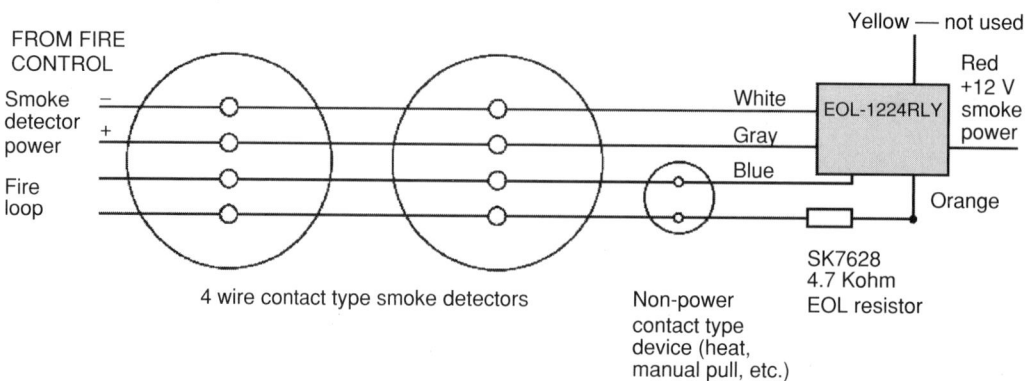

FIGURE 16.101
Typical installation of end-of-line relay of the EOL-1224RLY type, for supervision of a four-wire smoke detector voltage (Silent Knight, U.S.A.).

(a)

(b)

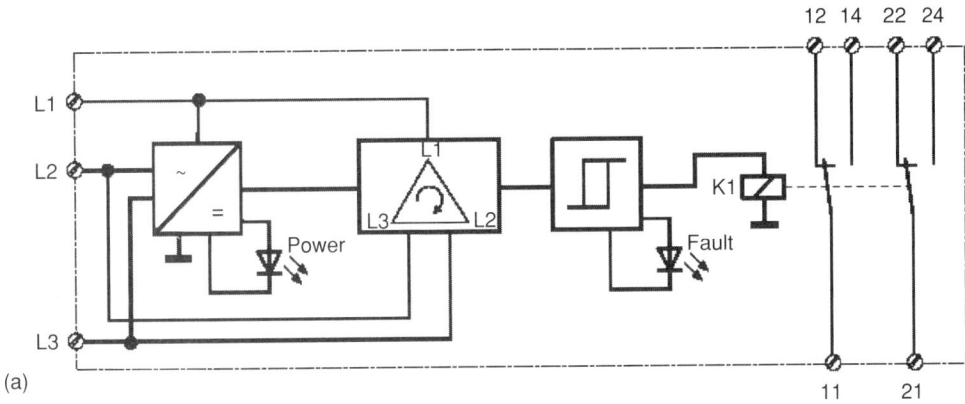

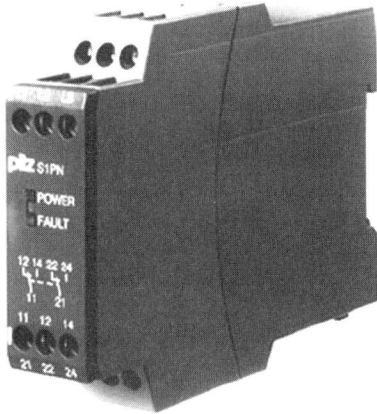

FIGURE 16.102
A phase sequence supervision relay of the S1PN type (Pilz GmbH & Co.)

FIGURE 16.103
Unbalance Supervision Relay of the EMR4-A type. (Klokner–Moeller 2004 online catalog.)

The relay has its own power supply, that is, an additional supply voltage is not required. As it is possible to see from the above descriptions of supervision relays, they are very simple devices and are intended for fulfillment of very simple functions. Nevertheless, even in this area, the attempts of usage of microprocessors are undertaken.

Naturally, such relays deserve a serious view (Figure 16.104 — compare this to the relay in Figure 16.95). For whom is it necessary? Perhaps the readers themselves know the answer to this question (if no, see chapter 16).

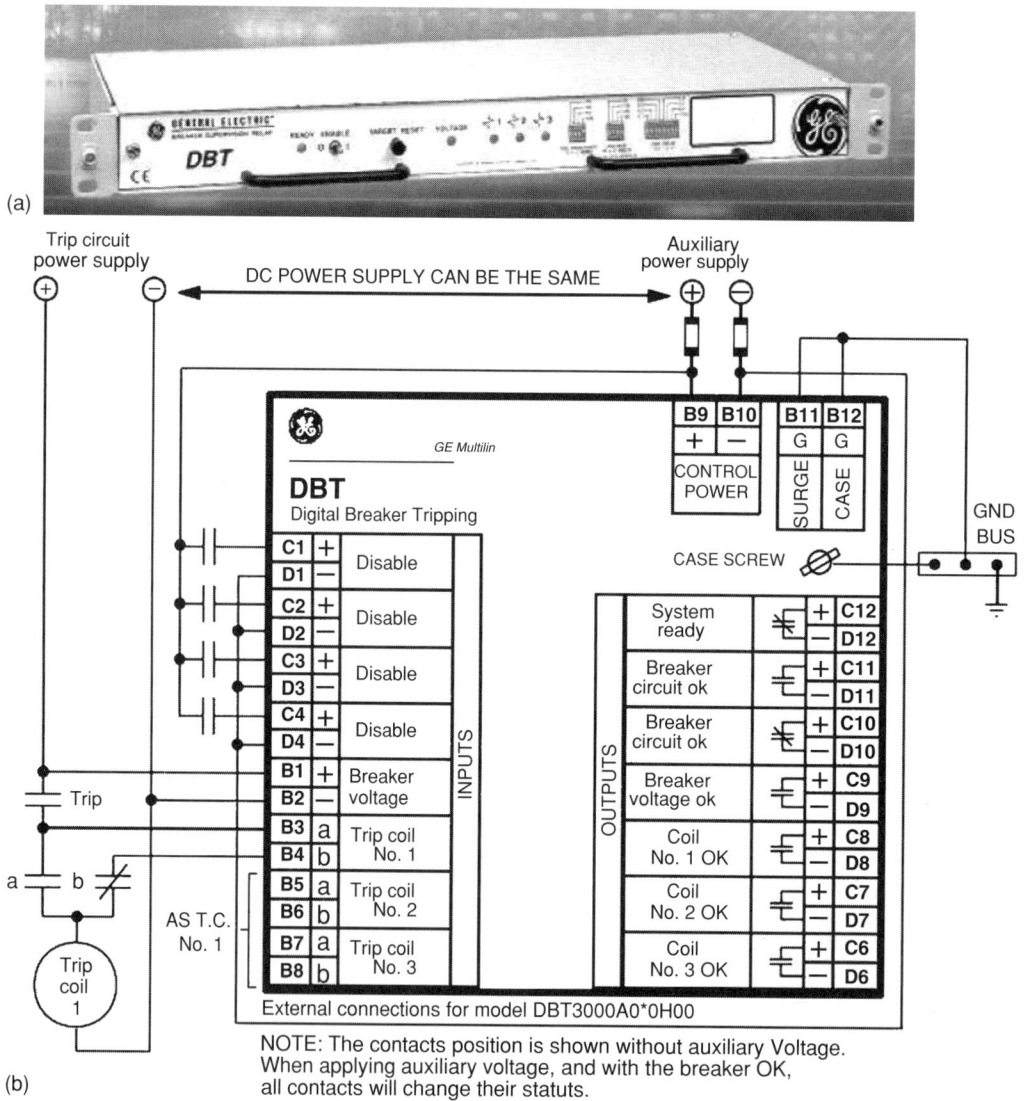

FIGURE 16.104

(a) The DBT type digital circuit breaker supervision relay (General Electric Co.). (b) Connection diagram of the DBT type supervision relay. (General Electric online catalog.)

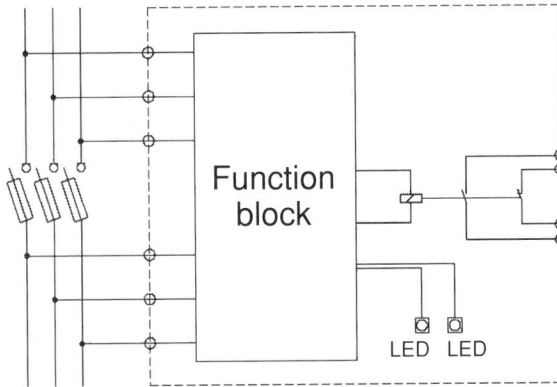

FIGURE 16.105
Principal of fuse supervision.

Fuse supervision relays are for detecting one or more blown fuses in important electrical equipment (Figure 16.105). In a normal condition of fuses, the voltage drop on them makes parts of volt even at high currents. When a fuse blows, on the input of the relay there is a voltage, causing actuation. Certainly the input circuit of such supervision relay should have high resistance and not influence load.

Many fuse supervision relays have similar principles of operation, for example, the static RXBA4 type three-phase relay (Figure 16.106). Often the fuse supervision relay is just enough to supervise only the voltage after the fuses (on part of the load), with actuation of the relay occurring at a loss of voltage in one of phases. Some companies build supervision relays on this principle (Figure 16.107).

The sort of strange solution, utilized in supervision relays manufactured by Siemens (in the Sirius-3R type), Rittal (United Kingdom), and some other companies (Figure 16.108), is based on the principle of automatic circuit breakers.

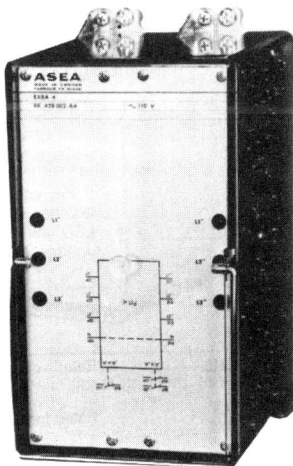

FIGURE 16.106
Fuse supervision relay of the RXBA4 type (ASEA).

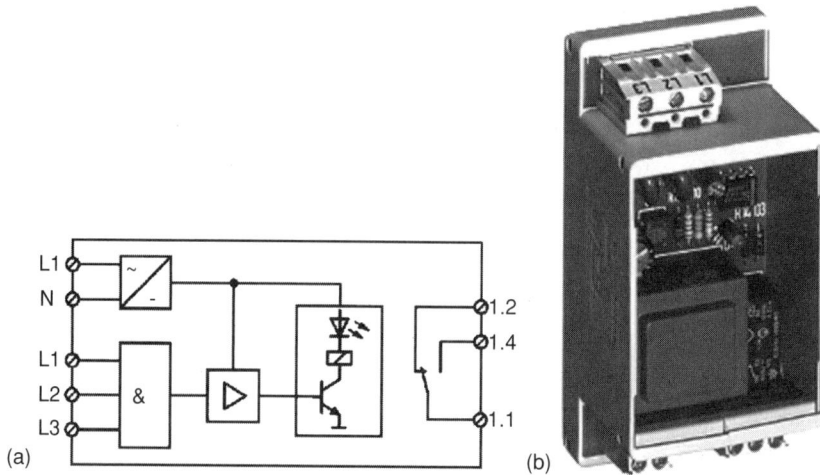

FIGURE 16.107
Fuse supervision relay produced by Weiland, Inc.

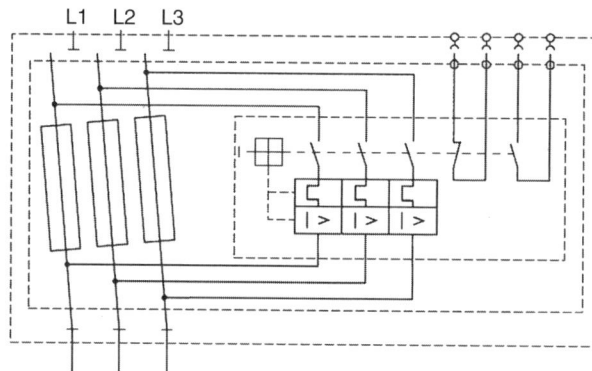

FIGURE 16.108
Fuse supervision relay based on the principle of the automatic circuit breaker.

It is designed so that at blowing of a fuse, all current will pass through these automatic circuit breakers, causing actuation and disconnecting the load. The auxiliary contacts of the automatic circuit breakers may also be used for the signaling (supervision) system.

16.14   Hydraulic-Magnetic Circuit Breakers

Hydraulic-magnetic circuit breakers (HMCB) is a relatively new type of a circuit breakers now produced by many companies for wide range of currents.

A HMCB consists (Fig. 16.109) of a magnetic frame comprising a series-connected solenoid coil wound around a hermetically sealed tube containing an iron core, a spring, and hydraulic fluid. Mounted onto this magnetic frame is an armature, which when activated unlatches the trip latch of the circuit breaker. As the electrical current passing

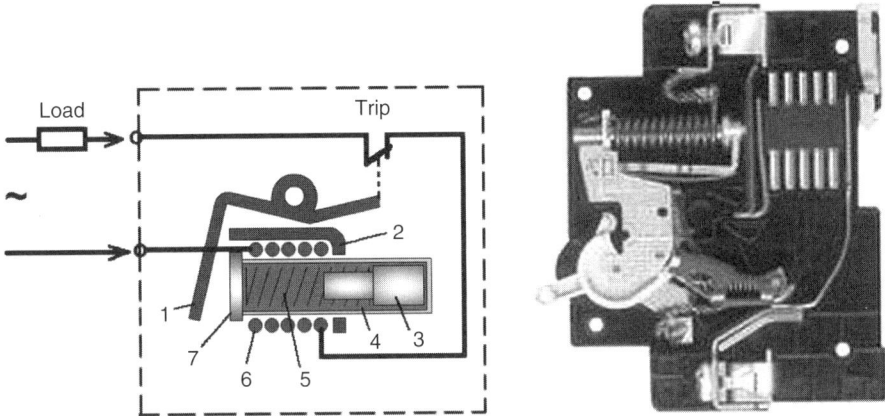

FIGURE 16.109
Construction of hydraulic-magnetic circuit breaker. 1 — Armature; 2 — frame; 3 — moving ferromagnetic core; 4 — temperature stable fluid with a special viscosity; 5 — spring; 6 — coil; 7 — pole piece.

through the coil increases, the strength of the magnetic field around the magnetic frame also increases. As the current approaches the circuit breaker's rating, the magnetic flux in the coil produces sufficient pull on the core to overcome the spring tension and start it moving towards the pole piece. During this movement, the hydraulic fluid regulates the core's speed of travel, thereby creating a controlled time delay, which is inversely proportional to the magnitude of the current.

As the normal operating or "rated" current flows through the sensing coil, a magnetic field is created around that coil. When the current flow increases, the strength of the magnetic field increases, drawing the spring-biased, movable magnetic core toward the pole piece. As the core moves inward, the efficiency of the magnetic circuit is increased, creating an even greater electromagnetic force, Fig. 16.110. When the core is fully "in," maximum electromagnetic force is attained. The armature is attracted to the pole piece, unlatching a trip mechanism thereby opening the contacts.

Under short-circuit conditions (Fig. 16.111) the resultant increase in electromagnetic energy is so rapid, that the armature is attracted without core movement, allowing the breaker to trip without induced delay. This is called "instantaneous trip." It is a safety feature that results in a very fast trip response when needed most.

The trip time delay is the length of time it takes for the moving metal core inside the current sensing coil to move to the fully "in" position, thereby tripping the circuit breaker. The time delay should be long enough to avoid nuisance tripping caused by harmless transients, yet fast enough to open the circuit when a hazard exists.

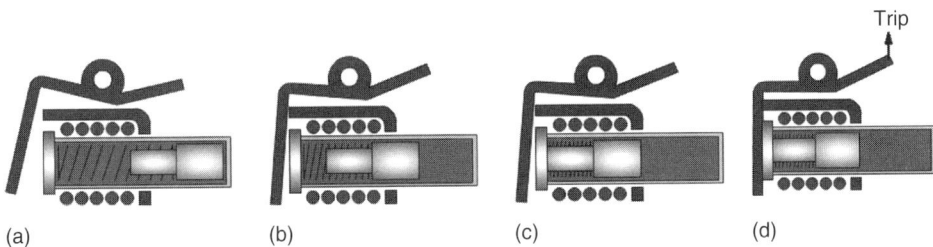

FIGURE 16.110
Phases of pick-up process under overload condition (trip time delay).

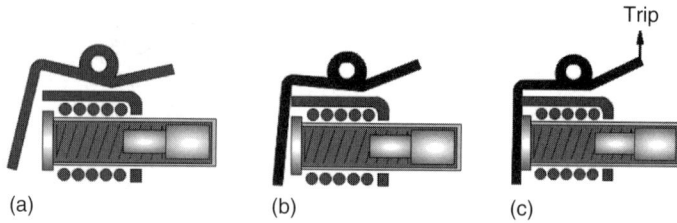

FIGURE 16.111
Phases of pick-up process under short-circuit condition (instantaneous trip).

Many hydraulic-magnetic breakers are available with a selection of delay curves to fit particular applications.

If the delay tube is filled with air, the core will move rather quickly, and the breaker will trip quickly. This is characteristic of the *ultrashort* trip time delay. Solid-state devices, which cannot tolerate even short periods of current overload, should use the *instantaneous* trip time delay, which have no intentional time delay.

When the delay tube is filled with a light viscosity, temperature stable fluid, the core's travel to the full "in" position will be intentionally delayed. This results in the slightly longer *medium* trip time delays that are used from general purpose applications.

When a heavy viscosity fluid is used, the result will be a very *Long* trip time delay. These delays are commonly used in motor applications to minimize the potential for nuisance tripping during lengthy motor start-ups.

Unlike the thermal-magnetic circuit breakers, the HMCB not affected by ambient temperature, and also it has no warm up period to slow down its response to overload and no cool down (thermal memory) period after overload event. After tripping, the HMCB may be reclosed immediately since there is no cooling down time necessary.

Fixed rating thermal-magnetic breakers are often limited in availability to ampere ratings not lower than about 15 amperes. The reason for this limitation is that the very light bi-metals that are necessary for the lower ampere ratings are often incapable of withstanding the stresses of the high let through or short-circuit currents for which they are normally designed. One way of solving this problem is through the use of HMCB's.

Hydraulic-magnetic technology also scores on their flexibility in regards to technical issues such as the required trip point, time delay curves, and inrush current handling capacity, and also on their ability to be tailored to suit a customer's particular application requirements. Such flexibility makes HMCB products a very responsive solution to any design problems or harsh environmental criteria.

HMCB tend to be sensitive to rotational position. These breakers should be mounted in a vertical position to prevent gravity from influencing the movement of the solenoid.

Basic Relay Terms and Definitions — Glossary

There are several sources for definitions: the C37 subcommittee on protective relays, C83 committee on Components for Electronic Equipment, and IEEE Standards Coordinating Committee on Definitions (SCC10) which as published ANSI/IEEE Std 100–1984 Dictionary of Electrical and Electronics Terms. IEC has Technical Committee 41 on all types of relays and publication 50 (446) is a chapter of definitions on electrical relays.

Addition sources: British Standard (BS) 142 Electrical Protective Relays; BS 4727 Relay and Measurement Terminology; C37–90 American National Standard for Relays and Relay System; IEC 255 International Electrotechnical Commission Standards for Electrical Relays; IEC 50 International Electrotechnical Vocabulary: Electrical Relays; IEEE Standard 37.2 Electrical Power System Device Function Numbers and other sources of terms, used in technical literature.

1. Relay Classes

Relay

An electric device that is designed to interpret input conditions in a prescribed manner and after specified conditions are met to respond to cause contact operation or similar abrupt change in an associated electric control circuit. Inputs are usually electric, but may be mechanical, thermal, or other quantities. A relay may consist of several units, when responsive to specified inputs, the combination providing the desired performance characteristic. The relays can be electrical, thermal, pneumatic, hydraulic, and other.

Electrical Relay

A device designed to produce sudden, predetermined changes in one or more electrical output circuits, when certain conditions are fulfilled in the electrical input circuits controlling the device.

Electromechanical Relay

An electrical relay in which the designed response is developed by the relative movement of mechanical elements under the action of a current in the input circuits.

Electromagnetic Relay

A relay whose operation depends upon the electromagnetic effects of current flowing in an energizing winding. There are two basic types of electromagnetic relays: attraction relay and induction relay.

Electromagnetic *attraction relay* operates by virtue of a plunger being drawn into a solenoid, or an armature being attracted to the poles of an electromagnet. Such relays may be actuated by DC or by AC quantities (current, voltage, power).

Electromagnetic *induction relay* uses the principle of the induction motor whereby torque is developed by induction in a rotor; this operating principle applies only to relays actuated by alternating current, and in dealing with those relays we shall call them simply "induction-type" relays. Induction relays are basically induction motors. The moving element, or rotor, is usually a metal disk, although it sometimes may be a metal cylinder or cup. The stator is one or more electromagnets with current or potential coils that induce currents in the disk, causing it to rotate. The disk motion is restrained by a spring until the rotational forces are sufficient to turn the disk and bring its moving contact against the stationary contact, thus closing the circuit the relay is controlling. The greater the fault being sensed, the greater the current in the coils, and the faster the disk rotates.

Electronic Relay

Electrical relays whose functions are achieved by means of electronic components (vacuum or gas filled valves, semiconductor elements) without mechanical motion.

Hybrid Relay

A relay in which electromechanical and electronic devices are combined to perform a switching function.

Static Relay

An electrical relay in which the designed response is developed by electronic, magnetic, optical, or other components without mechanical motion.

It should be noted though that few static relays have a fully static output stage, to trip directly from thyristors, for example. By far the majority of static relays have attracted armature output elements to provide metal-to-metal contacts, which remain the preferred output medium in general.

Semiconductor Relay

Electronic relay based on semiconductor elements (transistors, thyristors, SCR, opto-couplers).

Solid-State Relay

Electronic relay, designed as single epoxy molded solid-state module (usually, opto-coupled).

2. Relay Types

AC Directional Relay

Are used most extensively to recognize the difference between current being supplied in one direction or the other in an AC circuit, and the term "directional" is derived from this usage. Basically, an AC directional relay can recognize certain differences in phase angle between two quantities, just as a DC directional relay recognizes differences in polarity. This recognition, as reflected in the contact action, is limited to differences in phase angle exceeding 90° from the phase angle at which maximum torque is developed.

AC Time Overcurrent Relay

A relay that operates when its AC input current exceeds a predetermined value, and in which the input current and operating time are inversely related through a substantial portion of the performance range.

Annunciator Relay

(target relay, signal relay, flag relay)
A non-automatically reset device that gives a number of separate visual indications upon the functioning of protective relays, indicates the present or former state of a circuit or circuits and which may also be arranged to perform a lockout function.

Auxiliary Relay

A relay that operates to assist another relay or device in the performance of a function.

Bar Relay

A relay so designed that a bar actuates several contact simultaneously.

Bimetal Relay

A form of thermal relay using a bimetallic element to activate contacts when heated electrically.

Bistable Relay

An electrical relay which, having responded to an input energizing quantity (or characteristic quantity) and having changed its condition remains in that condition after the quantity has been removed. Another appropriate further energization is required to make it change over.

Crystal Can Relay

A term used to identify a relay housed in a hermetically sealed enclosure that was originally used to enclose a frequency-control type of quartz crystal.

Current-Balance Relay

A relay that operates when the magnitude of one current exceeds the magnitude of another current by a predetermined degree.

Current-Sensing Relay

A relay that functions as a predetermined value of current; an overcurrent or an undercurrent relay, or a combination of both.

Dashpot Relay

A relay employing the dashpot principle to effect a time delay.

Dependent-Time Measuring Relay

A specified-time measuring relay for which times depend, in a specified manner, on the value of the characteristic quantity.

Differential Relay

A relay with multiple inputs that actuating when the voltage, current, or power difference between the inputs reaches a predetermined value.

Distance Relay

A relay that functions when the circuit admittance, impedance, or reactance increases or decreases beyond predetermined limits.

Dry Reed Relay

Relay with hermetically sealed, magnetically actuated reed contact. No mercury or other wetting material is used. Typical atmosphere inside the enclosure is nitrogen or a specially dried air.

Enclosed Relay

Hermetically sealed — a relay contained within an enclosure that is sealed by fusion or other comparable means to ensure a low rate of gas leakage (generally metal-to-metal or metal-to-glass sealing is employed)
Encapsulated — a relay embedded in a suitable potting compound
Sealed — a relay contained in a sealed housing
Covered — a relay contained in an unsealed housing
Dustproof — a relay with a case to protect against dust penetration

Ferreed Relay

Combined name for a special form of dry reed switch having a return magnetic paths (reeds) of high remanence material (ferrite) that provides a bistable contact.

Flasher Relay

(*flashing-light relays*)
A self-interrupting relay, usually by thermo-magnetic or electronic type.

Field Relay

A relay that functions on a given or abnormally low value or failure of a machine field current, or on excessive value of the reactive component of armature current in an AC machine indicating abnormally low field excitation.

Frequency Relay

A relay that functions on a predetermined value of frequency (either under or over or on normal system frequency) or rate of change of frequency.

Hermetically Sealed Relay

Relay with the highest degree of sealing (particularly relays with metal cases and glass insulators).

Impulse Relay

A relay that follows and repeats pulses, as from a telephone dial; or a relay that operates on the stored energy of a short pulses after the pulse ends; or a relay that discriminates between length and strength of pulses, operating on long or strong pulses and not operating on short or weak ones; or a relay that alternately assumes one of two positions as pulsed.

Independent-Time Measuring Relay

A specified-time measuring relay the specified time for which can be considered as being independent, within specified limits, of the value of the characteristic quantity

Integrating Relay — rare usage

A relay that operates on the energy stored from a long pulse or a series of pulses of the same or varying magnitude, for example, a thermal relay.

Instantaneous Overcurrent Relay (Rate-of-Rise Relay)

A relay that functions instantaneously on an excessive value of current or on an excessive rate of current rise, thus indicating a fault in the apparatus or circuit being protected.

Instruments Relay (Meter Relay) — rare usage

A sensitive relay in which the principle of operation is similar to that of instruments such as the electrodynamometer, iron vane, galvanometer, and moving magnet.

Interlock Relay

A relay with two or more armatures having a mechanical linkage or an electrical interconnection, or both, whereby the position of one armature permits, prevents, or causes motion of another.

Inverse-Time Relay

A relay in which the input quantity and operating time are inversely related throughout at least a substantial portion of the performance range. Types of inverse-time relays are frequently identified by such modifying adjectives as *definite minimum time, moderately, very,* and *extremely* to identify relative degree *of* inverseness of the operating characteristics.

Latching Relay (Lockup Relay)

A relay that maintains its contacts in the last position assumed without the need of maintaining coil energization.

Magnetic latching

A relay that remains operated, held either by remanent magnetism in the structure or by the influence of a permanent magnet, until reset.

Mechanical latching

A relay in which the armature or contacts may be latched mechanically in the operated or unoperated position until reset manually or electrically.

Lockout Relay

A variety of latching relay with manual resetting. Lockout relays are applicable where several tripping functions need to be performed by the same relay. Typical applications for these relays include: line breaker tripping and lockout, lockout of all the line breakers in the same busbar, etc. One of the most important applications of lockout relays is the combination with differential relays, where the lockout relay needs to be reset manually for avoiding accidental reclosings, when an internal fault has activated the differential relay.

Measuring Relay (Protective Relay)

1. A device used to detect defective or dangerous conditions and initiate suitable switching or give warning when its characteristic quantity, under specified conditions and with a specified accuracy, attains its operating value (current relay, voltage relay, power relay, etc.)

2. A relay designed to initiate disconnection of a part of an electrical installation or to operate a warning signal, in the case of a fault or other abnormal condition in the installation. A protective relay may include more than one unit electrical relay and accessories.

The IEEE Standard 37.2 Electrical Power System Device Function Numbers assign device numbers (1 to 100) to various types of protective relays:

Some Often Used *Device Function Numbers*

2	Time Delay Starting Relay	52	AC Circuit Breaker
3	Checking or Interlocking Relay	53	Exciter Relay
21	Distance Relay	55	Power Factor Relay
23	Temperature Control Device	56	Field Applications Relay
25	Synchronizing Device	57	Short-Circuiting or Grounding Device
26	Apparatus Thermal Device	59	Overvoltage Relay
27	Undervoltage Relay	60	Voltage or Current Balance Relay
29	Isolating Contactor	63	Liquid or Gas Pressure Relay
30	Annunciator Relay	64	Ground Protective Relay
32	Directional Power Relay	68	Blocking Relay
36	Polarity Voltage Device	74	Alarm Relay
37	Undercurrent and Underpower Relay	76	DC Overcurrent Relay
38	Bearing Protective Device	78	Phase-Angle Measuring Relay
40	Field Relay	81	Frequency Relay
44	Unit Sequence Starting Relay	82	DC Reclosing Relay
46	Reverse Phase or Phase Balance Current Relay	85	Carrier or Pilot-Wire Receiver Relay
47	Phase-Sequence Relay	86	Lockout Relay
48	Incomplete Sequence Relay	87	Differential Relay
49	Machine or Transformer Thermal Relay	91	Voltage Directional Relay
50	Instantaneous Overcurrent or Rate-of-Rise Relay	92	Voltage and Power Directional Relay
51	AC Time Overcurrent Relay	94	Trip-Free Relay

Monostable Relay

An electrical relay, which, having responded to an input energizing quantity (or characteristic quantity) and having changed its conditions, returns to its previous.

Motor-Driven Relay

A relay whose contacts are actuated by the rotation of a motor shaft.

Neutral Relay

A relay whose operation is independent of the direction of the coil current, in contrast to a polarized relay.

Open Relay

An unenclosed relay.

Over Current Relay

Protection relay that is specifically designed to operate when its coil current reaches or exceeds a predetermined value.

Over Voltage Relay

Protection relay that is specifically designed to operate when its coil voltage reaches or exceeds a predetermined value.

Phase-Sequence Voltage Relay

A relay that functions upon a predetermined value of polyphase voltage in the desired phase sequence.

Plunger Relay

A relay operated by a movable core or plunger through solenoid action.

Polarized Relay

A relay whose operation is dependent upon the polarity of the energizing current.
Monostable (Biased) — a two-position relay that requires current of a predetermined polarity for operation and returns to the off position when the operating winding is deenergized or is energized with reversed polarity.
Bistable (Double-Biased) — a two-position relay that will remain in its last operated position keeping the operated contacts closed after the operating winding is deenergized.
Center-Stable (Un-Biased) — a polarized relay that is operated in one of two energized positions, depending on the polarity of the energizing current, and that returns to a third, off position, when the operating winding is deenergized.

Power Direction Relay

A protective relay used in protective circuits as a unit determining by the direction of power passing through the protected power line where the damage occurred: on the protected line or on some other outgoing lines adjoined to this substation.

Power Factor Relay

A relay that operates when the power factor in an AC circuit rises above or falls below a predetermined value.

Power Relay

1. A relay with heavy-duty contacts (output circuits), usually rated 10 to 25 A or higher. Sometimes called a contactor.
2. The protective relay are commonly used (as overpower relays) for protection against excess electric power flow in a predetermined direction.

Reed Relay

A relay using hermetic enclosed magnetic reeds as the contact members and coil or permanent magnet as source of operated magnetic field.

Remanent Relay (Remanence Relay)

A remanent, bistable relay adopts a particular switching position at an energizing direct current in any direction and is held in this position by the remanence in the magnetic circuit, that is, through the magnetization of parts of the magnetic circuit. The contacts shift to the other switching position on a small energizing current of limited amplitude in the opposite direction. This demagnetizes the magnetic circuit again.

RF Relay

A relay designed to switch electrical AC energy with frequencies higher than audio range (radio frequency).
Antenna Switching Relay — A special RF relay used to switch antenna circuits.
Coaxial Relay — A special RF relay that opens or closes a coaxial cable or line. It is generally a low impedance device.

Sealed Relay

Relays which are sealed against the penetration of specified PCB cleaners or lacquers.

Immersion Cleanable

Relays that can be cleaned, lacquered or cast-in together with the printed circuit board after soldering without anything penetrating the relay. The washing requires a suitable solvent. In ultrasonic washing processes, the limiting values for temperature, duration, and frequency must be observed.

Note: Relays sealed against washing can also be used for environments with aggressive atmospheres.

Starting Relay

A unit relay which responds to abnormal conditions and initiates the operation of other elements of the protective system.

Stepping Relay (Ratchet, Multiposition, Rotary)

A relay having many rotary positions, ratchet actuated, moving from one step to the next in successive operations, and usually operating its contacts by means of cams.

Telephone-Type Relay

A term sometimes applied to relay with an end on armature, an L-shaped heel piece, and contact springs mounted parallel to the long axis of the relay coil which are originally used in old telephone systems.

Thermal Relay

A relay that is actuated by the heating effects of an electrical current.

Time Delay Relay

A relay in which the actuation of the output circuit (operation or release) is delayed internally (coil slugs or sleeves), mechanically (clockwork, bellows, dashpot, etc.), or by an accompanying electronic timing circuit.

Tripping Relay (Trip-Free Relay)

A relay that functions to trip a circuit breaker, contactor or equipment, or to permit immediate tripping by other devices; or to prevent immediate reclosure of a circuit interrupter if it should open automatically even though its closing circuit is maintained closed.

Under Current Relay

A relay specifically designed to function when its energizing current falls below a predetermined value.

Under Voltage Relay

A relay specifically designed to function when its energizing voltage falls below a predetermined value.

Zero-Voltage-Turn-On (Off) Relay

A relay with isolated input and output in which added control circuitry delays the output turn-on (turn-off) until a zero voltage transition of the AC sine wave is detected. Construction may be all solid-state or hybrid with a solid-state output.

---

3. Parameters of Relay

Ampere-Turn

A unit of magnetizing force. The product of current flowing, measured in amperes, multiplied by the number of turns in a coil or winding.

Breakdown Voltage

Threshold value at which breakdown does not occur when AC voltage is applied between pins, similar to insulation resistance. Usually, the breakdown voltage is tested for 1 min and the current value that defines breakdown is 1 mA. The minimum value is specified.

Coil Resistance

DC resistance of the coil. Usually measured at 25°C. A tolerance of $\pm 10\%$ usually applies.

Coil Temperature Rise

Rise of the coil temperature at a given input (power or voltage).

Degree of Protection

Ratings, for example, defined in IEC 60 529, indicating how completely a cover, seal, etc., protect against water, humidity, dust, direct contact, etc.

Dielectric Strength

The maximum allowable AC rms voltage (50 or 60 Hz) which may be applied between two test points, such as the coil and case or current carrying and noncurrent carrying points, without a leakage current in excess of 1 mA.

Duty Cycle

The ratio between the switch *on* time and total cycle time during periodical switching. Fifty per cent duty cycle means the switch on time equals the switch off time.

Eddy Current (Foucault Current)

Circulating currents in magnetic field conductive materials caused by alternating magnetic fields. They represent power losses in relay core.

Electrical Life

Switching life of the contacts expressed as the number of operations measured when the rated voltage is applied to the relay and the relay is operated at the rated operating frequency with the rated load applied to the contacts.

Electromotive Force (EMF)

The force which causes current to flow in a conductor; in other words, the voltage or potential.

Flux

Magnetic lines of force.

Flux Density

Magnetic lines of force per unit of area.

Fritting

Electrical breakdown which can occur under special conditions (voltage, current) whenever thin contact films prevent electrical conductivity between closed contacts. Fritting is a process which generates (A-fritting) and/or widens (B-fritting) a conducting path through such a semi-conducting film on a contact surface. During A-fritting, electrons are injected into the undamaged film. The electron current alters the condition of the film producing a "conductive channel." During the following B-fritting, the current widens the channel increasing the conductivity.

Inrush Current

The inrush current of a machine or apparatus is the maximum current which flows after being suddenly and fully energized.

Insulation Resistance

The minimum allowable DC resistance between two parts electrically independent of each other, case at a specified voltage, usually 500 V. Usually, this specifies the insulation resistances between the coil and contact pin, between open contact pins, and between adjacent contact pins (if the relay has two or more contacts). In addition, the insulation resistance between the pins of the contacts that are open in the operate state is also specified.

Leakage Current

Error current that can degrade sensitive measurements. Even high resistance paths between low current conductors and nearby voltage sources can generate significant leakage currents. Leakage in insulating material, micro-contamination on insulating surfaces, and moisture (humidity) can have catastrophic effects on picoamp and sub-picoamp (femtoamp) measurements.

Limiting Continuous Current (Steady State Current Limit)

The highest value of the current (effective value for alternating current) that the previously closed output circuit can permanently carry under specified conditions.

Load Life

The minimum number of cycles the relay will make, carry, and break the specified load without contact sticking or welding, and without exceeding the electrical specifications of the device.
Load life is established using various methods including Weibull probability methods.

Maximum Carry Current

Maximum current that can flow between contacts when the contacts are closed.

Maximum Coil Voltage

The maximum voltage that can be applied to the coil. Usually, the ambient temperature is specified as a condition.

Maximum Switching Current

Maximum current switchable with relay contact.

Maximum Switching Power

Maximum load power switchable with relay contact.
The value under DC load is expressed in W, and that under AC load is expressed in VA.

Maximum Voltage

The highest permissible input (coil) voltage at the reference temperature at which the relay, with continuous energization, heats up to its maximum permissible coil temperature.

Maximum Switching Voltage

Maximum voltage switchable with relay contact.
The peak value is indicated in the catalog under DC load. The effective value (rms) is indicated under AC load.

Mechanical Life

Life expressed as the number of operations that can be performed when the nominal coil voltage is applied to the relay with the contacts not loaded and the relay is operated at the rated operating frequency.

Mechanical Shock, Nonoperating

That mechanical shock level (amplitude, duration, and wave shape) to which the relay may be subjected without permanent electrical or mechanical damage (during storage or transportation).

Minimum Energization Time

The minimum impulse length at the height of the nominal voltage that is required to change the switching position of a bistable relay.

Minimum Switching Power

Minimum load power through relay contact necessary for normal operation. Expressed as the minimum values of voltage and current.

Minimum Voltage

The lowest permissible input voltage at which the relay operates reliably at the reference temperature even after continuous energization (preenergizing) and brief deenergizing.

Must Operate Voltage

Minimum voltage required to place the make contact in operate state from the release state. Normally, the contact should be driven by a rectangular waveform voltage. The maximum value is specified. In the case of a latching relay, this term means a voltage (set voltage) that is required to place the relay in the set state from the reset state.

Must Release Voltage

Maximum voltage to place the relay in the release state (the break contact is closed) from the operate state. The minimum value is specified. In the case of a latching relay, the maximum value necessary for placing the relay in the reset state from the set state, and is expressed as a reset voltage. The maximum value is specified.

Nominal Coil Voltage

A standard voltage applied to the coil to use the relay.

Operate Time

The time interval from input (coil) energization to the functioning of the last output (contact) to function. This includes time for the coil to build up its magnetic field (a significant limiting factor) and transfer time of the moveable contact between stationary contact(s), and bounce time after the initial contact make. For a solid-state or hybrid relay in a nonoperated state, the time from the application of the pickup voltage to the change of state of the output.

Operating State

Switch position of a monostable relay in the energized state. For bistable relays the switch position specified by the manufacturer.

Operating Temperature Range

The ambient temperature range over which an unmounted relay is specified to operate.

Operating Voltage Range

Permissible range of the input voltage depending on the ambient temperature.

Pickup Value

The minimum input that will cause a device to complete contact operation or similar designated action.

Pickup Voltage (or Current)

The voltage (or current) at which the device starts to operate when its operating coil is energized under conditions of normal operating temperatures.

Power Dissipation Rating of Coil

A product of the coil voltage rating and coil current. Normal power dissipation to operate the relay.

Protection Classes for Relays (according to IEC 60. 529)

Class IP 54: nonsealed relays which are protected against flux by their base plate and cover (dust-proof).
Class IP 67: describes sealed (immersion cleanable) relays.

Rated Burden

The power of burden (*watts* if DC or *volt–amperes* if AC) absorbed under the reference conditions by a given energizing circuit of a relay and determined under specified conditions.

Rated Coil Voltage

The coil voltage at which the relay is intended to operate for the prescribed duty cycle. Note: The use of any coil voltage less than rated may compromise the performance of the relay.

Rating

The nominal value of an energizing quantity, which appears in the designation of a relay.

Release

Process in which a monostable relay shifts from the operating state back to the rest state.

Release Time

The time interval from input (coil) deenergization to the functioning of the last output (contact) to function. This includes time for the coil to dropout its magnetic field (a significant limiting factor) and transfer time of the moveable contact between stationary contact(s), and bounce time after the initial contact brake. For a solid-state or hybrid relay in an operated state, the time from the application of the drop-out voltage to the change of state of the output.

Response

Process in which a relay shifts from the rest state to the operating state.

Reset

Process in which a bistable relay returns from the operating state back to the rest state.

Rest State (Release state)

Switch position of a monostable relay in the unenergized state. In bistable relays this is the switch position specified by the manufacturer.

Reset Time

The time interval that elapses from the point of time at which a bistable relay in the operating state has the nominal voltage applied in the opposite direction to the point of time at which the last output circuit has closed or opened (not including the bounce time).

Return

A relay returns when sequentially: it disengages; it passes from an operated condition towards the prescribed initial condition; and it resets.

Returning Ratio

The ratio of the returning value to the operating value.

Setting

The limiting value of a 'characteristic' or 'energizing' quantity at which the relay is designed to operate under specified conditions.

Such values are usually marked on the relay and may be expressed as direct values, percentages of rated values, or multiples.

Shock Resistivity

Threshold value indicating that no abnormality occurs even when semisine wave pulsating mechanical shock has been applied to the relay. Even after the shock has been applied, the contacts that have been opened do not close or the contacts that have been closed are not opened.

Vibration Resistivity

In the same manner as shock, threshold value when sine-wave vibration has been repeatedly applied to the relay.

---

4. Contact Systems and Other Relay Components

According to the different switching functions of the relay contacts, a difference is made between the various contact configurations whose design and description are specified in DIN 41020, ANSI C83.16.

SP, single pole; **DP**, double pole; **ST**, single throw; **NO**, normally open; **NC**, normally closed; **C**, changeover; **B**, break; **M**, make; **DM**, double make; **DB**, double break; **DT**, double throw

Some additional forms:

2C, DPDT; **4C**, 4PDT; **P**, SPST-Latching; **R**, SPDT-Latching.

Armature

The moving magnetic member of an electromagnetic relay structure.

Armature Balanced

A relay armature that rotates about its center of mass and is therefore approximately in balance with both gravitational (static) and accelerative (dynamic) forces.

Armature End-On

A relay armature whose principal motion is parallel to the longitudinal axis of a core having a pole face at one end.

Armature Lever

The distance through which the armature buffer moves divided by the armature travel. Also, the ratio of the distance from the armature bearing pin (or fulcrum) to the armature buffer in relation to the distance from the bearing pin (or fulcrum) to the center of the pole face.

Armature Chatter

The undesired vibration of the armature due to inadequate AC performance or external shock and vibration.

Auxiliary Contact

A contact combination used to operate a visual or audible signal to indicate the position of the main contacts, establish interlocking circuits, or hold a relay operated when the original operating circuit is opened.

Bobbin

A spool or structure upon which a coil is wound.

Bias Electrical

An electrically produced force tending to move the armature towards a given position.

Bias Magnetic

A steady magnetic field (permanent magnet) applied to the magnetic circuit of a relay to aid or impede operations of the armature.

Bias Mechanical

A mechanical force tending to move the armature towards a given position.

Blowout Magnet

A device that establishes a magnetic field in the contact gap to help extinguish the arc by displacing it.

Break Contact (NC Contact)

A contact that is closed in the release (rest) state of a monostable relay and opens (breaks) when the relay coil is energized (operating state).

Bridging Make Contact

Compound contact with two simultaneously operating make contacts connected in series.

Changeover Contact

A combination of two contact circuits including three contact members: a make contact and a break contact with a common terminal. When one of these contact circuits is open, the other is closed and vice versa. On changing the switch position, the contact previously closed opens first followed by the closing of the contact that was previously open.

Coil

An assembly consisting of one or more windings, usually wound over an insulated iron core on a bobbin or spool. May be self-supporting, with terminals and any other required parts such as a sleeve or slugs.

Concentrically Wound — A coil with two or more insulated windings wound one over the other.

Double Wound

A coil consisting of two windings wound on the same core.

Parallel Wound

A coil having multiple windings wound simultaneously, with the turns of each winding being contiguous (see *winding, bifilar*).

Sandwich Wound

A coil consisting of three concentric windings in which the first and third windings are connected series aiding to match the impedance of the second winding. The combination is used to maintain transmission balance.

Tandem Wound

A coil having two or more windings, one behind the other, along the longitudinal axis. Also referred to as a two-, three-, or four-section coil, etc.

Clapper

Sometimes used for an armature that is hinged or pivoted.

Cold Switching

Closing the relay contacts before applying voltage and current, plus removing voltage and current before opening the contacts. (Contacts do not make or break current.) Also see Dry Circuit Switching. Larger currents may be carried through the contacts without damage to the contact area since contacts will not "arc" when closed or opened.

Contact Bifurcated

A forked, or branched, contacting member so formed or arranged as to provide some degree of independent dual contacting.

Contact Bounce

An unintentional phenomenon that can occur during the making or breaking of a contact circuit when the contact elements touch successively and separate again before they have reached their final position.

Caused by one or more of the following: impingement of mating contacts; impact of the armature against the coil core on pickup or against the backstop on dropout; momentary hesitation or reversal of the armature motion during the pickup or drop-out stroke.

Contact bounce period depends upon the type of relay and varies from 0.1 to 0.5 ms for small reed relays up to 5 to 10 ms for larger solenoid types. Solid-state or mercury wetted contacts (Hg) do not have a contact bounce characteristic.

Contact Bounce Time

The time from the first to the last closing or opening of a relay contact.

Contact Break-Before-Make

A contact combination in which one contact opens its connection to another contact and then closes its connection to a third contact.

Contact Carrier

Conductive metal part of the relay where the contact is attached to.

Continuous Current

The maximum current that can be carried by the closed contacts of the relay for a sustained time period. This specification is determined by measuring the resistance heating effect on critical relay components.

Contact Erosion

Material loss at the contact surfaces, for example, due to material evaporation by an arc.

Contact Force

The force which two contact tips (points) exert against each other in the closed position under specified conditions.

Contact Gap

The distance between a pair of mating relay contacts when the contacts are open.

Contact Chatter

Externally caused, undesired vibration of mating contacts during which there may or may not be actual physical contact opening. If there is no actual opening but only a change in resistance, it is referred to as dynamic resistance.

Contact Late

A contact combination that is adjusted to function after other contact combinations when the relay operates.

Contact Member

A conductive part of a contact assembly which is electrically isolated from other such parts when the contact circuit is open.

Contact Potential

A voltage produced between contact terminals due to the temperature gradient across the relay contacts, and the reed-to-terminal junctions of dissimilar metals. (The temperature gradient is typically caused by the power dissipated by the energized coil.) Also known as contact offset voltage, thermal EMF, and thermal offset (in special contact metal combinations a thermal induced voltage of a few 100 μV is possible). This is a major consideration when measuring voltages in the microvolt range. There are special low thermal relay contacts available to address this need. Special contacts are not required if the relay is closed for a short period of time where the coil has no time to vary the temperature of the contact or connecting materials (welds or leads).

Current Rated Contact

The current which the contacts are designed to handle for their rated life.

Current Surge Limiting

The circuitry necessary to protect relay contacts from excessive and possibly damaging current caused by capacitive loads or loads which have a higher current consumption on switch *on* than in subsequent continuous operation (e.g., light lamps).

Contact Rating

The electrical load-handling capability of relay contacts (voltage, current, and power capacities) under specified environmental conditions and for a prescribed number of operations.

Contact Resistance

The resistance of closed contacts is measured as voltage drop across contacts carrying 1 A at 6 VDC for power relays and smaller carrying current for miniature relays. Actually, this is the sum of the contact resistance and conductor resistance.
   The maximum initial value (on delivery) is usually set forth on the catalog.

Contact Roll

When a contact is making, the relative rolling movement of the contact tips (points) after they have just touched.

Contact Tip

That part of a contact member at which the contact circuit closes or opens.

Contact Wipe

When a contact is making, the relative nibbling movement of contact tips (points) after they have just touched.

Contact Weld

A contact failure due to fusing of contacting surfaces to the extent that the contacts fail to separate when intended.

Double Pole (Single Throw Version)

A double pole relay switches two electrically not connected common lines with two electrically independent load lines (like two separate make relays).

Double Throw (Single Pole Version)

A double throw (single pole) relay switches one common line between two stationary contacts, for example, between a NO contact and a NC contact (like changeover relay or form C).

Drop-Out Voltage

The voltage at which all contacts return to their "normal," unoperated positions. (Applicable only to nonlatching relays.)

Dry Switching

Switching below specified levels of voltage and current (usually: <1 mA, <100 mV) to minimize any physical and electrical changes in the contact junction.

Duty Cycle

The ratio between the switch on time and total cycle time during periodical switching. Fifty per cent duty cycle means the switch on time equals the switch off time.

Dynamic Contact Resistance

Variation in contact resistance due to changes in contact pressure during the period in which contacts are motion, before opening or after closing.

Make Contact (NO Contact)

A contact that is open in the release (rest) state of a monostable relay and closes (makes) when the relay coil is energized (operating state).

Maximal Break Current

The highest value of current that can switch an output circuit *off* under specified conditions (voltage, switch off rate, power factor, time constants, etc.).

Maximal Make Current

The highest value of current that can switch an output circuit *on* under specified conditions (voltage, switch on rate, power factor, time constants, etc.). Loads can frequently have a higher current consumption on switch on than in subsequent continuous operation (e.g., light lamps).

Maximal Switching Current

The maximum current that can switch a relay contact *on* and *off*.

Maximum Switching Power

Maximum permissible product of switching current and switching voltage (in W for direct current, in VA for alternating current).

Minimum Switching Capacity

Due to slight corrosion of contacts, a minimum current or voltage is needed to allow fritting to keep the contact resistance low.

Minimum Switching Power

Product of the switching current and switching voltage that should not be undercut to ensure switching.

Movable Contact

The member of a contact combination that is moved directly by the actuating system. This member is also referred to as the armature contact or swinger contact. The moveable contact is mounted on the armature or spring system.

Normally Closed Contact (NC)

A contact combination which is closed when the armature is in its unoperated position.

Normally Open Contact (NO)

A contact combination that is open when the armature is in its unoperated position (generally applies to monostable relays).

Plunger or Solenoid Armature

A relay armature that moves within a tubular core in a direction parallel to its longitudinal axis.

Premake Contact

Twin contact electrically connected where one contact always closes first (premake) and opens last. The premake contact, for example, out of tungsten (high resistive and very resistant to contact erosion), switches the current, while the second, like a low resistive silver contact, carries the load.

Reed Contact

A hermetically enclosed, magnetically operated contact using thin, flexible, magnetic conducting strips as the contacting members which are moved directly by a magnetic force.

Settle Time

The time required for establishing relay connections and stabilizing user circuits. For relay contacts, this includes contact bounce.

Shading Ring

A shorted turn surrounding a portion of the pole of an alternating-current electromagnet that delays the change of the magnetic field in that part, thereby tending to prevent chatter and reduce hum.

Single Contact

Contact configuration with a single stationary and moveable contact pair on the make and/or the break side (compare twin or double contacts).

Single Pole (Single Throw Version)

A single pole (single throw) relay connects one common line (moveable contact) to one load line (stationary contact).

Single Throw (Single Pole Version)

A single throw (single pole) relay connects one common line (moveable contact) to one load line (stationary contact).

Stagger Time

The time interval between the functioning of contacts on the same relay. For example, the time difference between the opening of
   normally closed contacts on pickup.

Stationary Contact

Nonmoveable contact, mounted on a contact carrier which is directly connected to a relay pin or faston blade.

Switching current

Current that can switch a relay contact *on* and *off*.

5. Specified Terms for Solid-State Relays

Critical Rate of Rise of Off-State Voltage (Critical dv/dt)

The minimum value of the rate of rise of the forward voltage which will cause switching from the off-state to the on-state.

Critical Rate of Rise of On-State Current (Critical di/dt)

The maximum value of the rate of rise of on-state current which a thyristor can withstand without deleterious effect.

DIP

Dual inline package.

di/dt

Rate of rise of current.

dv/dt

Rate of rise of voltage.

FET

Field effect transistor. A device in which the gate voltage (not current) controls the ability of the device to conduct or block current flows.

Gate-Controlled Turn-On Time (t_{gt})

The time interval between a specified point at the beginning of the gate pulse and the instant when the forward voltage (current) has dropped (risen) to a specific, low (high) value during switching of a thyristor from the off-state to the on-state by a gate pulse.

Gate Trigger Current, Input Current, Control Current (I_{GT})

The minimum gate current required to switch a thyristor from the off-state to the on-state.

Gate Trigger Voltage (V_{GT})

The gate voltage required to produce the gate trigger current.

GTO

Gate turn-off thyristor. A thyristor which can be turned on and off by control of its gate current.

Holding current (I_H)

The minimum forward current required to maintain the thyristor in the on-state without control current after its opening.

IGBT

Insulated gate bipolar thyristor. A bipolar power transistor whose gate is voltage-charge controlled in similar manner to the MOSFET.

Latching Current (I_L)

The minimum forward current required to maintain the thyristor in the on-state immediately after switching from the off-state to the on-state has occurred and the triggering signal has been removed.

MOSFET

Metal-oxide-semiconductor field effect transistor. Variety of FET transistor (see FET).

MOV

Metal oxide varistor. Used for transient suppression.

Off-State dv/dt

The rate of rise of voltage, expressed in volts per microsecond (V/msec), that the SRR output switching device can withstand without turning on.

Off-State Voltage

The maximum effective steady state voltage that the output is capable of withstanding when in off-state without breakover or damage.

On-State Resistance

In a power-FET relay, this is the intrinsic resistance of the output circuit in the on-state.

On-State Voltage or Voltage Drop (V_T, V_{DROP})

The voltage at maximum load current developed across the output switching element (thyristor, triac, etc.) when the relay is in the ON state.

Over-Voltage Rating

The guaranteed transient peak blocking (or breakdown) voltage rating of the SSR.

Peak Gate Power Dissipation

The maximum power which may be dissipated between the gate and main terminal (or cathode) for a specified time duration.

Power Dissipation

The maximum power dissipated by the SSR for a given load current.

Repetitive Overload Current

The maximum allowable repetitive RMS overload current that may be applied to the output for a specific duration and duty cycle while still maintaining output control.

Repetitive Peak Forward Voltage of an SCR (V_{FRM})

The maximum instantaneous cyclic voltage occurs across a thyristor in off-state which it can withstand without turn-on (without control signal).

Repetitive Peak Off-State Current (I_{DRM})

The maximum instantaneous value of the off-state current that results from the application of repetitive peak off-state voltage.

Repetitive Peak Off-State Voltage (V_{DRM})

The maximum instantaneous value of the off-state voltage which occurs across a thyristor, including all repetitive transient voltages, but excluding all nonrepetitive transient voltages.

Repetitive Peak Reverse Current of an SCR (I_{RRM})

The maximum instantaneous value of the reverse current that results from the application of repetitive peak reverse voltage.

Repetitive Peak Reverse Voltage of an SCR (V_{RRM})

The maximum instantaneous value of the reverse voltage which occurs across a thyristor, including all repetitive transient voltages, but excluding all nonrepetitive transient voltages.

SCR

Silicon controlled rectifier. Synonym of word "thyristor."

SIP

Single inline package.

SMD

Surface mounted device.

Snubber

RC-circuit placed in parallel with a solid-state commutation device to protect against overvoltage transients.

Surge (Nonrepetitive) On-State Current (I_{TSM})

The maximum nonrepetitive surge (or overload) on-state current of short-time duration and specified waveshape that the SCR can safely withstand without causing permanent damage or degradation.

Thermal Resistance, Junction to Ambient

The temperature difference between the thyristor junction and the ambient divided by the power dissipation causing the temperature difference under conditions of thermal equilibrium. Note: Ambient is defined as the point where the temperature does not change as a result of the dissipation.

Triac

Variety of SCR (more complex) especially intended for use in AC circuits.

Voltage Reverse Polarity

The maximum allowable reverse voltage which may be applied to the input of a solid-state relay without permanent damage.

Zero-Voltage Turn On Relay

A relay with isolated input and output in which added control circuitry delays the output turn-on until a zero-voltage transition of the AC sine wave is detected.

References

General Catalogs

Protective Devices. Catalog NS 1–89. Siemens.
Schutzeintrichtungen und Relais. Catalog. Preisliste R, 1972. Siemens (Germ.).
HV Protection and Protection Systems. Buyer's Guide 1989–1990. ABB Relays.
Relay Units and Components. Buyer's Guide 1989–1990. ABB Relays.
ASEA Relays. Buyer's Guide B03-0011E 1985–1986. ASEA.
Mebrelais Schutzrelais Regler elektrischer Groben. Lieferprogramme 1974/75. AEG (Germ.).
G. E. C. Catalogue of Electrical Installation Material. The General Electric Co. Ltd, 1935.

General Monographs

Davies, T. *Protection of Industrial Power Systems*. Newnes, 1998.
Sterl, N. *Power Relays. EH-Schrack Components AG*. Vienna, Austria, 1997.
Engineers Relay Handbook. Fifth Edition. National Association of Relay Manufacturers. Milwaukee, Wisconcin, 1996.
Protective Relays. Application Guide. GEC ALSTHOM T & D, 1995.
Titarenko, M. and Noskov-Dukelsky, I. *Protective Relaying in Electric Power Systems*. Moscow: Foreign Languages Publishing House.
Schleiecher, M. *Die Moderne Selektivschutztechnik und die Methoden zur Fehlerortung in Hochspannung-sanlagen*. Verlag von Julius Springer: Berlin, 1936 (Germ.).
Vitenberg, M. I. *Calculation of Electromagnetic Relays*. Leningrad, Energya, 1975 (Rus.).
Chunyhin, A. A. *Electric Apparatus*. Moscow: Energoatomizdat, 1988 (Rus.).
Alexeyev, V. S., Varganov, G. P., and Panphilov, B. I. *Protective Relays*. Moscow: Energya, 1976 (Rus.).
Bul, B. K. et al. *Electromechanical Apparatus of Automatics*. Moscow: Wishaya Shchola, 1988 (Rus.).
Chernobrovov, N. V. *Relaying Protection*. Moscow: Energya, 1974 (Rus.).
Berckovich, M. A., Molchanov, V. V., and Simeonov, V. A. *Basic Technology of Relaying Protection*. Moscow: Energoatomizdat, 1984 (Rus.).
Fabrikan, T. V. L., Gluhov, V. P., and Paperno, L. B. *Elements of Protective Relaying Devices and Electric-Power Automatic and it Designing*. Moscow: Wishaya Shchola, 1974 (Rus.).
Stupel, F. A. *Electromechanical Relays. Basic of Theory, Designing and Calculations*. Kharkov: Issue of Kharkov University, 1956 (Rus.).
Dorogunzov, V. G. and Ovcharenko, N. I. *Elements of Devices for Automatic of Power Systems*. Moscow: Energya, 1974 (Rus.).

Chapter 1

The Papers of Joseph Henry, Vol. 1. Washington: Smithsonian Institution Press, 1972, pp. 132–133.
Albert, E. M. and Joseph, H. *The Rise of an American Scientist*. Washington, D.C.: Smithsonian Institution Press, 1997, pp. 26 and 50–51.

Joseph, H. On some modifications of the electro-magnetic apparatus. *Transactions of the Albany Institute*, Vol. 1, pp. 22–24.

Chipman, R. A. The Earliest Electromagnetic Instruments, Contributions from the Museum of History and Technology, Paper 38, Smithsonian Institution, United States National Museum Bulletin 240. Washington: Smithsonian Institution, 1966, pp. 127–131.

Henry, J. On the application of the principle of the galvanic multiplier to electro-magnetic apparatus, and also to the development of great magnetic power in soft iron, with a small galvanic element. *American Journal of Science and Arts,* Vol. 19, January 1831.

King, W. J. The Development of Electrical Technology in the 19th Century: 1. The Electrochemical Cell and the Electromagnet, Contributions from the Museum of History and Technology, Paper 28, United States National Museum Bulletin 228, Washington: Smithsonian Institution, 1962.

Henry, J. On the application of the principle of the galvanic multiplier. *Scientific Writings,* Vol. 1, pp. 39–45.

Henry, J. An account of a large electro-magnet, made for the Laboratory of Yale College. *American Journal of Science and Arts,* Vol. 20, April 1831.

Statement of Prof. Henry, in relation to the History of the Electro-magnetic Telegraph, Annual Report of the Board of Regents of the Smithsonian Institution for the Year 1857. Washington, D.C.: William A. Harris, Printer, 1858, p. 105.

Gee, B. Electromagnetic engines: pre-technology and development immediately following Faraday's discovery of electromagnetic rotations. *History of Technology,* Vol. 13, 1991, pp. 41–72.

Malcolm, M. L. *The Rise of the Electrical Industry During the Nineteenth Century Princeton.* Princeton, NJ: Princeton University Press, 1943.

Ritchie, W. Experimental Researches in Electro-Magnetism and Magneto-Electricity, *Philosophical Transactions of the Royal Society of London,* Vol. 123, 1833, pp. 313–321.

Davis, D. Jr. *Davis's Descriptive Catalogue of Apparatus and Experiments.* Boston, 1838, pp. 27, 30.

Page, C. G. Experiments in electro-magnetism. *American Journal of Science and Arts,* Vol. 33, No. 1, January 1838.

Page, C. G. *History of Induction: The American Claim to The Induction Coil and its Electrostatic Developments,* Washington, 1867.

Sherman, R. Joseph Henry's contributions to the electromagnet and the electric motor, *National Museum of American History,* 1999.

Chapter 3

Kunath, H. Funkenloschung in Induktiv Belasteten Stromkreisen. *Electrotechnik,* 1988, No. 1 (Germ.).
Industrial Relays. Magnecraft & Struthers-Dunn Catalogue, Edition 101.
Midtex Relays, Inc. Catalogue, 1988.
3TF AC Contactors. Catalogue Siemens, 1989.

Chapter 4

Crystal Can Relays. Meder Electronic GmbH Catalogue, 1998.
Electromechanical Relay. Data Book. Teledyne Relays Catalogue, 1999.
Yellow Relays: General Purpose Relays. Idec Izumi Corp. Catalogue.
Hartman Product Brochure, 1999.
Instantaneous Auxiliary Relay Type HGA11A, G, H, J, K, P, S, T, V, W, X. Instruction GEH-1793E. General Electric Co.

Chapter 5

Reed Relays. Hamlin, Inc. Catalogue.

Reed Sensors & Reed Switches. Meder Electronic Catalogue, 1999.

Radio Frequency, High Voltage, DIP & SIP Reed Relays. Crydom Corp. Catalogue.

Reed Switches. ALEPH Corp. Catalogue, 1994.

Kharazov, K. I. *Automatic Devices with Reed Switches*. Moscow: Energya, 1990.

Koblenz, M. G. *Power Reed Switches*. Moscow: Energya, 1979 (Rus.).

Koblenz, M. G. *Sealed Commutation Devices on Power Reed Switches*. Moscow: Energoatomizdat, 1986 (Rus.).

Chapter 6

Gurevich, V. I. *High-Voltage Automatic Devices with Reed Switches*. Haifa, 2000.

Gurevich, V. I. *Protection Devices and Systems for High-Voltage Applications*. Marcel Dekker, Inc.: NewYork, 2003.

High Voltage Vacuum and Gas Filled Relays. Joslyn Jennings Corp. Catalogue REL-103, 1994.

High Voltage Relays and DC Power Switching Devices. Kilovac Corp. Catalogues, 1993.

Chapter 7

Eccles, W. H. and Jordan, F. W. A trigger relay utilizing three-electrode thermionic vacuum tubes. *Radio Review*, Vol. 1, No. 3, December, 1919, pp. 143–146.

Solid State Relays. Crouzet Automatismes Catalogue, 1993.

Solid State Relays for Motor Control. Celduc Relais Catalogue.

Solid State Relays & Power Modules. Crydom Corp. Catalogue.

Solid State Relays. Data Book. Teledyne Relays Catalogue, 1999.

Solid-State Relays & Solid-State Power Controllers. Teledyne Relays Catalogue, 2002.

Fast High-Voltage Solid-State Switches. Behlke Electronic GmbH Catalogue, 1998.

Voronin, P. A. *High-Power Solid-State Switches*. Moscow: Dodeka-XXI, 2001 (Rus.).

Isakov, U. A., Platonov, A. P., and Rudenko, V. S. *Basics of Industrial Electronics*. Kiev, Technika, 1976 (Rus.).

Kaganov, I. L. *Industrial Electronics*. Moscow: Wishaya Shchola, 1968 (Rus.).

Micklashewsky, S. P. *Industrial Electronics*. Moscow: Wishaya Shchola, 1973 (Rus.).

Semikron: Innovation + Service. Power Electronics (Headquarter). Nurnberg, 1999.

Gentry, F. E., Gutzwiller, F. W., and Holonyak, N. *Semiconductor Controlled Rectifiers: Principles and Applications of p-n-p-n-Devices*. Englewood Cliffs, NJ: Prentice-Hall, 1964.

Coughlin, R. F. and Driscoll, F. F. *Operational Amplifiers and Linear Integrated Circuits*. Englewood Cliffs, NJ: Prentice-Hall, 1977.

Tokheim, R. *Digital Electronics*. McGraw-Hill, New York, 1984.

High-Side Driver Switch MC33038. Data Sheet. Motorola, Inc.

Static Var Compensator. Catalog. Toshiba Corp., Japan.

300 MW, DC 125 kV, 2400 A — Catalog of Toshiba Corp. by Japan.

Directly Light-Triggered Thyristor (LTT) Valve. Catalog of Toshiba Corp.

High-Power Semiconductors. Catalog of Toshiba Corp.

Odegard, B. and Ernst, R. Applying IGCT Gate Units — ABB Application Note 5SYA 2031-01, December, 2002.

Chapter 8

Type RXKF-1 Pickup Time-Delay Relay. B03-1614. Relay Description. ASEA, 1983.
Programmable Timer MC14541B. Semiconductor Technical Data. Motorola Inc., 1999.
Type SSX 120, CSX 700 Time-Lag Relays Static with Delayed Pick-Up. CH-ES 64-41.5 Relay Description. ABB.
Timing Circuit MC1455, MC1555. Specifications and Applications Information. Motorola, Inc.
Elektronische Zeitschalter Typ ZSg and RZSg. Behandlungvorschrift. AEG, 1966 (Germ.).
Time-Lag Relay type RXKE-1 with Pick-Up Delay. Catalogue RK 33-12E. ASEA, 1979.
Timing Relays Types SAM11, SAM13, SAM18, SAM999. Instruction GEK-7393D. General Electric Co.
Transistor-Zeitrelais RZyb1. Schaltbild. AEG, 1996 (Germ.).
DC Operated Timing Relay Types: SAM14A12A AND UP and SAM14B15A AND UP. Instruction GEK-7398A. General Electric Co.
Elektronische Zeitschalter Typ ZSg 05 Ws, 10 Ws und 20 Ws mit Kaltkanthodenrohren. Badeinungsanweisung. AEG, 1958 (Germ.).
Type RXIDG-2H Time-Overcurrent Relay. Catalogue B03-2212. ABB, 1989.
Electro-mechanical High-precision AC Time Delay Relays Series MZ, MZA and MZJ. Catalogue. Schleicher AG.
Type RXMS-1 High-Speed Tripping Relay. Catalogue B03-1213. ABB, 1989.
CR5481-33 Normally Closed Time-Delay Dropout Contactor. Instruction GEH-1492A. General Electric Co.
Definite-Time Control Relay Type MC-13. Instructions GEH-1167A. General Electric Co.

Chapter 9

Morel, R. LV Circuit-Breaker Breaking Techniques — Cahier Technique No. 154 — Schneider Electric, 2000.
Circuit-Breakers for Overcurrent Protection for Household and Similar Installation. Part 1: Circuit-Breakers for AC Operation. International Standard IEC 60898-1, 2003.
AC or DC Temperature Overload Relays CR 2824-41C and -41H. Instruction GEI-19295. General Electric Co.
Cera-Mite Capacitor and Thermistor Full Line Catalog. Cera-Mite Corporation.
Thermal Relay Type GW. Instructions GEI-28027. General Electric Co.
Thermal Overcurrent Relay Type TMC11A. Instructions GEI-28826A. General Electric Co.

Chapter 10

Bushing-type Current Transformers. Instructions GEI-70372D. General Electric Co.
Bushing Current Transformers Types BR-B and BR-C. Instructions GEH-2020B.
Instrument Potential Transformers Type EU Oil Immersed Cascade Construction for Line-to-Neutral Operation on Power Circuits of 92 kV and Above. Instructions GEH-1629B. General Electric Co.
Indoor Current Transformers Dry Type, Type JS-1. Instructions GEC-333B. General Electric Co.

Types CO and COH Overcurrent Relays. Instructions I.L. 41-280G. Westinghouse, 1950.

Instruction Manual for Capacitor Voltage Transformers and Coupling Capacitors. Bulletin IM300–05. Trench Electric, 1988.

Inverse Time Current Relays. Instruction Bulletin 810. A-B, 1977.

Three-Phase Undervoltage Relay Type RXOB-23. Catalogue RK 43-12E. ASEA, 1977.

Three-Phase Overvoltage and Undervoltage Relay SPAU 1340 C. 34 SPAU 15 EN1 B, ABB, 1995.

Instantaneous Current and Voltage Relays Type RXIL and RXEL. Catalogue RK 41-11E. ASEA, 1971.

Over and Undervoltage Relays Type RXEB. Catalogue RK 42-12E. ASEA, 1971.

Overvoltage and Undervoltage Relay Type RXEF-2. Catalogue RK 42-16E. ASEA, 1978.

Small Over-Current and Over-Voltage Relays RXIC-1, RXID-1, RXEC-1, RXED-1. Catalogue RK 42-10E. ASEA, 1972.

Distribution System Feeder Overcurrent Protection. GET-6450. General Electric Co.

Gelfand, Y. S. *Relaying Protection of Electric-Power Networks*. Moscow: Energya, 1975 (Rus.).

Specktor, S. A. *Electrical Measurements of Physical Values*. Leningrad: Energoatomizdat, 1987 (Rus.).

Afanasyev, V. V., Adonyev, N. M., and Kibel, V. M. *Current Transformers*. Leningrad: Energoatomizdat, 1989 (Rus.).

Instantaneous Overcurrent Relays Type PJC11, PJC12, PJC13, PJC14. Instruction GEI-28803B. General Electric Co.

Harmonic Restraint Overcurrent Relay Type RAISA. Catalogue RK 65-50 E. ASEA, 1975.

Types IC91-1, IC91-3 modures[R] Inverse Time-Overcurrent Relays. CH-ES 65-11-1 Relay Description. ABB, 1983.

Short-Time Overcurrent Relays IAC55A — IAC55T, IAC56A, IAC56B. Instruction GEI-3101C. General Electric Co.

Time Overcurrent Relays Types IAC51A — IAC51R, IAC52A, IAC52B. Instruction GEH-1753E. General Electric Co.

Single-Phase Overcurrent Protective Relays. Catalogue RK 65-16E. ASEA, 1976.

Type RXIDF-2H AC Time-Overcurrent Relay. Catalogue B03-2210. ABB, 1987.

Overcurrent Relay Type RXIK-1. Catalogue RK 41-12E. ASEA, 1974.

Instantaneous Current Relays Type PJC. Instruction GEH-1790A. General Electric Co.

Type IJCV Time-Overcurrent Relays with Voltage Restraint. Renewal Parts GEF-4040A. General Electric Co.

Time Overcurrent Relays Types IAC66, IAC67. Instruction GEI-28818D. General Electric Co.

Type IFC Time Overcurrent Relays. Renewal Parts GEF-4533C. General Electric Co.

Unabhangige Strom-und Spannungs-Zeitrelais mit Transistor-Zeitglied. Technische Beschreibung. AEG-Telefunken, 1981 (Germ.).

Type RXIG-21 Instantaneous AC Over- and Undercurrent Relays. Catalogue B03-2032. ABB, 1982.

Primar-Relais MUT1, MU, MT1. Sprecher + Schuh GmbH (Germ.).

Instantaneous Voltage Relay Types PJV11A, PJV11B and PJV12A. Instruction GEI-30971F. General Electric Co.

Chapter 11

Power Relays Types RPB, RP 1 and RPF. Catalogue RK 51-1E, ASEA, 1968.

Type ICW Overpower Relays. Instruction GEA-3417D. General Electric Co.

Instructions for Maintenance of Relaying Protection, Electrical Automation and Another Electrical Equipment. Vol. 1. Irkutsk, 1968 (Rus.).

Directional Relay Type RXPE-4. Catalogue RK 51-10E. ASEA, 1976.

Type GGP Power Directional Relays. Renewal Parts. GEF-3956A. General Electric Co.

Type PPX 105b Power Relay. Catalogue CH-ES 62-51.10. ABB, 1987.

Chapter 12

Transformer Differential Relay with Percentage & Harmonic Restraint Types BDD15B, BDD16B. Instruction Manual GEH-2057A. General Electric Co.

Triple-Pole Differential Relays Type D2. Catalogue AB 309 c. Brown Bowery Co., 1967.

Transformer Differential Relay RQ4. AEG, 1959.

Transformer Differential Relays with Percentage and Harmonic Restraint Types STD15B, STD16B. Instruction Manual GEK-7362B. General Electric Co.

High-Speed Differential Relays Type CFD. Instruction Manual GEK-34124. General Electric Co.

Types IJD53C and IFD51A Percentage-Differential Relays. Instruction GEA-3236B. General Electric Co.

Pilot-Wire Differential Protection Type RYDHL for Feeders. Description of Relay Protection RK 60-324E. ASEA, 1969.

AC Pilot Wire Relaying System SPD and SPA Relays. Application Guide GET 6462A. General Electric Co.

Type HCB Pilot-Wire Relaying and Pilot-Wire Supervision. Catalog 41–658. Westinghouse, 1942.

Pilot-Wire Differential Relay Type RADHL for cables and Overhead Lines. Catalogue RK 61-10E. ASEA, 1978.

Transformer Differential Type BDD. Description, Construction, Applications GET-7278. General Electric Co.

Type PVD Differential-Voltage Relay. Instructions GEA-5449A. General Electric Co.

Type MFAC High Impedance Differential Relay. ALSOM.

High-Impedance Differential Relaying. GER-3184. General Electric Co.

Chapter 13

Impedance Relay Type RXZF. RK 556-300E Relay Description. ASEA, 1970.

Distance Relay Type RAZOG. Info-No. RK 615-302E, ASEA, 1975.

Type RANZP Power-Swing-Blocking Relay. B03-7111 Relay Description. ABB, 1989.

Type RAKZB Three-Phase Impedance Relay. B03-3213 Relay Description. ABB, 1989.

Distance Relays Application Guide. GER-3199. General Electric Co.

Type LZ95 modures® Solid-State Distance Relay. CH-ES 63-95.10 Relay Description. ABB, 1987.

Distance Relay Type RAZOA. RK 61-29E Relay Description. ASEA, 1981.

Fabrikant, V. L. Distance Protection. Moscow: Wishaya Shchola, 1978 (Rus.).

Operating Principles of GCX and GCY Relays. GET-6658. General Electric Co.

The Art of Protective Relaying. GET-7206A. General Electric Co.

Use of the R–X Diagram in Relay Work. GET-2230B. General Electric Co.

Three-Phase Impedance-Measuring Protective Relay Type RAKZA. Catalogue RK 65-52 E. ASEA, 1977.

Angle-Impedance Relay Types: CEX57D, CEX57E, CEX57F. Instruction GEK-49778B. General Electric Co.

Chapter 14

Type CF-1 Under and Over Frequency Relay. Descriptive Bulletin 41–500. Westinghouse, 1963.

Frequency Relay RFA. RK 50–302 E Relay Description. ASEA, 1966.

Frequency Relay Type IJF. GEI-19008. General Electric Co.

Static Frequency Relay Type SFF 31A, C, SFF 32A, C. Instruction GEC-49923A, General Electric Co.

New, W. C. Load Conservation by Means of Under-Frequency Relays. GER-2398, General Electric Co.

Berdy, J. Load Shedding — An Application Guide. Electric Utility Engineering Operation. General Electric Co., Schenectady, NY, 1968.

Under-Frequency Relay Type CFF13A. Instruction GEI-44246C. General Electric Co.

Frequency Relay Type FCX103. Catalog CH-ES 62-01 E. Brown Bowery Co., 1973, 1981.

Over- and under-Frequency Relay Assemblies (TFF). B03-2920 Relay Description. ASEA, 1985.

Type RXFE-4 Frequency Relay. B03-2910 Relay Description. ASEA, 1982.

Types FC95, FCN950 modures® Frequency Relay. CH-EC 62-01.60 Relay Description. ABB, 1987.

Chapter 15

Matsuda, T. et al. Experience with Maintenance and Improvement in Reliability of Microprocessor-Based Digital Protection Equipment for Power Transmission Systems. Report 34–104. SIGRE Session, 30 August–5 September 1992, Paris.

Information Notice No. 94–20. *Common-Cause Failures Due to Inadequate Design Control and Dedication*, Nuclear Regulatory Commission, March 17, 1994.

The Power Protection Handbook. APC, 1994.

Djohov, B. D. Features of electrical power supply of means of computer networks. *Industrial Power Engineering*, 1996, No. 2, pp. 17–24 (Rus.).

Reason, J. Realistic relay tests need fault reconstruction. *Electrical World*, 1991, Vol. 205, No. 5, pp. 41–42.

Ianoz, M. and Wipf, H. Modeling and simulation methods to assess EM terrorism effects. *Proceedings of 13th International Zurich Symposium and Technical Exhibition on Electromagnetic Compatibility*, February 16–18, 1999, pp. 191–194.

Stanley, E. and Zocholl, G. B. How microprocessor relays respond to harmonics, saturation, and other wave distortions. *Neta World*, Schweitzer Engineering Laboratories, Inc., Summer 2003.

Horak, J. Pitfalls and benefits of commissioning numerical relays. *Neta World*, Summer 2003.

Manuel, W. W. Electromagnetic terrorism — what are the risks? What can be done? *International Product Compliance Magazine*, 1977.

IEEE Power Engineering Society. IEEE Standard 1402–2000: *IEEE Guide for Electric Power Substation Physical and Electronic Security*, New York: IEEE, April 4, 2000.

Prishchepenko, A. B., Kiseljov, V. V., and Kudimov, I. S. Radio frequency weapon at the future battlefield, Electromagnetic environment and consequences, *Proceedings of the EUROEM94, Bordeaux, France*, May 30–June 3, 1994, part 1, pp. 266–271.

Wood, A. Microprocessors: Your Questions Answered. Sevenoaks: Butterworths & Co., 1982.

Aspects of Digital Protective Relaying. Report RE-626, 1991. *Research and Development Division of the IEC*.

Johnson G., Thomson M. Reliability Considerations of Multifunction Protection, *Basler Electric Co.*

Gurevich, V. The Hazards of Electro-Magnetic Terrorism, *Public Utilities Fortnightly*, June 2005.

Gurevich, V. Nonconformance in Electromechanical Output Relays of Microprocessor-Based Protection Devices under Actual Operating Conditions, *Electrical Engineering & Electromechanics*, 2006, No. 1.

Gurevich, V. Electromagnetic Terrorism: New Hazards, *Electrical Engineering & Electromechanics*, 2005, No. 4.

Chapter 16

Latching Relays. Gruner AG Catalogue, 2000.

Remanence Relay Type RXMVE-1. Catalogue RK 25-11 E. ASEA, 1975 (B03-1512 ABB, 1989).

Type RXMVB-2 Heavy-Duty Latching Relay, 8 Contacts. Technical Description B03-1510. ABB, 1987.

Type RXMVB-4 Heavy-Duty Latching Relay, 14 Contacts. Technical Description B03-1511. ABB, 1986.

Throw-Over Relays type RXMVB-2 and RXMVB-4. Catalogue RK 25-10 E. ASEA, 1978.

Haftrelais RHH-110. Schaltplan RS 237-63. Hartmann & Braun Meb-und Regeltechnik.

Type RXIK1 Highly Sensitive Instantaneous AC and DC Overcurrent Relay. B03-2031. ABB, 1987.

Handbook for adjustment of secondary circuits on power stations and substations. Editor E. Musaelyan. Moscow: Energoatomizdat, 1989 (Rus.).

Flashing-Light Relays Type RXSU. Catalogue RK 27-13 E. ASEA, 1974.

Signal Relay Type RXSL-1. Catalogue RK 27-11E. ASEA, 1974.

Type RXSF-1 Dual Target Relay. Technical Description B03-1215. ABB, 1987.

Target and Contact Data. Data Sheet GEP-662. General Electric Co.

Signal Systems Type RSF-1 with Flags. Catalogue RK 28-11E. ASEA, 1973.

Type HEA Auxiliary Relays. Renewal Parts. GEF-3325E. General Electric Co.

Gas Detector Relay. Instructions GEK-4817C. General Electric Co.

Gas Detector Relay and Fault Pressure Relay, Type E1 or E3 for Oil-Filled Transformer. Instructions GEI-22525B. General Electric Co.

Twin-Float Buchholz Relays with Change-Over Contacts acc. DIN 42 566. Technical Description. Siemens.

Buchholz Relays (Single and Double-Float Relays Models A and B). Technical Data, Design and Mode of Operation. Siemens.

Gas-Operated Relay Type TVGC. Instruction 4609 022 Ea. ASEA, 1966.

Index